Methods and Applications of Geospatial Technology in Sustainable Urbanism

José António Tenedório
Universidade NOVA de Lisboa, Portugal

Rossana Estanqueiro
Universidade NOVA de Lisboa, Portugal

Cristina Delgado Henriques
Universidade de Lisboa, Portugal

IGI Global
PUBLISHER of TIMELY KNOWLEDGE

A volume in the Advances in Geospatial Technologies (AGT) Book Series

Published in the United States of America by
IGI Global
Engineering Science Reference (an imprint of IGI Global)
701 E. Chocolate Avenue
Hershey PA, USA 17033
Tel: 717-533-8845
Fax: 717-533-8661
E-mail: cust@igi-global.com
Web site: http://www.igi-global.com

Library of Congress Cataloging-in-Publication Data

Names: Tenedório, José António, 1961- editor. | Estanqueiro, Rossana, 1980- editor. |Henriques, Cristina Delgado, 1968-editor.
Title: Methods and applications of geospatial technology in sustainable urbanism / José António Tenedório, Rossana Estanqueiro and Cristina Delgado Henriques, editors.
Description: Hershey, PA : Engineering Science Reference, 2020. | Includes bibliographical references. | Summary: "This book examines the use of geospatial technology in sustainable urbanism"-- Provided by publisher.
Identifiers: LCCN 2019037712 (print) | LCCN 2019037713 (ebook) | ISBN 9781799822493 (h/c) | ISBN 9781799822509 (s/c) | ISBN 9781799822516 (eISBN)
Subjects: LCSH: City planning--Geographic information systems. | Sustainable urban development--Geographic information systems | Land use--Planning--Geographic information systems.
Classification: LCC HT166 .M4455 2021 (print) | LCC HT166 (ebook) | DDC 307.1/2160285--dc23
LC record available at https://lccn.loc.gov/2019037712
LC ebook record available at https://lccn.loc.gov/2019037713

This book is published in the IGI Global book series Advances in Geospatial Technologies (AGT) (ISSN: 2327-5715; eISSN: 2327-5723)

British Cataloguing in Publication Data
A Cataloguing in Publication record for this book is available from the British Library.

All work contributed to this book is new, previously-unpublished material.
The views expressed in this book are those of the authors, but not necessarily of the publisher.

For electronic access to this publication, please contact: eresources@igi-global.com.

Advances in Geospatial Technologies (AGT) Book Series

ISSN:2327-5715
EISSN:2327-5723

Editor-in-Chief: Nilanjan Dey Techno India College of Technology, India

MISSION

The geospatial technology field is a fast-paced, high growth industry that is involved in a variety of fields, including military planning, public health, land use, environmental protection, and Google Earth mapping. With such a diverse body of applications, the research in geospatial technologies is always evolving and new theories, methodologies, tools, and applications are being developed. **Advances in Geospatial Technologies (AGT) Book Series** is a reference source and outlet for research that discusses all aspects of geographic information, including areas such as geomatics, geodesy, GIS, cartography, remote sensing, and other areas. Because geospatial technologies are so pervasive in such a variety of areas, AGT also includes books that address interdisciplinary applications of the technologies.

COVERAGE

- Geomatics
- Land Surveying
- Geospatial Analysis
- Global Positioning Systems
- Cartography
- Disaster Informatics
- Public Sector Use of Geospatial Technologies
- Public Health Applications of Geospatial Technologies
- Surveying Technologies
- Spatial Reference Systems

IGI Global is currently accepting manuscripts for publication within this series. To submit a proposal for a volume in this series, please contact our Acquisition Editors at Acquisitions@igi-global.com or visit: http://www.igi-global.com/publish/.

The Advances in Geospatial Technologies (AGT) Book Series (ISSN 2327-5715) is published by IGI Global, 701 E. Chocolate Avenue, Hershey, PA 17033-1240, USA, www.igi-global.com. This series is composed of titles available for purchase individually; each title is edited to be contextually exclusive from any other title within the series. For pricing and ordering information please visit http://www.igi-global.com/book-series/advances-geospatial-technologies/73686. Postmaster: Send all address changes to above address.

Titles in this Series

For a list of additional titles in this series, please visit:
http://www.igi-global.com/book-series/advances-geospatial-technologies/73686

Interdisciplinary Approaches to Spatial Optimization Issues
Sami Faiz (University of Tunis El Manar Tunis, Tunisia) and Soumaya Elhosni (University of Tunis El Manar Tunis, Tuisia)
Engineering Science Reference • © 2021 • 300pp • H/C (ISBN: 9781799819547) • US $195.00

Geospatial Technologies for Effective Land Governance
Moha El-Ayachi (Institut Agronomique et Vétérinaire Hassan II, Morocco) and Loubna El Mansouri (Institut Agronomique et Vétérinaire Hassan II, Moocco)
Engineering Science Reference • © 2019 • 302pp • H/C (ISBN: 9781522559399) • US $205.00

Emerging Trends in Open Source Geographic Information Systems
Naveenchandra N. Srivastava (Bhaskaracharya Institute for Space Applications and Geo-Informatics, India)
Engineering Science Reference • © 2018 • 302pp • H/C (ISBN: 9781522550396) • US $195.00

Geospatial Technologies in Urban System Development Emerging Research and Opportunities
Alok Bhushan Mukherjee (North-Eastern Hill University Shillong, India) Akhouri Pramod Krishna (Birla Institute of Technology Mesra, India) and Nilanchal Patel (Birla Institute of Technology, India)
Engineering Science Reference • © 2018 • 179pp • H/C (ISBN: 9781522536833) • US $145.00

Spatial Analysis Techniques Using MyGeoffice®
João Garrott Marques Negreiros (University of Saint Joseph, China & Macau Polytechnic Institute, China)
Engineering Science Reference • © 2018 • 336pp • H/C (ISBN: 9781522532705) • US $225.00

For an entire list of titles in this series, please visit:
http://www.igi-global.com/book-series/advances-geospatial-technologies/73686

IGI Global
PUBLISHER of TIMELY KNOWLEDGE

701 East Chocolate Avenue, Hershey, PA 17033, USA
Tel: 717-533-8845 x100 • Fax: 717-533-8661
E-Mail: cust@igi-global.com • www.igi-global.com

Editorial Advisory Board

Table of Contents

Detailed Table of Contents

The emergence of alienated patch in the periphery of the city or fragmentation of the main city are the results of irresponsible and poor planning. This global problem of sprawl is strengthening even more with the hasty pace of urbanization. Despite the existing policies and regulations, it is a huge failure to control the sprawl. Hence, city planners and policy makers need to be more efficient in designing the cities to achieve sustainable development goals. For that purpose, adequate and informative data of the urban morphology, growth pattern, sprawl characteristics are required. Geospatial technology is a cost-effective measure and best among currently available techniques for collecting real-time/near real-time geographical data of the entire globe. The geographic information system (GIS) provides numerous tools for assessment of multidimensionality of urban sprawl. This chapter discusses various urban models, different forms of urban expansion, and a few existing methods to quantify sprawl.

To understand the territory of fast-growing cities, where there are multiple stakeholders involved, the observation of such dynamics seems indispensable to formulate and implement policies and actions based on a better understanding of these territorial systems. This chapter offers a perspective on how urban territories should be observed through geoinformation technologies that can provide a means for creating monitoring indicators concerning land use of fast-growing cities. The city of Maputo was used as an experimental laboratory for the use of geoinformation technologies in the observation, discussion, and reflection on methods for sustainable urbanism. The discussion includes the implications of the case study and possible developments to take a step forward in land use planning processes to achieve the desired socio-spatial equality.

In this chapter, a hybrid approach integrating cellular automata (CA), fuzzy logic, logistic regression, and Markov chains for modelling and prediction of land-use and land-cover (LULC) change at the local scale, using geographic information with fine spatial resolution is presented. A spatial logistic regression model was applied to determine the transition rules that were used by a conventional CA model. The overall dimension of LULC change was estimated using a Markov chain model. The proposed CA-based model (termed CAMLucc) in combination with physical variables and land-use planning data was applied to simulate LULC change in Portimão, Portugal between 1947 and 2010 and to predict its future spatial patterns for 2020 and 2025. The main results of this research show that Portimão has been facing massive growth in artificial surfaces, particularly near the main urban settlements and along the coastal area, and reveal an early and intensive urban sprawl over time.

This chapter discusses urban mobility considering two main analyses approaches. Based on the relationship between mobility and vulnerability, the first approach analyzed commuter's vulnerability using basin as unit of analysis. The second one analyzes variables related to land use such as population density and its relation with job offer in the city and people's income using traffic zones as unit of analysis. The two scales dialogue and can be used concurrently. The municipality of São José dos Campos (Brazil) was used as a case study. Origin-destination research was the main database used in the analyses. Authors used geospatial tools, like spatial join operation and thematic maps, which enable the in-depth analysis of important data for urban studies or transport planning and can be replicated in any study area. The analysis of mobility data aggregated by basin contributed to an understanding of the implications of the urban configuration, with its displacement patterns related to water courses if any flooding or landslide occurs and interrupts people's flow.

Spatial modeling always involves choices. The existence of constraints, the uncertainty and even the reliability of the data, the purposes and the applications of the studies make these reflections a kind of guiding compass for GIS analysts. Building on a previous exercise of data acquisition (check-ins) based on two digital social networks (DSN – Facebook and Foursquare) and on the awareness of the use of volunteered geographic information (VGI) generated by tourists through DSN, this work aims to evaluate the contribution of spatial analysis applied to urban tourism in the "Alta and University of Coimbra" area. Concepts and procedural tasks related to density determination, cluster analysis, and identification of patterns have thus been implemented with the purpose of evaluating and comparing the results obtained through the application of two techniques of spatial analysis, kernel density estimation (KDE) and optimized hot spot analysis (OHSA) and inverse distance weighting (IDW) interpolation.

Urban transition is an unstoppable process. Globally, several planning measures are taken by the city and country administration to control the sprawling process. Despite all the planning, most of the cities experience appreciable impact of urbanization on the localized weather parameters. This chapter summarizes the understanding relating to urban modification of localized weather, that is, changes in precipitation, temperature, and wind speed in the form of increase or decrease, their spatio-temportal distribution, urban heat island (UHI), and urban wind island (UWI). The impacts of the urbanization are primarily because of changes in land-surface characteristics due to the alteration of land use in a city. The urbanization effects on local or mesoscale weather could be studied both through observations and/or numerical modeling. The purpose of this chapter is to provide a review of most of the relevant studies carried out globally and with a special emphasis on India.

Swagata Ghosh, Amity Institute of Geoinformatics and Remote Sensing (AIGIRS), Amity University, Noida, India
Krishna Vidhata N., Amity Institute of Geoinformatics and Remote Sensing (AIGIRS), Amity University, Noida, India
Sunil Kumar, Amity Institute of Geoinformatics and Remote Sensing (AIGIRS), Amity University, Noida, India
Kousik Midya, Department of Electronics and Communication Engineering, Ajay Kumar Garg Engineering College, Ghaziabad, India

The chapter has highlighted the adverse impact of conversion of natural land cover into urban concrete over inter-seasonal variation of land surface temperature (LST) in Faridabad district which is a major threat for sustainable urbanism. Apart from high LST in the dense urban area built-up in Faridabad city, inter-seasonal variation of LST has been observed in dry deciduous forested areas due to defoliation, fallow land, and over-grazed land in rural surrounding areas. Compared to NDVI, NDBI has significant positive and stable correlation with LST in all seasons (Pearson index: 0.35 to 0.60). Weaker correlation (Pearson index: 0.02 to 0.48) between NDVI and LST accounts for the seasonal impact over NDVI due to defoliation and agricultural practices over the study area. Overall, it can be remarked that image-based spectral indices and thermal band can be used for the evaluating thermal environmental contrast across seasons. Use of in situ measurement with good network of meteorological stations can validate satellite-derived LST better and increase the accuracy of the study.

Urban greening is an essential dimension of sustainable urban development. Through green planning, cities can develop strategies towards climate change and mitigation actions and reduce emissions. This chapter aims to evaluate the combined use of remote sensing data and GIS to produce different levels of sustainable indicators, starting from the city level to the neighborhood and the building levels. Green roofs can be part of the solution through the ecosystem services provided. Its benefits are evaluated for the different urban scales through scenario analysis. The impact in the outdoor thermal comfort is assessed at the neighborhood level, while the overall potential to improve energy efficiency is evaluated at the building level. The methodology is implemented in the city of Lisbon, Portugal through a pilot case study. The results can provide rationales for the city when formulating new incentives to encourage the uptake of green roofs in Lisbon.

Vegetation is an excellent strategy to tackle global warming and climate change. In this context, the green wall is a kind of green infrastructure that is used as a bioclimatic strategy. Also, it can be used as an approach to insert green areas in an urban context, contributing to urban areas to become more sustainable. This chapter aims to evaluate the thermal performance of green walls using three different native climbing vegetation from the Cerrado biome. Thus, the authors used a computational tool named ENVI-met (v.3) as a method to analyze and evaluate which plants would bring the best performance in terms of improving thermal comfort. The database the authors obtained from the simulations will serve as a comparison of the thermal performance between walls without vegetation cover, as well as clarify the morphological and physiological characteristics that influence its thermal performance. The results demonstrate that Arrabidaea pulchra species enabled

This chapter outlines the experience of the Épura Research and Extension Group of the Federal University of Mato Grosso for the identification, characterization, and demarcation of informal settlements in Cuiabá (Brazil). This involved cooperating with the local residents' associations and public authorities in recognizing local demands and the need to open up urban land regularization processes. The cartographic issue is central to the recognition and legitimization of these territories and the use of geoinformation technologies, namely geographic information systems and unmanned aerial vehicles, can increase the speed and accuracy of the mapping of the neighborhoods. Additionally, the demands for urban and housing improvements and provision of space for everyday activities can be highlighted through the visualization and measuring tools that are embodied in these technologies. Furthermore, the experience requires the university to play a role in seeking more participatory and proactive practices within urban planning processes.

The creation of 3D models of urban elements is extremely relevant for urbanists constituting digital archives and being especially useful for enriching maps and databases or reconstructing and analyzing objects/areas through time, building/recreating scenarios and implementing intuitive methods of interaction. The widespread data available online offer new opportunities to generate realistic 3D models without the need to go physically to the place. This chapter aims to demonstrate the potential 3D modeling and visualization/interaction of urban elements in the city for multiple purposes, and it is organized in four main topics: The first deals with the theoretical framework regarding the bases of the human perception of the spatial environment and the importance of 3D modelling. The second and third deal with technical procedures on terrestrial/aerial data acquisition and demonstrate alternatively data gathered online to generate 3D models for the visualization of urban elements of the city, and the fourth introduces 3D model visualization within an augmented reality environment.

In this chapter, the authors are presenting opportunities for the use of unmanned aerial vehicles (UAV) in town. Methods for the optimization of flight routes of UAVs in the dependence of target tasks in the city are presented, for example, area monitoring; search and rescue operations; retransmission of communication (in places, where the antenna coverage cannot be set due to terrain specifications); organization of logistics as the safe, cheap, and fast transportation method of goods; for aerial photography, for controlling traffic; for the provision of the first aid to people in emergencies; unmanned taxi. It is done using air navigation information and mathematical methods. Authors suggest dynamic programming methods, GRID analyses, expert judgment method, and fuzzy-logic methods for estimation of risk/safety of flights in the city. Optimization of flows and flexible redistribution of UAV routes in multilevel airspace is provided according to air navigation requirements and standards.

UAV photogrammetry has simplified the measurement of the terrain's surface; however, the accuracy of the method is still a matter of research as it depends on multiple factors such as the UAV employed and the decisions made by the user during information acquisition and processing. This chapter analyses the accuracy of 27 digital elevation models (DEMs) built from the data acquired with a DJI Phantom 3 drone, which was flown at different heights and trajectories in a case study in the Colombian Department of Antioquia. The statistical analysis confirmed that the use of ground control points (GCPs) is fundamental to increase the accuracy of the map products. It also showed that accuracy, understood as the comparison between the coordinates of points measured in the field (CPs) with their corresponding values estimated in the orthophotomosaics, depends on the number of CPs and their location. An insufficient number of CPs or an inadequate distribution, for instance all or most of them located in a single land cover, will cause the accuracy to be overestimated.

A case study drone that constitutes a shock-resistant aerial vehicle is discussed in the chapter. The aerial motor platform is placed in gimbal joints of the exclusive framework (shell). The platform is a helicopter type aerial vehicle powered with two coaxial rotors of contra rotation. Mathematical model of the platform spatial dynamics bases Lagrange's equations to bring reliable solutions so that advanced model-based control law design techniques can be used. Though the case study implies utilizing an automatic flight mode of the aerial vehicle, it can be piloted remotely on radio. The on-board video cameras and other sensors are used to bring about both navigational duties and surveillance missions such as building constructions monitoring.

Urbanism has mainly used 2D data for both the urban analysis and diagnosis and the presentation of proposals for changes in the whole city or parts of the city. Even regarding the production of urban indicators, using, for example, the quantification of the existing green area in relation to the resident population, this practice is regularly based on the area and rarely on volume. This situation is mainly justified by the sluggishness and costs associated with obtaining 3D data. The recent development of data collection technology by unmanned aerial vehicles has triggered a change in this scenario. This chapter presents the UAV data acquisition and processing chain, analyses the positional accuracy of UAV data processing performed with GCP measurements obtained from GNSS, demonstrates how positional accuracy assessment and UAV workflow's quality control are relevant for ensuring the accuracy of derived UAV geospatial products, and demonstrates the usability of 3D models in a theoretical 3D urbanism context.

Foreword

Technology as a Tool to Support Sustainable Urban Development

Along with industrialization, the urbanization process has become a dominant trend across the globe at different levels and paces. Different factors have driven urbanization. Across many decades, new technologies have been essential for economic specialization, better living and working conditions, and improved mobility, health, and communication. The quality of life and public health has improved in general but often accompanied by social polarization and socio-spatial fragmentation in cities and metropolises. For some decades, emerging technologies have been accelerating urbanization processes through economic restructuring and new agglomeration effects. Nowadays, more than half of the global population lives in urbanized areas, and this share will further increase to about 68% of the world's population by 2050, according to UN forecasts.

This global urbanization process is one of the driving forces endangering sustainable development as it is rather strongly connected to economic growth and the growing consumption of energy, as well as the production of emissions. Thus, urban agglomerations, cities, and metropolises have become an essential driver of emissions with their problematic effects on global temperature and climate change. In view of these realities, we can conclude that our way of working and living, our development model is still too attached to emissions as they keep on increasing globally. However, several countries, even those with higher shares of urbanization rates, have shown some success in decoupling socio-economic development from the production of emissions because of their climate-friendly policies that have strengthened sustainable development. Obviously, specific policies supported by new technologies do have a positive impact. Very recently, Ursula von der Leyen, the President of the EU-Commission, announced the "Green Deal", emphasizing that climate-friendly industries and new technologies will displace old growth models based on fossil, non-renewable energies.

This assumption that technologies will become crucial for sustainable development is not new. There are some reasons why technology provides new chances but also carries new/old risks. On the one hand, new technologies have improved significantly (remote) sensing, measuring and monitoring urbanization trends at different scales. Advanced tools and methods allow the analysis of complex system dynamics, enable decision-making in real-time conditions, support decision-making processes through sophisticated modelling and simulation, and finally provide good possibilities for the visualization and communication of results. Thus, modern technologies enable the exploration and explanation of new and actual urbanization trends, identify complex dynamics of socio-ecological systems, allow the monitoring of impacts, and support learning processes. In particular, new and smart devices enable the construction of new and precise indicators to support sustainable development strategies. However, on the other hand, new technologies have created new and more activities, changed lifestyles, and enforced the so-called rebound effects, which are clearly jeopardizing sustainability.

Therefore, new technologies may have a positive or negative impact, but they have improved the quality of scientific research on sustainable development. Similar to the smart city debate, it remains the question of which policies to implement and how such tools and results should be used predominantly. Is it only to increase the efficiency of existing systems that have rebound effects? Do new technologies have a socially polarizing effect because of their exclusive character? Who will be the owner of the new data, and for whom is that kind of evidence produced?

To cope with these challenges, the scientific community has not only been asked to use new technologies to obtain more precise information but also to use them in an inclusive way enabling learning processes and participatory decision-making across all groups and interests. Only in that way, technology will become the fourth pillar of sustainability.

Rudolf Giffinger
TU Wien, Austria
Vienna, 14th March 2021

Rudolf Giffinger *is a Full Professor of Regional Science and the Head of the Centre of Urban and Regional Studies, which is part of the Institute of Spatial Planning at the TU Wien, Austria. He is an expert in urban and regional development research under globalized conditions and addresses his strategic advice activities on concepts of smartness, sustainability, and resilience of cities.*

Preface

Urban sustainability is on the global political and scientific agenda. In 2015, the United Nations (UN) member states adopted 17 Sustainable Development Goals (SDGs). These goals convey a global purpose: achieving a more sustainable future for everyone. So much so that the UN has translated this great purpose into the sentence, "Take Action for the Sustainable Development Goals.". "Take Action" seems to indicate a willingness to steer political discourse into action progressively. Eradicating poverty and protecting the planet, achieving peace and prosperity for everyone are the SDGs' fundamental pillars for the 2030 horizon based on science, technology, creativity, and global financial resources.

Goal 11 is geared towards "Sustainable cities and communities". According to the UN, it is an ambitious goal, but absolutely justifiable considering that by 2050 the world urban population will have reached 6.5 billion people. The problems arising from urbanization globally are now more than ever at the centre of the debate in prominent international organizations. The management of future cities will be guided by inclusion, security, citizen participation, resilience, and sustainability. Cities will have to go through processes of replacing polluting transport with more environmentally friendly options. The improvement of living conditions in cities will have to be coupled with nature-based solutions, including increased green spaces and the progressive improvement of energetic efficiency. Urban planning should enlist technology to design a better future on a global, regional, and local scale.

More and more countries are including the SDGs in their urban policy agendas. Meanwhile, an alternative economic model based on the circular economy is being developed (albeit very timidly). Urban management can and indeed must increase the coordination of different production and consumption systems in closed circuits from which new business models will emerge. Urban metabolism stands to gain from the circular economy because it creates new business opportunities, promotes wealth, maintains the added value of urban products for more extended periods, and encourages new forms of production and consumption. The city of the future, both a consumer and producer of goods and services, may have a more balanced metabolism if the international political agenda is implemented with all UN member states' combined efforts.

Many countries have been incorporating the 17 Sustainable Development Goals into publicly funded research agendas. In competitive public tenders, tender documents require researchers to identify the SDGs that can directly benefit from the funded research. For example, the European Commission has broad investment guidelines for research and innovation inspired by the SDG agenda. It acknowledges that universities can be the driving force behind the European strategic plan in this regard. In this context, universities, which are places of innovation par excellence, should remain connected to the region in which they are located, contributing to meet the goal of "Sustainable cities and communities" while also promoting "smart specialization and innovation ecosystems". Therefore, universities as centres of knowledge production must develop strategies to become more engaged with the territories and local communities to meet the goals of sustainable development progressively.

Sustainable urbanism reflects the global concerns briefly described here.

The concept of *sustainable urbanism* originated in Cambridge University (United Kingdom) in the 1990s, and its creation has been attributed to Susan Owens. Despite being a budding scientific field and a recent praxis, *sustainable urbanism* is currently on the agenda of university teaching and scientific research. Texas A&M University is credited with having devised the first undergraduate degree officially called "Sustainable Urbanism" in 2002. Since then, there has been a proliferation of conferences, symposia, and workshops dedicated to this wide-ranging subject, extending far beyond the "classic" design of the urban form. The large number of books and papers that have been published in scientific journals shows just how important the topic has become. This has helped to consolidate an idea of urbanism with multiple dimensions and a central concern: sustainability. In 2007, Douglas Farr published a book entitled *Sustainable Urbanism: Urban Design With Nature* and in the following years, several critical reviews of the work were published. This is one of the references for those that wish to learn about an idea of urban life on Earth, of how to build cities, which has progressively inspired the planning praxis.

Green Urbanism was the concept used by Timothy Beatley in 2000, embedding it in the title *Green Urbanism. Learning from European Cities*. This book presents the theoretical principles, discusses the methodologies, and reveals the European case studies. This concept bears similarity to the idea of *sustainable urbanism* but clearly prioritizes the trilogy "zero fossil-fuel energy use, zero waste, and zero emissions". Steffen Lehmann (2010) uses the concept in a fascinating article entitled *Green Urbanism: Formulating a Series of Holistic Principles*. In 2017, Puay Yok Tan and Chi Yung Jim edited *Greening Cities. Forms and Functions* using the term *urban greenery* but the concept of *sustainable urbanism* appears to be marginal.

In all these cases, the relationship between technology and urbanism is very little discussed.

The relationship between urbanism and technology has not been much explored, although some indirect references to this relationship can be found in the literature. What is lacking in the scientific literature is the conceptual discussion of the relationship between the modern designation of *sustainable urbanism* and geospatial technologies and methods (remote sensing, geographic information systems, web mapping technologies, 3D point cloud, spatial modelling, etc.). But why is it necessary to discuss this relationship? Because megacities are a reality, much like the environmental pollution they cause (pollution: sanitation, air, and water). These disturbances can be modelled in computational environments with data obtained by modern technology (e.g. by UAV) to monitor cities almost in real-time and simulate risk situations, suggesting solutions for the future. These solutions can be inspired by the theoretical principles of *sustainable urbanism* as developed by the various Schools. We can state that urban praxis, namely urban planning and urbanism, cannot exist without the support of geospatial technologies. Geospatial technologies allow the operationalization of the concept of *sustainable urbanism*. This is the thesis defended in this book.

Present and future urbanism consider the physical, ecological, economic, social, health, and equity factors as critical factors for constructing the city from a sustainable and circular economy perspective. We believe that this construction should be based on the opportunities brought to us by the progress of geospatial technology.

This book's title may be defined as follows: *sustainable urbanism* is the modern practice of planning the city based on principles that can be operationalized using geospatial technology. This technology consists not only of "artefacts" created by human beings (satellites, UAVs, computers, parallel processing, computing, among others) but considers each person as a sensor that produces and/or shares data in real-time (e.g. in social networks). These data, in some cases big data, can help design the city of the future based on sustainable urban design principles (compact urban growth, transport-driven urban development, mixed land uses, appropriate sizing of urban blocks, greener public spaces, soft mobility, public transport, regulation of motor vehicles, green buildings, renewable energy, efficient waste management, efficient water management).

As mentioned above, the book's overview is guided by the fusion of ideas between *sustainable urbanism* and geospatial technology, highlighting some methods and applications. Thus, the book focuses on demonstrating the effectiveness of technology in implementing this critical concept: urban sustainability. The authors of the 18 chapters that comprise this book come from very diverse scientific and technical areas: urbanism (with a vision of *sustainable urbanism*), urban planning, architecture, geospatial technology, engineering, geography, computer science, and social sciences and humanities.

The contents of the book's chapters can be summarized in this cloud of terms, originated from the abstracts. As expected, terms related to geospatial technology (UAV, GIS, etc.) and methods (analysis, models, modelling, etc.) coupled with data underpin the fundamental concepts of this book (sustainable urbanism, planning, urban, green solutions).

Figure 1.

As editors of this book, which has contributions from authors from several parts of the world, we are convinced that it will increase the knowledge of the cross-relationship between urbanism, geospatial technology, urban spatial data and urban sustainability. Methods and applications are present in most of the chapters. Reading them will provide urbanists, urban planners, local government agencies, public administration, companies, urban designers, researchers, professors, and students in urbanism, urban planning, computer science, geography, engineering, among others, with increased knowledge and insights into new forms of looking at the cities of the future from a technological standpoint.

Chapter 1, written by Srutisudha Mohanty, Jagabandhu Panda and Sudhansu Rath, is an example of the relationship between urban growth and GIS technology. As the authors point out "the emergence of alienated patch in the periphery of the city or fragmentation of the main city are the results of irresponsible and poor planning. This global problem of sprawl is strengthening even more with the hasty pace of urbanization. Despite the existing policies and regulations, it is a huge failure to control the sprawl. Hence, city planners and policy makers need to be more efficient in designing the cities to achieve sustainable development goals. For that purpose, adequate and informative data of the urban morphology, growth pattern, and sprawl characteristics are required. Geospatial technology is a cost-effective measure and best among currently available techniques for collecting real-time/near real-time geographical data of the entire globe. Geographic information systems (GIS) provide numerous tools for assessment of multidimensionality of urban sprawl. "

Chapter 2, written by Cristina Delgado Henriques, conveys a significant concern: linking research to action in the field of sustainable urban development. Her thesis explicitly addresses the role she assigns to geotechnologies and spatial data to implement the urban observation of Maputo, Mozambique. Indeed, according to the author, "this chapter offers a perspective on how urban territories should be observed through geoinformation technologies that can provide a means for creating monitoring indicators concerning land use of fast-growing cities. The city of Maputo was used as an experimental laboratory for the use of geoinformation technologies in the observation, discussion, and reflection on methods for sustainable urbanism. The discussion includes the implications of the case study and possible developments to take a step forward in land use planning processes to achieve the desired socio-spatial equality."

Chapter 3, written by Raquel Faria de Deus, José António Tenedório and Jorge Rocha, presents a proposal for modelling land use and land cover change in a geographical coastal area in Portugal. According to the authors, "in this chapter, a hybrid approach integrating Cellular Automata (CA), fuzzy logic, logistic regression and Markov chains for modelling and prediction of land-use and land-cover (LULC) change at the local scale, using geographic information with fine spatial resolution is presented. A spatial logistic regression model was applied to determine the transition rules that were used by a conventional CA model. The overall dimension of LULC change was estimated using a Markov chain model. The proposed CA-based model (termed CAMLucc) in combination with physical variables and land-use planning data was applied to simulate LULC change in Portimão, Portugal, between 1947 and 2010 and to predict its future spatial patterns for 2020 and 2025. The main results of this research show that Portimão has been facing massive growth in artificial surfaces, particularly near the main urban settlements and along the coastal area, and reveal an early and intensive urban sprawl over time."

Chapter 4, authored by Rui Pedro Juliao, Amilton Amorim, João Paulo Hespanha, Guilherme Barros de Souza, Ronaldo Messias Correia and Rafael Destro, is dedicated to revealing new methodological trends in the acquisition and management of geographic data in small municipalities. According to the chapter's authors, "promoting and managing sustainable territorial development requires adequate tools that enable territorial managers to formulate appropriate choices. Technological solutions have emerged, and the paradigm has shifted from isolated GIS to more collaborative production and dissemination of geographic data using Spatial Data Infrastructures (SDI). A critical dataset for municipal land administration is cadastre. ISO 19152 standard of the Land Administration Domain Model (LADM) was published in 2012. Also, technology evolution, namely Unmanned Aerial Vehicles (UAV), has changed data acquisition for cadastre purposes. These are the three pillars of modern territorial management: Openness, co-production and data sharing (SDI); Models (LADM); Affordable technology (UAV). This chapter presents how municipalities can develop an SDI project incorporating LADM guidelines and UAV data acquisition. The case study is based on a group of 32 small municipalities from São Paulo state, Brazil, known as UNIPONTAL."

Chapter 5, written by Lívia Tomás, Maria Carolina Jurema, Janaina dos Santos, Luciana Londe, Regina Reani, Claudia de Linhares and Leonardo Santos, focuses on urban mobility in the municipality of São José dos Campos (Brazil) following an interesting change of geographical scale. The authors state that this chapter "discusses urban mobility considering two main analysing approaches. Based on the relationship between mobility and vulnerability, the first approach analysed commuter's vulnerability using basin as a unit of analysis. The second one analysed variables related to land use, such as population density and its relation with job offers in the city and people's income using Traffic Zones as a unit of analysis. (...) The authors used geospatial tools, such as spatial join operation and thematic maps, which enable the in-depth analysis of important data for urban studies or transport planning and can be replicated in any study area."

In Chapter 6, José Gomes dos Santos, Liliana Azevedo and Luís Leitão, use social network data and spatial statistics to reflect on urban tourism in the *Alta* and University of Coimbra. The chapter abstract says "spatial modeling always involves choices. The existence of constraints, the uncertainty and even the reliability of the data, the purposes and the applications of the studies, make these reflections a kind of guiding compass for GIS analysts. Building on a previous exercise of data acquisition (check-ins) based on two Digital Social Networks (DSN — Facebook and Foursquare) and on the awareness of the use of Volunteered Geographic Information (VGI) generated by tourists through DSN, this work aims to evaluate the contribution of spatial analysis applied to urban tourism (…). Concepts and procedural tasks related to density determination, cluster analysis and identification of patterns have thus been implemented with the purpose of evaluating and comparing the results obtained through the application of two techniques of spatial analysis, Kernel Density Estimation (KDE) and Optimized Hot Spot Analysis (OHSA) & Inverse Distance Weighting (IDW) Interpolation."

Chapter 7, authored by Sudhansu Rath, Jagabandhu Panda and Srutisudha Mohanty, reviews and discusses urban transitions' impacts. According to the authors, "this chapter summarizes the understanding relating to urban modification of localized weather that is changes in precipitation, temperature and wind speed in the form of increase or decrease, their spatiotemporal distribution, Urban Heat Island (UHI) and Urban Wind Island (UWI). The impacts of the urbanization are primarily because of changes in land-surface characteristics due to the alteration of land use in a city. The urbanization effects on local or mesoscale weather could be studied both through observations and/or numerical modeling."

Chapter 8, written by Swagata Ghosh, Krishna Vidhata N., Sunil Kumar and Kousik Midya, uses remotely sensed data for seasonal contrasts of land surface temperature in a district in India, noting that the chapter has "highlighted the adverse impact of conversion of natural land cover into urban concrete over inter-seasonal variation of land surface temperature (LST) in Faridabad district which is major threat for sustainable urbanism. Apart from high LST in dense urban built-up in Faridabad city, inter-seasonal variation of LST has been observed in dry deciduous forested part due to defoliation, fallow land, and over-grazed land in rural surrounding. Compared to NDVI, NDBI has significant positive and stable correlation with LST in all seasons (Pearson index: 0.35 to 0.60). Weaker correlation (Pearson index: 0.02 to 0.48) between NDVI and LST accounts for the seasonal impact over NDVI due to defoliation and agricultural practices over the study area. Overall, it can be remarked that image-based spectral indices and thermal band can be used for evaluating thermal environmental contrast across seasons."

Chapter 9, authored by Teresa Santos, Caio Silva, José António Tenedório and Thiago Montenegro Góes, focuses on the demonstration of the usefulness of GIS spatial analysis and modelling to identify potential green roofs in the city of Lisbon. The authors contextualize the method within the urban greening framework by referring that "this chapter aims to evaluate the combined use of remote sensing data and GIS to produce different levels of sustainable indicators, starting from the city level to the neighborhood and the building levels. Green roofs can be part of the solution through the ecosystem services provided. Its benefits are evaluated for the different urban scales through scenario analysis. The impact on the outdoor thermal comfort is assessed at the neighborhood level, while the overall potential to improve energy efficiency is evaluated at the building level. The results can provide rationales for the city when formulating new incentives to encourage the uptake of green roofs in Lisbon."

Chapter 10, written by Pedro Sorte, Caio Silva, Abner Calixter and Marta Romero, emphasizes the vegetation's role within urban environmental strategies. The authors argue that "vegetation is an excellent strategy to tackle global warming and climate change. In this context, the green wall is a kind of green infrastructure used as a bioclimatic strategy. This chapter aims to evaluate the thermal performance of green walls using three different native climbing vegetation from the Cerrado biome. Thus, the authors used a computational tool named ENVI-met (v.3) as a method to analyze and evaluate which plants would bring the best performance in terms of improving thermal comfort. The results demonstrate that Arrabidaea pulchra enabled green walls with lower temperatures and higher air humidity rate in most situations."

In Chapter 11, David Green, Billy Gregory, Jason Hagon, Alex Karachok, Jakob Larsen and Alastair Skitmore demonstrate UAV's potential to study built environments. The authors produced an overview which they summarize as follows: "the 'built environment' includes rural, urban, and underwater environments. Together with low-cost image processing and softcopy photogrammetry, fixed-wing and multi-rotor UAVs can collect a wide range of imagery for generating 3D models of individual buildings, record and analyse architecture and infrastructure, and terrain models. Consideration is given to non-imaging sensors carried on UAVs. Input to GIS provides a basis to create visually realistic models of the 'built-environment' for urban and rural planning and decision-making for sustainable urbanisation. 3D visualisation software, Virtual, and Augmented Reality will allow public engagement with the spatial planning process. Safety and operational considerations are needed for UAV flights. The chapter concludes by examining how this technology will develop in the future to play a role in sustainable urbanization."

In Chapter 12, Josep Roca and Blanca Arellano apply UAV data to the identification and classification of dense green spaces in an area of Barcelona, stating that "the objective of this chapter is to show the usefulness of conventional UAVs for the identification, inventory and classification of trees in the context of dense green spaces. The aim is to demonstrate the potential of low-cost drones (with traditional red, green, blue [RGB] sensors) to identify and classify trees in public parks. A case study is discussed on Turó Parc in Barcelona, in which a 3D model was developed and an exercise to identify and classify the vegetation was carried out using the information provided by a UAV. The example confirms that conventional drones could be useful for studying green urban spaces characterized by a high density of plant species. Non-professional UAVs have a potential that should not be undervalued, as they enable three-dimensional point clouds to be obtained of high spatial density."

In Chapter 13, Andrea Figueiredo Arruda, Cristina Delgado Henriques and Claudio Miranda start with a group experience and fieldwork to discuss the importance of counter-mapping. In short, according to the authors, "although advances have been made in policies and programs, as well as in Brazilian legislation with regard to informal settlements in urban areas, there remain serious gaps in our understanding. The cartographic issue is central to the recognition and legitimization of these territories and the use of geoinformation technologies, namely Geographic Information Systems and Unmanned Aerial Vehicles, can increase the speed and accuracy of the mapping of the neighborhoods."

Chapter 14, written by Luis Marques and Josep Roca, deals with 3D modelling of photographs taken by tourists of historical monuments in some European cities and shared on social networks. According to the authors, "the creation of 3D models of urban elements is extremely relevant for urbanists constituting digital archives and being especially useful for enriching maps and databases or reconstructing and analyzing objects/areas through time, building/recreating scenarios, and implementing intuitive methods of interaction. The widespread data available online offer new opportunities to generate realistic 3D models without the need to go physically to the place."

Chapter 15, written by Tetiana Shmelova, Vitalii Lazorenko and Oleksandr Burlaka, discusses the relationship between UAVs and smart cities. According to the abstract, in this chapter, "the authors are presenting opportunities for the use of Unmanned Aerial Vehicles (UAV)'s in town. Methods for the optimization of UAV flight routes in the dependence of target tasks in the city are presented, for example, area monitoring; search and rescue operations; retransmission of communication (in places where the antenna coverage cannot be set due to terrain specifications); organization of logistics as the safe, cheap and fast transportation method of goods; for aerial photography, for controlling traffic, for the provision of the first aid to people in emergencies; unmanned taxi. It is done using air navigation information and mathematical methods."

Chapter 16, authored by Engelberth Soto-Estrada, Ann Wellens, Andrés Pava-Restrepo and Camilo Hernández Mejía, focuses on the accuracy of UAV-based digital elevation models. According to the authors, "this chapter analyses the accuracy of 27 digital elevation models (DEMs) built from the data acquired with a DJI Phantom 3 drone, which was flown at different heights and trajectories in a case study in the Colombian department of Antioquia. The statistical analysis confirmed that the use of ground control points (GCPs) is fundamental to increase the accuracy of the map products."

Chapter 17, written by Konstantin Dergachov and Anatolii Kulik, deals with impact-resistant flying platforms to be used in urban construction monitoring, as the title states. The authors summarize their work as follows: "a case study drone that constitutes a shock-resistant aerial vehicle is discussed in the chapter. The mathematical model of the platform spatial dynamics bases Lagrange's equations to bring reliable solutions to use advanced model-based control law design techniques. Though the case study implies utilizing an automatic flight mode of the aerial vehicle, it can be piloted remotely on the radio."

Chapter 18, written by Rossana Estanqueiro, José António Tenedório, Carla Rebelo and João Marques, discusses the concept of 3D urbanism theoretically and suggests that this concept may be fundamental in the theory and practice of future urbanism, associated with other dimensions (4D, 5D). As stated in the abstract, "urbanism has mainly used 2D data for both the urban analysis and diagnosis and the presentation of proposals for changes in the whole city or parts of the city. Even regarding the production of urban indicators, using, e.g., the quantification of the existing green area in relation to the resident population – this practice is regularly based on the area and rarely on volume. This chapter presents the UAV data acquisition and processing chain; analyses the positional accuracy of UAV data processing performed with GCP measurements obtained from GNSS; demonstrates how positional accuracy assessment and UAV workflow's quality control are relevant for ensuring the accuracy of derived UAV geospatial products; and demonstrates the usability of 3D models in a theoretical 3D urbanism context."

This book brings together contributions from many experts engaged in research and university education in the cross-cutting field of *sustainable urbanism* and technology. In addition to demonstrating the fundamental importance of geospatial technology in urbanism as a tool for obtaining and analyzing spatial information, this book contributes to the progress of research methods that place scientific evidence in the place it deserves to influence policy-making.

José António Tenedório
Universidade NOVA de Lisboa, Portugal

Rossana Estanqueiro
Universidade NOVA de Lisboa, Portugal

Cristina Delgado Henriques
Universidade de Lisboa, Portugal

Acknowledgment

The editors are most grateful to all the people involved in this project. Without their support, this book would not have been possible. Our sincere gratitude goes to each one of the chapter's authors who contributed their time and knowledge to this book. The editors also wish to acknowledge the valuable contributions of the reviewers regarding the improvement of quality, coherence, and content presentation of chapters. Some of the authors also served as referees; we highly appreciate their double task. The editors would like to express their thankfulness to the Editorial Advisory Board for taking the time to share their expertise. Additionally, the editors would like to express their gratitude to all that provide valuable assistance such as technical editing and revising the manuscripts for grammar and syntax.

José António Tenedório
Universidade NOVA de Lisboa, Portugal

Rossana Estanqueiro
Universidade NOVA de Lisboa, Portugal

Cristina Delgado Henriques
Universidade de Lisboa, Portugal

Chapter 1
Geospatial Technology in Urban Sprawl Assessment:
A Review

Srutisudha Mohanty
https://orcid.org/0000-0002-8642-6234
National Institute of Technology, Rourkela, India

Jagabandhu Panda
https://orcid.org/0000-0002-4238-1820
National Institute of Technology, Rourkela, India

Sudhansu S. Rath
https://orcid.org/0000-0003-4529-3719
National Institute of Technology, Rourkela, India

ABSTRACT

The emergence of alienated patch in the periphery of the city or fragmentation of the main city are the results of irresponsible and poor planning. This global problem of sprawl is strengthening even more with the hasty pace of urbanization. Despite the existing policies and regulations, it is a huge failure to control the sprawl. Hence, city planners and policy makers need to be more efficient in designing the cities to achieve sustainable development goals. For that purpose, adequate and informative data of the urban morphology, growth pattern, sprawl characteristics are required. Geospatial technology is a cost-effective measure and best among currently available techniques for collecting real-time/near real-time geographical data of the entire globe. The geographic information system (GIS) provides numerous tools for assessment of multidimensionality of urban sprawl. This chapter discusses various urban models, different forms of urban expansion, and a few existing methods to quantify sprawl.

DOI: 10.4018/978-1-7998-2249-3.ch001

INTRODUCTION

Urbanization can be defined as the migration or shifting of rural populations towards the urban areas. The major driving factors for rural to urban transition are industrialization, employment, better standard of living, and social security benefits (Aithal & Ramachandra, 2016; Bhatta, 2009). In the 21st century, urbanization blooms as one of the biggest problems for many developing as well as developed countries all over the world (Jaeger et al., 2010; Muñoz, 2003). It is taking place at an unprecedented scale by creating profound impacts on the environment, economy, as well as the society (Sun et al., 2013). The foreseeable consequences of this process are the unplanned and uncontrolled spatial expansion of urban cores beyond their administrative boundaries in order to accommodate the growing population, which results in scattered patches or fringes (Marshall & Dolley, 2019). The sprawled cities are facing serious issues with respect to basic amenities, such as electricity, water, sewerage channels, transportation, and communication, along with a dramatic increase in pollution, poor health facilities, deteriorating educational standards, unemployment, and poverty. To overcome such issues, urban planners have to understand the dynamics of a particular urban area and the current urbanization process, which will solidify the basis of future predictions and guide in preparedness for urban sustainability (Anas et al., 1998).

There are three popular models for urban development, such as the concentric zone model (Burgess Model), the sector model (Hoyt Model), and the multiple nuclei model (Ullman Harris model). In the concentric zone model-based cities (e.g. Chicago, United State), multiple zones, such as industrial corridors, working-class housing, middle-class housing, and commuter zones with suburbs of higher-income-class housing are usually designed around a central business district (CBD). Whereas, the sector model-based cities (e.g. Sunderland, United Kingdom) are divided into multiple wedge-shaped (not ring shaped) sectors of transportation, industries, and low-medium-high-class residential areas designed around a central circular CBD. In both of these models, the CBD plays a vital role, however, in contrast, a less significant CBD is presented for the multiple nuclei model (e.g. Los Angeles and Houston). Multiple centres attract a particular kind of population, which is imitated by neighbourhood land use. Apart from these conceptual models, different countries develop their own theories and policies for urban development (Figure 1).

Figure 1. Global trends of urban morphology

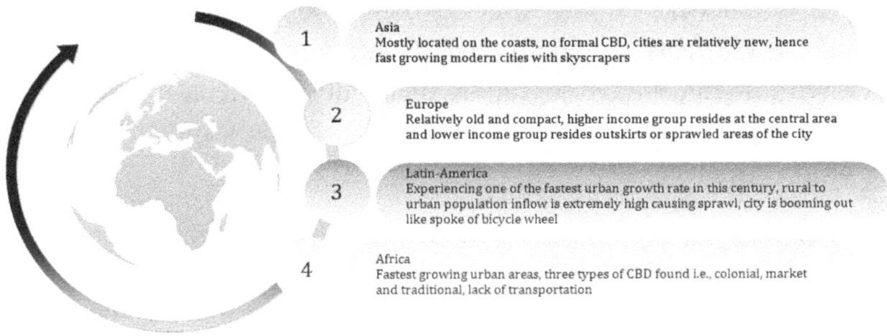

Several studies have been conducted to investigate the spatial heterogeneity and temporal distinctions of urbanizing trends in the small, as well as the megacities of the world collectively, and as individual continents i.e. for Asia (Enqvist et al., 2019, Friedmann & Sorensen, 2019; Masoudi et al., 2019; Lehner et al., 2016; Deuskar et al., 2015; Dutta, 2012; Gupta et al., 2012; Mohammad & Sidaway, 2012; Van Dijk, 2011; Griffiths et al., 2010; Kundu, 2011; 2006), Africa (Murphy & Carmody, 2019; Epstein, 1967; Little, 1960), and Europe, Latin America, the Caribbean, and North America (Dong et al., 2019; González & Medina, 2004). Hackenberg (1980) reported diffused and patchy patterns of urban expansion in South-Asian cities. However, Asian megacities in recent decades are experiencing a faster growth rate than the rest of the developed world. As a result, the peri-urban space around these cities of sprawl is of major concern due to irregular distribution of resources (Marshall & Dolley, 2019). Study of 129 European cities revealed very peculiar findings due to the process of re-urbanization and showed that economically revitalizing cities are experiencing disperse and thin centres with polycentric development (Salvati et al., 2019). On the other hand, smaller cities in Africa are comparatively less compact and more dispersed (Xu et al., 2019) due to physical constraints, poor urban policies, and irrepressible demography leading to sprawled urbanization and reduction in land use efficiency of the cities and their neighbouring area (Todes et al., 2010).

The major forms of urban expansion are of three types, such as development of newer patches adjacent to the edge of older ones, infilling the open spaces present in an existing urban area, or a completely new patch blooming out separately from the main city (Figure 2). Besides above generic forms, there are several other aspects such as land use, density of built-up area, shape etc. can be used for characterizing urbanization. Figure 3 represents the classification of spatial representation of urban expansion (Angel et al., 2005). Out of all these categories, decentralized leapfrogging is considered as sprawled development by most of the researchers. Various forms of urban expansions are mainly governed by the physical features, terrain, weather, transportation network, master plans and land use policies of the concerned city (Schneider & Woodcock, 2008). Angel et al. (2005) defined six driving forces, which directly or indirectly attribute towards swaying the urban expansion, effect of natural environment, the demographic distribution, effect of transport system, effects of economy, effects of consumer preferences for proximity to other amenities and effect of governance.

Figure 2. Different forms of urbanization (darker shades with dashed outline represents newer patches and solid line encircles older urban patches)

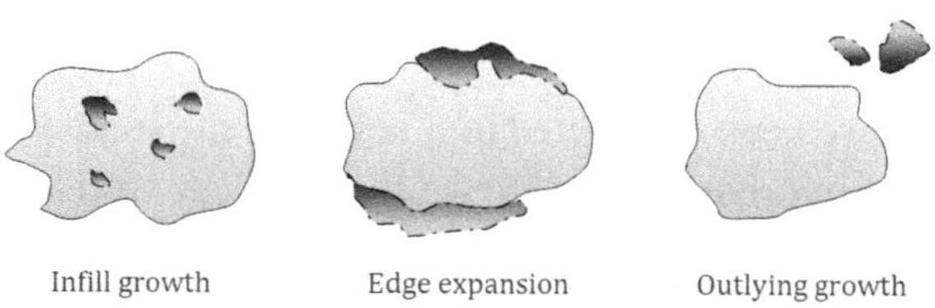

Primitive cities are mostly established near coasts (e.g., Kolkata, London, New York), which have vast space to grow (first linearly followed by edge expansion and infill growth), unlike many mountainous cities, which cannot expand unlimitedly. This is one of the consequences of urban expansion, which can be either a constructive type or an adverse type (Jenks et al., 1996). There are primarily two different views among researchers working on urban sprawl. One group that supports sprawling argues that low-density sprawl improves the economy of the city and creates new job opportunities (Hall, 2001; Newman & Kenworthy, 1992). Whereas, the anti-sprawl policymakers state that leapfrog development will decrease the open space and lengthen the transport or commuting network, resulting in more vehicular GHGs, more energy consumption, reduced socio-cultural interaction, and will lead to fragmentation of society (Bosch et al., 2019; Chen et al., 2010).

Figure 3. Classification of spatial forms of urbanization

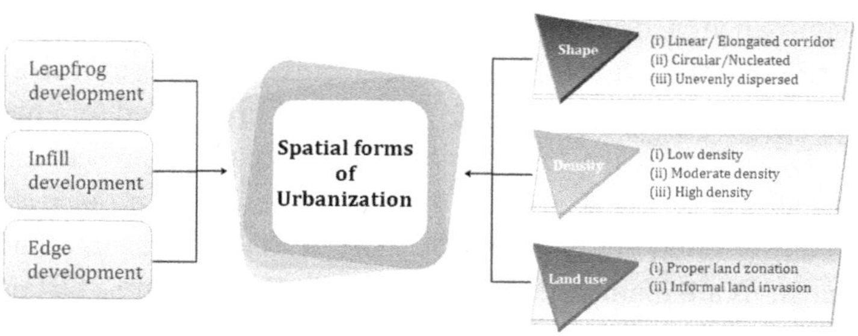

Urban landscape is a spatial entity, which integrates physical, ecological, socio-economic, and environmental factors of a city. With the pace of rapid urbanization, it is a challenging task to maintain the balance between these factors for the planners. Transition of land use/ land cover (LULC) is mandatory in the process of urbanization. In order to achieve sustainability, urban sprawl and its dynamics have to be understood carefully. For optimal use of urban land, it is not only necessary to prepare a database on existing land use, but also to monitor the dynamics of land triggering from the new demands of the growing population. According to Olorunfemi (1983), it is difficult and time-consuming to monitor the LULC changes using traditional methods. Remote sensing (RS) and Geographic Information System (GIS) provide cost-effective tools to obtain information over a very large area (Dewan & Yamaguchi, 2009; Pathan et al., 1993).

LULC distribution is a major challenge for urban planners to minimize the negative effect of land transformation (Esbah, 2007). The primary concerns include planning for proper green space, open urban space, and monitoring the construction of high-rise/mid-rise buildings, which replace the arable agriculture and barren land within the judicial area of the city. Various studies have been conducted to emphasize the cause and effect of LULC changes in different cities (Chen & Yu, 2016; Kantakumar et al., 2016; Taubenböck et al., 2012; Epstein et al., 2002). For instance, Yang & Lo (2002) performed a detailed LULC analysis of Atlanta (Georgia) metropolitan area using Landsat MSS and TM data and an unsupervised clustering method; their study revealed the loss of vegetation and occurrence of urban sprawl in the city. Weber et al. (2005) conducted the research considering SPOT (Satellite Pour l'Observation de la Terre) imagery and adopted parallelepiped classification along with a brightness index and a Normalized Difference Vegetation Index (NDVI) to identify the urban green space of Athens metropolitan. The multi-temporal trends of urban expansion have been characterized and its impact on alteration of land transformation has been studied by Xiao et al. (2006). Similar studies are found for Tegucigalpa, Honduras (Son et al., 2016), Washington DC (Sexton et al., 2013; Masek et al., 2000), Chongming Island, China (Shen et al., 2017), Ludhiana, India (Sharma et al., 2005), Delhi, India (Jain et al., 2016; Mohan et al., 2011), Kemer, Turkey (Onur et al., 2009), Jakarta, Bankok, and Manila (Murakami et al., 2005), Nairobi, Kenya (Mundia & Aniya, 2005), Sanghai, China (Li et al., 2010), Dhaka, Bangladesh (Dewan & Yamaguchi, 2009), Tabriz, Iran (Dadashpoor et al., 2019), Riverside County, California (Chen et al., 2010), Nagpur, India (Ansari et al., 2016), and Jigjiga, Ethiopia (Barow et al., 2019). The spatiotemporal patterns of urbanization in major agglomerations of India are well documented by Taubenböck et al. (2009).

All of the aforementioned studies, conducted in last few decades with the evolution of geo-informatics and satellite sensing technologies, helped to understand the prevailing issues associated with urban sprawling. Geo-technology, a broader technological spectrum comprising of RS, GIS, Global Positioning System (GPS), artificial intelligence, machine learning to Internet of Things (IoT) help significantly to create impactful platforms for the urban related studies. RS plays a vital role in the preparation of cadastral map, boundary delineation/extraction, create databases for transportation network, hydrological connectivity, gas pipelines and all other amenities. It provides very high-resolution imagery capturing each minute details of the terrain, which is cost-effective, relatively less time consuming and with minimized manual-error. These properties push the accuracy of the final thematic map (e.g. LULC) and helps in real-time visualization and decision making. For temporal-basis applications like monitoring urban green cover, traffic congestion, daily/weekly pollution assessment, the near-real time data can be accessed through the RS technology. High resolution imageries are acquired by sensors named GeoEye, IKONOS, RapidEye, QuickBird, Cartosat 2 series etc (Jensen 2015). Various algorithms such as Maximum likelihood classifier, random forest, decision tree, artificial neural network, and support vector machine are available for classifying the images. The classified LULC layer acts as a baseline for sprawl analysis (Banai & DePriest, 2010). Besides satellite imagery, aerial photographs, LiDAR point cloud, historical maps, and toposheets are also used for this purpose. Also, RS coupled with GIS can help in identifying ideal location for residential apartments, commercial complex, industrial corridor as well as overall urban planning, traffic surveillance, utility mapping and management such as hospitals, gas stations, water supply, electricity, waste management etc.

Studies show that the "low density", "unplanned" single use type automobile dependent growth has large scale adverse impacts on quality of living, health and economy of a city. But such sprawled development can be supressed with stringent policies and smart growth pertaining to new urbanism, which aims to create an organized city with mixed use, walkable, transit-friendly and compact (Bibri 2019; Knaap & Talen 2005). A research program entitled "Gulf sustainable urbanism" has been launched comprising of eight countries that border the Persian Gulf, which include Oman, the United Arab Emirates, Qatar, Bahrain, Saudi Arabia, Kuwait, Iran and Iraq, to monitor urban settlements and proposes a holistic sustainability model to solve the current challenges (Ardalan, 2014). Such kind of implications are classic examples of the urge to create an environmentally safe and economical strong city under new urbanist principles, which can't be possible without advanced geo-technologies (Alabi, 2020; Aurigi & Odendaal 2020; Lee et al., 2020; Benton-Short et al., 2019; Chouhan et al., 2018; Oktay 2012; Vall-Casas et al., 2011). Effective city management in general and sprawl management in particular requires the proper understanding of urban morphology, sprawl characteristics, geographical limitations, the socio-economic status of the city, and adequate sprawl-regulating policies (as illustrated in Figure 4).

The objective of this review is to emphasize upon the significance of geo spatial technology in urban sprawl monitoring and assessment in particular and city planning in general. Accordingly, the chapter starts with the general introduction followed by urban sprawl analysis. The next part emphasizes upon available methods of sprawl quantification. It includes descriptions about 'Shanon's entropy method' and 'spatial metrics'. Then a case study of a port city named Visakhapatnam of India was presented to discuss about the sprawl analysis over it as an example. Then a comparison of good and bad sprawls is presented before suggesting the possible solutions and putting forward the recommendations in this regard. Then the future research directions were discussed before concluding the review.

Figure 4. Complex feedback system of urbanization and its major components

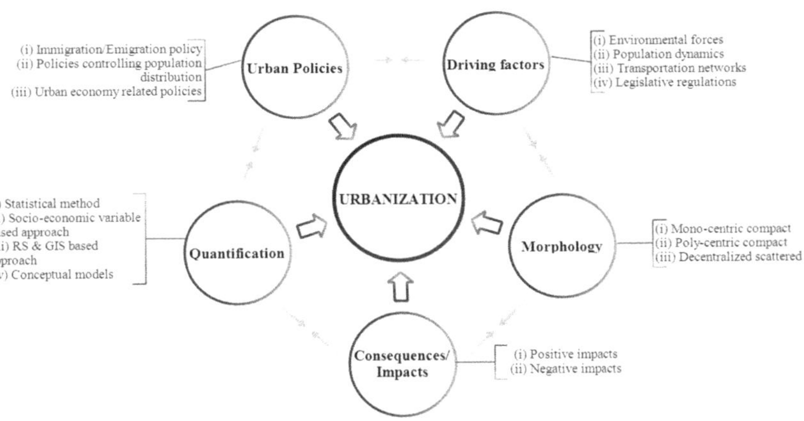

SPRAWL ANALYSIS

Urban sprawl denotes the spatial expansion of a city towards its peripheral areas, which can open up its enormous space for scattered, fringe/ribbon, or secondary city core development (Gómez-Antonio et al., 2016; Tewolde & Cabral, 2011; Galster et al., 2001; Zhang, 2001). Since World War II, urban sprawl became one of the dominant patterns all over the world with variability in time, causes, and consequences (Mosammam et al., 2017). Due to a rush of massive population inflow, the outburst of an existing urban core's judicial boundary begins. To avail a better living space, mainly the high-income group starts to colonize in the outskirt areas of the city. This random colonization predominantly occurs along the major transportation networks, which ignites the occurrence of sprawls. Since the poorly planned scattered patches rise in the sub-urban area, it encourages several socioeconomic, environmental, and political complications (Cobbinah et al., 2015). Sprawl positively destroys the open green space, increases energy consumption, traffic congestion, and automobile dependence that leads to air pollution, indirectly boosts poor health, reduces overall productivity, and enhances crime rates (Barnes et al., 2001). In developing countries like India, the adverse impacts of sprawl contribute to the depletion of fragile natural resources at an alarming rate (Sudhira et al., 2004). Hence, there is a definite urgency to quantify the urban sprawl so that policies can be modified for the designing of sustainable cities.

Figure 5. Built-up density in the city core and periphery

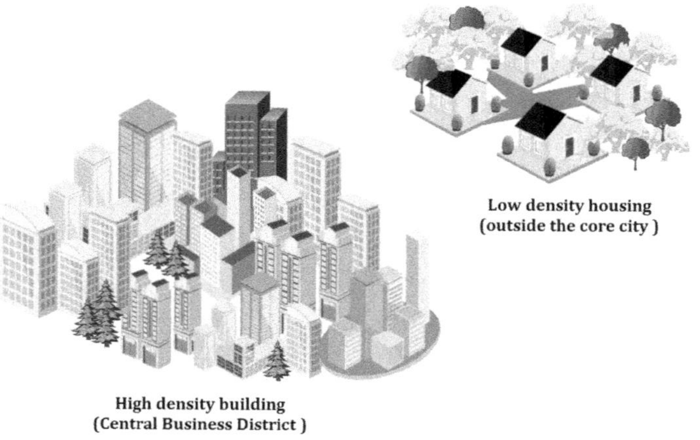

To establish a linkage between urbanization and sustainability, the concept of a "compact city" has been put forward (Jenks et al., 1996). It suggests that a sustainable city must possess appropriate "urban morphology", efficient public transport, and a compactness that encourages social interaction. Many historical European cities are designed on the basis of this compact development theory. Artmann et al. (2019) supported the compact city theory for densified urban development to achieve a multidimensional and multiscale compact green city. This may seem simple theoretically, but it is enormously complex for practical applications. In contrast, western cities (mainly in the US) are growing star-shaped or more or less radially. The CBD concentrates the economic and administrative focal point and the periphery constitutes mostly residential areas and open spaces (Figure 5) while the edge of the city serves as industrial corridors or slums. This kind of urban planning is facing severe issues of sprawling and associated problems, including high crime rates, improper electrification, inadequate sewer lines, and lack of basic amenities.

The existing policies are not strict enough to control the sprawling; hence it could be considered as a global problem.

Angel et al. (2005) categorized the urban expansion policies into three major classes: 1.) controlling the rural-urban or international migration; 2.) distribution of the urban population within the city; 3.) process of urban development for individual cities. "Smart growth" and "new urbanism" policies in North America and "compact city" and "multifunctional land use" policies in the European Union have the potential to curb the urban sprawl problem, at least in the bigger agglomerations (Dieleman & Wegener, 2004).

Patterns and dynamics of sprawl can be perceived by evaluation of the spatial data using RS and GIS techniques. Time-series based study of LULC is the foremost constituent for sprawl analysis. The eight dimensions of sprawl such as density, continuity, concentration, clustering, centrality, nuclearity, mixed uses, and proximity given by Galster et al. (2001) presented in Figure 6. All those combined can capture the characteristics of sprawled growth for an urban agglomeration (Song & Knaap, 2004). Considering the wide applicability of these dimensions, Cutsinger et al. (2005) conducted an extensive research by taking up 14 LU indices for 50 major metropolitans of the US, and found that density, continuity, and mixed-use indices were highly correlated with one another and the analysis showcased the pattern of sprawl and interaction with housing and job. Shift-share analysis can be conducted considering the population, average commuting distance, and employment data for a specific zonal time-series. Socio-economic and biophysical variables can be considered for understanding the trend of sprawl dynamics to strengthen the prediction of future urban development (Zhang, 2001).

SPRAWL QUANTIFICATION METHODS

The various quantification methods adopted for sprawl analysis are illustrated in Table 1. In the absolute method developed by Bhatta et al. (2010), a direct black and white dimension is created to distinguish between compact and sprawled urban core. The disparities in individual assessment to set a threshold for sprawling and non-sprawling makes the quantification a tough task. Further, the relative method deals with the measurement of sprawls in terms of comparative analysis, i.e., among different cities, different time frames for a city or different zones within the city. Several studies conducted to measure sprawl by adopting a multi-indices method using GIS (Inostroza et al., 2013) or descriptive statistical analysis using growth rate, density, spatial configuration, and LU efficiency. Angel et al. (2007) demonstrated sprawling using five "sprawl metrics" (viz. main urban core, secondary urban core, urban fringe, ribbon development, and scatter development), five "suburbanization metrics" (viz. cohesion, decentralization, city centre shift, minimum average distance, and density gradient), density metrics, contiguity, openness, and compactness metrics.

Figure 6. Eight dimensions of sprawled development. The examples are adopted from Google Earth.

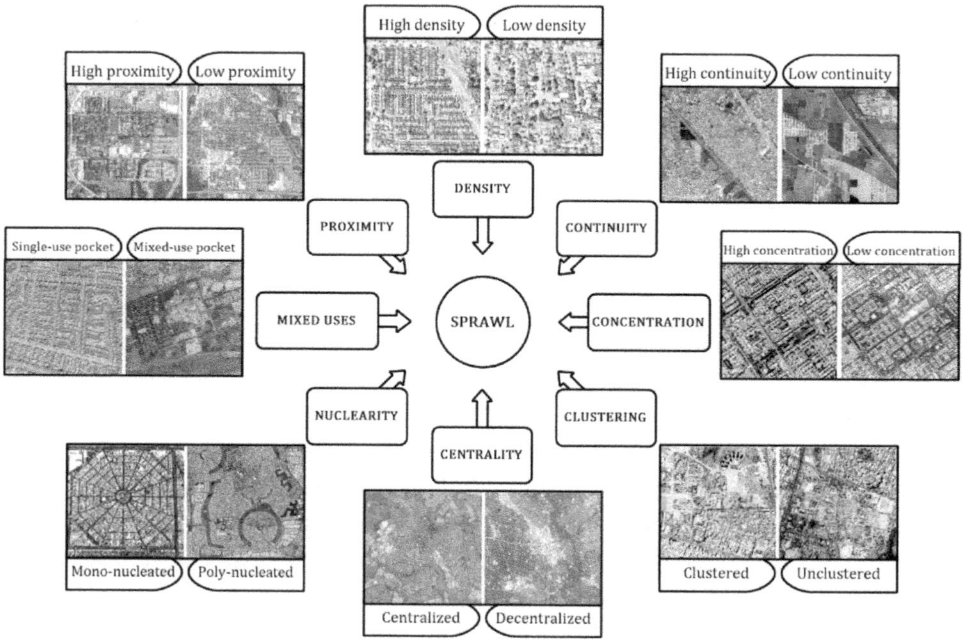

Apostolos (2008) adopted the fractal analysis method to characterize the degree of sprawl in the peri-urban areas of the Greek metropolitan Thessaloniki, and highlighted the complexity of decentralized urban expansion by considering changes in area and boundary fractal dimension. Urban growth patterns have been examined using spatial metrics and gradient approach for the city of Chennai by Aithal & Ramachandra (2016). Aithal & Ramachandra (2013) adopted a combined approach to investigate the urban growth and sprawl in the Tier II cities of Karnataka, India, by considering LU change using multi-temporal, remotely-sensed, satellite imageries and land cover using NDVI, followed by density gradient, zonal analysis, and the computation of Shannon's entropy, along with spatial metrics. Zeng et al. (2005) proposed "Sprawl signature curve" approach, by integrating the concept of a "two-dimensional urban field" and linear gradient analysis with GIS to characterize the urban sprawl in all directions of a city.

Table 1. Sprawl quantification techniques

References	Adopted method
Bhatta et al. (2010)	Absolute/ Relative method
Angel et al. (2007)	Sprawl metrics, sub-urbanization metrics, density metrics, contiguity, openness metrics, and compactness metrics
Fan et al. (2009); Taubenböck et al. (2009)	Density gradient method
Apostolos (2008)	Fractal analysis method
Tewolde & Cabral (2011); Bhatta et al. (2010); Sudhira et al. (2003)	Shannon's entropy method
Zeng et al. (2005)	Sprawl signature curve
Weilenmann et al. (2017)	Composite sprawl metrics
Hasse & Lathrop (2003)	Land Resources Impact (LRI) indicators
Berling-Wolff & Wu (2004)	Phoenix Urban Growth Model (PHX-UGM)
Xu et al. (2019)	Proximity Expansion Index (PEI)
Zhou et al. (2019); Zeng et al. (2015)	Spatial autocorrelation index (Moran's I)
Sutton (2003)	Night-time light imagery
Zeng et al. (2015)	Multivariate Regression Model
Zeng et al. (2005)	Urban field index (UFI)
Li et al. (2016)	Expansion intensity index (EII)
Ciommi et al. (2018)	Principal component analysis (PCA)+Linear regression+ Descriptive statistics

Weilenmann et al. (2017), adopted a multidimensional method by selecting four metrics for sprawl characterization i.e., percentage of built-up area, dispersion of settlements, utilisation density, and composite sprawl metrics. Hasse & Lathrop (2003) proposed LRI indicators (density, farmland forfeiture, loss of natural wetlands, deterioration of forest habitat, and increased impervious surface) to examine the impact of sprawled land expansion on critical land resources of different municipalities. A combined LRI indicator was also developed to identify the collective impact of sprawl over the LULC of a city.

Berling-Wolff & Wu (2004) adopted a Human Induced Land Transformation model and developed PHX-UGM for performance optimization by adding agriculture-urban area land conversion besides original urban and non-urban land and discovered that the leap-frogged sprawl can be measured successfully. To understand the macro-dynamics of existing urban patches, Xu et al. (2019) applied a concentric ring method, followed by an estimation of PEI for the new patches, considering the distance and boundary-sharing rate between new and old patches. The spatial autocorrelation index is useful to measure patch fragmentation and gravity centre migration, which is used for tracking spatial evolution of urban land (Zhou et al., 2019; Zeng et al., 2015).

Shannon's Entropy Method

The concept of entropy is widely accepted to compute the periodic dispersion of urban sprawl. The entropy approach quantifies the fragmentation of natural entity by invasion of new urban patches. It provides a straightforward implication of sprawled zone and rate of sprawl growth, which seems to be significantly useful for urban planners and policy makers to restrict the fragmentation of the city.

The original concept of entropy put forward by Rudolf (1867) in thermodynamics was later adopted by the Max Planck (Jaynes, 1957) and proposed the formulation of Boltzmann entropy i.e.,

$$S = K \ln W$$

where 'S' represents entropy, 'K' is the Boltzmann's constant, and 'W' is the maximum number of possibilities for the system. Shannon (1948) defined entropy as the logarithm of the possibilities of arrangement:

$$H = -\sum_{i=1}^{n} p_i \log p_i$$

where, 'H' is entropy, 'p' is the probability distribution of the variable i, and 'n' is the number of observations. An experiment with higher number of observations, statistically, logarithmic scale offers better interpretability due to its ability for linear comparison.

The application of entropy in urban studies was introduced by Wilson (1967) who proposed a framework for spatial interaction and associated location model (Wilson, 2010). Medvedkov (1967) divided a certain settlement into two components i.e., a uniform and random component. In real-world, there can be infinite combinations of these components. If the entropy for the uniform component is zero, it will vary for the random depending upon the density and amount of scatter.

For the computation of entropy, the area of interest is divided into 'n' number of zones/ rings of constant interval (Kiran & Joshi 2013; Punia & Singh, 2012; Sarvestani et al. 2011). The degree of spatial dispersion or concentration for the urban agglomeration can be calculated as:

$$H_n = \sum_{i=1}^{n} (P_i * \ln \frac{1}{P_i}).$$

Here, H_n = Shannon's Entropy, P_i = Probability/proportion of the land use in the ith spatial zone, and n = total number of zones/rings. The *ln(n)* gives the maximum entropy possible for the urban area. The sprawl analysis is done temporally with respect to *ln(n)* value. Hence, the value of absolute entropy ranges from zero to *ln(n)*. If the sprawl is maximally concentrated in a certain zone, the value will be nearly "zero", and it keeps increasing with the rise in dispersed growth zones. Thus, lower entropy value is desirable for sustainable and healthy environment. For simpler interpretation of the spatial dispersion, another concept called "Relative Entropy" was introduced by Thomas (1981), which scaled the entropy from zero to one. It can be calculated as follows:

$$H_n = \sum_{i=1}^{n} \frac{P_i * \ln \frac{1}{P_i}}{ln(n)}.$$

If the sprawl distribution is even and compact, entropy will be nearly "zero" and it will be ~1 for the maximally dispersed built-ups. This computation is also called as Shannon's equitability index as it has potential to measure the diversity and evenness of the landscape (Cabral et al., 2013).

Shanon's entropy is usually adopted to study spatiotemporal changes in an urban core and its buffer area. It is a unique approach to statistically summarize the urban growth pattern (Jat et al., 2008). It indicates the degree of compactness in the city centre or transportation network and the spread of urban sprawl in all directions. Sudhira et al. (2003) investigated the zone of prevailing and potential sprawl by conducting Shannon's entropy analysis for the Bengaluru-Mysore highway corridor to identify compact and dispersed growth. Kiran & Joshi (2013) took up multi-buffer rings for Vadodara City, India, to analyse entropy for different time periods and found evenly dispersed patches in the city area. An entropy analysis conducted for Jaipur at the city and ward levels, adopting multivariate regression techniques to quantify the spatial patterns of urbanization and the impact of variables on the urban growth (Punia & Singh, 2012). Jha et al. (2008) considered 1-km buffer rings for entropy analysis to evaluate the spatial extent of urban development from the city center and from the highway for Haridwar, India, to prepare future LU plans. For assessing the spatial and temporal variability of the impervious surface and sprawled built-up of Ajmer, India, Jat et al. (2008) adopted multivariate statistical techniques, landscape metrics and Shannon's entropy approach. Shen et al. (2015) proposed a hybrid

Entropy-McKinsey matrix derived from development and coordinating index to quantify the sustainable urbanization considering a set of social (education, safety, health, livelihood, transportation, entertainment and harmony), economic (GDP, consumption, industry structure and science and information) and environmental (population, resource, pollution and governance) indicators. While this kind of study strengthens the detailed understanding of the city's status to achieve sustainability and may prove to be extremely helpful to the city planners for designing suitable development plans, its weakness lies in the spectrum of data it requires, which is not available for all cities.

Spatial Metrics

The LU dynamics can be investigated to understand the sprawling over a city. Spatial metrics consider LULC data as comprehending landscape features (Ramachandra et al., 2012; Herold et al., 2005; 2002). It quantifies the spatial patterning of individual patches, classes, or the whole landscape mosaic as a single geographical unit. It serves as an important tool to assess the fragility of the sprawl over a certain period and assist with evaluation of the composition and configuration of a city.

The spatial heterogeneity of land cover patch / class can be identified at a specific user-defined resolution by spatial metrics. The accuracy of such metrics is directly dependent upon the accuracy of thematic layers. Thus, these are used extensively to study urban growth, pattern, sprawl, fragmentation, and certain model applications (Bhatta et al., 2010; Dasgupta et al., 2009). Spatial metrics can be broadly divided into the following categories: area/edge, shape, core area, isolation/proximity, diversity, evenness, connectivity, and dominance metrics (McGarigal et al., 2002). The detailed formulation of some widely used metrics are presented in Table 2. A single metric cannot identify the physical characteristics of urban sprawl but requires a set of metrics for this purpose. It can be selected considering the available literature and the geographical location to avoid redundancy. Globally, landscape metrics are widely accepted to quantify the composition, dynamics and configuration of cities (Terfa et al., 2019; Sahana et al., 2018; Ramachandra et al., 2015; Toosi et al., 2012; Dasgupta et al., 2009; Hai & Yamaguchi, 2008; Shi & Xiao, 2008; Dibari, 2007; Huang et al., 2007). For instance, Aithal & Ramachandra (2016) adopted landscape metrics like PLAND, NP, PD, LPI, NLSI, the interspersion and juxtaposition index, the clumpiness index, AI, the percentage of like adjacencies, and the patch cohesion index (COHESION) to illustrate the spatial characteristics of Chennai.

Table 2. Spatial metrics

Sprawl metrics	Formulae
Number of patches (NP)	$NP = n_i$ n_i= number of patches in landscape of class i
Largest patch index (LPI)	$\frac{\max^{n}_{j=1}}{A}(100)$
Percentage of landscape (PLAND)	$P_i = \frac{\sum_{j=1}^{n} a_{ij}}{\text{Total area}}$ P_i= no. of patches of class i P_i= proportion of landscape occupied by class i a_{ij} =area of patch ij
Patch density (PD)	$\frac{n_i}{\text{Total area}} * (10{,}000)(100)$
Aggregation index (AI)	$\left[\frac{g_{ii}}{\max g_{ii}}\right](100)$ g_{ii} = no. of like adjacencies between pixels of class i using single count method
Edge density (ED)	$\frac{\sum_{k=1}^{m} e_{ik}}{\text{Total area}} * (10{,}000)$ e_i= total length (m) of edge of the patch belongs to class i
Fractal dimension coefficient of variation (FRAC_CV)	$\text{Fractal dimension}\left[\frac{\text{SD}}{\text{Mean}}\right] * (100)$
Area-weighted mean nearest neighbour (AWMNN)	ENN= h_{ij} e_i= distance (m) from patch ij to nearest neighboring patch of class i
Normalized landscape shape index (NLSI)	$\frac{e_i - e_{\min}}{e_{\max} - e_{\min}}$ e_i= perimeter of edge of class i
Contagion Index (CONTAG)	$\left[1 + \frac{\sum_{i=1}^{m}\sum_{k=1}^{m}\left[P_i \frac{g_{ik}}{\sum_{k=1}^{m} g_{ik}}\right]\left[\ln\left(P_i \frac{g_{ik}}{\sum_{k=1}^{m} g_{ik}}\right)\right]}{2\ln(m)}\right](100)$
Splitting Index (SPLIT)	$\frac{(\text{Total area})^2}{\sum_{i=1}^{m}\sum_{j=1}^{n} a_{ij}^2}$ a_{ij} =area of patch ij
Modified Simpson's diversity index (MSIDI)	$-\ln \sum_{i=1}^{m} P_i^2$ P_i = proportion of the landscape occupied by class i

Figure 7. Sprawled (leapfrog, fringe and scattered) urbanization in Visakhapatnam city, India

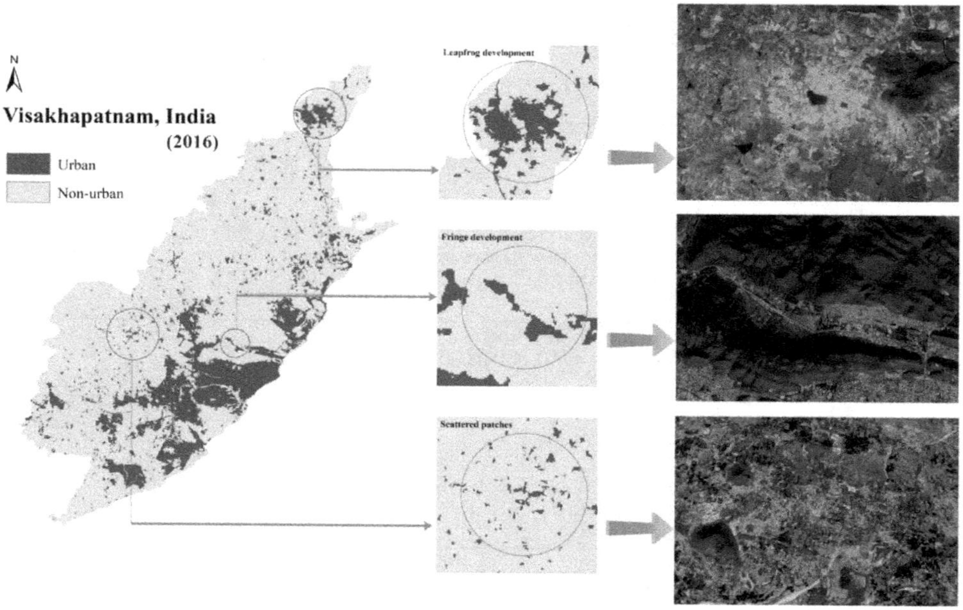

CASE STUDY OF VISAKHAPATNAM, INDIA

A case study is taken up for the eastern Indian port city of Visakhapatnam, to assess the sprawled growth of the urban agglomeration. A unique geographical location, connected freight corridor and presence of a major port helped Visakhapatnam to grow as a large industrial and commercial centre as well as major tourist attraction. The city inhabits more than a million people and the rate of population growth is skyrocketing due to its huge pool of employment opportunities. The coastal landscape and moderately pleasant weather conditions, plus a strong economy, attracts even more people from all around the country. The current planning area of Visakhapatnam Urban Development Authority is spread over ~1700 km^2 from 17°29' to 18°11'N latitudes and 82°56' to 83°31'E longitude. To highlight the urban land transformation, ISRO-BHUVAN classified images for the year 2005 and 2015 are pre-processed and reclassified into Urban and Non-Urban class. The "urban" class comprises of all the urban land categories resulted as pre-existing impervious surfaces such as residential, commercial, industrial, public use land cover etc. Using the LULC layer as the baseline, the spatio-temporal expansion of the urban cover is evaluated and for detailed understanding, the existing sprawl morphology has been considered (Figure 7). Further, Shannon's entropy is computed to understand the overall degree of concentration/ dispersion of the city for the selected years and evaluation of spatial metrics has thrown some light upon the relative distribution of urban patches in the city landscape. The selected spatial metrics, NP, and AI are computed using 500 m × 500 m moving window of FRAGSTATS v4.2 (Figure 8).

Figure 8. Moving window analysis considering 500m×500m grid) NP and AI metric (unit less) for 2005 and 2015 over Visakhapatnam

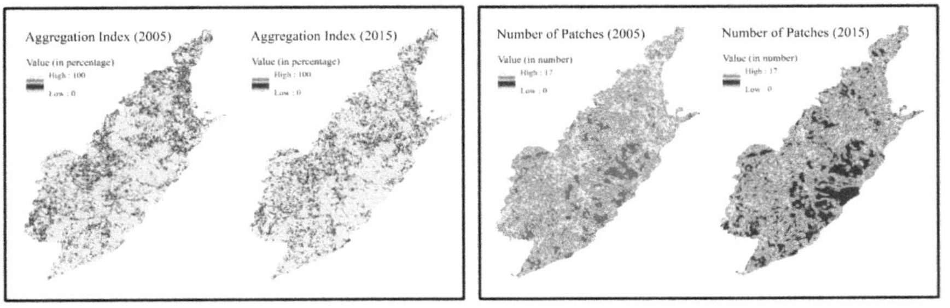

The topographical variations and geographical constraint hinder the city development process to a great extent. The vicinity of CBD area is gradually followed the infill growth and exposed to edge expansion in recent years with coming up of new residential plots, commercial complexes, other amenities such as hospitals, recreational centres. The study shows an overall pattern of sprawled development with higher degree of fragmentation. In Figure 7, the encircled areas represent outlying, fringe, and dispersed types of sprawl development in the Visakhapatnam metropolitan area. Further, Shannon's entropy analysis, using a multi-ring buffer method of 2 km intervals depicts a consistently high value for the last decade. It supports the recent findings regarding the city's higher degree of fragmented growth. However, the selected spatial metrics NP and AI provide a straight forward and versatile understanding of the city landscape. NP seems to persist with higher values all around the city boundary except for the CBD and its peripheral area. It is supported by the high AI value in the compact patches of the city core and relatively lower values found in the low-density scattered patches. This city sets a classic example of decentralized, low-continuity, multi-nucleated growth which may further affect the city's economy and the compromise the environment. The current study revealed the expeditious occurrence of sprawl around the city core, wiping out the lush greenery, especially in the hilly neighbourhood of the city. With tremendous developable area, the city has great potential to be designed beautifully keeping the integrity of the metropolis to achieve sustainable development goals

GOOD SPRAWL VERSUS BAD SPRAWL: A CONCEPTUAL AMBIGUITY

The general concept of sprawl has been critically vague among different research groups for a very long period of time. The literature suggests that sprawl can be a unique pattern of colonization besides traditional models, or an invasion of a new LU due to human-intervention. Irrespective of reasons, policymakers and planners have failed to control the sprawl, even in developed countries. Hence, it is inevitable, as are its consequences. It brings both positive outcomes in terms of "good sprawl," and/or negative ones such as "bad sprawl." For instance, development of scattered patches decreases the open spaces and encroaches on fertile land or fragile ecosystems and demands new commute lines; but it curtails the housing gap between different communities or income groups, which may lead to an increase in social interaction. The fringed sprawl areas are generally confined to the adjacent transportation networks, but they lie far away from the city centres and miss out on the basic amenities. The compact alienated urban patch remaining outside the city creates surplus job opportunities and may have the potential to grow as a new independent agglomeration, but initially it lacks all facilities and becomes a burden for the existing city. Due to judiciary limitations, sprawled patches usually face a lack of proper attention from city administrators, which result in an unfavourable economy and poor productivity. To combat these problems, new policies are formulated to achieve sustainable development goals for the cities by supporting them through technological advancement of new automobile technologies (electric vehicles, solar-charged buses, attractive public transport, such as high-speed rail, etc.), green architecture (LEED, GRIHA rating system), as well as conservation of green belts to protect the environment.

SOLUTION AND RECOMMENDATIONS

In the process of urbanization, the first and foremost step is industrialization, which stimulates the existing economy of the country. It increases the productivity of the city and creates job opportunities that attract the population to move into the city (Turok & McGranahan, 2013). Hence, in order to achieve higher per capita income, highly productive vibrant cities need to be established, which cannot be boosted up without industrialization. It is an inevitable step, which comes with several constraints, such as sprawl. Sprawling itself is a potential threat for the growth and development of a smart sustainable city. To overcome this problem, city planners can start identifying locations with a robust potential for the establishment of a CBD. This kind of approach may result in preparing a productive blueprint that can help lay the foundation for a new city to tackle a target population and their demands. This will help in sharing the load with the existing crowded cities and also with better management of natural resources. More numbers of special economic zones (SEZ) and inter-city industrial corridors should be designed, especially within developing countries. Micro-urbanizations, which are experiencing population inflow, even if at a slow rate, should be taken into considerations for sustainable futuristic planning to combat climate extremes.

FUTURE RESEARCH DIRECTIONS

Geospatial technology plays a vital role in city planning. High resolution, panchromatic, and multispectral imageries, provided by several commercial and government agencies (Cartosat series, IKONOS, GeoEye, QuickBird, Worldview, KOMPSAT, etc.) and dedicated for cartographic applications can be used for effective urban planning through a precise understanding of the changes in the spatial extent of urbanized and semi-urbanized land areas. LIDAR mapping and 3-D modelling of a city enhances the accuracy of the transport network, hydrological channels, as well as designing the city landscape. Appropriate use of RS and GIS techniques, including those of spatial metrics, modelling can be quite helpful in the LULC mapping and associated applications to establish how the urbanization is taking place both qualitatively and quantitatively. Further, relevant tools and resulting information from these processes can be used in the urban planning process.

CONCLUDING REMARKS

According to the World Urbanization Prospects (UNDESA, 2018; 2015) report, half of the world's population would reside in cities. This trend does not appear to be to slowing down in near future, rather it is increasing at a rapid pace, especially in Africa and Asia. With proper management and planning, positive urban growth can be attained by reducing sprawled colonization (Alsharif & Pradhan, 2014). This requires an integrated approach to achieve the sustainable goals.

Although there are a bunch of methods to evaluate urban sprawling, each method has some computational limitations (Nazarnia et al., 2019). Urban sprawl is a very complex phenomenon with intermingled cause-and-effect feedback mechanisms. Thus, only one technique may not be able to highlight the peculiarity of the sprawl, or the same method may not work for all cities, as each possesses its unique pattern of growth. The use of a single method may not able to differentiate the shape and forms of urban growth, and may lead to a misinterpretation of the results. Appropriate suitability criteria (intuitive interpretation, mathematical sensitivity, etc.) can be adopted to understand the impact of sprawl.

Development of small scattered pocket of sprawl, hinders the process of sustainable urbanism or smart growth. Due to lack of administrative and institutional support, sprawled areas are rather becoming a burden to the city governance. In such a scenario, the "area concerned" becomes the connecting corridor between the mainland and the sprawled patch. Here, the question arises, what the best suitable land cover for that transitional area would be and whether that should be urbanized or converted to a green buffer or a recreational centre. That is a real challenge for the planners, because if this issue is left unattended within the correct time, it may affect the neighbourhood society and its economy. The leap frogged patch may act as a seed nuclei and random disorganized properties/buildings may come up, which will further amplify sprawl-related problems exponentially. Another possibility is establishment of slums in the buffer areas.

To avoid such chaotic consequences, the cities should be planned in such a way that it can provide a healthy environment to its existing population and the design should be adequate enough to handle the migrants too. The sustainable urbanism promotes walkable neighbourhood, a harmony between job and housing, accessibility for transit and amenities, and mixed-use land parcels to enable social connectivity. To achieve this, there has to be fine synchronization between urban policies, planning, management, governance and public participation. It can't be possible without the advanced geo-technologies.

REFERENCES

Aithal, B. H., & Ramachandra, T. V. (2013). Measuring urban sprawl in Tier II cities of Karnataka, India. *2013 IEEE Global Humanitarian Technology Conference: South Asia Satellite (GHTC-SAS)*, 321-329. 10.1109/GHTC-SAS.2013.6629939

Aithal, B. H., & Ramachandra, T. V. (2016). Visualization of urban growth pattern in Chennai using geoinformatics and spatial metrics. *Photonirvachak (Dehra Dun)*, *44*(4), 617–633. doi:10.100712524-015-0482-0

Alabi, M. O. (2020). Sustainable Urban form and challenges of open space utilization, Akure, Nigeria as a case study. *International Journal of Urban Sustainable Development*, *12*(3), 1–12. doi:10.1080/19463138.2020.1728691

Alsharif, A. A., & Pradhan, B. (2014). Urban sprawl analysis of Tripoli Metropolitan city (Libya) using remote sensing data and multivariate logistic regression model. *Photonirvachak (Dehra Dun)*, *42*(1), 149–163. doi:10.100712524-013-0299-7

Anas, A., Arnott, R., & Small, K. A. (1998). Urban spatial structure. *Journal of Economic Literature*, *36*(3), 1426–1464.

Angel, S., Parent, J., & Civco, D. (2007). Urban sprawl metrics: an analysis of global urban expansion using GIS. *Proceedings of ASPRS 2007 Annual Conference*, *7*(11).

Angel, S., Sheppard, S., Civco, D. L., Buckley, R., Chabaeva, A., Gitlin, L., Kraley, A., Parent, J., & Perlin, M. (2005). *The dynamics of global urban expansion*. World Bank, Transport and Urban Development Department.

Ansari, T. A., Katpatal, Y. B., & Vasudeo, A. D. (2016). Spatial evaluation of impacts of increase in impervious surface area on SCS-CN and runoff in Nagpur urban watersheds, India. *Arabian Journal of Geosciences*, *9*(18), 702. doi:10.100712517-016-2702-5

Apostolos, L. (2008). Evaluating Urban Sprawl Patterns through Fractal Analysis: The Case of Greek Metropolitan Areas and Issues of Sustainable Development. *Scienze regionali*.

Ardalan, N. (2014). Towards Sustainable Urbanism in the Persian Gulf: Analysis of the Past. *International Journal of Islamic Architecture*, *3*(1), 171–186. doi:10.1386/ijia.3.1.171_1

Artmann, M., Inostroza, L., & Fan, P. (2019). Urban sprawl, compact urban development and green cities. How much do we know, how much do we agree? *Ecological Indicators*, *96*(2), 3–9. doi:10.1016/j.ecolind.2018.10.059

Aurigi, A., & Odendaal, N. (2020). From "Smart in the Box" to "Smart in the City": Rethinking the Socially Sustainable Smart City in Context. *Journal of Urban Technology*, 1–16. doi:10.1080/10630732.2019.1704203

Banai, R., & DePriest, T. (2010). Urban sprawl: Definitions, data, methods of measurement, and environmental consequences. *Journal of Sustainability Education, 7.*

Barnes, K. B., Morgan, J. M. III, Roberge, M. C., & Lowe, S. (2001). *Sprawl development: Its patterns, consequences, and measurement*. Towson University.

Barow, I., Megenta, M., & Megento, T. (2019). Spatiotemporal analysis of urban expansion using GIS and remote sensing in Jigjiga town of Ethiopia. *Applied Geomatics*, *11*(2), 121–127. doi:10.100712518-018-0245-z

Benton-Short, L., Keeley, M., & Rowland, J. (2019). Green infrastructure, green space, and sustainable urbanism: Geography's important role. *Urban Geography*, *40*(3), 330–351. doi:10.1080/02723638.2017.1360105

Berling-Wolff, S., & Wu, J. (2004). Modeling urban landscape dynamics: A case study in Phoenix, USA. *Urban Ecosystems*, *7*(3), 215–240. doi:10.1023/B:UECO.0000044037.23965.45

Bhatta, B. (2009). Analysis of urban growth pattern using remote sensing and GIS: A case study of Kolkata, India. *International Journal of Remote Sensing*, *30*(18), 4733–4746. doi:10.1080/01431160802651967

Bhatta, B., Saraswati, S., & Bandyopadhyay, D. (2010). Urban sprawl measurement from remote sensing data. *Applied Geography (Sevenoaks, England)*, *30*(4), 731–740. doi:10.1016/j.apgeog.2010.02.002

Bibri, S. E. (2019). Data-driven smart sustainable urbanism: The intertwined societal factors underlying its materialization, success, expansion, and evolution. *GeoJournal*, 1–26. doi:10.100710708-019-10061-x

Bosch, M., Chenal, J., & Joost, S. (2019). Addressing urban sprawl from the complexity sciences. *Urban Science*, *3*(2), 60. doi:10.3390/urbansci3020060

Cabral, P., Augusto, G., Tewolde, M., & Araya, Y. (2013). Entropy in urban systems. *Entropy (Basel, Switzerland)*, *15*(12), 5223–5236. doi:10.3390/e15125223

Chen, X., Li, B. L., & Allen, M. F. (2010). Characterizing urbanization, and agricultural and conservation land-use change in Riverside County, California, USA. *Annals of the New York Academy of Sciences*, *1195*(1), E164–E176. doi:10.1111/j.1749-6632.2009.05403.x PMID:20586769

Chen, Y., & Yu, S. (2016). Assessment of urban growth in Guangzhou using multi-temporal, multi-sensor Landsat data to quantify and map impervious surfaces. *International Journal of Remote Sensing*, *37*(24), 5936–5952. doi:10.1080/01431161.2016.1252473

Chouhan, H. A., Parthasarathy, D., & Pattanaik, S. (2018). Urban at the Edges: Mumbai's Coastline Urbanisms. Sustainable Urbanization in India, 279-293.

Ciommi, M., Chelli, F. M., Carlucci, M., & Salvati, L. (2018). Urban Growth and Demographic Dynamics in Southern Europe: Toward a New Statistical Approach to Regional Science. *Sustainability*, *10*(8), 2765. doi:10.3390u10082765

Cobbinah, P. B., Erdiaw-Kwasie, M. O., & Amoateng, P. (2015). Rethinking sustainable development within the framework of poverty and urbanisation in developing countries. *Environmental Development*, *13*, 18–32. doi:10.1016/j.envdev.2014.11.001

Cutsinger, J., Galster, G., Wolman, H., Hanson, R., & Towns, D. (2005). Verifying the multi-dimensional nature of metropolitan land use: Advancing the understanding and measurement of sprawl. *Journal of Urban Affairs*, *27*(3), 235–259. doi:10.1111/j.0735-2166.2005.00235.x

Dadashpoor, H., Azizi, P., & Moghadasi, M. (2019). Land use change, urbanization, and change in landscape pattern in a metropolitan area. *The Science of the Total Environment*, *655*, 707–719. doi:10.1016/j.scitotenv.2018.11.267 PMID:30476851

Dasgupta, A., Kumar, U., & Ramachandra, T. V. (2009). Urban Landscape analysis through spatial metrics. *Proceedings of International Conference on Infrastructure, Sustainable Transportation and Urban Planning,(CISTUP@ CiSTUP),* 18-20.

Deuskar, C., Baker, J. L., & Mason, D. (2015). *East Asia's changing urban landscape: Measuring a decade of spatial growth*. Academic Press.

Dewan, A. M., & Yamaguchi, Y. (2009). Using remote sensing and GIS to detect and monitor land use and land cover change in Dhaka Metropolitan of Bangladesh during 1960–2005. *Environmental Monitoring and Assessment*, *150*(1-4), 237–249. doi:10.100710661-008-0226-5 PMID:18317939

DiBari, J. N. (2007). Evaluation of five landscape-level metrics for measuring the effects of urbanization on landscape structure: The case of Tucson, Arizona, USA. *Landscape and Urban Planning*, *79*(3-4), 308–313. doi:10.1016/j.landurbplan.2006.04.004

Dieleman, F., & Wegener, M. (2004). Compact city and urban sprawl. *Built Environment*, *30*(4), 308–323. doi:10.2148/benv.30.4.308.57151

Dong, T., Jiao, L., Xu, G., Yang, L., & Liu, J. (2019). Towards sustainability? Analyzing changing urban form patterns in the United States, Europe, and China. *The Science of the Total Environment*, *671*, 632–643. doi:10.1016/j.scitotenv.2019.03.269 PMID:30939316

Dutta, V. (2012). Land use dynamics and peri-urban growth characteristics: Reflections on master plan and urban suitability from a sprawling north Indian city. *Environment & Urbanization Asia*, *3*(2), 277–301. doi:10.1177/0975425312473226

Enqvist, J. P., Tengö, M., & Bodin, Ö. (2019). Are bottom-up approaches good for promoting social–ecological fit in urban landscapes? *Ambio*, 1–13. PMID:30879271

Epstein, J., Payne, K., & Kramer, E. (2002). Techniques for mapping suburban sprawl. *Photogrammetric Engineering and Remote Sensing*, *68*(9), 913–918.

Epstein, L. A., Bruner, E. M., Horowitz, M. M., Little, K. L., McCall, D. F., Mayer, P., Miner, H., Plotnicov, L., Schwab, W. B., & Schack, W. A. (1967). Urbanization and Social Change in Africa. *Current Anthropology*, *8*(4), 275–28410. doi:10.1086/200808

Esbah, H. (2007). Land use trends during rapid urbanization of the city of Aydin, Turkey. *Environmental Management*, *39*(4), 443–459. doi:10.100700267-005-0331-y PMID:17265112

Fan, F., Wang, Y., Qiu, M., & Wang, Z. (2009). Evaluating the temporal and spatial urban expansion patterns of Guangzhou from 1979 to 2003 by remote sensing and GIS methods. *International Journal of Geographical Information Science*, *23*(11), 1371–1388. doi:10.1080/13658810802443432

Friedmann, J., & Sorensen, A. (2019). City unbound: Emerging mega-conurbations in Asia. *International Planning Studies*, *24*(1), 1–12. doi:10.1080/13563475.2019.1555314

Galster, G., Hanson, R., Ratcliffe, M. R., Wolman, H., Coleman, S., & Freihage, J. (2001). Wrestling sprawl to the ground: Defining and measuring an elusive concept. *Housing Policy Debate*, *12*(4), 681–717. doi:10.1080/10511482.2001.9521426

Gómez-Antonio, M., Hortas-Rico, M., & Li, L. (2016). The causes of urban sprawl in Spanish urban areas: A spatial approach. *Spatial Economic Analysis*, *11*(2), 219–247. doi:10.1080/17421772.2016.1126674

González, R. R., & Medina, J. S. (2004). Theories, models and urban realities. From New York to Kathmandu. *Dela*, (21), 69–81. doi:10.4312/dela.21.5.69-81

Griffiths, P., Hostert, P., Gruebner, O., & van der Linden, S. (2010). Mapping megacity growth with multi-sensor data. *Remote Sensing of Environment*, *114*(2), 426–439. doi:10.1016/j.rse.2009.09.012

Gupta, R., Joshi, R., & Gupta, N. (2012). Urbanization and air environment in Jaipur. *Environment & Urbanization Asia*, *3*(2), 353–358. doi:10.1177/0975425312473231

Hackenberg, R. A. (1980). New patterns of urbanization in Southeast Asia: An assessment. *Population and Development Review*, *6*(3), 391–419. doi:10.2307/1972408

Hai, P. M., & Yamaguchi, Y. (2008). Characterizing the urban growth of Hanoi, Nagoya, and Shanghai city using remote sensing and spatial metrics. In *IGARSS 2008-2008 IEEE International Geoscience and Remote Sensing Symposium* (Vol. 5, pp. V-9). IEEE.

Hall, P. (2001). Sustainable cities or town cramming. *Planning for a Sustainable Future*, 101-114.

Hasse, J. E., & Lathrop, R. G. (2003). Land resource impact indicators of urban sprawl. *Applied Geography (Sevenoaks, England)*, *23*(2-3), 159–175. doi:10.1016/j.apgeog.2003.08.002

Herold, M., Couclelis, H., & Clarke, K. C. (2005). The role of spatial metrics in the analysis and modeling of urban land use change. *Computers, Environment and Urban Systems*, *29*(4), 369–399. doi:10.1016/j.compenvurbsys.2003.12.001

Herold, M., Scepan, J., & Clarke, K. C. (2002). The use of remote sensing and landscape metrics to describe structures and changes in urban land uses. *Environment & Planning A*, *34*(8), 1443–1458. doi:10.1068/a3496

Huang, J., Lu, X. X., & Sellers, J. M. (2007). A global comparative analysis of urban form: Applying spatial metrics and remote sensing. *Landscape and Urban Planning*, *82*(4), 184–197. doi:10.1016/j.landurbplan.2007.02.010

Inostroza, L., Baur, R., & Csaplovics, E. (2013). Urban sprawl and fragmentation in Latin America: A dynamic quantification and characterization of spatial patterns. *Journal of Environmental Management*, *115*, 87–97. doi:10.1016/j.jenvman.2012.11.007 PMID:23246769

Jaeger, J. A., Bertiller, R., Schwick, C., Cavens, D., & Kienast, F. (2010). Urban permeation of landscapes and sprawl per capita: New measures of urban sprawl. *Ecological Indicators*, *10*(2), 427–441. doi:10.1016/j.ecolind.2009.07.010

Jain, M., Dimri, A. P., & Niyogi, D. (2016). Urban sprawl patterns and processes in Delhi from 1977 to 2014 based on remote sensing and spatial metrics approaches. *Earth Interactions*, *20*(14), 1–29. doi:10.1175/EI-D-15-0040.1

Jat, M. K., Garg, P. K., & Khare, D. (2008). Monitoring and modelling of urban sprawl using remote sensing and GIS techniques. *International Journal of Applied Earth Observation and Geoinformation*, *10*(1), 26–43. doi:10.1016/j.jag.2007.04.002

Jaynes, E. T. (1957). Information theory and statistical mechanics. *Physical Review*, *106*(4), 620–630. doi:10.1103/PhysRev.106.620

Jenks, M., Burton, E., & Williams, K. (1996). Compact cities and sustainability: an introduction. *The Compact City: A Sustainable Urban Form*, 11-12.

Jensen, J. R. (2015). Introductory digital image processing: a remote sensing perspective (2nd ed.). Prentice-Hall Inc.

Jha, R., Singh, V. P., & Vatsa, V. (2008). Analysis of urban development of Haridwar, India, using entropy approach. *KSCE Journal of Civil Engineering*, *12*(4), 281–288. doi:10.100712205-008-0281-z

Kantakumar, L. N., Kumar, S., & Schneider, K. (2016). Spatiotemporal urban expansion in Pune metropolis, India using remote sensing. *Habitat International*, *51*, 11–22. doi:10.1016/j.habitatint.2015.10.007

Kiran, G., & Joshi, U. B. (2013). Estimation of variables explaining urbanization concomitant with land-use change: A spatial approach. *International Journal of Remote Sensing*, *34*(3), 824–847. doi:10.1080/01431161.2012.720738

Knaap, G., & Talen, E. (2005). New urbanism and smart growth: A few words from the academy. *International Regional Science Review*, *28*(2), 107–118. doi:10.1177/0160017604273621

Kundu, A. (2006). Trends and patterns of urbanization and their economic implications. *India Infrastructure Report, 2006*, 27-41.

Kundu, A. (2011). *Trends and Processes of Urbanisation in India, Urbanisation and Emerging Population Issues 6*. International Institute for Environment and Development (IIED). https://pubs.iied.org/10597IIED/

Lee, A., Mackenzie, A., Smith, G. J., & Box, P. (2020). Mapping platform urbanism: Charting the nuance of the platform pivot. *Urban Planning*, *5*(1), 116–128. doi:10.17645/up.v5i1.2545

Lehner, A., Kraus, V., & Steinnocher, K. (2016). Urban growth scenarios of a future Mega City: Case study Ahmedabad. *ISPRS Annals of the Photogrammetry, Remote Sensing and Spatial Information Sciences*, *3*, 165.

Li, M., Zhang, Z., Lo Seen, D., Sun, J., & Zhao, X. (2016). Spatiotemporal characteristics of urban sprawl in Chinese port cities from 1979 to 2013. *Sustainability*, *8*(11), 1138. doi:10.3390u8111138

Li, X., Zhang, L., & Liang, C. (2010). A GIS-based buffer gradient analysis on spatiotemporal dynamics of urban expansion in Shanghai and its major satellite cities. *Procedia Environmental Sciences*, *2*, 1139–1156. doi:10.1016/j.proenv.2010.10.123

Little, K. (1960). West African urbanization as a social process. *Cahiers d'etudes Africaines, 1*(3), 90-102.

Marshall, F., & Dolley, J. (2019). Transformative innovation in peri-urban Asia. *Research Policy*, *48*(4), 983–992. doi:10.1016/j.respol.2018.10.007

Masek, J. G., Lindsay, F. E., & Goward, S. N. (2000). Dynamics of urban growth in the Washington DC metropolitan area, 1973-1996, from Landsat observations. *International Journal of Remote Sensing*, *21*(18), 3473–3486. doi:10.1080/014311600750037507

Masoudi, M., Tan, P. Y., & Liew, S. C. (2019). Multi-city comparison of the relationships between spatial pattern and cooling effect of urban green spaces in four major Asian cities. *Ecological Indicators*, *98*, 200–213. doi:10.1016/j.ecolind.2018.09.058

McGarigal, Cushman, Neel, & Ene. (2002). *FRAGSTATS: spatial pattern analysis program for categorical maps*. Academic Press.

Medvedkov, Y. V. (1967). The concept of entropy in settlement pattern analysis. *Papers / Regional Science Association. Regional Science Association. Meeting*, *18*(1), 165–168. doi:10.1007/BF01940320

Mohammad, R., & Sidaway, J. D. (2012). Spectacular urbanization amidst variegated geographies of globalization: Learning from Abu Dhabi's trajectory through the lives of South Asian men. *International Journal of Urban and Regional Research*, *36*(3), 606–627. doi:10.1111/j.1468-2427.2011.01099.x

Mohan, M., Pathan, S. K., Narendrareddy, K., Kandya, A., & Pandey, S. (2011). Dynamics of urbanization and its impact on land-use/land-cover: A case study of megacity Delhi. *Journal of Environmental Protection*, *2*(09), 1274–1283. doi:10.4236/jep.2011.29147

Mosammam, H. M., Nia, J. T., Khani, H., Teymouri, A., & Kazemi, M. (2017). Monitoring land use change and measuring urban sprawl based on its spatial forms: The case of Qom city. *The Egyptian Journal of Remote Sensing and Space Sciences*, *20*(1), 103–116. doi:10.1016/j.ejrs.2016.08.002

Mundia, C. N., & Aniya, M. (2005). Analysis of land use/cover changes and urban expansion of Nairobi city using remote sensing and GIS. *International Journal of Remote Sensing*, *26*(13), 2831–2849. doi:10.1080/01431160500117865

Muñoz, F. (2003). Lock living: Urban sprawl in Mediterranean cities. *Cities (London, England)*, *20*(6), 381–385. doi:10.1016/j.cities.2003.08.003

Murakami, A., Zain, A. M., Takeuchi, K., Tsunekawa, A., & Yokota, S. (2005). Trends in urbanization and patterns of land use in the Asian mega cities Jakarta, Bangkok, and Metro Manila. *Landscape and Urban Planning*, *70*(3-4), 251–259. doi:10.1016/j.landurbplan.2003.10.021

Murphy, J. T., & Carmody, P. R. (2019). Generative urbanization in Africa? A sociotechnical systems view of Tanzania's urban transition. *Urban Geography, 40*(1), 128–157. doi:10.1080/02723638.2018.1500249

Nazarnia, N., Harding, C., & Jaeger, J. A. (2019). How suitable is entropy as a measure of urban sprawl? *Landscape and Urban Planning, 184*, 32–43. doi:10.1016/j.landurbplan.2018.09.025

Newman, P. W., & Kenworthy, J. R. (1992). Is there a role for physical planners? *Journal of the American Planning Association, 58*(3), 353–362. doi:10.1080/01944369208975814

Oktay, D. (2012). Human sustainable urbanism: In pursuit of ecological and social-cultural sustainability. *Procedia: Social and Behavioral Sciences, 36*, 16–27. doi:10.1016/j.sbspro.2012.03.003

Olorunfemi, J. F. (1983). Monitoring urban land use in developing countries—An aerial photographic approach. *Environment International, 9*(1), 27–32. doi:10.1016/0160-4120(83)90111-3

Onur, I., Maktav, D., Sari, M., & Kemal Sönmez, N. (2009). Change detection of land cover and land use using remote sensing and GIS: A case study in Kemer, Turkey. *International Journal of Remote Sensing, 30*(7), 1749–1757. doi:10.1080/01431160802639665

Pathan, S. K., Sastry, S. V. C., Dhinwa, P. S., Rao, M., Majumdar, K. L., Sampat Kumar, D., Patkar, V. N., & Phatak, V. N. (1993). Urban growth trend analysis using GIS techniques—A case study of the Bombay metropolitan region. *International Journal of Remote Sensing, 14*(17), 3169–3179. doi:10.1080/01431169308904431

Punia, M., & Singh, L. (2012). Entropy approach for assessment of urban growth: A case study of Jaipur, India. *Photonirvachak (Dehra Dun), 40*(2), 231–244. doi:10.100712524-011-0141-z

Ramachandra, T. V., Aithal, B. H., & Sanna, D. D. (2012). Insights to urban dynamics through landscape spatial pattern analysis. *International Journal of Applied Earth Observation and Geoinformation, 18*, 329–343. doi:10.1016/j.jag.2012.03.005

Ramachandra, T. V., Bharath, A. H., & Sowmyashree, M. V. (2015). Monitoring urbanization and its implications in a mega city from space: Spatiotemporal patterns and its indicators. *Journal of Environmental Management, 148*, 67–81. doi:10.1016/j.jenvman.2014.02.015 PMID:24768450

Rudolf, C. (1867). *The Mechanical Theory of Heat: With its Applications to the Steam-engine and to the Physical Properties of Bodies*. Academic Press.

Sahana, M., Hong, H., & Sajjad, H. (2018). Analyzing urban spatial patterns and trend of urban growth using urban sprawl matrix: A study on Kolkata urban agglomeration, India. *The Science of the Total Environment*, *628*, 1557–1566. doi:10.1016/j.scitotenv.2018.02.170 PMID:30045573

Salvati, L., Serra, P., Bencardino, M., & Carlucci, M. (2019). Re-urbanizing the European city: A multivariate analysis of population dynamics during expansion and recession times. *European Journal of Population*, *35*(1), 1–28. doi:10.100710680-017-9462-0 PMID:30976266

Sarvestani, M. S., Ibrahim, A. L., & Kanaroglou, P. (2011). Three decades of urban growth in the city of Shiraz, Iran: A remote sensing and geographic information systems application. *Cities (London, England)*, *28*(4), 320–329. doi:10.1016/j.cities.2011.03.002

Schneider, A., & Woodcock, C. E. (2008). Compact, dispersed, fragmented, extensive? A comparison of urban growth in twenty-five global cities using remotely sensed data, pattern metrics and census information. *Urban Studies (Edinburgh, Scotland)*, *45*(3), 659–692. doi:10.1177/0042098007087340

Sexton, J. O., Song, X. P., Huang, C., Channan, S., Baker, M. E., & Townshend, J. R. (2013). Urban growth of the Washington, DC–Baltimore, MD metropolitan region from 1984 to 2010 by annual, Landsat-based estimates of impervious cover. *Remote Sensing of Environment*, *129*, 42–53. doi:10.1016/j.rse.2012.10.025

Shannon, C. E. (1948). A mathematical theory of communication. *The Bell System Technical Journal*, *27*(3), 379–423. doi:10.1002/j.1538-7305.1948.tb01338.x

Sharma, M. P., Kaur, A., & Shalley, V. (2005). Satellite based study of land transformation in Ludhiana district, Punjab. *Photonirvachak (Dehra Dun)*, *33*(1), 63. doi:10.1007/BF02989991

Shen, G., Ibrahim Abdoul, N., Zhu, Y., Wang, Z., & Gong, J. (2017). Remote sensing of urban growth and landscape pattern changes in response to the expansion of Chongming Island in Shanghai, China. *Geocarto International*, *32*(5), 488–502. doi:10.1080/10106049.2016.1156166

Shen, L., Zhou, J., Skitmore, M., & Xia, B. (2015). Application of a hybrid Entropy–McKinsey Matrix method in evaluating sustainable urbanization: A China case study. *Cities (London, England)*, *42*, 186–194. doi:10.1016/j.cities.2014.06.006

Shi, Y., & Xiao, J. (2008). Evaluating landscape changing due to urbanization using remote sensing data: A case study of Shijiazhuang, China. In *Education Technology and Training, 2008. and 2008 International Workshop on Geoscience and Remote Sensing. ETT and GRS 2008. International Workshop on*, (vol. 1, pp. 559-562). IEEE. 10.1109/ETTandGRS.2008.183

Son, N. T., Chen, C. F., Chen, C. R., & Chiang, S. H. (2016). Mapping urban growth of the capital city of Honduras from Landsat data using the impervious surface fraction algorithm. *Geocarto International*, *31*(3), 328–341. doi:10.1080/10106049.2015.1047469

Song, Y., & Knaap, G. J. (2004). Measuring urban form: Is Portland winning the war on sprawl? *Journal of the American Planning Association*, *70*(2), 210–225. doi:10.1080/01944360408976371

Sudhira, H. S., Ramachandra, T. V., & Jagadish, K. S. (2004). Urban sprawl: Metrics, dynamics and modelling using GIS. *International Journal of Applied Earth Observation and Geoinformation*, *5*(1), 29–39. doi:10.1016/j.jag.2003.08.002

Sudhira, H. S., Ramachandra, T. V., Raj, K. S., & Jagadish, K. S. (2003). Urban growth analysis using spatial and temporal data. *Photonirvachak (Dehra Dun)*, *31*(4), 299–311. doi:10.1007/BF03007350

Sun, C., Wu, Z. F., Lv, Z. Q., Yao, N., & Wei, J. B. (2013). Quantifying different types of urban growth and the change dynamic in Guangzhou using multi-temporal remote sensing data. *International Journal of Applied Earth Observation and Geoinformation*, *21*, 409–417. doi:10.1016/j.jag.2011.12.012

Sutton, P. C. (2003). A scale-adjusted measure of "urban sprawl" using nighttime satellite imagery. *Remote Sensing of Environment*, *86*(3), 353–369. doi:10.1016/S0034-4257(03)00078-6

Taubenböck, H., Esch, T., Felbier, A., Wiesner, M., Roth, A., & Dech, S. (2012). Monitoring urbanization in mega cities from space. *Remote Sensing of Environment*, *117*, 162–176. doi:10.1016/j.rse.2011.09.015

Taubenböck, H., Wegmann, M., Roth, A., Mehl, H., & Dech, S. (2009). Urbanization in India–Spatiotemporal analysis using remote sensing data. *Computers, Environment and Urban Systems*, *33*(3), 179–188. doi:10.1016/j.compenvurbsys.2008.09.003

Terfa, B. K., Chen, N., Liu, D., Zhang, X., & Niyogi, D. (2019). Urban Expansion in Ethiopia from 1987 to 2017: Characteristics, Spatial Patterns, and Driving Forces. *Sustainability*, *11*(10), 2973. doi:10.3390u11102973

Tewolde, M. G., & Cabral, P. (2011). Urban sprawl analysis and modeling in Asmara, Eritrea. *Remote Sensing*, *3*(10), 2148–2165. doi:10.3390/rs3102148

Thomas, R. W. (1981). *Information statistics in geography, geo abstracts*. University of East Anglia.

Todes, A., Kok, P., Wentzel, M., Van Zyl, J., & Cross, C. (2010). Contemporary South African urbanization dynamics. *Urban Forum, 21(3), 331-348*. doi:10.100712132-010-9094-5

Toosi, N. B., Fakheran, S., & Soffianian, A. (2012). Analysis of landscape pattern changes in Isfahan city during the last two decades. In *International Conference on Applied Life Sciences*. IntechOpen.

Turok, I., & McGranahan, G. (2013). Urbanization and economic growth: The arguments and evidence for Africa and Asia. *Environment and Urbanization*, *25*(2), 465–482. doi:10.1177/0956247813490908

UNDESA. (2015). *World urbanization prospects: The 2014 revision*. United Nations Department of Economics and Social Affairs, Population Division.

UNDESA. (2018). *Revision of world urbanization prospects*. United Nations Department of Economics and Social Affairs, Population Division.

Vall-Casas, P., Koschinsky, J., & Mendoza, C. (2011). Retrofitting suburbia through pre-urban patterns: Introducing a European perspective. *URBAN DESIGN International*, *16*(3), 171–187. doi:10.1057/udi.2011.9

Van Dijk, T. (2011). Networks of urbanization in two Indian cities. *Environment & Urbanization Asia*, *2*(2), 303–319. doi:10.1177/097542531100200211

Weber, C., Petropoulou, C., & Hirsch, J. (2005). Urban development in the Athens metropolitan area using remote sensing data with supervised analysis and GIS. *International Journal of Remote Sensing*, *26*(4), 785–796. doi:10.1080/01431160512331316856

Weilenmann, B., Seidl, I., & Schulz, T. (2017). The socio-economic determinants of urban sprawl between 1980 and 2010 in Switzerland. *Landscape and Urban Planning*, *157*, 468–482. doi:10.1016/j.landurbplan.2016.08.002

Wilson, A. (2010). Entropy in Urban and Regional Modelling: Retrospect and Prospect. *Geographical Analysis*, *42*(4), 364–394. doi:10.1111/j.1538-4632.2010.00799.x

Wilson, A. G. (1967). 1967: A statistical theory of spatial distribution models. *Transportation Research*, *1*(3), 253–269. doi:10.1016/0041-1647(67)90035-4

Xiao, J., Shen, Y., Ge, J., Tateishi, R., Tang, C., Liang, Y., & Huang, Z. (2006). Evaluating urban expansion and land use change in Shijiazhuang, China, by using GIS and remote sensing. *Landscape and Urban Planning*, *75*(1-2), 69–80. doi:10.1016/j.landurbplan.2004.12.005

Xu, G., Dong, T., Cobbinah, P. B., Jiao, L., Sumari, N. S., Chai, B., & Liu, Y. (2019). Urban expansion and form changes across African cities with a global outlook: Spatiotemporal analysis of urban land densities. *Journal of Cleaner Production*, *224*, 802–810. doi:10.1016/j.jclepro.2019.03.276

Yang, X., & Lo, C. P. (2002). Using a time series of satellite imagery to detect land use and land cover changes in the Atlanta, Georgia metropolitan area. *International Journal of Remote Sensing*, *23*(9), 1775–1798. doi:10.1080/01431160110075802

Zeng, C., Liu, Y., Stein, A., & Jiao, L. (2015). Characterization and spatial modeling of urban sprawl in the Wuhan Metropolitan Area, China. *International Journal of Applied Earth Observation and Geoinformation*, *34*, 10–24. doi:10.1016/j.jag.2014.06.012

Zeng, H., Sui, D. Z., & Li, S. (2005). Linking urban field theory with GIS and remote sensing to detect signatures of rapid urbanization on the landscape: Toward a new approach for characterizing urban sprawl. *Urban Geography*, *26*(5), 410–434. doi:10.2747/0272-3638.26.5.410

Zhang, T. (2001). Community features and urban sprawl: The case of the Chicago metropolitan region. *Land Use Policy*, *18*(3), 221–232. doi:10.1016/S0264-8377(01)00018-7

Zhou, W., Jiao, M., Yu, W., & Wang, J. (2019). Urban sprawl in a megaregion: A multiple spatial and temporal perspective. *Ecological Indicators*, *96*, 54–66. doi:10.1016/j.ecolind.2017.10.035

KEY TERMS AND DEFINITIONS

Central Business District: The focal point of a city where most of the commercial activities are concentrated.

Fringe: A linear corridor of the urbanized patch.

Image Classification: Extracting information from the given remotely sensed imagery categorically.

Land Use: Alteration in the natural land cover through anthropogenic means.

Leapfrog: Development of urban patches beyond the boundary of an existing city/town due to the availability of resources at a significantly lower cost than the city/town.

Sprawl: The unplanned and unrestricted growth in urban and peri-urban areas characterized by low-density housing, single LU based zoning, and dependence on private vehicles.

Sustainable: Fulfilling the requirement of the current population without exploiting the needs of the future generation.

Urbanism: Interaction of the city dwellers with their niche in an urban ecosystem.

Chapter 2
Geoinformation Technologies in Land Use Monitoring of Fast-Growing Cities for Sustainable Urban Development:
Maputo as a Laboratory for Research and Action

Cristina Delgado Henriques
https://orcid.org/0000-0003-0100-5586
CIAUD, Faculdade de Arquitetura, Universidade de Lisboa, Portugal

ABSTRACT

To understand the territory of fast-growing cities, where there are multiple stakeholders involved, the observation of such dynamics seems indispensable to formulate and implement policies and actions based on a better understanding of these territorial systems. This chapter offers a perspective on how urban territories should be observed through geoinformation technologies that can provide a means for creating monitoring indicators concerning land use of fast-growing cities. The city of Maputo was used as an experimental laboratory for the use of geoinformation technologies in the observation, discussion, and reflection on methods for sustainable urbanism. The discussion includes the implications of the case study and possible developments to take a step forward in land use planning processes to achieve the desired socio-spatial equality.

DOI: 10.4018/978-1-7998-2249-3.ch002

INTRODUCTION

Despite countless good intentions and urban development agendas, there are still many persistent issues playing an important role in the urban expansion process in developing countries, which can lead to spatial inequities and social deprivation.

At the same time, emerging issues such as climate change and migration are extending this issue and hampering the process of reconciling urbanization and sustainable development.

The United Nations World Cities Report 2016 (United Nations Human Settlements Programme, 2016) refers to specific policies and actions that can drive transformation change to leverage a New Urban Agenda. Of particular note among these is a global monitoring framework to increase the availability and usefulness of data to support decision-making, accountability mechanisms, and the capacity of countries/cities to deliver and report on the New Urban Agenda and SDGs (Sustainable Development Goals).

Moreover, the New Urban Agenda supports

"the role and enhanced capacity of national, subnational and local governments in data collection, mapping, analysis and dissemination and in promoting evidence-based governance, building on a shared knowledge base using both globally comparable as well as locally generated data, including through censuses, household surveys, population registers, community-based monitoring processes and other relevant sources."

In addition, it fosters

"the creation, promotion and enhancement of open, user-friendly and participatory data platforms using technological and social tools available to transfer and share knowledge among national, subnational and local governments and relevant stakeholders, including non-State actors and people, to enhance effective urban planning and management, efficiency and transparency through e-governance, approaches assisted by information and communications technologies, and geospatial information management." (United Nations, 2017: 40).

Concerning urban territorial planning and policymaking in general, and in fast-growing cities in particular, spatial information management through geoinformation technologies is of vital importance in the monitoring and evaluation of the planning process. The determination of ratios and indicators through spatial data makes it possible, among other operations, to quantify needs and to understand the transformation or evolution of certain phenomena, thereby contributing to the construction of an informed planning system and building evidence-based governance.

To understand the urban territory, where the stakeholders of its transformations are multiple, the observation of its dynamics seems indispensable to formulate and implement policies and actions based on a better understanding of these territorial systems. In short, observing to understand and understanding to act is an elementary element for promoting the sustainable development of cities.

Maputo, Mozambique's capital city, located in a coastal area of Southeast Africa, struggles from climate change effects, observable particularly during floods, land pressure, both from real estate investments and low-income families looking to settle where they can find a means of survival, and from the lack of effective integrated planning, based on a monitoring framework enabling public authorities to obtain periodic assessments of the various dimensions of urbanization and their impacts.

In this context, this chapter provides a perspective on how urban territories should be observed through geoinformation technologies that can provide the means for creating monitoring indicators concerning the land use of fast-growing cities. The city of Maputo has been used here as an experimental laboratory for the use of geoinformation technologies in the observation, discussion, and reflection on methods for sustainable urbanism.

The chapter is structured into two sections, as follows. The first section provides an overview on geoinformation technologies in urban land use monitoring processes, discussing: i) the concepts of geoinformation technologies, land use, and urban sustainability, as well as the methods that presently provide multitemporal information about their dynamics and the repercussions that these dynamics have on sustainable urban development; ii) the evolution of technology and data to monitor dynamics; iii) the most used methods to analyze and model the current land use situation and to simulate and/or create scenarios for future evolution. It also discusses monitoring and assessment as phases of the planning process, supported by geoinformation technologies, by providing an overview of progress in the field of technology and data acquisition. To this end, examples of the use of geoinformation technologies in monitoring and evaluation processes in very different development contexts will be analyzed. Special reference is made to the notion that geoinformation technologies facilitate the transparency of processes that should distinguish democratic societies. The second section uses Maputo city as a case study to illustrate the idea that geoinformation technologies are indispensable for the analysis of many spatial processes, and are fundamental in monitoring the rapid change of land use in urban areas. Through urban planning storytelling, the new land use planning paradigm that Mozambique is aiming to achieve does not deviate from public interest, social and spatial justice, environmental sustainability, and the transparency of processes. However, it needs to take full advantage of the opportunities created by geoinformation technologies for sustainable urban development, for more informed urban policies, and ultimately to support a more just and resilient city.

The need for new evidence on urban territorial development to support the design of urban policies has long been advocated and the observation of land use dynamics has been one of the objects of research in many cities around the world, although this is less frequent for African cities. Framed by these ideas, the chapter concludes with a discussion of the implications of the case study, and possible developments to take a step forward in land use planning processes to achieve the desired socio-spatial equality.

GEOINFORMATION TECHNOLOGIES IN URBAN LAND USE MONITORING

Concepts and Methods

Urban areas are more than just a set of buildings, streets, and people living therein. They are systems of networks and flows (Batty, 2013) and, therefore, they are spatial systems characterized fundamentally by their structure and their functioning.

Regarding structure, urban areas and cities are systems which operate on two fundamental elements: i) networks, including roads and other infrastructures (basic sanitation, water supply, and energy), which guarantee flows and, therefore, allow for a balanced urban metabolism when they are sized to respond well to demand; ii) physical distribution of functions, that is, in the planned or self-organized allocation of land use, which, as the designation implies, associates a morphology and an arrangement to each area, territory or city, resulting from a social and economic vision at a given time. Regarding the function of each land parcel, this derives from the social use and is economically referenced by an agent or group of stakeholders that transform the territory and the city. These transformations (land use, social, housing, etc.) result in dynamics, and the functioning of the systems results from these dynamics.

Understanding the urban territory, at the same time both a living space and a managed space, mainly in terms of its functioning and the inequalities generated by growth, is of the utmost importance to present and support sustainable policies for spatial and social equity.

Land use planning of urban areas, besides being a process for regulating land use decisions and providing a framework to promote sustainable urbanism, also involves the systematic assessment of the physical, social, and economic transformations arising from the implementation of a plan and emerging urban development issues.

The transition from one land use status to another is an essential indicator of urban change. Focusing the evolution of a city on the geometric and semantic changes of the smallest parts that compose it, and that are (re)produced over time, it is possible to understand, albeit partially, the dominant model of this evolution. If it is true that the evolution of a city should not be reduced to just one of its tangible aspects (the land use), it is also true that ignorance of the speed of transitions, in an explanatory framework for the change that only considers the less tangible aspects (such as culture, policies, the degree of human development, information flows, capital flows, or migratory flows, among other factors), has consequences. Most developing countries, regardless of their individual histories, which necessarily have created their differences, are commonly characterized by high levels of poverty, low indexes of education and technical knowledge, fast-growing cities, lack of investment in social needs, considerable pressure on land and natural resources due to high population growth rates, and absence of physical planning – all of this coupled with other negative discourses that place almost all these countries at the bottom of any human development ranking. In short, there is a vast portion of territories in the world with basic social requirements and an absence of planning and management procedures for rational spatial organization and justice.

Cities in developing countries are diverse and difficult to conceptualize, like any other cities. However, most of them are undergoing a rapid pace of urbanization and land use change that imperatively need meaningful, available and up-to-date data and tools to understand trends, mitigate human impacts on the natural environment and simultaneously preserve social and economic structures, and ensure quality of life.

Rabinovitch (1996: 51) argues that the ultimate goal of any land use policy is to

"affect social environmental and economic aspects so as to benefit as many people as possible, especially the most vulnerable groups – which in the context of most developing countries, constitute the majority of the population anyway."

Land use planning is one of the basic instruments that leaders of fast-growing cities have to improve living conditions – by setting the terms of urbanization, especially policies targeted at using urban land and expanding basic infrastructures and public services – as well as managing the city's physical shape.

In a context of low financial resources, and, in most cases, a conflict between the interests and strategies of the private sector, local communities, and public administration, and the difficulties in harmonizing these interests, how can the spatial discourse be changed to respond affirmatively and critically and to achieve a desirable sustainable development regarding land use?

The monitoring and qualitative and quantitative assessment of spatial-temporal transformations of land use is a fundamental step in understanding the dynamics of the processes that lead to the alteration of urban landscapes and consequently the geographical framework of the planning actions that are intended to be triggered (Henriques, 2008).

The use of geoinformation technologies "is, for many years now, embedded in many of the stages of the planning process, in particular for data collection, data visualization (in its many forms) and modeling and simulation of the spatial objects" (Pinto et al., 2014: xiv). They are, therefore, essential tools in the planning and monitoring of land use changes in any city in general and especially in those where such fast transformations are laborious to follow.

Data from remote sensing has been used for several decades now in the systematization and analysis of land use transformation processes. Whether by the use of conventional aerial photography, or by images from satellites, or more recently from unmanned aerial vehicles (UAVs), these types of location-based data play a critical role in the production of fundamental diachronic data sets and therefore in generating developing information. Achieving sustainable development goals goes beyond the representation of land use change. Measurable indicators based on geodata can provide a relational basis for the assessment of land use planning policies and in accomplishing progress and defined targets, and can also provide inputs for the modeling and simulation of future scenarios.

The utilization of land use transformation models dates back to the 1970s, having been developed with the aim of mathematically reproducing the nonlinear dynamic relationships that characterize the spatial evolution of cities. Its use made it possible to measure, recognize, interpret, and anticipate some of the sustainability problems in cities. The need to use these models remains today and their relevance in fast-growing urban areas is considerable, as they serve as an instrument to support integrated planning strategies that provide ways of understanding urban dynamics and their consequences at the human dimensional level (Henriques, 2008).

There are many approaches to the simulation of land use changes, most of them integrated, nowadays, in software packages. Cellular Automata, Markov chain algorithm, and Artificial Neural Networks, by themselves or combined, are common approaches for modeling land use changes. Mas et al. (2014) review and discuss several methods of some of these software packages.

Geoinformation technologies are widely used in contemporary decision-making processes and are unavoidable in planning and managing urban spatial development. Monitoring and evaluation through appropriate and affordable tools are ingredients for an effective and integrated perspective on the planning and management of fast-growing urban territories.

Monitoring and Assessment

Firstly, if the purpose of planning is to seek balance among competing interests in the use of land and guarantee a sustainable future, mechanisms to provide evidence on the progress of achieving stated goals should be assured. Secondly, if geoinformation technologies are appropriate tools for data collection, data visualization, and modeling, supporting an integrated view of the territory and the human activities that occur therein, advantage must be taken of them in providing such evidence.

Indicator frameworks based on geographic data and technologies can provide clear and direct information to support urban planning and policy-making activities. Various types of indicators are recommended in a process of planning and managing land use changes in fast-growing cities, but commonly they must ensure: i) monitoring of transformation trends; ii) benchmarking through the supplying of a set of time-lapse "images" of the territory; iii) referencing different life conditions to allocate the necessary infrastructures, social amenities and utilities in a balanced way; iv) evaluation of the success and impacts of the policies, measuring the distance between the existing situation and the stated goals.

Statistical capacity is an essential component of the monitoring and evaluation processes. Lack of reliable data on the territory and its inhabitants compromises planning and the evaluation of the planning process and undermines the transparency desirable in democratic systems. However, in many developing countries, statistical and cartographic data are not so readily available. Most of it is restricted to census data and to outdated maps undertaken as part of specific development projects which were not maintained after the project was completed.

Data from remote sensing, since the launch of the Landsat program, are an important source for building indicators of land cover status and the changes that occur therein. The ready availability, the convenient temporal resolutions, and the recently improved spatial resolution of several recent sensors, combined with free and open-source software packages for data processing, have enabled a turnaround in the acquisition of land cover data. Taking the example of the Sentinel 2 MultiSpectral Instrument (MSI) data, which are open data associated with ESA open software, can be a strong argument for deciding to use Vegetation, Soil, and Water indicators, as well as land use as basic information to observe the cities at a macro-scale. These indicators are biophysical but may be used with indicators of a different nature to improve the analysis of urban territories. At the scale of the neighborhoods and of the urban fabric, the ability to discern urban features (that may include soil, concrete, asphalt, plastic, metal, wood, water, vegetation…) and their use (residential, industrial, agricultural, educational…) from satellite images requires high spatial and spectral resolution and image processing software that accounts for spectral and contextual information. However, remote sensors from satellites that supply images with sub-meter spatial resolutions and spectral resolutions beyond the panchromatic, the three bands of visible, and near infrared, such as WorldView-3, are commercial programs and still have high acquisition costs. Images freely provided by Google, Bing, ESRI, Maxar, among others, are not suited for automated remote sensing analysis of any kind. What is extracted from the satellite image is a mere photo, which does not contain the various spectral bands of multispectral resolution satellite data. Nonetheless, when incorporated as a base map in several Geographic Information Systems (GIS) packages, this can contribute with highly detailed information on the complexity of the urban landscape inferred by visual image interpretation. Several crowdsourcing and Volunteer Geographic Information initiatives have contributed to enhancing geospatial databases around the world based on these images where OpenStreetMap and Collect Earth stand out. In Maputo, "#MapeandoMeuBairro" is an example of a non-governmental organization (NGO) that has been contributing to acquiring data on self-produced neighborhoods.

The widespread use of UAVs, due to their affordable cost and ease of handling, is also becoming an important source of urban data. "… data obtained using UAVs are useful for 3D urban modeling, i.e., buildings and urban areas, with a view intending to support large scale urbanism and urban planning studies" (Tenedório et al, 2016). Although more limited in the extent of the covered area, the assessment of the impact of a given occurrence or the obtaining of 3D information for a better understanding of a specific urban context, have much to benefit from the information collected through this technology.

According to Sawicki and Flynn (1996), the development and use of urban indicators is not always a consensual process. They argue that on the planners' and policy-makers' side, the idea of "straightforward, easily understood, and applicable to a specific program or policy" indicators opposes the academics' will for "innovative methodologies, models, theories, and complex indices for publication" not in line with the "repetitive exercise of updating yearly reports." However, if an urban observatory is preconized for a city, within the context of the functions of a university (building capacity, research, and social extension), the scope and applicability of urban research are considered and both types of indicators can be incorporated thereby contributing to the improvement of the quality of life.

Land Use Change Observatories: Tools For Spatial Planning

The need to record, analyze, and interpret the cumulative results of human action, translated into the physical expansion of cities and the emergence of new forms of urbanization and urbanity, has long been advocated (Atkinson, 2005; United Nations Development Program, 2016) and recently emphasized by the emergence of climate change challenges (Bai et al., 2018). Means for the effective implementation of the New Urban Agenda also refer to the

"use of digital platforms and tools, including geospatial information systems (…) to improve long-term integrated urban and territorial planning and design, land administration and management, and access to urban and metropolitan services"

and which fosters

"the creation, promotion and enhancement of open, user-friendly and participatory data platforms using technological and social tools available to transfer and share knowledge among national, subnational and local governments and relevant stakeholders, including non-State actors and people, to enhance effective urban planning and management, efficiency and transparency through e-governance, approaches assisted by information and communications technologies, and geospatial information management." (United Nations, 2017)

Developing the city dynamics of countries often leads to irreversible consequences such as the geographical and social impact of many construction projects carried out in unsuitable areas or the generation of conflicts concerning sustainable nature between urban land uses and the various stakeholders (Jenkins, 2013). The development of observation methods, structured around geographic information, either to map the global change of territories and human settlements or to carry out spatial analysis on land use oriented towards the management of urban territories, are necessary and desirable, but should be tailored to each local reality: the availability of data and equipment, and technical capacities must be checked, in order to guarantee its usefulness, continuity, and appropriation by users.

An urban observatory implemented through the modeling of phenomena and guided by a set of indicators can show the evolution trends of the urban system over time, space, and in relation to specific objectives, thereby guiding local development (*Figure 1*).

Figure 1. Urban land use observatory model (adapted from Henriques, 2008)

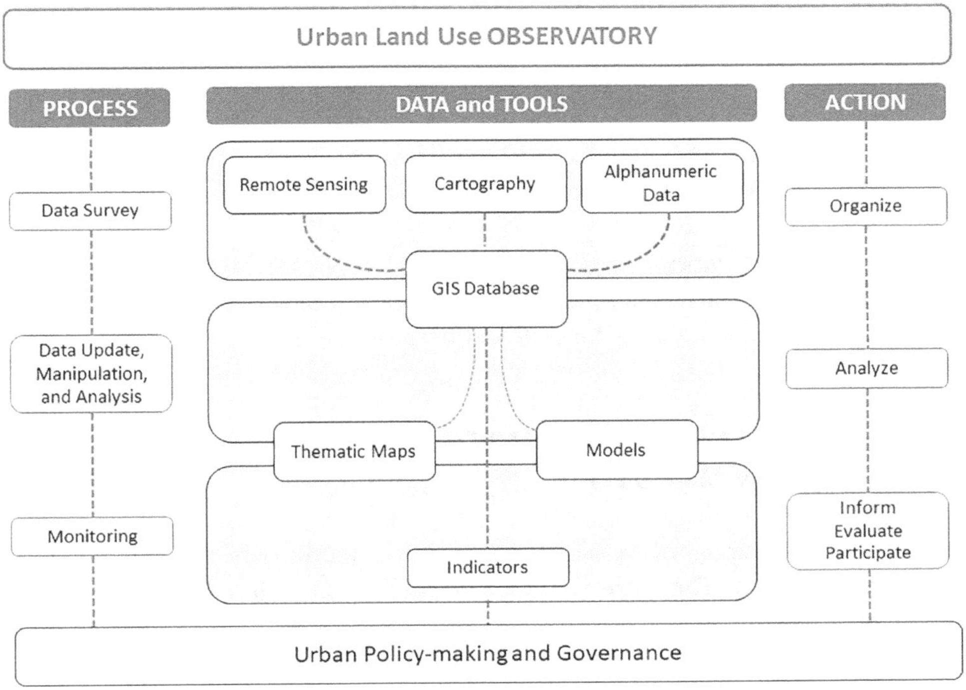

The process and means of gathering and analyzing geographical information within a monitoring frame can be guided by the following questions: 1) observe why?; 2) observe what?; 3) observe how? (Henriques, 2008).

1. The urban territory is a space with different levels of organization. Understanding the city, at the same time a lived in and an administered space, mainly with regard to its functioning and the inequalities generated by growth, is an element of the greatest importance to show and support democratic policies involving spatial and social equity, in this way observing to understand and understanding to act;
2. The question of "observe what" leads us to land use as a tangible part of urban growth and the basis for deciding on the preferential guidelines for future expansions, of the conditions to be established, and of the actors within the urban economy to be elected, considering that they will be promoters of the sustainable development of the territory;
3. Urban observation currently requires technically simple means to implement, but difficult to maintain in view of the scarce resources of many developing countries. Geoinformation technologies are barely replaceable in urban contexts

where the rates of change in land use are large and very often irreversible. Local policymakers and organizations from civil society (Universities, local NGOs, national and international institutions, and other stakeholders), if working together, can share costs and knowledge to develop such an observatory. The promotion and enhancement of open, user-friendly, and participatory data platforms using technological and social tools available to transfer and share knowledge should also be considered in this context.

An urban observatory can be a cross-disciplinary project strengthening ties between different urban stakeholders (researchers, policymakers, planners, residents), producing knowledge together for different purposes.

LAND USE PLANNING AND PATTERNS OF CHANGE IN MAPUTO

Maputo city, the capital of Mozambique, is one of the fast-changing urban areas of Southeast Africa. The extremely high population growth and the consequent accelerated variation in land use, are problems of great importance in this city, due to the human and environmental repercussions that they entail.

The surrounding territory of the colonial city's core is an enormous coalescent urban area based on self-produced residential land use, which originally was not structured by any master plan. The urban growth was 'organic' and very dependent on the social and economic population conditions (migration from rural areas to Maputo) and on the political system both before and after Independence in 1975. According to Henriques & Tenedório (2010), the most important transition of land use between 1964 and 2008 was from Natural, Semi-natural, and Leisure Areas to Residential Areas (Periphery).

The territory of Maputo has been recorded by remote sensing images (aerial photographs and satellite images) since the 1960s. This has enabled the acquisition of knowledge on developing trends at a macro scale, the observation of the different urban fabrics produced over time (planed, organic, self-produced, imposed), and the modeling of the urban expansion for the near future. The spatial and temporal aspects of land use throughout five decades, as well as the model of urban development for 2011, was carried out by Henriques (2008) and was used for drawing up the municipality's master plan.

From Urban Planning In The Colonial Period To The Explosion of Self-Produced Settlements

In the early 1940s, the Municipal Council of Lourenço Marques (the former name of Maputo) started to face an effective occupation of the city surroundings with self-produced settlements compromising the expansion of the colonial city. Therefore, the idea of having a Plan for the city and its suburbs began to emerge. However, it was only in 1947 that the process of drawing up an urbanization plan started, under the coordination of João Aguiar, and in 1952 that plan was approved. Notwithstanding the distance between its proposals and real needs, the "General Urbanization Plan for Lourenço Marques" (*Plano Geral de Urbanização de Lourenço Marques*) had some of its proposals implemented, especially concerning the road network, and stressed the differences between the solutions presented and the actual situation. This plan was used until 1969 when a new plan was proposed.

During the 1960s, the city was confronted with a process of fast growth and strong pressure from the private sector, and the Urbanization Plan of 1952 was inadequate to regulate those forces. In an attempt to fill this gap, two documents were produced: an urbanization study in 1965, and the Regulation Plan of Land Occupation for Lourenço Marques Surroundings (*Plano Regulador da Ocupação do Solo nos Arredores de Lourenço Marques* – PROSALM) in 1966. Both studies tried to zone the city's expansion by defining the location for different land uses.

Under the coordination of Mário de Azevedo, the works leading to the Master Urbanization Plan of Lourenço Marques (*Plano Director de Urbanização de Lourenço Marques*) started in 1967 and were delivered in 1969. Only in 1972 was it formally approved by the Municipal Council. This plan, consisting of twenty-three volumes, was intended to be a tool for the framing and coordination of land use in the urban area and its surroundings in a malleable and dynamic manner according to the urban reality and to the perspectives that would progressively be defined, but always supported by permanent natural constraints and by the main options determining the development framework proposed (Azevedo, 1969). In respect to the definition of the land use, the Master Plan defined and mapped six main classes: i) Residential, ii) Traditional Residential, iii) Industry, iv) Retail and Services, v) Regional Equipment, vi) Tourism and Recreational Equipment, Regeneration Areas and Green Areas. Additionally, it identified twenty-two subclasses.

The Master Urbanization Plan of Lourenço Marques has been the only legal document able to regulate the urbanization processes of the whole region under the influence of the city almost down to the present day. In fact, after Independence, in 1975, several attempts were made to address urban land issues. Between 1979 and 1985, the National Institute of Physical Planning carried out a planning action that culminated in the Maputo Structure Plan. Facing serious difficulties with the ongoing civil war, high levels of demographic growth (immigration and high natural growth rates), and a shortage of financial resources, such physical planning activity was urgent. The Structure Plan intended to be a legal tool for regulating the location of human activities and stopping informal land supply activity. It proposed three different scenarios for the city's development. However, its implementation was doomed from the start. At that time, the Maputo Council had very little awareness of the importance of urban planning. The political and administrative culture of centralization did not help. Moreover, there was no legal framework for its approval by the municipal council. This was aggravated by the absence of mechanisms for the implementation and control of development proposals. The Structure Plan was therefore never put into practice.

Later, in 1999, another attempt was made concerning land use planning in the Maputo metropolitan area. The Greater Maputo Metropolitan Structure

"covered the two municipalities of Maputo and its satellite Matola, and areas between these and the two nearest towns in Maputo province to the north and south. It hence involved three administrations, namely, two autonomous municipalities, newly elected during the course of the project, and one appointed provincial government of long standing. (…) Hence, while the structure plan proposal called for metropolitan functions to be developed for certain activities (e.g., land management and development, infrastructure planning and environmental protection), there was no way that this could be implemented for all three administrations, and only possibly for the two municipalities if they choose to form an association. In practice, it is quite likely that they will compete rather than collaborate." (Jenkins, 2000)

These failed attempts to have an instrument for urban land use management during a long period, when urban demographic growth and its consequent demand for land was critical, led to an explosion of self-produced settlements on the peripheries. In fact, this 13% of residential areas of the Maputo municipality in 1973, two years before Independence, corresponded to 41% in 2008, when the new structure plan was approved (Henriques, 2008).

The Urban Structure Plan of Maputo Municipality of 2008 was the first practical consequence of the approval of a set of regulations concerning the definition of land use strategies in urban areas of Mozambique. For the first time after Independence, the citizens of Maputo had an instrument which had been subjected to a democratic process of approval, debate, and public participation which was able to support procedures for rational spatial organization. This plan intended to be a document that illustrated the locations and the relations between geographical, natural, and infrastructural features in order to manage the evolution of the city as required and to provide a framework for decisions regarding its future development. It was not only a necessary document but a mandatory one since the legislation at that time concerning territorial planning required the preparation and the approval of spatial planning instruments by the local administration under municipal jurisdiction.

The Urban Structure Plan of Maputo was designed with the following principles of urban development in mind (consistent with the Constitution): i) the right to the city; ii) the right to urbanized land; iii) the right to adequate housing; iv) the right to environmental sanitation; v) the right to infrastructures, utilities, and urban amenities for education, health, information, and culture, sport, leisure, and safety; vi) the right to participation. In addition, with the establishment of the following priorities to achieve those principles: i) reordering of informal neighborhoods; ii) recuperation of land for social activities, services, and public amenities; iii) adequate city occupation densification; iv) reservation of areas to restore the ecological balance and to ensure environmental quality; v) establishment of the necessary conditions for mobility; vi) support for social housing construction; vii) encouragement of urban agricultural activities; viii) carrying out the diversification of urban activities to avoid suburbanization and spatial and environmental segregation of low-income households; ix) reduction of inequalities and privileges concerning the selection of sites for network infrastructures and social amenities (Conselho Municipal de Maputo, 2008).

The principles and priorities listed above aimed at achieving three main objectives: qualification of land, the control of environmental impacts, and the correction of social inequalities manifested through the existing urban land use. Urban qualification involves land use classification and the definition of its indexes and densities for present and future urban development. The control of environmental impacts is supported by the definition of *non aedificandi* areas in relation to the ecological structure of the city, such as water bodies, mangrove and humid areas, beaches, green spots and others, and the definition of partially protected areas with specific regulation concerning their use. The correction of social inequalities implies, among other interventions, the renewal of neighborhoods and the improvement of accessibility from each neighborhood to the city center and within the geographical region.

Maputo's growth during recent decades has been an extensive one with a densification process only in the neighborhoods close to the city center and therefore leading to a nearly complete occupation of the municipal territory. The exceptions are: the green belt along the river Infulene and along the Eastern coast (consisting here of urban agriculture, mangrove, and other flooded areas), Catembe on the south bank of the estuary, and Inhaca island (*Figure 2*).

Figure 2. Main land use classes in Maputo Municipality in 2018

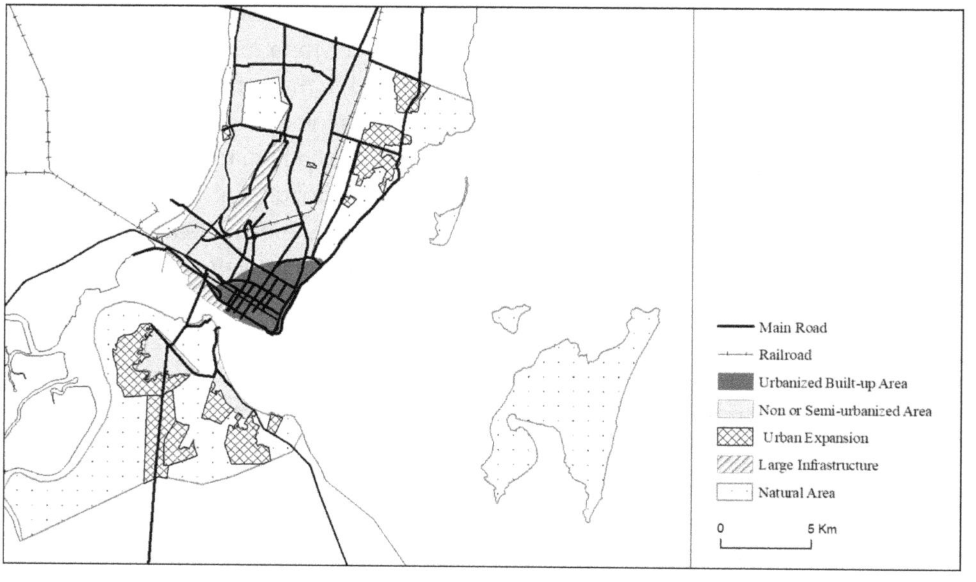

Given this scenario, the Urban Structure Plan proposed a densification of the more peripheral areas as well as some new areas for urban expansion (30% more than the current residential areas), either along the coastal area or in Catembe. These new areas for urban expansion have meanwhile been subjected to Partial Urbanization Plans. As Catembe is located on the south bank, improvement of fluvial transportation is foreseen as well as the construction of a bridge. Besides the expansion and densification of the residential areas, the Urban Structure Plan also proposes eleven multipurpose Urban Centers throughout the municipal territory and therefore reversing the polarization of activities in the prior city center, structuring the territory in a balanced way and with principles of spatial justice.

To achieve the inclusion of the different parts of the municipality and to incorporate the new multipurpose centers, a new accessibility and mobility structure is planned. This Plan proposes a consolidation of the existing road network through improving its infrastructure and the creation of new major circulation infrastructures.

From what was said, it is clear that Mozambique has made significant progress regarding the practice and culture of land use planning and has the legal framework necessary to implement different planning processes. However, some problems remain and still pose some challenges when it comes to enhancing the promotion of the territory as a more attractive, sustainable, and fair one. The great pressure on land has led to the need to establish new areas for expansion. However, suitable land use practices should be tailored, especially in places sensitive to flooding and erosion. The experience of the past (notably in 2000) proved that the occupation of these sites, although in that case involving unplanned and precarious situations, can be catastrophic.

Regarding technical issues and in order to track and monitor the implementation, and now to review the Urban Structure Plan, a different organizational structure is needed. However, in spite of the major investment made by the Municipal Council concerning the creation of a georeferenced database and tools for spatial analysis, the Urban Structure Plan of Maputo was not designed to be manipulated and tracked by a Geographical Information System. Its regulation also sustains the need to disseminate and to ensure public consultation, but there is no platform to ensure this process. The technical departments of the municipality, namely the ones related to urban management and planning, still work in a gated way. Most of the information produced therein is not shared with other institutions that produce relevant data for the planning process and therefore duplicate and generate inconsistent datasets. A need to take a step forward in the land use planning processes is obvious and necessary to achieve the desired spatial and social justice, but it will have to face the game of competing interests from the different stakeholders that play a major role in this process. The creation of open, user-friendly, and participatory platforms supported by geoinformation and technological tools for sharing knowledge among different stakeholders can enhance effective urban planning and management and promote a sustainable future for the Maputo territory.

Land Use Change In Maputo Between 1964 and 2018: Changing Patterns and Their Consequences

Analysis of the change in land use between 1964 and 1973, before Independence, provides for the identification of the main change: almost 30% of the total land use change corresponds to an alteration from "Other Natural, semi-natural and Leisure Areas" to "Peripheral Residential Area," leading to an occupation that is mainly residential made through the appropriation of "Bush Areas" (Table 1). The period coincided with the implementation of strategic policies intended to maintain the Portuguese presence in the colonies as a response to nationalistic movements, resulting in a migration wave from the metropolis to the colonized territories and a certain level of investment in industry and infrastructures.

The years between 1973 and 1982 were years of political change in Mozambique. The Independence brought changes to the territorial administration systems, particularly due to the nationalization of land and built property. The rural exodus had reflexes in the urban landscape especially in the growth of the periphery. The incentives given to agricultural production and the stimulus to create cooperatives had its main impact in the landscape of the utilization of the best soil areas for agricultural production.

Table 1. Land use rate of change (%) between 1964 and 2010 in Maputo City[1]

Land Use Classes	1964 -1973	1973 -1982	1982 -1991	1991 -2000	2000 -2009	2009-2018
Residential Areas (Urbanized)	17.1	-0.9	1.6	13.2	16.1	10.1
Residential Areas (non- or semi-urbanized)	70.0	45.6	48.0	53.2	13.0	0.8
Social Amenities, Infrastructure and Public Services Areas	24.3	12.0	6.2	2.7	14.0	3.8
Agricultural and other Economic Activity Areas	5.3	81.9	80.3	-23.8	26.6	-4.1
Vacant and Derelict Urban Areas	24.9	-10.4	-0.9	-41.3	38.8	-9.3
Water Bodies and Floodable Areas	13.2	-13.8	-45.5	86.9	-56.4	-7.1
Other Natural, Semi-natural and Leisure Areas	-18.5	-24.5	-41.1	-60.5	-72.3	-0.5

[1] Does not include Catembe and Inhaca.

Source: Author's aerial image interpretation and field work.

Figure 3. Residential Area in Maputo between 1964 and 2018

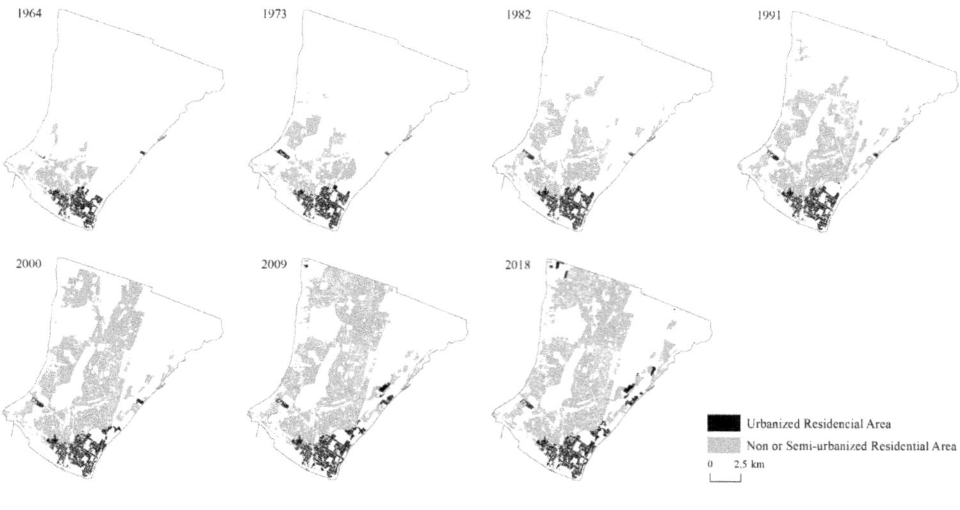

Between 1982 and 1991, the civil war was still devastating Mozambique. As a consequence, the rural exodus to Maputo city intensified, stressing the accelerated growth of the periphery. Many areas that were rejected for residential purposes were occupied with houses during this period.

During the next decade, the Peace Agreement and the democratization of the political system influenced the process of urban development. In terms of land use change, these years were clearly marked by the expansion of the periphery to remote areas (Figure 3). Until 2008 there was no legal framework and no planning instruments to control urban expansion. Several *ad-hoc* measures were taken by different authorities and carried out in specific areas (such as the Maxaquene Pilot-Project), in an attempt to eliminate the contrasts between the formal colonial city and the self-produced neighborhoods through reordering its precarious residential areas, but still lacking a global policy for land use within the municipal territory of Maputo that could change the pattern of spatial decisions.

The considerable rate of change in the territory, in a short time followed by the outgrowing of financial and management capacity, left scars that were not possible to heal and which actually compromised the future: vast areas with precarious settlements, lack of security of tenure, considerable environmental sensitivity are obvious in the landscape of the city of Maputo. Desirable resilience should imply, among other things, political strategies based on thorough knowledge about the functioning of the interactions that take place among the various stakeholders that shape the territory.

The great pressure on land has led to the need to establish new areas for expansion. However, suitable land use practices should be tailor made, especially in places sensitive to flooding and erosion. The experience of the past (notably with the floods of 2000 and 2001) showed that the occupation of these sites, although in that case referring to informal and precarious situations, can be catastrophic. Moreover, the lack of an environmental impact assessment, concerning the new large infrastructures already under construction, has been pointed out and criticized by various groups of citizens concerned with the sustainable development of the region.

At the social level, the issues relating to the resettlement of the population as a consequence of the construction of the new infrastructures foreseen in the plan are worthy of note. Despite the recent regulation on the resettlement process resulting from economic activities (Decree-law 31/2012 of 8 August) trying to protect the rights of affected populations, in practice, the displacement of families to more peripheral areas, most of which being without basic infrastructures was a widespread occurrence. A conflict of interests by stakeholders and the high illiteracy level in communities offers a difficult scenario for the effective performance and transparency of a praxis process.

Many other examples of political, technical, environmental, and social issues could be addressed in this context. Perhaps what is most important to emphasize here is the importance that the new land use planning paradigm that Mozambique is aiming to draw up does not deviate from public interest, social, and spatial justice, environmental sustainability and transparency of processes, taking full advantage of the opportunities created by the growing demand for this territory.

Census data from 1997, 2007, and 2017 provide a demographic and socio-economic picture of the city and combining this with geometric and semantic data supplied by aerial images are basic elements for the creation of monitoring and evaluation indicators for an urban development model, as well as supporting the definition of policies that should guide the future of the territory. The idea of "socializing the pixel" and "pixelizing the social", as discussed by Geoghegan, et al. (1998) is a potential area of research and action here if framed by an urban observatory laboratory.

FUTURE RESEARCH DIRECTIONS

Land use monitoring for planning and management purposes in fast-growing cities can greatly benefit from currently available geoinformation technologies that provide an integrated view of the environment and the agents that intervene in it.

However, to enhance the advantages that these technologies may bring in improving the quality of life and sustainability in adverse contexts with a strong transformational dynamic, certain steps forward should be considered and evaluated:

1. Developing and integrating effective methodologies for obtaining and providing data from remote sensors within planning and management processes;
2. Providing technological means to add and show significant geographic information to promote a debate around urban issues among researchers, practitioners, decision-makers, citizens, and other stakeholders, and to strengthen partnerships, while improving knowledge and action methodologies;
3. Revitalizing urban planning education, introducing diverse forms of reading the territory, particularly through the use of geoinformation technologies, and encouraging universities to embrace long-term research programs concerning emerging urban issues of their cities;

CONCLUSION

Land is the resource that populations can use to express their culture. Planning land use – by understanding and clarifying the complex relations between the different components that produce space – is an indispensable step towards providing a framework for decisions regarding the future development of any country.

Urban planning processes should incorporate monitoring and assessment phases that can be supported by geoinformation technologies. Geographic data supported by these technologies enables the definition and use of indicator systems to evaluate contexts, realities, and trends; in other words, permanent monitoring by detecting deviations or variations in previous situations and by identifying the progress already achieved compared to the intended or expected final results.

The city of Maputo was set up as a laboratory to show the relation between territories with a large population concentration and functions and the technology that describes and should monitor this concentration in urbanistic, economic, social, and geographic terms. At the technical level, there are already several conditions in place for it to become a methodological case study for action. If framed by a strong link between research and urban policy and practice (as was already started through collaborating with the municipality in the developing of land use plans), it can have a stronger role in urban planning activities.

Availability of up-to-date data is the most important bottleneck in preventing geoinformation technologies from being used effectively in developing countries, but current free or low-cost data from remote sensing platforms can help to overcome these difficulties. Moreover, several crowdsourcing programs led by civil society organizations have begun to play a significant role in the construction of information on the territory. Geoinformation technologies anchored within a land use observatory of fast-growing cities could be a groundbreaking project to engage researchers, planners, policymakers in promoting a better understanding of the urban phenomena and their impacts and, therefore, in achieving the desired sustainability goals.

ACKNOWLEDGMENT

This research was funded by CIAUD-Research Centre for Architecture, Urbanism and Design, grant number UIDB/04008/2020.

The author would also like to acknowledge the contribution of José António Tenedório.

REFERENCES

Atkinson, A. (2005). Urban development: Reviving and activating Utopian strategies. *City*, *9*(3), 279–295. doi:10.1080/13604810500392548

Azevedo, M. (1969). *O Plano Director de Urbanização de Lourenço Marques*. Separata do Boletim n. 7. Câmara Municipal de Lourenço Marques.

Bai, X., Dawson, R. J., Ürge-Vorsatz, D., Delgado, G. C., Barau, A. S., Dhakal, S., Dodman, D., Leonardsen, L., Masson-Delmotte, V., Roberts, D., & Schultz, S. (2018). Six research priorities for cities and climate change. *Nature*, *555*(7694), 23–25. doi:10.1038/d41586-018-02409-z

Batty, M. (2013). *The New Science of Cities*. The MIT Press. doi:10.7551/mitpress/9399.001.0001

Conselho Municipal de Maputo. (2008). *Plano de Estrutura Urbana do Município de Maputo (PEUMM)*. Conselho Municipal de Maputo.

Geoghegan, J., Pritchard, L., Jr., Ogneva-Himmelberger, Y., Chowdhury, R. R., Sanderson, S., & Turner, B. L. (1998). "Socializing the Pixel" and "Pixelizing the Social in Land-Use Land-Cover Change". In D. Liverman, E. Moran, R. Rindfuss, & P. Stern (Eds.), People and Pixels. Linking Remote Sensing and Social Science (pp. 51-69). National Academic Press.

Henriques, C. D. (2008). *Maputo. Cinco Décadas de Mudança Territorial. O uso do solo observado por tecnologias de informação geográfica*. Instituto Português de Apoio ao Desenvolvimento.

Henriques, C. D., & Tenedório, J. A. (2010). Mapping Urban Change to Plan the Future: Maputo City Change Model and the Municipal Structure Plan. In *Remote Sensing for Science, Education, and Natural and Cultural Heritage: Proceedings of the 30th EARSeL Symposium*. Paris: EARSeL.

Jenkins, P. (2000). Urban management, urban poverty and urban governance: Planning and land management in Maputo. *Environment and Urbanization*, *12*(1), 137–152. doi:10.1177/095624780001200110

Jenkins, P. (2013). *Urbanization, Urbanism, and Urbanity in an African City. Home Spaces and House Cultures*. Palgrave Macmillan. doi:10.1057/9781137380173

Mas, J. F., Kolb, M., Paegelow, M., Olmedo, M. T., & Houet, T. (2014). Inductive pattern-based land use/cover change models: A comparison of four software packages. *Environmental Modelling & Software*, *51*, 94–111. doi:10.1016/j.envsoft.2013.09.010

Pinto, N. N., Tenedório, J. A., Antunes, A. P., & Cladera, J. R. (2014). *Technologies for Urban and Spatial Planning: Virtual Cities and Territories*. IGI Global. doi:10.4018/978-1-4666-4349-9

Rabinovitch, J. (1996). Innovative land use and public transport policy: The case of Curitiba, Brazil. *Land Use Policy*, *13*(1), 51–67. doi:10.1016/0264-8377(95)00023-2

Sawicki, D. S., & Flynn, P. (1996). Neighborhood Indicators: A Review of the Literature and an Assessment of Conceptual and Methodological Issues. *Journal of the American Planning Association*, *62*(2), 165–183. doi:10.1080/01944369608975683

Tenedório, J. A., Estanqueiro, R., Matos Lima, A., & Marques, J. (2016). Remote sensing from unmanned aerial vehicles for 3D urban modelling: case study of Loulé, Portugal. In *Back to the Sense of the City: International Monograph Book* (pp. 1118–1128). Centre de Política de Sòl i Valoracions.

United Nations. (2017). *New Urban Agenda*. https://www.habitat3.org/

United Nations Development Programme. (2016). *Sustainable Urbanization Strategy. UNDP's support to sustainable, inclusive and resilient cities in the developing world*. https://www.undp.org/content/undp/en/home/librarypage/poverty-reduction/sustainable-urbanization-strategy.html

United Nations Human Settlements Programme. (2016). *Urbanization and development: emerging Futures. World Cities Report 2016*. https://unhabitat.org/sites/default/files/download-manager-files/WCR-2016-WEB.pdf

ADDITIONAL READING

Babere, N. (2015). Social Production of Space: "Lived Space" of Informal Livelihood Operators; the Case of Dares Salaam City Tanzania. *Current Urban Studies*, *3*(04), 286–299. doi:10.4236/cus.2015.34024

Henriques, C. D., & Tenedório, J. A. (2009). Remote Sensing, GIS Application and Simulation of Coastal Land Use Changes Based on Cellular Automata: A Case Study of Maputo, Mozambique. *Journal of Coastal Research*, *56*, 1518–1521.

Melo, V. (2013). Urbanismo português na cidade de Maputo: passado, presente e futuro. *URBE - Revista Brasileira de Gestão Urbana (Brazilian Journal of Urban Management)*. *5*(1), 71-88.

Schindler, S. (2017). Towards a paradigm of Southern urbanism. *City*, *21*(1), 47–64. doi:10.1080/13604813.2016.1263494

Simwanda, M., & Murayama, Y. (2017). Integrating Geospatial Techniques for Urban Land Use Classification in the Developing Sub-Saharan African City of Lusaka, Zambia. *ISPRS International Journal of Geo-Information*, *6*(4), 102. doi:10.3390/ijgi6040102

Todes, A., Karam, A., Klug, N., & Malaza, N. (2010). Beyond master planning? New approaches to spatial planning in Ekurhuleni, South Africa. *Habitat International*, *34*(4), 414–420. doi:10.1016/j.habitatint.2009.11.012

Ulbrich, P., Porto de Albuquerque, J., & Coaffee, J. (2019). The Impact of Urban Inequalities on Monitoring Progress towards the Sustainable Development Goals: Methodological Considerations. *ISPRS International Journal of Geo-Information*, *8*(1), 6. doi:10.3390/ijgi8010006

United Nations Economic Commission for Africa. (2018). *Geospatial information for sustainable development in Africa: African Action Plan on Global Geospatial Information Management - 2016-2030*. https://repository.uneca.org/handle/10855/23834

KEY TERMS AND DEFINITIONS

Geoinformation Technologies: Technological tools used for measuring, mapping, modeling, and visualizing features from human and natural environments.

Land Use: The functional use of land units based on human activity.

Self-Produced Settlements: Non- or semi-urbanized areas produced more or less autonomously in relation to state order and control by the resident population.

Spatial Justice: When social resources are adequately distributed in space promoting equitable opportunities to benefit from them.

Urban Indicators: Tools that can provide a comprehensive means for the monitoring and assessment of urban conditions, trends, and policies.

Urban Observatory: Infrastructure element to record, analyze and interpret the cumulative results of human action, translated into the physical transformation of cities and enabling the monitoring and assessing of changes and policies, through a set of dedicated indicators.

Urban Structure Plan: Land use normative planning instrument, also known as master plan.

Chapter 3
Modelling Land-Use and Land-Cover Changes:
A Hybrid Approach to a Coastal Area

Raquel Faria de Deus
https://orcid.org/0000-0002-9274-2049
Interdisciplinary Centre of Social Sciences (CICS.NOVA), NOVA FCSH, Universidade NOVA de Lisboa, Portugal

José António Tenedório
https://orcid.org/0000-0003-3047-0807
Interdisciplinary Centre of Social Sciences (CICS.NOVA), NOVA FCSH, Universidade NOVA de Lisboa, Portugal

Jorge Rocha
https://orcid.org/0000-0002-7228-6330
Institute of Geography and Spatial Planning, Universidade de Lisboa, Portugal

ABSTRACT

In this chapter, a hybrid approach integrating cellular automata (CA), fuzzy logic, logistic regression, and Markov chains for modelling and prediction of land-use and land-cover (LULC) change at the local scale, using geographic information with fine spatial resolution is presented. A spatial logistic regression model was applied to determine the transition rules that were used by a conventional CA model. The overall dimension of LULC change was estimated using a Markov chain model. The proposed CA-based model (termed CAMLucc) in combination with physical variables and land-use planning data was applied to simulate LULC change in Portimão, Portugal between 1947 and 2010 and to predict its future spatial patterns for 2020 and 2025. The main results of this research show that Portimão has been facing massive growth in artificial surfaces, particularly near the main urban settlements and along the coastal area, and reveal an early and intensive urban sprawl over time.

DOI: 10.4018/978-1-7998-2249-3.ch003

INTRODUCTION

Sustainable land-use and land-cover (LULC) development requires increasingly more reliable and high-quality spatial information to guide smart spatial planning and land-use decision-making (Magarotto, Faria-De-Deus, Ferreira Costa, & Masanet, 2017; Rocha, 2019). This applies in particular to the geographical context of coastal areas, which have been undergoing rapid and strong urbanization dynamics in the past few decades. The fast population and economic growth along coastal areas presents both an opportunity and a challenge for the sustainability of coastal ecosystems and built environments (Freire, Santos, & Tenedório, 2009; Martinez et al., 2017). Presently, 2.5 billion people, i.e., 40% of the world's population, live within 100 km of the coast (Burke, Reytar, Spalding, & Perry, 2011) adding great pressure to these productive and valuable coastal ecosystems. As coastal populations increase and natural coastal protection is reduced or lost, the sea level rise and local impacts of land-based stress factors, namely LULC change, occur together with global and regional stressors, such as climate change (Burke et al., 2011), resulting in altered dynamics and the subsequent reduced resilience of territories (Rocha, Ferreira, Simões, & Tenedório, 2007; Rocha, Gutierres, Gomes, & Teodoro, 2018).

In Portugal, as in the rest of the world, this has posed the challenge of rethinking the types of innovative actions that may contribute to more efficient LULC planning policies. Moving in this direction implies the application of state-of-the-art technologies to the studies targeting the analysis, monitoring, assessing and modelling of the process of LULC change in coastal areas (Boavida-Portugal, Rocha, & Cardoso Ferreira, 2016). The scientific advances that have been made in the field of urban growth and LULC change Cellular Automata (CA)-based simulation since the late 1990s have confirmed CA-based simulation as one of the most remarkable and promising geographic modelling approaches to interpret and represent the spatiotemporal complexity of LULC change, mainly due to their natural ability to capture and reproduce complex bottom-up processes through simple rules (Bone, Dragicevic, & White, 2011; Verburg et al., 2015; White, Engelen, & Uljee, 2015).

Research on LULC change CA-based modelling has since evolved a great deal, moving from its main focus on understanding the drivers and impacts of LULC change towards novel approaches (see, for example, Li & Gong, 2016; Gomes, Abrantes, Banos, & Rocha, 2019), which use this knowledge to properly develop sustainable land-use management practices and spatial policies. Even though CA-based models have already proven to be very effective at modelling and predicting large-scale LULC change, research on the development of fine-scale and long-term series LULC datasets to evaluate, quantify, and model LULC change, particularly at the local level, is still lacking. In order to produce more realistic urban phenomena simulations, CA-based models should improve their ability to simulate at a high level of spatial resolution. At the same time, increasingly complex urban spatial models that are very time-dependent require the use of physical and socioeconomic data from several time periods, since previous research has shown that the calibration process over time improves the results of simulation models (Liu, 2009). Thus, objectively approaching the analysis and modelling of LULC change at a fine spatial scale should be central for urban planning teams to efficiently define and conduct forthcoming land-use planning decisions.

This Chapter looks to address the issue of LULC change in coastal areas through the application of a CA-based model (CAMLucc) to the municipality of Portimão, which has one of the fastest developing urban coastal areas in Portugal. In this Chapter, a new hybrid framework integrating Geographical Information Systems, fuzzy logic, logistic regression, Markov chain analysis, and CA is introduced and implemented for the analysis, modelling, and prediction of LULC change from 1947 to 2025. The innovative nature of this research comes from the analysis and modelling of LULC change at a fine spatial scale together with the use of highly detailed LULC categorization and long-term series of LULC maps, thus advancing our understanding of LULC change patterns, processes, and structural trends. The Chapter begins with background information on CA LULC change modelling to specify both the principles and the terminology involved. Section 2 is devoted to the methodological framework that was tested in the municipality of Portimão. Section 3 describes the study area and the spatial information necessary for this research. In section 4, the implementation of the CAMLucc model is presented and the results of the calibration, simulation, and validation process are discussed. The conclusion and future research directions are provided in Section 5.

BACKGROUND: CELLULAR AUTOMATA FOR LAND-USE AND LAND-COVER CHANGE AND URBAN GROWTH MODELLING

CA models have been one of the most productive approaches within the urban Geosimulation field of studies in the recent past (Benenson & Torrens, 2004). Although many studies have been dedicated to the addition of new components to perfect the functionality of formal CA first introduced by Ulam (1952), von Neumann (1966) and Wolfram (1983, 2002), at the most basic level, a CA is most often an n-dimensional regular matrix (usually one or bi-dimensional raster) but can also be an irregular (vector) lattice (Semboloni, 2000; Pinto, Antunes, & Roca, 2017). Each cell is represented by a discrete state – which is iteratively and synchronously changed over time *t* based on its previous states, the characteristics of its neighbouring cells, and the recursive application of a given set of transition rules (Torrens & O'Sullivan, 2000).

CA models revolutionised urban and spatial modelling (Couclelis, 2005). These start from a microscopic approach in which the structure of urban space is defined by spatial relations between interacting automata in order to easily reproduce the geographic reality. Thus, a CA model may be defined by the following five components: an *n* matrix of cells, the state of the cells, a neighbourhood that defines the extension to which spatial interactions between cells are calculated according to their state (Figure 1), the transition rules that change the state of the cells at a given $t+1$ moment simulating their spatial transitions, and time, which gives CA models their dynamic character (Torrens & Benenson, 2005). Clarke, Parks, & Crane (2002) also stress that these essential elements provide a large capacity to capture new, emergent, self-organized phenomena and not only to extrapolate future trends.

The improvements in the field of urban growth and LULC change CA-based simulation that have been made since the late 1990s have re-affirmed CA modelling as one of the most remarkable and promising modelling approaches to interpret and represent the spatial and temporal complexity of LULC change (White & Engelen, 1994, Portugali, 2002, Xia Li & Yeh, 2004; Pontius, Shusas, & McEachern, 2004; Petrov, Lavalle, & Kasanko, 2009; Caruso, Peeters, Cavailhes, & Rounsevell, 2009; Bone et al., 2011). In order to provide an overview of the recent trend towards the applications of CA-based models to modelling LULC change and urban growth, we used B-on and Google Scholar to search for articles containing any of the terms "cellular automata+land-use and land-cover change" and "cellular automata+urban growth". Table 1 lists the selected CA applications along with the cell size, time frame, and simulated LULC category information.

Even though CA-based models have been widely used for the analysis of urban phenomena, research on CA land-use models employing a high level of spatial resolution is limited. This could be due to the cost associated with acquiring high-resolution imagery, the sheer data size and/or technological constraints related to the use of image processing software and GIS. However, Table 1 shows that temporal resolution has been increasing in recent years. Thus, a trend can be envisaged towards wider more comprehensive spatial studies with coarser spatial resolution and long-term datasets (both in calibration and simulation), addressing the concern about global issues/challenges, e.g., climate change, food/water security, smart and green growth, etc.

There are, nonetheless, a few remarkable examples. Stanilov & Batty (2011) addressed the challenge of improving the representation of the morphology of the land-use patterns of West London, England, by adapting the METRONAMICA modelling system to simulate its historical urban growth from 1875 to 2005 on a 20-year basis. The model incorporates nine land-use categories and uses a cell size of 25 metres. The land-use data layers were extracted from historical OS maps (1:2 500 scale) using the visual image interpretation technique. The research carried out by Stanilov & Batty (2011) focused on the search for more effective CA modelling. Their improvements on the METRONAMICA modelling system were an important contribution to the analysis of urban phenomena and estimation of long-term forecasts. Rocha (2012) has developed a CA-based modelling framework named UrbGeoSim that combines fuzzy logic, predictive sub-models, and CA in a GIS environment in order to predict the urban growth of the Almada municipality in Portugal (a peri-urban area integrated in the Lisbon metropolitan area) for the year 2021. UrbGeoSim uses a cell size of 5 metres and incorporates existing maps for that urban area (years 1995, 1998, 2000, 2006, and 2008) that were derived from historical cartography (1:28 000 scale), aerial photos (1:25 000 scale) and orthophotomaps (1:10 000 scale). His work aims to re-address and re-evaluate several predictive models already recognized as robust tools and relate them in order to provide a better understanding of urban growth dynamics, as well as suggest future directions for research and point out the challenges that may lie ahead.

Figure 1. Classical 2-dimensional CA neighbourhoods: a) von Neumann's 3x3 neighbourhood, b) Moore's 3x3 neighbourhood, and c) von Neumann's 5x5 neighbourhood
Source: adapted from Benenson & Torrens (2004).

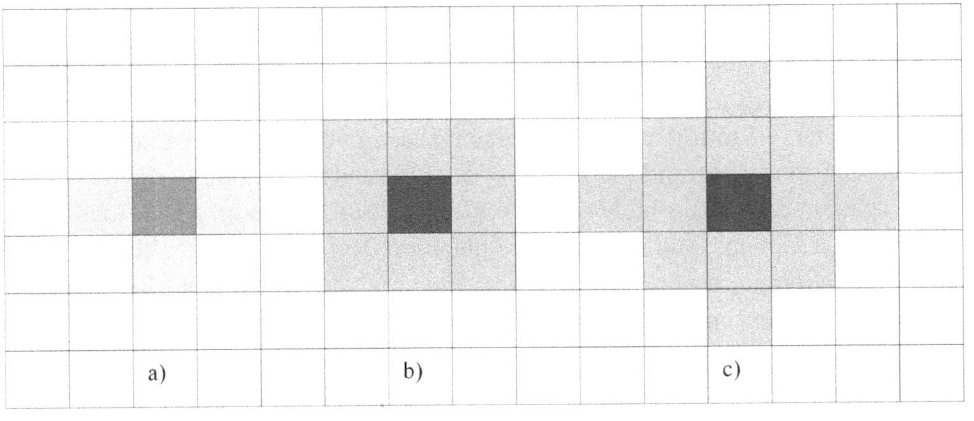

Table 1. Selected studies on applying CA-based models for LULC change and urban growth

Paper	Model	Cell size	Time frame	Simulated LULC categories	Study area
Agyemang & Silva (2019)	SLEUTH	200x200m	Calibration: 1990, 2000, 2010, 2015; Simulation: 2020, 2025, 2030, 2035, 2040.	Built-up; Non built-up	Accra city-region, Ghana
Liang, Liu, Li, Zhao, & Chen (2018)	FLUS	100 x100 m	Calibration: 2000, 2005, 2008, 2010, 2013; Simulation: 2019, 2031, 2040, 2052.	Urban; Non-urban, Water	Pearl River delta region, China
Ustaoglu, Williams, Petrov, Shahumyan, & Van Delden (2018)	MOLAND	100 x100 m (urban classes); 3 ha (others)	Calibration: 1990, 2000 and 2006 Simulation: 2026, 2040, 2045, 2050, 2055.	24 land-use classes (9 are urban)	Dublin, Ireland
Naghibi, Delavar, & Pijanowski (2016)	ABC-CA (Artificial Bee Colony)	30x30 m	Calibration: 1997, 2006, 2015; Simulation: 2016.	Urban, agricultural lands, wastelands	Urmia, Iran
Rocha (2012)	UrbGeoSim	5x5m	Calibration:1995, 1998, 2000, 2006, 2008; Simulation: 2021.	Urban, Non-urban	Almada, Portugal
Stanilov & Batty (2011)	Metronamica	25x25 m	Calibration:1875, 1895, 1915, Simulation: 1875 to 2005.	9 land-use classes	West London, England

As shown in Table 1, previous CA-based models have favoured one resolution over the others (e.g., temporal, cell size and number of simulated LULC categories). In this Chapter, we introduce and implement a new hybrid modelling approach for the analysis, modelling and prediction of LULC change combining the use of (i) a fine spatial scale; (ii) a highly detailed LULC categorization; (iii) a long-term series of LULC maps; and (iv) land-use planning data both in the calibration and in the simulation processes, which can advance our understanding of LULC change patterns, processes, and structural trends.

METHODOLOGICAL FRAMEWORK

In this study, the CAMLucc model is introduced to model LULC change at a fine resolution (10 m) from 1947 to 2010 and to predict its future spatial patterns for two periods: from 2010 to 2020 (10-year period) and to 2025 (15-year period).

Figure 2. Conceptual CAMLucc model

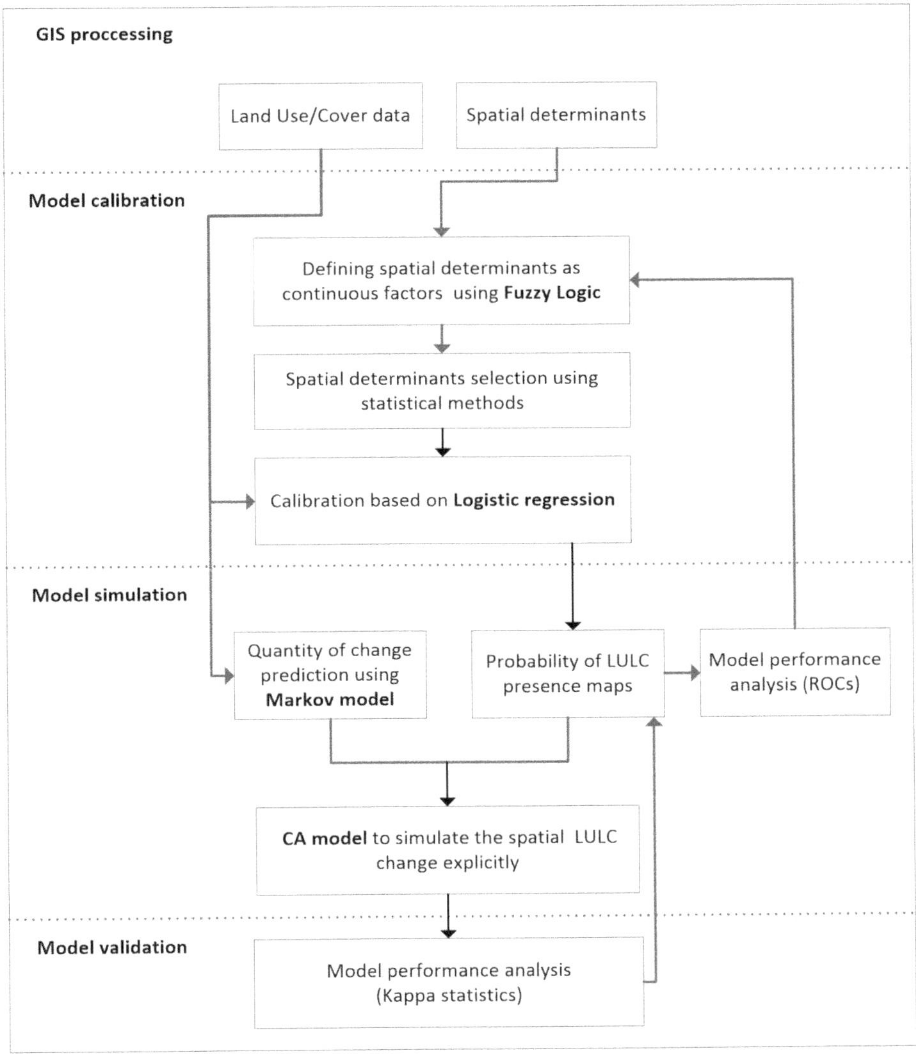

CAMLucc Model Framework

The CAMLucc model is implemented by combining fuzzy logic, logistic regression, Markov chain analysis, and a CA model into a hybrid framework, in which LULC changes emerge through local interaction according to transition rules that reflect the effect of key determinants on LULC change. In this study, besides physical determinants, we have also incorporated institutional determinants that together govern LULC change in varied ways. Figure 2 provides an overview of the conceptual CAMLucc model framework that consists of three core components: model calibration, simulation, and validation.

This hybrid methodology comprises a step-by-step model design to (i) deal with the spatial and temporal complexity of LULC change and (ii) assess and validate each increment to the CAMLucc model. The application of this gradual design of the CAMLucc model should provide insights into the spatial evolution patterns of Portimão's LULC change and their temporalities. The CAMLucc framework is structured around three main concepts or dimensions: transition, location, and temporality, which we believe that together may effectively provide a good understanding of LULC change. Firstly, the main principle of the proposed approach is that LULC changes emerge by means of the transition of the states of cells, which in turn is governed by the spatial interaction between each cell and its surroundings according to specific transition rules. Secondly, as our analysis of LULC change and modelling will concern both its spatial and temporal dynamics, location-based data are used to generate transition rules in order to obtain more realistic simulations of land-use changes. Additionally, a logistic regression is used in the calibration process of the CAMLucc model. The combined effect of different spatial determinants on LULC configuration was quantified from the logistic regression model that was further applied to estimate preferred locations for each LULC to occur. The technique used to quantify the temporal change in each land-use transition is Markov models, which provide robust and flexible techniques in modelling the temporality of several phenomena, including LULC change. There is one drawback in addressing the LULC change prediction by means of Markov models, however. There is no spatial component in the Markov model's final results. A CA model is therefore applied to make the dynamics of LULC change spatially explicit.

Calibration Using Logistic Regression

Several CA models have been established through empirical estimation methods using statistical techniques in order to determine their transition rules and parameters. Logistic regression as an empirical estimation method is one of the preferred approaches for CA modelling mainly due to the transparency of the parameters and straightforward analysis of the results (Al-sharif & Pradhan, 2013). This method is well suited to combine the influence of different independent variables (either categorical or continuous variables) on the dependent variable (binary categorical variable, e.g., presence/absence, 1/0) resulting in spatially explicit models of probability for a single dependent variable.

The transition rules of the CAMLucc model were configured using a binomial logistic regression approach (BLR). The integration of the continuous independent variables defined in section 3 into a single land-use suitability map per each land-use category was specified by the following general BLR formula:

$$p\left(y_{il}=1|x_j\right)=\frac{exp\left(\sum_{j=1}^{k}\beta_{jl}x_{ji}\right)}{1+exp\left(\sum_{j=1}^{k}\beta_{jl}x_{ji}\right)} \tag{1}$$

and

$$logitP\left(Y_{il}=1|x_j\right)=ln\left(\frac{p}{1-p}\right)=\beta_0+\beta_1x_1+\beta_2x_2+\ldots+\beta_kx_k+\varepsilon \tag{2}$$

where $p\ (y_{il}=1|x_i)$ is the probability of the occurrence of land-use category l at grid cell i (y_{il} being 1 if grid cell i pertains to land-use category l; y_{il} being 0 otherwise), k is the number of independent variables, x_{ji} is the value of grid cell i for the independent variable j and β_{jl} is the estimated parameter of independent variable j for land-use category l. The regression parameter β_{jl} represents the independent variable function (Al-sharif & Pradhan, 2013). The logistic transformation was performed on equation (1) through equation (2) where ε is the error term and β_0, β_1, β_2, β_k are the parameters of independent variables that were estimated by means of Maximum Likelihood Estimation method. A 10% sample of the dataset was used via a stratified random sampling in order to account for spatial autocorrelation on error terms while reducing processing time considerably.

Temporal LULC Change Simulation Using Markov Modelling

The Markov chain model is currently one of the most popular methods to simulate past LULC change and to forecast its stability quantitatively (Dadhich & Hanaoka, 2011) chiefly due to its inter-complementarity with both remote sensing data and GIS technology. The basic assumption of this kind of stochastic models that are grounded on the notion of a Markov process is that the state of a cell at any given time is merely a function of its preceding state (Muller & Middleton, 1994). The structure of Markov chains when applied to land-use modelling is grounded on two elements: *(i)* a transition matrix (P_{ij}) that defines the probability of a cell converting from state i to state j (in this case for each land-use category) in a discrete time period; and *(ii)* a vector (S_t) that represents the distribution of each LULC at time t (Pastor & Johnson, 1992). Another vector (S_{t+1}) describing LULC distribution for a given time period in the future can be calculated by multiplying S_t and P_{ij} (Peña, Nilsson, Björkegren, & Tegnér, 2007). Accordingly:

$$S_{t+1} = P_{ij} \times S_t \tag{3}$$

The transition probability matrix (P_{ij}) can be represented as (Hou, Change, & Yu, 2004):

$$P_{ij} = \begin{bmatrix} P_{11} & P_{12} & \cdots P_{1n} \\ P_{21} & P_{22} & \cdots P_{2n} \\ \cdots & \cdots & \cdots & \cdots \\ P_{n1} & P_{n2} & \cdots P_{nn} \end{bmatrix}. \tag{4}$$

Given that transitions between states are probabilities, this implies that:

$$0 \le P_{ij} < 1 \, and \sum_{j=1}^{N} P_{ij} = 1, \left(i, j = 1, 2, \ldots, n\right) \tag{5}$$

This means that must be a non-negative quantity, ranging from 0 to 1, and the row sums of must be 1. The probability of a transition between i and j (P_{ij}) may be estimated by means of the maximum likelihood method of the form (Anderson & Goodman, 1957):

$$\hat{P}_{ij} = n_{ij} / \sum_{j=1}^{n} n_{ij} \tag{6}$$

where n_{ij} represents the number of transitions occurred from i to j.

Rather than a simple linear projection of future LULC changes, as the Markov chain progresses in sufficient time frames, the probability of the Markov chain being at the subsequent time period in state j becomes independent from the former state of the chain. Thus, the transition probabilities of Markov models change over time as the different transitions achieve a *steady state distribution* (Weng, 2002) or an *equilibrium state* (Eastman, 2009), which may effectively specify the trend of the LULC change process. Another basic assumption of Markov chain models relates to *stationarity* or *temporal homogeneity* (JRC & ESA, 1994). As Weng (2002) already stated, it is unlikely that LULC change is stationary when LULC change itself reproduces the dynamics and interaction of several spatial determinants over time. However, this assumption has frequently been relaxed in research papers. It is generally assumed that the LULC change phenomenon is *reasonably stationary* in the short and medium terms (Weng, 2002). The importance of the stationarity assumption of Markov chain models lies in the fact that it will allow the prediction of LULC change for a given point in time using current LULC distribution and the change trend of the system. In this study, it was assumed that: *(i)* the Markov process holds LULC data; *(ii)* LULC transitions are treated as a stochastic process; and *(iii)* LULC categories are considered chain states.

STUDY AREA AND DATASETS

The Study Area: Portimão Municipality, Portugal

The proposed CA-based modelling approach was applied to the modelling and prediction of LUC change in the municipality of Portimão, located in the Western coast of the Algarve region, Portugal (Figure 3). According to the results of the Portuguese 2011 Census (INE, 2011), the largest growth in population (14%) and in the number of households (24.8%) occurred in the region of the Algarve. Since 2001, the number of dwellings and buildings has also increased significantly in this region to 36.9% and 24.9%, respectively. Indeed, the most meaningful increase in households in mainland Portugal occurred in the municipality of Portimão (54%). Such results show that the largest population growth continued to occur in coastal areas. The municipality of Portimão covers an area of approximately 183 sq. km and a total population of around 55 614 inhabitants. Population density in 2011 was 305.5 people per sq. km and the population growth rate was 43% between 1991 and 2011 (INE, 2011). Portimão is nowadays a reference municipality in the Algarve, it is one of the main tourist destinations both for national and foreign guests. It is currently the third most important cruise port in Portugal (following Lisbon and Funchal), which receives a wide number of passengers from cruise ships, thus strengthening the economy of tourism and services (the structuring activities of the municipality's economy). Regarding land-use/land-cover and planning, urban pressure for tourist purposes had induced considerable and continuous changes for the past 40 years, such as irreversible land-uses even in areas with strong physical restrictions, loss of agricultural land-use in favour of residential areas, infrastructures and leisure facilities (EEA, 2017), urban sprawl and related social-environment vulnerabilities. By 2015, around 14% of the study area was occupied by artificial areas (quantified using COS - *Carta de ocupação do Solo de Portugal*, 2015), mostly located by the coastline. Thus, in order to study this phenomenon with greater accuracy, we need to use higher-resolution spatial data. Furthermore, there is a current need for analysing, modelling, and predicting LULC change in order to assist both urban and coastal management.

Figure 3. Study area of this research. The figure contains elevation data available in Gonçalves (2011)

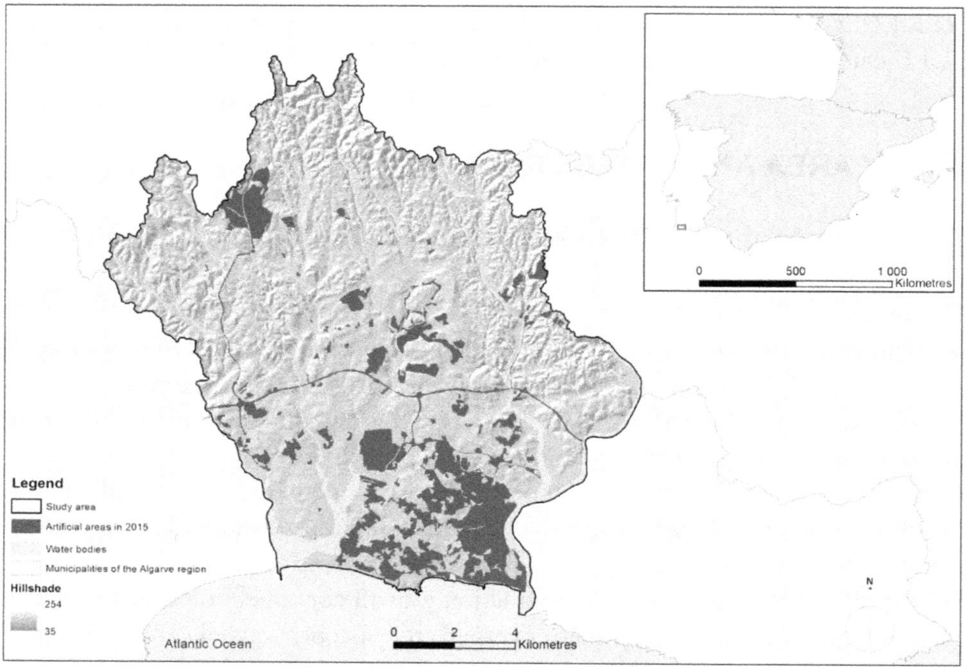

Data and Pre-Processing

Five main groups of variables were assumed to have impact on LULC change processes. These included state/transition variables, density variables, distance variables, topographic characteristics, and land-use planning policies. All spatial information was georeferenced to the ETRS89/PT-TM06 reference system (the official reference system for Portugal) and processed as raster maps using ArcGIS 10 – advanced – software (©ESRI, Environmental Systems Research Institute). A spatial scale of 10 m was chosen for this research that resulted in a total of 3,375,526 cells comprising the whole municipality of Portimão. Such a choice represents a trade-off between the size of the study area, the level of detail of LULC data and computer processing capacity (Hengl, 2006). These raster maps were further converted to ASCII format. The spatial information needs for this research are presented in Table 2.

Table 2. Description of the model's spatial information

Category	Variable	Year	Data Source	Description
State/ transition	Land-use/ Land-cover	1947 - 2010	Aerial photos and orthoimagery datasets provided by the DGT and IGeoE	State of cell
	Spatial trend of change surface	1947 - 2010	Derived from LULC maps	3rd order polynomial trend surfaces
Distance	Distance from land-use/land-cover categories	1947 - 2010	Extracted from LULC maps	Euclidean distances (metres)
	Distance from agricultural and shrubland areas	1947 - 2010	Extracted from LULC maps	
	Distance from coastal line	2010	Extracted from the 2010 Portuguese Administrative Map (https://www.dgterritorio.pt/)	
Density	Transportation Network attractiveness surface	1947 - 2010	Derived from LULC maps	(metres)
Topographic	Slopes	2004	Computed from 25-m resolution Digital Elevation Model (DEM) provided by the DGT	(%)
Land-use planning	Urban planning schemes	1942 - 2011	Historical and recent land-use plans, Municipal Master Plan, Planning permits provided by the DGT and the Local Planning Department of Portimão's City Hall	

Land-Use and Land-Cover Data

A major issue regarding the use of nationwide LULC maps provided by the *Direção-Geral do Território* (DGT) and European research projects (e.g. Corine Land Cover) for LULC change modelling relates to the difference between scales of analysis, levels of information disaggregation, dates, and data sources. In fact, the minimum mapping unit (MMU) of the referred thematic LULC maps ranges from 1 to 25 ha, which involves a major abstraction of urban landscape and may mask the existing spatial variability of land-use changes, such as sprawl patterns that are better seen on a micro scale. Considering these shortcomings, unprecedented high-precision and highly detailed thematic LULC maps were created to provide insights into the space-time nature of LULC change over the past 63 years in the Portimão municipality. For this study, eight thematic LULC maps at a nominal scale of 1:10 000 with an MMU of 1000 sq. m covering the entire study area were generated for the years 1947, 1958, 1968, 1972, 1987, 1995, 2000, and 2010. The nominal scales of the aerial photographs were mostly well-suited for the production of LULC maps at a scale of 1:10 000, as the basemap scale ranged between 17 000 and 26 000 (Redweik, 2007).

The historical aerial photographs were georeferenced to the ETRS89/PT-TM06 reference system using 2007 vector cartographic data as geometric reference. The integration of data and image processing techniques was carried out using Autodesk® AutoCAD Map 3D 2012 software and Autodesk® AutoCAD Raster Design 2011 software. In order to georeference the aerial photographs, 7 to 20 ground control points (GCP) were used in an affine (first-order) polynomial transformation to a root mean square error (RMSE) ranging between 3.2 and 10.3 m. Following the standard values recommended by Conolly & Lake (2006), these geometric accuracy measures were considered acceptable. Subsequently, LULC maps were derived by visual image interpretation based on the technical specifications of *Carta de Ocupação do Solo de Portugal* 2007 (DGT, 2018). These LULC maps contain five levels of thematic detail, with 45 categories at the most detailed level (Figure 4). Figures 5 and 6 illustrate the final LULC maps produced for the municipality of Portimão. The classification methodology followed a downdating procedure. A thematic accuracy assessment was performed both to evaluate the 2010 orthoimagery interpretation and to guarantee a good quality 2010 LULC. This was a crucial step since the 2010 LULC was used to derive the earlier LULC maps. In this study, collecting ground-truthing data with both higher resolution and time intervals that are similar to those of historical images proved impossible to accomplish. Therefore, for the other seven LULC maps no thematic accuracy assessment was performed. In these cases, ancillary data proved to be essential for a proper interpretation of the historical images. Google Earth images were used to evaluate the accuracy of land-use and land-cover categories. To assess the global thematic quality of the 2010 LULC map both Overall Accuracy Index and K_{hat} statistic were computed. The results indicate an overall accuracy of 96.9% and a K_{hat} of 96.8%, meaning that these categories are very well mapped as the value of thematic accuracy resulting from photo-interpretation based on remote sensing should be ³ 85% (Anderson, Hardy, Roach, & Witmer, 1976) and thus greatly represent the land-cover on the study area.

Spatial Determinants

Several spatial determinants were used to guide the simulation of LULC change processes and patterns considering the physical characteristics and detailed land-use planning data of the study area. These included spreading and constraint factors, such as the spatial trend of change surfaces; the distances from *(i)* the coastal line, *(ii)* existing agricultural and shrubland areas, and *(iii)* existing urban land-use categories; the slopes; the areas proposed for urban development and the transportation network factor, which we refer to as transportation network attractiveness surface. This is given as a density index computed combining a simple line density surface weighted by the road network hierarchy and a kernel point density surface weighted by the linear distance from the railway stations and highway exits to the closest urban areas. Policies and natural constrains (e.g., land excluded from development defined by urban planning schemes, agroforestry and ecological protected areas, water bodies) were also considered during the simulation experiments that enabled us to illustrate the land-use planning actions taken by the local government within the study area. In Table 2, all spatial determinants and their main technical specifications are presented.

CAMLucc MODEL IMPLEMENTATION AND RESULTS

In this study, both physical determinants and land-use planning data were selected to calibrate the CAMLucc model over a period of near seven decades and simulate LULC change in Portimão for 2020 and 2025 at a fine spatial resolution. Six land-use and land-cover categories were considered to define the state variable of the CAMLucc model. These included vertical urban fabric (TUV); horizontal urban fabric (TUH); discontinuous urban fabric (TUD); sparsely discontinuous urban fabric (TUDE); other artificial surfaces (OAA), and non-built areas (NB). Their selection was based on the analysis of the main trends of LULC transition in the study area. The proposed sub-categorization of urban land-use, which is rarely used in studies for both LULC change and urban growth modelling, together with the high-resolution spatial data of the model should improve the geo-representation of LULC change and urban growth patterns as they offer highly-detailed information on urban land-uses and their change patterns.

The CAMLucc model is a regular two-dimensional CA used to represent the municipality of Portimão, consisting of a grid of cells. Each cell in the grid can only be in one of the specified states. The temporal behaviour of LULC change was modelled synchronously. This means that every cell in the CAMLucc model changes its state simultaneously depending on its own characteristics and its neighbouring cells in each discrete time period. For this study, a cell area of 100 sq. m was defined, which represents the smallest discernible feature that can be identified. This modelling framework was constructed based on two approaches: a retrospective simulation method that intended to calibrate and validate the CAMLucc model by replicating the historical LULC change process of the study area, and a prospective method that led to the prediction of future LULC changes for 2020 and 2025. Years 1958, 1968, 1972, 1987, 1995, and 2000 were assigned as starting points to perform both calibration and validation processes, while the year 2010 was used to forecast the LULC change for 2020 and 2025.

Figure 4. Illustration of the adjusted COS 2007 legend for the study area (DGT, 2018)

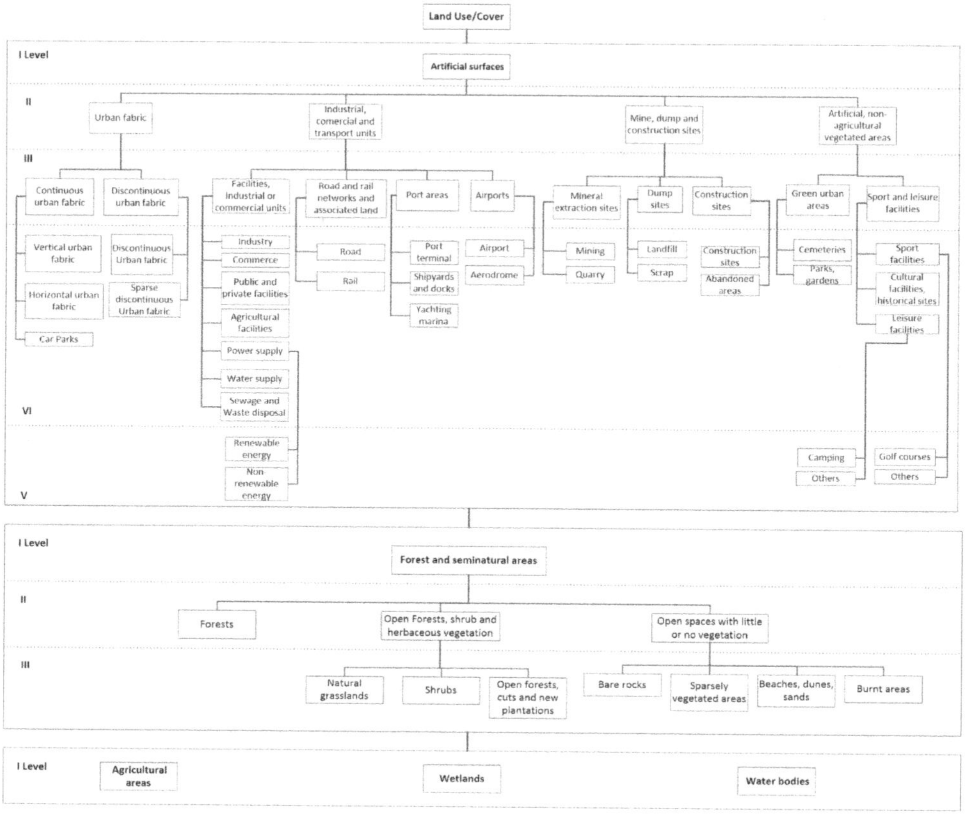

The CAMLucc model was implemented using the functionalities of the IDRISI Taiga software and is composed of three main elements: a spatially explicit transition rule configuration, an non-spatial LULC demand computation, and a spatially explicit LULC change allocation procedure. In the spatially explicit transition rule configuration, the suitability maps produced by means of logistic regression were used to establish the probability of a single category of land-use to occur for each spatially explicit cell, and functioned as the transition rules to execute the CAMLucc model simulations (the specific combination of these two models is well-known as suitability-based CA models). A standard von Neumann 5 x 5 neighbourhood configuration was used as the CA contiguity filter. According to research carried out by Mondal, Sharma, Kappas, & Garg (2012), the 5 x 5 contiguity filter generates spatially distributed improved outcomes when compared with other contiguity filters, such as 3 x 3 or 7 x 7. The transition area matrices produced by the non-spatial Markov Chain model were used in the non-spatial LULC demand computation

Figure 5. Historical LULC maps of Portimão of (a) 1947, (b) 1958, (c) 1968, and (d) 1972 classified at the most detailed level

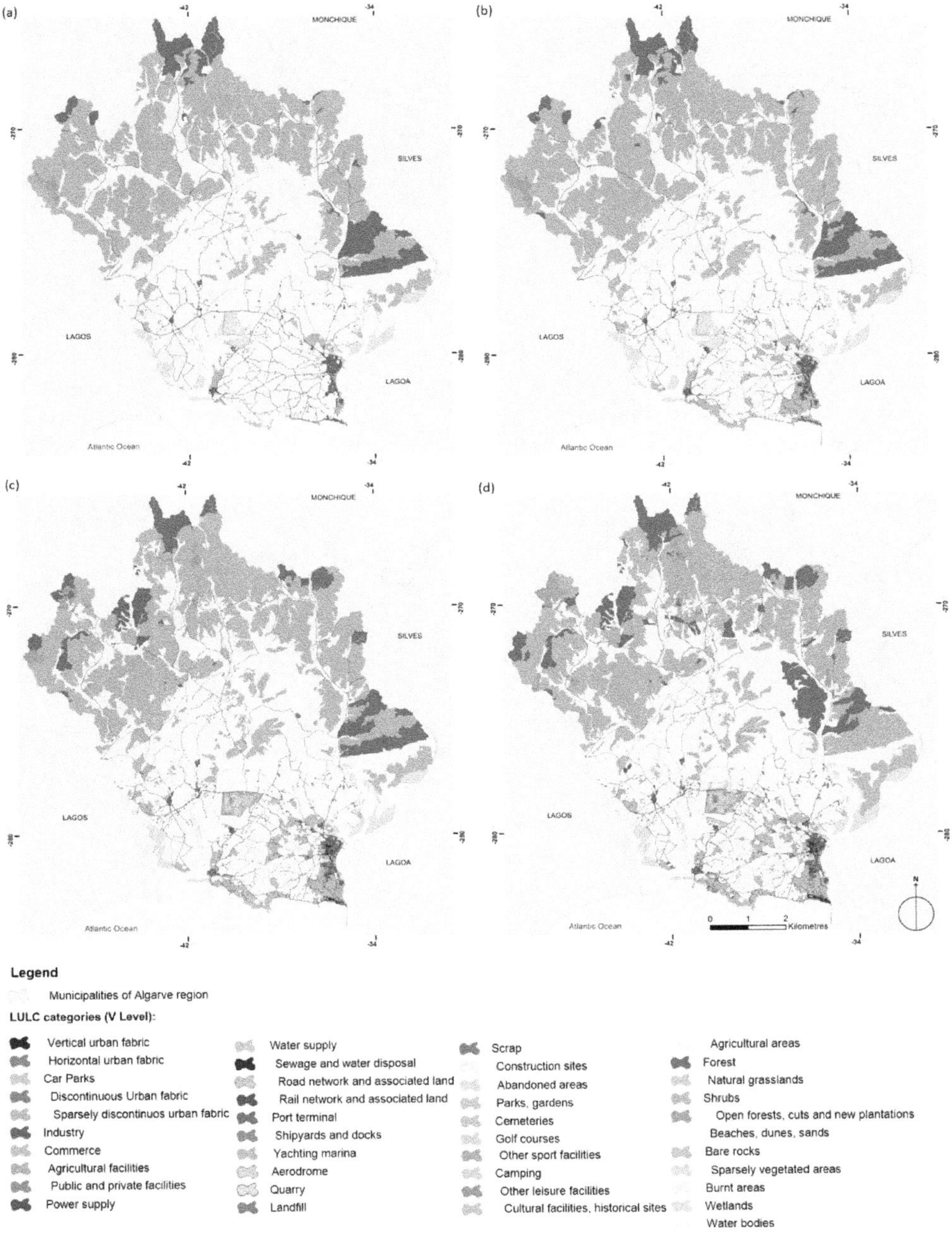

Figure 6. LULC maps of Portimão of (e) 1987, (f) 1995, (g) 2000, and (h) 2010 classified at the most detailed level

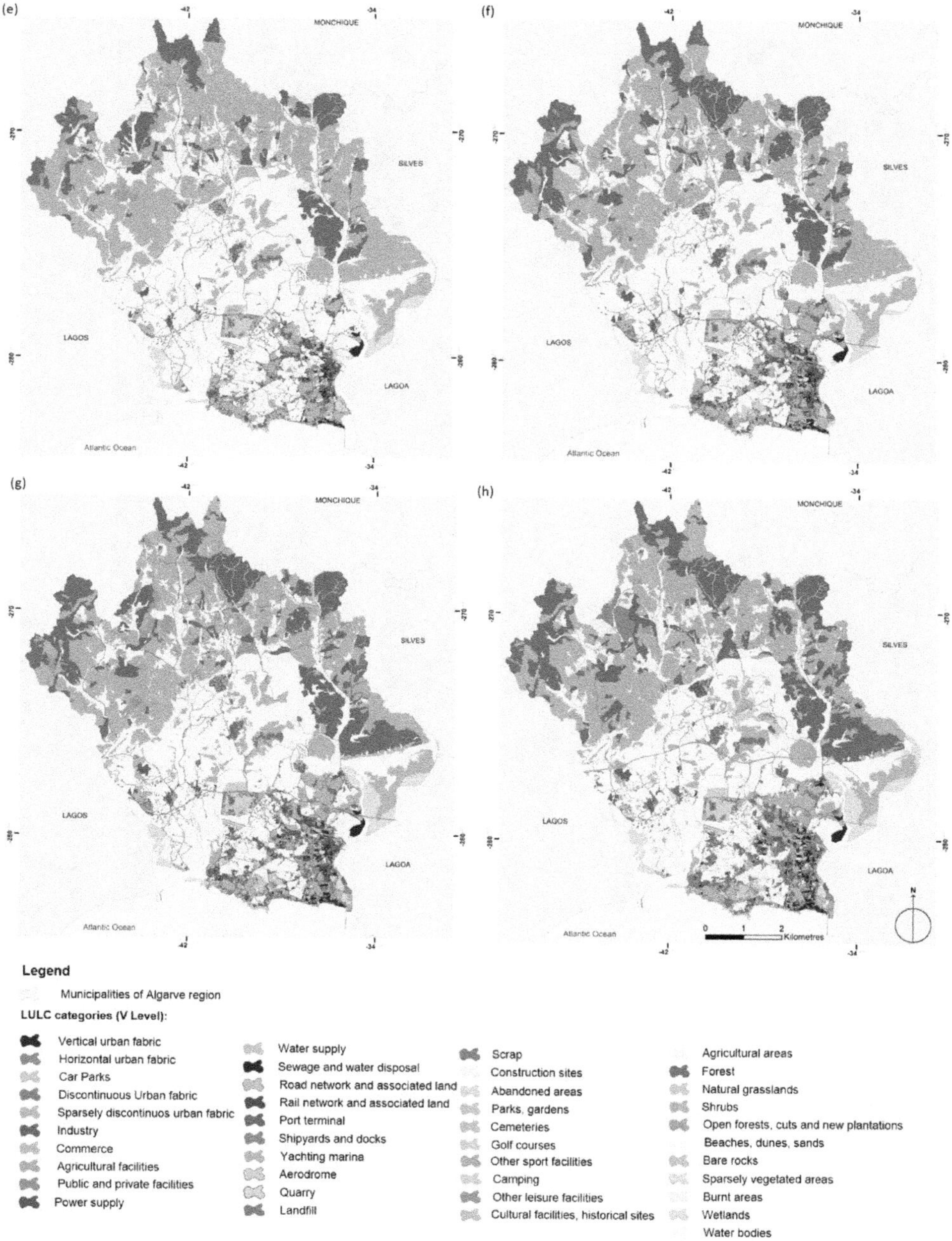

process. These transition area matrices provided the area change for every LULC category. They were then used by the CA that iterated the allocation of LULC until the estimated quantity of change was achieved. The CAMLucc model was established using different iteration numbers since they were governed by the number of years used for each projection. The spatially explicit LULC change allocation procedure stems from the dynamic simulation of competition or conflicts among the six land-use categories. The allocation of the LULC change procedure implemented in the CAMLucc is based on a multi-objective land allocation analysis (MOLA) (Eastman, Kyem, & Toledano, 1993). MOLA is a heuristic iterative procedure that allocates LULC change until it reaches the quantity of cells that are expected to change into each category and simultaneously conflicting location. The CA component emerged from such iterative process of LULC change allocation that was run for the entire study area, along with the application of the CA contiguity filter in each iteration. The latter decreases the suitability of cells further away from existing locations of the same LULC category, so that each cell changes into the corresponding category based on what is happening in its neighbourhood (Eastman, 2009). Accordingly, LULC change evolves as a diffusion process in highly suitable sites closer to existing areas of the same category.

Model Calibration

Four main types of transition rules reproducing the influence of spatial determinants on LULC change in Portimão from 1947 to 2010 were considered. These included neighbourhood, site-specific characteristics, distance, and constraints. Site-specific characteristic rules serve to relate the value of a location according to local biophysical and socioeconomic characteristics with LULC change (namely locations with high transportation network density and locations of high-intensity transitions in terms of land-use). Distance rules quantify the influence of Euclidean distance between each location and the nearest spatial determinant. Constraint rules serve to represent locations where LULC change may not occur based on both policies and natural constraints. All spatial variables (excluding land-use planning data and policies and natural constrains) were rescaled to a common continuous scale between 0 and 1 by means of fuzzy logic in order to allow their comparison and further combination through logistic regression. Therefore, in the CAMLucc model, the evaluation of suitable areas for LULC change was performed based on a fuzzy set of spatial determinants that allowed us to cope with fuzziness and uncertainty in the development of transition rules. The theory of fuzzy sets (Zadeh, 1965) allows us to attribute degrees of membership to a given fuzzy set. The fuzzy membership degree varies from 0 to 1, displaying a continuous increment from non-membership to full membership (Zadeh, 1965; Eastman, 2009). Accordingly, the logic of fuzzy sets gives us the possibility of thinking about the degree to which a location is suitable for a specific LULC change rather than merely whether a location is suitable or not. In this study, both sigmoidal monotonically increasing and decreasing functions were used as membership functions to develop the continuous spatial variables for each land-use category. The sigmoidal membership function has been preferred in several studies (Batty, 2005; Eastman, 2009) mostly due to its simplicity and superiority in fuzzifying variables for urban modelling. The settings of the fuzzy transformation were established through experiments and literature review to define how the suitability of a particular land-use category changes according to each spatial variable individually.

In order to represent the impact of land-use planning actions taken by the local government regarding LULC change, a number of areas proposed for urban development in the urban planning schemes were introduced in the model as spatial determinants promoting more urban transition in specific areas within a given time period. To reclassify the aforementioned locations into an ordinal map of suitability for urban land-use change they were coded with different values indicating the suitability associated with the nature of each urban planning scheme. In this case, we used "1.0" to codify all the areas planned for urban development in mandatory plans, "0.8" for all the areas planned for urban development in guiding plans, and "0.0" for the remaining study area. This was necessary because here the assumption is that mandatory plans are edificability commitments and therefore more likely to promote urban change within their boundaries than in areas bounded by guiding plans (most historical plans were unapproved plans or if approved not homologated). We have also used some constraint rules in this model. A diachronic analysis of the process of Portimão's LULC change allows us to verify that urban development has been occurring within valuable and protected areas. Therefore, ecologically and agroforestry valuable and protected areas are considered as constraints limiting rather than prohibiting LULC change. These constraints were realized by coding the ecological and agroforestry locations so that they have a low level of suitability for LULC change (all ecological and agroforestry locations were assigned a 0.2 value) and those where LULC change can occur were assigned a value of 1. Land excluded from development defined by urban planning schemes over the modelling period along with the "water bodies" land-use category was expressed as Boolean maps so that these locations were coded with 0 (locations unavailable for LULC change) and those where LULC change may occur were coded with 1.

Before estimating the final suitability maps for each land-use category using logistic regression, an exploratory data analysis was performed using statistical tests to cope with two important properties of spatial variables, namely multicollinearity and spatial dependence. Both Pearson´s correlation and variation inflation factor (VIF) were applied to identify multicollinearity. Presence of multicollinearity among explanatory variables was prevented by selecting a single variable from strongly correlated pairs, i.e., r > -0.85 or r < 0.85 (Berry & Feldman, 1985). Presence of multicollinearity was further prevented by excluding explanatory variables with values of VIFk ³ 5 (Rogerson, 2001), which suggests severe multicollinearity effects. In order to verify the presence of spatial autocorrelation on explanatory variables, the first-lag autocorrelation of explanatory variables was computed via global Moran's I spatial autocorrelation index throughout the modelling period. In order to establish the group of explanatory variables that adequately grasp the intrinsic spatial pattern of simulated land-uses, preventing pseudoreplication in the proposed model, only highly autocorrelated variables, i.e., I ³ 0.9, were retained.

Later, logistic regression was performed to test the relationship between each land-use category independently using a raster binary map and the defined set of independent variables resulting in six spatially-explicit models of suitability of the land-use presence at each location. Figure 7 presents an example of such a suitability map. The relative operating characteristic (ROC) measure was used to assesses the goodness of fit of logistic regression models. The ROC values range from 0.5 to 1, where a value of 0.5 indicates a random fit and 1 a perfect fit. Thus, the greater the ROC value (above the 0.5 threshold) the better the ability of the model to properly specify locations. The validation process was carried out by comparing the suitability maps computed by a logistic regression model with the real LULC maps from 1968 to 2010. The fairly high ROC values obtained ranging between 0.771 and 0.987 suggest that the spatial pattern of the six land-use categories may be adequately determined by the selected set of explanatory variables (Table 6, Appendix 1).

Additionally, all suitability maps were multiplied by the constraint layer to produce a unique suitability map and then integrated into the CAMLucc model as a prediction of LULC spatial patterns. Such maps (so-called regression covers, Kok & Veldkamp (2001)) represent the probability of a single land-use category to occur in each location and were applied to model LULC change in the subsequent time periods.

Model Simulation and Validation

The Markov chain model in this study was first employed to simulate transitions from the past to the present using the observed LULC data of Portimão. The significance of the calibration process was evaluating the simulation precision and then forecasting the quantity of LULC change for 2020 and 2025. The transition matrices between LULC maps in 1947-1958, 1958-1968, 1968-1972, 1972-1987, 1987-1995, and 1995-2000 were used to backcast LULC changes quantitatively in 1968, 1972, 1987, 1995, 2000, and 2010, respectively. Furthermore, LULC maps of 2000 and 2010 were used as the base year matrices to produce forecasts for the years

2020 and 2025. Additionally, the transition area matrices describing the quantity of LULC transitions to each land-use category were computed. Both absolute and relative errors were computed in order to assess the simulation precision of the Markov chain model's final outcomes. The formula and results are presented in Table 7 (Appendix 2).

The comparison between real LULC areas (from LULC maps) and estimated LULC areas (from Markovian transition area matrices) shows acceptable forecast precisions during the simulation periods. Both 2000 and 1972 simulation years show higher forecast precision with average relative errors of less than -13%, followed by 1968 and 1995 simulation years with average relative errors of -16% and -18%, respectively. These results suggest that the Markov chain model is an effective estimator for short to medium term LULC change trends.

After the determination of suitability maps and transition area matrices, the CAMLucc model simulated the LULC change in Portimão throughout the study period (Figures 8, 9, and 10). Actual LULC data were used as reference maps to evaluate the model's performance. Several simulations were carried out during the calibration process so that only the best results produced by the CAMLucc model were applied in the validation of the model.

Compared to reference LULC maps, the CAMLucc model generates the basic LULC change spatial patterns in the study area. The results indicate that the model efficiently generates both significant compact LULC spatial patterns similar to those found in the reference maps, and several large and isolated areas of urban land-use categories (Figure 8 (c), (d); Figure 9 (c), (d); and Figure 10 (c), (d)). This suggests that the CAMLucc model that detects potential changes based on spatial determinants including the transportation network gravity factor, topographic and distance constraints, and urban planning controls is an adequate solution in terms of CA-based modelling, providing high reliability regarding the spatial heterogeneity represented in the reference LULC maps.

This might be attributed to the ability of the CAMLucc model to detect non-linear and complex relationships and interactions over time among the different LULC categories. However, we can see in the CAMLucc-simulated LULC maps some areas that were missed by the model. For example, in 1968 (Figure 8 (c)), the urban changes northwest to the core/central area of Portimão were entirely missed by the model. This was mainly due to the fact that the transition from water bodies to urban land-use categories was not allowed in the calibration process. Consequently, since such urban areas were previously occupied by water bodies and wetlands their LULC category was maintained.

Figure 7. Example of estimated land-use suitability surfaces for the 2000-2010 modelling period

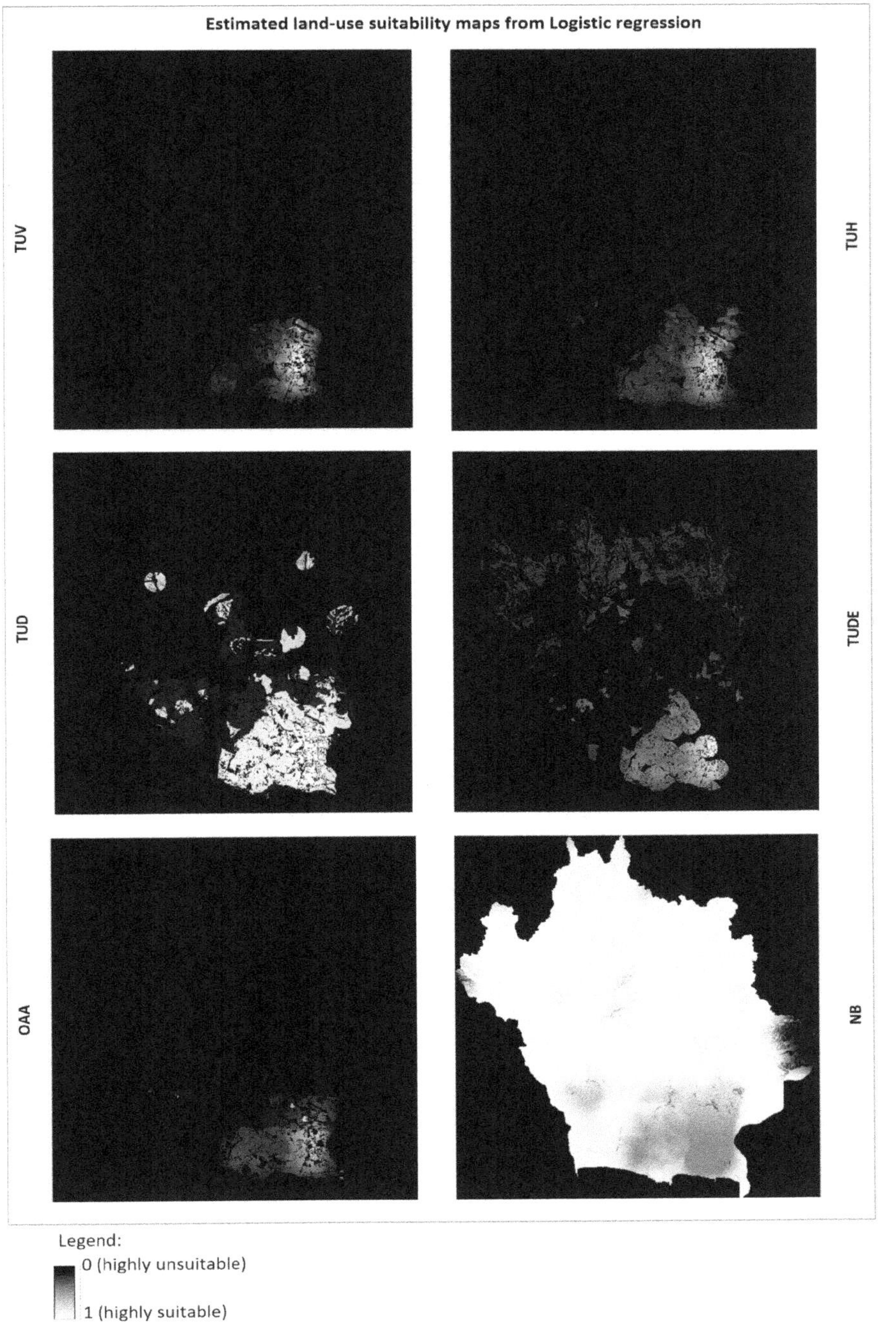

Figure 8. Historical reference LULC map of 1968 (a) and 1972 (b) versus the CAMLucc simulated LULC maps of 1968 (c) and 1972 (d)

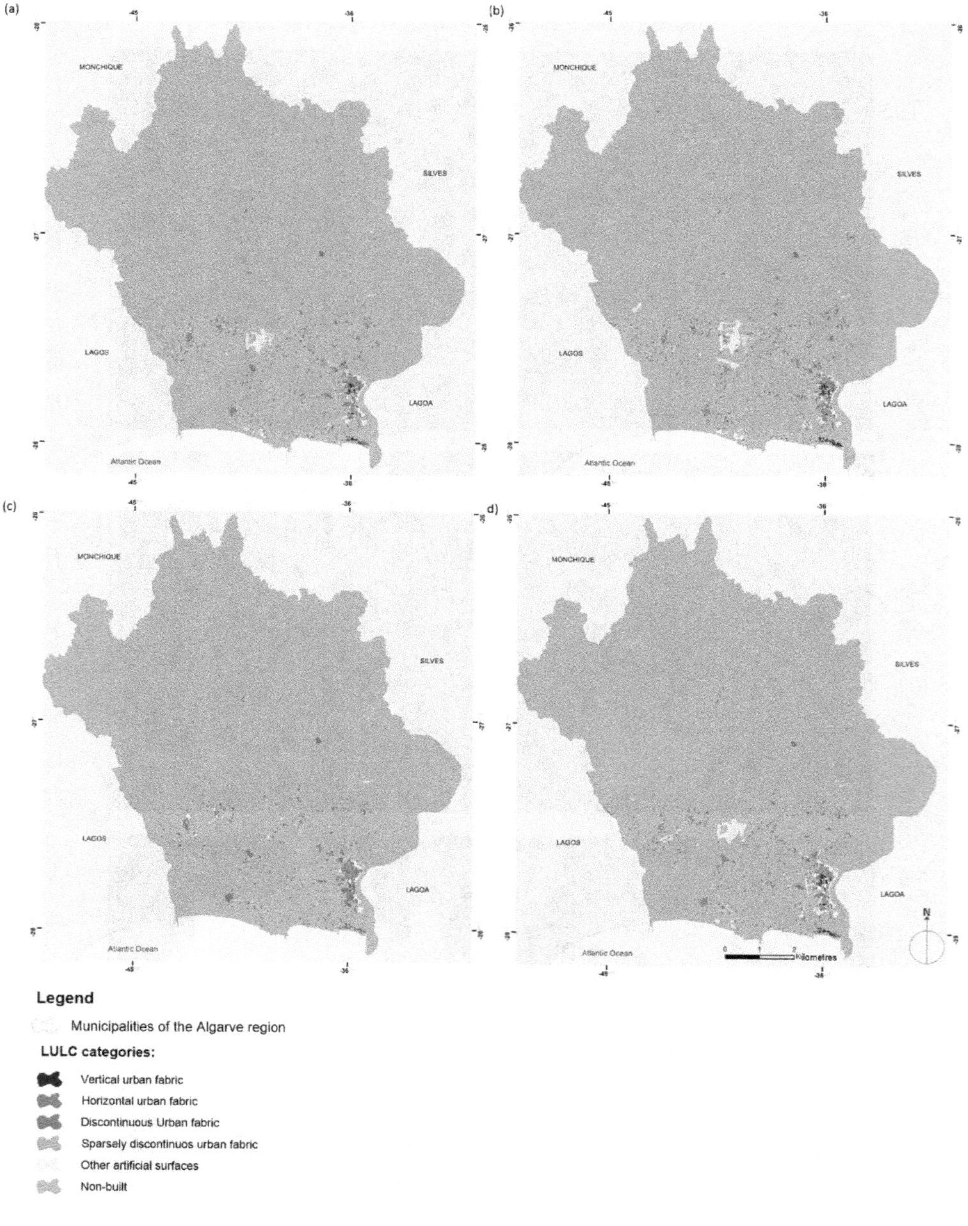

Figure 9. Historical reference LULC map of 1987 (a) and 1995 (b) versus the CAMLucc simulated LULC maps of 1987 (c) and 1995 (d)

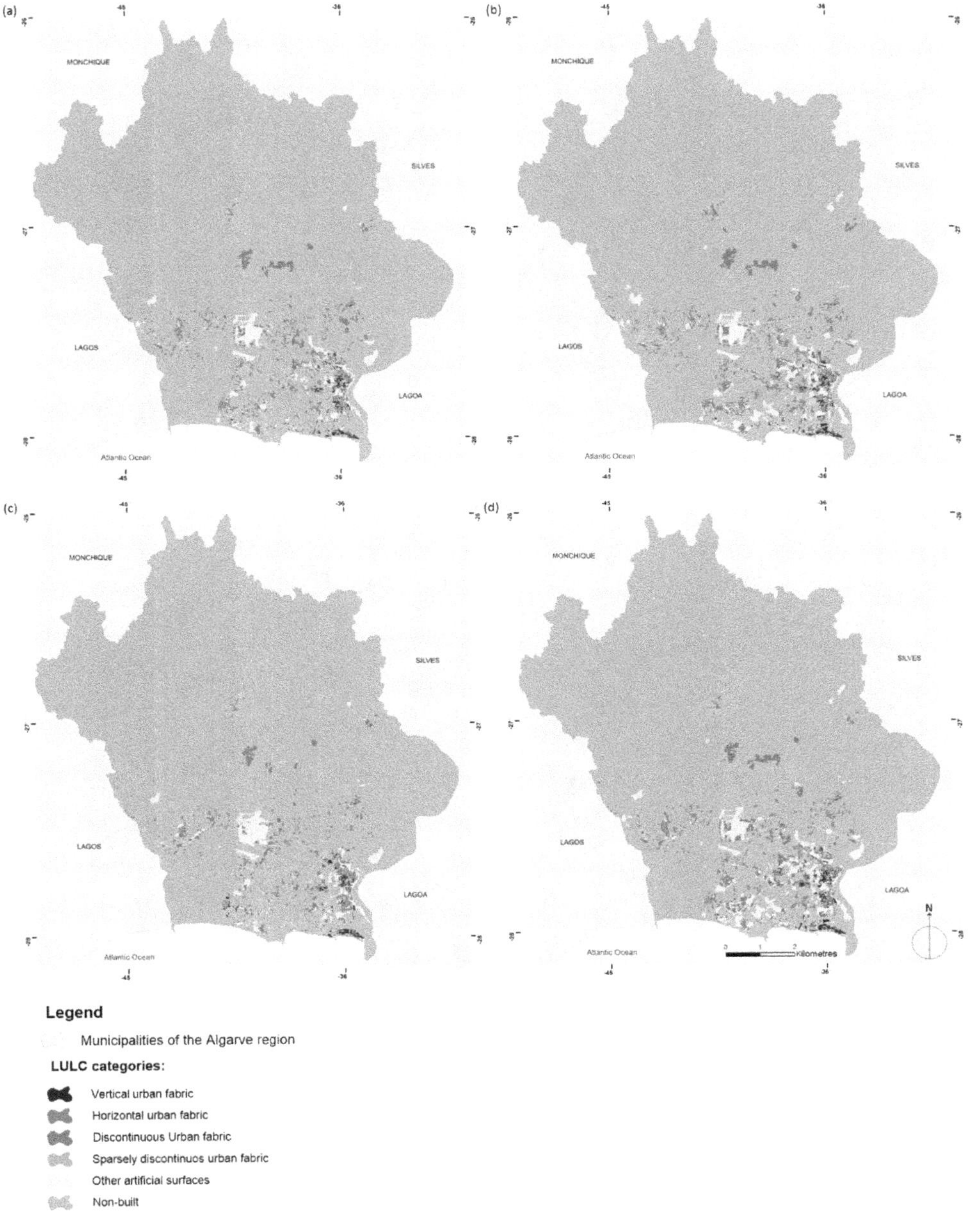

Figure 10. Historical reference LULC map of 2000 (a) and 2010 (b) versus the CAMLucc simulated LULC maps of 2000 (c) and 2010(d)

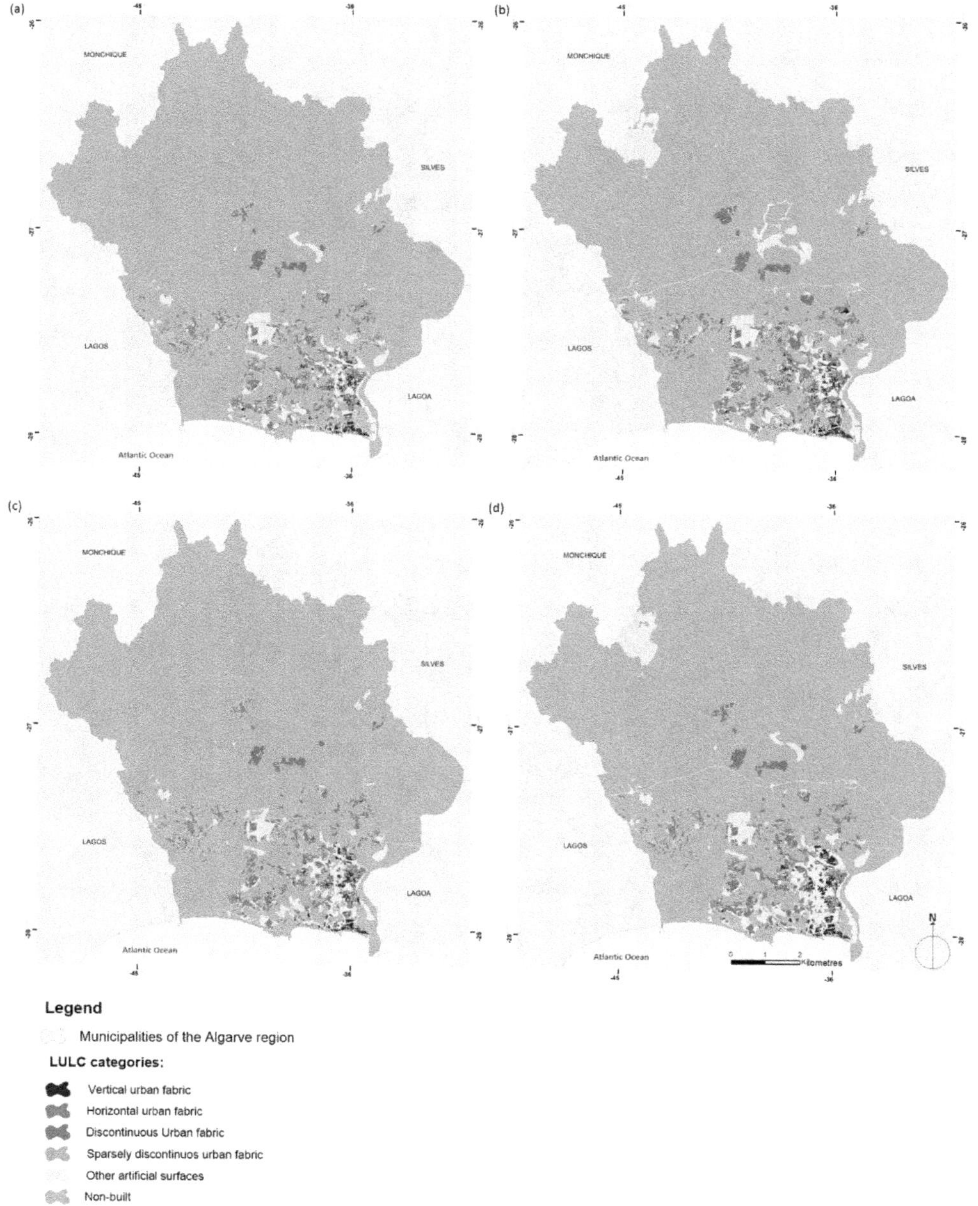

CAMLucc Model's Goodness-of-Fit Results: 1942-2010

The CAMLucc simulation results were validated using four Kappa statistics including the traditional Kappa index of agreement, KIA (Cohen, 1960), the Kappa for no ability (Kno), Kappa for quantity (Kquantity) and Kappa for location (Klocation). To validate the model, LULC maps derived from the visual image interpretation were used as reference data. The historical and actual reference data are independent from the calibration/simulation data used. The Kno statistic (Pontius, 2002) stems from general KIA statistics and specifies the overall accuracy of a model by performing a pairwise comparison. The main advantage of Kno statistics over the KIA statistics is that the former effectively accounts for both quantification and location errors (Wang & Mountrakis, 2011). Kquantity and Klocation (Pontius, 2002) are special forms of the general KIA statistics as well, and effectively discriminate accuracies in quantity and location, respectively. Klocation measures the location accuracy by analysing how well cells of the same category are located in the grid space, and Kquantity measures the accuracy in terms of the quantity of correct cells in each category (Eastman, 2009). All four indices of agreement range from 0 to 1, where a value of 1 indicates a perfect agreement. Additionally, an index of agreement due to chance (Pontius, 2002) was computed. Such index measures the agreement that can be reached with no information regarding both the quantity and location of the categories and was further compared to the overall accuracy of the simulated LULC maps. Table 3 shows the CAMLucc model simulation accuracies.

Results from the simulation accuracies indicate that the CAMLucc model simulations show great consistency with the spatial patterns of the reference LULC maps. The results summarized in Table 3 illustrate the continuous improvement of the KIA index over time. For the years 2000 and 2010, the CAMLucc model produced a KIA index of 0.97 and 0.95, and a Kno index of 0.98 and 0.97, respectively.

Additionally, Table 3 shows that there is a substantial difference between KIA index and Agreement Chance index values suggesting strong agreement between the CAMLucc-simulated LULC maps and the reference LULC maps since the actual agreement is better than chance. These results justify the use of both the logistic regression model and Markov chains as the basis of the CAMLucc model that generated LULC change models with great accuracy by incorporating few spatial determinants that are only partly responsible for the LULC change in Portimão.

Table 3. CAMLucc model simulation accuracies and comparison between the actual area and simulated area for each category over the study period

Percent of area in each category (%)	CAMLucc model											
	1968		1972		1987		1995		2000		2010	
	Actual area	**Simulated area**	**Actual area**	**Simulated area**	**Actual area**	**Simulated area**	**Actual area**	**Simulated area**	**Actual area**	**Simulated area**	**Actual area**	**Simulated area**
Tuv	0.005	0.008	0.069	0.095	0.234	0.368	0.277	0.376	0.396	0.475	0.493	0.688
Tuh	0.124	0.150	0.314	0.387	0.360	0.561	0.540	0.650	0.675	0.771	0.700	0.762
Tud	0.647	0.754	0.565	0.539	0.616	0.812	1.392	1.752	1.808	2.025	2.026	2.442
Tude	0.392	0.415	0.483	0.517	0.613	0.975	0.764	0.829	0.773	0.778	0.864	1.041
OAA	1.178	1.226	1.702	1.903	2.038	3.202	3.216	3.813	4.181	4.765	5.163	6.972
Simulation Accuracies												
KIA index	0.757		0.797		0.801		0.844		0.969		0.954	
Kno index	0.984		0.983		0.974		0.973		0.981		0.970	
Klocation index	0.872		0.837		0.836		0.853		0.974		0.963	
Kquantity index	0.799		0.786		0.743		0.710		0.972		0.959	
AgreementChance index	0.143		0.143		0.143		0.143		0.143		0.143	

Figure 11. Comparison between simulated and actual areas for each LULC category over the study period

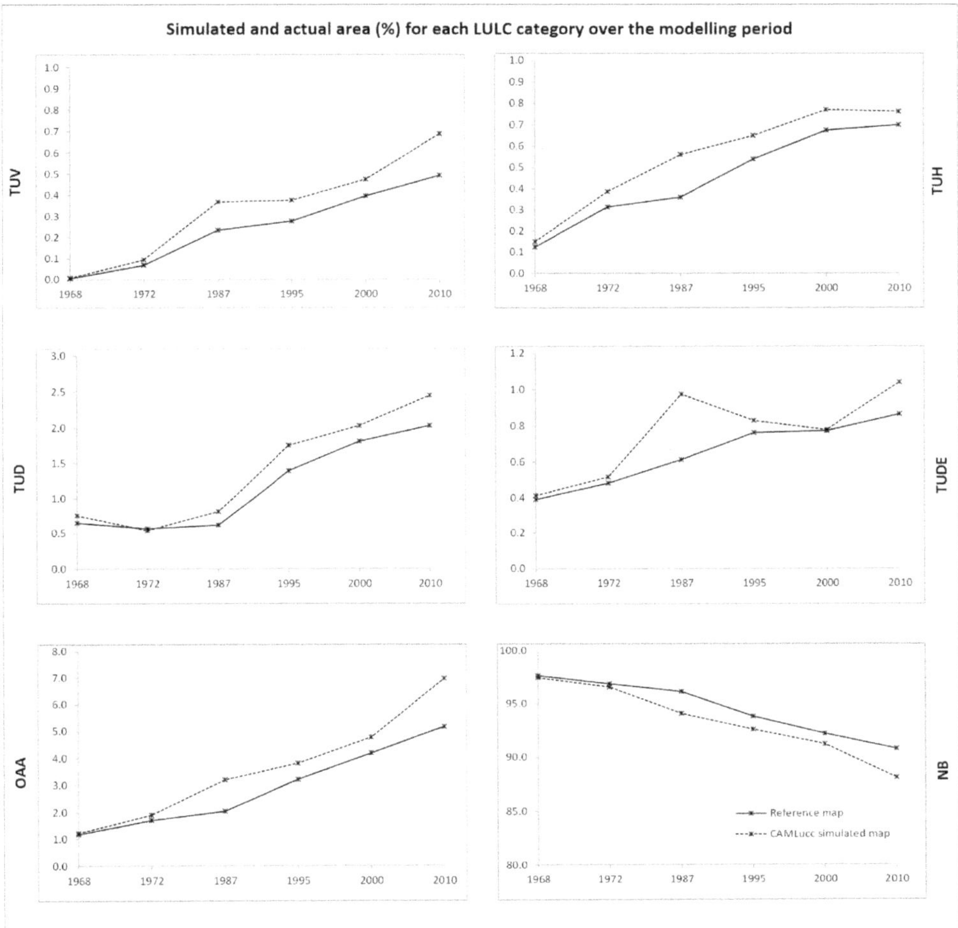

Figure 12. Projected LULC maps for the years 2020 (a) and 2025 (b)

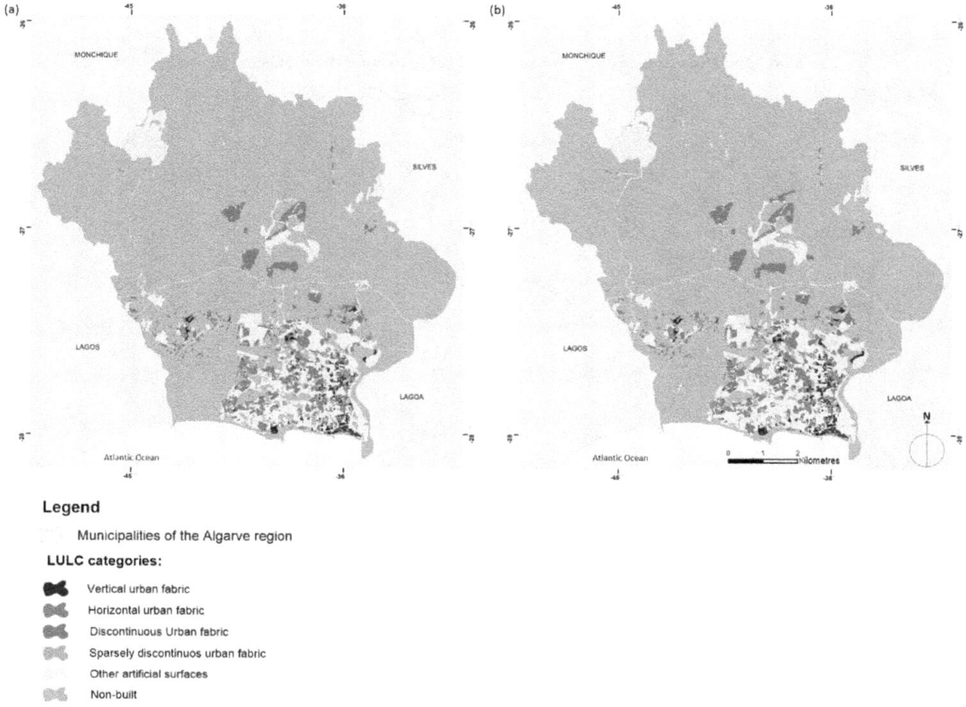

Table 4. Projected LULC composition for the study area from 2010 to 2020 and 2025

LULC category	Expected LULC composition (%)		
	2010 (actual)	2020	2025
TUV	0.49	0.93	1.04
TUH	0.70	0.91	0.98
TUD	2.03	3.33	3.63
TUDE	0.86	1.24	1.33
OAA	5.16	11.45	12.85
NB	90.75	82.18	80.21

Figure 13. Overview of expected changing areas from 2010 to 2020 (a) and 2025 (b)

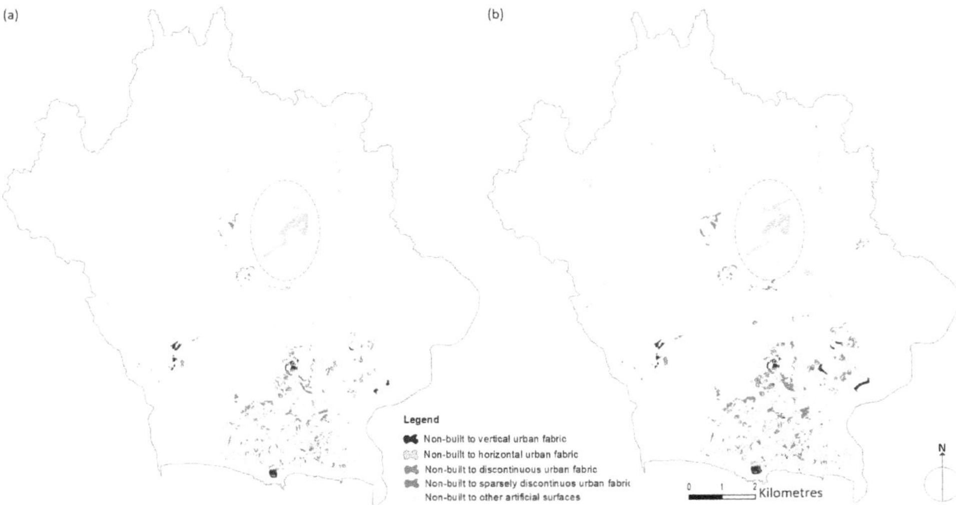

The Klocation indices (Table 3) reveal that the CAMLucc model produces LULC maps that are in strong agreement with the reference LULC maps in terms of the grid cell-level location of each category. Every Klocation index value is higher than 0.80, which according to Landis & Koch (1977) represents a nearly perfect fit. Indeed, there is significant improvement in the years 2000 and 2010 with values of K location index of 0.97 and 0.96, respectively. Regarding the estimated quantity of change, Table 3 shows that Kquantity values are slightly lower than the values of the Klocation index, however, especially for the years 1987 and 1995 with values of 0.74 and 0.71, respectively. This is also revealed in the percentage of area in each category for both CAMLucc simulations and reference data (Table 3) showing greater dissimilarities between simulated and actual areas during those years for all LULC categories. These results indicate that although the CAMLucc model was capable of capturing the basic trends in LULC dynamics in terms of quantity of LULC changes in the study period as seen in more detail in Figure 11, it has a tendency to overestimate the amount of area that is expected to change in every LULC category, particularly for the years 1987 and 1995. This may be attributed to the combination of two factors. First, because it was for these years that most discrepant time periods between the earlier and later LULC maps (4 years for 1987 and 15 for 1995) and the projection of LULC change forward (15 years for 1987 and 8 for 1987) were used to estimate the quantity of change by Markov chains. Secondly, due to the Markov chains' strong dependence on past LULC change trends – e.g., the quantity of change estimation for the year 1995 was based on the 1972-1987 transition area matrices. This was one of the most remarkable periods of population growth and economic activities' expansion, mainly due to the independence movements in Portugal's African colonies in 1975. The return of Portuguese people residing in the former colonies (over half a million) resulted in a significant increase in population that triggered a strong urban pressure on the Portuguese territory.

CAMLucc Model Results: Forecasts For 2020 and 2025

Based on the initial LULC map of the year 2010, the CAMLucc model was applied to forecast medium-term LULC change in Portimão for the next 10 and 15 years. The LULC forecast results for the years 2020 and 2025 are provided in Figure 12. The projected LULC composition for the study area from 2010 to 2020 and 2025 is listed in Table 4.

Results show once more that the CAMLucc model has generated compact LULC spatial patterns in Portimão's municipality similar to those observed in the reference LULC maps. From the visual analysis of Figure 12, we can see that urban growth will be dominated by a coalescence process in which nearby urban land-use categories shall become progressively interconnected. However, we can also clearly see the gradual increase of discontinuous urban fabric over the entire study area and sparsely discontinuous urban fabric along the transportation network, which are evidences of urban sprawl into the surrounding agricultural land and shrublands. According to the results, LULC categories – e.g., other artificial surfaces, discontinuous urban fabric, and sparsely discontinuous urban fabric – will have a significant increase in terms of area in 2025 when compared with 2010. By 2020 and 2025, discontinuous urban fabric will cover 3.3% and 3.6% of the study area, respectively (Table 4). The sparsely discontinuous urban fabric will occupy 1.2% by 2020 and 1.3% by 2025, whilst non-built areas will decrease in the two periods and will occupy 82% and 80% of the total area by 2020 and 2025, respectively (Table 4). In the two predicted LULC maps, changes into other artificial surfaces follow both infilling and edge-expansion growth types and we can expect an increase in other artificial surfaces by approximately 11 sq. km from 2010 to 2025, which will cover 12.9% of the entire study area. Based on the assumption that past LULC change trends will continue into the future, Figure 13 displays the dominant LULC changes in 2020 and 2025. The LULC maps indicate that the greatest LULC changes predicted by the CAMLucc model will occur along the coastal area, to the west (gently sloping areas), north and northwest areas of the core/central area of Portimão.

As we can gather from Table 5, most of the predicted horizontal urban fabric for 2020 and 2025 is expected to be converted from discontinuous urban fabric and non-built areas and located in large-scale projects planned by the municipality – such as the urban land-use plan of "Morgado do Reguengo" – and linearly along the transportation network towards the northern part of the municipality (Figure

13). Most of the vertical urban fabric will come from other artificial surfaces and non-built areas (Table 5) and will be interspersed among them. The vertical urban fabric will increase by about 2 sq. km from 2010 to 2025, which means that changes will occur vertically more often. The discontinuous urban fabric and sparsely discontinuous urban fabric will be generally converted from non-built areas (Table 5). These land-use categories will spread towards the northwest of the core/central area of Portimão and cluster in the southwest part of the municipality. Lastly, the CAMLucc model predicted clusters of other artificial surfaces (which will be converted mostly from non-built areas) in the southeast area of the municipality and some in the municipality-planned areas (namely, the urban land-use plan of "Morgado do Reguengo") located in the northern part of the city. Generally, the most dynamic areas of Portimão are found along the coastal area, in the west, north, and northwest parts of the municipality.

Table 5. Expected LULC transition matrix from 2010 to 2020 and 2025 (unit: sq. km)

2010	TUV		TUH		TUD		TUDE		OAA		NB	
	2020	2025	2020	2025	2020	2025	2020	2025	2020	2025	2020	2025
TUV	1.25	1.24	0.00	0.00	0.00	0.01	0.00	0.00	0.01	0.02	0.01	0.01
TUH	0.01	0.01	1.42	1.40	0.02	0.03	0.00	0.00	0.00	0.01	0.00	0.00
TUD	0.05	0.07	0.12	0.17	4.46	4.26	0.07	0.10	0.19	0.27	0.02	0.03
TUDE	0.00	0.00	0.01	0.01	0.03	0.04	1.82	1.77	0.01	0.01	0.05	0.07
OAA	0.10	0.15	0.01	0.02	0.25	0.36	0.03	0.05	14.64	14.30	0.33	0.49
NB	0.27	0.42	0.10	0.17	1.29	1.90	0.33	0.49	5.96	8.76	148.97	145.19
TOTAL	1.68	1.90	1.66	1.77	6.05	6.60	2.25	2.42	20.81	23.36	149.37	145.78

CONCLUSION AND FUTURE RESEARCH DIRECTIONS

In this chapter, we present a high spatial resolution CA-based model that is associated with GIS at the local scale to address past and current LULC changes and future change trends within the municipality of Portimão. The main strength of this modelling approach lies on its hybrid framework – the CAMLucc model – that incorporates the advantages of fuzzy logic, logistic regression, Markov chain analysis, and CA for LULC change modelling and prediction. One of the original contributions of the modelling approach is in the fine-scale spatiotemporal analysis of LULC transition/change at the local government level, since coupling Markov chain models and regression logistic models with CA is not novel in LULC change research. Thus, the first challenge involved collecting high-resolution spatial datasets on a municipal scale and producing long-term and fine-scale LULC maps with the aim of better capturing the past and current undetected emergent LULC structures. Main data sources were derived from multi-resolution imagery, specifically, aerial photographs and orthorectified digital aerial images. Aerial photographs still prove to be good and rich sources for mapping LULC, as seen in the years 1947, 1958, 1968, 1972, and 1987. However, collecting and processing such long-term imagery (from 1947 to 2010) along with the production of eight unprecedented high-precision and highly detailed thematic LULC maps (45 categories at the most detailed level) by means of visual interpretation required a substantial initial investment. In this research, we have considered a very fine minimum mapping unit (1000 sq. m) and multi-temporal LULC datasets to obtain detailed LULC data, providing extremely rich information on the LULC structural evolution of Portimão, which may assist urban planners and policy-makers to perform further analysis apart from modelling and predicting LULC change.

This research has provided an important contribution to LULC change analysis by incorporating both physical variables (LULC maps, distance and density variables, transportation network, and slope) and detailed land-use planning data into the CAMLucc model. Community behaviours and policies are major factors, since they are affected by and/or affect the environment and the economy. However, at this stage, our model did not integrate socioeconomic variables (e.g., economic growth rate, population growth) or the so-called 'fast variables' such as climate events and economic shocks. The capability to integrate data related to such extreme events, if available in the future, provides grounds for future work in LULC change research.

The CAMLucc model has been successfully applied to modelling and predicting LULC change even in the geographical context of a coastal area that has been undergoing rapid and strong coastal urbanization dynamics in the past decades. We provide a modelling approach with a spatially explicit transition rule configuration, followed by a non-spatial LULC demand computation, and a spatially explicit LULC change allocation procedure, complemented by accuracy analysis at different stages of the process. The model's calibration and validation results indicated that the simulation accuracy of LULC changes is highly significant. Therefore, the advanced methods used in this hybrid framework proved to be complementary and performed well together.

Our model contributes with several insights into the factor combinations required to meet sustainability goals, mainly by overtaking the constraints of traditional models that were normally applied to understand territorial sustainability resilience. The model's ability to simulate with a remarkable level of reality the dynamics of urban growth and LULC change patterns of Portimão's coastal area suggests that the proposed CA-based model could be used as a decision support tool in the regular practice of the urban planning process.

Although the CAMLucc model appears in this research as a simple and helpful tool to explain and cope with the complexity of LULC change using few spatial determinants that are only partly responsible for LULC change, some future developments are still required. The Markov chains' forecasting precision decayed severely specifically when long time periods were used *(i)* between the earlier and later LULC maps, and *(ii)* to project LULC change forward. These results suggest that the Markov chain model is an effective estimator for at least short to medium term LULC change trends. Consequently, to produce more accurate predictions of LULC change, future developments should incorporate population changes as a complementary variable to assist in the prediction of LULC transitions. Another limitation of this model is its strong dependence on past LULC change trends. This implies that Markov chains become quite vulnerable whenever time periods where there has been an extreme or unexpected change in land-use are used. However, by identifying and isolating the areas where extreme LULC changes have occurred and by further breaking down LULC into sub-categories this problem is potentially overcome.

Furthermore, we used a multi-objective land allocation (MOLA) procedure to allocate LULC change to land units where its suitability is highest. Even though the application of this procedure is considerably affected by the type and quantity of the available data, we consider this procedure could be practical and appropriate for LULC change allocation where cases of competition or conflicts occur among LULC categories.

Still, the CAMLucc model seems to be sufficiently versatile to be replicated and used in other geographic contexts when special features are considered. Any replication of the model presented in this research inevitably involves adjustments that rely on the specific characteristics (e.g., LULC spatial pattern, urban and demographic dimension) and on the legal and juridical planning framework of each geographic context.

ACKNOWLEDGMENT

This work is financed by national funds through FCT - Foundation for Science and Technology, I.P., within the scope of the project «UIDB/04647/2020» of CICS. NOVA - Centro Interdisciplinar de Ciências Sociais da Universidade Nova de Lisboa. The first author was financed by national funds through the FCT, under the Norma Transitória - DL 57/2016 and by a grant number SFRH/BD/61544/2009. We would like to thank the reviewers for the insightful comments on an earlier version of this chapter.

REFERENCES

Agyemang, F. S. K., & Silva, E. (2019). Simulating the urban growth of a predominantly informal Ghanaian city-region with a cellular automata model: Implications for urban planning and policy. *Applied Geography (Sevenoaks, England)*, *105*, 15–24. doi:10.1016/j.apgeog.2019.02.011

Al-sharif, A., & Pradhan, B. (2013). Monitoring and predicting land use change in Tripoli Metropolitan City using an integrated Markov chain and cellular automata models in GIS. *Arabian Journal of Geosciences*, *7*. doi:10.100712517-013-1119-7

Anderson, J. R., Hardy, E. E., Roach, J. T., & Witmer, R. E. (1976). *A land use and land cover classification system for use with remote sensor data*. Professional Paper., doi:10.3133/pp964

Anderson, T. W., & Goodman, L. A. (1957). Statistical Inference about Markov Chains. *Annals of Mathematical Statistics*, *28*(1), 89–110. doi:10.1214/aoms/1177707039

Batty, M. (2005). *Cities and Complexity: Understanding Cities with Cellular Automata, Agent- Based Models, and Fractals*. The MIT Press.

Benenson, I., & M Torrens, P. (2004). *Geosimulation: Automata-Based Modeling of Urban Phenomena*. doi:10.1002/0470020997

Berry, W., & Feldman, S. (1985). *Multiple Regression in Practice*. doi:10.4135/9781412985208

Boavida-Portugal, I., Rocha, J., & Cardoso Ferreira, C. (2016). Exploring the impacts of future tourism development on land use/cover changes. *Applied Geography* (Vol. 77). doi:10.1016/j.apgeog.2016.10.009

Bone, C., Dragicevic, S., & White, R. (2011). Modeling-in-the-middle: Bridging the gap between agent-based modeling and multi-objective decision-making for land use change. *International Journal of Geographical Information Science*, *25*(5), 717–737. doi:10.1080/13658816.2010.495076

Burke, L., Reytar, K., Spalding, M., & Perry, A. (2011). Reefs at Risk Revisited. Washington, DC: World Resources Institute (WRI).

Caruso, G., Peeters, D., Cavailhes, J., & Rounsevell, M. (2009). Space–Time Patterns of Urban Sprawl, a 1D Cellular Automata and Microeconomic Approach. *Environment and Planning. B, Planning & Design*, *36*(6), 968–988. doi:10.1068/b34009

Clarke, K., Parks, B., & Crane, M. (2002). *Geographic Information Systems and Environmental Modeling*. Prentice Hall.

Cohen, J. (1960). A Coefficient of Agreement for Nominal Scales. *Educational and Psychological Measurement*, *20*(1), 37–46. doi:10.1177/001316446002000104

Conolly, J., & Lake, M. (2006). Geographical Information Systems in Archaeology. *Cambridge Manuals in Archaeology*. doi:10.1017/CBO9780511807459

Couclelis, H. (2005). "Where has the future gone?" Rethinking the role of integrated land-use models in spatial planning. *Environment & Planning A*, *37*(8), 1353–1371. doi:10.1068/a3785

Dadhich, P. N., & Hanaoka, S. (2011). Spatio-temporal Urban Growth Modeling of Jaipur, India. *Journal of Urban Technology*, *18*(3), 45–65. doi:10.1080/10630732.2011.615567

DGT. (2018). *Especificações Técnicas da Carta de Uso e Ocupação do Solo de Portugal Continental para 1995, 2007, 2010 e 2015*. Retrieved from http://mapas.dgterritorio.pt/atom-dgt/pdf-cous/COS2015/ET-COS-1995-2007-2010-2015.pdf

Eastman, J. (2009). *IDRISI Taiga Tutorial*. Graduate School of Geography, Clark University.

Eastman, J. R., Kyem, P., & Toledano, J. (1993). A procedure for multi-objective decision making in GIS under conditions of conflicting objectives. *European Conference on Geographical Information Systems*.

EEA. (2017). *Landscapes in transition, An account of 25 years of land cover change in Europe. EEA Report No /10/2017*. Publications Office of the European Union.

Freire, S., Santos, T., & Tenedório, J. A. (2009). Recent urbanization and land use/land cover change in Portugal— The influence of coastline and coastal urban centers. *Journal of Coastal Research*, 1499–1503. http://www.jstor.org/stable/25738039

Gomes, E., Abrantes, P., Banos, A., & Rocha, J. (2019). Modelling future land use scenarios based on farmers' intentions and a cellular automata approach. *Land Use Policy* (Vol. 85). doi:10.1016/j.landusepol.2019.03.027

Gonçalves, J. (2011). *DEM SRTM*. Retrieved July 2, 2019, from FCUP-University of Oporto website: https://www.fc.up.pt/pessoas/jagoncal/srtm

Hengl, T. (2006). Finding the Right Pixel Size. *Computers & Geosciences*, *32*(9), 1283–1298. doi:10.1016/j.cageo.2005.11.008

Hou, X., Change, B., & Yu, X. (2004). Land use change in Hexi corridor based on CA Markov methods. *Nongye Gongcheng Xuebao (Beijing)*, *20*(5), 286–291.

INE. (2011). *Censos 2011 – Resultados Definitivos*. Retrieved from https://censos.ine.pt/xportal/xmain?xpid=CENSOS&xpgid=ine_censos_publicacao_det&contexto=pu&PUBLICACOESpub_boui=73212469&PUBLICACOESmodo=2&selTab=tab1&pcensos=61969554

Jianping, L., Bai, Z., & Feng, G. (2005). RS-and-GIS-supported forecast of grassland degradation in southwest Songnen plain by Markov model. *Geo-Spatial Information Science*, *8*(2), 104–109. doi:10.1007/BF02826848

JRC & ESA. (1994). *Modelling Deforestation Processes—A Review*. Luxembourg: ECSC-EC-EAEC.

Kok, K., & Veldkamp, A. (2001). Evaluating Impact of Spatial Scales on Land Use Pattern Analysis in Central America. *Agriculture, Ecosystems & Environment* (Vol. 85). doi:10.1016/S0167-8809(01)00185-2

Landis, J. R., & Koch, G. G. (1977). The Measurement of Observer Agreement for Categorical Data. *Biometrics*, *33*(1), 159–174. doi:10.2307/2529310 PMID:843571

Li, X., & Gong, P. (2016). Urban growth models: progress and perspective. *Science Bulletin*, *61*(21), 1637–1650. doi:10.100711434-016-1111-1

Li, X., & Yeh, A. G.-O. (2004). Analyzing spatial restructuring of land use patterns in a fast growing region using remote sensing and GIS. *Landscape and Urban Planning*, *69*(4), 335–354. doi:10.1016/j.landurbplan.2003.10.033

Liang, X., Liu, X., Li, D., Zhao, H., & Chen, G. (2018). Urban growth simulation by incorporating planning policies into a CA-based future land-use simulation model. *International Journal of Geographical Information Science*, *32*(11), 2294–2316. doi:10.1080/13658816.2018.1502441

Liu, Y. (2009). *Modelling Urban Development with Geographical Information Systems and Cellular Automata*. doi:10.1201/9781420059908

Magarotto, M., Faria-De-Deus, R., Ferreira Costa, M., & Masanet, É. (2017). Green areas in coastal cities – conflict of interests or stakeholders' perspectives? *International Journal of Sustainable Development and Planning*, *12*(08), 1260–1271. doi:10.2495/SDP-V12-N8-1260-1271

Martinez, M. L., Silva, R., Lithgow, D., Mendoza, E., Flores, P., Martínez, R., & Cruz, C. (2017). Human Impact on Coastal Resilience along the Coast of Veracruz, Mexico. *Journal of Coastal Research*, *77*(SI), 143–153. doi:10.2112/SI77-015.1

Mondal, D. M. S., Sharma, N., Kappas, M., & Garg, P. (2012). Modeling of spatio-temporal dynamics of land use land cover - a review and assessment. *Journal of Geomatics*, *6*.

Muller, M. R., & Middleton, J. (1994). A Markov model of land-use change dynamics in the Niagara Region, Ontario, Canada. *Landscape Ecology*, *9*(2), 151–157. doi:10.1007/BF00124382

Naghibi, F., Delavar, M. R., & Pijanowski, B. (2016). Urban Growth Modeling Using Cellular Automata with Multi-Temporal Remote Sensing Images Calibrated by the Artificial Bee Colony Optimization Algorithm. *Sensors (Basel)*, *16*(12), 2122. Advance online publication. doi:10.339016122122 PMID:27983633

Pastor, J., & Johnson, C. A. (1992). *Using Simulation Models and Geographic Information Systems to Integrate Ecosystem and Landscape Ecology*. doi:10.1007/978-1-4612-4382-3_11

Peña, J. M., Nilsson, R., Björkegren, J., & Tegnér, J. (2007). Towards scalable and data efficient learning of Markov boundaries. *International Journal of Approximate Reasoning*, *45*(2), 211–232. doi:10.1016/j.ijar.2006.06.008

Petrov, L., Lavalle, C., & Kasanko, M. (2009). Urban land use scenarios for a tourist region in Europe: Applying the MOLAND model to Algarve, Portugal. *Landscape and Urban Planning* (Vol. 92). doi:10.1016/j.landurbplan.2009.01.011

Pinto, N., Antunes, A. P., & Roca, J. (2017). Applicability and calibration of an irregular cellular automata model for land use change. *Computers, Environment and Urban Systems*, *65*, 93–102. doi:10.1016/j.compenvurbsys.2017.05.005

Pontius, R. (2002). Statistical Methods to Partition Effects of Quantity and Location During Comparison of Categorical Maps at Multiple Resolutions. *Photogrammetric Engineering and Remote Sensing*, *68*.

Pontius, R. G. Jr, Shusas, E., & McEachern, M. (2004). Detecting important categorical land changes while accounting for persistence. *Agriculture, Ecosystems & Environment*, *101*(2), 251–268. doi:10.1016/j.agee.2003.09.008

Portugali, J. (2002). Self-Organization and the City. *J. Artificial Societies and Social Simulation* (Vol. 5). doi:10.1007/978-3-662-04099-7

Redweik, P. (2007). *Fotogrametria Aérea*. Departamento de Engenharia Geográfica, Geofísica e Energia, Faculdade de Ciências da Universidade de Lisboa.

Rocha, J. (2012). *Sistemas complexos, modelação e geosimulação da evolução de padrões de uso e ocupação do solo.* Universidade de Lisboa. Retrieved from http://hdl.handle.net/10451/6772

Rocha, J. (2019). Simulation and Sustainability BT - Encyclopedia of Sustainability in Higher Education. doi:10.1007/978-3-319-63951-2_413-1

Rocha, J., Ferreira, J., Simões, J., & Tenedório, J. (2007). Modelling Coastal and Land Use Evolution Patterns through Neural Network and Cellular Automata Integration. *Journal of Coastal Research SI, 50.*

Rocha, J., Gutierres, F., Gomes, P., & Teodoro, A. C. (2018). *A Hybrid CA-ANN-Fuzzy Model for Simulating Coastal Changing Patterns BT - Beach Management Tools - Concepts, Methodologies and Case Studies* (C. M. Botero, O. Cervantes, & C. W. Finkl, Eds.). doi:10.1007/978-3-319-58304-4_10

Rogerson, P. (2001). *Statistical Methods for Geography*. doi:10.4135/9781849209953

Semboloni, F. (2000). The Growth of an Urban Cluster into a Dynamic Self-Modifying Spatial Pattern. *Environment and Planning. B, Planning & Design*, *27*(4), 549–564. doi:10.1068/b2673

Stanilov, K., & Batty, M. (2011). Exploring the Historical Determinants of Urban Growth Patterns through Cellular Automata. *Transactions in GIS*, *15*(3), 253–271. doi:10.1111/j.1467-9671.2011.01254.x

Torrens, P. M., & Benenson, I. (2005). Geographic Automata Systems. *International Journal of Geographical Information Science*, *19*(4), 385–412. doi:10.1080/13658810512331325139

Torrens, P. M., & O'Sullivan, D. (2000). Cities, cells and complexity: developing a research agenda for urban geocomputation. *Geocomputation 2000*. Retrieved from http://www.geosimulation.org/geosim/gc044.htm

Ulam, S. (1952). Random processes and transformations. In *International Congress of Mathematicians*, (pp. 264–75). Providence, RI: American Mathematical Society.

Ustaoglu, E., Williams, B., Petrov, O. L., Shahumyan, H., & Van Delden, H. (2018). Developing and Assessing Alternative Land-Use Scenarios from the MOLAND Model: A Scenario-Based Impact Analysis Approach for the Evaluation of Rapid Rail Provisions and Urban Development in the Greater Dublin Region. *Sustainability*, *10*(2), 61. Advance online publication. doi:10.3390u10010061

Verburg, P. H., Crossman, N., Ellis, E. C., Heinimann, A., Hostert, P., Mertz, O., Nagendra, H., Sikor, T., Erb, K.-H., Golubiewski, N., Grau, R., Grove, M., Konaté, S., Meyfroidt, P., Parker, D. C., Chowdhury, R. R., Shibata, H., Thomson, A., & Zhen, L. (2015). Land system science and sustainable development of the earth system: A global land project perspective. *Anthropocene*, *12*, 29–41. doi:10.1016/j.ancene.2015.09.004

Von Neumann, J. (1966). The theory of automata: Construction, reproduction, homogeneity. In J. Von Neumann & A. W. Burks (Eds.), *Theory of Self-Reproducing Automata* (pp. 89–250). University of Illinois Press.

Wang, J., & Mountrakis, G. (2011). Developing a multi-network urbanization model: A case study of urban growth in Denver, Colorado. *International Journal of Geographical Information Science*, *25*(2), 229–253. doi:10.1080/13658810903473213

Weng, Q. (2002). Land use change analysis in the Zhujiang Delta of China using satellite remote sensing, GIS and stochastic modelling. *Journal of Environmental Management*, *64*(3), 273–284. doi:10.1006/jema.2001.0509 PMID:12040960

White, R., & Engelen, G. (1994). Urban systems dynamics and cellular automata: Fractal structures between order and chaos. *Chaos, Solitons, and Fractals*, *4*(4), 563–583. doi:10.1016/0960-0779(94)90066-3

White, R., Engelen, G., & Uljee, I. (2015). *Modeling Cities and Regions as Complex Systems: From Theory to Planning Applications*. MIT Press. doi:10.7551/mitpress/9780262029568.001.0001

Wolfram, S. (1983). Statistical mechanics of cellular automata. *Reviews of Modern Physics*, *55*(3), 601–644. doi:10.1103/RevModPhys.55.601

Wolfram, S. (2002). *A New Kind of Science*. Wolfrom Media.

Zadeh, L. A. (1965). Fuzzy sets. *Information and Control*, *8*(3), 338–353. doi:10.1016/S0019-9958(65)90241-X

ADDITIONAL READING

Batty, M. (2013). *The New Science of Cities*. The MIT Press. doi:10.7551/mitpress/9399.001.0001

Brunet, R. (2001). Models in geography? A sense to research. *Cybergeo: European Journal of Geography*.

Dutta, V. (2012). Land Use Dynamics and Peri-urban Growth Characteristics Reflections on Master Plan and Urban Suitability from a Sprawling North Indian City. *Environment and Urbanization Asia* (Vol. 3). doi:10.1177/0975425312473226

Guermond, Y. (2008). *The Modeling Process in Geography, From Determinism to Complexity*. John Wiley & Sons. doi:10.1002/9780470611722

Herold, M., Goldstein, N., & Clarke, K. (2003). The spatiotemporal form of urban growth: Measurement, analysis and modeling. *Remote Sensing of Environment, 86*(3), 286–302. Advance online publication. doi:10.1016/S0034-4257(03)00075-0

Pinto, N., Tenedório, J. A., Antunes, A. P., & Cladera, J. R. (2014). *Technologies for Urban and Spatial Planning: Virtual Cities and Territories*. IGI Global. doi:10.4018/978-1-4666-4349-9

Pumain, D. (2004). An Evolutionary approach to Settlement Systems. In T. Champion & G. Hugo (Eds.), *New Forms of Urbanization, Beyond the Urban-Rural Dichotomy* (pp. 231–247). Ashgate.

Rocha, J., & Tenedório, J. A. (Eds.). (2018). *Spatial Analysis, Modelling and Planning*. IntechOpen. doi:10.5772/intechopen.74452

KEY TERMS AND DEFINITIONS

Coastal Area: The land area that extends landward of the coastline and that is directly or indirectly influenced by coastal processes. This is a more general term than coastal zone, which refers to the geographic area defined by the legislation as a zone for coastal management purposes.

Diffusion Cellular Automata Model: A model that reproduces the capacity of a cell to spread or propagate information over space through neighbouring cells.

Hybrid Modelling: A method that integrates the advantages of different techniques into one approach, while preserving accuracy.

Land-Use and Land-Cover: The physical transposition of human activity on the earth's surface, i.e., the territorial inscription of human activity.

Land-Use and Land-Cover Change: The physical expression of land surface change.

Model: A simplified and approximate representation of a phenomenon, structure, or system of the real world.

Neighbourhood: In cellular automata, the neighbourhood of a cell is the set of nearby cells with which a specific cell interacts.

Spatial Determinant: A variable that may increase or detract from the suitability of a location for the presence of a particular land-use or land-cover category.

Transition Rule: Expresses the parameters and weights related with the spatial variables involved in a simulation and governs the result of its spatial evolution.

Urban Phenomena Modelling: The approach concerned with developing and applying models in order to explore specific issues in urban studies by means of spatial computation techniques.

APPENDIX 1

Table 6. Results of logistic regression for the six land-use categories in each modelling period

Land-use category	Variables	Modelling periods							
		1958 - 1968	1968 - 1972	1972 - 1987	1987 - 1995	1995 - 2000	2000 - 2010	2010 - 2020/25	Average ROC
TUV	Intercept (β_0)	-10 307	-82 500	-87 690	-25 489	-11 953	23 999	33.074	
	Dist_CL (x_1)[a]	0.924	2 917	2 277	2.39	1 892	2 355	0.875	
	IDRV (x_2)[b]	2 140	0.254	0.421	0.349	2 714	0.139	0.6790	
	Slope (x_3)	2 944	0.402	1 326	0.229	0.690	0.46	1.700	
	STC_TUV (x_4)[c]	-	-	-	-	-	-	8.585	
	STC_TUH (x_5)	-	-	-	-	-	-	-	
	STC_TUD (x_6)	-	-	-	-	-	-	-	
	STC_TUDE (x_7)	-	-	-	-	-	-	-	
	STC_OAA (x_8)	-	-	-	-	-	-	-	
	Dist_ existing_TUV (x_9)[d]	2 589	80 683	85 102	15 523	10 530	21 449	39.757	
	Dist_ existing_TUH (x_{10})	-	-	-	-	-	-	-	
	Dist_ existing_TUD (x_{11})	-	-	-	-	-	-	-	
	Dist_ existing_TUDE (x_{12})	-	-	-	-	-	-	-	
	Dist_existing_OAA (x_{13})	-	-	-	-	-	-	-	
	Dist_AUA (x_{14})[e]	-	-	-	-	-	-	-	
	Dist_AS (x_{15})[f]	-	-	-	-	-	-	-	
	ROC	0.932	0.741	0.971	0.965	0.978	0.965	0.968	0.931
TUH	Intercept (β_0)	-13 614	-71 744	-58 571	-17 110	-52 910	12 345	18 848	
	Dist_CL (x_1)	0.675	2 167	0.922	0.378	0.111	0.286	0.072	
	IDRV (x_2)	5.110	0.226	2.843	2 415	2 142	0.102	2 012	
	Slope (x_3)	11 288	0.958	1 133	1 092	1 133	1 091	0.797	
	STC_TUV (x_4)	-	-	-	-	-	-	-	
	STC_TUH (x_5)	-	-	-	-	-	0.950	-	
	STC_TUD (x_6)	-	-	-	-	-	-	-	
	STC_TUDE (x_7)	-	-	-	-	-	-	-	
	STC_OAA (x_8)	-	-	-	-	-	-	-	
	Dist_ existing_TUV (x_9)	-	-	-	-	-	-	-	
	Dist_ existing_TUH (x_{10})	19 614	71 135	38 808	11 587	51 912	14 710	19 054	
	Dist_ existing_TUD (x_{11})	-	-	-	-	-	-	-	
	Dist_ existing_TUDE (x_{12})	-	-	-	-	-	-	-	
	Dist_existing_OAA (x_{13})	-	-	-	-	-	-	-	
	Dist_AUA (x_{14})	-	-	-	-	-	-	-	
	Dist_AS (x_{15})	-	-	-	-	-	-	-	
	ROC	0.943	0.976	0.965	0.920	0.964	0.935	0.931	0.948
TUD	Intercept (β_0)	-8 185	-33 047	-41 211	-50 126	-47 569	11 937	60 117	
	Dist_CL (x_1)	0.49	0.261	0.029	-0.132	-0.254	-0.201	-0.632	
	IDRV (x_2)	1 288	0.977	1.096	0.52	0.490	0.023	0.914	
	Slope (x_3)	1 621	1 032	0.574	1 060	0.469	1 146	0.664	
	STC_TUV (x_4)	-	-	-	-	-	-	-	
	STC_TUH (x_5)	-	-	-	-	-	-	-	
	STC_TUD (x_6)	-	-	-	-	0.580	-	0.553	
	STC_TUDE (x_7)	-	-	-	-	-	-	-	
	STC_OAA (x_8)	-	-	-	-	-	-	-	
	Dist_ existing_TUV (x_9)	-	-	-	-	-	-	-	
	Dist_ existing_TUH (x_{10})	-	-	-	-	-	-	-	
	Dist_ existing_TUD (x_{11})	1 775	30 472	37 421	41 076	51 569	11 244	68 441	
	Dist_ existing_TUDE (x_{12})	-	-	-	-	-	-	-	
	Dist_existing_OAA (x_{13})	-	-	-	-	-	-	-	
	Dist_AUA (x_{14})	-		-	-	-	-	-	
	Dist_AS (x_{15})	0.048	-	-	-	0.886	0.984	1 369	
	ROC	0.881	0.934	0.874	0.856	0.921	-	0.871	0.890

continued on following page

Table 6. Continued

Land-use category	Variables	Modelling periods							
		1958 - 1968	1968 - 1972	1972 - 1987	1987 - 1995	1995 - 2000	2000 - 2010	2010 - 2020/25	Average ROC
TUDE	Intercept (β_0)	-48 300	-59 234	-10 511	-51 093	-41 332	39 497	15.155	
	Dist_CL (x_1)	0.089	0.052	-0.521	-0.463	-0.504	-0.426	-0.175	
	IDRV (x_2)	0.421	0.329	0.522	0.914	0.572	0.049	0.716	
	Slope (x_3)	0.670	0.915	1 218	0.814	0.919	0.937	0.833	
	STC_TUV (x_4)	-	-	-	-	-	-	-	
	STC_TUH (x_5)	-	-	-	-	-	-	-	
	STC_TUD (x_6)	-	-	-	-	-	-	-	
	STC_TUDE (x_7)	-	-	-	-	-	0.033		
	STC_OAA (x_8)	-	-	-	-	-	-	-	
	Dist_ existing_TUV (x_9)	-	-	-	-	-	-	-	
	Dist_ existing_TUH (x_{10})	-	-	-	-	-	-	-	
	Dist_ existing_TUD (x_{11})	-	-	-	-	-	-	-	
	Dist_ existing_TUDE (x_{12})	48 382	58 766	15 115	50 809	40 133	49 467	18.544	
	Dist_existing_OAA (x_{13})	-	-	-	-	-	-	-	
	Dist_AUA (x_{14})	-	-	-	-	-	-	-	
	Dist_AS (x_{15})	0.302	-	1 056	-	-	-	0.137	
	ROC	0.883	0.876	0.792	0.772	0.876	0.876	0.880	0.851
OAA	Intercept (β_0)	-24 048	-31 116	-8 598	-13 163	-63 402	-46 871	-12 338	
	Dist_CL (x_1)	0.745	0.337	0.240	0.752	1 013	0.859	0.193	
	IDRV (x_2)	1 511	2 766	3 163	2 315	1 923	0.119	1 793	
	Slope (x_3)	0.932	1 115	1 010	1 248	0.897	0.956	0.966	
	STC_TUV (x_4)	-	-	-	-	-	-	-	
	STC_TUH (x_5)	-	-	-	-	-	-	-	
	STC_TUD (x_6)	-	-	-	-	-	-	-	
	STC_TUDE (x_7)	-	-	-	-	-	-	-	
	STC_OAA (x_8)	-	-	-	-	-	-	0.140	
	Dist_ existing_TUV (x_9)	-	-	-	-	-	-	-	
	Dist_ existing_TUH (x_{10})	-	-	-	-	-	-	-	
	Dist_ existing_TUD (x_{11})	-	-	-	-	-	-	-	
	Dist_ existing_TUDE (x_{12})	-	-	-	-	-	-	-	
	Dist_existing_OAA (x_{13})	20 862	29 849	5 870	7 310	60 155	37 224	8 282	
	Dist_AUA (x_{14})	-	-	-	-	-	-	-	
	Dist_AS (x_{15})	-	-	-	-	-	-	-	
	ROC	0.896	0.917	0.842	0.840	0.922	0.914	0.771	0.872
NB	Intercept (β_0)	49 654	-3 956	16 662	15 806	-3 725	5 826	10 121	
	Dist_CL (x_1)	-3 009	-2 429	-0.555	-0.738	-1 720	-1 955	-0.374	
	IDRV (x_2)	4 268	-4 459	-2 551	-2 334	-3 519	0.001	-2 495	
	Slope (x_3)	-5 531	-3 502	-1 078	-1 309	-1 770	-1 324	-1 085	
	STC_TUV (x_4)	-	-	-	-	-	-	-	
	STC_TUH (x_5)	-	-	-	-	-	-	-	
	STC_TUD (x_6)	-	-	-	-	-	-	-	
	STC_TUDE (x_7)	-	-	-	-	-	-	-	
	STC_OAA (x_8)	-	-	-	-	-	-	-	
	Dist_ existing_TUV (x_9)	-	-	-	-	-	-	-	
	Dist_ existing_TUH (x_{10})	-	-	-	-	-	-	-	
	Dist_ existing_TUD (x_{11})	-	-	-	-	-	-	-	
	Dist_ existing_TUDE (x_{12})	-	-	-	-	-	-	-	
	Dist_existing_OAA (x_{13})	-	-	-	-	-	-	-	
	Dist_AUA (x_{14})	-52 325	-8 224	-11 158	-10 669	-7 136	-1 890	-6 177	
	Dist_AS (x_{15})	-	-	-	-	-	-	-	
	ROC	0.987	0.981	0.807	0.823	0.970	0.979	0.812	0.908

Notes: " – " indicates that a variable was not selected. [a] Distance to coastal line; [b] transportation network density variable; [c] spatial trend of change surfaces; [d] distance to existing similar LULC category; [e] distance to all urban areas; and [f] distance to agricultural and shrubland areas.

APPENDIX 2

The absolute and relative errors were calculated with the following expressions (Jianping, Bai, & Feng, 2005):

$$'' N = N - N'$$

and

$$\varepsilon = \frac{N - N'}{N} \times 100\%$$

Where ΔN represents the absolute error, ε the relative error, N is the real area of LULC categories, and N' the estimated LULC areas.

Table 7. Simulation precision of LULC categories

Simulation year		TUV	TUH	TUD	TUDE	OAA	NB	Base year matrix	Year increments projected forward	Average relative error (%)
1968	Absolute error/sq. km	-0.0047	-0.0469	-0.1939	-0.0403	-0.0872	0.3723	1947 - 1958 (11)	10	-16
	Relative error (%)	-51.6484	-20.7798	-16.4797	-5.6474	-4.0702	0.2097			
1972	Absolute error/ sq. km	-0.0469	-0.1321	0.048	-0.0618	-0.3664	0.5593	1958 - 1968 (10)	4	-12
	Relative error (%)	-37.4601	-23.1430	4.6724	-7.0347	-11.8423	0.3175			
1987	Absolute error/ sq. km	-0.2432	-0.3646	-0.3558	-0.6567	-2.1131	3.9335	1968 - 1972 (4)	15	-43
	Relative error (%)	-57.1160	-55.5623	-31.7423	-58.8125	-56.9738	2.2477			
1995	Absolute error/ sq. km	-0.1805	-0.2004	-0.6529	-0.1188	-1.0858	2.2387	1972 - 1987 (15)	8	-18
	Relative error (%)	-35.8990	-20.4094	-25.7870	-8.5560	-18.5664	1.3124			
2000	Absolute error/ sq. km	-0.1435	-0.175	-0.397	-0.0105	-1.0647	1.7241	1987 - 1995 (8)	5	-10
	Relative error (%)	-19.9416	-14.2682	-12.0819	-0.7477	-14.0111	1.0291			
2010	Absolute error/ sq. km	-0.3553	-0.1137	-0.7562	-0.3221	-3.2887	4.8358	1995 - 2000 (5)	10	-20
	Relative error (%)	-39.6806	-8.9373	-20.5372	-20.5081	-35.0451	2.9315			

N.B.: Numbers in brackets represent the number of time periods between the earlier and the later LULC maps.

Chapter 4
New Solutions and Methodologies for Data Acquisition and Management in Small Municipalities

Rui Pedro Juliao
Interdisciplinary Centre of Social Sciences (CICS.NOVA), NOVA FCSH, Universidade NOVA de Lisboa, Portugal

Amilton Amorim
https://orcid.org/0000-0002-4012-7079
UNESP, Brazil

João Paulo Hespanha
https://orcid.org/0000-0002-5506-8868
ESTGA, Universidade de Aveiro, Portugal

Guilherme Henrique Barros de Souza
UNESP, Brazil

Ronaldo Celso Messias Correia
UNESP, Brazil

Rafael Delli Colli Destro
UNESP, Brazil

ABSTRACT

Promoting and managing sustainable territorial development require adequate tools that enable territorial managers to formulate appropriate choices. Technological solutions have emerged, and the paradigm has shifted from isolated GIS to a more collaborative production and dissemination of geographic data using spatial data infrastructures (SDI). A critical dataset for municipal land administration is cadastre. ISO 19152 standard of the Land Administration Domain Model (LADM) was published in 2012. Also, technology evolution, namely unmanned aerial vehicles (UAV), has changed data acquisition for cadastre. These are three pillars of modern territorial management: openness, co-production, and data sharing (SDI); models

DOI: 10.4018/978-1-7998-2249-3.ch004

(LADM); affordable technology (UAV). This chapter presents how municipalities can develop an SDI project, incorporating LADM guidelines and UAV data acquisition. The case study is based on a group of 32 small municipalities from São Paulo state, in Brazil, known as UNIPONTAL.

INTRODUCTION

Territorial managers are facing huge challenges due to the increase of pace of spatial pressure and transformations. There are multiple and complex problems that require wise and informed decisions, and for that accurate spatial datasets and decision support models are mandatory. This is not an exclusive requirement of major urban and metropolitan areas, as in rural context economic agents do create expectations and put pressure on the governing organizations. Thus, promoting and managing sustainable spatial planning and territorial development of small size and income municipalities, namely in rural context, does also require adequate tools that enable decision makers to formulate appropriate choices.

During the last decades a large set of technological based solutions have been developed by the scientific community and business companies linking the ability to acquire, store, process and publish spatial datasets (basically, traditional GIS and Remote Sensing Applications) with the capability of doing it based on web services, taking advantage of Internet and the significant cost reduction of connectivity.

Figure 1. Paradigm change – From GIS to SDI
Source: Julião, 2010

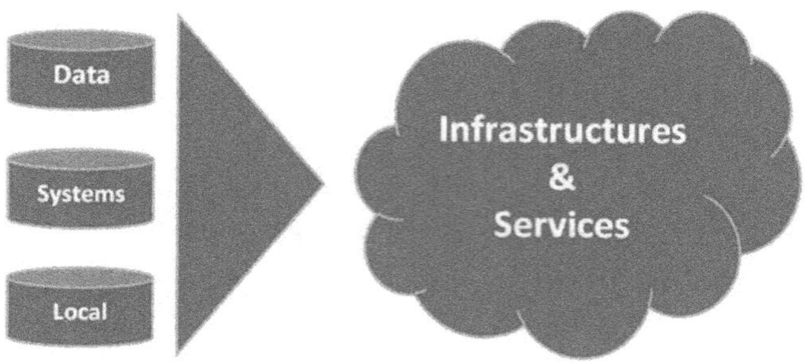

Therefore, today, more than the simple representation of land features through cartography, it is important to collect, organize, store, retrieve and explore spatial data finding the necessary knowledge for action. More than data repositories, it is important to have dynamic data flowing through the Information Society channels. In fact, the actual paradigm has shifted from an isolated GIS implementation approach to a more collaborative production and dissemination of Geographic Information, namely through the Internet, users groups and mobile technologies. It is an infrastructure approach, usually known by the designation of Spatial Data Infrastructures (SDI), which is shown in Figure 1.

Since SDI early days, and especially during the last years, there was a significant growing of interest about them acknowledging their important role in what concerns the formulation, implementation and monitoring of spatial public policies and private activities that spread over the territory. Technological growth and improvement, in particular the so-called Information and Communication Technologies (ICT), has helped in smart cities initiatives. Thus, Santana et al., 2016 pointed out four emerging technologies within systems for smart cities:

- **Internet of Things (IoT)**: It is the connection of everyday objects and sensors, transforming an entire environment into a network. Such objects may be lamps, refrigerators, microwaves, televisions, and other sensors and devices. For the operation of the network, each of the connected objects must be uniquely identiðed so that their access within the network is possible. Because large numbers of these objects are connected in a large urban center, smart cities systems are "forced" to use IoT-related technologies;
- **Big Data**: Strongly related to IoT, as objects that structure a smart city network are sending information continuously, producing massive amounts of data;
- **Cloud Computing**: Local physical storage has become a problem for complex smart city systems with a very high number of collected and processed data. Therefore, cloud computing can be a very useful technology for such systems because it offers a large, elastic and highly available infrastructure for both storage and processing of this data. Santana et al. (2016) also introduced the concept of "Cloud of Things," which corresponds to the junction of IoT and Cloud Computing, where the second technology would process and store data from the first;
- **Participatory Mobile Sensing**: In a smart city environment, citizens' own devices are used for data generation. Such devices may be, for example, smartphones and smartwatches, and data generated may be their location and travel speed. Areas such as health, transportation, safety and urban planning can take advantage of such data to improve their development. This is directly linked with the concept of Volunteered Geographic Information presented by Goodchild (2007), where he discussed the role of citizens becoming active environmental sensors.

On the other hand, concerning specific data, a critical dataset for municipal land administration is cadastre. Cadastral data is unique in its richness enabling a true and sound knowledge of the territorial context, as it can be used as a reference model to link and integrate other datasets (Amorim et al., 2018). Fully aware of such enabling power, the scientific community pushed forward for the establishment of a framework integrating key elements related to property. This led to ISO 19152 standard of the Land Administration Domain Model (LADM) published in 2012.

Finally, technology evolution has empowered a revolution in what concerns easy and fast data acquisition for cadastre. Indeed, Unmanned Aerial Vehicles (UAV) are growing as base technology for small area surveying, despite some limitations common to all airborne solutions.

These are three pillars of modern territorial management:

- Openness, co-production and data sharing (SDI)
- Models (LADM)
- Light and affordable technology (UAV)

However, the application of the conjunction of these solutions, mandatory for the full implementation of smart cities principles, is mainly restricted to wealthy organizations, mostly referring to large cities or municipalities within a metropolitan context. In fact, because of resource shortage, many small municipalities never hire an aerial survey by conventional photogrammetry. Consequently, the use of UAV and modern photogrammetric processing software have made possible the modernization of territorial management procedures.

This chapter presents how a small size municipality can still develop an affordable, and truly useful for public administration and citizens, SDI project compliant with INSPIRE, LADM, and other main standards, based on geodata acquisition through UAV. Thus, the methodological options can be replicated in different geographical contexts (different regions, countries or continents) with similar political, administrative and socioeconomic conditions. Our case study is based on a group of 32 small size municipalities from São Paulo state, in Brasil, known as UNIPONTAL (União dos municípios do Pontal do Paranapanema).

This chapter is structured as follows. It starts with a brief presentation of the case study region, then there are three blocks concerning data acquisition, management and sharing, and ends up with some conclusions.

UNIPONTAL REGION

The study area region is known as Pontal do Paranapanema and it is located in the western part of São Paulo State, spreading across an area of 18.442 km^2, representing more or less 7,5% of the state and contributing with 0.89% of its GNP (PEDEPP,

2015). It has a population of 600.000 inhabitants mostly resident in urban areas (urbanization rate of 91%).

This region is managed and represented by an association of the comprising 32 municipalities called Associação dos Municípios do Pontal do Paranapanema – UNIPONTAL, as shown in Figure 2. These municipalities have similar physical and social characteristics.

Figure 2. Study area location

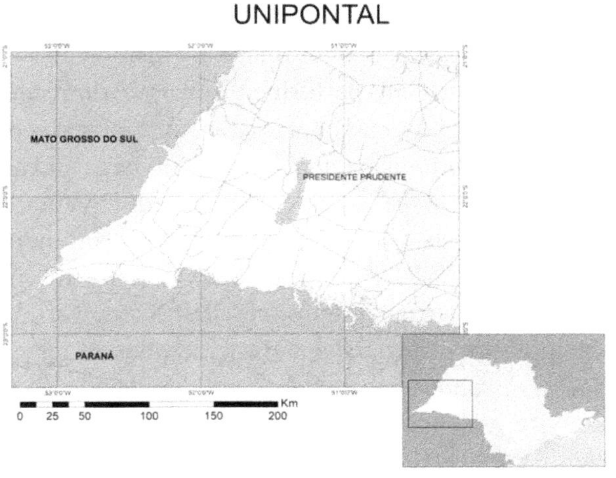

This region has a strategic location in São Paulo state, as it is directly linked to São Paulo city by SP 270 highway, fluvial channel Tietê-Paraná, and a foreseen expansion of the railroad north-south that shall cross it.

Pontal do Paranapanema region has several possibilities to expand its economic activities, considering the touristic and agronomic vocation, its diversified industrial poles, and the human capital and technological development with several universities. There is a large potential for leisure and tourism activities due to major rivers and lakes, but limited by a single active commercial airport located in Presidente Prudente, the main city of the region.

During the last years, this region has gone through several changes, supported by the enlargement of its logistics infrastructure and road duplication, as for instance, the modernization and expansion of sugar cane production and industry, growth in cattle and meat industry, and increase in industrial entrepreneurship. Nevertheless, Pontal do Paranapanema region is still considered as one of the less developed regions of São Paulo state.

Furthermore, the 32 municipalities of the region are of small size, and concerning cadastral data they have different management and governance models with prevalence of tax purpose applications.

DATA ACQUISITION SOLUTIONS

UAV are one of last major innovations for geographic data acquisition with a wide and diversified range of application, resulting from a strong technological evolution. There are several typologies with different configurations, weight and flight autonomy.

Nowadays, UAV devices developed for image acquisition allow applications for tasks that, until recently, were only performed by large companies specialized in photogrammetry procedures and techniques, given the complexity of maintaining an aircraft equipped with conventional cameras, which are large and heavy equipment. With constant improvements in the development of both UAV and sensors, as well as in photogrammetric processes, it is now possible to board increasingly smaller and more powerful sensors on UAV devices necessary for the photogrammetric mapping, such as cameras, inertial systems and GNSS receivers.

Systems for obtaining small-format camera aerial photographs, such as cameras mounted on micro-UAV, cannot be compared with conventional photogrammetry in terms of productivity, as there are restrictions on its use in urban areas and also by the limited size of the overflown area. On the other hand, UAV equipment is already used by some companies for photogrammetric mapping, with specific objectives of small area mapping, which can be used even for the generation of cadastral cartographic products, either in rural or urban areas.

Several studies have been carried out with the objective of enabling the creation of cartographic products compliant with the mandatory levels of precision established by legislation and required by the bodies responsible for land regularization. Current results are satisfactory. Thus, the use of these equipment has been significantly disseminated, since it speeds up the process of data acquisition, maintaining the levels of precision, and enabling the achievement of a good cost-benefit ratio.

Bearing in mind the use of UAV images for urban land registry and cadastre applications, some technical issues should be considered. Although a very good geometrical accuracy can be achieved, and even precise calculations of built area, it is necessary to consider that the built area must be calculated by measuring the external walls of a building and not by projecting the roof on the ground. This is a limitation that should be considered regardless of the method used for area calculation by aerial images, since objects in the aerial image are roofs and not the building outer walls.

Thus, it must be understood that the use of aerial images is very useful for the production of various cartographic documents used in registration, and can become a useful tool to detect changes even at the level of built-up area, but field work with on-site measurement cannot be dismissed.

Furthermore, current image processing software makes it easily possible to produce three dimensional urban models that are so important for spatial planning and territorial management.

DATA MANAGEMENT MODEL

The development of an international standard for the domain of land administration is mainly a response of the worldwide professional and academic community to the challenges identified (initially) in the Rio "Earth Summit" of 1992. It is one answer, although partial, to the pressing problems of land management and environmental protection.

A joint United Nations – Habitat and "Fédération Internationale Géomètres" (FIG) workshop held in Bogor in 1996 announced an initial vision for the future of cadastre, with the aim of efficient land and property markets, protecting land rights and support sustainable development, among other related goals (UN-FIG, 1996). It also claimed that a common cadastral vision is possible and essential to the development of modern cadastral systems worldwide.

A second joint UN and FIG workshop held in Bathurst in 1999 also produced an important declaration reinforcing the need for development of land administration infrastructures in face of the rapid change brought by growing population pressures. It identified a set of major issues, namely, the land administration role in sustainable development and the need to re-engineering land administrations systems (UN-FIG, 1999). It concludes with the sentence "Sustainable development is not attainable without sound land administration".

In between these two declarations, a clear vision for the future of cadastral systems was presented by two experts of the Work Group 7.1 – Modern Cadastral Systems of FIG. It included a review of the strengths and weaknesses of (then) current cadastral systems (Kaufmann & Steudler, 1998). Its vision for a future cadastre system, called Cadastre 2014 is condensed around six statements. The following aspects are especially relevant for the domain model:

- Cadastre 2014 will show the complete legal situation of land, including public rights and restrictions;
- The separation between maps and registries will be abolished;
- The cadastral mapping will be dead, long live modelling.

Considering the above declarations and statements as a background, a proposal to develop a Core Cadastral Domain Model (CCDM) was presented in a FIG Congress in Washington in 2002. The model was perfected in the following years, with contributions from several organizations and academia. It assumed two main goals (Lemmen & von Oosterom, 2006):

1. To avoid reinventing and re-implementing the same functionality over and over again, but provide an extensible basis for efficient and effective cadastral system development based on a model driven architecture (MDA);
2. To enable involved parties, both within one country and between different countries, to communicate based on the shared ontology implied by the model.

From the development of the CCDM resulted a first mature version, submitted by FIG to the standardization procedure at ISO, as a New Working Item Proposal, in 2008. The now called Land Administration Domain Model (LADM) addresses the following issues:

- Defines a reference model covering all information related to components of land administration;
- Provides a conceptual schema with five basic packages concerning Parties, Spatial Units, RRR (Rights, Restrictions and Responsibilities), Surveying and Geometry and Topology;
- Provides a terminology allowing a shared description of practice in various jurisdictions;
- Is a basis to develop national or regional profiles;
- enables the coherent integration of land administration information from various sources.

LADM should be able to accommodate any legal framework, supporting the reform of land administration systems at national or other jurisdictional level (Lemmen et al., 2009).

The domain of land administration can be summarized through the definition of its four basic functions, as described in the fundamental work of (Williamson et al., 2009):

- **Land Tenure**: Titles, mortgages, servitudes and other land rights, usually recorded at a Land Registry. It provides security of legal rights on land and is the legal component of a cadastre;
- **Land Value**: Definition of the value of the land, determining a tax on immovable property. It is the fiscal component of a cadastre;

- **Land Use**: Can be regulated through land administration and management policies. It provides support for the control of land use and is the land management component of a cadastre;
- **Land Development**: Can be promoted through urbanization plans or land consolidation projects. Development control can be implemented locally through the issue of licenses. It is the development component of a cadastre.

A correct implementation of relationships integrating the land tenure and land value functions contributes to an efficient market, while the integration of the land use and development functions contributes to an efficient land management. At the core of the four basic functions should lay the Land Information Infrastructure, supplying the basic spatial data which works as the "glue" integrating land tenure, value, use and development.

The IT infrastructure component in a land reform integrating the above basic functions, using the capabilities of the LADM standard, can be achieved through a methodology described in (Hespanha, 2012). The main contributions to the methodology were, in brief:

- A systems approach to land registries, including different aspects, namely, technical, legal, administrative and institutional;
- A combination of the Unified Process for software development life cycle and the Model Driven Architecture (MDA);
- The "blueprint" for the reformed infrastructure shall be supplied by the LADM standard, supplemented by the Legal Model published in its informative annexes.

The method to achieve an implemented IT solution starting from the domain model has a total of 12 phases and 33 activities within these phases (Hespanha, 2012, p. 130-132). The following enumeration summarizes key aspects along the Unified Process stages:

1. **Inception**: Starting from LADM as a highly conceptual description of the domain, critical requirements should be captured by UML Use Cases, and complemented by a business and iteration and development plans;
2. **Elaboration**: should be achieved through a series of iterations. The initial consider cadastral components (geometry, legal and administrative), concluding with an integration in a single country specific model. These iterations include new Use Cases and Activity Diagrams, as well as a specialized LADM model, incorporating different geometric and legal profiles. The final result is still a platform independent class model;

3. **Construction**: starting with a new iteration and development plan, the final model is implemented with the help of code generation and persistence tools to a database and web based interface. In these last phases, a map based edition layer shall be achieved through existing API, and preferably, a cartographic data model including feature portrayal as in ISO 19117.

In the years following the publication of the LADM standard, a number of workshops were promoted by FIG with the collaboration of other entities like ISO and the World Bank. One of the mandatory discussions is about the implementation of country profiles, from proposals to implementation plans, including prototypes with real data.

The following paragraphs summarize the presentations for Croatia (Vučić, 2017) and Greece (Dimopoulou, 2017), concerning their country profiles.

In the case of Croatia, the spatial topological profile was adopted, and specialized classes and code lists were considered for the parcels and its boundaries, and for the vertices. Land use and building types are very detailed, with tenths of entries in the code lists. Besides the specialization of buildings, the utility networks have also a number of country specific code lists.

In the case of Greece, the approach is more comprehensive and includes all the major packages of LADM, starting with a legacy model translated into a UML class model. The legal and administrative components received close attention, considering specializations for servitudes and administrative source documents. The LA_Party class is also specialized, including new code lists. A separate thematic model was also derived from LADM, for the implementation of a maritime cadastre. This considers the laws defining interests in a marine environment and the demarcation of marine parcels (with a 3D spatial unit profile).

In the specific case of the Unipontal Region SDI development, initial requirements were gathered based on previous proposals for the adoption of LADM in Brazil and further activities carried out in the “Pontal Transparente” research project at UNESP.

The proposal described in Paixão et al. (2015) focused on the indigenous tribes’ land rights, while having a larger scope in the inception stage, for the context Use Case. The existing land administration for public lands in Brazil is described as a set of thematic cadastral systems which are not integrated and do not share data. The conclusion is that the management and control of land use and tenure arrangements in public lands is impaired. The proposal included a jurisdiction specific country model with both dynamic (Use Case and Activity) and static (Class) UML diagrams. Correspondence between entities in LADM and laws and regulations in Brazil started with the LA_Party package and moved to spatial units and then to the people-to-land relations through the LA_RRR package (for Rights, Restrictions and Responsibilities).

Given its specific approach for indigenous lands, the proposal did not consider private property land parcels in the rural and urban cadastral systems. Uses specialized spatial unit group classes to identify tribal lands and other indigenous federal lands (Reserve, Park, Agricultural Colony or Federal Territory).

"Pontal Transparente" project assumed that land administration theme is a fundamental spatial data layer in the development of the SDI. The specialized country model is being derived from LADM, including both rural and urban land parcels. In Brazil, rural parcels are administered at federal level by INCRA (Agrarian Land Reform National Institute), while urban parcels are administered at municipal level by each individual local government ("Prefeitura"). Rural parcels are in a process of georeferencing, establishing a digital cadastral database at national level, CNIR (Rural Immovables National Cadastre). For the urban cadastre, given the diversity of implementations existing nowadays, a set of guidelines issued by the ministry of cities gave the initial set of requirements and concepts (CONFEA, 2017). Furthermore, the first guidelines for the proposed integrated cadastre, SINTER (Land Information and Management National System), was also considered at the inception stage. The existing cadastral systems are not integrated and have different purposes. While rural cadastre is mostly an agricultural land parcel and (federal level) tax collection system, the various urban cadastres are mainly fiscal cadastre with various degrees of land use and land development functions, considering namely data required for administering municipal master plans and issuing of licenses. Usually, urban cadastral systems in Brazil include also a set of elements outside the scope of LADM, which typically belong to a large scale topographic dataset. The least developed component is the legal one, giving that there is no federal level land registry code or regulation, and land registry offices are not integrated and many still work with analogue records.

"Pontal Transparente" project will establish a land administration theme derived from LADM, including geometric, legal, fiscal, and land use and development components. Thus, forming a multipurpose cadastral set of interrelated layers at its core. The method starts with the conceptual model expressed in UML (LADM) and uses available open source and proprietary modeling tools in order to assist for the definition of a physical database structure and an object-oriented programming layer. The process is summarized in the following list:

- Importing of the publicly available UML model project with ISO 19152 LADM into modeling software. Visual Paradigm (proprietary) and Modelio (open source) are being used to test code and database generation, besides basic UML diagramming;
- Specific country requirements gathering, including above referred laws and regulations for the urban and rural cadastres. To be complemented

with questionnaires and team visits to local governments in the "Pontal do Paranapanema" region;

- Derivation of a country specific model, BR_LADM, with required specialized classes and code lists. This will be the platform independent model for the following MDA processes;
- Use of the code and database generation capabilities of the modeling software, to obtain a programming layer and database structuring scripts. Both software have independent code generation and database logical and physical model, and also Hibernate tools or plugins.

The database server to be used in the project is postgreSQL with PostGIS extension to manage spatial data layers. The existent SDI prototype uses PHP code for the WebGIS programming component.

The schematic diagram in figure 3 shows the main stages of the method, as above described.

DATA SHARING PROPOSAL

Nowadays, SDI are the key-paradigm for data sharing and are considered to be indispensable at different administration levels. Being fully based on technologic platforms, they involve many other essential components for its proper functioning. As defined by the Brazilian SDI – INDE, they are an "Integrated set of technologies; policies; coordination and monitoring mechanisms and procedures; standards, and agreements necessary to facilitate and manage the generation, storage, access, sharing, dissemination, and use of geospatial data from federal, state, district, and municipal sources." (https://www.inde.gov.br/Inde/Apresentacao). That is why SDI are the current model of sharing, access and exploitation of geographic data, ensuring that they are exploited to the fullest of their potential. It is only in this way from an institutional, academic and commercial perspective that the investment in geographical data acquisition and management is maximized and it effectively contributes to the creation of solutions for intelligent territories and the promotion of development.

First SDI projects started in 1990 (Portugal), USA and The Netherlands (1994), Canada (1996), Colombia (1999), but only in the last decade they spread through the world and assumed a critical role (Masser, 2005). INSPIRE Directive, approved in May 2007, has stressed the importance and set a new framework for the development of SDI in Europe, as tools for good and modern governance, through the establishment of a coordinated network of National SDI (NSDI). It also sets a new model for the development of SDI projects across the world, by the pushing forward ISO 19000 series and OGC standards.

Figure 3. Land administration SDI theme data modelling process

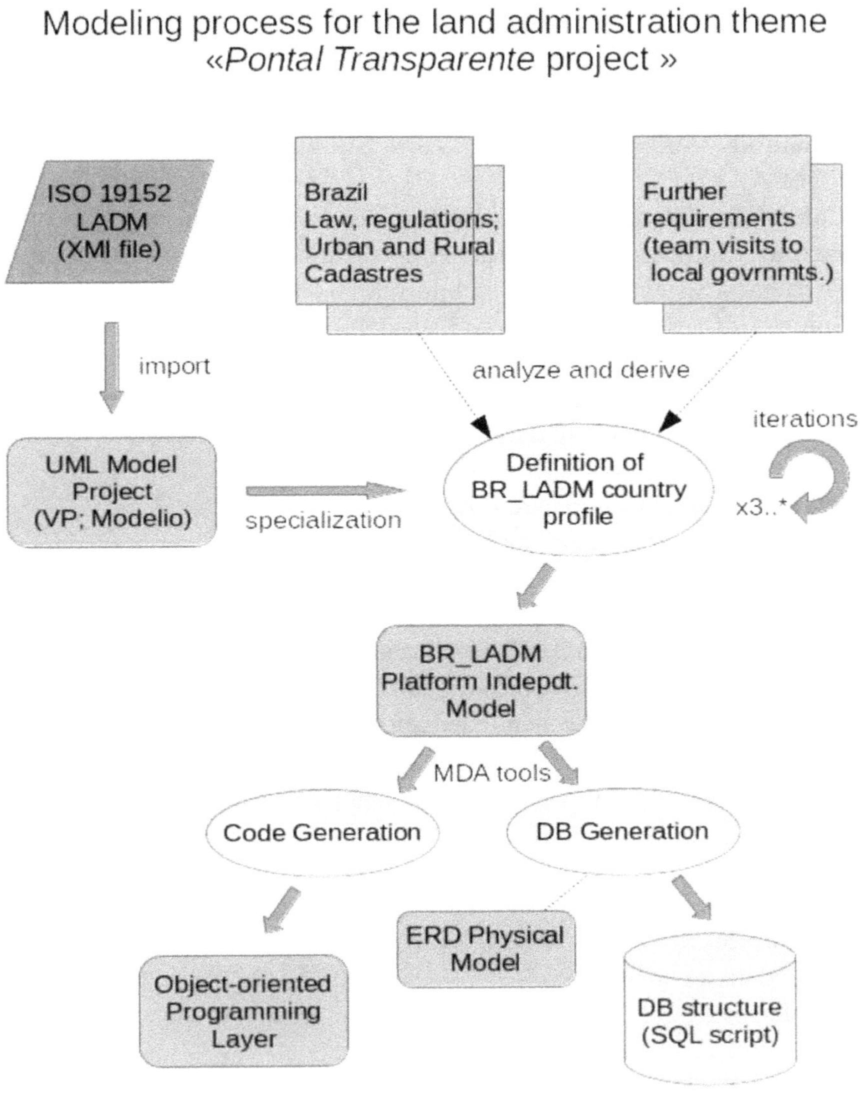

However, no SDI can be copied without adjustments to fit other geographical and/or institutional background. Moreover, small municipalities do not have the same power, resources and opportunities of larger ones. However, with a proper framework, based upon reference projects, it is possible to scope and develop useful projects.

There are now several examples of SDI projects with different spatial and thematic scopes that are set out as examples to inspire UNIPONTAL SDI draft, and some of them are even mandatory reference elements (marked with *), namely:

- **Brazil** – Infraestrutura Nacional de Dados Espaciais (INDE)*
- **Brazil** – Infraestrutura de Dados Espaciais para o Estado de São Paulo (IDE-SP)*
- **Colombia** – Infraestructura de Datos Espaciales para el Distrito Capital (IDECA)
- **Colombia** – Infraestructura Colombiana de Datos Espaciales (ICDE)
- **Europe** – Infrastructure for spatial information in Europe (INSPIRE)
- **Portugal** – Sistema Nacional de Informação Geográfica (SNIG)
- **Spain** – Infraestructura de Datos Espaciales de España (IDEE)

The standards of the following organizations are also reference elements, many of which are reflected in the previous mentioned SDI examples:

- Open Geospatial Consortium (OGC)
- International Organization for Standardization (ISO)

Building a SDI core can be an affordable investment, even in the context of budget restraints. This is of great value for small organizations, like rural municipalities, which do not have the same resource availability as bigger ones. Additionally, a SDI project, as demonstrated in the mentioned reference projects, is a process more than just setting up technology and filing in databases. Taking this under consideration, UNIPONTAL SDI project has 4 main stages combined with each other that will be later described, namely:

1. Inception
2. Pilot-project
3. General dissemination
4. Maintenance

As mentioned before, INSPIRE – Infrastructure for Spatial Information in Europe – (Directive 2007/2/EC of the European Parliament and of the Council of 14 March 2007) has set a new background for SDI development in Europe and across the world. It is a legal document binding for all 27 European member states with the main objective of creating an "Infrastructure for Spatial Information in the

European Community (...), for the purposes of Community environmental policies and policies or activities which may have an impact on the environment" (Article 1). INSPIRE is based on a set of key principles:

- Collected once and use many, across different scales;
- To combine spatial data from different sources and share it;
- Data for good governance should be readily and transparently available;
- Data should be easy to find, as well as their characteristic and terms of usage.

Of course everything should be based on network services: discovery services; view services; download services; transformation services; and, services allowing spatial data services to be invoked (Article 11).

To make this effective, INSPIRE Directive is supported by a set of Implementing Rules (also compulsory for Member States to follow) that specify to very detailed level different topics, as:

- Metadata
- Data Specifications
- Network services
- Data and Service sharing
- Monitoring and Reporting

Most of the specifications are supported by ISO and OGC standards and that is also why INSPIRE assumed a reference role concerning SDI implementation.

Coming back to UNIPONTAL SDI project the four stages are now briefly detailed.

Inception

The initial preparation phase is fundamental for stabilizing the concept, objectives and model of SDI to be implemented. UNIPONTAL SDI is meant to be a support infrastructure for the development of the region, providing access to seamless data from the 32 municipalities, under the umbrella coordination of UNIPONTAL.

Given the territorial reality of the Pontal do Paranapanema region and the institutional structure of the 32 municipalities that comprise it, the SDI proposal aims at creating a centralized geoportal architecture at UNIPONTAL headquarters with the possibility of integrating web services from the municipalities that are able to provide them. Figure 4 presents a generic view of UNIPONTAL SDI connections.

Figure 4. UNIPONTAL SDI connections

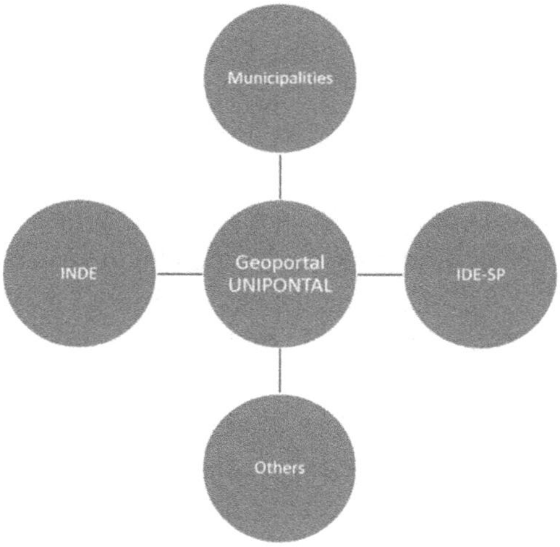

Likewise, through the use of standard services (OGC and ISO), UNIPONTAL SDI must be able to articulate with IDE-SP (http://www.idesp.sp.gov.br/), and INDE (https://www.inde.gov.br/), to provide services to other stakeholders and to integrate external sources.

By the end of the Inception stage, it will be possible to have a clear definition of the geographical data sets and web-based services definitions and availability, a user's assessment regarding functionalities that supports its definitions and priorities, and a detailed budget with funding sources identified.

As a starting point, some indispensable functionalities are identified, like:

- **Catalogue**
  - Free text, thematic and geographical searches
  - Viewer Integration
  - CSW service (Catalog Service for the Web is an OGC standard for sharing metadata)
- **Viewer**
  - Free text, thematic and geographical searches
  - Viewer Integration
  - CSW service (Catalog Service for the Web is an OGC standard for sharing metadata)
- **Catalogue**
  - Catalogue integration
  - Multiple layers support

- External web services integration
- Key-tools (search, query, measure, sketch, print)

Concerning data requirements, main points to take under consideration are:

- **Geopositioning parameters** (projection and coordinate systems, positional accuracy, etc.)
- **Themes** (topics to consider)
- **Access availability** (viewing / downloading, restrictions, costs, etc.)

There are some initial technological options that must be considered in order to insure a 365/7/24 availability of the system, with adequate response time and load stress handle. At the same time, taking under consideration that an overall solution must be affordable, the use of FOSS4GEO components is mandatory. So the main hub of the UNIPONTAL SDI is to be based on Geonetwork, a well-known open source catalog application able to manage spatial data with a powerful metadata editor and search tool, and integration with a map viewer.

Let us remind of the SDI definition by GSDI (2012) as a "relevant base collection of technologies, policies and institutional arrangements that facilitate the availability of and access to spatial data". Thus, it is critical for UNIPONTAL SDI the creation of a set of governing rules that establish the basis for its development, namely:

- **General rules**
    - Concepts/definitions
    - System architecture
    - Data and data services
    - Responsibilities
    - Funding
- **Technical rules**
    - Geoportal
    - Metadada publication
    - Data publication

Pilot-Project

The second stage of UNIPONTAL SDI project is the setting up of a pilot-project that will consist of:

1. **UNIPONTAL SDI geoportal**. Setting up of the central infrastructure of the UNIPONTAL SDI based on Geonetwork with metadata catalog upload and

viewer, and definition of initial available themes through view (WMS) and download (WFS) services.

2. **Local nodes.** Setting up of UNIPONTAL SDI nodes in major municipalities and testing of integration of available themes through view (WMS) and download (WFS) services at the geoportal.
3. **Training and dissemination.** Awareness raising and capacity building activities for different user's profiles, including general public.

General Dissemination and Maintenance

General dissemination and maintenance are the following stages. It is foreseen that in the second year of the project, all 32 municipalities (some due to personal and budget constraints may still opt for a dependent service on the central geoportal) will be connected to UNIPONTAL SDI geoportal, providing through it access to data and services.

Also during this dissemination stage activities will continue in order to further engage more stakeholders.

CONCLUSION

SDI are the actual paradigm for data sharing and promotion of a better articulation between public bodies and stakeholders, based on a common geographical base. This approach is already consolidated at central and regional level bodies and some municipal project are emerging. Some of the critical barriers are: 1) access to technological resources; 2) human resources capacitation; 3) institutional acknowledgement.

The term "smart city" does not yet have a unique and consensual meaning (Albino et al., 2015). Even though, two main domains can be identified. The "hard" one, concerning buildings, natural resources, mobility and logistics, and the "soft" one, related to areas such as education, culture, political innovations, social inclusion and the like. In addition, Neirotti et al. (2014) also states that the consensus and diffusion of a single meaning for the term is hampered by the fact that there are several initiatives around the globe, with different backgrounds and objectives. In addition to these two great domains that generalize smart cities, Albino et al. (2015) and Santana et al. (2016), have set out some application areas of smart city initiatives:

- **Intelligent Economy**: Evaluates the economic situation of the city, taking into consideration the companies operating in its territory and its entrepreneurship environment;

- **Smart People**: Assesses how assisted is the development of the population by the city environment in areas such as education, employment and income;
- **Intelligent Governance**: Services provided by public agencies, its transparency, ease of service to citizens and investors;
- **Smart Mobility**: Entire transportation network and services provided to the population, such as buses, subways, highways and bike paths;
- **Intelligent Environment**: Evaluates the environment of the city and its sustainability with regard to providing welfare to its inhabitants;
- **Smart Living**: Evaluates the quality of life of city residents regarding their lives and happiness, considering for example entertainment, culture and safety.

UNIPONTAL SDI project seeks to provide a base infrastructure for the creation of a smart territory and has addressed, among others, three challenges providing an affordable solution, making it a reference for small scale projects of less favored regions/municipalities. The highlights of this approach are as follows:

- A robust data model, following LADM principles and adjusted to local reality;
- Usage of light and flexible data acquisition methods based on UAV;
- A technological solution, integrated with state and national SDI, fully based on open source components, making it adjustable and affordable;
- A capacity building plan to engage all 32 municipalities;
- A dissemination plan to raise awareness of the project among other stakeholders.

Futures prospects include the incorporation of Volunteered Geographic Information data sources and the development of a SDI toolkit that will empower other municipalities to establish their local SDI compliant with the requirements of São Paulo SDI and INDE.

ACKNOWLEDGMENT

This work is financed by Portuguese funds through FCT - Foundation for Science and Technology, I.P., within the scope of the project «UIDB/04647/2020» of CICS. NOVA - Centro Interdisciplinar de Ciências Sociais da Universidade Nova de Lisboa, and also by Brasilian funds through Fundação de Amparo à Pesquisa do Estado de São Paulo (FAPESP) within the scope of project «18/23155-3 – Pontal Transparente».

REFERENCES

Albino, V.; Berardi, U.; Dangelico, R. M. (2015). Smart Cities: Definitions, Dimensions, Performance, and Initiatives. *Journal of Urban Technology, 22*(1), 3-21.

Amorim, A., Pelegrina, M. A., & Julião, R. P. (2018). *Cadastro e gestão territorial. Uma visão luso-brasileira para a implementação de sistemas de informação cadastral nos municípios*. Editora UNESP Digital.

Amorim, A.; Pessoa, G. G.; Lemes, I. R.; Komazaki, J. M.; Jorge, L. C.; Galo, M.; Camargo, P. O. (2016). *Utilização de VANT para o georreferenciamento de imóveis rurais*. I Seminário Internacional UAV, Lisboa, Portugal.

Associação dos Municípios do Pontal do Paranapanema – UNIPONTAL. (2015). *PDEPP – Plano de Desenvolvimento Regional e Local dos Municípios do Pontal do Paranapanema*. Presidente Prudente.

CONFEA, CREA, ABNT, MINISTÉRIO DAS CIDADES (2017). *Grupo Técnico Operacional. Norma de Engenharia para Georreferenciamento Urbano no Brasil.* Normas e Procedimentos de Engenharia para Cadastro Urbano no Brasil.

Dimopoulou, E. (2017). *GR_LADM*. The 6th Land Administration Domain Model (LADM) Workshop, Delft, The Netherlands.

Goodchild, M. F. (2007). Citizens as sensors: The world of volunteered geography. *GeoJournal, 69*(4), 211–221. doi:10.100710708-007-9111-y

GSDI. (2012). *Spatial Data Infrastructure Cookbook 2012 Update*. Global Spatial Data Infrastructure Association. Retrieved from http://gsdiassociation.org/images/publications/cookbooks/SDI_Cookbook_from_Wiki_2012_update.pdf

Hespanha, J. P. (2012). Development Methodology for an Integrated Legal Cadastre. ed. Delft, The Netherlands: TU Delft. Repository of Delft University.

Julião, R. P. (2010). Low-Cost SDI. The Portuguese example of building a SDI for small countries. *FIG Congress 2010.*

Kaufmann, J. & Steudler, D. (1998). Cadastre 2014 – A vision for a future cadastral system. *FIG*.

Lemmen, C., & van Oosterom, P. (2006). *Version 1.0 of the FIG Core Cadastral Domain Model. XXIII FIG Congress*, Munich, Germany.

Lemmen, C.; van Oosterom, P.; Uitermark, H.; Thompson, R.; Hespanha, J. (2009). *Transforming the Land Administration Domain Model (LADM) into an ISO Standard (ISO19152).* FIG Working Week 2009, Eilat, Israel.

Masser, I. (2005). *GIS Worlds. Creating Spatial Data Infrastructures*. ESRI Press.

Neirotti, P., deMarco, A., Cagliano, A. C., Mangano, G., & Scorrano, F. (2014). Current Trends in Smart City Initiatives: Some Stylised Facts. *Cities (London, England)*, *38*, 25–36. doi:10.1016/j.cities.2013.12.010

Paixão, S., Hespanha, J. P., Ghawana, T., Carneiro, A. F. T., Zevenbergen, J., & Frederico, L. N. (2015). Modeling indigenous tribes' land rights with ISO 19152 LADM: A case from Brazil. *Land Use Policy*, *49*, 587–597. doi:10.1016/j.landusepol.2014.12.001

Santana, V. D., Silva, J. S., Ribeiro, H. L. M., Arantes, A. J. R., Silva, O. C., & Alves, L. L. (2016). InfoDengue – Ferramenta Colaborativa de Combate e Prevenção à Dengue. *Computer on the Beach*, (9), 1-4. Retrieved from https://siaiap32.univali.br/seer/index.php/acotb/article/view/10787

UN-FIG (1996). *The Bogor Declaration*. United Nations Interregional Meeting of Experts on the Cadastre, Bogor, Indonesia.

UN-FIG. (1999). *Executive Summary of the Bathurst Declaration. Workshop on Land Tenure and Cadastral Infrastructures for Sustainable Development*, Bathurst, Australia.

Vučić, N. (2017). *LADM country profile Croatia*. The 6th Land Administration Domain Model (LADM) Workshop, Delft, The Netherlands.

Williamson, I., Enemark, S., Wallace, J., & Rajabifard, A. (2009). *Land Administration for Sustainable Development*. ESRI Press.

KEY TERMS AND DEFINITIONS

Cadastre: Spatial dataset regarding the localization, spatial configuration, and rights of a parcel

Catalogue: Specific website or its component that enables the user to search and evaluate existing geographic datasets or services through its metadata records.

Data Acquisition: Process of retrieving data regarding features of the real world to feed databases and information systems.

Infrastructure: Set of facilities that supports the provision of a service.

Metadata: Data that describes the characteristics of a dataset or service.

Viewer: Specific website or its component that enables the user to view and explore geographic datasets.

Web Services: Access to data or services between devices through the internet.

Chapter 5
From Macro to Micro: Two Approaches to Study Urban Mobility in a Brazilian Municipality

Lívia Rodrigues Tomás
https://orcid.org/0000-0002-2464-1879
National Center for Monitoring and Early Warning of Natural Disasters (CEMADEN), Brazil

Maria Carolina Barbosa Jurema
São Paulo State University (UNESP), Brazil

Janaina Cassiano dos Santos
Federal Rural University of Rio de Janeiro (UFRRJ), Brazil

Luciana de Resende Londe
National Center for Monitoring and Early Warning of Natural Disasters (CEMADEN), Brazil

Regina Tortorella Reani
National Center for Monitoring and Early Warning of Natural Disasters (CEMADEN), Brazil

Claudia de Albuquerque Linhares
National Center for Monitoring and Early Warning of Natural Disasters (CEMADEN), Brazil

Leonardo B. L. Santos
National Center for Monitoring and Early Warning of Natural Disasters (CEMADEN), Brazil

ABSTRACT

This chapter discusses urban mobility considering two main analyses approaches. Based on the relationship between mobility and vulnerability, the first approach analyzed commuter's vulnerability using basin as unit of analysis. The second one analyzes variables related to land use such as population density and its relation with job offer in the city and people's income using traffic zones as unit of analysis. The two scales dialogue and can be used concurrently. The municipality of São José dos Campos (Brazil) was used as a case study. Origin-destination research was the main database used in the analyses. Authors used geospatial tools, like spatial join

DOI: 10.4018/978-1-7998-2249-3.ch005

operation and thematic maps, which enable the in-depth analysis of important data for urban studies or transport planning and can be replicated in any study area. The analysis of mobility data aggregated by basin contributed to an understanding of the implications of the urban configuration, with its displacement patterns related to water courses if any flooding or landslide occurs and interrupts people's flow.

INTRODUCTION

Historically, early civilizations emerged near the banks of rivers in search of greater agricultural productivity and to facilitate transportation. Over the years, the relationships between transport infrastructures and hydrological processes became complex, especially because they happen at the same geographic location.

Relations between water resources and transport are explored in the literature (Santos, Bacelar, & Santos, 2017; Tucci, 2007), and some works deal with the impact of hydrological extremes on highways and in urban mobility. These works also highlight the possibility of using urban mobility data to infer exposure (time dynamics and heterogeneity in space) of different population groups at different hydrological risks (Cavion & Lombardo, 2014; Doll et al., 2014; Eidsvig, Kristensen, & Vangelsten, 2017; Londe, Santos, Soriano, Tomás, & Carvalho, 2015; Pregnolato, Ford, Wilkinson, & Dawson, 2017; Santos, Londe, Soriano, Souza, & Coelho, 2015; Santos et al., 2017).

However, in a scenario of climate change, with increasing frequency and intensity of extreme rainfall, it is important to develop scientific research at the interface between water resources and urban mobility from the perspective of sustainable development.

This chapter considered two main analyses approaches. Considering environmental topics, the first approach presents a hydrographic scale to analyze data on urban mobility. The second one discusses and analyzes variables related to land use such as population density and its relation with job offer in the city and people's income, using Traffic Zones (TZs) as unit of analysis. The municipality of São José dos Campos (Brazil) was used as a case study. This chapter also emphasizes the use of geoprocessing as an auxiliary tool in urban planning and shows the matter of occupation in environmentally fragile areas.

BACKGROUND

Urban mobility, "the whole of trips generated daily by the inhabitants of a city, and the methods and conditions associated with such trips (modes of transport selected, length of trip, time spent in transport, etc.)" (Shang, Doulet, & Keane, 2009), is directly related to the city habitant's lifestyle.

A range of studies have been conducted on the relationship between urban density, automobile dependence, and energy consumption (McLoughlin, 1991; Kenworthy & Laube, 1999; Newman, & Kenworthy, 1999; Mees, 2010a; Mees, 2010b). Studies on the relationship between urban structures and transport energy consumption are mostly based on private car travel, but the level of service (LOS) of public transport is also a considerable factor. Urban compactness will increase travel demand density, allowing a higher LOS and modal share of public transport. Modal shift from private to public transport is expected as a mitigation measure for the problem of global warming, but depends on travel density and the efficiency of public transport. Except in very large cities, private cars are the dominant transport mode in most developed cities. This reflects the lower profitability and LOS of public transport that comes with lower travel density. If an administration forces increased public transport service in a region with low travel demand density, it could increase CO2 emissions due to the higher energy intensity of public transport at lower occupancy ratios (Doi & Kii, 2012; Kii, Hirota, & Minato, 2005; Kii & Hanaoka, 2003).

According to Brazilian Public Transport Association (ANTP, 2011) the choice of a mode of transport has a direct impact on energy consumption, the emission of particulate pollutants, the figures of traffic accidents, and the track area consumed. According to the ANTP's Environmental Impacts Simulator, if 20% of the transfers by automobiles were walking in São José dos Campos (Brazil), there would be a reduction of 16% of track space consumption, 17% of energy consumption, 14% of local pollutants emission, and 15% of CO2 (carbon dioxide) emission, beside an increase of 38% on the travel time.

As to the environmental impact, urban expansions on environmentally fragile areas, either with edifications or road mesh, associated with changes in land use increase the exposure of the population that lives, works or moves along the boundaries of those areas, consequently increasing the risk of social and environmental disasters. Damage caused by floods and landslides impairs the mobility and accessibility of people that use the route to travel.

The expansion of paves - impermeable areas, which prevent rain from being absorbed by the soil - increases the risk of flood, particularly in low-lying areas (UNDRR, 2015). Land use types (together with soils and topography) influence the hydrologic and physical nature of the basin (U.S. Environmental Protection Agency, 2008). As a result, the peak discharge, volume, and frequency of floods increase in nearby streams (Konrad, 2003).

Social and environmental disasters cause great human and material losses throughout the world, and their increasing risk is a global matter. The large Brazilian territorial extension and growing population, allied with its social and economic conditions, contribute to the country's susceptibility to some kinds of disasters, and their impacts affect thousands of millions of people, often recurrently, each year. Among the deflagrating processes of disasters that affect Brazil, mass movements, droughts, and floods are noteworthy. According to the *Atlas Brasileiro de Desastres Naturais* (*Centro de Estudos e Pesquisas em Engenharia e Defesa Civil*, 2012), between 1991 and 2010, floods and flash floods were 50% of the disasters related to natural hazards.

As to the impact on land use, settlement patterns are important to understand the basin conditions and source dynamics (U.S. Environmental Protection Agency, 2008), and also help to determine what kind of transport is used. Single-family homes on large lots increase the need for cars, whereas high-rise apartments (with limited parking) create the conditions for people to choose to use public transport (subways, buses) or taxis (Bouton, Knupfer, Mihov, & Swartz, 2015).

Land use affects mobility and has a huge impact on the sustainability of cities. Jobs or study places far away from residential areas are inaccessible for pedestrians, cyclists, and, sometimes, for public transport users. In other words, a mixed zoning that allows the existence of homes, educational institutions, and jobs is more sustainable.

In previous times, individuals were only mobile or active within a limited radius, thus could get around walking or by slow vehicles. Nowadays, mobility patterns have changed. Dispersed and low-density settlement patterns imply longer distances between living areas and the various service facilities. This also increases the distances that people have travel each day, and, in many cases, these mobility patterns are only feasible using a private car (Mobility and Residential Costs Project, 2014).

In the United States, for example, the rise of the automobile during the 1950s, 1960s, and 1970s allowed people to move out of cities into bigger houses in the suburbs. In addition, zoning codes separated residential, commercial, and industrial areas, meaning that people needed to drive to work and to shop. This model makes it difficult to offer public-transit cost-effectively, and large parts of the United States are public-transit deserts. No wonder the United States has the highest ownership rate in the world, at 1.93 vehicles per household (Bouton, Knupfer, Mihov, & Swartz, 2015).

In contrast, many European cities developed their urban cores before the invention of the automobile. Hence, they are denser, more walkable, and more likely to have mixed uses. As a result, there are lower rates of vehicle ownership and fewer miles traveled by car (Bouton, Knupfer, Mihov, & Swartz, 2015).

Disperse settlement structures increase the length for daily trips (i.e., shopping, work, and services). On the other hand, the length of streets and infrastructure lines (i.e., gas, electricity, and sewage) increase substantially. Costs for infrastructure decrease with higher housing densities. While in densely populated towns or cities, only 20 m of road per housing unit are needed, this figure rises significantly in disperse settlements which are made up of detached or single-family houses, where up to 150 m of road length per housing unit can be necessary (Mobility and Residential Costs Project, 2014).

The role of infrastructure as a factor of economic development has led to a permanent demand to expand networks and to increase their complexity. The paradox lies in the fact that the indiscriminate expansion of infrastructure networks turns into more dispersion of the city, which generates, among other things, a model of unsustainable mobility, high energy consumption, and increasing internalization of costs for broader layers of the population, according to Vallejo (as cited in Cruz & Fonseca, 2018, p. 563).

It becomes a vicious circle, beginning with thoroughfare widening, which generates (for a period) faster travel to more distant locations. Access to remote lands causes owners to request expansion of the urban perimeter. Under political and economic pressure, new areas are incorporated into municipal urban zoning. New subdivisions and businesses develop, and the population moves to cheaper and more distant residences. Higher flows to more distant areas increase road congestion. Once again, it is necessary to create or expand roads (Cruz & Fonseca, 2018).

In order to understand the transformations and implications of the new urban configurations, with their new spatial and sociodemographic patterns, it is essential to also consider the vulnerability of places and commutes. The relationship between mobility and vulnerability is evident, whether in spatial or social terms (Marandola Jr., 2012).

The United Nations Office for Disaster Risk Reduction (UNDRR, 2019) defines vulnerability as "the conditions determined by physical, social, economic, and environmental factors or processes which increase the susceptibility of an individual, a community, assets or systems to the impacts of hazards" (n.p.).

Among the most relevant migratory phenomena in the metropolitan or urban agglomeration context is the commuting between the cities of the region, also known as pendulum migration (Marandola Jr., 2012). The search for a service, mainly study and work, in another municipality creates a new urban configuration, where the municipality of origin becomes a dormitory town (or commuter town) and the municipality of destination becomes a centrality, attracting a large number of trips. This commuting pattern results in a particular picture of vulnerability, with variation in space and time. The consequences of a disaster will vary in size, if it occurs on a typical work day or on a weekend or holiday, for example.

If it was a normal working day, roads would have a higher traffic, streets would be much more crowded, and people would be at schools, work or other public buildings, instead of their own homes. Legeard (as cited in Veyret, 2007, p. 43) argued that, once this dynamic nature of mobility processes is well known, it is possible to produce vulnerability maps by time slots, as those in a thesis, about the city of Nimes in France, which established the following slots:

- **Daily Vulnerability:** Commercial period, except for peak hours.
- **Peak Vulnerability:** Period of hard movement in transportation net (roads, collective transportation, stations).
- **Night Vulnerability:** When population is concentrated in residential areas.

Legeard (as cited in Veyret, 2007, p. 43) uses a scale with three vulnerability levels matching the three time slots; the scale is calculated for each sector of the land occupation plan, according to the kind of activity.

In Brazil, Carmo and Anazawa (2014) analyze time slots concerning landslides in the municipality Nova Friburgo: The time when landslide happened, around 4:00 a.m., suggests that most people were sleeping at home. Mortality patterns were related to age and gender, considering people with less ability to escape.

Londe et al. (2015) analyzed the dynamic vulnerability and time slots for floods in São José dos Campos, and pointed three main peaks referring to three main moments: Early morning (home to workplaces paths), lunch time, and evening (return home). Especially for Traffic Zone (TZ) 2, which is central and one of the most crowded, floods would severely impact mobility during those three peak periods of the day. TZ 2, TZ 3, and TZ 5 remain vulnerable from 21:00 h to 23:00 h, while TZ 40 an TZ 54 are not vulnerable from 19:00 h to 05:00 h a.m. (see Figure 2 for reference of each TZ location).

Mobility is one of the most important phenomena of contemporary society. It requires readings that transcend the idea of mere measurement of flows or commutes (Marandola Jr., 2012) and must be studied in an integrated way to territorial planning.

STUDY AREA

São José dos Campos is located in the interior of the state of São Paulo, Brazil (Figure 1). The city is located 93 km from São Paulo, the state capital, and its initial point has the coordinates 23°10'47'' S and 45° 53' 14'' W. The municipality has a population of 673,255 inhabitants and occupies an area of 1,100 km^2 (*Instituto Brasileiro de Geografia e Estatística*, 2018), of which nearly 356 km^2 are urban.

The gross domestic product (GDP) is in the order of R$ 37.3 billion (Brazilian currency), corresponding to 1.83% of the GDP of the State of São Paulo, being the eighth largest state GDP. GDP per capita is R$ 54,875.51, while the state average is R$ 47,003.04 per inhabitant (*Fundação Sistema Estadual de Análise de Dados*, 2016).

Figure 1. Location of the municipality of São José dos Campos

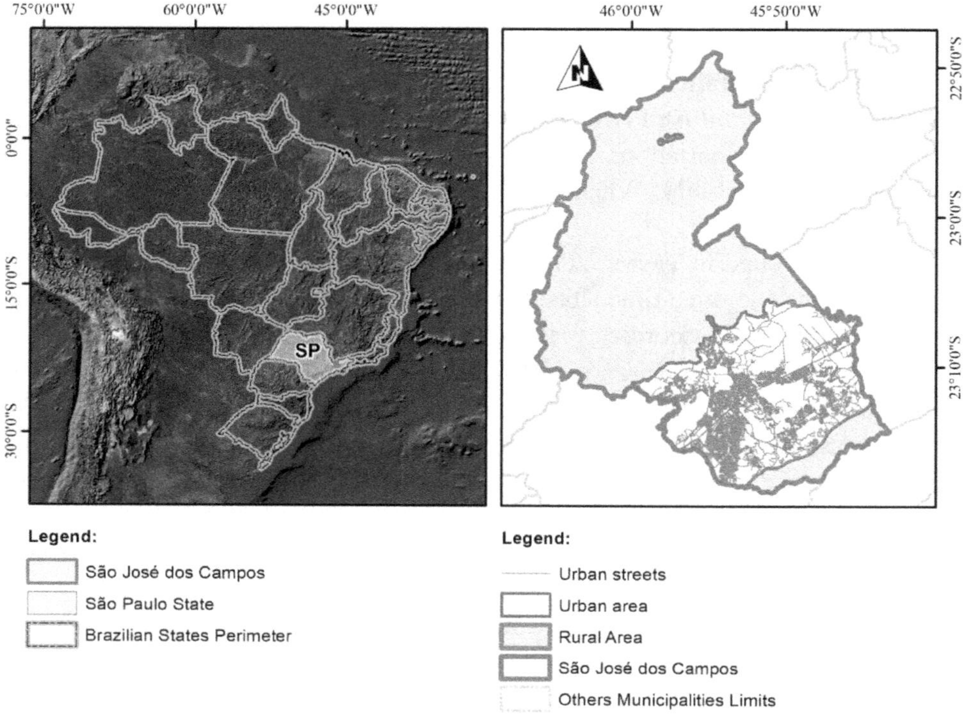

The Paraíba do Sul River is the main water course of the city, permeating with some of its tributaries the entire urban area. The altitude varies from 550 m, in the alluvial plain of the Paraíba do Sul River, to 2,082 m, in the Pico do Selado mountain, at the north of the municipality. At the main urban area the altitude varies from 550 m to 690 m above sea level. It is one of the municipalities in the region of the Paraíba Valley which is more affected by floods.

The main urban area of São José dos Campos is located in the southeast portion of the municipality and is cut by the Paraíba do Sul River, which constitutes an important barrier of spatial segregation. In addition, due to a series of constraints of the city occupation, which the authors will illustrate in the following section, the area is characterized by a spread population distribution, forming discontinuous densification nuclei.

According to Vendrame, Senaubar, & Scofield (2000), most of the urban population of the municipality is concentrated in the southern part of its territory, which presents soft relief, which is formed by an immense plateau. In this region, urban densification is concentrated along the banks of the rivers.

Urbanization and Conditions of Occupation

The main aspects that determine the urban occupation of São José dos Campos are related to the existing barriers, which derive from physical characteristics or from the use and occupation of the land along the historical process of urbanization of the city. The main elements that condition the occupied urban areas and interfere with the mobility of people are: Paraíba do Sul river, with its tributaries and valley areas (e.g., Banhado); Dutra (BR-116) and Carvalho Pinto highways; railway line; electric power transmission line; environmental protection areas - APAs; industrial areas (mainly REVAP and General Motors - GM); aeronautics center (CTA) (Figure 2).

Observing maps that represent the conditioning factors of the urban occupation, from the 1920s, it is possible to note that São José dos Campos was built on a grid form, but following topology, courses of rivers, rail line, and roads. At the beginning of the city development, there were 3 kernels: Eugenio de Mello and Santana districts and Downtown, which occupied the central plateau Lavapés and Vidoca rivers and Banhado shore had formed. The estimated population at the begging of 1920s was 30,681 inhabitants, and there were two small factories (a ceramic factory and a textile industry) and two large sanatoriums to treat tuberculosis (Santos, 2010).

From 1940, the central nucleus of the city grew, with installation of new industries integrated to the urban area. In 1946, the urban area was cut by a high-voltage electricity transmission line, with an extension of approximately 22 km. In 1947, there was the donation of a large territorial area to the CTA. At that time, this area was at the edges of the urban nucleus, not interfering in the urban mesh.

In 1951, the construction of an important highway (BR-116/Dutra) to connect the cities of São Paulo and Rio de Janeiro changed the morphology of São José dos Campos. The route of the BR-116, a line nearly straight and flat across the municipality, ignored the particularities of the city relief, passing through farms and plateaus that were grouped into industrial and residential parcels (Santos, 2010).

An intense process of industrialization took place, with the arrival of large factories, such as Johnson & Johnson (1954), Ericsson (1954), Kanebo (1956), GM (1957), Eaton (1957), Alpargatas (1959), Bendix/Bundy (1961), Avibras (1961), FSRA (1962), Matarazzo - IRFM (1962), KODAK (1969), Panasonic (1970), Embraer (1970), Hitachi (1972), Engesa (1973), Ericsson II (1973), Philips (1974), Granja Itambi (1974), Monsanto (1974), Santista (1974), Kone (1976), Minisa (1976), and *O Valeparaibano* Journal (1977). In 1961, the Brazilian Institute for Space Research was created beside the CTA (Santos, 2010).

The choice of areas where the municipality allowed for industrialization, together with the implementation of the CTA, the electricity transmission line, and BR-116 highway were determinants in the constitution of the current urban network of São José dos Campos, creating great physical barriers, which impede people's transfers.

Figure 2. Urban occupation and location of the popular housing program

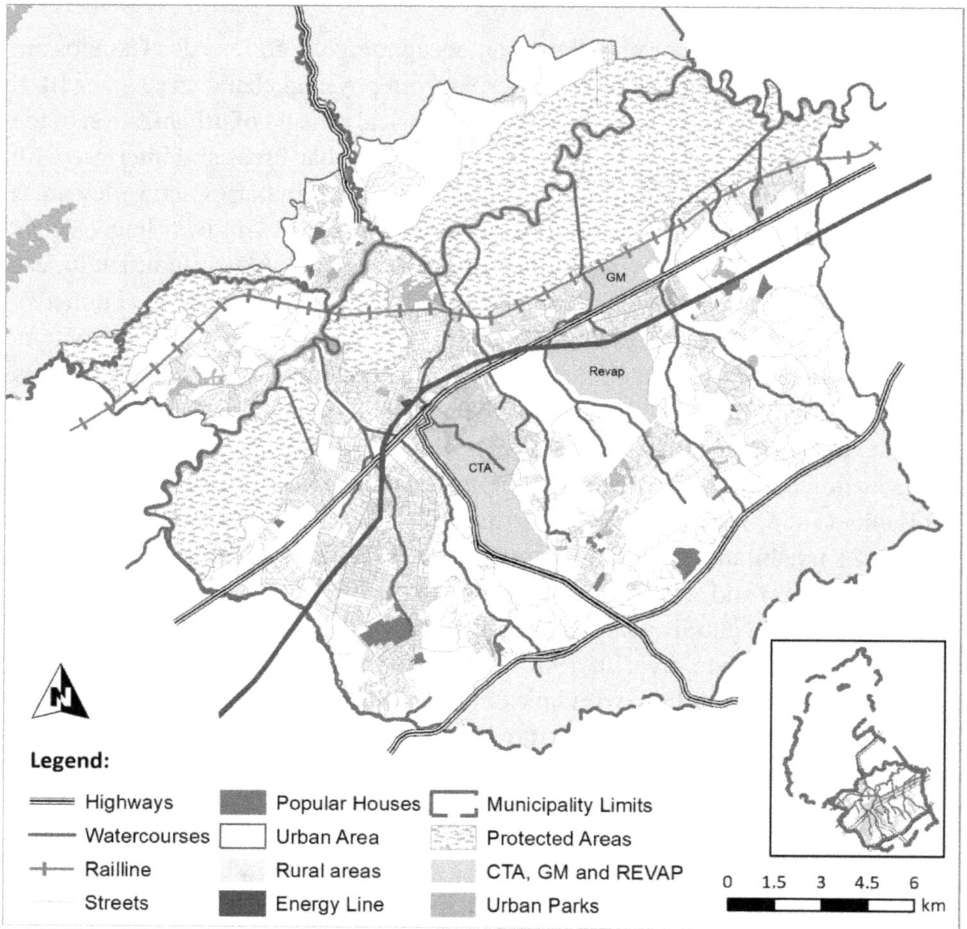

Road System

The internal road system to the municipality has a total extension of approximately 2,900 km, of which about 2,150 km are inserted in the urban perimeter. Urban roads are composed of express, arterial, collector, and local paths, as well as important regional highway links, which, constitute barriers, but are also widely used for urban traffic (Prefeitura de São José dos Campos, 2015).

Table 1. Roads extension by hierarchy

Hierarchy	Total Extension (km)	Urban Extension (km)
Local	2,231.15	1,608.77
Collector	185.48	185.48
Arterial	210.83	210.83
Express	47.55	47.55
State highway	61.94	31.13
Federal highway	47.86	47.86
Municipal highway	114.54	16.8
Total	2,899.35	2,148.42

The road hierarchy is defined by law (Municipal Law 156/2016) and aims to define the road functions, establish adequate capacity and speed in the road operation, and differentiate the flows. The expressways have the function of connecting regions, the arterial roads connect neighborhoods, the collecting roads distribute the flow within the neighborhood, and the local roads are used to access the allotments.

There is a railway section which is operated only for the transport of loads. Public passenger transport consists of buses and vans. There are 103 buses lines in operation, which cover an average of 3,089 km per month, carrying an average of 2.19 passengers per km (Prefeitura de São José dos Campos, 2019).

The fleet of motor vehicles, especially automobiles, have had a significant increase, following the trend in Brazil. In December 2018, São José dos Campos had a total of 435,309 motor vehicles, 69% of which were automobiles and 13% motorcycles. In the period from 2001 to 2018, there was a 155% increase in the total fleet, and the inhabitant occupancy rate per vehicle fell from 3.21 to 1.64.

Figure 3 presents the fleet of motor vehicles, where the y-axis shows the absolute number of vehicles and the x-axis shows the years of reference (from 2001 to 2018).

Figure 3. Evolution of the fleet of motor vehicles (Departamento Nacional de Trânsito, 2019)

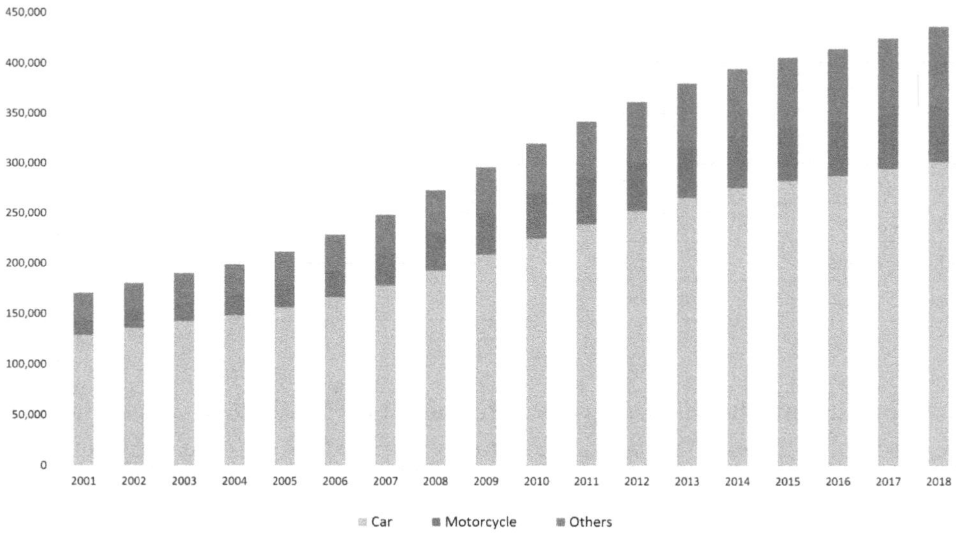

METHODOLOGY

Origin-Destination (OD) Data

In 2011, São José dos Campos conducted an OD research on a sample of 4,185 houses in the urban and rural zones of the municipality. In order to survey and analyze the collected data, the city was divided in 55 areas named TZs (Figure 6, Figure 7 and Figure 8). These zones were obtained by dividing the research areas into smaller units, according to some criteria that considerate existing TZs, census districts from *Instituto Brasileiro de Geografia e Estatística* - IBGE, the road system and other natural physical barriers, as well as homogeneous characteristics in the use and occupation of the land.

The OD research is a survey of the volume and characteristics of the population trips in their daily activities, establishing quantitative relations between the travels made and many other variables, such as socioeconomic attributes and physical elements of human occupation. The questionnaire allows to know the reason of the commute (e.g., work and studies), the mean of transport used, the time waste, and the origin and the destination of each travel (Prefeitura de São José dos Campos & Instituto de Pesquisa, Administração e Planejamento, 2014).

The data bank of the OD research allows to create a matrix with the number of commuters or the number of commutings between TZs, forming then a matrix with

55 lines and 55 columns, in which each line represents the origin and each column represents a destination. The connection between an origin to a destination forms an OD pair.

The OD was the main data used in this work. Authors used tables (in xls format) and polygons (in shapefile). This layer represents the TZs perimeter and it has an attribute table with information about zone number, zone name and area in km^2.

Other Data

The other variables used in this work fell into three categories: hydrographic data, socioeconomic data from demographic census (*Instituto Brasileiro de Geografia e Estatística*, 2010) and cadastral data provided by the PSJC.

In the first approach, authors used two hydrographic data in shapefile format: basin (polygons) and rivers (lines); and auxiliary geographic data for visual detection of nonurbanized areas, such as satellite images and road mesh. The rivers shapefile was provided by PSJC.

The analysis of mobility data aggregated by basin will contribute to an understanding of the implications of the urban configuration, with its commuting patterns related to water courses, if any flooding or landslide occurs and interrupts people's flow.

According to Carvalho et al. (2016), basin is the set of all points in space which are responsible for draining water to a specific point, which is named outlet. For example, the entire region which contributes to send any amount of water to the final water volume in outlet "A" belongs to the basin of/associated to "A."

The variables population, income and jobs by census district (IBGE, 2010) were used in *xls* format. It was also used a geographic layer containing the delimitation of the census districts (polygons) in shapefile format.

The PSJC provided two shapefiles with geometry type: point, representing the location of vertical buildings; and polygon, representing the housing programs for low-income population. The attribute table of the layer vertical buildings contains these columns: cadastral code, number of apartments, number of floors, allotment area (m^2). This layer was used to visualize the most vertical areas of the city.

Brazil has some housing programs for the low-income population. Usually, the municipality donates an area for housing construction, which is funded largely by the federal government. However, the cost of infrastructure (i.e., water, sewage, electricity, bus lines, and schools) is the responsibility of the municipality. The PSJC afforded the footprint of these houses with these information: code, typology of the building (number of floors), number of dwellings and year of implementation.

Analyses Approaches

Urban form can be viewed from various geographical scales. Some variables operate only at certain levels, such as the jobs–housing balance. Density, for instance, may carry different meanings at different levels and may differently affect human activities, such as travel behavior (Tsai, 2005). This research focuses on two scales: basin and TZ.

The analysis per basin is justified because from a hydrological point of view this unit is the most appropriate to study hydrological processes and the water balance. The TZ, in turn, is the unit adopted in the OD Survey. It is the smallest unit to study travel's patterns and urban mobility.

Areas susceptible to the occurrence of flooding or mass movement can be mapped locally, but they may be result of processes that occur upstream. Thus, the basin must be considered.

Basin based plans also make use of demographic data, including information on the people in the basin, such as the number of persons or families, commuting patterns, household structure, age, gender, race, economic conditions, employment, and educational information to help determine future trends and needs of the populations (U.S. Environmental Protection Agency, 2008).

In view of this scenario, the authors propose the calculation of affected people in case of occurrence of disasters per basin (1) and also analyze socioeconomic and structural data per TZ (2).

The two scales dialogue and can be used concurrently. The first gives the big picture of the spatiotemporal processes that occur along the basin; the other provides details regarding the characteristics of the city and the population.

In the first approach authors used an original set of basins, generated by means of altimetric information of the SRTM (Shuttle Radar Topography Mission) digital elevation model (DEM). The computational tools used are part of the ArcHydro extension at AcGIS, which are:

- Fill Sinks;
- Flow Direction;
- Flow Accumulation;
- Stream Definition;
- Stream Segmentation;
- Catchment Grid Delineation;
- Catchment Polygon Processing;
- Drainage Line Processing;
- Adjoint Catchment Processing;
- Drainage Point Processing;

- Batch Point Generation;
- Watershed Delineation.

ArcHydro is a model developed for building hydrologic information systems to synthesize geospatial and temporal water resources data that support hydrologic modeling and analysis. The model is developed as an Add-on to ArcGIS software.

The last output of these steps, wathershed delineation, was used in this work. The analysis considered seven hydrographic basins, whose main rivers are Alambari, Cambuí, Comprido, Lavapés, Pararangaba, Senhorinha, and Vidoca.

The hydrography data, including basin (polygons) and rivers (lines), and TZs (polygons) were inserted into ArcGIS, a Geographic Information System (GIS), to check intersection between them. Then, TZs that intercept the watercourses and the hydrographic basins were selected to compute the number of commuters that would be affected in their mobility.

The TZs that have an insignificant territorial area in the intercepted area were excluded from the analysis. Later, a more detailed analysis of the remaining TZs was made, using auxiliary geographical data, such as satellite images and road mesh, in order to exclude the TZs that have only urban emptiness or nonurbanized areas in the intersection with hydrological data. By this way, only the TZs which were relevant for the analysis were kept. Then, all commuters with origin or destination in one of these zones were summed up.

For each basin, a matrix of origin and destination was created, with the number of people that move along the selected TZs. Only then, the internal force (IF) of the watercourse and basin was calculated. The IF is the number of people that move along in a certain unit of analysis.

São José dos Campos is part of the national register of municipalities with areas susceptible to landslides of great impact, sudden floods or related geological or hydrological processes, because of the risks and their history of occurrences. According to the report Evaluation of Risk Areas of São José dos Campos Municipality (Prefeitura de São José dos Campos, 2016), it has 35 areas which are considered at risk of landslides and 20 areas considered at risk of floods. Figure 4 shows the location of these areas.

Risk areas, in shapefile format, were aggregated per basin in ArcGIS using 'Spatial Join' operation. It affixes data from one feature layer's to other based on spatial location. The geometry type of the source feature class and the join feature class can be polygons to polygons or points to polygons. It is possible to summarize the data in the original table before join it to a layer, creating a new table containing summary statistics derived from the table. Quantitative information per area was summed, resulting in number of risk sectors and number of houses at risk per basin.

Total population and total occupied houses per basin were obtained by aggregating census information by basin.

The population living or working in risk areas is directly affected in cases of occurrences of floods or landslides; however, people who transit through these points are also affected indirectly. The breakdown of mobility data by basin will allow to calculate the number of people that move along this unit of analysis, giving a more realistic dimension of affected people.

In the second approach, demographic density, verticalization, location of popular housing program, job density, and income stratum were analyzed. Subsequently, the authors analyzed OD data, including crossings on mode of transport, time of travel, mobility index by mode of transportation, and income. The analysis of the data of population per km^2 in level aggregated by TZ allows to observe the distribution of the population in the territory and the different levels of densification.

Variables were aggregated in ArcGIS using the abovementioned 'Spatial Join' operation. The geometry type of the source feature class and the join feature class were polygons to polygons (the case of census district to ZT and house program to ZT) and points to polygons (buildings do ZT). For calculate population, employees and number of vertical buildings and apartments per TZ it was used sum in summary statistics; to calculate income, number of floors and allotment area it was used average. Thus, the TZ shapefile has the variables population, average income, number of vertical buildings and number of apartments, average number of floors, average allotment area (m^2) and the total of popular dwellings.

Using the territorial area column of attribute table (km^2) of the shapefile, population density (Eq. 1) and job density (Eq. 2) were computed in new columns.

- Gross Population Density: Gross population density measured the entire population or number of residents within a designated geographic area divided by the size of the designated area in km^2, which was the TZ.

$$\boldsymbol{Pop.Density} = \frac{\boldsymbol{Population\ in\ TZ_i}}{\boldsymbol{Area\left(km^2\right) of\ the\ TZ_i}}, \tag{1}$$

- Gross Jobs Density: Gross employment density measured the number of employees within a designated geographic area divided by the size of the designated area, the TZ.

$$\textit{Job Density} = \frac{\textit{Number of Jobs in } TZ_i}{\textit{Area}\left(km^2\right)\textit{of the } TZ_i}, \tag{2}$$

The variable Job Density is important because the reason 'work' has great representativeness in the number of trips, and according to Barnes (2005) public transit ridership in a city is more heavily influenced by the number of jobs in the city's downtown area than by almost any other factor.

Household income is an explanatory variable of demand of trips, since, in addition to allowing the assessment of the population's living conditions, it provides important subsidies for the analysis of the travel behavior of different strata (Petersen & Vovsha, 2008). Income is correlated with land use characteristics and which is independently affect travel (Small & Verhoef, 2007).

Due to the variability observed in the distribution of income in each zone, a statistical stratification of the TZ was carried out, which consists of the decomposition of the population into six homogeneous strata, seeking to identify representative segments of the population. The cut-off range adopted for stratification considers the concentration of households with an income of up to 5 minimum wages over the total households in each zone (Eq. 3). Stratum 1 represents the highest income concentration, while stratum 6 is the lowest income concentration.

$$\textit{Income Stratum} = \frac{\textit{Number of household with up to } 5\,M.W.}{\textit{Total household in } TZ}, \tag{3}$$

The typology of buildings (high-rise or low-rise) permits detect compactness or sprawl areas, complementing density variable. It is also relevant to calculate infrastructure costs and helps to understand the relationship between the demand for transport services and the capacity of the system. As already mentioned in the item background, the length of road per housing unit for disperse settlements, which are made up of detached or single-family houses, can be 7.5 higher than the length for higher housing densities (Mobility and Residential Costs Project, 2014).

The location of housing programs for the low-income population shows how municipality is dealing with demands for land use and transport and real estate market. In this case, the location of these enterprises is more relevant than the typology to calculate the infrastructure costs.

Population density, household income and job density are traditional socioeconomic variables used in travel demand modeling in which it is necessary to represent the most important components of urban system, such as the location and interaction of activities, the real estate market and the transport system. These variables are interrelated. The phenomenon of the circulation of people and goods is explained as the results of the economic and spatial interaction between the activities the transport system and real estate market. In turn, the accessibility that results from the transport system influences the location and interaction between activities, also affecting land rent (Barra, 1998; National Academies of Sciences, Engineering, and Medicine, 2012).

The OD indexes used in the analyses took into account total commutes, commutes per mode, commutes per income and mode together, total commutes per mode and time spent, and population. These information were obtained in OD data.

The ratio between the total internal commutes and the total population forms the mobility index (Eq. 4). It is also possible to decompose this index by mode of transport (Eq. 5), income, age or gender. In that cases, the numerator changes according to the desired clipping. For example, to decompose per mode of transport, numerator is the number of commutes per mode in analysis.

Equations 4 to 7 show the calculations performed:

$$Mobility\ Index = \frac{Total\ commutes}{Total\ population}, \tag{4}$$

$$Mobility\ Index\ per\ Mode_i = \frac{Number\ of\ commutes\ per\ mode_i}{Total\ population}, \tag{5}$$

$$Mobility\ Index\ per\ income_i\ and\ mode_i = \frac{Total\ commutes\ per\ income_i\ and\ mode_i}{Total\ population}, \tag{6}$$

The modal split is calculated by the ratio of the number of trips per mode under analysis by the total of trips:

$$Mode(\%) = \frac{Number\ of\ commutes\ per\ mode_i}{Total\ commutes}, \tag{7}$$

Trips were also tabulated by mode of transport and time slots to identify oscillations throughout the day. The presented chart highlights peak times for each mode.

Note that, unlike first approach, in which the number of commuters was used, in this approach (2), the number of trips of OD research was used in the analysis.

RESULTS AND DISCUSSION

Hydrographic Scale

The Alambari river runs through a low population density area with low-medium income, passing through GM and REVAP industries, and cuts eight TZs (18, 20, 21, 40, 46, 49, 50, and 51). Its basin is approximately 45 km^2.

The Cambuí river, also known as Ribeirão dos Putins, intersects six TZs (4, 10, 22, 23, 40, and 46). Its basin is approximately 48 km^2.

The Comprido river separates the municipalities of São José dos Campos and Jacareí, passing through three TZs (31, 33, and 47) and its basin occupies an area of 57 km^2. The basin has a peripheral location in relation to the two municipalities and has suffered environmental impacts resulting from the urban expansion with an accelerated process of socioenvironmental degradation. It has a varied occupation pattern with presence of leisure farms, low-income precarious housing, popular housing complexes, and a small portion of high construction standard (Makinodan & Costa, 2004).

The Lavapés river runs through a very urbanized part of the city, passing through six TZs (1, 2, 3, 5, 11, and 54). Most of its route was channeled in the open air mode. It reliably triggers flooding problems that can seriously affect the mobility of people who travel through these areas, especially during peak flows (Londe et al., 2015). Its basin has an area of approximately 12 km^2.

The Pararangaba river has the biggest basin, with almost 95 km^2, in the most eastern part of the municipality. Its watercourse intersects six TZs (19, 41, 42, 43, 44, and 45). With the exception of TZ 19, the other TZs have low population density with low income.

The Senhorinha river is a tributary of the Vidoca river. The researchers highlighted it because it passes through the more densely populated areas of the city, where there is a project of a linear park. Its watercourse cuts five TZs (13, 14, 25, 28, and 30), and its basin is 18 km^2.

The Vidoca river cuts six TZs (7, 12, 13, 14, 15, and 24) and its basin is 58 km^2 (including the Senhorinha basin).

Figure 4 shows the main rivers, the analyzed basins, and the risk sectors. The risk sectors were scaled to improve visibility in the figure.

These 55 mapped areas resulted in 131 risk sectors, of which 109 were landslides and 22 flood sectors. Each sector has a rating in relation to risk, ranging from "no risk" to "very high risk." Table 2 summarizes data per basin the authors analyzed.

Figure 4. Main rivers, analyzed basins, and risk sectors

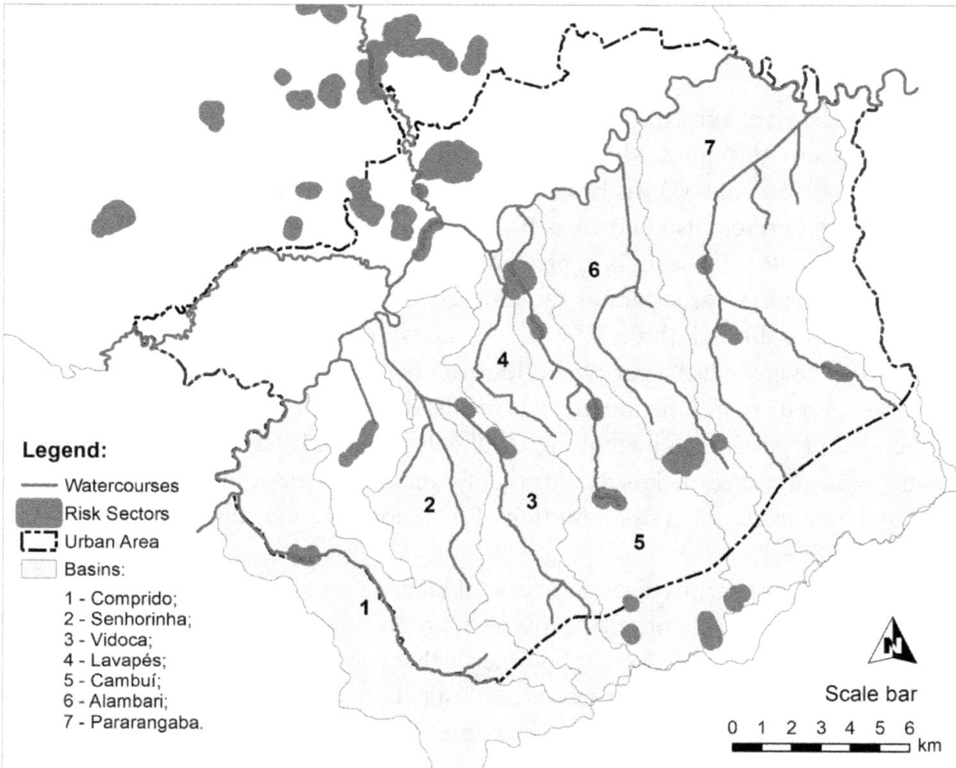

Table 2. Number of risk sectors at hydrographic basins

ID	Basin	Total Population	Total Occupied Houses	Number of Risk Sectors	Number of Houses at Risk	Risk Type
6	Alambari	42,124	12,515	6	354	Landslide
5	Cambuí	56,296	15,712	10	297	Landslide/flood
1	Comprido	41,821	11,290	2	93	Landslide/flood
4	Lavapés	34,463	11,345	2	43	Flood
7	Pararangaba	101,252	27,549	9	182	Landslide/flood
3	Vidoca	204,438	62,796	4	287	Landslide
	Total	480,394	141,207	30	1,256	Landslide/flood

The occupation of lowland areas, with inappropriate topography and soil type for this purpose, is one of the main reasons that leads to classification as an area of risk. These sectors are occupied, for the most part, by the low-income population. Considering an average of 3.4 people per household (ration between total population and total occupied houses), there are 4,273 people living in risk areas. This estimate takes into account only people who live in houses at risk, excluding commuters who enter or leave the area.

Although there is no risk area in its basin, the Senhorinha stream basin is going through fast urbanization, which turns the area impermeable increasing runoff and sediment transport; as a consequence, the microdrainage system became undersized. Frequent floods occurred and also hydraulic and urban transport system problems happened in the last two decades (Vendrame, Mota, & Reis, 2016, p. 1).

The analysis of the IFs shows the total of people who would be affected in case of flooding or landslide. Table 3 and Figure 5 show the IF of each watercourse and of each basin.

Table 3. Internal force by watercourse and basin

Name	Watercourse IF	Basin IF
Alambari	36,785	63,339
Cambuí	42,917	86,773
Comprido	20,159	65,602
Lavapés	43,779	145,894
Pararangaba	118,562	166,544
Senhorinha	202,575	361,078
Vidoca	94,170	555,978

As to the watercourse, the Pararangaba river and the Senhorinha brook are the ones with greatest IFs. Although the Pararangaba river may be located in the East geographical region, with lower population density and great urban emptiness, it cuts the Presidente Dutra road (BR-116) and goes through many TZs. The Senhorinha brook has a much shorter extension and intercepts less TZs. However, it is located in a very densely packed region and with a high travel attraction power.

Observing the IF of the basins, the Vidoca river and the Senhorinha brook are the ones that present the highest values. The Senhorinha brook is an affluent of the Vidoca river. Together, they have a great territorial extension and permeate an area with a strong appeal for travel attraction and of high population density. The IF of the Vidoca river encompasses the Senhorinhas IF. The number of people that move along this area represents 73% of the city's total population. This means that the partial or total interruption of the passage in this area would affect an expressive percentage of the residents in São José dos Campos.

Figure 5. Basin and watercourse internal force

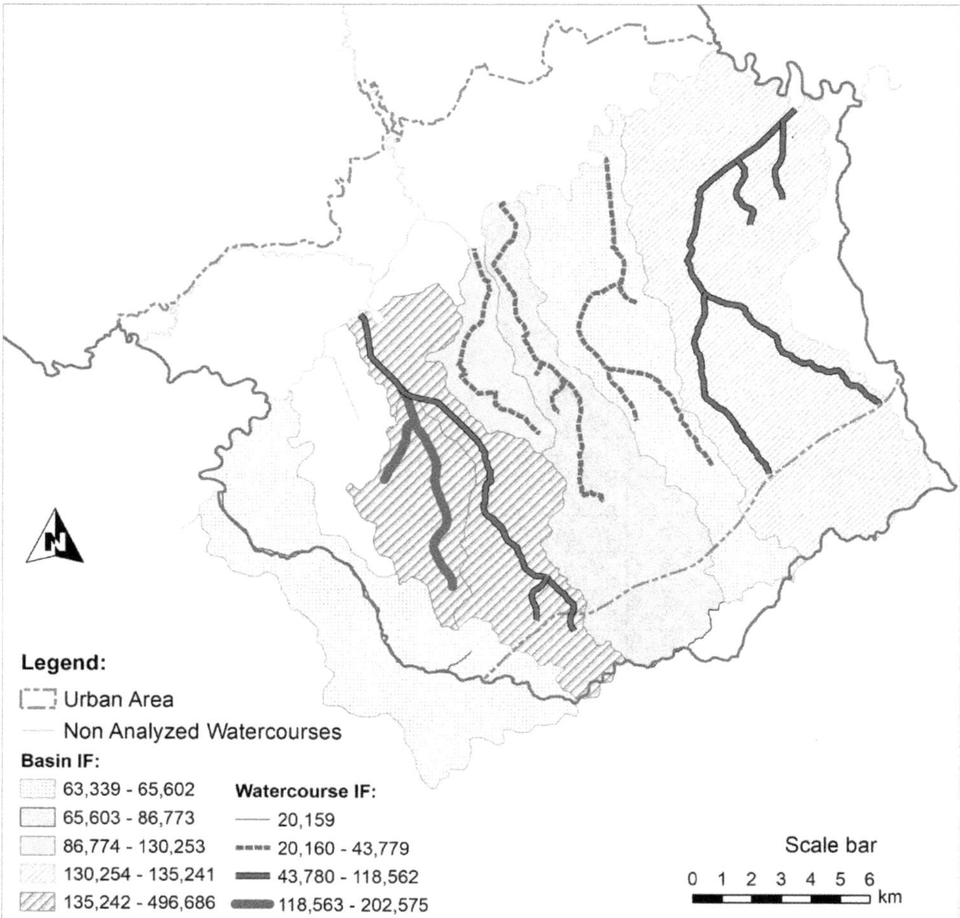

The results can provide decision makers with different information for analyzing maps, elaborating strategies, and developing specialized planning, and can mitigate the negative impacts of hydrologic extremes in transport infrastructures and urban mobility. This scale indicated the area to be considered in spatiotemporal process and estimated the number of vulnerable people in case of occurrence of disaster.

Analysis Per Traffic Zone

The largest densities occur in the southern geographic region at TZs 26, 27, 28, and 29, respectively. In the North region, TZ 16 stands out with 9,746 inh./km². Eight TZs have density between 6 and 9 thousands inhabitants per km² (TZs 3, 6, 10, 4, 9, 13, 2, and 19, respectively), four of them are in the Central region.

TZs 8, 15, 33, 35, 36, 40, 41, and 54 are almost unhabited, since they are APAs, some of them are totally protected (TZs 15 and 35), and others are partially protected.

Figure 6. Demographic density per traffic zone

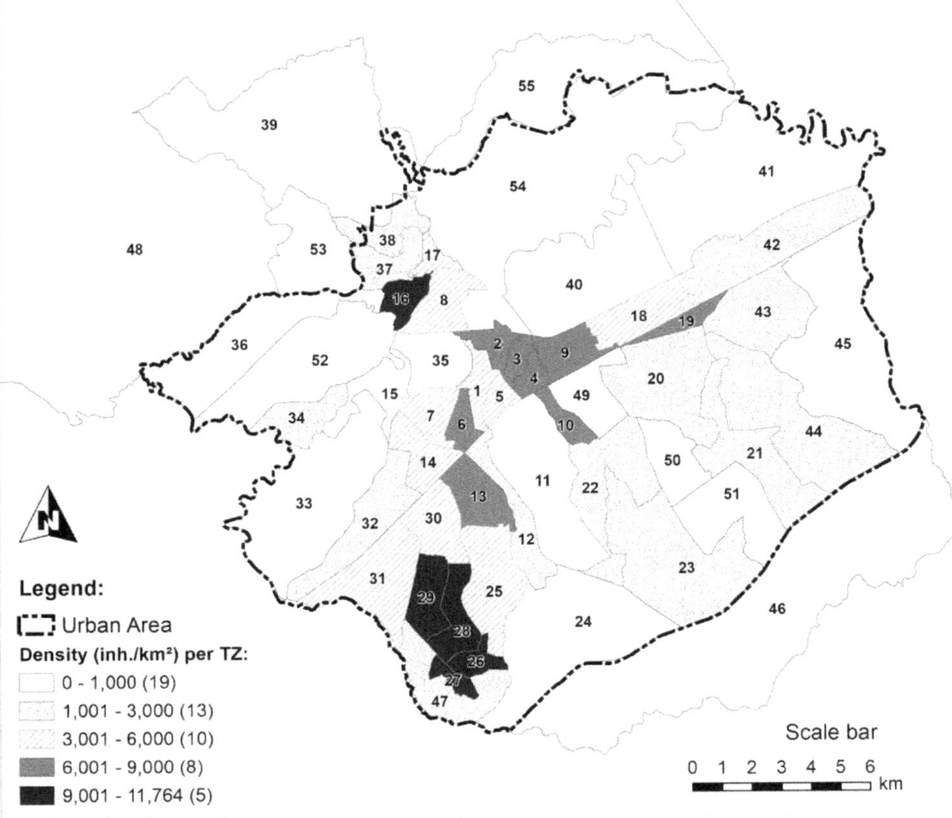

TZ 1 is downtown, which is the oldest part of the city. It has an intense occupation by commerce and services, an important concentration of institutions, and some residential activity. It attracts a considerable number of trips during business hours, but at night it is underutilized.

TZ 34 has an intense occupation by a gated community with a single road access. It has an occupation pattern of low population density, away from downtown, with a high cost of infrastructure per capita (or per dwelling).

The comparison of the denser TZs with the location of the vertical residential buildings highlights that TZs 26 and 27 have no vertical buildings, and that TZs 28 and 29 do not have a building number proportional to their densification. In other words, these TZs have a horizontal typology. TZ 16 also follows this typology, having only two low-rise buildings with a total of 72 apartments. Table 4 shows the 10 TZs with the highest number of apartments, the number of buildings, the average number of floors, and the area of the lot.

Some TZs with a density between 6 and 9 thousand inhabitants per km^2 are among the 10 most verticalized, but it is noted that, in general, the denser areas of the city have a horizontal typology, with a small lot size and most probably a larger number of inhabitants per residence.

Table 4. The top ten traffic zones in number of apartments in vertical buildings (Prefeitura de São José dos Campos, 2013)

TZ	Number of Building	Number of Apartments	Average Number of Floors	Average Allotment Area (m^2)
14	84	7,322	17	2,229
7	91	5,714	14	1,693
31	34	5,040	8	3,504
13	72	4,664	10	1,377
6	106	4,061	12	480
30	80	3,893	7	1,582
9	26	3,053	6	1,401
25	49	2,782	7	1,289
32	64	2,727	7	1,567
1	48	2,492	13	384

Figure 2 shows the location of housing programs (1997 to 2017). It is possible to observe that, as a rule, the implantation occurs in the periphery, in areas with low supply of jobs and with difficult access by public transport. The enterprise with the largest number of dwellings has 1,461 households and was implanted along the Carvalho Pinto highway (in TZ 23), which is the southern boundary of the urban area. At the time of the implantation, this area did not have the necessary infrastructure to house this high number of dwellings. The whole network of water, light, sewage, and public transport had to be extended to that fringe of the city. Also, no school places and enough health units were available to serve the new population.

Figure 7. Job density per traffic zone (Prefeitura de São José dos Campos, 2015)

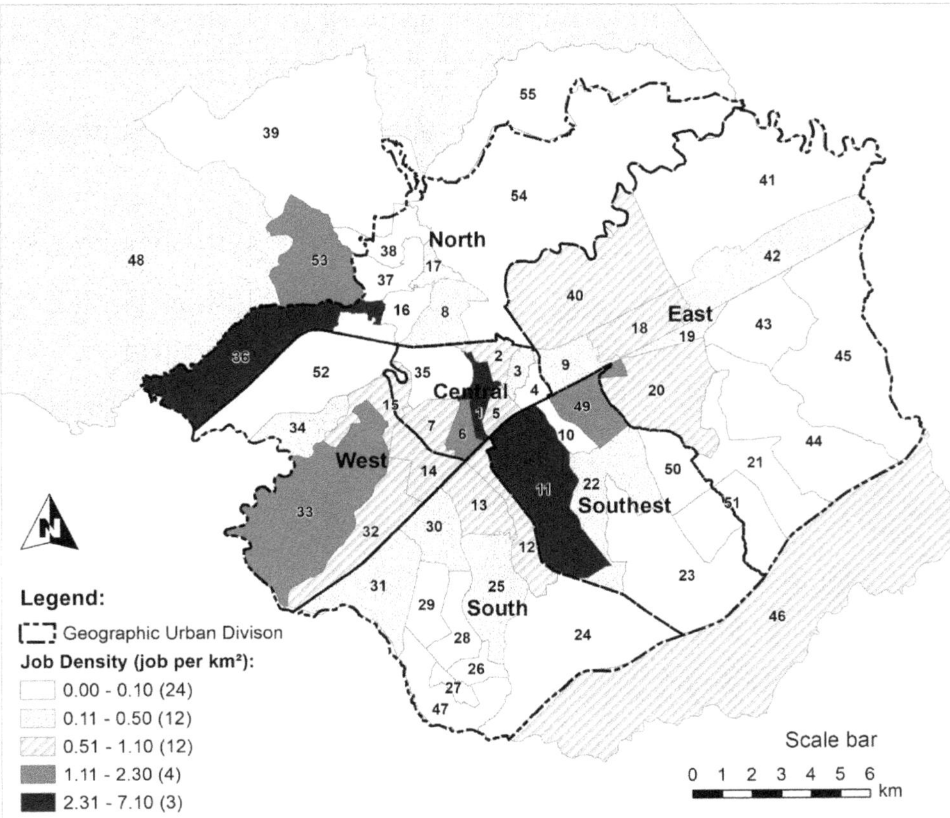

The aforementioned enterprise is a typical example of the standard the municipality adopted for the supply of housing programs for the low-income population. Typically, sites with low land costs or even areas belonging to the public authority are sought; however, when considering all the investments to provide the area with the necessary infrastructure, the cost becomes very high. Central areas have land with high venal value, but already have the entire infrastructure for implementation, which equalizes the account, and allow the population to be closer to the places of interest, which decreases travel time due to work and study.

The municipality of São José dos Campos is one of the main development centers of the State of São Paulo and concentrates an important industrial cluster, which is composed of industries mainly of the aeronautical, aerospace, and petrochemical sectors. The city is considered a center of technological reference, concentrating institutes of higher education and research, which also support the development of regional industry.

Also noteworthy are the tertiary activities which are linked to trade and services in the center area and the jobs linked to industry in the areas along the axis of the Dutra highway. Figure 7 illustrates the density of jobs per inhabitant per TZ for the year 2014 (Prefeitura de São José dos Campos, 2015).

Figure 8 shows population income information in six homogeneous strata, in order to identify representative segments of the population.

Figure 8. Income stratum (Prefeitura de São José dos Campos, 2015)

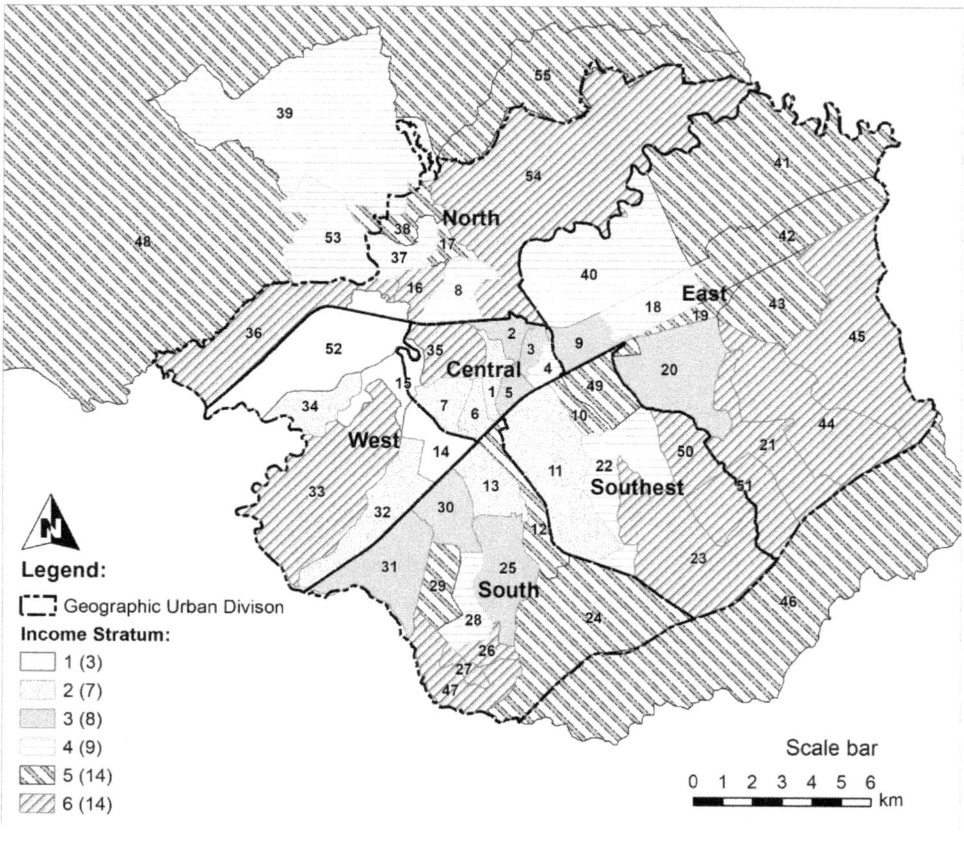

Figure 9. Mobility index by income and transportation mode
Source: Prefeitura de São José dos Campos and Instituto de Pesquisa, Administração e Planejamento (2014)

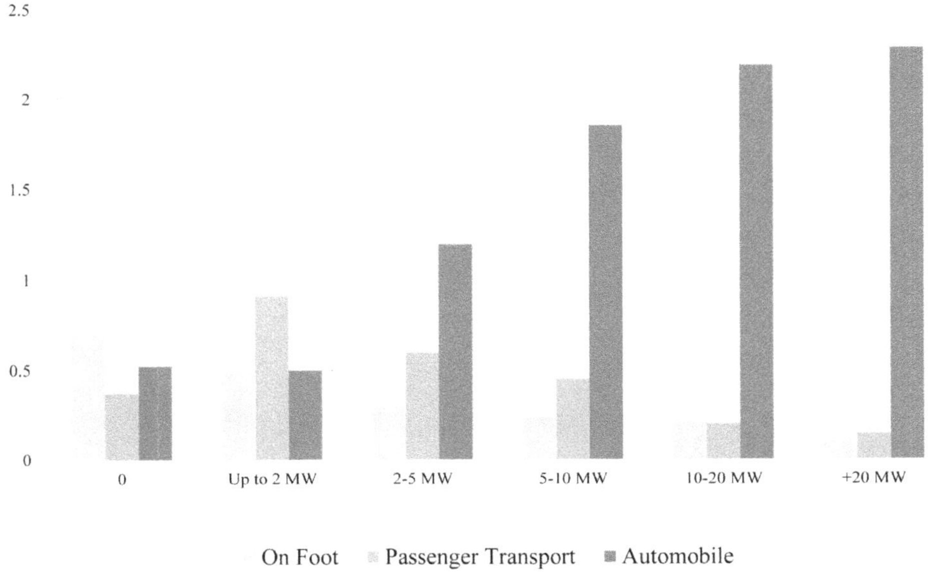

Figure 10. Transport mode and type
Source: Prefeitura de São José dos Campos and Instituto de Pesquisa, Administração e Planejamento (2014)

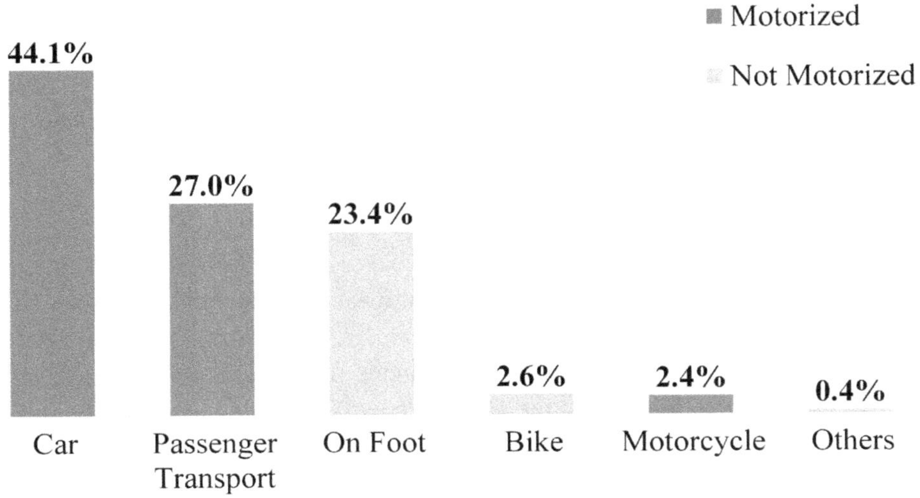

Figure 11. Number of trips by mean of transportation and time slots

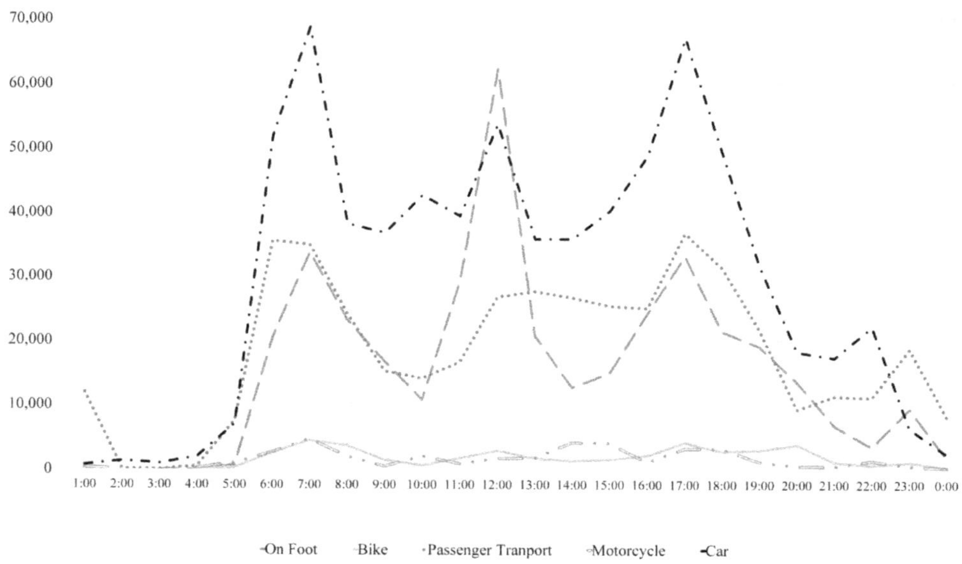

As Figure 6 illustrates, the highest population densities are in TZs 26 and 27, precisely in areas belonging to the lowest income stratum (Figure 8), and among those with lower density jobs per capita (Figure 7).

The highest employment densities per inhabitant are in TZ 11 (where the CTA and the Brazilian Institute for Space Research are located), followed by the center area. It is important to highlight that the CTA specifically shows low population density, since it has a special land-use, while the center has higher density levels, which means that it has a greater supply of jobs.

The highest average incomes are in TZs of the western region (TZs 14, 34, and 52) and the center (TZs 6 and 7), belonging to strata 1 and 2, of higher income.

OD Data Analysis

The mobility index at São José dos Campos was equal to 2.54. It means that each inhabitant does an average of 2 trips per day. The mean index for other similar Brazilian cities is 1.65, according to *Associação Nacional de Transportes Públicos* (2011).

When analyzing by mode, the mobility index for car in São José dos Campos is 1.11; this is twice bigger than the national mean (0.45) (Associação Nacional de Transportes Públicos, 2011). The index by income and mode together (Figure 9) indicates that walking is the main mode of transportation of the population with no income, transport of passengers is the main mode of transportation of the population with income up to 2 MWs, and the car is the main mode of transportation of the population with income from 2 MWs. The graph below also shows that the higher the income, the higher the mobility index.

In the case of the population that does not have income, the second option is the mode car, since, in most cases, they travel by car as passengers (i.e., children, adolescents, and the elderly) (Prefeitura de São José dos Campos & Instituto de Pesquisa, Administração e Planejamento, 2014).

The modal split in São José dos Campos also shows the prevalence of the car over other modes of transportation. Travel by car represents 44% of total travel, followed by passenger transport with 27% and walking with 23% (Figure 10). Motorized modes represent almost 74% of travel together.

Although passenger transport is a motorized mode, it can be categorized as sustainable, since its impact is divided by the number of users. Considering this question, it is also possible to categorize trips into sustainable and unsustainable ones.

According to the *Atlas of Origin and Destination Research* (Prefeitura de São José dos Campos & Instituto de Pesquisa, Administração e Planejamento, 2014), 53.3% of the trips in São José dos Campos are made by "sustainable" transport and 46.7% by "not sustainable" transport. Trips by individual and motorized transport were classified as "not sustainable." The category "sustainable" represents individual not motorized transport and transport of passengers. The mode "passenger transport" groups different types of trips in which the traveler behaves as a passenger in front of a contracted and motorized service, be it of a collective nature, such as buses, or individual (taxi).

Regarding the number of trips per hour, it is possible to notice that São José dos Campos there has three typical peak times: Between 6:00 and 7:00, at noon, and 17:00, as Figure 11 shows.

The walk mode has a characteristic that differs from the others: the highest peak occurs at noon, as the population walks small distances near the workplace to have lunch and/or make small purchases.

Both car mode and passenger transport maintain a steady flow of travel throughout the day, with three marked peaks. The morning and afternoon peaks stand out from the lunch peak. It is possible to notice that the public transport user leaves home earlier than car users. The morning peak of passenger transportation starts at 6:00 and stays high until 7:00, while the peak of the car is at 7:00.

During the night, the number of trips increases, because in São José dos Campos, as in Brazil in general, a large part of the population studies at night.

As the number of trips of bicycle and motorcycle is much smaller in relation to the other modes, the visualization of the profiles was compromised in the graph. Bicycle trips have a peak at 7:00, which decreases a little until 8:00, and a peak at 17:00. Then, there is a nocturnal peak between 19:00 and 21:00, higher than noon.

Motorcycle trips have a peak at 7:00, but then differ from other modes, as the number of trips does not increase between 14:00 and 16:00.

CONCLUSION

Mobility challenges differ from city to city and vary according to geographical, ecological, economic and political factors. Mobility policies need to go beyond the implementation of infrastructure or transport services, they should guarantee the right of more equitable and efficient access to urban spaces and facilities.

This chapter explored the analysis and discussion of important themes and variables for the organization of the urban space, the patterns of distribution of the activities, the densities (of jobs and residents), and urban design. The geospatial tools used, although simple, enable the in-depth analysis of important data for urban studies or transport planning and can be replicated in any study area.

It also contributed to the relationship between urban mobility and vulnerability to the risks of socio-environmental disasters. There are a significant number of areas at risk of flooding and mass movements in the municipalities, and, in generally, only the vulnerability of the population residing in these locations is measured. Authors presented the vulnerability of commuters, who enter and leave basin unit. This vulnerability is little considered in the indices that quantify urban risks.

Floods and landslides influence mobility patterns and therefore, the inclusion of the hydrological scale in the agenda of Urban Planning is essential to understand the processes that trigger disasters. The analysis of mobility at hydrographic scale chipped in the integration of the environmental component and urban planning.

Although the basin is the most appropriate spatial unit for hydrological studies, it is not suitable to visualize urban dynamics, such as the variables the authors analyzed per TZs. This is due to the big territorial area of the basins and the diversity of land use in it. The basins have significant urban voids and a large variation of population density. As a result, the authors recommend to use different territorial units, according to the variables which are analyzed.

Sectorial studies are important; however, it is not possible to separate urban planning from urban mobility and from land use. Territorial planning must be integrated, so that it becomes possible to build sustainable cities, balancing the social, environmental, and urbanistic dimensions of mobility.

The analysis per basin contributes to such understanding of integrated planning, considering the rivers and their impact on urban mobility. Based on the study and diagnosis of how rivers relate to mobility, decision makers are able to better define investments and actions for transport of people and goods during the following years.

ACKNOWLEDGMENT

Funding: São Paulo Research Foundation (FAPESP), Grant Number 2015/50122-0 and DFG-IRTG Grant Number 1740/2; FAPESP Grant Number 2018/06205-7; CNPq Grant Number 420338/2018-7.

REFERENCES

Associação Nacional de Transportes Públicos - ANTP. (2011). *Sistema de Informações da Mobilidade Urbana - Relatório Geral 2011*. São Paulo: ANTP. Retrieved from http://www.antp.org.br/

Associação Nacional de Transportes Públicos - ANTP. (2019). *Environmental Impacts Simulator.* Retrieved from http://www.antp.org.br/simulador-de-impactos-ambientais/sobre-os-resultados.html

Barnes, G. (2005). The importance of trip destination in determining transit share. *Journal of Public Transportation Record*, *8*(2), 1–15. doi:10.5038/2375-0901.8.2.1

Barra, T. (1989). *Integrated Land Use and Transport Modelling: Decision Chains and Hierarchies (Cambridge Urban and Architectural Studies)*. Cambridge University Press. doi:10.1017/CBO9780511552359

Bouton, S., Knuper, S. M., Mihov, I., & Swartz, S. (2015). *Urban mobility at a tipping point. Insights on Sustainability*. McKinsey & Company. Retrieved from https://www.mckinsey.com/business-functions/sustainability/our-insights/urban-mobility-at-a-tipping-point

Carmo, R. L., & Anazawa, T. M. (2014). Mortalidade por desastres no Brasil: O que mostram os dados. *Ciencia & Saude Coletiva*, *19*(9), 3669–3681. doi:10.1590/1413-81232014199.07432014 PMID:25184574

Carvalho, T., Santos, L. B. L., Luz, E. F. P., Ishibashi, R., Jorge, A. A. S., & Londe, L. R. (2016). An Open Source Approach for Watershed's Delimitation to Support Flood Monitoring. *International Journal of Distributed Systems and Technologies*, *7*(4), 77–88. doi:10.4018/IJDST.2016100105

Cavion, R., & Lombardo, M. A. (2014). Floods and Urban Design Performance. In *Delft University of Technology, Faculty of Architecture and the Built Environment (Org.), New Urban Configurations - Delta Urbanism: Living with water in the urban deltas (pp. 1030-1035)*. IOS Press.

Centro de Estudos e Pesquisas em Engenharia e Defesa Civil – CEPED. (2012). *Atlas Brasileiro de Desastres Naturais 1991 a 2010: volume Brasil. Centro Universitário de Estudos e Pesquisas sobre Desastres. Universidade Federal de Santa Catarina*. CEPED/UFSC.

Cruz, M. F., & Fonseca, F. C. P. (2018). Vetores em contradição: Planejamento da mobilidade urbana, uso do solo e dinâmicas do capitalismo contemporâneo. *Cadernos Metrópole*, *20*(42), 553–576. doi:10.1590/2236-9996.2018-4212

Departamento Nacional de Trânsito - DENATRAN. (2019). *Frota de veículos automotores licenciados por município*. Retrieved from http://www.denatran.gov.br/index.php/estatistica/237-frota-veiculos

Doi, K., & Kii, M. (2012). Looking at sustainable urban mobility through a cross-assessment model within the framework of land-use and transport integration. *International Association of Traffic and Safety Sciences -. IATSS Research*, *35*(2), 62–70. doi:10.1016/j.iatssr.2012.02.004

Doll, C., Trinks, C., Sedlacek, N., Pelikan, V., Comes, T., & Schultmann, F. (2014). Adapting rail and road networks to weather extremes: Case studies for southern Germany and Austria. *Natural Hazards*, *72*(1), 63–85. doi:10.100711069-013-0969-3

Eidsvig, U. M. K., Kristensen, K., & Vangelsten, B. V. (2017). Assessing the risk posed by natural hazards to infrastructures. *Natural Hazards and Earth System Sciences*, *17*(3), 481–504. doi:10.5194/nhess-17-481-2017

Fundação Sistema Estadual de Análise de Dados – SEADE. (2016). *Gross Domestic Product (GDP) from municipalities of São Paulo, Brazil (XLS table)*. Retrieved from https://www.seade.gov.br/produtos/pib-municipal/

IBGE – Instituto Brasileiro de Geografia e Estatística. (2010). *Demographic Census* [2010]. Retrieved from http://www.sidra.ibge.gov.br/cd/defaultcd2010.asp?0=4&i=P

IBGE – Instituto Brasileiro de Geografia e Estatística. (2018). *Population Estimate* [2018]. Retrieved from https://sidra.ibge.gov.br/tabela/6579

Kenworthy, J., & Laube, F. (1999). An International Sourcebook of Automobile Dependence in Cities, 1960–1990. *Transportation Research Part A, Policy and Practice*, *33*, 691–723. doi:10.1016/S0965-8564(99)00006-3

Kii, M., & Hanaoka, S. (2003). Comparison of efficiency between private and public transport considering urban structure. *IATSS Research*, *27*(2), 6–15. doi:10.1016/S0386-1112(14)60139-4

Kii, M., Hirota, K., & Minato, K. (2005). A study on modal shift potential considering public transport operation. *Compendium of Papers of the 9th International conference on Computers in Urban Planning and Urban Management.*

Konrad, C. P. (2003). *Effects of Urban Development on Floods.* USGS Fact Sheet FS-076-03. Retrieved from https://pubs.usgs.gov/fs/fs07603/pdf/fs07603.pdf

Londe, L. R., Santos, L. B. L., Soriano, E., Tomás, L. R., & Carvalho, T. (2015). Urban mobility data to support the assessment of dynamic vulnerability to disasters. In *Proceedings of the Second International Conference on Information and Communication Technologies for Disaster Management.* Rennes, Paris: IEEE. 10.1109/ICT-DM.2015.7402036

Makinodan, D. Y., & Costa, S. M. F. (2004). Estudo das características socioeconômicas e ambientais da microbacia do Rio Comprido. In *Proceedings of XIV Encontro Nacional de Estudos Populacionais – ABEP.* Caxambú, Brazil: ABEP.

Marandola, E. Jr. (2012). Mobilidades contemporâneas: distribuição espacial da população, vulnerabilidade e espaços de vida nas aglomerações urbanas. In J. M. P. C. Cunha (Ed.), *Mobilidade Espacial da População: Desafios teóricos e metodológicos para o seu estudo* (pp. 95–115). Nepo/Unicamp.

McLoughlin, L. (1991). Urban consolidation and urban sprawl: A question of density. *Urban Policy and Research, 9*(3), 148–156. doi:10.1080/08111149108551499

Mees, P. (2010a). *Density and sustainable transport in US, Canadian and Australian cities, another look at the data.* Compendium of Papers, the 12th WCTR conference, Lisbon, Portugal.

Mees, P. (2010b). *Transport for Suburbia: Beyond the Automobile Age.* Earthscan.

Mobility and Residential Costs Project – MORECO. (2014). *Why is MORECO of interest to you?* Retrieved from http://www.moreco-project.eu/

National Academies of Sciences, Engineering, and Medicine. (2012). *Travel Demand Forecasting: Parameters and Techniques.* Washington, DC: The National Academies Press. . doi:10.17226/14665

Newman, P., & Kenworthy, J. (1999). *Sustainability and Cities: Overcoming Automobile Dependence.* Island Press.

Petersen, E. R., & Vovsha, P. (2008). Directions for Coordinated Improvement of Travel Surveys and Models. In *Proceedings of Innovations in Travel Demand Modeling* (Vol. 2, pp. 85–88). Transportation Research Board of the National Academies.

Prefeitura de São José dos Campos - PSJC & Instituto de Pesquisa, Administração e Planejamento - IPPLAN. (2014). *Atlas da Pesquisa Origem e Destino: Panorama da Mobilidade em São José dos Campos*. São Carlos, Brazil: Editora Cubo Prefeitura de São José dos Campos - PSJC.

Prefeitura de São José dos Campos - PSJC. (2015). *Urban mobility report - Diagnosis and prognosis*. Retrieved from https://www.sjc.sp.gov.br/media/56153/diagnostico_planmob.pdf

Prefeitura de São José dos Campos - PSJC. (2016). *Evaluation of Risk Areas of São José dos Campos Municipality*. Retrieved from https://www.sjc.sp.gov.br/media/45004/pmrr_final_i_e_ii.pdf

Prefeitura de São José dos Campos - PSJC. (2019). *Public transport operational data (XLS table)*. Retrieved from https://www.sjc.sp.gov.br/servicos/governanca/portal-da-transparencia/dados-do-transporte-coletivo/dados-operacionais/

Pregnolato, M., Ford, A., Wilkinson, S. M., & Dawson, R. J. (2017). The impact of flooding on road transport: A depth disruption function. *Transportation Research Part D, Transport and Environment*, *55*, 67–81. doi:10.1016/j.trd.2017.06.020

Santos, A. P. (2010). Urbanização e Indústria: Interpretação Cartográfica da Evolução Urbana de São José dos Campos, São Paulo, 1920-2000. XI Seminário da História da Cidade e do Urbanismo - Seminário da História da Cidade e do Urbanismo, 11, 1-22.

Santos, J. D., Bacelar, R. B., & Santos, L. B. L. (2017). Desastres e mobilidade urbana - um recorte da literatura. In *Proceedings of XXII Simpósio Brasileiro de Recursos Hídricos (SBRH)*. Florianópolis, Brazil: ABRHidro.

Santos, L. B. L., Carvalho, T., Anderson, L. O., Rudorff, C. M., Marchezini, V., Londe, L. R., & Saito, S. M. (2017). A RS-GIS-based Comprehensive Impact Assessment of Floods - a Case Study in Madeira River, Western Brazilian Amazon. *IEEE Geoscience and Remote Sensing Letters*, *14*(9), 1–4. doi:10.1109/LGRS.2017.2726524

Santos, L. B. L., Londe, L. R., Soriano, E., Souza, A., & Coelho, A. F. (2015). Potential flood-related daily urban mobility problems in Rio de Janeiro (Brazil). *Revista do Departamento de Geografia*, *29*, 175–190.

Shang, D., Doulet, J.-F., & Keane, M. (2009). Urban Informatics in China: Exploring the Emergence of the Chinese City 2.0. In M. Foth (Ed.), *Handbook of Research on Urban Informatics: The Practice and Promise of the Real-Time City* (pp. 379–389). IGI Global. doi:10.4018/978-1-60566-152-0.ch026

Small, K. A., & Verhoef, E. T. (2007). *The Economics of Urban Transportation*. Routledge. doi:10.4324/9780203642306

Tsai, Y.-H. (2005). Quantifying Urban Form: Compactness versus 'Sprawl'. *Urban Studies (Edinburgh, Scotland)*, *42*(1), 141–161. doi:10.1080/0042098042000309748

Tucci, C. E. M. (2007). *Inundações Urbanas*. ABRH.

UNDRR - United Nations Office for Disaster Risk Reduction. (2015). *Poorly planned and managed urban development. PreventionWeb, the knowledge plataform for disaster risk reduction*. UNDRR.

UNDRR - United Nations Office for Disaster Risk Reduction. (2019). *Terminology on Disaster Risk Reduction.* Retrieved from https://www.unisdr.org/we/inform/terminology

U.S. Environmental Protection Agency. (2008). *Handbook for Developing Watershed Plans to Restore and Protect Our Waters. EPA 841-B-08-002*. Nonpoint Source Control Branch.

Vendrame, I. F., Mota, M. S., & Reis, D. (2016). Inundações no Córrego Senhorinha e suas Consequências na Mobilidade Urbana. In *Proceedings of II Encontro sobre Impactos Potenciais de Desastres Naturais em Infraestrutura de Transporte e Mobilidade Urbana (IPTMU)*. São José dos Campos, Brazil: UNESP.

Vendrame, I. F., Senaubar, M. A., & Scofield, G. B. (2000). Aspectos da Evolução da Urbanização e dos Problemas de Inundação em São José dos Campos. *Proceedings of XXVII Congresso Interamericano de Engenharia Sanitária e Ambiental*, *27*, 1-9.

Veyret, Y. (2007). *Os riscos: o homem como agressor e vítima do meio ambiente*. Contexto.

KEY TERMS AND DEFINITIONS

Accessibility: The quality of places that are easily reached or entered by any person, even if the person has reduced mobility.

Hydrographic Basin: It is any surface area from which runoff resulting from rainfall is collected and drained through a common point. It is synonymous with a watershed or catchment area.

Land Use: The way the geographic space is being occupied by the human being and its activities in it.

Sustainable Urban Mobility: The result of a set of transport and circulation policies aimed at providing broad and democratic access to urban space, through the prioritization of non-motorized and collective modes of transport, in an effective, socially inclusive and ecologically sustainable way, based on people and not in vehicles.

Traffic: Is the use of roads by motor vehicles, non-motorized vehicles, and pedestrians for purposes of movement, temporary stop or parking.

Transport: A system or means of conveying people or goods from place to place by means of a vehicle, aircraft, or ship.

Urbanization: Is the process of occupation of an area of the city with implantation of some urban infrastructure.

Chapter 6
Spotting Premium Hot Spots for Urban Tourism Based on Facebook and Foursquare Data Using VGI and GIS

José Gomes dos Santos
University of Coimbra, Portugal

Liliana Raquel Simões Azevedo
University of Coimbra, Portugal

Luís Carlos Roseiro Leitão
University of Coimbra, Portugal

ABSTRACT

Spatial modeling always involves choices. The existence of constraints, the uncertainty and even the reliability of the data, the purposes and the applications of the studies make these reflections a kind of guiding compass for GIS analysts. Building on a previous exercise of data acquisition (check-ins) based on two digital social networks (DSN – Facebook and Foursquare) and on the awareness of the use of volunteered geographic information (VGI) generated by tourists through DSN, this work aims to evaluate the contribution of spatial analysis applied to urban tourism in the "Alta and University of Coimbra" area. Concepts and procedural tasks related to density determination, cluster analysis, and identification of patterns have thus been implemented with the purpose of evaluating and comparing the results obtained through the application of two techniques of spatial analysis, kernel density estimation (KDE) and optimized hot spot analysis (OHSA) and inverse distance weighting (IDW) interpolation.

DOI: 10.4018/978-1-7998-2249-3.ch006

INTRODUCTION

Working with spatial models and Geographical Information Systems (GIS) involves choices. There are always constraints, either of a financial nature or related to the specifics of the software itself, to the algorithms being used, to the level of uncertainty or to the contingencies that affect the consistency and accuracy of the data. Spatial modeling encompasses all these components, and the researchers should be aware of this fact if they wish to be able to discuss the best solutions for each case. To these initial thoughts one should add the words of George E. P. Box and Norman Richard Draper when they wrote in 1987: "Remember that all models are wrong; the practical question is how wrong they have to be to not be useful. Essentially, all models are wrong, but some are useful" (Rocha, 2012, 323).

Geostatistical methodologies, modeling and spatial analysis are currently used in such varied fields as the communication strategies of numerous economic and scientific/research activities, both in the state and private sector, and applied to many different branches of knowledge, from the biomedical to the spatial sciences and, of course, the geosciences. They resort to a set of tools that allow for the construction of geospatial models and the inclusion of predictive capabilities conducive to the shaping of a system aimed at explaining both natural and social phenomena.

The present study builds on a previous broader research project that sought to compare and assess the reliability of three geostatistical algorithms, two of which in association (complement), applied to the study of urban tourism in the "Alta and University of Coimbra", the uptown area and University of Coimbra main campus, in Portugal's Centre region.

Background

Regionalization as something that is reflected in the structured nature of phenomena will thus be best articulated in the language of Random Functions. However, one must point out that these concepts got it wrong in initially assuming an overly naturalistic bent, because today's knowledge in the area of Geostatistics lends itself to a treatment similar to that of other types of anthropic variables, as is the case with the areas of Health, Criminology, Road Accident Monitoring, Tourism, Natural or Technological Hazards and Risks, among others. This work seeks to correlate variables, attributes and their respective interactions, and the authors concur with Getis's notion (Getis, 1991), that the family of spatial interaction models is a special instance of the broader models used in spatial autocorrelation. With the necessary adaptations, the present study applies Getis & Ord's paper (Getis & Ord, 1992) on "sudden death syndrome by county in North Carolina" and its potential correlation with the "price of housing units sold by zip-code district in the San Diego metropolitan region." It is also based on a kind of first corollary of the "First Law of Geography" (Tobbler, 1970), according to which the behavior of a regionalized variable is more likely to show similarities in "the neighborhood" of a given point than in a geographically distant location. What this postulate means is that, in accordance with Probability Theory (Andriotti, 1988), the values of two spatially close samples must be correlated. Therefore, it will be impossible to study numerical values as independent of each other, which is to say, considering the classical statistical methods alone.

According to Matheron (*apud* Andriotti, 1988) who is usually credited with the founding of Geostatistics and the development of related concepts, "Geostatistics is the application of the formalism of Random Functions to the reconnaissance and estimation of natural phenomena". Based on the conceptual contributions to Geostatistics made by Matheron in the mid-1960s and developed by other authors (Önsoy, & Bocquillon, C.,1980; Andriotti, 1988) this area of knowledge can be faced as a practical application of the Regionalized Variable Theory (Matheron, 1965, 1970).

A "regionalized variable" is a dipole of (seemingly contradictory) features and formed by variables whose behavior is at once random and structured. It is random in that the values of one's measurements can vary considerably within the same sample, yet the behavior pattern of the variable cannot fail to evince a regionalized dimension structured according to spatial law, assuming that (as can be easily understood) no sample values are entirely independent from their respective geographic location.

The relevance of these techniques in urban planning, based on tourism demand of sites of great historical and scenic value is attested by other recent studies, notably by García-Palomares et al. (2015) and Zhihui et al. (2016). As far as the former is concerned, the researchers would like to single out the data acquisition processes, as they bear close similarities to the methodological procedures of the present study (check-in data mining), namely with regard to the use of GIS-processed data on urban tourist destinations collected through photo-sharing services. But the work carried out by Zhihui et al. (2016) also shares procedural and methodological affinities with those used in the present study, namely with regard to the comparative analysis of patterns that identify potential places of tourist interest and their application in European metropolises based on the use of data obtained from of digital social networks.

MAIN FOCUS OF THE CHAPTER

Although the main goal of this comparative analysis applied to urban tourism in the area of the "Alta and University of Coimbra" is not to interpret data in light of the classic conceptions of descriptive statistics, the authors sought to maintain some data that characterize the sample universe and allow the anticipation of a number of properties among the variables at play, whose dominant feature is a strong independence and, as a consequence, a considerable degree of randomness. The availability of modern spatial analysis techniques, such as density determination, cluster analysis, or pattern identification for turning native vector data into intelligent raster images, led us to test the data resulting from comparing two methods of density analysis, one of which in combination with a geostatistical interpolator – Kernel Density Estimation (KDE) and Optimized Hot Spot Analysis (OHSA) & Inverse Distance Weighting (IDW) Interpolation. This is definitely the domain of "Geostatistics" and the exploratory and confirmatory techniques that lie at the "core of a spatial analysis module for a GIS" (Anselin & Getis, 1992), and on that basis the researchers should be able to identify patterns as well as to estimate, delimit and possibly predict areas with homogeneous characteristics in terms of behavior.

Geodata, Geostatistics and Spatial Analysis – Stage I: Point Density, Cluster and Outlier Analysis (Anselin Local Moran's I)

The starting point consisted of the inventory and classification of the 86 Points of Interest (POI) corresponding to the check-ins provided by tourists on Digital Social Networks (DSN) - Facebook and Foursquare - that allowed us to collect and process geodata. These 86 identified POIs were the subject of thousands of notifications (336 037 check-ins) made available by tourists on the referred DSN (Appendix I), constituting our database that we designated as "check-in data" or, for short, only "check-ins". Due to the relative dispersion of POI typologies, categories were defined. Once organized into a table (Appendix I), these data were grouped and classified into 11 derived thematic categories: Accommodation; Shopping center; Leisure and Culture; Exploratory - Space Science House; Historical green space; Railway station; Bus station; Historical monument; Campsite; Square or Park; Restoration (Figure 1).

The second stage of the present study consisted in identifying areas with a higher density of points (incidents shared by tourists in DSN - check-ins), in a geo-collaboration process that only became possible with the advent of the web 2.0 environment (the web as a platform), in which the user can also be a data producer and provider. This new human-machine interaction was at the origin of a new paradigm – The Voluntary Geographic Information (VGI). Given that some of the POI are spatially very close, the authors decided that it would be useful for the work to identify the areas with the highest density (point density). The option for a pixel value of 10m was selected and the search radius was set to 140m, rounding up the proposed default value. The sample was later reclassified into 25m classes and the result is shown in Figure 2.

A cluster of points can easily be seen to stand out from the entire neighborhood, forming a spatial unit that can be defined as the whole area comprising the various (and major) sites of the University of Coimbra and Museums, plus the historical monuments surrounding both the Old and New Cathedrals ("Sé Velha" and "Sé Nova"). It should be pointed out that the authors focused on 86 point (POIs) extent, locations checked in by DSN users (Azevedo, 2015), even if in the case at hand, and given that the methodology only includes "Falls inside" points, two of them ("Choupal" and Coimbra-B train station) ended up excluded from the cartographic representation.

Proceeding with the analysis in order to try to identify spatial patterns and association between POI, the next stage lead the authors to perform the algorithm Anselin Local Moran's I statistic for cluster-specific mapping[1], according to a workflow (made available by ESRI in ArcGIS 10.6)[2] that followed the path, **... \ Arctoolbox \ Spatial Statistics tools \ Mapping clusters \ Cluster and Outlier Analysis (Anselin Local Moran's I)**. The configuration Wizard required to create a "Checkin_Final" layer (point theme), with the attribute *Sum_FF* (number of check-ins in each POI). The options "Inverse distance for the Conceptualization of Spatial Relationships field" and "Euclidean Distance for the Distance method field", were selected. As can be readily gathered from the results shown in Figure 3, identification of objects remains very limited, and so the influence of location and distance on the decision to visit the remaining POI at the "Alta and University of Coimbra" in the Historical Centre of Coimbra is left unclear. When opening the attribute table of the layer "Checkin_final_Clusters_Outlier" one can see that new fields with "z-scores" and "p-values" have been created[3] (Figure 4). The results thus obtained show the existence of a considerable number of points whose check-in attribute value, their location and physical proximity notwithstanding, are not indicative of cluster formation.

It is probable that the reason lies in such factors as neighborhood, proximity, and iconographic standing, and it seems to be reasonable to conclude that if it is not possible to identify actual urban tourist hot spots in the "Alta and University of Coimbra" area, this is due to the fact that this whole area operates as a cluster in which no POI stands out from the neighborhood but is rather very much a part of it. This is explained by the fact that, as we have already mentioned, the neighborhood is very similar in terms of POI and check-ins, and this translates in the absence of high z-score for those POI, which means that they are not part of a similar neighborhood. Due to the high prestige and patrimonial interest of the monuments being visited and the short distance that separates them, tourists end up establishing visiting routes that allow them to identify the "Alta and University of Coimbra" POI as a strong cluster.

Figure 1. Categories of Points Of Interest (POI) in Coimbra Historical Centre. Data source: Tourist Check-ins (incident reports) on Facebook and Foursquare.

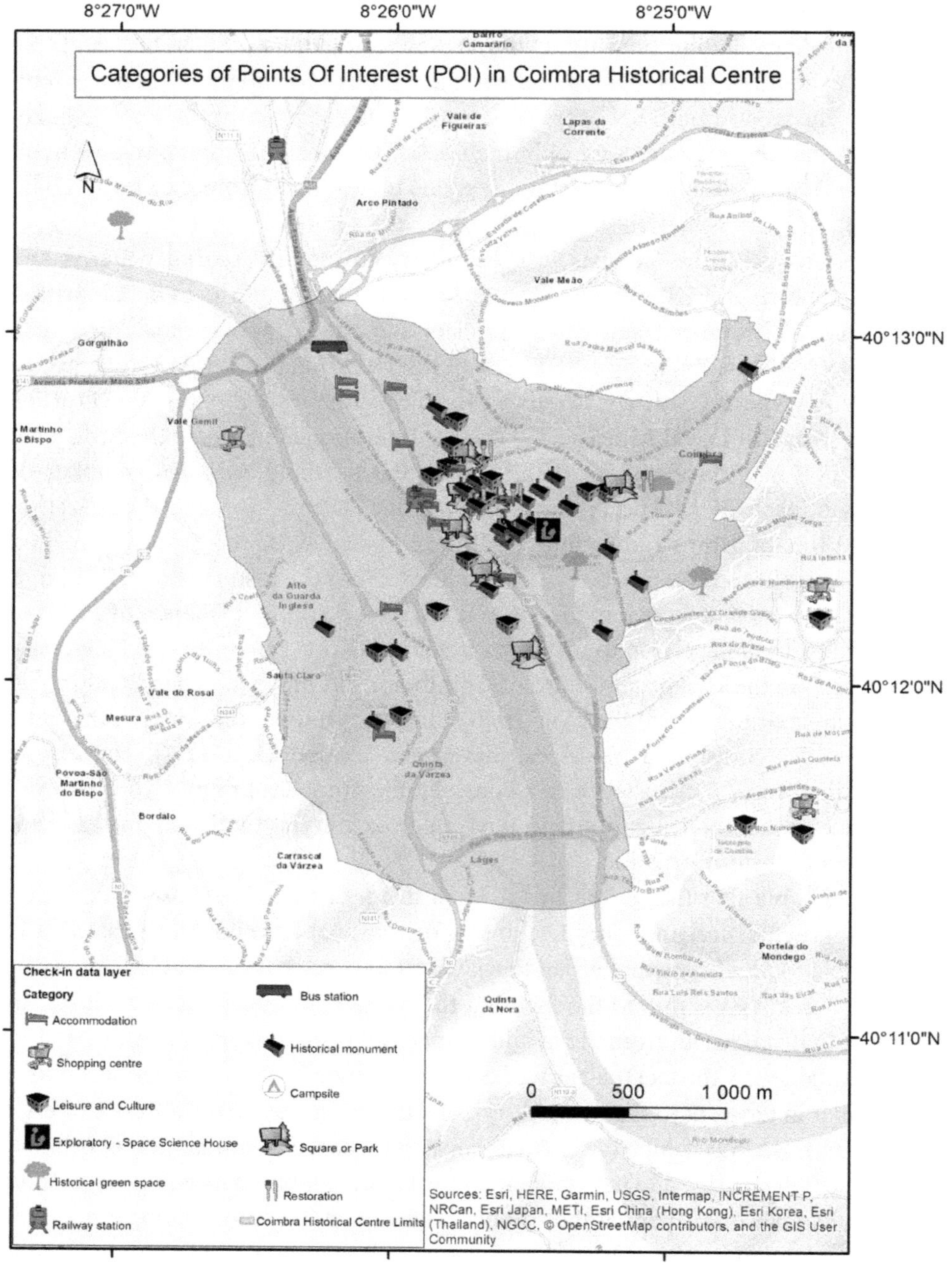

Figure 2. Point Density map of POI at the "Alta and University of Coimbra", Portugal.

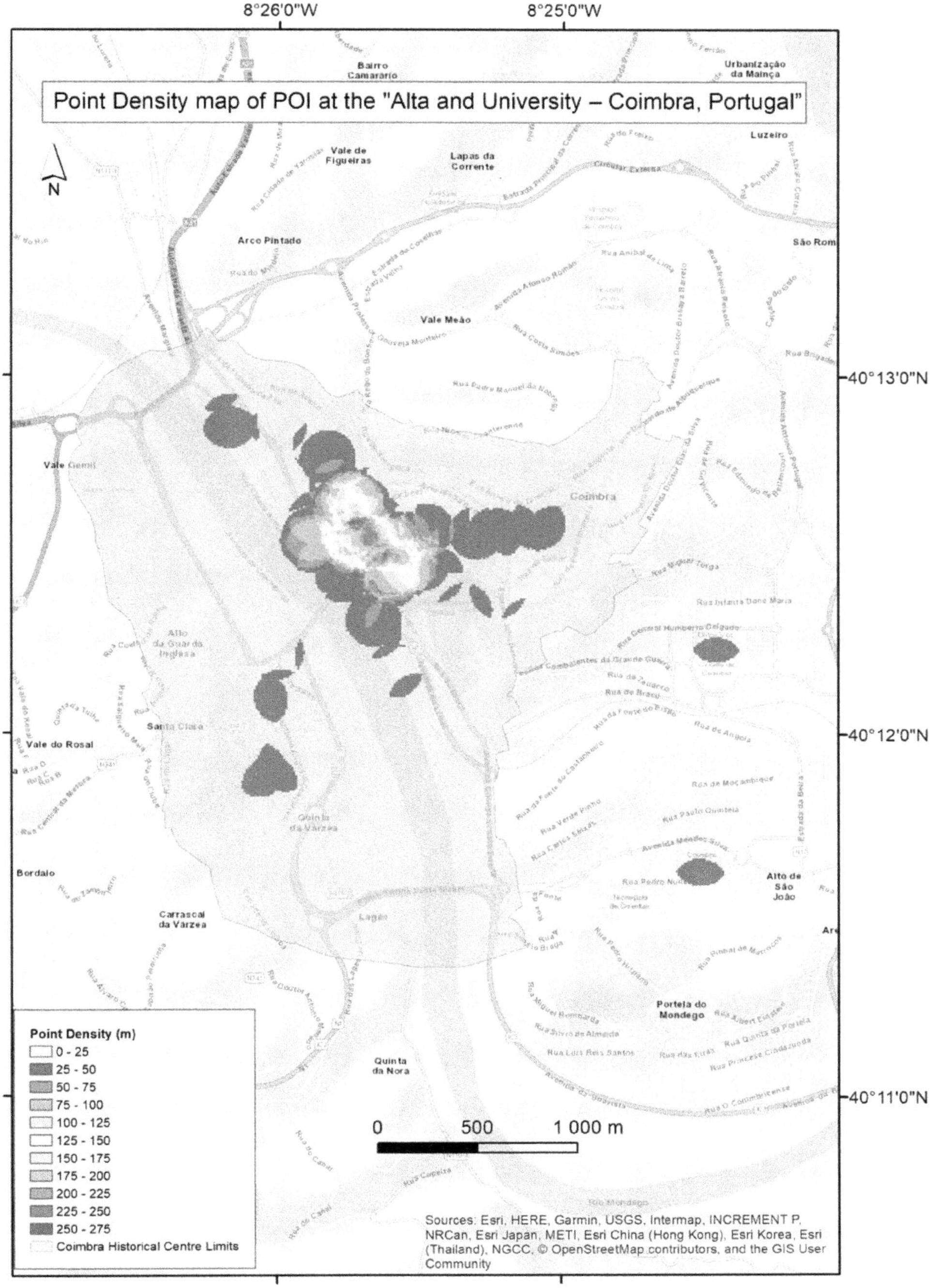

Figure 3. Cluster and outlier analysis (Anselin Local Moran's I).

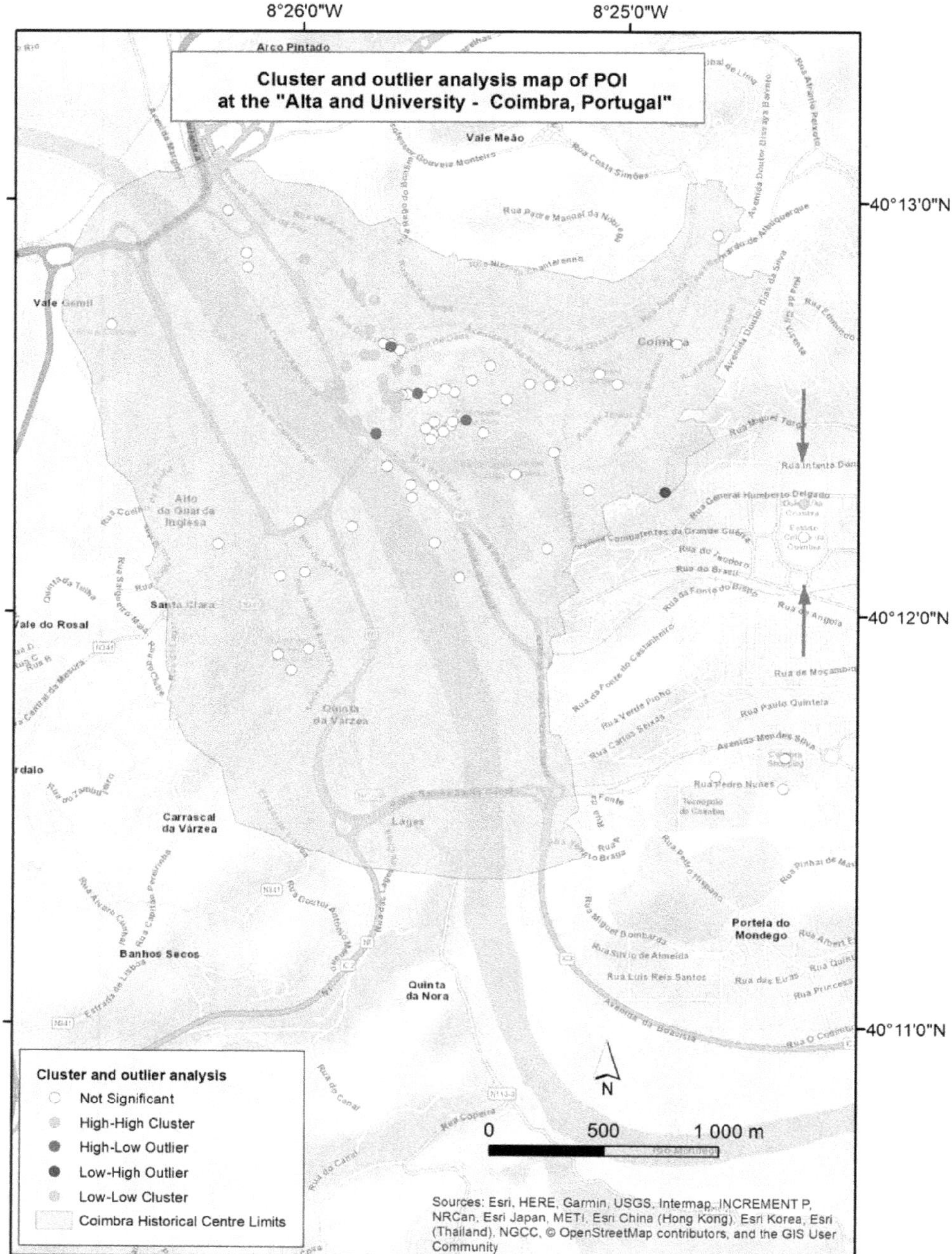

Figure 4. Results (table and map) from the application of the algorithm Cluster and outlier analysis statistic (Ansel-in Local Moran's I).

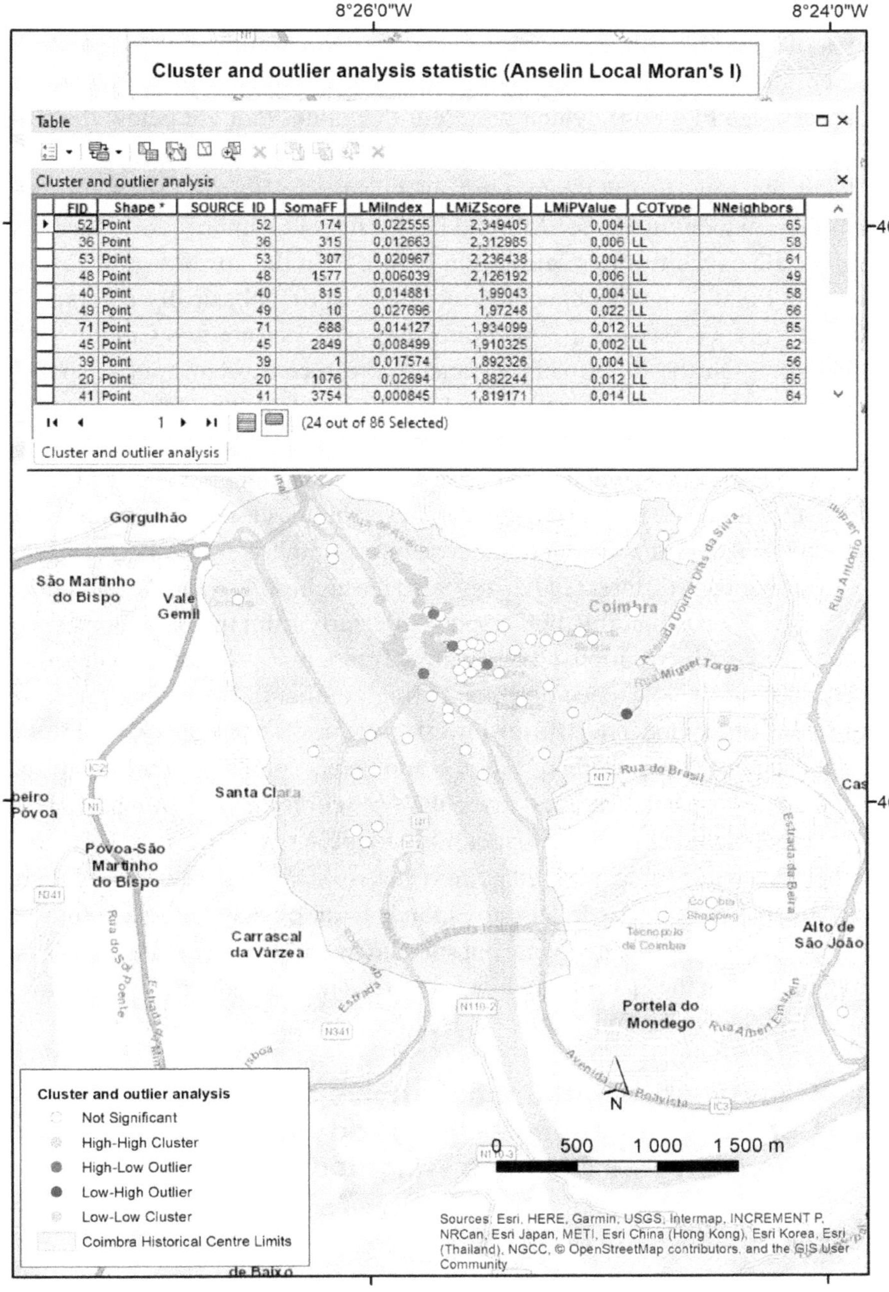

FID	Shape *	SOURCE_ID	SomaFF	LMiIndex	LMiZScore	LMiPValue	COType	NNeighbors
52	Point	52	174	0,022555	2,349405	0,004	LL	65
36	Point	36	315	0,012663	2,312985	0,006	LL	58
53	Point	53	307	0,020967	2,236438	0,004	LL	61
48	Point	48	1577	0,006039	2,126192	0,006	LL	49
40	Point	40	815	0,014881	1,99043	0,004	LL	58
49	Point	49	10	0,027696	1,97248	0,022	LL	66
71	Point	71	688	0,014127	1,934099	0,012	LL	65
45	Point	45	2849	0,008499	1,910325	0,002	LL	62
39	Point	39	1	0,017574	1,892326	0,004	LL	56
20	Point	20	1076	0,02694	1,882244	0,012	LL	65
41	Point	41	3754	0,000845	1,819171	0,014	LL	64

The whole "Alta and University of Coimbra" functions as a somewhat indistinguishable POI unit that are only a few meters away from each other. This assumption could even justify the absence of high-high cluster values, but also, the high number of undefined situations (represented by "Not Significant", in Figure 3) and the scarce number of outliers, namely, high-low and low-high outliers. Regarding the low outliers (see Figure 4), which represent the values that are below the first quartile of the statistical distribution, it seems to be clear that there is an appreciable amount of data that represent POI that stand out, negatively, from the neighborhood where the premium monuments of "Alta and University of Coimbra" are placed. By identifying the ID of the features selected in Figure 4, in the attribute table of the layer "Checkin_Final", one can conclude that these outliers basically correspond to the POI categories classified as "Accommodation", "Leisure and Culture" and "Square or Park". On the other hand, a much more recent urban tourist spot in Coimbra stands out, constituted by a complex unit in which the Alma Shopping Centre - City of Coimbra Stadium, Multipurpose Pavilion and Municipal Swimming Pools are placed. This is a POI that reveals a considerable number of check-ins with high "z-scores2. Subjected to geostatistical analysis, in this case, to the application of an algorithm to assess the association between the number of check-ins and the proximity effect between different POI, it revealed the highest "z-scores"of the entire sample set, which means that this POI stands out clearly from the neighborhood.

Given the check-in distribution (and the relative randomness of that distribution), attention is drawn to the existence of 4 points (high outliers). How significant are they? These data correspond indeed to the lowest "p-values" and negative "z-scores", showing that each of the objects in question could be part of a neighborhood if the values of the nearest objects were less irregular ("z-scores" and "p-values"). That not being the case, these values are statistically atypical. A text (HL) field (COType) was nevertheless created for these 4 red points (*please confront* Figure 3 and Figure 4). This means that each of the latter corresponds to an object characterized by a large number of occurrences (check-ins) but surrounded (in what can be called "first buffer") by points with lower values, while further away there are other points with higher values. Such distribution cannot be described as cluster based.

Geodata, Geostatistics and Spatial Analysis – Stage II; Kernel Density Estimation *vs* Optimized Hot Spot Analysis & Inverse Distance Weighted

Kernel Density Estimation (KDE)

In some open-source GIS software (such as QGIS), the KDE function is – mistakenly, in our view – referred to as "heatmaps". Without going into too much detail, as that would fall outside our present scope, it can be attributed to the fact that its closest equivalent might be confused with Hot Spot map, which is different from KDE. The authors should point out that KDE is particularly useful in terms of reading, interpreting and analyzing thematic maps, and that it is the result of converting a vector theme to a raster theme which in turn can be changed back to vector (if deemed helpful). Kernel estimation can be very useful and Medeiros (2019) highlights two advantages in the use of Kernel maps:

- Whenever there is an excessive concentration of points, visual analysis can end up being hindered. Thus, for example, a given point in a given area may in fact represent more than one occurrence (as can be the case with illnesses, crime, among others);
- Representation is not limited to predefined areas, as is the case with neighborhood or county polygons.

Building a KDE map requires a point mesh, and for each point it is necessary to determine if the distance to their location is greater or lesser than the radius of the predefined circle. One of the problems to be tackled, and for which the authors offer a solution, is to determine the radius value. It becomes necessary to estimate the Kernel function for each interior point, as well as the accumulated value of point concentration, with the result being converted to a raster data model. Density will tell us about the existence of clusters, but not whether they carry any statistical significance.

Another frequently asked question is to know the number of points required (after aggregation) to achieve reliable results. The statistical representativeness of the data depends on the work scale, among other factors, the general idea being that "The more points you have, the more reliable your results will tend to be." In the present case, using ArcGIS 10.6, the researchers followed the path, **...\ Arctoolbox \ Spatial Analyst tools \ Density \ Kernel density**. In the configuration Wizard (Figure 5) they entered the theme under analysis and the attribute *Sum_FF*. The default value of the cell size resolution (16.6530470571816) was modified and the value 10m was selected to obtain a less pixelated image, but, on the contrary, a visually smoother output surface. In the next field the researchers set the radius value for searching neighborhood. The final cartographic result will depend greatly on the value entered in this field. The authors performed an impressive number of tests, which led to the finding that there exist major differences in the final layouts.

Figure 5. Configuration Wizard of KDE in ArcGIS.

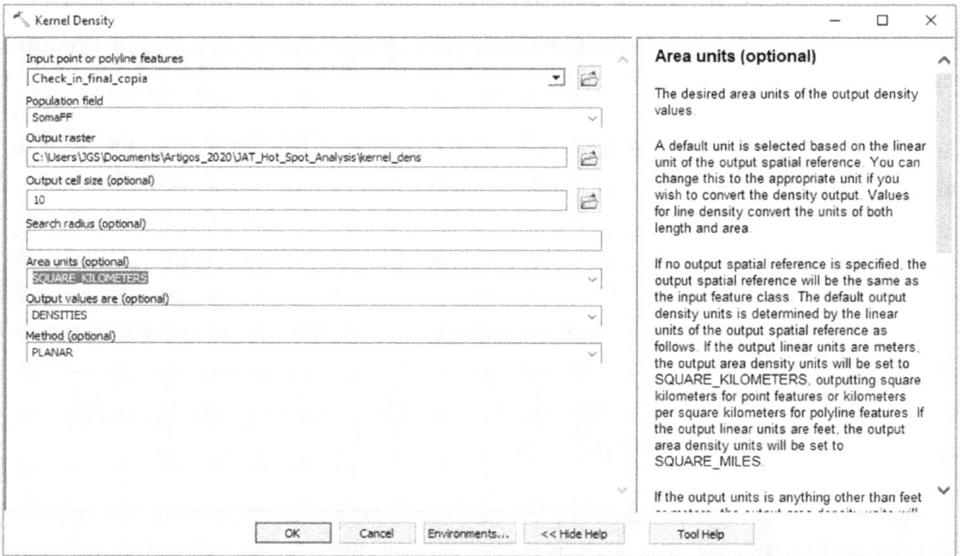

Figure 6. Calculate Distance Band from Neighbor Count tool wizard (Source: ArcGIS-ESRI).

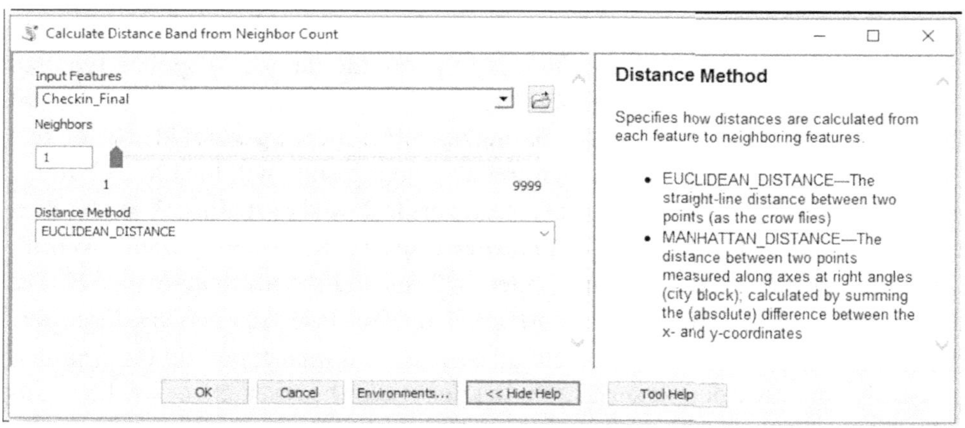

Once the Wizard was set up, a surface was generated, as shown in Figure 7. This is a continuous surface, where notifications reported by tourist activity on Facebook and Foursquare (check-ins) seem to gain relevance in terms of identifying clusters and hot spots. As it was to be expected the "Alta and University of Coimbra" area

– the University and the city's most symbolic historical and religious monuments – includes the highest density of ckeck-ins. This cartographic information and analysis confirmed the importance of the premium monuments of "Alta and University of Coimbra", which together represent approximately 33% of the total check-ins made available by tourists on digital social networks Facebook and Foursquare. It should be noted that this type of analysis and its correlative cartographic representation are not restricted to administrative or tourism purposes but allow the representation of several other types of events, such as crime analysis, urban traffic accidents, outbreaks of infectious diseases, among others.

Optimized Hot Spot Analysis (OHSA)

When using the Hot Spot Analysis algorithm (HSA) the researchers are faced with processes leading to the aggregation of data and values (the presence of clusters) that can be either high or low, and certainly are not lacking in statistical significance (measured by confidence intervals). It requires to deal with the probability that distribution is not entirely random but follows some spatial pattern instead, and to consider the proximity and neighborhood factors. These concepts become especially relevant for the implementation of this analytical technique and for spatial modeling, because they contribute with more and better information, at least in visual terms, than other methodologies. But in spite of making it possible to overcome the difficulties resulting from classic choropleth mapping (think of a sample with thousands of points, each representing thousands of occurrences, as is the case with calls to the Medical Emergency Number), the representation of such an amount and variety of information can prove a highly entropic exercise and thus turn into yet another obstacle instead of a decision support tool. On the other hand, it is possible to complement the OHSA processed vector data (points or polygons) with interpolation-based surfaces whose smoothness depends on the value assigned to the pixel, thereby generating cartographic information for analyzing patterns capable of explaining the behavior of such phenomena as the association proposed by the authors with regard to the total number of check-ins (336 037) and the 86 tourist POI classified in 11 categories, with the effects of proximity and neighborhood.

The Inverse Distance Weighted (IDW) is one of the most popular interpolators that is frequently used as a complement of HSA. But, instead of the HSA algorithm, in the present work it was performed the Optimized Hot Spot Analysis (OHSA) algorithm (Figure 8), because it allows the less experienced user or the one less knowledgeable in the field of Geostatistics to execute a workflow with preconfigured parameters, thereby reducing error and uncertainty.

Figure 7. Kernel Density Estimation map.

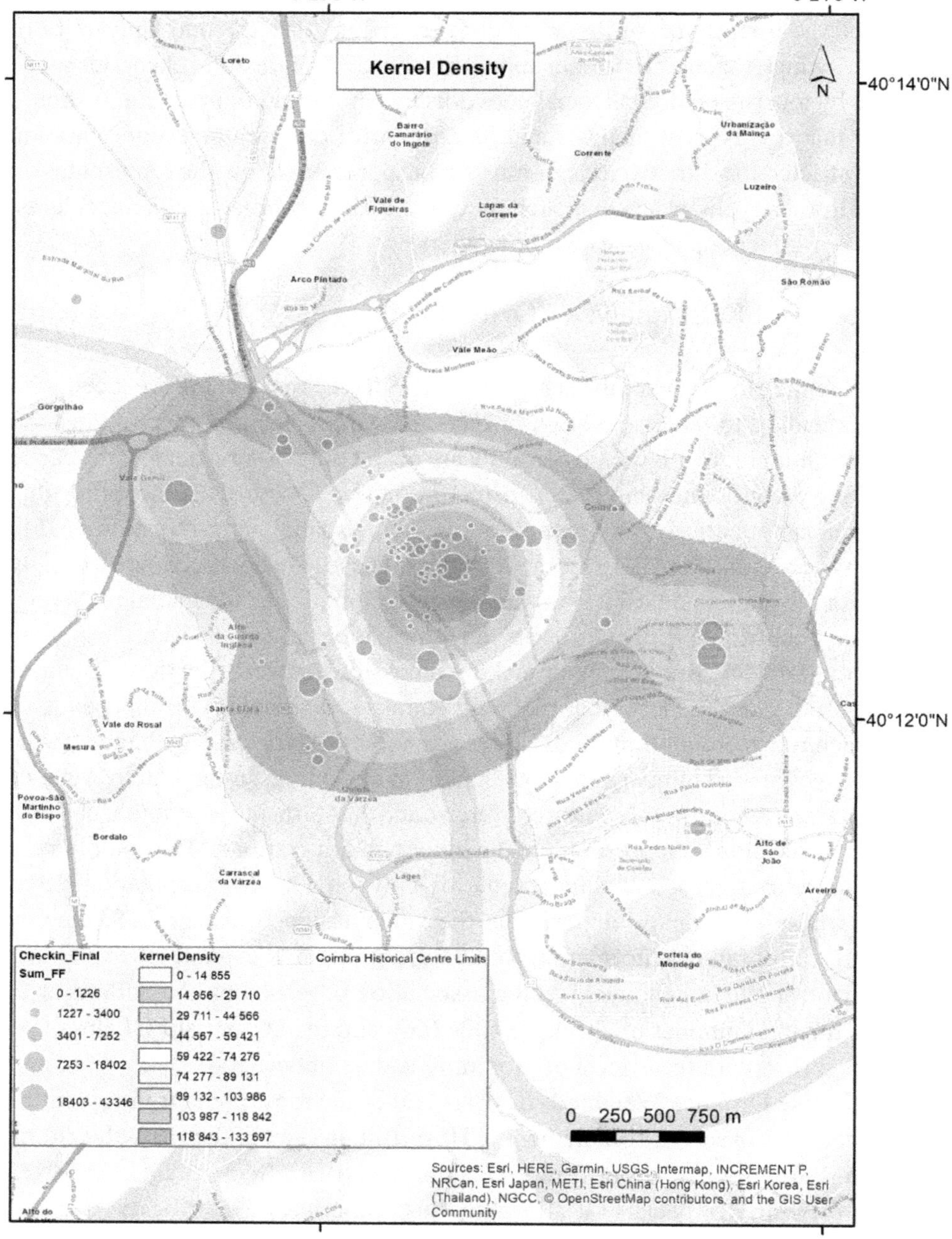

Figure 8. Incident cluster map of the "Alta and University of Coimbra" - OHSA algorithm.

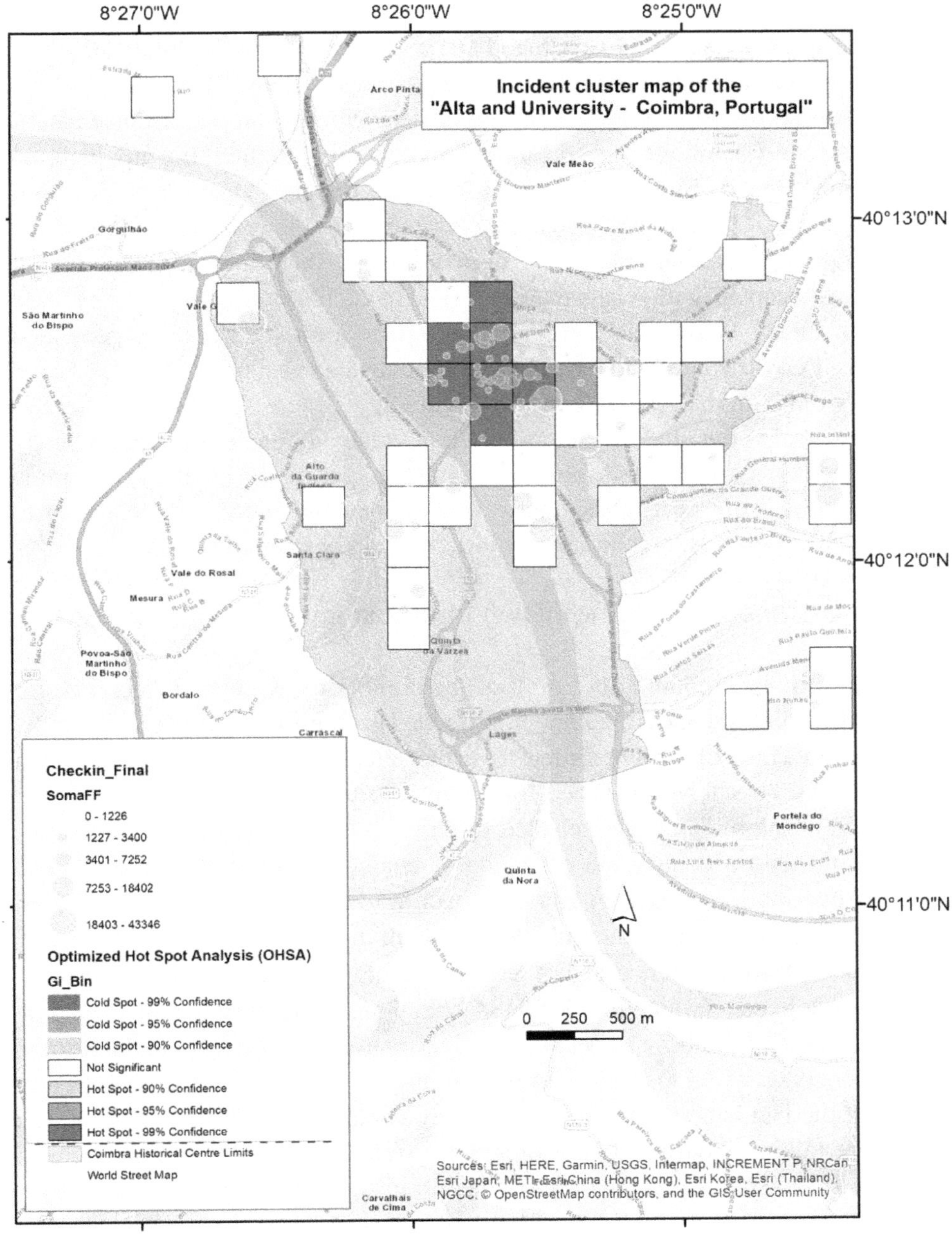

This tool is based on the Getis-Ord GI* statistical estimation procedure (where

GI* should be pronounced as "G Eye star". As it was mentioned before, the qualifier "Optimized" is an allusion to the fact that the tool itself preconfigures some of the parameters of the data being processed. Bearing in mind that the researchers have been working with one variable (POI) and with the spatial implications of one of its quantitative attributes (occurrences – check-ins) and given also that it is our aim to establish a correlation between those attributes and the distance factor, the OHSA tool was initially tested for the POI according to the following path, **...\ Arctoolbox \ Spatial Statistical tools \ Mapping clusters \ Optimized Hot Spot Analysis**. In the ArcMap OHSA wizard configuration, the researchers placed the "Checkin_Final" layer and instead of inserting an attribute the suggested default method was accepted, without attribute data, i.e., Incident Data Aggregation Method (Count_Incidents_Within_Fishnet_Polygons). As a result, the cells thus generated indicated a POI distribution that confirms the existence of a cluster in the "Alta and University of Coimbra" (*please confront* Figure 8).

The next step consisted in using a different value aggregation method "Snap_Nearby_Incidents_to_Create_Weighted_Points", to permit the creation of a single weighted point from nearby incidents. Thus, it becomes necessary to return to working with vector data (geometry: point), as it can be seen in Figure 9.

Inverse Distance Weighting (IDW) Interpolation

The weight for each point is the result of the number of incidents (check-ins) for each nearby point in the defined neighborhood. The existence of the cluster in the "Alta and University of Coimbra" area, as further attested to by the number of related check-ins, can be safely confirmed. In a return to the raster data model (Figure 10), the IDW interpolator was selected to perform the assessment of continuous surfaces and try to identify the existence of tendencies and patterns associated with this distribution. It is a common procedure to allude to the fact that a high number of occurrences is commonly equated with the presence of a Hot Spot. For clarification on the existence of hot and cold spots it is required the analysis of the Gi-Bin field data (which in this OHSA technique is executed and displayed by default, as mentioned above), because it assigns confidence intervals to the results. The authors understand the reason of this common mistake is so often and hastily made, but the fact is that the relation is not always that straightforward. As it was said before, talking about a Hot Spot forces you to talk about neighborhood relations; there is no Hot Spot unless a particular feature is aggregated with other features with similar occurrences (incidents), but that stand out from a set of neighboring features. If all the features have high values (incidents) and look similar, they will not stand out from the neighborhood and therefore will not constitute real hot spots. To give an example, (*please confront* Figure 9) there are 25 POI that once aggregated

Figure 9. The "Alta and University of Coimbra" in vector data (geometry: point) - OHSA algorithm.

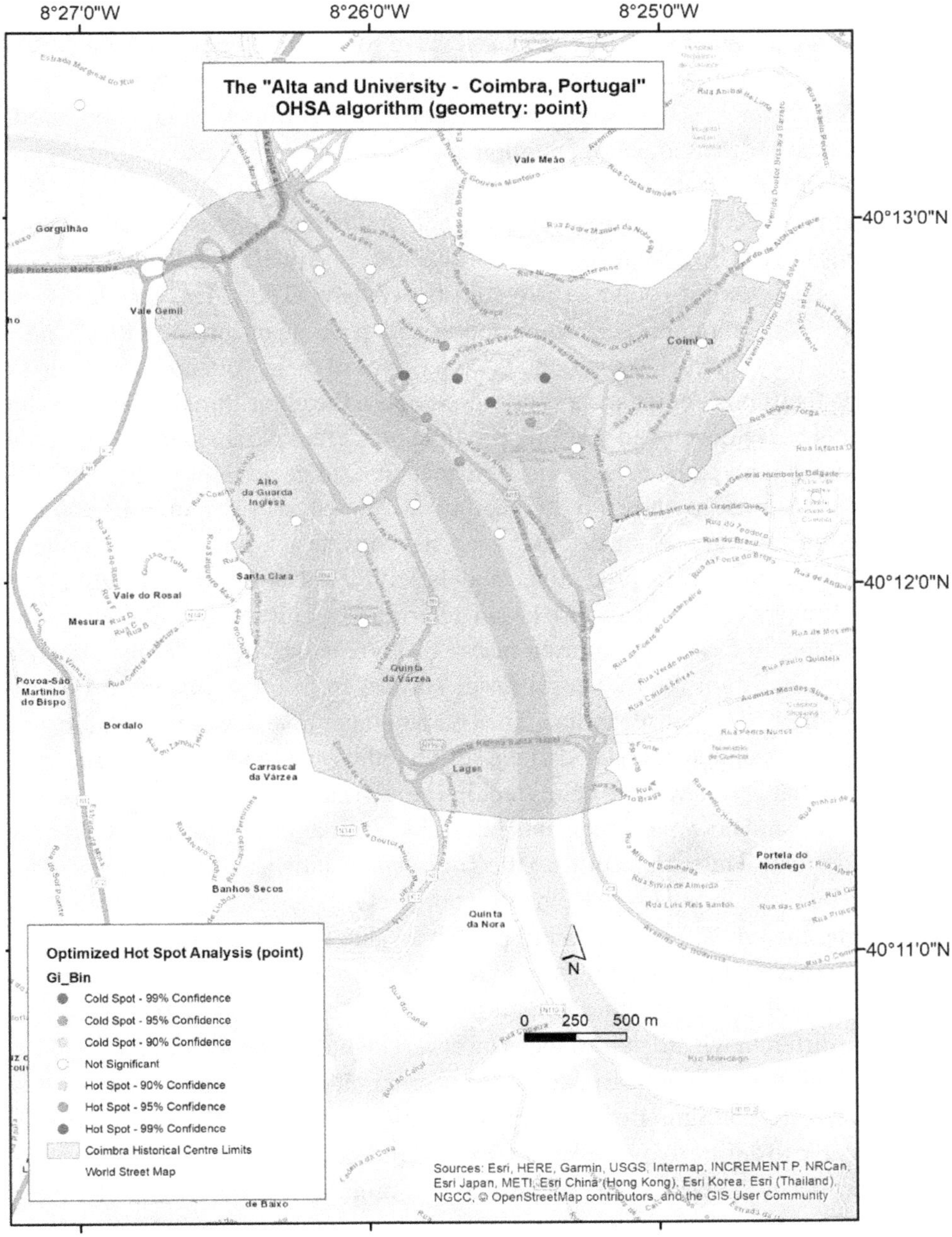

form clusters characterized by a high level of confidence (95%) but which, after the performing of the IDW interpolation, fall within areas whose level of confidence in terms of identifying the existence of hot spots is not necessarily the same; in the case at hand, what the results show is a value somewhere between 1.8 and 2.1, which is still far from the maximum required for being able to say with certainty that one is in the presence of Hot Spots – as they stand out from their neighborhood. Mutatis mutandis, the same interpretation applies to cold spots, absent from Figure 10 and Figure 11.

A brief note recommends justifying the authors' choice to present both figures mentioned above. The IDW interpolator is available in ArcGIS ArcToolbox, in two toolsets: Spatial Analyst Tools and Geo-statistical Analyst Tools. This double option for the same algorithm may pose some doubts and generate uncertainty for the GIS operators, as they should keep in mind the objectives of their work and according to the criteria they consider most appropriate to choose the algorithm that best fits the study that is being developed. Despite providing different parameters, the wizards available from both paths reveal a performance that, in general, presents identical cartographic results, as shown by the comparative analysis of figures 10 and 11. However, for the purposes of detailed analysis, considering that these techniques allow the development of predictive models with a focus on spatialization, the rigorous design of the areas must be an important requirement when choosing the algorithms. In case the path used in ArcToolbox is that of the Toolset Spatial Analyst the algorithm interpolates a raster surface from points using an Inverse Distance Weighted (IDW) technique. The operator takes note that the algorithm will favor Euclidean parameters related to the search radius and the number of incidents recorded in that action radius, being required to define the search radius value to analyze which and how many incidents fall inside an imaginary circle whose radius has been defined. The path leading to the Toolset Geostatistical Analyst leads to an algorithm that (according to the Help toolset of ArcGIS) "Uses the measured values surrounding the prediction location to predict a value for any unsampled location, based on the assumption that things that are close to one another are more alike than those that are farther apart", therefore, it follows Tobler's Law (1970). This option provides a different wizard, with a focus on criteria which also involves the distance factor, but through derived variables such as proximity and neighborhood. In this case, the wizard configuration leads us to configure the neighbor-hood settings, being possible to define two search radii (maximum and minimum) in which records (incidents) occur, as well as the maximum and minimum numbers of neighbors. Referring to some other aspects that require configuration of the IDW algorithm when accessed by one of the two above mentioned paths available in ArcGIS, it should be added that, in both cases, a layer with "z-scores" and "p-values" must be used, for that this interpolator interprets the values and returns the desired results in the form of a cartographic image in a raster data model. This is one of the main reasons that justify the use of this interpolator only after following the procedural path which led us to perform the OHSA algorithm.

Figure 10. OHSA & IDW surface map (Spatial Analyst).

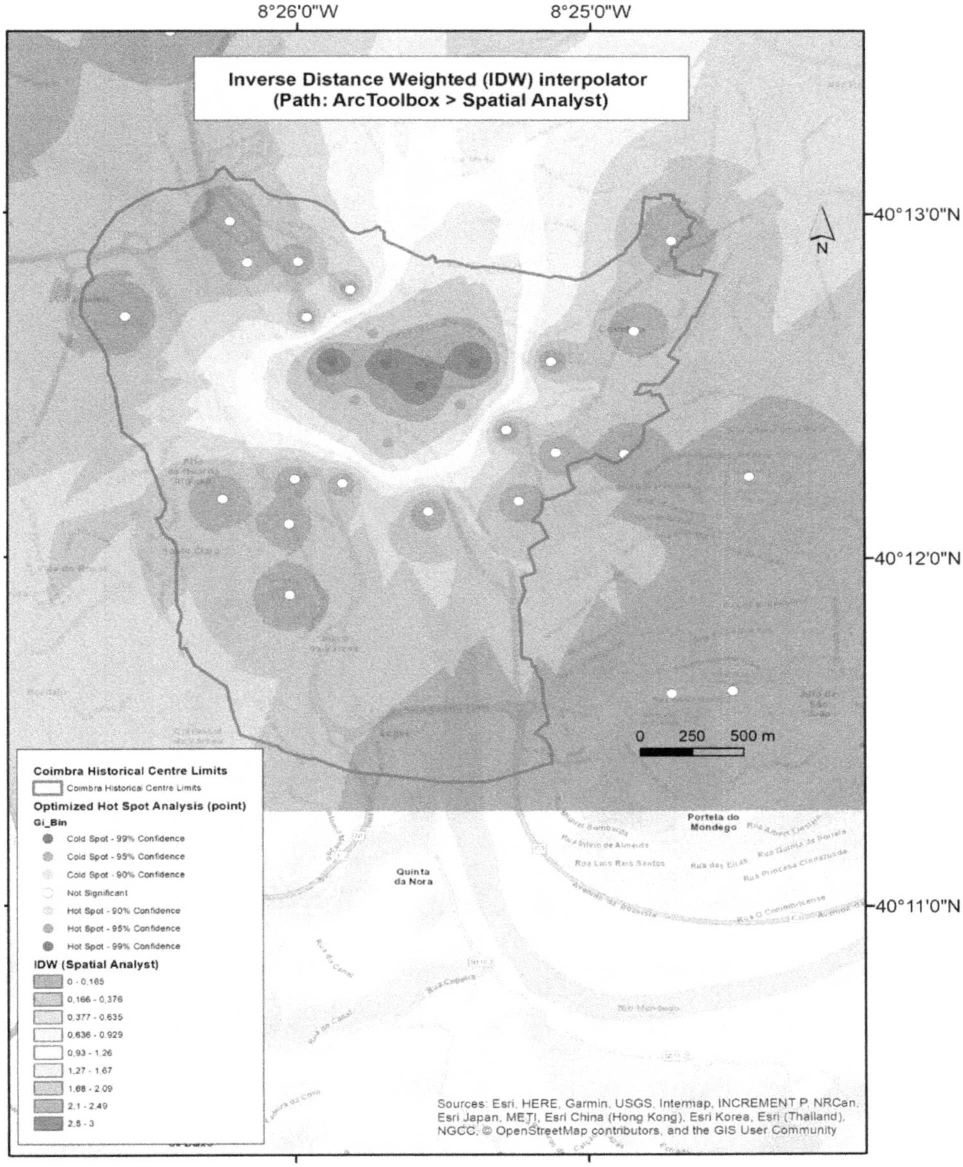

Figure 11. OHSA & IDW surface map (Geostatistical Analyst tool).

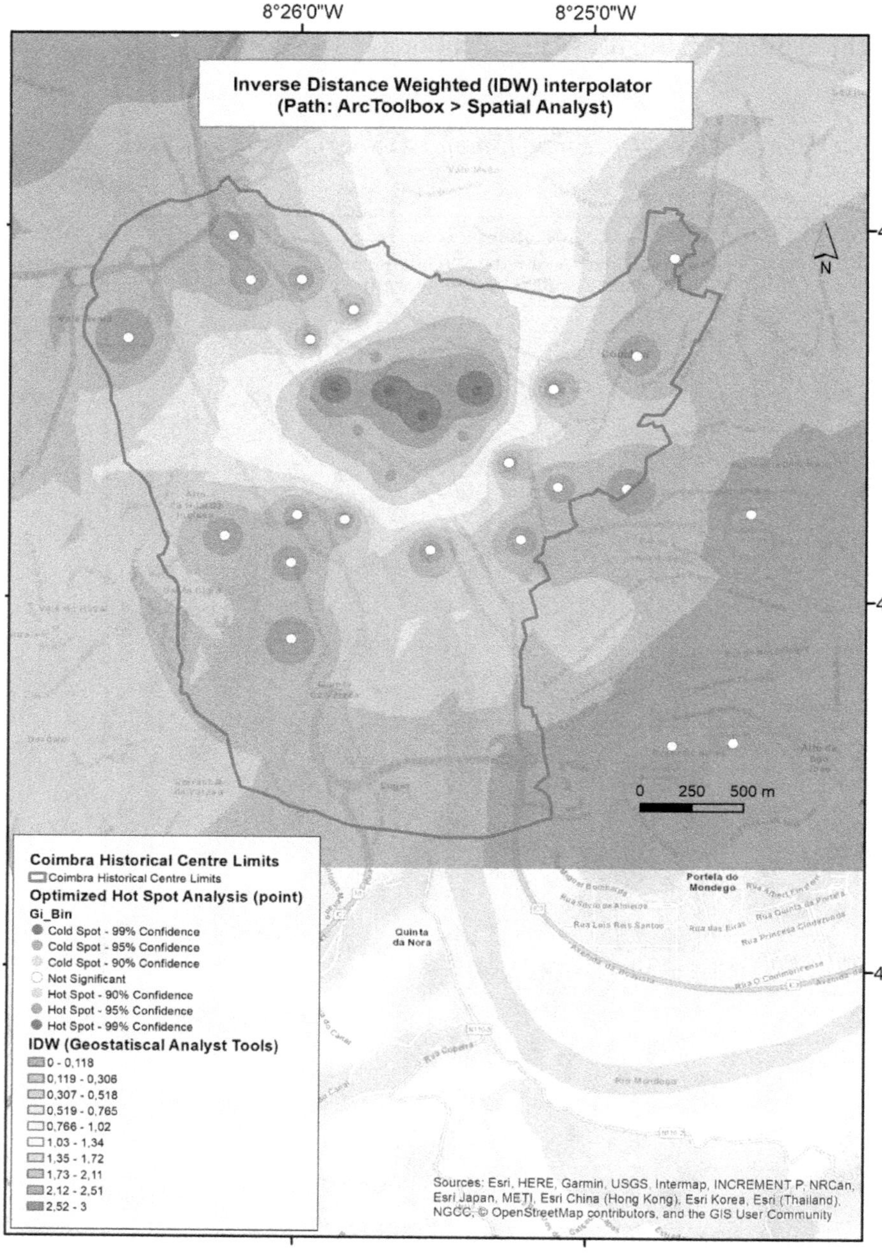

The comparative analysis of Figure 10 and Figure 11 shows the strength of the POI cluster located at the "Alta and University of Coimbra", a POI that includes the premium architectural monuments that have been mentioned (University, Machado de Castro National Museum - former Roman Forum, Old Cathedral Church, Joanina Library and Chapel Hall). But it is also noticeable, in both figures, the design of areas with high aptitude for the development of strategies for tourism purposes, namely, the extension to WNW of the main spot of the identified hot-spot identified in the "Alta and University of Coimbra" defining a region that is located between the University and the "Santa Cruz Monastery - National Pantheon". Another area with similar characteristics in terms of aptitude and potential for tourism development can be identified, which includes premium architectural monuments such as the Science Museum and goes in the direction of ENE, to "Caldeiras House" and the surroundings of the "República Square".

SOLUTIONS AND RECOMMENDATIONS

An analysis of the data and of the results obtained by implementing the modeling processes at our disposal shows the existence of optimal tourist areas in the "University – Quebra-Costas" axis as well as in the city's Green Spaces, namely in the "Parque Verde" (Green Park) area and in "República Square" area. However, in view of the check-ins (86 POIs) and the 336 037 incidents reported by tourists on Facebook and Foursquare – the two DSN under consideration - the present comparative methodological exercise allows us to formulate some important conclusions.

The KDE interpolation method performs an algorithm that leads to estimate the intensity of an event/occurrence/incident in an area where no real occurrence has been generated. It is a probabilistic, non-parametric estimator, based on the ranking of data, whose only premise is that, when the occurrence of events in space has been evaluated, its location must be considered random. For this reason, and as the occurrences do not follow a heterogeneous distribution, it does not reflect its spatial distribution in the best possible way, but it indicates, without the required precision, therefore, without reliability, the areas where the occurrences show greater concentration. It means that this method suggests the existence of a cluster identified by the number of POI and the related incidents reported. The KDE method showed the existence of a cluster in the University area, but it did not allow us to accurately determine the radius value of a polygon with statistical significance for the occurrences (incidents), and that means that the surface thus generated must vary from according to the value set in the wizard.

This research recommended that the analysis should involve other dimensions, as it was essential to adequately assess the probable relationship between the number of

POI/incidents and the distance factor, both in terms of Euclidean distance and, also, in terms of neighborhood and proximity effect. This conviction justifies testing the implementation of a chain of procedures and techniques for spatial and geostatistical analysis that define a hybrid methodology that integrates the performance of the OHSA algorithm and the IDW interpolator to estimate an appropriate radius value to find POI-incidents that fit within an imaginary circle with a defined radius and analyze the neighborhood effect. The OHSA technique showed with a high degree of confidence the existence of a spatial distribution pattern very similar to that evidenced by the KDE method, allowing to identify the presence of a tourist hot spot in the "Alta and University of Coimbra" and the existence of spatial patterns that reveal high aptitude for the development of tourist activities motivated, in good measure, by the effect of proximity and neighborhood. This observation provides adequate information for those responsible for the dynamics of urban tourism and invites a more correct management of the areas that recommend priority intervention for attracting investments.

Both techniques used (KDE and OHSA/IDW) seem to produce very similar spatial distributions. It means that it is possible to discern real tourist points of interest marked by the presence of premium historical monuments in urban areas. Given that the whole "Alta and the University of Coimbra" is a vast cluster of POI that cannot be distinguished from the neighborhood, the authors sought other solutions designed to represent, cartographically, if possible, a different type of situation, in which a given POI is part of a neighborhood instead of moving away from it. Therefore, it seems more appropriate to refer to a Region Of Interest (ROI). However, it should also be noted that the IDW interpolator is available in two different toolsets of ArcGIS and this fact requires an appropriate decision by the researchers. Although the selected results are very similar in terms of cartographic reproduction, the parameters are not the same and their configuration is also different. This fact requires attention from the researcher and the GIS operator who measures the decision by one or another toolset, because in selected contexts this can have a notable influence on the spatial delimitation and on the morpho-functional classification of the urban areas under analysis.

CONCLUSION

As a first conclusion, mention must be made to the fact that geospatial information is proving to be a paradigm in the new social, economic and demographic dynamics, a fact that has been largely amplified by the services made available by Digital Social Networks (DSN) and Geographic Volunteered Information (VGI). They were both born from the evolution of the web as a platform that since the evolution to the web

2.0, started to allow the user to also be a data producer and provider. Voluntary Geographic Information would constitute a solid and reliable way for data acquisition and availability that, in addition to allowing a considerable reduction in costs for research itself, introduced a concept of generalized democracy to all citizens. In the field of geographic information, Digital Social Networks combined with GIS and VGI proved to be a powerful association, namely, for the purposes that involve geospatial and geostatistical data and analysis.

As a second conclusion, we can assume the premise that the effects of neighborhood and proximity influence the tourist when choosing the POI to visit and in deciding the routes to be defined, a fact that reminds us of the importance of Tobbler's ideas (1970). The methods and techniques of analysis and spatial modeling to be used must adapt as best as possible to reality and must allow the identification of areas with the greatest potential for the development of tourist activities as well as the identification of spatial patterns and trends in terms of land use and occupation, territorial organization and urban planning.

As a third conclusion, however, it seems clear that, whatever the methodology used in a geospatial modeling process, it is necessary to carefully consider the list of algorithms available for execution. The choice of the most appropriate algorithm, the correct configuration of the algorithm and the appropriate choice of the numerical and qualitative parameters provided by the configuration wizard are factors that can influence the results, which, in turn, can translate into significant differences at the level of cartographic representation itself. On the other hand, reliability, completeness, consistency and accuracy are fundamental attributes required for the input data, but they are also essential factors to guarantee the quality of the work.

Another conclusion consists of the need for awareness to the fact that the greater the number of incidents associated with tourist check-ins published by users of digital social networks, the greater the likelihood that spatial and geostatistical analysis will reveal a closer approximation to reality. On the other hand, the identification of spatial patterns and the possibility of classifying urban areas according to their tourist potential is a task that can receive a notable contribution to this type of study. Modern geotechnologies, including the use of GIS and Geostatistical analysis, provide a set of robust solutions for analyzing spatial patterns, dynamic analysis involving "time" and, also, for the development of predictive models. The identification of hot spots and the concepts of proximity and neighborhood effect can be invaluable in terms of applying GIS and spatial modeling based on geostatistical operators to the analysis of urban tourism. The authors are very aware of the capabilities of these methodologies and believe that they should be used as an analytical tool to support tourism management because determining the "where?" can make a difference in terms of urban and tourism planning and management. Understanding tourist behavior and preferences are good indicators for finding the right solutions to provide tourists with better conditions to prepare their visits and ensure a greater number of potential visitors.

Finally, the analysis of the distribution of check-ins/incidents provided by tourists on Digital Social Networks after their travels and journeys for the purpose of tourism will allow the development of better strategies to expand the range of tourist offers but it will also allow tour operators "to be where tourists are". On the other hand, where there is no tourist demand, one should try to seek the reason behind this fact, so that it is possible to find creative solutions to promote tourism activity in those specific locations in a city.

ACKNOWLEDGMENT

This work was supported by the European Regional Development Funds, through the COMPETE 2020 – Operational Program 'Competitiveness and Internationalization', under Grant POCI-01-0145- FEDER-006891; and by National Funds through the Portuguese Foundation for Science and Technology (FCT) under Grant UID/GEO/04084/2013.

REFERENCES

Andriotti, J. L. (1988). Introdução à Geoestatística. *Estudos Tecnológicos – Acta Geológica Leopoldensia, São Leopoldo, 11*(27), 5-81. http://rigeo.cprm.gov.br/jspui/handle/doc/14835

Anselin, L. (1995). Local Indicators of Spatial Association-LISA. Geographical Analysis, 27(2), 93-115. doi:10.1111/j.1538-4632.1995.tb00338.x

Anselin, L., & Getis, A. (1992). Spatial Statistical Analysis and Geographic Information Systems. *The Annals of Regional Science*, *26*(1), 19–33. doi:10.1007/BF01581478

Azevedo, L. (2015). *Geolocalização, Geocolaboração e Redes Sociais. Estudo de caso aplicado ao turismo no Centro Histórico de Coimbra. tokusei* [Master Thesis]. Universidade de Coimbra, Portugal.

García-Palomares, J. C., Gutierrez, J., & Mínguez, C. (2015). Identification of tourist hot spots based on social networks: A comparative analysis of European metropolises using photo-sharing services and GIS. *Applied Geography, Elsevier, 63*, 408–417. doi:10.1016/j.apgeog.2015.08.002

Getis, A. (1991). Spatial Interaction and Spatial Autocorrelation: A Cross-Product Approach. *Environment and Planning. SAGE Journals, 23*(9), 1269–1277. doi:10.1068/a231269

Getis, A., & Ord, K. (1992). The Analysis of Spatial Association by Use of Distance Statistics. Geographical Analysis, 24, 189-206. doi:10.1111/j.1538-4632.1992.tb00261.x

Matheron, G. (1965). *Les variables régionalisées et leur estimation: Une application de la théorie des fonctions aléatoires aux sciences de la nature*. Masson.

Matheron, G. (1970). *La théorie des variables régionalisées, et ses applications*. Cahiers du Centre de Morphologie Mathématique de Fontainebleau. Fascicule 5. École des Mines de Paris.

Medeiros, A. (2019). *Evento gratuito mostra como fazer análise espacial com Mapas de Kernel em software livre*. https://mundogeo.com/blog/2015/07/02/evento-gratuito-mostra-como-fazer-analise-espacial-com-mapas-de-kernel-em-software-livre/

Önsoy, H., & Bocquillon, C. (2009). Interpolation de variables régionales et cartographie automathique: Modèle triangulaire hydrologique (MTH)/Interpolation of regional variables and automatic drawing of isoplets. *Hydrological Sciences Bulletin, 25*(3), 297–306. doi:10.1080/02626668009491936

Rocha, F. J. (2012). *Sistemas complexos, modelação e geossimulação da evolução de padrões de uso e ocupação do solo* [PhD Thesis]. IGOT, Universidade de Lisboa.

Tobler, W. R. (1970). Computer Movie Simulating Urban Growth in the Detroit Region. *Economic Geography, 46*, 234-240. https://www.tandfonline.com/doi/abs/10.2307/143141

Zhihui, G., Yan, Z., Yu, C., & Xiaomeng, C. (2016). Analysis of Attraction Features of Tourism Destinations in a Mega-City Based on Check-in Data Mining - A Case Study of Shenzhen, China. *ISPRS, International Journal of Geo-Information, 5*(11), 1–13. doi:10.3390/ijgi5110210

ADDITIONAL READING

Martí, P., García-Mayor, C., & Serrano-Estrada, L. (2021). Taking the urban tourist activity pulse through digital footprints. *Current Issues in Tourism*, *24*(2), 157–176. doi:10.1080/13683500.2019.1706458

Timothy, D. J. (2018). Geography: The substance of tourism. *Tourism Geographies*, *20*(1), 166–169. doi:10.1080/14616688.2017.1402948

KEY TERMS AND DEFINITIONS

Digital Social Network: Online platform-based services which people use to communicate and to exchange various types of content, from Data and Information to Knowledge and Intelligence. They are multi-purpose platforms, which can be categorized according to the main purposes of sharing information.

Geostatistics: Is a branch of Statistic Sciences focusing on spatial relations and analysis and/or in the spatiotemporal datasets. Geostatistics involves regionalized variables and some algorithms are incorporated in Geographical Information Systems (GIS). Formally, according to Matheron (1965) "Geostatistics is the application of the formalism of Random Functions to the reconnaissance and estimation of natural phenomena."

Geographic Information Systems: Is a framework for gathering, managing, and analyzing data. Rooted in the science of geography, GIS integrates many types of data. It analyzes spatial location and organizes layers of information into visualizations using maps and 3D scenes. With this unique capability, GIS reveals deeper insights into data, such as patterns, relationships, and situations - helping users make smarter decisions (Source: ESRI.com).

Inverse Distance Weighted: Is an interpolation method which "explicitly, makes the assumption that things that are close to one another are more alike than those that are farther apart. To predict a value for any unmeasured location, IDW uses the measured values surrounding the prediction location. The measured values closest to the prediction location have more influence on the predicted value than those farther away. IDW assumes that each measured point has a local influence that diminishes with distance. It gives greater weights to points closest to the prediction location, and the weights diminish as a function of distance, hence the name inverse distance weighted" (Source: ESRI.com).

Kernel Density Estimation: Is a mathematic process of finding an estimate probability density function of a random variable. The estimation attempts to infer characteristics of a population, based on a finite data set. It is a powerful way to estimate probability density (Source: deepai.org).

Optimized Hot Spot Analysis: Given incident points or weighted features (points or polygons), Optimized Hot Spot Analysis creates a map of statistically significant hot and cold spots using the Getis-Ord Gi* statistic (Source: ArcGIS-ESRI).

Spatial Analysis: Refers to statistical analysis based on patterns and spatial relations between variables, human and/or natural.

Urban Tourism: According to the World Tourism Organization (UNWTO), urban tourism is defined as "a type of tourism activity which takes place in an urban space with its inherent attributes characterized by non-agricultural based economy such as administration, manufacturing, trade and services and by being nodal points of transport (Source: wikipedia.org).

Volunteered Geographical Information: The term was coined by Michael Goodchild (2007) to describe "an explosion of interest in using the Web to create, assemble, and disseminate geographic information provided voluntarily by individuals." VGI is the spatial subset of user-generated content appearing at that time tied to the advent of Web 2.0 technologies (Source: Encyclopedia of GIS, 2017).

ENDNOTES

1 (...) new general class of Local Indicators of Spatial Association – LISA (ANSELIN, 1995).

2 According to the *ESRI Help Toolset*, "Given a set of weighted features, [cluster and outlier analysis] identifies statistically significant hot spots, cold spots, and spatial outliers using the Anselin Local Moran's I statistic."

3 """z-scores" are simply standard deviations. If, for example, a tool returns a z-score of +2.5, you would say that the result is 2.5 standard deviations. Both "z-scores" and "p-values" are associated with the standard normal distribution. The p-value is a probability. For the pattern analysis tools, it is the probability that the observed spatial pattern was created by some random process. When the p-value is very small, it means it is very unlikely (small probability) that the observed spatial pattern is the result of random processes, so you can reject the null hypothesis." (Source: ArcGIS Help Desktop 10.6).

APPENDIX

Collection of Geodata related to check-ins reported by tourists on Facebook and Foursquare.

Figure 12.

OBJECTID	Name	Foursquare	Facebook	Sum_FF	Inquiry	Category	POINT_X	POINT_Y	Category Order
1	A Capela	74	1059	1133	0	Diversão e Cultura	-25134.51135170000	60267.42872270000	3
2	Arcos dos Jardim	553	836	1389	0	Monumento histórico	-24465.98664130000	59814.88451570000	8
3	Jardim Botânico	707	14178	14885	34	Espaço verde histórico	-24628.91804390000	59716.77681420000	5
4	Torre da Universidade	119	0	119	0	Monumento histórico	-24983.10764940000	59946.26941780000	8
5	Capela de São Miguel	130	366	496	0	Monumento histórico	-24989.19304950000	59893.71721690000	8
6	Biblioteca Joanina	377	99	476	27	Monumento histórico	-24997.39514960000	59871.22761660000	8
7	Sala dos Capelos	114	301	415	0	Monumento histórico	-25017.79814990000	59918.05691730000	8
8	Reitoria da UC	83	173	256	0	Monumento histórico	-24980.08244830000	59931.55281750000	8
9	Universidade de Coimbra	2665	40681	43346	31	Monumento histórico	-24842.08444720000	59954.15901790000	8
10	Porta Férrea	241	1554	1795	0	Monumento histórico	-24903.73244820000	59948.48611780000	8
11	Igreja da Sé Nova	431	1450	1881	13	Monumento histórico	-24823.85394690000	60133.21862060000	8
12	Museu da Ciência	126	493	619	14	Monumento histórico	-24740.00794570000	60198.89532170000	8
13	Colégio de São Jerónimo	92	386	478	10	Monumento histórico	-24867.51204450000	60047.49341930000	8
14	TAGV	520	3382	3902	5	Diversão e Cultura	-24478.59914160000	60113.11022030000	3
15	Praça da República	529	17873	18402	0	Praça ou Parque	-24397.42484040000	60137.29312070000	10
16	Casa das Caldeiras	941	2065	3006	0	Diversão e Cultura	-24567.28764300000	60117.44932040000	3
17	Museu da Cidade - Chiado	11	128	139	2	Monumento histórico	-25189.05475260000	60044.68881930000	8
18	Torre da Almedina	230	710	940	7	Monumento histórico	-25176.35965240000	60065.06171960000	8
19	Fado ao Centro	133	1093	1226	0	Diversão e Cultura	-25135.61375180000	60062.68051960000	3
20	Fangas Mercearia bar	266	722	988	0	Restauração	-25150.95965200000	60008.97001870000	11
21	Praça do Comércio	529	547	1076	0	Praça ou Parque	-25206.25765290000	60095.22432010000	10
22	Igreja da Sé Velha	1049	427	1476	31	Monumento histórico	-25020.78435000000	60056.33051960000	8
23	Largo da Sé Velha	34	17879	17913	0	Praça ou Parque	-25054.91565060000	60073.52841970000	10
24	Largo da Portagem	1973	4817	6790	0	Praça ou Parque	-25236.42025330000	59895.86041700000	10
25	Igreja de São Tiago	27	58	85	0	Monumento histórico	-25209.82955290000	60138.36152070000	8
26	Serenata Hostel	313	493	806	0	Alojamento	-24992.73844960000	60078.82011980000	1
27	Auditório da Reitoria	60	250	310	0	Espaço Ciência	-24768.90054610000	59901.01971710000	4
28	Parque Dr. Manuel Braga	245	440	685	6	Praça ou Parque	-25084.91595100000	59666.18591350000	10
29	Museu da água	7	24	31	4	Monumento histórico	-25079.65425090000	59609.84521260000	8
30	Parque Verde do Mondego	2191	23457	25648	6	Praça ou Parque	-24869.70704770000	59248.71470700000	10
31	Docas de Coimbra	2210	9647	11857	6	Diversão e Cultura	-24977.38284890000	59404.95140940000	3
32	Hotel Ibis	433	568	1001	0	Alojamento	-24983.34574940000	59663.31761340000	1
33	Hotel Vitória	44	39	83	0	Alojamento	-25386.85655550000	60047.96961930000	1
34	Hotel Bragança	79	398	477	0	Alojamento	-25409.59038620000	60074.48091870000	1
35	Estação de Coimbra - A	867	463	1330	5	Estação Ferroviária	-25454.35795670000	60062.41591950000	6
36	Hotel Oslo	288	688	976	0	Alojamento	-25397.08875580000	60113.90392130000	1
37	Almedina Coimbra Hotel	254	61	315	0	Alojamento	-25510.06605750000	60372.93152430000	1
38	Terminal Rodoviário de Coimbra	2576	77	2653	5	Estação Rodoviária	-25882.60016330000	60889.92843230000	7
39	Igreja do Carmo	0	0	0	2	Monumento histórico	-25310.96475450000	60511.22072650000	8
40	Igreja da Graça	1	0	1	1	Monumento histórico	-25348.00845500000	60566.25412730000	8
41	Teatro da Cerca de São Bernardo	43	772	815	5	Diversão e Cultura	-25243.69185340000	60488.39582610000	3
42	Jardim da Manga / Restaurante	484	3270	3754	0	Restauração	-25086.42315100000	60322.66062360000	11
43	Mosteiro de Santa Cruz	336	1024	1360	22	Monumento histórico	-25171.93675230000	60296.62562320000	8
44	Café Santa Cruz	1109	4847	5956	0	Diversão e Cultura	-25172.78335230000	60282.65652290000	3
45	Praça 8 de Maio	1159	0	1159	0	Praça ou Parque	-25208.35365280000	60296.61232320000	10
46	Salão Brazil	296	2553	2849	0	Diversão e Cultura	-25285.68135410000	60244.54222240000	3
47	Hotel D. Inês	381	2129	2510	0	Alojamento	-25801.26926210000	60700.26142940000	1
48	Hotel Vila Galé	842	4400	5242	0	Alojamento	-25797.43946200000	60635.96762840000	1
49	Hotel Tivoli	530	1047	1577	0	Alojamento	-25652.17515820000	60672.74472900000	1
50	Torre do Anto	10	0	10	0	Monumento histórico	-25143.65275190000	60178.81962130000	8
51	Convento de Santa Ana	5	44	49	0	Monumento histórico	-24307.03853900000	59644.36021310000	8
52	Seminário maior	23	35	58	0	Monumento histórico	-24487.48414180000	59381.36380900000	8
53	Vintage Loft	13	161	174	0	Diversão e Cultura	-25369.47945640000	60195.11352160000	3
54	Diligência Bar	90	217	307	0	Diversão e Cultura	-25266.60925380000	60358.52052410000	3
55	Casa da Escrita	1	440	441	0	Diversão e Cultura	-25063.62055060000	60177.75682130000	3
56	Bacchus	40	683	723	6	Diversão e Cultura	-25186.75786260000	59749.42201470000	3
57	Hotel Quinta das lágrimas	268	1359	1627	0	Alojamento	-25600.69925890000	58823.90770040000	1
58	Fonte dos Amores	111	365	476	0	Monumento histórico	-25657.06558880000	58895.34530150000	8
59	Quinta das Lágrimas	716	4570	5286	0	Diversão e Cultura	-25526.08664780000	58922.72970190000	3
60	Fórum Coimbra	1115	34413	35528	6	Centro Comercial	-26390.87897110000	60378.40402440000	2
61	Mosteiro de Santa Clara-a-Nova	130	104	234	12	Monumento histórico	-25821.24266380000	59396.13640930000	8
62	Portugal dos Pequenitos	802	16858	17660	29	Diversão e Cultura	-25650.04415970000	59254.58400710000	3
63	Mosteiro de Santa Clara-a-Velha	390	2689	3079	31	Monumento histórico	-25542.88775810000	59273.16890740000	8
64	Praça da Canção	532	4331	4863	0	Diversão e Cultura	-25339.15815480000	59478.16741050000	3
65	Pátio das Escolas	132	412	544	0	Monumento histórico	-24943.25244880000	59606.29821710000	8
66	Museu nacional Machado de Castro	445	359	804	35	Monumento histórico	-24894.56904800000	60080.39441980000	8
67	Jardim da Sereia	334	3418	3752	25	Espaço verde histórico	-24183.79093710000	60117.11862040000	5
68	Casa de Chá do Jardim da Sereia	127	363	490	0	Restauração	-24263.16613830000	60164.21452110000	11
69	Pousada da juventude de Coimbra	78	0	78	0	Alojamento	-23926.60033310000	60298.36702320000	1
70	Mosteiro de Celas	9	0	9	2	Monumento histórico	-23747.49053030000	60783.26993070000	8
71	Hostel Be Coimbra	120	48	168	0	Alojamento	-25244.57195340000	60244.02682240000	1
72	Hotel Astória	171	517	688	0	Alojamento	-25322.67705480000	59962.13791780000	1
73	Loggia Restaurante	201	1434	1635	0	Restauração	-24934.69124870000	60093.10822000000	11
74	Penedo da Saudade	318	3082	3400	6	Espaço verde histórico	-23975.16023380000	59637.46391330000	5
75	Quebra Bar	371	1171	1542	0	Diversão e Cultura	-25097.53165120000	60069.74101970000	3
76	Tapas nas Costas	40	340	380	0	Restauração	-25122.27025160000	60066.83061960000	11
77	Quebra Costas	318	281	599	0	Monumento histórico	-25143.17235190000	60054.13061940000	8
78	Fever Suite Hotel	107	176	283	0	Alojamento	-25568.94065850000	59500.51531090000	1
79	Estação de Coimbra - B	4359	2893	7252	5	Estação Ferroviária	-26170.99656770000	61903.54914800000	6
80	Estádio Cidade de Coimbra	3140	24198	27338	22	Diversão e Cultura	-23071.52732450000	59437.92740990000	3
81	Dolce Vita	9728	2157	11885	6	Centro Comercial	-23372.05652450000	59586.62301220000	2
82	Coimbra Shopping	4201	0	4201	6	Centro Comercial	-23446.72202570000	58436.68919440000	2
83	Conservatório de Música de Coimbra	243	1750	1993	5	Diversão e Cultura	-23456.24712580000	58299.10559230000	3
84	Mata do Choupal	505	2453	2958	24	Espaço verde histórico	-26973.56318010000	61505.97224180000	5
85	Parque de Campismo	307	346	653	3	Parque de campismo	-22634.42561310000	57740.28738370000	9
86	Oficina Municipal de teatro	16	228	244	5	Diversão e Cultura	-23749.24973040000	58352.26989310000	3

Chapter 7
Urban Transition and Its Impact on Mesoscale Weather:
A Review

Sudhansu S. Rath
https://orcid.org/0000-0003-4529-3719
National Institute of Technology, Rourkela, India

Jagabandhu Panda
https://orcid.org/0000-0002-4238-1820
National Institute of Technology, Rourkela, India

Srutisudha Mohanty
https://orcid.org/0000-0002-8642-6234
National Institute of Technology, Rourkela, India

ABSTRACT

Urban transition is an unstoppable process. Globally, several planning measures are taken by the city and country administration to control the sprawling process. Despite all the planning, most of the cities experience appreciable impact of urbanization on the localized weather parameters. This chapter summarizes the understanding relating to urban modification of localized weather, that is, changes in precipitation, temperature, and wind speed in the form of increase or decrease, their spatio-temportal distribution, urban heat island (UHI), and urban wind island (UWI). The impacts of the urbanization are primarily because of changes in land-surface characteristics due to the alteration of land use in a city. The urbanization effects on local or mesoscale weather could be studied both through observations and/or numerical modeling. The purpose of this chapter is to provide a review of most of the relevant studies carried out globally and with a special emphasis on India.

DOI: 10.4018/978-1-7998-2249-3.ch007

INTRODUCTION

Urban transition is broadly the shift from rural to urban land use (LU) and economically from agricultural to industrial, commercial, or service employment (Pannel, 1995). According to the United Nations (UN), globally, 55% percent of the population currently resides in cities, and this figure projected to be two-thirds of the world population by 2050 as the growing population concentrates in cities and towns in search of better livelihoods and opportunities (DESA, 2018). This process of urbanization gives rise to the expansion of existing cities to form mega cities and evolution of new cities, which accelerates the changes in existing land use and land cover (LULC). Studies suggest that the major causes of land cover (LC) changes are neither population nor poverty alone; rather, it is people's response to economic opportunities. Mostly, global factors are the main determinants of LU change as they are weakened by the local factors (Lambin et al., 2001). Besides these causes of urban transition, the development of an area as urban land takes place only when the post-development value of the land and the percentage growth rate of rent increases. Urban transition in the outskirts of core cities is directly affected by the emerging economy and due to foreign investments.

Figure 1. Global trends and rate of change of urbanization.

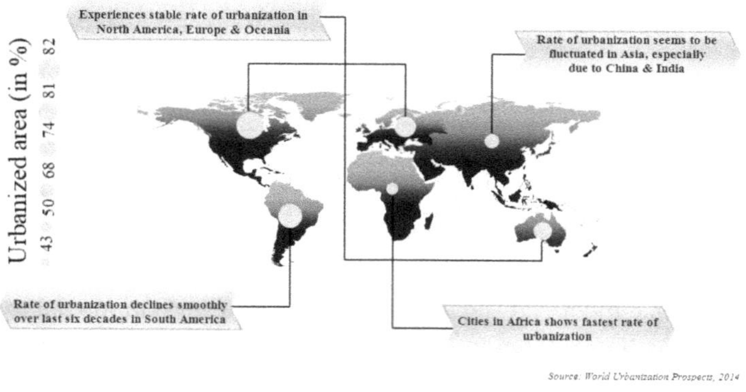

The global trends and rate of urbanization over different parts of the world illustrated in figure 1. The rate of urbanization found to be either stagnant or declining in other parts of the world except for Asia and Africa, where the rural to urban transition mostly occurs. However, the urbanization in African countries is driven by global historical process like population dynamics and rural-urban migration, both of which are stimulated by technological and institutional changes.

In Asia, the southeastern region, including India and China, are in a continuous state of development. Chinese cities are proceeding toward rapid urbanization with a dense population and a structural shift in employment. Economic reforms and development-friendly policies by the local governments are primarily the causes for the rapid shift (Pannell, 1995; Han, 2010). In contrast to popular beliefs, cities in China are becoming more and more like their first-world counterparts.

The policymakers exclude rapidly growing areas outside the boundary of large urban agglomerations, which often misleads their conclusions regarding the growth rate (Zheng et al., 2009). Thus the focus of the urban related studies need to be diverted toward the developing countries like India, where the diversified low density development in the peri-urban area is seen. In India, both Gibrat's Law and Zipf's Law failed to explain the spatial equilibrium of urbanization (Chauvin et al., 2017). Urbanization in India started close to ports during the British colonization era. There are similarities between urban developments in India and in European countries, which deteriorated severely because they lacked proper expansion plans and-passed the focus over other smaller cities after independence (Harari, 2016). India also witnessed the development of corporate urbanism in forms of company towns to industrial cities to the present-day urban agglomerations (Sood, 2015). Growth of Indian cities is found to be driven by the initial population, capital city status, and proximity to larger cities (Abhishek et al., 2017).

Although the process of urbanization is relatively localized, it has regional and global impacts as well. Human activities influence the fluxes, and the amount of gases present between the land surface and atmosphere (Guo et al., 2016). Anthropogenic emissions from urban areas also alter the chemical composition of the atmosphere and the radiative balance in the earth (Lindberg et al., 2013). The resulting feedback of land-atmospheric interaction consequently influences the hydrological cycle (Brubaker & Entekhabi, 1996). Urbanization can influence the rainfall occurrence and distribution over a city and its neighboring regions (Li et al., 2016) too.

India, as an emerging economy, has mega cities like Chennai, Delhi, Mumbai, and Kolkata, and building numerous tier-II cities throughout the country. The tier-II cities eventually grew and became the hub for economic and industrial activities mostly for particular areas of the country. Therefore, the topic of urban transition and its consequences needs to be emphasized by researchers, government bodies, and planning personnel. The impacts on water, energy, and atmospheric composition can be analyzed, and appropriate measures can be taken for the cities.

This review compiles the works done related to urban expansion and its impact on mesoscale weather over different parts of the world and mostly emphasizing the Indian scenario. It consists of a general introduction followed by the concepts of land and atmospheric interaction. Observational and numerical modeling studies related to the impact on temperature, and precipitation is discussed subsequently.

A novel concept of urban wind anomaly is found in several pieces of literature and discussed afterward. The future scopes are mentioned in the section before putting forward an overall concluding remark.

INFLUENCE ON LAND ATMOSPHERIC INTERACTIONS

The earth is a highly complex system that consists of interlinked components like land, ocean, and atmosphere. They interact with each other in various spatial and temporal scales. Fluxes of moisture and heat from the underlying land surface determine the overlying distributions of atmospheric water vapor, temperature, cloud properties, precipitation, and therefore the downward radiative fluxes at the surface. Changes in LULC due to urbanization affect these processes through changes in the surface layer properties (Dickinson, 1995). For instance, the incoming solar radiation is same for both rural vegetated locations and highly urbanized locations. However, urban and rural areas have contrasting surface properties, like albedo and heat capacity, and thus, they exhibit different energy budgets.

A diagram for energy balance on a typical sunny day for a city and surrounding rural location is illustrated in Figure 2, which is adopted from Shepherd (2005). Both rural and urban locations receive energy due to radiative processes from the sun and lose energy back to the atmosphere. The incoming short-wave radiation warms the earth's surface during the day and cools at night by emitting long-wave radiation. For instance, the illustration in Figure 2 indicates the differential radiative fluxes for a city in the United States, where the incident solar radiation, Q_I, is 7.6 kilowatt-hours per square meter per day ($kWh/m^2/day$) for both rural and urban locations. With an albedo of 0.25, which is typical of a rural vegetation cover, the reflected solar radiation, Q_R, is 1.9kWh/m^2/day in rural areas. Urban locations may have an average albedo as low as 0.05 because of the high density of low-reflecting buildings and roads, resulting in an urban reflected energy of only 0.4kWh/m^2/day. Thus, the net energy obtained from the sun in the rural area is 5.7 kWh/m^2/day, while that in the urban area is 7.2 kWh/m^2/day, which is 1.5 kWh/m^2/day more than any average rural location. The differences mentioned above give rise to different fluxes over urban and rural locations. Urban areas also emit anthropogenic heat flux, which is entirely absent in rural areas with non-human activity.

Figure 2. Typical energy balance values for an urban and rural location (unit W/ m2/day, Adopted and modified after http://www.ruf.rice.edu/~sass/UHI).

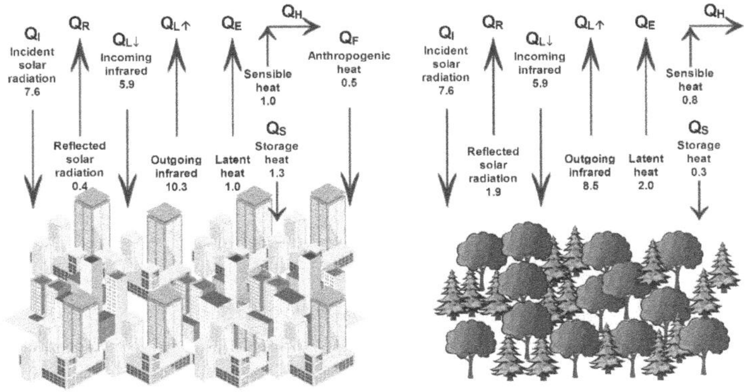

Anomalous moisture and temperature states affect each other and create mutual positive feedback in the atmosphere (Brubaker & Entekhabi, 1996). These changes in the energy budget eventually impact the local atmospheric conditions, such as enhancing the urban heat island (UHI) effect and altering the amount and distribution of precipitation (Barros & Hwu, 2002). Field experiments have proved that the urban circulations are affected by the temperature, surface flux gradient, and surface roughness (Guo et al., 2016).

Urbanization influences the local weather and climate of a city. In this regard, cities need to be prepared at the planning level. Master Plans usually do not include climatic aspects, and very few studies have addressed climate change at the urban scale (Viegas et al., 2013). Observational evidence suggests that sprawled areas are at a greater risk of extreme weather events than that of a compact planned metropolitan region (Stone et al., 2010). The impact of urbanization on weather and climate has been studied since the 1970s using observational analysis and numerical modelling in different parts of the globe but lacks for India. For both urbanization and sustainable city to survive, urban climate research outcomes need to be applied in the design and development of cities (Grimmond et al., 2010; Hurlimann & March, 2012). Recently, the government of India is trying to incorporate climate change as an active component in development goals (Rajasekar et al., 2018). Studies over Indian metropolises reveal how the local development authorities are grappling with large deficit in infrastructure and services and not prioritizing climate change adaptation plans (Sharma & Tomar, 2010; Joerin et al., 2012 Prashar et al., 2012; Kumar et al., 2016). There is a severe need for enhancement of disaster resilience in mainstream adaptation for urban planning in India (Surjan & Shaw, 2009; Prashar et al., 2012).

IMPACT ON TEMPERATURE

Boundary layer processes are modified by the urban impervious layers in various ways. One of the foremost process is UHI effect. The impervious surfaces in urban areas have different thermal properties than the natural LC (Christen & Vogt, 2004), which typically stores more heat energy and contributes to the sensible heat production. The increased sensible heat contributes to increases in the surface air temperature. Various studies suggest that the UHI effect can increase the temperature of urban areas up to 10°C (Gedzelman et al., 2003; Chang et al., 2007; Wilby, 2008; Sharifi & Lehmann, 2015) than the surrounding rural areas due to their difference in land surface properties. The UHI effect can have diurnal and seasonal cycles that are similar to that of temperature. The spread of the city, wind patterns, and cloud coverage are few of the important factors that can modulate the UHI impact. Human activities like vehicular fossil fuel burning and cooling and heating processes of buildings can enhance the anthropogenic heat emissions in urban than rural areas contributing more to the UHI effect (Allen et al., 2011).

Luke Howard conducted the first study related to UHI for London during the eighteenth century. Afterwards, several studies were conducted worldwide, which reveal that the UHI is the clear example of the impact of urban-induced LULC changes in cities (Landsberg, 1970; Oke et al., 1991; Arnfield, 2003; Gedzelman et al. 2003; Yow 2007; Pielke Sr et al. 2011; Chun & Guldmann 2014; Chapman et al. 2017). Most of the preliminary investigations are based on any particular urban region using observational networks. Basel Urban Boundary Layer Experiment (BUBBLE) helped in understanding the energy and radiation balance over urban areas, including the variability in temperature (Christen & Vogt, 2004). Studies over Sydney confirm that road networks and open spaces in cities are more prone to higher surface UHI than other parts of the city (Sharifi & Lehmann, 2015). Significant differences in minimum temperature between dense cities and suburbs are revealed by long-term studies of the metropolitan areas of Hamburg and Turkey (Tayanç et al., 1997; Schlünzen et al., 2010). Most of the results suggest that urban atmosphere is always warmer than its surroundings throughout the year.

A large number of studies are conducted for several Chinese and Indian cities, where approximately 40 percent of the world's populations resides. This high density of population makes these places vulnerable due to heat stress hazards. Although long-term assessment requires quality data, studies using both traditional and modern approach agreed with the impact of UHI on the cities resulting from urban–rural temperature differences (e.g., Wang & Yan, 2016). In China, eastern and northern cities are more affected than other parts of the country. Coastal cities experience weaker UHI than inland cities (Hua et al., 2008; Li et al., 2010; Yang et al., 2011). Analysis suggests UHI increases with the expansion of population and the rapid increase of gross domestic product. Additionally, UHI causes hot days and heat-waves, which directly cause adverse health effects on urban dwellers of Shanghai (Zhang et al., 2010; Tan et al., 2010; Bian & Ren, 2019).

Micrometeorological observations, remote sensing data analysis, and improved LULC classification are few of the methods adopted for analyzing surface and canopy layer UHI impacts on Indian cities (Mohan et al., 2012; Ramachandra & Kumar, 2010; Mohan et al., 2011; Pandey et al., 2012; Mohan et al., 2013; Mallick et al., 2013; Borbora & Das, 2014; Thomas et al., 2014; Mohan & Kandya, 2015; Joshi et al., 2015; Kikon et al.; 2016; Mathew et al., 2016; Kotharkar & Surawar, 2016; Singh et al., 2017; Mathew et al., 2018; Kotharkar & Bagade, 2018). Dimri (2019) well-documented the regional changes in daily surface temperature extremes over India. Shastri et al. (2017) studied the contrasting behavior of surface UHI over India during pre-monsoon due to the low levels of vegetation that were reported over non-urban regions.

The National Capital Region (NCR) is more extensively studied than other regions in India. Professor Manju Mohan and her research group (at Indian Institute of Technology, Delhi) carefully observed the diurnal temperature range, LULC specific UHI, and temperature trends using *in situ* as well as satellite observations (Mohan et al., 2011; Mohan et al., 2012; Mohan et al., 2013; Mohan & Kandya, 2015). Pandey et al. (2012) studied the association of particulate matter with UHI over Delhi. Similar studies were also performed over other metropolitans like Bengaluru & Chennai (Ramachandra & Kumar, 2010; Amirtham, 2016) and for relatively smaller cities like Guwahati, Lucknow, Nagpur, Chandigarh & Ahmedabad (Borbora & Das, 2014; Joshi et al., 2015; Kikon et al., 2016; Singh et al., 2017; Mathew et al., 2016; Kotharkar & Surawar, 2016). These studies reveal that the cause of UHI can be attributed to population density and poor design of cities. Incorporation of local climatic zone (LCZ) classification methods improved the UHI accuracy in cities like Nagpur and Kochi (Thomas et al., 2014; Kotharkar & Bagade, 2018). By incorporating relatively newer LULC within WRF modeling framework, the 2015 heavy rainfall event in Chennai was simulated (Rath & Panda, 2019). The results from the studies indicate increased near surface temperature while using recent LU with larger impervious layers (Figure 3). The impact of urban impervious layers in localized heating of the near surface was realized before the heavy rainfall event. The urban induced LULC changes encouraged the underlying land-surface to absorb and emit more heat resulting in a relatively warmer air and this enhanced near surface temperature. A recent study by Gogoi et al. (2019) also highlighted the possible role of urban-induced LULC changes and anthropogenic activities in accelerating the near surface air temperature rise over the eastern Indian state of Odisha.

Figure 3. Temperature (2m) difference simulated using WRF-ARW (1.2 km horizontal grid resolution for the inner-most domain with 1:3 parent to nest grid ratio with 3-domain configuration) over Chennai, India, using recent ISRO LU and older USGS LU. The black boundary represents the Chennai metropolitan area and the other one is for the coast-line.

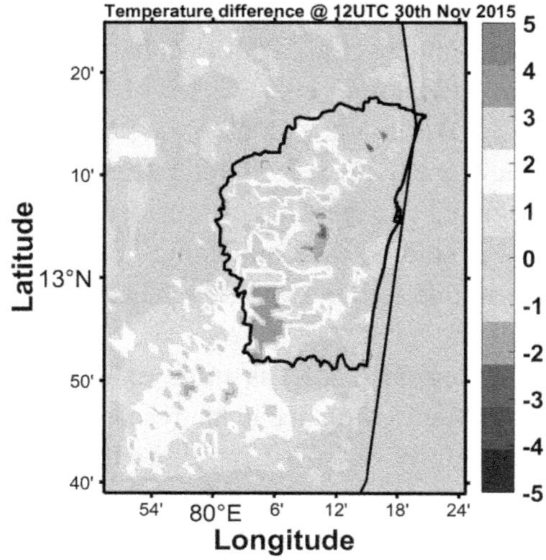

Several efforts using numerical modelling have been made in the past to understand the urban impact on the UHI phenomenon and boundary layer structures. Urban canopy variables were found to be better simulated/predicted by the numerical models than the conventional variables at several instances (Miao et al. 2009). Regression analysis using model re-analyses and downscaled WRF model results are found to be quite helpful in UHI assessment (Doan et al., 2016; Goddard & Tett, 2019; Jahangir & Moghim, 2019). Numerical modelling of UHI has shown that the simulation results are very much sensitive to the representation of LULC in the numerical modelling framework (Hale et al., 2008; Salamanca et al., 2012; Bhati & Mohan, 2016; Göndöcs et al., 2017; Li et al., 2018a; Vitanova et al., 2019). Numerical modelling in combination with high-resolution surface modelling significantly increases the accuracy in temperature predictions (Liu et al., 2019). Future projections of extreme urban heating using representative concentration pathway scenarios also predicted the impact of urbanization on temperature trends and severity of heatwaves (Ortiz et al., 2019). Also, some of the studies found that, urbanization contributes appreciably the boundary layer processes, but has a negligible impact on the intensification of heatwaves (Wang & Li, 2019; Kumar & Mishra, 2019).

IMPACT ON PRECIPITATION

A built-up area usually influences localized and mesoscale convective processes, which was known even in the 1920s. Later, researchers, including Atkinson (1968) illustrated that urbanization can impact the distribution of thunder rainfall over a city like London. The majority of the studies published before 1970 relating to urban impacts on precipitation focused on the United States. The Metropolitan Meteorological Experiment (METROMEX) is one of the early studies that identify the urban modification of precipitation over major urbanized areas during the 1970s (Changnon, 1981). Balling and Brazel (1987) reported enhancement in afternoon thunderstorms over urban areas during the peak population growth period over Arizona. Moreover, Jáuregui & Romales (1996) found that the daytime UHI seems to be associated with the intensification of rainfall over urban stations of Mexico City, where as rainfall over non-urban stations remains unaffected. Additionally, Bronstein & Lin (2000) confirmed that during nearly calm regional flow conditions, the UHI over urban area initiates convective activity and further, the building barrier effect bifurcates the clouds to move around the area.

The second-generation studies after the METROMEX experiment focused on validating the findings and extended the same study to more urban locations inside the United States. Lowry (1998) suggested that the methods followed to study urban impacts on temperature are inappropriate for precipitation analysis because the temperature is continuous in both time and space, whereas precipitation is continuous in neither. More recent studies (during the last three decades) are focusing on urban areas of other parts of the globe. The rapid expansion of Asian and European cities aided the research related to urbanization impacts. In addition, the remote sensing technique helps in quantifying the urban LULC changes and global analysis of precipitation and temperature in the absence of sufficient *in situ* data. There could be a diverse role of urban-induced LULC changes on precipitation processes (Pielke Sr et al., 2007). Recent studies suggested the influence of the city and the presence of inland water bodies on the atmosphere, which causes an increase in heavy rains over Chicago (Changnon & Westcott, 2002). Observations using space-borne RADAR reveals an increase in rainfall rate within few kilometers downwind of the metropolis (Shepherd et al., 2002). Dixon & Mote (2003) found that UHI over Atlanta could initiate and enhance the thunderstorm developments while in some US cities, they are more frequent over moisture-rich areas rather than those with higher UHI. However, freezing rain events in US cities found to be decreasing due to UHI effect (Changnon, 2003).

Additionally, Burian & Shepherd (2005) reported that the diurnal cycle of rainfall distribution experienced a shift from pre to post urban periods. Kaufmann et al. (2007) highlighted that urbanization could be responsible for a reduction of precipitation over the Pearl River Delta region, which may be due to changes in surface hydrology that extend beyond the UHI effect and energy-related aerosol emissions. In addition, analysis for spatial precipitation patterns for Tokyo, using long-term hourly data showed a positive anomaly in the afternoon of the warm season rainfall, supporting to the reality of UHI effect on the precipitation (Fujibe et al., 2009). Using long-term RADAR reflectivity and lightning data, Ashley et al. (2012) found positive urban amplification of thunderstorm frequency and intensity over major cities in the United States, while the degree of urban influence depends on the geometry of the cities. Furthermore, a relatively recent study by Ganeshan et al. (2013) found that the UHI has a dominant influence on precipitation for inland cities, especially during the afternoon and nocturnal hours. While propagating storms appear to converge over the urban regions, the rainfall intensification occurs downwind of the city. Haberlie et al. (2015) supported the same over the southeastern United States, indicating that human-induced changes in LC in moist tropical environments lead to greater initiation of thunderstorms over the developed areas.

Urban impact studies in Indian context are in a pre-mature stage. The majority of the studies are based on numerical modelling, while very few are observation-based. Studies like Kishtawal et al. (2010) identified the urban signature in heavy rainfall climatology over the India. Additionally, Mitra et al. (2012) studied the relationship between pre-monsoonal rainfall climatology and urban LC dynamics over Kolkata. The analysis shows a significant increase in observed rainfall with an increase in urban land cover. The observational studies over India is limited possibly because of the scarcity of well distributed weather networks.

A number of modelling studies investigated the urban influence on local weather. First simulations using primitive equations revealed that the UHI intensity, urban roughness, and boundary layer instability has a role in UHI circulation and convection. Over coastal cities, the interaction of UHI and sea breeze depends upon the intensity of UHI, the size of the urban area, and distance from the coast, which was studied using 2-D and 3-D mesoscale models (Kusaka et al., 2000; Ohashi & Kida, 2002). For instance, Thielen et al. (2000) used a 2-D model to describe the importance of surface parameters, such as sensible heat flux, on convective precipitation over Paris. Additionally, Baik et al. (2001) used a similar 2-D mesoscale model to understand the role of atmospheric heating due to UHI impacting the dry and moist convection process. Also, these studies delineated the process of precipitation occurrence in the downwind direction of the urban area. Using a storm-resolving version of regional atmospheric modelling system (RAMS), Rozoff et al. (2003) investigated the role of the urban atmosphere on deep moist convections over St. Louis. The UHI was found to play a vital role in initiating moist convection in the downwind of the city, as emphasized previously by Baik et al. (2001). The simulated storms have a greater sensitivity to the details of urban surface characteristics.

Urban growth, release of moisture, aerosol, and heat significantly affect the downwind precipitation over higher-altitude cities like Fairbanks, Alaska (Mölders & Olson, 2004). In addition, the prediction efficiency of precipitation depends on grid sizes because sub-grid processes are resolved on the finer grid. This process is very well-connected to changes in urban induced LULC, which significantly alters the spatial variability of surface heat, moisture, and momentum (e.g., Li et al., 2013; 2016). Along with the increase in size of urban areas, the effects are more pronounced over the surrounding areas rather than only over downwind areas (Lin et al., 2008a).

Furthermore, using WRF model along with Noah land surface and Urban Canopy Model (UCM), several studies emphasized the significance of anthropogenic heat (AH) in determining the boundary layer processes and UHI intensity (e.g., Li et al., 2013) especially during night-time and early morning hours (Lin et al., 2008b). Shepherd et al. (2010) used an urban LU prediction model to predict the future LU of Huston and incorporated the result in numerical modelling to suggest that convective processes might be more vigorous in response to the future urban land-use scenario, possibly indicating the urban impacts. Niyogi et al. (2011) found that the storm of interest could not be simulated without the urban area in the smaller domain. Removing the Indianapolis urban region from the simulation caused distinct differences in the regional convergence, convection, and surface energy balance. However, the synoptic-scale rainfall patterns may not be sensitive to different urban LU representations inside the model (Yang et al., 2014).

Several studies used numerical models, different LU scenarios, and UCM to perform simulations over cities in China. For example, Miao et al. (2011) showed that simulation using early urbanization prior to 1980 decreases the maximum rainfall, whereas high-rise urban growth in Beijing is conducive to bifurcating the path of rainfall. Similar to Baik et al. (2001) and Rozoff et al. (2003), increases in precipitation over the downwind direction are reported by Wan et al. (2013). Similarly, Zhang (2014), Zhong & Yang (2015a; 2015b) investigated urban impacts over Beijing for cold frontal and summertime rainfall events using urban and non-urban LU scenarios with UCM coupled to a WRF model. The robust upward motion along with higher convergence and abundant moisture transportation is caused by the thermodynamics and dynamics of urbanization, which directly contributes to changes in surface energy balance and planetary boundary layer characteristics and the production of rainfall. Wan et al. (2015) did the same as Zhong & Yang (2015a; 2015b) for a short-term rain event using ensemble WRF simulations for Yangtze River Delta (YRD) in China and found significant urban modifications to rainfall. On the contrary, reduction in rainfall is seen in the case of urban simulations with moisture deficits; however, increases in temperature are observed over urban agglomeration of China (Wang et al., 2012). Additionally, Zhong et al. (2015; 2017) evaluated the impacts of UHI versus aerosol on precipitation over YRD and Beijing using WRF-Chem and UCM coupling. Additional studies focusing on Asian urban agglomerations suggested that more realistic urban LU representation affects the location of thunderstorms and precipitation, leeward movement of precipitation and enhanced UHI (Li et al., 2016).

Figure 4. WRF-ARW simulations of multiple thunderstorm events over Bhubaneswar-Cuttack urban complex of Eastern India. The near-surface wind vectors are represented by arrows and difference of accumulated rainfall between recent ISRO LU and older USGS LU simulations are represented in blue (-ve) to red (+ve) shades.

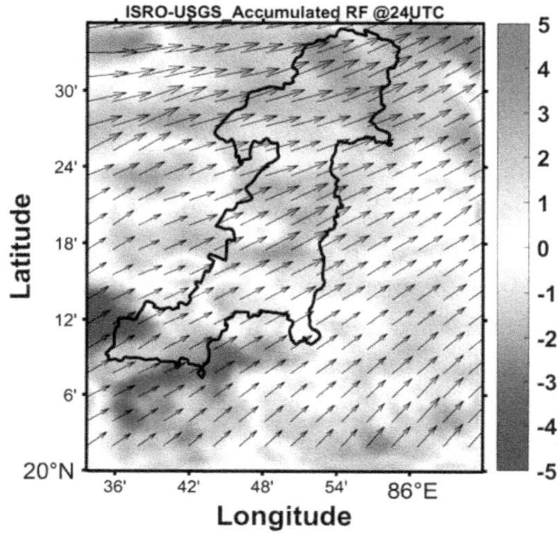

India has also adopted numerical model-based studies in the last three decades. Kumar et al. (2008) found that mesoscale land-surface feedback modulated the location and intensity of the rain by changing the wind flow, regional moisture convergence, and formation of mesoscale vortices during a heavy rainfall event over Mumbai. Additionally, Lei et al. (2008) and Chang et al. (2009) found the influence of urbanization during the same rainfall event over Mumbai using RAMS and WRF model. Based on their research, they suggested that the explicit urban land surface representation within the modeling framework improves the model performance for extreme precipitation studies. Furthermore, Goswami et al. (2010) used two urban scenarios in MM5 mesoscale model to simulate rainfall events over Mumbai, Bengaluru, and Chennai. Based on this study, the authors suggested that urbanization increases the diurnal temperature range and affects both the intensity and spatial distribution of rain. Li et al. (2018b) studied the impact on microclimate over Kolkata city using a WRF model with improved LU representation. Moreover, a recent study of near-surface boundary layer characteristics during the 2015 Chennai flood in the context of urban-induced LU changes shows the dominance of localized effects due to urbanization (Rath & Panda, 2019). Noticeably higher values of the near-surface variables such as air temperature, wind speed, relative humidity (RH), moisture flux, Convective Available Potential Energy (CAPE), and BL height are attributed to the increased impervious layers inside the city boundary due to urbanization and its growth.

Although there are several studies available for India, these studies are mainly concentrated over metropolitan cities like Mumbai, New Delhi, and Kolkata. More studies like this are required for other Indian cities to understand the impacts of urbanization in different geographical conditions. One such attempt is a study that was conducted over Bhubaneswar-Cuttack, an urban complex of eastern India, which revealed similar results as other global studies (Wan et al., 2015). The ensemble analysis of multiple thunderstorm simulations using different LU scenarios shows a significant impact of urbanization on precipitation distribution. The increase in rainfall observed over the downwind direction of the city, as shown in Figure 4, is one of the important consequences of urbanization and well emphasized by several researchers (Shem & Shepherd, 2009; Wan et al., 2013; Wan et al., 2015). The increase in rainfall in the downwind direction is seen within Bhubaneswar-Cuttack urban complex (eastern part of India) and in the neighboring areas (Figure 4).

To summarize, the various possible mechanisms for urban environments to impact precipitation or convection can be understood as: 1) enhanced convergence due to urban-induced surface roughness (Changnon et al., 1981; Bornstein & Lin, 2000; Thielen et al., 2000). The impact of urban-induced surface roughness is found to be higher than the orographic effect; 2) unstable urban boundary layers due to UHI and energy budget alteration to enhance the urban afternoon thunderstorms (Balling & Brazel, 1987; Jáuregui & Romales, 1996; Shepherd et al., 2002; Shepherd & Burian, 2003; Dixon & Mote, 2003; Fujibe et al., 2009; Ganeshan et al., 2013); 3) aerosol enhanced/diminished cloud condensation nuclei (CCN) sources in the urban environment (Diem & Brown 2003; Mölders & Olson 2004; Kaufmann et al. 2007); 4) bifurcation or diversion of precipitating systems by the urban canopy or related processes (Bornstein & Lin 2000; Loose & Bornstein 1977); 5) shifting of precipitation towards the downwind direction of the city belt (Shepherd et al., 2002; Ganeshan et al., 2013; Haberlie et al., 2015). Several authors have also hypothesized that urban areas serve as sources of moisture convergence, which is essential for convective processes (Dixon & Mote, 2003). Aforementioned mechanisms either work alone or can have their integrated effect depending on the urban area.

URBAN WIND ISLAND

The wind is undoubtedly a key component in urban meteorology due to its relevance in transporting the air pollutants, moisture, and momentum. The wind is also key for its help in temperature advection and in the partial change of the precipitation distribution. The wind flow variation is created by the temperature and pressure gradient and by dynamics of the atmospheric boundary layer (ABL). Studies related to the wind speeds within urban locations are very limited to the observed differences between urban and nearby rural areas. Previous studies over New York City show both increase and decrease in wind speed are observed within the city boundary compared with sites outside of the city (Bronstein & Johnson, 1977). Bronstein & Johnson (1977) proposed a critical wind speed of approximately four meter per second (ms^{-1}), a decrease to which is attributed to increased values of the surface roughness over the city. This increase is associated with temperature gradient due to well-developed UHI. The studies carried out over the city of London (Lee, 1979) agreed with Bronstein & Johnson (1977) about the concept of critical wind speed within the city environment. Lee (1979), further suggested that under the conditions of light regional airflow and intense UHI, strong rural atmospheric stability and the unstable city atmosphere allows a greater flux of momentum to the surface and thus maintains higher wind speeds within the city than the rural area.

Additionally, another study using tower measured wind speed over St. Louis raised concerns about the previously formulated critical wind speed of approximately four ms^{-1} and found slightly higher wind speeds over the city center under nearly calm conditions, weak UHI, and conditional instability (Shreffler; 1979). A recent study over the Netherlands used a straightforward mixed-layer bulk model to identify differences in the wind under equal geostrophic forcing and found that urban winds are surprisingly higher than rural winds and termed the phenomenon as Urban Wind Island (UWI) effect, which is similar to UHI (Droste et al., 2018). The occurrence of UWI is highly dependent upon boundary-layer depth, initial wind speed, and geostrophic winds along with the driving factor of the Coriolis force. The most favorable conditions for UWI to occur are high urban and low rural boundary-layer depths. The urban wind research and the concept of UWI need to be explored further over different cities all over the world in order to understand transport of pollutants, to complement wind-engineering studies, and to foster further research related to urban air circulation.

FUTURE RESEARCH DIRECTIONS

Observation alone or modelling attempts are not enough to understand the impact of urbanization on mesoscale weather. As the findings from major modelling studies all over the world suggest, it is still not clear how much urban induced LULC change is influencing the weather. Therefore, more sophisticated experiments with modern numerical models are required in different parts of the world to reach a certain conclusion on the impact of urbanization on mesoscale weather. Further, the impact assessment of urbanization on local weather conditions requires the proper representation of LU categories within the numerical model. For this purpose, the detailed LU classification with sub-pixel level accuracy is required. In addition, the coupling of UCM with the WRF model is shown to be beneficial in depicting the explicit urban impacts. This requires input of various urban canopy parameters (e.g., building height, road width) besides the sub-classification of urban LU into categories such as commercial, high residential, low residential, and industrial (e.g., Li et al., 2013; 2016), which is not available for all highly-urbanized regions of the world. For instance, some modelling studies could be carried out over urban areas of India as seen from the available literature, but the lack of urban-specific dense meteorological observations would not allow the validation of these results. The unavailability of appropriate satellite-derived LULC observations and urban-specific parameters has repressed the urban-specific modeling studies for Indian regions as well. Such criticisms may hold good for other parts of the developing world too, where the urban specific studies are not prioritized.

The need for extensive research in the direction of developing an urban database for cities of the world has been started under the World Urban Database and Access Portal Tools (http://www.wudapt.org/) using LCZ LU classification (Ching et al., 2018). In the first stage, few of the Indian cities are under consideration for developing urban parameter specific datasets, which can be extended to other cities. The development or customization of numerical weather prediction models according to the geography of the study region will be greatly helpful. There is a need for the establishment of a city specific dense network of weather stations too.

SUMMARY

The facts and research outcomes of urban transition and its impact over regional weather have been summarized in this chapter based on some of the relevant literatures. The focus is on urban growth and consequent modification of precipitation, UHI and UWI effects experienced over cities. The review of urbanization and its impacts on mesoscale weather may be summarized as follows:

- Urbanization found to be influencing the precipitation intensity and distribution in and around the urban area.
- Several studies found an increase in precipitation due to the increased urban area, while some also reported precipitation deficiencies.
- The role of urbanization is confirmed in the enhancement of UHI over urban core than the surrounding rural area.
- Few of the recent studies found an interesting phenomenon named as 'UWI effect' similar to that of UHI over highly dense urban areas.
- Improved LCZ based urban LULC classification using high-resolution satellite images and through ground truthing would be quite helpful always for each urban area. The information can then be incorporated in numerical weather prediction models for forecasting weather over a particular city.
- Homogenous distribution of high resolution measuring instruments must be put in all cities of India to develop a mesoscale weather network and record continuous measurements of weather parameters. It will help in model output validation too.
- Although some urban canopy models available for coupling within WRF framework, improved modelling systems to explicitly resolve urban canopy parameters, complex land surfaces and anthropogenic effects required and must be advanced further to understand the associated interactions and feedbacks exclusively.
- Local governing bodies or the city planning and administration need to think about the weather and climate resiliency while planning for cities, which can help in coping up with the adverse effects of extreme weather and climate change. It should be of utmost importance in the recent smart city initiative.

REFERENCES

Abhishek, N., Jenamani, M., & Mahanty, B. (2017). Urban growth in Indian cities: Are the driving forces really changing? *Habitat International*, *69*, 48–57. doi:10.1016/j.habitatint.2017.08.002

Allen, L., Lindberg, F., & Grimmond, C. S. B. (2011). Global to city scale urban anthropogenic heat flux: Model and variability. *International Journal of Climatology*, *31*(13), 1990–2005. doi:10.1002/joc.2210

Amirtham, L. R. (2016). Urbanization and its impact on urban heat island intensity in Chennai Metropolitan Area, India. *Indian Journal of Science and Technology*, *9*(5), 1–8. doi:10.17485/ijst/2016/v9i5/87201

Arnfield, A. J. (2003). Two decades of urban climate research: A review of turbulence, exchanges of energy and water, and the urban heat island. *International Journal of Climatology*, *23*(1), 1–26. doi:10.1002/joc.859

Ashley, W. S., Bentley, M. L., & Stallins, J. A. (2012). Urban-induced thunderstorm modification in the Southeast United States. *Climatic Change*, *113*(2), 481–498. doi:10.100710584-011-0324-1

Atkinson, B. W. (1968). A preliminary examination of the possible effect of London's urban area on the distribution of thunder rainfall 1951-60. *Transactions of the Institute of British Geographers*, (44), 97–118. doi:10.2307/621751

Baik, J. J., Kim, Y. H., & Chun, H. Y. (2001). Dry and moist convection forced by an urban heat island. *Journal of Applied Meteorology*, *40*(8), 1462–1475. doi:10.1175/1520-0450(2001)040<1462:DAMCFB>2.0.CO;2

Balling, R. C. Jr, & Brazel, S. W. (1987). Recent changes in Phoenix, Arizona summertime diurnal precipitation patterns. *Theoretical and Applied Climatology*, *38*(1), 50–54. doi:10.1007/BF00866253

Barros, A. P., & Hwu, W. (2002). A study of land-atmosphere interactions during summertime rainfall using a mesoscale model. *Journal of Geophysical Research, D, Atmospheres*, *107*(D14), 4227. doi:10.1029/2000JD000254

Bhati, S., & Mohan, M. (2016). WRF model evaluation for the urban heat island assessment under varying land use/land cover and reference site conditions. *Theoretical and Applied Climatology*, *126*(1-2), 385–400. doi:10.100700704-015-1589-5

Bian, T., & Ren, G. (2019). Reassessment of urbanization effect on surface air temperature trends at an urban station of North China. *Meteorology and Atmospheric Physics*, *131*(2), 237–250. doi:10.100700703-017-0568-z

Borbora, J., & Das, A. K. (2014). Summertime urban heat island study for Guwahati city, India. *Sustainable Cities and Society*, *11*, 61–66. doi:10.1016/j.scs.2013.12.001

Bornstein, R., & Lin, Q. (2000). Urban heat islands and summertime convective thunderstorms in Atlanta: Three case studies. *Atmospheric Environment*, *34*(3), 507–516. doi:10.1016/S1352-2310(99)00374-X

Bornstein, R. D., & Johnson, D. S. (1977). Urban-rural wind velocity differences. *Atmospheric Environment, 11*(7), 597-604.

Brubaker, K. L., & Entekhabi, D. (1996). Analysis of feedback mechanisms in land-atmosphere interaction. *Water Resources Research*, *32*(5), 1343–1357. doi:10.1029/96WR00005

Burian, S. J., & Shepherd, J. M. (2005). Effect of urbanization on the diurnal rainfall pattern in Houston. *Hydrological Processes: An International Journal, 19*(5), 1089–1103. doi:10.1002/hyp.5647

Chang, C. R., Li, M. H., & Chang, S. D. (2007). A preliminary study on the local cool-island intensity of Taipei city parks. *Landscape and Urban Planning, 80*(4), 386–395. doi:10.1016/j.landurbplan.2006.09.005

Chang, H. I., Kumar, A., Niyogi, D., Mohanty, U. C., Chen, F., & Dudhia, J. (2009). The role of land surface processes on the mesoscale simulation of the July 26, 2005 heavy rain event over Mumbai, India. *Global and Planetary Change, 67*(1-2), 87–103. doi:10.1016/j.gloplacha.2008.12.005

Changnon, S. A., Jr. (1981). METROMEX: A review and summary. Meteorological Monograph No. 40. American Meteorological Society.

Changnon, S. A. (2003). Urban modification of freezing-rain events. *Journal of Applied Meteorology, 42*(6), 863–870. doi:10.1175/1520-0450(2003)042<0863:UMOFE>2.0.CO;2

Changnon, S. A., & Westcott, N. E. (2002). Heavy rainstorms in chicago: Increasing frequency, altered impacts, and future implications 1. *Journal of the American Water Resources Association, 38*(5), 1467–1475. doi:10.1111/j.1752-1688.2002.tb04359.x

Chapman, S., Watson, J. E., Salazar, A., Thatcher, M., & McAlpine, C. A. (2017). The impact of urbanization and climate change on urban temperatures: A systematic review. *Landscape Ecology, 32*(10), 1921–1935. doi:10.100710980-017-0561-4

Chauvin, J. P., Glaeser, E., Ma, Y., & Tobio, K. (2017). What is different about urbanization in rich and poor countries? Cities in Brazil, China, India and the United States. *Journal of Urban Economics, 98*, 17–49. doi:10.1016/j.jue.2016.05.003

Ching, J., Mills, G., Bechtel, B., See, L., Feddema, J., Wang, X., ... Mouzourides, P. (2018). WUDAPT: An urban weather, climate, and environmental modeling infrastructure for the anthropocene. *Bulletin of the American Meteorological Society, 99*(9), 1907–1924. doi:10.1175/BAMS-D-16-0236.1

Christen, A., & Vogt, R. (2004). Energy and radiation balance of a central European city. *International Journal of Climatology: A Journal of the Royal Meteorological Society, 24*(11), 1395-1421.

Chun, B., & Guldmann, J. M. (2014). Spatial statistical analysis and simulation of the urban heat island in high-density central cities. *Landscape and Urban Planning, 125*, 76–88. doi:10.1016/j.landurbplan.2014.01.016

DESA. (2018). *Revision of world urbanization prospects*. Population Division of the UN Department of Economic and Social Affairs. https://population. un.org/wup

Dickinson, R. E. (1995). Land-atmosphere interaction. *Reviews of Geophysics*, *33*(S2), 917–922. doi:10.1029/95RG00284

Diem, J. E., & Brown, D. P. (2003). Anthropogenic impacts on summer precipitation in central Arizona, USA. *The Professional Geographer*, *55*(3), 343–355.

Dimri, A. P. (2019). Comparison of regional and seasonal changes and trends in daily surface temperature extremes over India and its subregions. *Theoretical and Applied Climatology*, *136*(1-2), 265–286. doi:10.100700704-018-2486-5

Dixon, P. G., & Mote, T. L. (2003). Patterns and causes of Atlanta's urban heat island–initiated precipitation. *Journal of Applied Meteorology*, *42*(9), 1273–1284. doi:10.1175/1520-0450(2003)042<1273:PACOAU>2.0.CO;2

Doan, Q. V., Kusaka, H., & Ho, Q. B. (2016). Impact of future urbanization on temperature and thermal comfort index in a developing tropical city: Ho Chi Minh City. *Urban Climate*, *17*, 20–31. doi:10.1016/j.uclim.2016.04.003

Droste, A. M., Steeneveld, G. J., & Holtslag, A. A. (2018). Introducing the urban wind island effect. *Environmental Research Letters*, *13*(9), 094007. doi:10.1088/1748-9326/aad8ef

Fujibe, F., Togawa, H., & Sakata, M. (2009). Long-term change and spatial anomaly of warm season afternoon precipitation in Tokyo. *Scientific Online Letters on the Atmosphere*, *5*, 17–20. doi:10.2151ola.2009-005

Ganeshan, M., Murtugudde, R., & Imhoff, M. L. (2013). A multi-city analysis of the UHI-influence on warm season rainfall. *Urban Climate*, *6*, 1–23. doi:10.1016/j. uclim.2013.09.004

Gedzelman, S. D., Austin, S., Cermak, R., Stefano, N., Partridge, S., Quesenberry, S., & Robinson, D. A. (2003). Mesoscale aspects of the urban heat island around New York City. *Theoretical and Applied Climatology*, *75*(1-2), 29–42. doi:10.100700704-002-0724-2

Goddard, I. L., & Tett, S. F. (2019). How much has urbanisation affected United Kingdom temperatures? *Atmospheric Science Letters*, *20*(5), 896. doi:10.1002/asl.896

Gogoi, P. P., Vinoj, V., Swain, D., Roberts, G., Dash, J., & Tripathy, S. (2019). Land use and land cover change effect on surface temperature over Eastern India. *Scientific Reports*, *9*(1), 8859. doi:10.103841598-019-45213-z PMID:31222135

Göndöcs, J., Breuer, H., Pongrácz, R., & Bartholy, J. (2017). Urban heat island mesoscale modelling study for the Budapest agglomeration area using the WRF model. *Urban Climate, 21*, 66–86. doi:10.1016/j.uclim.2017.05.005

Goswami, P., Shivappa, H., & Goud, B. S. (2010). Impact of urbanization on tropical mesoscale events: Investigation of three heavy rainfall events. *Meteorologische Zeitschrift (Berlin), 19*(4), 385–397. doi:10.1127/0941-2948/2010/0468

Grimmond, C. S. B., Roth, M., Oke, T. R., Au, Y. C., Best, M., Betts, R., ... Freitas, E. (2010). Climate and more sustainable cities: Climate information for improved planning and management of cities (producers/capabilities perspective). *Procedia Environmental Sciences, 1*, 247–274. doi:10.1016/j.proenv.2010.09.016

Guo, W., Wang, X., Sun, J., Ding, A., & Zou, J. (2016). Comparison of land–atmosphere interaction at different surface types in the mid-to lower reaches of the Yangtze River valley. *Atmospheric Chemistry and Physics, 16*(15), 9875–9890. doi:10.5194/acp-16-9875-2016

Haberlie, A. M., Ashley, W. S., & Pingel, T. J. (2015). The effect of urbanisation on the climatology of thunderstorm initiation. *Quarterly Journal of the Royal Meteorological Society, 141*(688), 663–675. doi:10.1002/qj.2499

Hale, R. C., Gallo, K. P., & Loveland, T. R. (2008). Influences of specific land use/land cover conversions on climatological normals of near-surface temperature. *Journal of Geophysical Research, D, Atmospheres, 113*(D14), D14113. doi:10.1029/2007JD009548

Han, S. S. (2010). Urban expansion in contemporary China: What can we learn from a small town? *Land Use Policy, 27*(3), 780–787. doi:10.1016/j.landusepol.2009.10.010

Harari, M. (2016). *Cities in bad shape: Urban geometry in India. Processed.* Wharton School of the University of Pennsylvania.

Hua, L. J., Ma, Z. G., & Guo, W. D. (2008). The impact of urbanization on air temperature across China. *Theoretical and Applied Climatology, 93*(3-4), 179–194. doi:10.100700704-007-0339-8

Hurlimann, A. C., & March, A. P. (2012). The role of spatial planning in adapting to climate change. *Wiley Interdisciplinary Reviews: Climate Change, 3*(5), 477–488. doi:10.1002/wcc.183

Jahangir, M. S., & Moghim, S. (2019). Assessment of the urban heat island in the city of Tehran using reliability methods. *Atmospheric Research, 225*, 144–156. doi:10.1016/j.atmosres.2019.03.038

Jáuregui, E., & Romales, E. (1996). Urban effects on convective precipitation in Mexico City. *Atmospheric Environment, 30*(20), 3383–3389. doi:10.1016/1352-2310(96)00041-6

Joerin, J., Shaw, R., Takeuchi, Y., & Krishnamurthy, R. (2012). Assessing community resilience to climate-related disasters in Chennai, India. *International Journal of Disaster Risk Reduction, 1*, 44–54. doi:10.1016/j.ijdrr.2012.05.006

Joshi, R., Raval, H., Pathak, M., Prajapati, S., Patel, A., Singh, V., & Kalubarme, M. H. (2015). Urban heat island characterization and isotherm mapping using geo-informatics technology in Ahmedabad city, Gujarat state, India. *International Journal of Geosciences, 6*(03), 274–285. doi:10.4236/ijg.2015.63021

Kaufmann, R. K., Seto, K. C., Schneider, A., Liu, Z., Zhou, L., & Wang, W. (2007). Climate response to rapid urban growth: Evidence of a human-induced precipitation deficit. *Journal of Climate, 20*(10), 2299–2306. doi:10.1175/JCLI4109.1

Kikon, N., Singh, P., Singh, S. K., & Vyas, A. (2016). Assessment of urban heat islands (UHI) of Noida City, India using multi-temporal satellite data. *Sustainable Cities and Society, 22*, 19–28. doi:10.1016/j.scs.2016.01.005

Kishtawal, C. M., Niyogi, D., Tewari, M., Pielke, R. A. Sr, & Shepherd, J. M. (2010). Urbanization signature in the observed heavy rainfall climatology over India. *International Journal of Climatology, 30*(13), 1908–1916. doi:10.1002/joc.2044

Kotharkar, R., & Bagade, A. (2018). Evaluating urban heat island in the critical local climate zones of an Indian city. *Landscape and Urban Planning, 169*, 92–104. doi:10.1016/j.landurbplan.2017.08.009

Kotharkar, R., & Surawar, M. (2016). Land use, land cover, and population density impact on the formation of canopy urban heat islands through traverse survey in the Nagpur urban area, India. *Journal of Urban Planning and Development, 142*(1), 04015003. doi:10.1061/(ASCE)UP.1943-5444.0000277

Kumar, A., Dudhia, J., Rotunno, R., Niyogi, D., & Mohanty, U. C. (2008). Analysis of the 26 July 2005 heavy rain event over Mumbai, India using the Weather Research and Forecasting (WRF) model. *Quarterly Journal of the Royal Meteorological Society, 134*(636), 1897–1910. doi:10.1002/qj.325

Kumar, P., Geneletti, D., & Nagendra, H. (2016). Spatial assessment of climate change vulnerability at city scale: A study in Bangalore, India. *Land Use Policy, 58*, 514–532. doi:10.1016/j.landusepol.2016.08.018

Kumar, R., & Mishra, V. (2019). Decline in surface urban heat island intensity in India during heatwaves. *Environmental Research Communications*, *1*(3), 031001. doi:10.1088/2515-7620/ab121d

Kusaka, H., Kimura, F., Hirakuchi, H., & Mizutori, M. (2000). The effects of land-use alteration on the sea breeze and daytime heat island in the Tokyo metropolitan area. *Kisho Shushi. Dai2shu*, *78*(4), 405–420. doi:10.2151/jmsj1965.78.4_405

Lambin, E. F., Turner, B. L., Geist, H. J., Agbola, S. B., Angelsen, A., Bruce, J. W., ... George, P. (2001). The causes of land-use and land-cover change: Moving beyond the myths. *Global Environmental Change*, *11*(4), 261–269. doi:10.1016/S0959-3780(01)00007-3

Landsberg, H. E. (1970). Man-Made Climatic Changes: Man's activities have altered the climate of urbanized areas and may affect global climate in the future. *Science*, *170*(3964), 1265–1274. doi:10.1126cience.170.3964.1265 PMID:17829423

Lee, D. O. (1979). The influence of atmospheric stability and the urban heat island on urban-rural wind speed differences. *Atmospheric Environment, 13*(8), 1175-1180.

Lei, M., Niyogi, D., Kishtawal, C., Pielke, R. A. Sr, Beltrán-Przekurat, A., Nobis, T. E., & Vaidya, S. S. (2008). Effect of explicit urban land surface representation on the simulation of the 26 July 2005 heavy rain event over Mumbai, India. *Atmospheric Chemistry and Physics*, *8*(20), 5975–5995. doi:10.5194/acp-8-5975-2008

Li, Q., Li, W., Si, P., Xiaorong, G., Dong, W., Jones, P., Huang, J., & Cao, L. (2010). Assessment of surface air warming in northeast China, with emphasis on the impacts of urbanization. *Theoretical and Applied Climatology*, *99*(3-4), 469–478. doi:10.100700704-009-0155-4

Li, X., Mitra, C., Dong, L., & Yang, Q. (2018b). Understanding land use change impacts on microclimate using Weather Research and Forecasting (WRF) model. *Physics and Chemistry of the Earth Parts A/B/C, 103*, 115–126. doi:10.1016/j.pce.2017.01.017

Li, X. X., Koh, T. Y., Entekhabi, D., Roth, M., Panda, J., & Norford, L. K. (2013). A multi-resolution ensemble study of a tropical urban environment and its interactions with the background regional atmosphere. *Journal of Geophysical Research, D, Atmospheres*, *118*(17), 9804–9818. doi:10.1002/jgrd.50795

Li, X. X., Koh, T. Y., Panda, J., & Norford, L. K. (2016). Impact of urbanization patterns on the local climate of a tropical city, Singapore: An ensemble study. *Journal of Geophysical Research, D, Atmospheres*, *121*(9), 4386–4403. doi:10.1002/2015JD024452

Li, Y., Zhao, C., Zhang, T., Wang, W., Duan, H., Liu, Y., Ren, Y., & Pu, Z. (2018a). Impacts of land-use data on the simulation of surface air temperature in Northwest China. *Journal of Meteorological Research*, *32*(6), 896–908. doi:10.100713351-018-7151-5

Lin, C. Y., Chen, F., Huang, J. C., Chen, W. C., Liou, Y. A., Chen, W. N., & Liu, S. C. (2008b). Urban heat island effect and its impact on boundary layer development and land–sea circulation over northern Taiwan. *Atmospheric Environment*, *42*(22), 5635–5649. doi:10.1016/j.atmosenv.2008.03.015

Lin, C. Y., Chen, W. C., Liu, S. C., Liou, Y. A., Liu, G. R., & Lin, T. H. (2008a). Numerical study of the impact of urbanization on the precipitation over Taiwan. *Atmospheric Environment*, *42*(13), 2934–2947. doi:10.1016/j.atmosenv.2007.12.054

Lindberg, F., Grimmond, C. S. B., Yogeswaran, N., Kotthaus, S., & Allen, L. (2013). Impact of city changes and weather on anthropogenic heat flux in Europe 1995–2015. *Urban Climate*, *4*, 1–15. doi:10.1016/j.uclim.2013.03.002

Liu, Y., Yue, T., Jiao, Y., Zhao, N., & Zhao, M. (2019). Temperature changes in the Heihe River Basin based on high accuracy surface modelling. *Meteorological Applications*, *26*(4), 720–732. doi:10.1002/met.1803

Loose, T., & Bornstein, R. D. (1977). Observations of mesoscale effects on frontal movement through an urban area. *Monthly Weather Review*, *105*(5), 563–571. doi:10.1175/1520-0493(1977)105<0563:OOMEOF>2.0.CO;2

Lowry, W. P. (1998). Urban effects on precipitation amount. *Progress in Physical Geography*, *22*(4), 477–520. doi:10.1177/030913339802200403

Mallick, J., Rahman, A., & Singh, C. K. (2013). Modeling urban heat islands in heterogeneous land surface and its correlation with impervious surface area by using night-time ASTER satellite data in highly urbanizing city, Delhi-India. *Advances in Space Research*, *52*(4), 639–655. doi:10.1016/j.asr.2013.04.025

Mathew, A., Khandelwal, S., & Kaul, N. (2016). Spatial and temporal variations of urban heat island effect and the effect of percentage impervious surface area and elevation on land surface temperature: Study of Chandigarh city, India. *Sustainable Cities and Society*, *26*, 264–277. doi:10.1016/j.scs.2016.06.018

Mathew, A., Khandelwal, S., & Kaul, N. (2018). Analysis of diurnal surface temperature variations for the assessment of surface urban heat island effect over Indian cities. *Energy and Building*, *159*, 271–295. doi:10.1016/j.enbuild.2017.10.062

Miao, S., Chen, F., LeMone, M. A., Tewari, M., Li, Q., & Wang, Y. (2009). An observational and modeling study of characteristics of urban heat island and boundary layer structures in Beijing. *Journal of Applied Meteorology and Climatology*, *48*(3), 484–501. doi:10.1175/2008JAMC1909.1

Miao, S., Chen, F., Li, Q., & Fan, S. (2011). Impacts of urban processes and urbanization on summer precipitation: A case study of heavy rainfall in Beijing on 1 August 2006. *Journal of Applied Meteorology and Climatology*, *50*(4), 806–825. doi:10.1175/2010JAMC2513.1

Mitra, C., Shepherd, J. M., & Jordan, T. (2012). On the relationship between the premonsoonal rainfall climatology and urban land cover dynamics in Kolkata city, India. *International Journal of Climatology*, *32*(9), 1443–1454. doi:10.1002/joc.2366

Mohan, M., & Kandya, A. (2015). Impact of urbanization and land-use/land-cover change on diurnal temperature range: A case study of tropical urban airshed of India using remote sensing data. *The Science of the Total Environment*, *506*, 453–465. doi:10.1016/j.scitotenv.2014.11.006 PMID:25437763

Mohan, M., Kandya, A., & Battiprolu, A. (2011). Urban heat island effect over national capital region of India: A study using the temperature trends. *Journal of Environmental Protection*, *2*(04), 465–472. doi:10.4236/jep.2011.24054

Mohan, M., Kikegawa, Y., Gurjar, B. R., Bhati, S., Kandya, A., & Ogawa, K. (2012). Urban heat island assessment for a tropical urban airshed in India. *Atmospheric and Climate Science*, *2*(2), 127–138. doi:10.4236/acs.2012.22014

Mohan, M., Kikegawa, Y., Gurjar, B. R., Bhati, S., & Kolli, N. R. (2013). Assessment of urban heat island effect for different land use–land cover from micrometeorological measurements and remote sensing data for megacity Delhi. *Theoretical and Applied Climatology*, *112*(3-4), 647–658. doi:10.100700704-012-0758-z

Mölders, N., & Olson, M. A. (2004). Impact of urban effects on precipitation in high latitudes. *Journal of Hydrometeorology*, *5*(3), 409–429. doi:10.1175/1525-7541(2004)005<0409:IOUEOP>2.0.CO;2

Niyogi, D., Pyle, P., Lei, M., Arya, S. P., Kishtawal, C. M., Shepherd, M., Chen, F., & Wolfe, B. (2011). Urban modification of thunderstorms: An observational storm climatology and model case study for the Indianapolis urban region. *Journal of Applied Meteorology and Climatology*, *50*(5), 1129–1144. doi:10.1175/2010JAMC1836.1

Ohashi, Y., & Kida, H. (2002). Local circulations developed in the vicinity of both coastal and inland urban areas: A numerical study with a mesoscale atmospheric model. *Journal of Applied Meteorology*, *41*(1), 30–45. doi:10.1175/1520-0450(2002)041<0030:LCDITV>2.0.CO;2

Oke, T. R., Johnson, G. T., Steyn, D. G., & Watson, I. D. (1991). Simulation of surface urban heat islands under 'ideal'conditions at night Part 2: Diagnosis of causation. *Boundary-Layer Meteorology*, *56*(4), 339–358. doi:10.1007/BF00119211

Ortiz, L. E., González, J. E., Horton, R., Lin, W., Wu, W., Ramamurthy, P., Arend, M., & Bornstein, R. D. (2019). High-resolution projections of extreme heat in New York City. *International Journal of Climatology*, *39*(12), 4721–4735. doi:10.1002/joc.6102

Pandey, P., Kumar, D., Prakash, A., Masih, J., Singh, M., Kumar, S., Jain, V. K., & Kumar, K. (2012). A study of urban heat island and its association with particulate matter during winter months over Delhi. *The Science of the Total Environment*, *414*, 494–507. doi:10.1016/j.scitotenv.2011.10.043 PMID:22154211

Pannell, C. (1995). China's urban transition. *The Journal of Geography*, *94*(3), 394–403. doi:10.1080/00221349508979342 PMID:12178548

Pielke, R. A., & Adegoke, J., BeltraáN-Przekurat, A., Hiemstra, C. A., Lin, J., Nair, U. S., ... Nobis, T. E. (2007). An overview of regional land-use and land-cover impacts on rainfall. Tellus B. *Chemical and Physical Meteorology*, *59*(3), 587–601.

Pielke, R. A. Sr, Pitman, A., Niyogi, D., Mahmood, R., McAlpine, C., Hossain, F., ... Reichstein, M. (2011). Land use/land cover changes and climate: Modeling analysis and observational evidence. *Wiley Interdisciplinary Reviews: Climate Change*, *2*(6), 828–850. doi:10.1002/wcc.144

Prashar, S., Shaw, R., & Takeuchi, Y. (2012). Assessing the resilience of Delhi to climate-related disasters: A comprehensive approach. *Natural Hazards*, *64*(2), 1609–1624. doi:10.100711069-012-0320-4

Rajasekar, U., Chakraborty, S., & Bhat, G. (2018). Climate resilient smart cities: opportunities for innovative solutions in India. In *Climate Change in Cities* (pp. 203–227). Springer. doi:10.1007/978-3-319-65003-6_11

Ramachandra, T. V., & Kumar, U. (2010). Greater Bangalore: Emerging urban heat island. *GIS Development, 14*(1), 86-104.

Rath, S. S., & Panda, J. (2019). A Study of Near-Surface Boundary Layer Characteristics During the 2015 Chennai Flood in the Context of Urban-Induced Land Use Changes. *Pure and Applied Geophysics*, *176*(6), 2607–2629. doi:10.100700024-018-2069-5

Rozoff, C. M., Cotton, W. R., & Adegoke, J. O. (2003). Simulation of St. Louis, Missouri, land use impacts on thunderstorms. *Journal of Applied Meteorology*, *42*(6), 716–738. doi:10.1175/1520-0450(2003)042<0716:SOSLML>2.0.CO;2

Salamanca, F., Martilli, A., & Yagüe, C. (2012). A numerical study of the Urban Heat Island over Madrid during the DESIREX (2008) campaign with WRF and an evaluation of simple mitigation strategies. *International Journal of Climatology*, *32*(15), 2372–2386. doi:10.1002/joc.3398

Schlünzen, K. H., Hoffmann, P., Rosenhagen, G., & Riecke, W. (2010). Long-term changes and regional differences in temperature and precipitation in the metropolitan area of Hamburg. *International Journal of Climatology*, *30*(8), 1121–1136. doi:10.1002/joc.1968

Sharifi, E., & Lehmann, S. (2015). Correlation analysis of surface temperature of rooftops, streetscapes and urban heat island effect: Case study of central Sydney. *Journal of Urban and Environmental Engineering*, *9*(1), 3–11. doi:10.4090/juee.2015.v9n1.3-11

Sharma, D., & Tomar, S. (2010). Mainstreaming climate change adaptation in Indian cities. *Environment and Urbanization*, *22*(2), 451–465. doi:10.1177/0956247810377390

Shastri, H., Barik, B., Ghosh, S., Venkataraman, C., & Sadavarte, P. (2017). Flip flop of day-night and summer-winter surface urban heat island intensity in India. *Scientific Reports*, *7*(1), 40178. doi:10.1038rep40178 PMID:28067276

Shem, W., & Shepherd, M. (2009). On the impact of urbanization on summertime thunderstorms in Atlanta: Two numerical model case studies. *Atmospheric Research*, *92*(2), 172–189. doi:10.1016/j.atmosres.2008.09.013

Shepherd, J. M. (2005). A review of current investigations of urban-induced rainfall and recommendations for the future. *Earth Interactions*, *9*(12), 1–27. doi:10.1175/EI156.1

Shepherd, J. M., & Burian, S. J. (2003). Detection of urban-induced rainfall anomalies in a major coastal city. *Earth Interactions*, *7*(4), 1–17. doi:10.1175/1087-3562(2003)007<0001:DOUIRA>2.0.CO;2

Shepherd, J. M., Carter, M., Manyin, M., Messen, D., & Burian, S. (2010). The impact of urbanization on current and future coastal precipitation: A case study for Houston. *Environment and Planning. B, Planning & Design*, *37*(2), 284–304. doi:10.1068/b34102t

Shepherd, J. M., Pierce, H., & Negri, A. J. (2002). Rainfall modification by major urban areas: Observations from spaceborne rain radar on the TRMM satellite. *Journal of Applied Meteorology*, *41*(7), 689–701. doi:10.1175/1520-0450(2002)041<0689:RMBMUA>2.0.CO;2

Shreffler, J. H. (1979). Urban-rural differences in tower-measured winds, St. Louis. *Journal of Applied Meteorology*, *18*(7), 829–835. doi:10.1175/1520-0450(1979)018<0829:URDITM>2.0.CO;2

Singh, P., Kikon, N., & Verma, P. (2017). Impact of land use change and urbanization on urban heat island in Lucknow city, Central India. A remote sensing based estimate. *Sustainable Cities and Society*, *32*, 100–114. doi:10.1016/j.scs.2017.02.018

Sood, A. (2015). Industrial townships and the policy facilitation of corporate urbanisation in India. *Urban Studies (Edinburgh, Scotland)*, *52*(8), 1359–1378. doi:10.1177/0042098014562318

Stone, B., Hess, J. J., & Frumkin, H. (2010). Urban form and extreme heat events: Are sprawling cities more vulnerable to climate change than compact cities? *Environmental Health Perspectives*, *118*(10), 1425–1428. doi:10.1289/ehp.0901879 PMID:21114000

Surjan, A., & Shaw, R. (2009). Enhancing disaster resilience through local environment management: Case of Mumbai, India. *Disaster Prevention and Management: An International Journal*, *18*(4), 418–433. doi:10.1108/09653560910984474

Tan, J., Zheng, Y., Tang, X., Guo, C., Li, L., Song, G., Zhen, X., Yuan, D., Kalkstein, A. J., Li, F., & Chen, H. (2010). The urban heat island and its impact on heat waves and human health in Shanghai. *International Journal of Biometeorology*, *54*(1), 75–84. doi:10.100700484-009-0256-x PMID:19727842

Tayanç, M., Karaca, M., & Yenigün, O. (1997). Annual and seasonal air temperature trend patterns of climate change and urbanization effects in relation to air pollutants in Turkey. *Journal of Geophysical Research, D, Atmospheres*, *102*(D2), 1909–1919. doi:10.1029/96JD02108

Thielen, J., Wobrock, W., Gadian, A., Mestayer, P. G., & Creutin, J. D. (2000). The possible influence of urban surfaces on rainfall development: A sensitivity study in 2D in the meso-γ-scale. *Atmospheric Research*, *54*(1), 15–39. doi:10.1016/S0169-8095(00)00041-7

Thomas, G., Sherin, A. P., Ansar, S., & Zachariah, E. J. (2014). Analysis of urban heat island in Kochi, India, using a modified local climate zone classification. *Procedia Environmental Sciences*, *21*, 3–13. doi:10.1016/j.proenv.2014.09.002

Viegas, C. V., Saldanha, D. L., Bond, A., Ribeiro, J. L. D., & Selig, P. M. (2013). Urban land planning: The role of a Master Plan in influencing local temperatures. *Cities (London, England)*, *35*, 1–13. doi:10.1016/j.cities.2013.05.006

Vitanova, L. L., Kusaka, H., Doan, V. Q., & Nishi, A. (2019). Numerical Study of the Urban Heat Island in Sendai City with Potential Natural Vegetation and the 1850s and 2000s Land-Use Data. *Journal of the Meteorological Society of Japan. Ser. II.*

Wan, H., Zhong, Z., Yang, X., & Li, X. (2013). Impact of city belt in Yangtze River Delta in China on a precipitation process in summer: A case study. *Atmospheric Research*, *125*, 63–75. doi:10.1016/j.atmosres.2013.02.004

Wan, H., Zhong, Z., Yang, X., & Li, X. (2015). Ensembles to model the impact of urbanization for a summertime rainstorm process in Yangtze River Delta, China. *Meteorological Applications*, *22*(1), 105–112. doi:10.1002/met.1360

Wang, J., Feng, J., Yan, Z., Hu, Y., & Jia, G. (2012). Nested high-resolution modeling of the impact of urbanization on regional climate in three vast urban agglomerations in China. *Journal of Geophysical Research, D, Atmospheres*, *117*(D21), n/a. doi:10.1029/2012JD018226

Wang, J., & Yan, Z. W. (2016). Urbanization-related warming in local temperature records: A review. *Atmospheric and Oceanic Science Letters*, *9*(2), 129–138. doi:10.1080/16742834.2016.1141658

Wang, L., & Li, D. (2019). Modulation of the urban boundary-layer heat budget by a heatwave. *Quarterly Journal of the Royal Meteorological Society*, *145*(722), 1814–1831. doi:10.1002/qj.3526

Wilby, R. L. (2008). Constructing climate change scenarios of urban heat island intensity and air quality. *Environment and Planning. B, Planning & Design*, *35*(5), 902–919. doi:10.1068/b33066t

Yang, L., Smith, J. A., Baeck, M. L., Bou-Zeid, E., Jessup, S. M., Tian, F., & Hu, H. (2014). Impact of urbanization on heavy convective precipitation under strong large-scale forcing: A case study over the Milwaukee–Lake Michigan region. *Journal of Hydrometeorology*, *15*(1), 261–278. doi:10.1175/JHM-D-13-020.1

Yang, X., Hou, Y., & Chen, B. (2011). Observed surface warming induced by urbanization in east China. *Journal of Geophysical Research, D, Atmospheres*, *116*(D14), D14113. doi:10.1029/2010JD015452

Yow, D. M. (2007). Urban heat islands: Observations, impacts, and adaptation. *Geography Compass*, *1*(6), 1227–1251. doi:10.1111/j.1749-8198.2007.00063.x

Zhang, K., Wang, R., Shen, C., & Da, L. (2010). Temporal and spatial characteristics of the urban heat island during rapid urbanization in Shanghai, China. *Environmental Monitoring and Assessment*, *169*(1-4), 101–112. doi:10.100710661-009-1154-8 PMID:19795217

Zhang, Y., Smith, J. A., Luo, L., Wang, Z., & Baeck, M. L. (2014). Urbanization and rainfall variability in the Beijing metropolitan region. *Journal of Hydrometeorology*, *15*(6), 2219–2235. doi:10.1175/JHM-D-13-0180.1

Zheng, Y., Chen, T., Cai, J., & Liu, S. (2009). Regional concentration and region-based urban transition: China's mega-urban region formation in the 1990s. *Urban Geography*, *30*(3), 312–333. doi:10.2747/0272-3638.30.3.312

Zhong, S., Qian, Y., Zhao, C., Leung, R., Wang, H., Yang, B., Fan, J., Yan, H., Yang, X.-Q., & Liu, D. (2017). Urbanization-induced urban heat island and aerosol effects on climate extremes in the Yangtze River Delta region of China. *Atmospheric Chemistry and Physics*, *17*(8), 5439–5457. doi:10.5194/acp-17-5439-2017

Zhong, S., Qian, Y., Zhao, C., Leung, R., & Yang, X. Q. (2015). A case study of urbanization impact on summer precipitation in the Greater Beijing Metropolitan Area: Urban heat island versus aerosol effects. *Journal of Geophysical Research, D, Atmospheres*, *120*(20), 10,903–10,914. doi:10.1002/2015JD023753

Zhong, S., & Yang, X. Q. (2015a). Ensemble simulations of the urban effect on a summer rainfall event in the Great Beijing Metropolitan Area. *Atmospheric Research*, *153*, 318–334. doi:10.1016/j.atmosres.2014.09.005

Zhong, S., & Yang, X. Q. (2015b). Mechanism of urbanization impact on a summer cold-frontal rainfall process in the greater Beijing metropolitan area. *Journal of Applied Meteorology and Climatology*, *54*(6), 1234–1247. doi:10.1175/JAMC-D-14-0264.1

KEY TERMS AND DEFINITIONS

Boundary-Layer: Also known as the 'Atmospheric Boundary Layer', is the lowest part of the atmosphere near the ground that is directly influenced by heat, moisture, or momentum transfer to or from the surface.

Hydrological Cycle: 'Hydrological cycle' or 'water cycle' is the continuous movement of water within the earth-atmosphere system. The movement involves various processes, including evaporation, transpiration, condensation, precipitation, and runoff. The total amount of water within the cycle remains constant, while phases and distribution of water change during these processes.

LCZ: LCZ or 'Local Climate Zones' is a systematic classification scheme for urban land use and land cover, which is used for zoning and categorization of the internal structure of urban areas. The main applications include studies related to urban climatology.

Micrometeorology: 'Micrometeorology' is the study of small-scale atmospheric phenomena ranging up to several kilometers in diameter and confined to the lower troposphere.

Thunderstorms: 'Thunderstorms' are generally accompanied by lightning and thunder with rain and gusty winds, sometimes with hail or snow, produced by cumulonimbus clouds.

Tier-I/II: Indian cities are classified into different tiers based on their population according to the 2011 census. Tier-I cities generally consist of metropolitan areas like New Delhi, Mumbai, Bengaluru, Chennai, Kolkata, and Hyderabad with developed infrastructure and high cost of living. Tier-II includes relatively smaller cities and state capitals, which are still more developed than tier-III (all other cities). Tier-I, II, and III are also known as X, Y, and Z categories of cities respectively.

UHI/UWI: UHI or 'Urban Heat Island' refers to a part or whole of the urban area, which is significantly warmer than the surrounding rural counterpart due to human activities. Similarly, UWI or 'Urban Wind Island' refers to the scenario when the wind speed is noticeably higher over urban areas.

Upwind/Downwind: In meteorology, a wind direction is a direction, the wind is coming from. 'Upwind' is in the direction towards the source of the wind, and 'Downwind' is in the direction away from the source of the wind.

Urbanization: 'Urbanization' is the act or fact of taking on the characteristics of a city.

Chapter 8
Seasonal Contrast of Land Surface Temperature in Faridabad:
An Urbanized District of Haryana, India

Swagata Ghosh
Amity Institute of Geoinformatics and Remote Sensing (AIGIRS), Amity University, Noida, India

Sunil Kumar
Amity Institute of Geoinformatics and Remote Sensing (AIGIRS), Amity University, Noida, India

Krishna Vidhata N.
Amity Institute of Geoinformatics and Remote Sensing (AIGIRS), Amity University, Noida, India

Kousik Midya
Department of Electronics and Communication Engineering, Ajay Kumar Garg Engineering College, Ghaziabad, India

ABSTRACT

The chapter has highlighted the adverse impact of conversion of natural land cover into urban concrete over inter-seasonal variation of land surface temperature (LST) in Faridabad district which is a major threat for sustainable urbanism. Apart from high LST in the dense urban area built-up in Faridabad city, inter-seasonal variation of LST has been observed in dry deciduous forested areas due to defoliation, fallow land, and over-grazed land in rural surrounding areas. Compared to NDVI, NDBI has significant positive and stable correlation with LST in all seasons (Pearson index: 0.35 to 0.60). Weaker correlation (Pearson index: 0.02 to 0.48) between NDVI and LST accounts for the seasonal impact over NDVI due to defoliation and agricultural practices over the study area. Overall, it can be remarked that image-based spectral indices and thermal band can be used for the evaluating thermal environmental contrast across seasons. Use of in situ measurement with good network of meteorological stations can validate satellite-derived LST better and increase the accuracy of the study.

DOI: 10.4018/978-1-7998-2249-3.ch008

INTRODUCTION

Urbanization, a dynamic process involving remarkable changes in the vast area of land cover associated with human population concentration (Ramachandra et al., 2012), has become a global issue (Newbold and Scott, 2013). Such ongoing urbanization accompanied by population growth has been predicted to contribute 2.5 billion people of the world urban population by 2050 which is about 2/3 of the total global population and 90% of that increase will concentrate over African and Asian countries. Between 2018 and 2050, India together with China and Nigeria expected contribute 35% of the projected growth of world urban population. Among such countries, India having an unprecedented urban population growth from 290 million to 340 million from 2001 to 2008 (Sankhe et al., 2010) solely has been predicted to add 416 million urban occupants. Although, many mega-cities have been formed since 1990 and by 2030, the world is expected to have 43 mega cities; yet, nearly half of the world's urban inhabitants dwell in comparatively small settlements (UN, 2018).

Planned and well-managed urbanization has greatly accelerated economic growth of a nation. However, such unprecedented urban expansion and urban sprawl has resulted into urban encroachment in green cover and agricultural land (Chadchan and Shankar, 2012; Pandey and Seto, 2015) leading to conversion of natural land-cover into urban concrete which gives rise to numerous socio-environmental problems. Some of them are: alteration of thermal properties of urban areas (Van and Bao, 2010), atmospheric pollution (Alves and Skole, 1996), undesirable hydrological impact in terms of supplying pollutants to surface water and causing erosion (Weng, 2001), ongoing downfall in the urban services apparent from the basic issues in slum, housing, lack of treated water supply, insufficient infrastructure, higher pollution levels, traffic congestion, degraded quality of life and many more (Ramachandra et al., 2012). In such a scenario, 'sustainable urbanism' is the key to achieve the benefit of urbanization for the human society. Sustainable urbanism has become a major target which the national, local governments and planners of any developing countries are keen to achieve.

Sustainable urbanism is the way to construct, plan and manage the city with the principles to maintain equity between urban growth and natural resources to prevent ecological imbalance and make the city climate resilient as well (Sharifi 2016). It is well-linked with the concept of Sustainable Development. Sustainable development is the combination of principles and approaches for the growth of the present society by judiciously using resources required to meet human development goals without compromising the stability of the ecosystem and conserve the resources for the future generations too (UN, 1987). Sustainable urbanism principally focuses on amalgamation of social, economic and environmental aspects of sustainable development. Such multi-faceted sustainable development can be achieved through proper planning and successful implementation of policies to manage the urban growth. Among social, economic and environmental dimensions of sustainable development, environmental degradation with respect to climate change grabs major attention. Further, the 2030 agenda for sustainable development puts special emphasis on adaptation and mitigation measures to combat climate change impact. Adaptation (to enhance the capability environmental, social, and economic systems to adjust the climate change impact by reducing their vulnerability towards actual and expected climate) and Mitigation (to reduce the rate of global warming effects by limits the factors responsible for it) are the two directions towards which scientific community are working to battle the climate change and protect the society form it's effect. Among 17 sustainable development goals (SDGs) to be accomplished by 2030 for sustainable development, goal 13 is to take necessary measures to combat adverse impact of climate change.

Planning of climate change adaptation and mitigation measures needs systematic information on urban thermal environmental problem occurred because of vast change in land use land cover (LULC) due to rapid urbanization, development of urban sprawl, population growth, their choice and way of life. Urban thermal environment can be evaluated by the detailed study of variation of land surface temperature (LST) along with land use (the changed situation of land after anthropogenic intrusion to occupy the land for human benefit, e.g. residential/ commercial/ industrial area) and land cover (the natural condition of the land, e.g. forest, water bodies) change (Gupta and Roy, 2012; Yang et al. 2018). LST is deðned by the International LST and Emissivity Working Group as a degree of hotness of the surface of the Earth when one touches it on any location (Liang 2018). LST is rapidly changing over space and time (Prata et al. 1995) and such changes are immensely controlled by land use, adjacent rural environs, building materials, length of the day, season etc. (Valsson and Bharat 2009). With the rapid rate of urbanization natural landscape has been converted into concrete material to meet the growing population. However, the use of construction material like concrete and asphalt in the urban residential, commercial and industrial areas and pavement roads make the soil impermeable which possesses higher LST as larger percentage of incident solar radiation is getting absorbed because of lower solar reflectance properties of urban surfaces. Such dynamic nature of LST demands detailed quantification with high spatial and temporal resolution which is challenging to accomplish using traditional ground measurement. Hence, satellite-derived thermal infrared (TIR) data has been used to calculate LST through the radiative transfer equation (Li et al. 2013) because of advantages of having short acquisition period, wide coverage and being economic for users (Weng 2009). Several studies have been utilized TIR band information of various satellites to compute LST using different approaches to apply in various field like climate change, hydrology, impact analysis for urban growth, environmental studies, urban air pollution research etc. (Duan et al., 2019; Wang et al. 2018; Kayet et al. 2016, Ding and Shi, 2013; Hansen et al. 2010, Mallick et al. 2008).

Such studies have documented the increase of urban land surface temperature due to long-term changes in LULC and thereby creation of urban heat island (UHI). UHI are the small urban pockets significantly warmer compared to the rural surrounding. Recent research has the reported the UHI has an impact over regional as well as urban micro-climate by modifying the amount and rate of rainfall (Liu and Niyogi 2019). More than 1000 cities are suffering from UHI all over the world (Stewart, 2011). In sustainable urbanism, different UHI mitigations strategies have been suggested viz. high albedo material, green cover on the surface and building roofs which influences the selective absorption and reflection of incident solar radiation and thus control the effect of UHI. Further, it helps in purifying the urban air (Sodoudi et al. 2014; Senanayake et al. 2013).

However, few earlier studies reported the effect of seasonal inconsistency in LST over urban thermal environment by influencing water vapour, cloudiness and precipitation (Liu and Weng 2008; Dai et al. 1999). UHI effect is more prominent during summer to late leaf-shed season (Oke, 1982). LST in urbanized area during summer can increase the energy demand and deteriorate the air quality level (Mathew et al. 2019). Despite the significant impact of seasonal irregularity in LST over urban thermal environment, research over inter-seasonal variation of land surface temperature is scarce. Moreover, it is worth to mention that, unlike the UHI phenomenon in temperate and sub-tropical region, the formation of daytime urban cool island (UCI) (UCI can be defined as cooler urban area compared to the non-urban/rural surroundings) has been reported in the arid and semi-arid climatic regions (Rasul et al. 2015; Lazzarini et al., 2013) and even in the neighbouring region of the present study area (Kumar et al. 2019).

Therefore, the objective of the present chapter is to study the inter-seasonal variation of land surface temperature of highly urbanized area Faridabad district, Haryana in the semi-arid zone of India which is a crucial component for the evaluation of urban microclimate.

BACKGROUND

Overview of Different Satellite Sensors Providing Thermal Information

With the advancement of sensor technology, imaging sensors employ thermal imaging systems which detect emitted energy from the earth's surface. Numerous thermal satellite sensors viz. Advanced Very High Resolution Radiometer (AVHRR), Geostationary Operational Environmental Satellite (GOES), Moderate Resolution Imaging Spectroradiometer (MODIS), Landsat Thematic Mapper (TM), Enhanced Thematic Mapper Plus (ETM+), Operational Land Imager (OLI), are able to provide thermal information of the surface of the earth.

Advances Space Borne Thermal Emission and Reflection Radiometer (ASTER)

ASTER collects thermal infrared information using five thermal bands (8.1 to 11.6 µm) with 90m spatial resolution. ASTER data can be downloaded from Earth Explorer.

Advanced Very High-Resolution Radiometer (AVHRR)

AVHRR launched by the National Oceanic and Atmospheric Administration (NOAA) collects data with thermal bands (10.3-11.3 µm and 11.3-12.5 µm) at 1.1 km. It can be obtained from Earth Explorer.

Geostationary Operational Environmental Satellite (GOES)

GOES launched by the United States' National Oceanic and Atmospheric Administration (NOAA) have two thermal band (10.2-11.2 µm and 11.5-12.5 µm) with spatial resolution 4km. GOES imagery can be obtained from https://www.star.nesdis.noaa.gov/goes/.

Moderate Resolution Imaging Spectroradiometer (MODIS)

MODIS carried by both NASA's Aqua and Terra Satellites provides thermal data with bands 31(10.78 – 11.28 µm) and 32 (11.77 – 12.27 µm) with 1000m spatial resolution. MODIS data can be downloaded from https://lpdaac.usgs.gov/.

Thermal Infrared Multispectral Scanner (TIMS)

TIMS developed by NASA and the Jet Propulsion Laboratory collects data with six different bands ranging from 8.2 to 12.2 µm with spatial resolution of 18m.

Thematic Mapper (TM) and Enhanced Thematic Mapper Plus (ETM+)

Landsat 4-5 TM and Landsat 7 ETM+ have thermal band at 10.40-12.50 µm with spatial resolution 120 m resampled into 30m are freely available from Earth Explorer.

Operational Land Image (OLI) and Thermal Infrared Sensor (TIRS)

Landsat 8 Operational Land Image (OLI) and Thermal Infrared Sensor (TIRS) have two thermal bands at 10.60 µm -11.19 µm and 11.50 µm -12.51 µm at 100m resolution resampled into 30m. It is also freely available from Earth Explorer.

Indian Thermal Sensor

INSAT-3D by the Indian Space Research Organization (ISRO) collects thermal information using TIR sensors 10.3 µm-11.3 µm and 11.5 µm-12.5 µm wavelength ranges. Very High-Resolution Radiometer (VHRR) of Kalpana-1 (K1) provides the information using 10.5 µm- 12.5 µm wavelength range with 30 minutes interval.

Theoretical Basis of Reflectance and LST Retrieval Using Satellite Images

This section has described the basic steps to convert digital number (DN) (the raw digital value of a pixel) of satellite images to physical units following the method given in the USGS website (https://landsat.usgs.gov/Landsat8_Using_Product.php). First, it has described the conversion of DN values in reflectance in case of optical bands, and then the conversion of DN values in LST in case of thermal band.

For optical bands, DN to Top of Atmospheric (TOA) reflectance conversion can be divided in two steps:

§ Conversion of DN to Radiance
§ Conversion of Radiance to TOA Reflectance

DN can be converted into spectral radiance at sensor (L_λ) following the Gain and Bias method using Equation 1

$$L_\lambda = Gain * DN + Bias \quad (1)$$

The gain can be calculated following the equation

$$\text{Gain}= \left(L_{MAX} - L_{MIN}\right) \Big/ \left(Q_{CALMAX} - Q_{CALMIN}\right) \tag{2}$$

So, radiance can be calculated following the equation 3

$$L_{\lambda} = \frac{\left(L_{MAX} - L_{MIN}\right)}{\left(Q_{CALMAX} - Q_{CALMIN}\right)} \times \left[Q_{CAL} - Q_{CALMIN}\right] + L_{MIN} \tag{3}$$

Where

Q_{CALMAX}: Maximum quantized pixel value (2^n for n bit system)
Q_{CALMIN}: Minimum quantized pixel value (typically 1)
L_{MAX}: Spectral radiance corresponding to Q_{CALMAX}
L_{MIN}: Spectral radiance corresponding to Q_{CALMIN}
Q_{CAL}=DN Value

Apparent reflectance or TOA reflectance can be computed for each band following the equation 4

$$\rho_{\lambda} = \frac{\pi \times L_{\lambda} \times d^2}{E_{\lambda} \operatorname{Cos}\theta} \tag{4}$$

Where,

ρ_{λ}: Apparent reflectance for specific band λ
d: Sun-Earth distance for the image acquisition day
E_{λ}: Mean exoatmospheric solar irradiance
θ: Solar zenith angle

Here it should be mentioned that Landsat 8 data is available with band specific multiplicative scaling factor and band specific additive rescaling factor which includes the value of $E_{\lambda}/(\pi d^2)$. So, for LANDSAT 8 image, DN can be converted to TOA reflectance without correction for solar angle at sensor directly using the equation

$$\rho'_{\lambda} = M_p Q_{CAL} + A_p \tag{5}$$

Where,

ρ'_λ : TOA planetary reflection without correction for solar angle
M_P: Band specific multiplicative rescaling factor from the metadata
A_P: Band specific additive rescaling factor from the metadata

TOA reflectance with solar zenith angle correction is obtained as

$$\rho_\lambda = \frac{\rho'_\lambda}{\text{Cos}\,\theta} \tag{6}$$

In case of thermal band, DN values are converted to radiance using the Equation 3. Radiance to LST conversion method is based on Plank's law. Every physical body, kept at temperature higher than 0 K, spontaneously emits electromagnetic radiation. The energy emitted per unit area of a black body, per unit solid angle, per unit wavelength, per unit time at temperature T can be obtained following the Plank's Law:

$$B_\lambda = \frac{2hc^2}{\lambda^5} \frac{1}{e^{\frac{hc}{\lambda k T_B}} - 1} \tag{7}$$

Where,

h: Plank constant
k: Boltzmann constant
c: Speed of light
λ: Wavelength
T_B: Temperature (K) of the blackbody

The expression can be rearranged as

$$e^{\frac{hc}{\lambda k T_B}} = \frac{2hc^2}{\lambda^5} \frac{1}{B_\lambda} + 1$$

Or,

$$\frac{hc}{\lambda k T_B} = \ln\left(\frac{2hc^2}{\lambda^5} \frac{1}{B_\lambda} + 1\right) = \ln\left(\frac{C_1}{B_\lambda} + 1\right)$$

Or,

$$T_B = \frac{hc/(\lambda k)}{\ln\left(\frac{C_1}{B_\lambda}+1\right)} = \frac{C_2}{\ln\left(\frac{C_1}{B_\lambda}+1\right)} \tag{8}$$

Or,

$$T_B\left({}^{o}C\right) = \frac{C_2}{\ln\left(\frac{C_1}{B_\lambda}+1\right)} - 273.15 \tag{9}$$

Where,

$$C_1 = \frac{2hc^2}{\lambda^5}$$

$$C_2 = hc/(\lambda k)$$

So, C_1 and C_2 are wavelength dependant constants.

Black body acts as a perfect emitter. A real material would radiate lesser energy at the same temperature. Emissivity (ε) of a material is defined as the ratio of amount of energy emitted by the material at temperature T to the energy emitted from a black body at the same temperature.

For thermal band, B_λ is equivalent to L_λ. Also, several other effects are included to obtain better accuracy in calculation of brightness temperature T_B. Hence, the values of constants, also called calibration constants, differ in each sensor.

For LANDSAT 8, L_λ has been used to calculate the At-Satellite Brightness Temperature by the following Equation 10

$$T_B = \frac{K_2}{\ln\left(\frac{K_1}{L_\lambda}+1\right)} \tag{10}$$

Where:

T_B = At-satellite brightness temperature
K_1= Calibration constant=774.8853 (For TIR1 in Landsat 8)
K_2=Calibration constant=1321.0789 (for TIR1 in Landsat 8)

Based on the theoretical basis described above, several authors developed algorithm for retrieving LST using thermal band of Landsat 4 and 5 TM, Landsat 7 ETM+ and Landsat 8 TIRS. Single-channel (SC) algorithm has been developed to derive LST from the single thermal band (10.40-12.50μm) available in TM, ETM+, Landsat 8 TIRS 1 (10.60-11.19 μm) (Qin et al. 2001). Using the two thermal bands (TIRS1:10.60-11.19 μm and TIRS2: 11.50-12.51 μm) available in Landsat 8, split-window algorithm has been developed also (Jiménez-Muñoz et al. 2008b). Further, modification, revision and improvements have been performed in those algorithms by eminent scientists (Du et al. 2015; Cristóbal et al. 2009; Jiménez-Muñoz et al. 2008a). Though, United States Geological Survey (USGS) advise not to use TIRS2 of Landsat 8 for the large calibration uncertainty. Moreover, comparison of different algorithms used for extracting LST using Landsat Thermal bands revealed that these algorithms required highly accurate input parametres (land surface emissivity (LSE), atmospheric transmittance, water vapor content, mean atmospheric temperature) to extract precise LST calculation which is a great challenge for the researcher (Wang et al. 2019; Jiménez-Muñoz et al. 2014; Yu et al. 2014; Li et al. 2013). Such input parametres are not available for the present study area. Therefore, for the present study, emissivity calculated through NDVI has been used to convert brightness temperature derived from Landsat 8 TIR1 to LST, which has been successfully used in the previous related research (Kumar et al. 2019; Kumari et al. 2018; Amirtham et al. 2009).

Role of Remote Sensing To Explore LST in The Context of Sustainable Urbanism

Introduction section of the present chapter highlights the adverse effect of unprecedented urbanization, urban encroachment in the green space, increased LST, formation of UHI. In the light of that discussion, it can be remarked that detailed analysis of spatio-temporal and inter-seasonal variation of LST is of critical importance to achieve the key components of sustainable urbanism. Although, traditional fixed thermometer networks and automatic weather station provides data with good temporal resolution, yet such methods are time and resource intensive and spatial coverage is relatively low (Qaid et al. 2016). On the contrary, current development in satellite remote sensing cost-effectively assesses LST using TIR bands and monitor

UHI. Moreover, multi-spectral, multi-temporal, high-resolution satellite data gather information about the changing urban landscape. Such evidences collectively can contribute as baseline information for the effective UHI management at the macro, meso and micro level to design mitigation and adaptation measures to the adverse impact of climate change. Previous researches successfully utilized LST data from thermal infrared sensors like AVHRR (Polydoros and Cartalis, 2014; Stathopoulou and Cartalis, 2009), MODIS (Keramitsoglou et al. 2011; Fabrizi et al. 2010), Landsat TM/ETM+, Landsat 8 TIRS (Sultana and Satyanarayana, 2018; Tsou et al. 2017; Sharma and Joshi, 2014) and ASTER (Mallick et al. 2013; Liu and Zhang, 2011).

To discover the linkage between LULC change and urban thermal environment, various researches have been conducted on inter-annual variation of LST over urbanized areas (Kumar et al., 2019; Hua and Ping 2018; Kumari et al. 2018; Rousta et. al. 2018; Kayet et al. 2016, Ding and Shi, 2013). Kumari et al. (2018) has reported continuous increase in LST from 2003 to 2017 because of rapid expansion of urban areas. Kayet et al. (2016) has evaluated the encroachment of mining area into the vegetation cover of Saranda forest of Jharkhand which increased the surface temperature. Amirtham et al. (2009) has identified thermal hotspot as an impact urban expansion in Chennai city. Unlike the general observation of UHI in Indian city area, few studies have reported the formation of UCI in urban area in India. Kumar et al. (2019) has correlated spectral indices Normalized Difference Vegetation Index (NDVI) and Normalized Difference Built-up Index (NDBI) with LST to investigate the relationship between built-up, green cover and LST in Hisar Tehsil in semi-arid climatic condition and reported contrasting scenario of cooler urban area compared to rural surrounding due to 'oasis effect' of green space in urban parks. Also, Bengaluru urban district in Southern India has experienced formation of UCI despite of notable conversion of barren and agricultural land in to urban land use (Govind and Ramesh, 2019).

Apart from the inter-annual variation of LST over urbanized areas, it is evident from past research that changes in LST are not only inter-annual but also seasonal, diurnal, and even abrupt (Verbesselt et. al. 2010; Zhan et al. 2014). Likewise, LULC also vary with seasons especially in those regions where agriculture is the prime source of livelihood. Sultana and Satyanarayana (2018) has identified higher UHI intensity during winter even due to prominent increase in urban built-up at the cost of vegetated lands and barren lands over Nagpur, Lucknow and Jaipur and Hyderabad. A strong correlation has been observed by Mallick et al (2013) between impervious surface area and relative surface temperature in central business district of by studying nighttime satellite imagery. Sharma and Joshi (2014) has investigated distribution and magnitude of UHI during winter, spring, summer, monsoon, and post-monsoon and identified commercial, industrial and airport as the thermal hotspot by computing mean annual LST anomaly. Highest and lowest seasonal

UHI intensities have been identified in summer and winter seasonal, respectively. Likewise, maximum variation of LST has been in observed in the month of June in Mumbai city (Sahana et al. 2019).

Based on the above observation in earlier research, it can be remarked that inter-seasonal variation of LST is an important phenomenon affecting the urban micro-climate. Therefore, to evaluate the role of seasonal variation on the relationship between LULC pattern and LST, present study has attempted to assess correlation between LST and spectral indices which has been successfully used in earlier studies (Mukherjee et al. 2017; Ranagalage et al. 2017; Pal and Ziaul 2017). As those spectral indices like NDVI and NDBI indicate the presence of vegetation cover and impervious surface, therefore the change analysis of NDVI and NDBI also designate the removal of vegetation cover by asphalt, concrete, metal in urban areas affecting the urban thermal environment.

STUDY AREA

The study area (Figure 1) covers Faridabad (28°13′N-28°30′N latitude 77°8′E-77°29′E longitude), a southeastern district in the state of Haryana in Gurgaon administrative division with headquarters in Faridabad. Faridabad district covers a total area 741 km^2 with 218.42 km^2 urban area. The study area lies in the semi-arid region where the average temperature ranges from 28°C to 35°C and 13°C to 20°C during summer and winter season, respectively. It is mainly occupied by flat terrain except Aravalli Hills in the north-western part of the district which acts like a prohibitive topography for urban expansion. Yamuna River, Badkhal and Surajkund lakes are some major water bodies located in the study area.

Faridabad is the fastest growing million plus metropolitan city of the state of Haryana and the share of which population in total population of the district has followed an increasing trend in 1971 (7.16%), 1981 (41.43%), 1991 (48.57%) and 2001 (77.80%) owing to the large number of migrants. Faridabad stands 1st in population size during 2011 with a total population of 1,809,733, of which 79.51% is urban population with decadal population growth of 32.54% during 2001-2011 (Census, 2011). Faridabad is the most industrialized and populated city in the whole state of Haryana due to migration into NCR from other regions of Northern India. Migrant are attracted to the several manufacturing industries providing employment opportunities. However, such industrialization and thereby population increase are accompanying with urban agglomeration and resultant extreme pressure on the environment. Although past studies have reported 38% increase in the built-up region at the cost of vegetation, water, open space during 1998-2017 (Ghosh and Ghosh 2017), deterioration of water quality, local ecology and biodiversity due to mining activities in Aravalli hills affect the (Nathalia et al. 2017; Wadhawan and Ahmad 2010), depletion of water from Barkhal lake (Rai and Nair 2013), the number of research related to environmental impact of urban growth particularly over Faridabad district is still limited. Therefore, the present study has attempted detailed analysis of the relationship of spatial pattern and characteristics of LST and spectral indices (NDVI and NDBI). Unlike the previous studies, the analysis has been conducted on district level.

DATA USED AND METHODOLOGY

To produce the spectral indices and LST maps, Landsat 8 Operational Land Imager

Figure 1. Location map of the study area showing Faridabad District (as viewed on Landsat 8, OLI image of 2018). Black lines (N–S and W–E) represent profiles across the study area. Dot represents the only meteorological station at Sector -16A Faridabad, Haryana State Pollution Control Board (HSPCB)

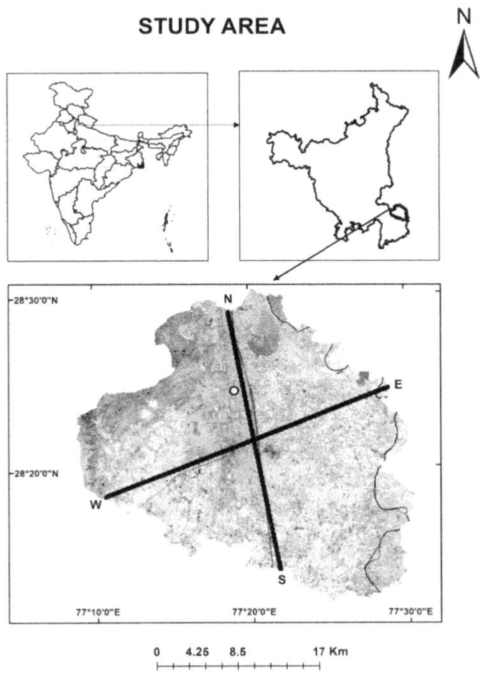

(OLI) and Thermal Infrared Sensor (TIRS) level-1 data acquired on January 19,

2018, April 25, 2018, September 16, 2018 and October 02, 2018 have been obtained from the United States Geological Survey (USGS) web server which represent the winter, pre-monsoon or summer, southwest (SW) monsoon and post-monsoon season in the present study. All the images for the same year representing different season has been selected for logical inter-seasonal comparison. All the cloud-free images have been acquired with approximately at the 10:30 am local time with +/- 15 minutes which makes the analysis more feasible. Specifications of the data used in the present study are given in Table 1.

All the data sets are terrain corrected with Universal Transverse Mercator (UTM) coordinate system, 43 North Zone and WGS 84 Datum. During pre-processing, DN values of optical bands (Visible Red and NIR, SWIR1) have been converted to TOA reflectance using Equation 5. TOA reflectance with solar zenith angle correction

Table 1. Specifications of the data used in the present study

Name	Band Description	Resolution		Date of Acquisition	Cloud Cover	Image ID	Source
		Spatial (m)	Spectral (µm)				
	Coastal / Aerosol	30	0.43 to 0.45				
	Visible blue	30	0.45 to 0.51				
	Visible green	30	0.53 to 0.59				
	Visible red	30	0.64 to 0.67	January 19, 2018	1%	LC08_L1TP_146040_20180119_20180206_01_T1	
	Near-infrared	30	0.85 to 0.88	April 25, 2018	Cloud free	LC08_L1TP_146040_20180425_20180502_01_T1	
Landsat 8 OLI & TIRS	Short wave infrared (SWIR1)	30	1.57 to 1.65	September 16, 2018	6%	LC08_L1TP_146040_20180916_20180928_01_T1	USGS NASA
	Short wave infrared (SWIR2)	30	2.11 to 2.29	October 02, 2018	Cloud free	LC08_L1TP_146040_20181002_20181010_01_T1	
	Panchromatic	15	0.50 to 0.68				
	Cirrus	30	1.36 to 1.38				
	Thermal infrared (TIR1)	100 (resampled to 30)	10.60 to 11.19				
	Thermal infrared (TIR2)	100 (resampled to 30)	11.50 to 12.51				

has been obtained using Equation 6.

NDVI has been computed by dividing the difference of reflectance between Red band and NIR band, by the sum of the two-reflectance using Equation 11:

$$\text{NDVI}=\frac{(\text{NIR-RED})}{(\text{NIR+RED})} \tag{11}$$

Likewise, NDBI has been computed by dividing the difference of reflectance between NIR band and SWIR1 band by the sum of the two-reflectance using Equation 12:

$$\text{NDBI}=\frac{\text{SWIR-NIR}}{\text{SWIR}+\text{NIR}} \tag{12}$$

In the present study TIR1 (10.60 to 11.19μm) band data has been used to calculate the LST. DN values of TIR1 have been converted to TOA spectral radiance using the Equation 13 given in the USGS website (https://landsat.usgs.gov/Landsat8_Using_Product.php).

$$L_\lambda = M_L Q_{CAL} + A_L \tag{13}$$

Where,

L_λ: Spectral radiance at the sensor's aperture
Q_{CAL}: Quantized calibrated pixel value [DN]
M_L: Band-specific multiplicative rescaling factor
A_L: Band-specific additive rescaling factor
L_λ: has been used to calculate the At-Satellite Brightness Temperature by using Equation 10.

Land surface emissivity (LSE) corrected LST has been computed using the Equation 14 following Artis and Carnahan (1982):

$$\text{LST}=\frac{T_B}{1+\left(\frac{\lambda T_B}{\rho}\right)\ln\varepsilon} \tag{14}$$

Where, λ= 10.9 μm for Band10 (the central wavelength of emitted radiance)

$$\rho=\frac{hc}{k}=1.438\times10^{-2}\ mK$$

(h denotes the Planck Constant equals to 6.626×10^{-34} Js; c denotes the velocity of light equals to 2.998×10^{8} m/s; k denotes the Boltzmann constant equals to 1.38×10^{-23} J/K)

ε: LSE

LSE (ε) has been estimated using the Equation15 following Estoque et al. (2017):

$$\varepsilon = mP_v + n \tag{15}$$

Where

m=0.004 and n=0.986 have been used (Sobrino et al. 2004)
P_V: Vegetation Proportion
P_V has been estimated using the Equation 16 (Carlson and Ripley, 1997)

$$P_V = \left(\frac{NDVI - NDVI_{min}}{NDVI_{max} - NDVI_{min}} \right)^2 \tag{16}$$

LST in Kelvin has been finally converted to degree Celsius by (Kelvin)-273.15.
For each season, considering N–S and E–W profiles (as marked in Figure 1), the pixel values of LST and spectral indices (NDVI and NDBI) have been selected to cover the spatial variation of surface temperature, NDVI and NDBI. Further, a regression analysis has been performed between LST and NDVI and LST and NDBI values to establish the correlation between them. Further, to investigate the inter-seasonal changes of spatial patterns of LST with reference to the major LULC (viz. water bodies, forest, agricultural land, vegetation, grass/grazing land, built-up, waste lands) 245 sample points randomly selected depending on the areal coverage of each land cover observed in this region. Overall insight of LULC information has been obtained through visual interpretation elements using False Colour Composite (FCC) image with combination NIR, Red and Green bands as RGB (as shown in figure 1), Google Earth imagery, preliminary survey in the study area and previous study (Ghosh and Ghosh 2017). Average LST of each land use/cover category for each season has been calculated. The flowchart of overall methodology has been given in figure 2.

ANALYSIS AND DISCUSSION

Figure 2. Flow chart of the methodology

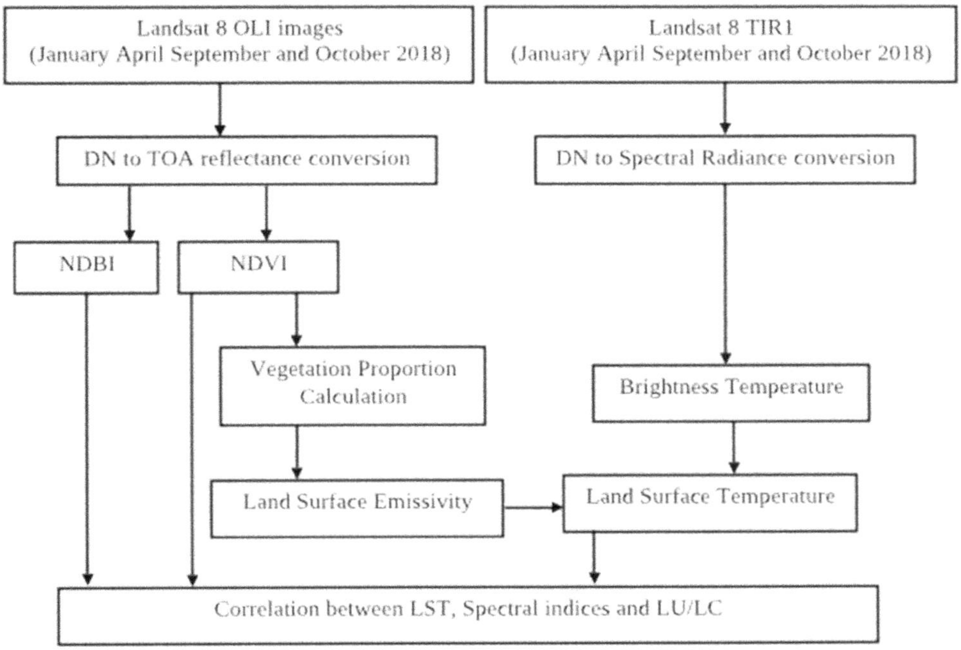

LST maps have revealed the spatial distribution of LST in Faridabad district of four dates representing four seasons in 2018 (Figure 3). Validation of LST has become difficult because only one meteorological station (at Sector-16A Faridabad, Haryana State Pollution Control Board (HSPCB)) is available in the study area. To validate the satellite image-derived, the average near-surface air temperature has been used from the meteorological station. First the pixel has been identified in LST maps where the meteorological station exists. A 200m × 200m square buffer has been prepared around the pixel. Arithmetic mean has been taken for the LST of all the pixels coming under the buffer. Such average LST has been compared with average near-surface air temperature. The difference between the image-derived LST and near-surface air temperature is within ±2°C. However, in case of pre-monsoon season the difference is about 5°C which is with the error limit reported in the earlier studies (Avdan and Jovanovska, 2016; Gallo et al. 2011). Compared to the other seasons, the difference between maximum and minimum temperature is higher in the pre-monsoon season (11.02°C, 20.62°C, 10.96°C and 16.98°C in winter, pre-

Figure 3. Spatial distribution of LST (°C) (a) January 19, 2018; (b) April 25, 2018; (c) September 16, 2018; (d) October 02, 2018

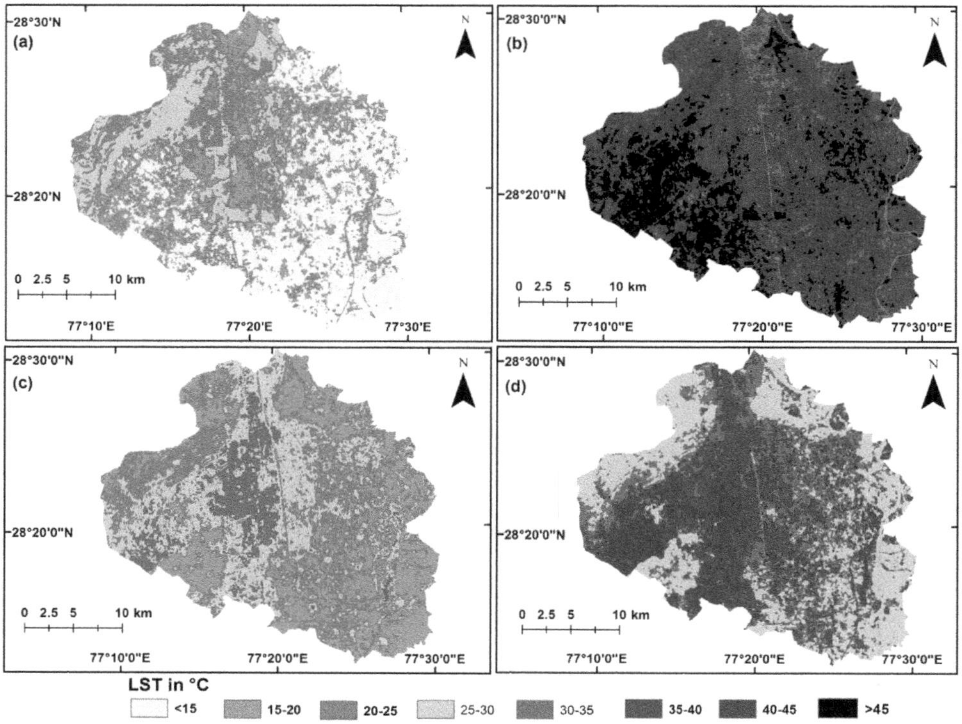

Table 2. Statistics of LST, NDVI and NDBI

	LST (°C)			NDVI			NDBI		
	Minimum	Maximum	Mean	Minimum	Maximum	Mean	Minimum	Maximum	Mean
Jan 19, 2018	16.85	27.87	20.20	-0.22	0.70	0.33	-0.92	0.66	-0.37
Apr 25, 2018	31.46	52.08	43.21	-0.08	0.44	0.12	-0.66	0.64	-0.07
Sept 16, 2018	24.62	35.58	28.14	-0.08	0.74	0.47	-0.78	0.16	-0.44
Oct 02, 2018	25.72	42.70	31.43	-0.22	0.74	0.43	-0.88	0.21	-0.39

monsoon, southwest monsoon, post-monsoon.) Highest (43°C) and lowest (20°C) mean temperature have been observed in the pre-monsoon and winter season, respectively (Table 2).

For all the four dates, lowest temperature has been observed in the water bodies. However, higher temperature has been observed in the Faridabad city developed

Table 3. Inter-seasonal variation of average temperature (°C) in different LULC types

LULC types	Jan 19, 2018	Apr 25, 2018	Sept 16, 2018	Oct 02, 2018
Water Bodies	18.54	34.57	26.53	28.02
Forest	22.09	43.39	27.38	29.69
Agricultural Land	19.20	43.64	27.74	31.32
Vegetation	20.66	40.84	28.52	31.43
Grass/Grazing Land	21.40	42.98	27.84	31.16
Built up	20.99	42.37	29.99	33.06
Waste Lands	22.03	43.53	28.45	31.36

along Delhi-Mathura National Highway and Agra canal where built-up region has encroached into vegetation, agricultural areas and open spaces in all the seasons which has been reported in earlier studies (Ghosh and Ghosh 2017, Saini et al. 2017). Being the 9th largest industrial estate in Asia, Faridabad has continuous industrial activities (Wadhawan and Ahmad 2010) in the form of multinational companies (MNC), ISO-based industries, medium, large and small-scale industries, which is one of the major contributors of higher temperature. Apart from that, higher temperature has also been contributed by the waste land, grass/grazing in all the seasons because of reduced evaporative cooling in the semi-arid climatic regions (Fan et al., 2017).

During the winter season, higher temperature has been observed in the vicinity of Aravalli hills which contain dry deciduous scrub forest which sheds the leaves during this season. Moreover, large scale removal of scrub forest and mining activities (Nathalia et al. 2017) contributes to the increase of temperature in this zone. Though, in the summer season, the temperature in the lower portion of the Aravalli hills and the sub urban part in the western part of Agra canal is 5-10°C higher than the city centre due to the presence of stone quarries and brick kilns. Also, agricultural area has possessed higher temperature in this season. During this season uncultivated or fallow agricultural land, overgrazed land and sandy area contribute to the higher temperature due to the low rainfall and thereby lower moisture content in this region.

During the monsoon season, the temperature is relatively lower compared to the other season over the Aravalli hills due to the rainfall and appearance of the leaves

Figure 4. NDVI maps of (a) January 19, 2018; (b) April 25, 2018; (c) September 16, 2018; (d) October 02, 2018

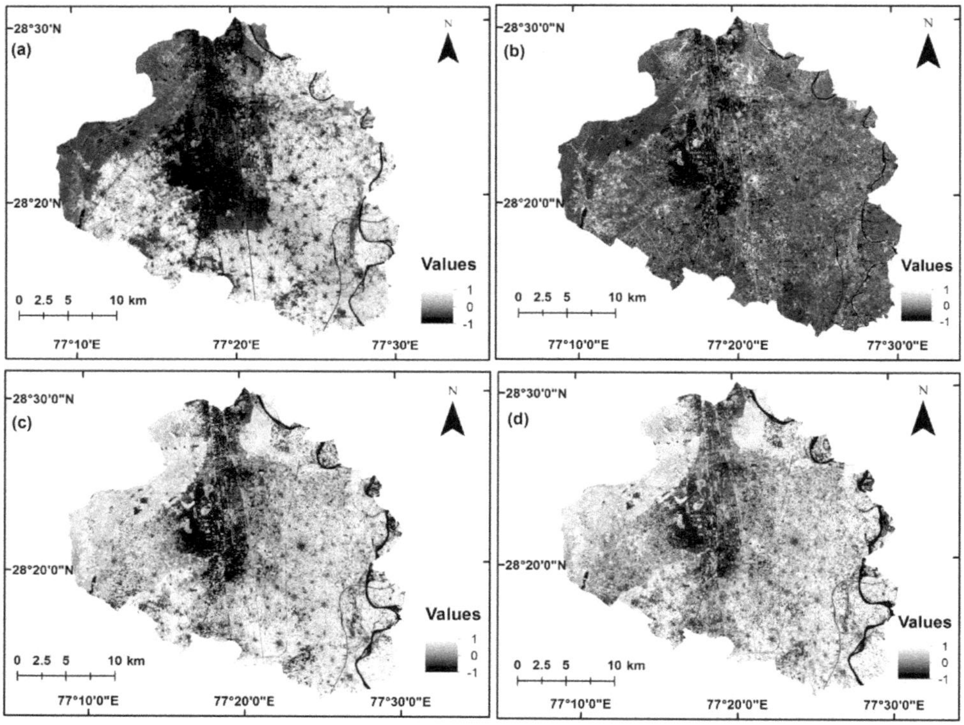

on different species of scrub forest which has increased the rate of evapotranspiration leading to sufficient amount of heat exchange between surface and atmosphere. Compared to the central and western portion of the district, the south eastern part of the district has lower temperature where mostly rural built-up and agricultural lands exist. In that rural region also, different parts have higher temperature because of the presence of grazing land near the rural built-up. During the monsoon and post-monsoon season, built-up area has higher temperature compared to any other LULC unit, whereas during the summer and winter season, surrounding rural region comprising, fallow land, grazing, wasteland have possessed higher temperature compared to built-up Overall, it can be mentioned that the spatial distribution of high LST is rather continuous and compact in the urbanized and industrialized city centre whereas irregular occurrence of high value of LST has been observed in the sub-urban and rural region (Figure 3, Table 3).

Spatial distribution of NDVI and NDBI in Faridabad district in 2018 have been presented in Figure 4 & 5 and Table 2.The ranges between the maximum and minimum NDVI values have been observed 0.91, 0.49, 0.82 and 0.96 for winter,

Figure 5. NDBI maps of (a) January 19, 2018; (b) April 25, 2018; (c) September 16, 2018; (d) October 02, 2018

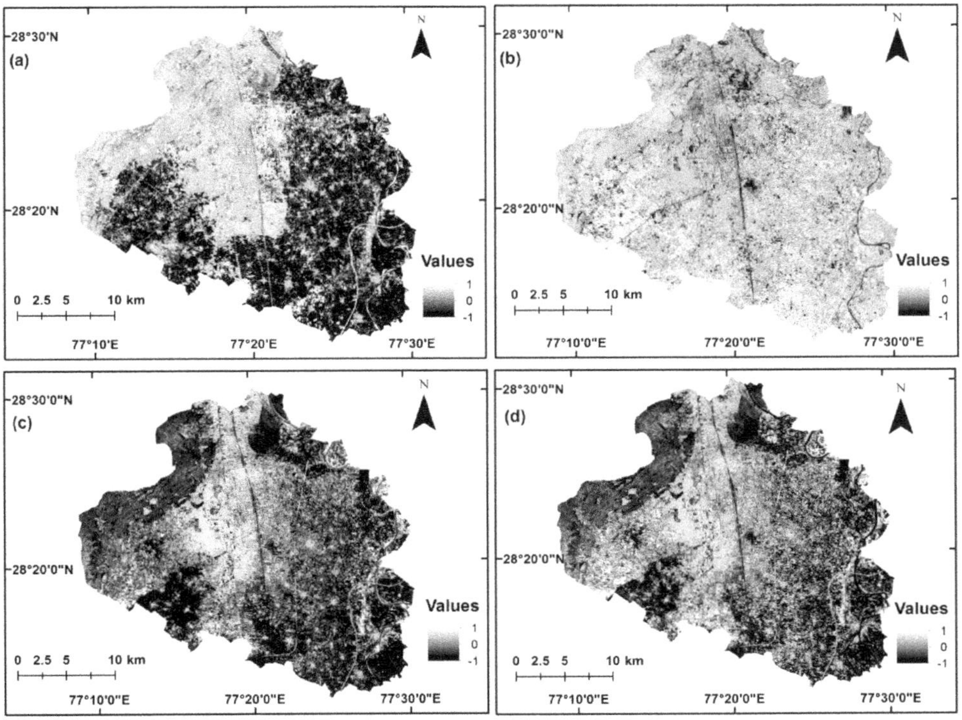

pre-monsoon, SW monsoon and post-monsoon respectively as shown as in Table 2. Higher ranges in NDVI values have been obtained for winter, SW monsoon and post-monsoon, whereas range in pre-monsoon was less. The lowest and highest mean NDVI has been recorded in the pre-monsoon and SW monsoon seasons, respectively. For all the seasons, low NDVI values have been found in the city centre because of the concrete buildings. Few high NDVI values were obtained because of the discontinuous urban green spaces. SW monsoon NDVI map displayed higher values (with high range of -0.08 to 0.74 and highest mean 0.47) as the foliation was complete. During the post-monsoon season also, the average NDVI is high because of the sowing of wheat and mustard and significant defoliation was yet to be started. During the month of April (pre-monsoon season), foliation was not complete and vast fallow land has contributed to the lower NDVI values due to harvest of Rabi crops. During winter, all the vegetation was defoliated which contributed the lower mean NDVI value of 0.33 (Figure 4).

Unlike NDVI, high NDBI values have been concentrated in the city area with relatively continuous dense built-up zone. The ranges between the maximum and minimum NDBI values have been observed 1.58, 1.30, 0.94 and 1.09 for winter, pre-monsoon, SW monsoon and post-monsoon respectively (Table 2). Although in previous studies (Chen et al. 2013; Li and Liu 2008), less seasonal variation of NDBI values were reported as they applied the index in smaller city area to analyse the UHI. However, in the present study NDBI has been applied in district level and it has been observed that NDBI is not only high in the dense urban built-up, but high values of NDBI has been found in vast rocky area, fallow land and over-grazed land. Therefore, considerable seasonal variation in NDBI values has been observed (Figure 5).

To examine the seasonal variation of relationship between LST and spectral indices (NDVI and NDBI) 1200 sample points have been selected to obtain the pixel values of LST and NDVI, NDBI along the profile lines (W–E and N–S) marked in figure 1.

In figure 6 and 7, each point represents mean LST and NDVI, NDBI values. Regression analysis has exhibited negative between mean LST and NDVI values and positive correlation between mean LST and NDBI values for all the seasons which has been reported in the earlier studies also (Malik et al. 219; Chen et al. 2013; Li

Figure 6. Regression analysis of NDVI and LST (a) January 19, 2018; (b) April 25, 2018; (c) September 16, 2018; (d) October 02, 2018

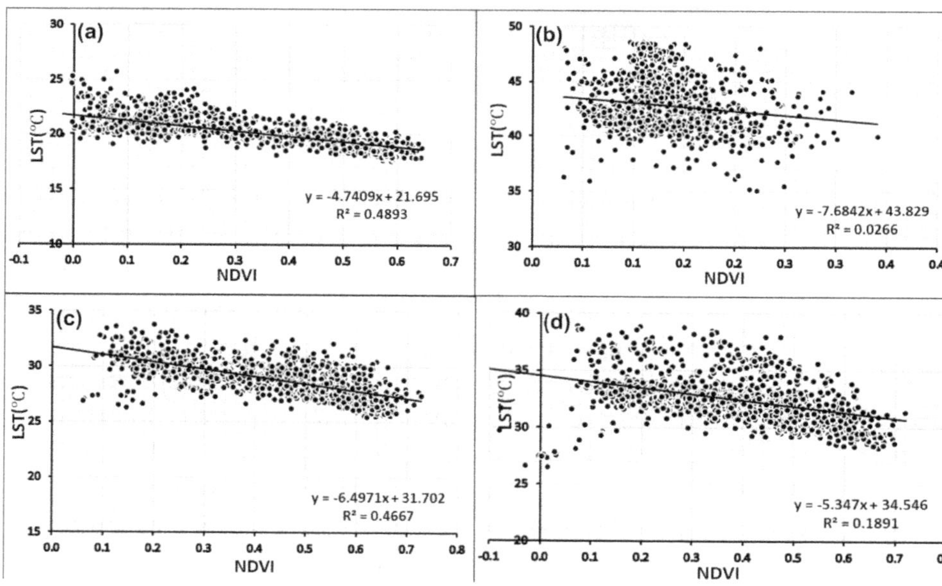

Figure 7. Regression analysis of NDBI and LST (a) January 19, 2018; (b) April 25, 2018; (c) September 16, 2018; (d) October 02, 2018

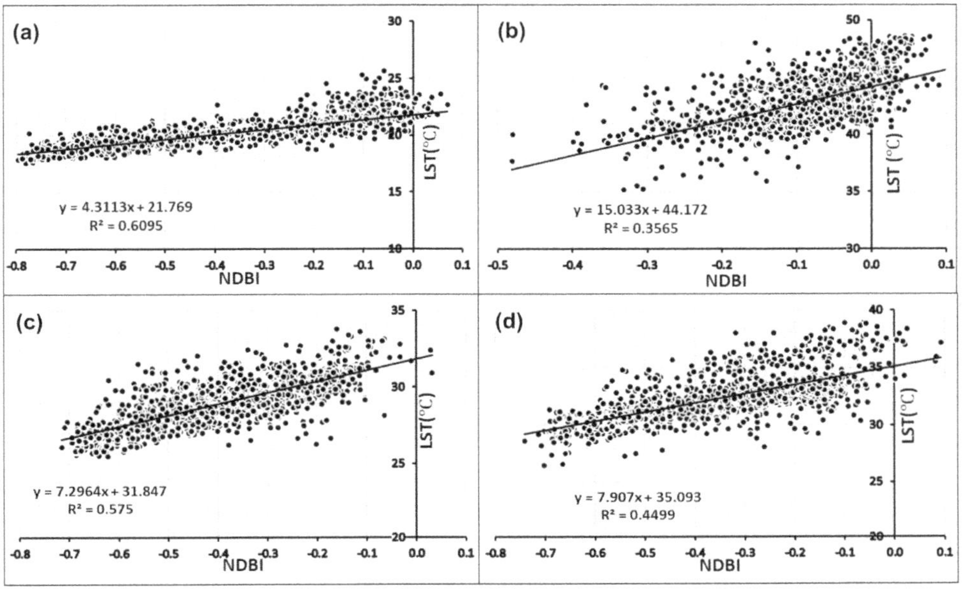

and Liu 2008; Yue et al. 2007). Comparing both the regression analysis, it can be remarked that NDBI and LST has a more stable relationship whereas the relation between NDVI and LST varied with seasons. Irregular patterns of NDVI values have been observed in the pre-monsoon and post-monsoon season in comparison to the other two seasons. Compared to the NDVI values, the pattern is regular in case of NDBI for all the seasons. Even though, it should also be mentioned the significant stability of the NDBI values in all the seasons cannot be obtained unlike the previous studies, as high NDBI values has not been obtained in a compact way when it is applied in the district level. Other than the dense built-up, high NDBI values have been found in other land covers (grazing land, fallow land, rocky areas etc.).

CONCLUSION

In the era of rapid urbanization, urban surface and its impact on urban climate is a major threat towards the well-being of the society. Therefore, development of climate-resilient cities by appropriate adaptation and mitigation strategies is major focus area of sustainable urbanism. To examine and govern the climate change in the urban context, this chapter provides a scientific purview to the policy makers and urban planners working towards sustainable urban development. This chapter has also described the role of thermal remote sensing which can complement and compensate the limitation of traditional method in measuring LST.

The results of the present research carried out in the data-scarce region like Faridabad, which was rarely examined despite the rapid change of LU/LC due to rapid urbanisation and industrialization, has exhibited significant spatial and inter-seasonal variation of LST not only in the urban but in the rural areas as well. Although weaker, negative and varying relation between NDVI and LST has been obtained in four seasons. The reason behind that is human interference in the form of concrete buildings, removal of scrub forest for domestic and industrial purpose and natural defoliation due to the dry deciduous nature of scrub forest in Aravalli range. Apart from the dense built-up region, high NDBI values have been observed in the Aravalli range due to mining activities, over-grazed land in the rural areas also. Relationship between LST and NDBI for all the four seasons, suggests that along with urban area, fallow land, rocky zone, overgrazing also accounts for variation in LST. Unlike the temperate climatic zone, areas under arid and semi-arid climatic condition has exhibited such kind of scenario as the soil condition, sandy areas, scanty vegetation, fallow land in the pre-monsoon season has reduced the rate of evaporation. Although, continuous high LST has been observed in the compact dense built-up zone of highly urbanised and industrialised city centre for all the seasons, but in the summer, rural areas exhibited higher temperature than city core. Urban green cover in the form of parks, and plantation here plays an important role as a contributor high evapotranspiration and thereby lower mean LST. Therefore, urban greening in the form of green roofs, green walls, and green parks has been widely practiced all over the world to achieve environmental sustainability. Also, check is required in over-grazing, illegal mining to control the increase in the temperature.

The unavailability of a good network of meteorological stations could be considered as limitation of the present study as it affects the validation of the satellite derived LST. Therefore, this study recommends the implementation of meteorological stations in such highly, urbanized, and industrialized area. For future studies field survey will be carried out for in situ measurements of LST using hand-held monitor.

ACKNOWLEDGMENT

This research was supported by Science & Engineering Research Board (SERB), Department of Science & Technology (DST), Government of India [File no. ECR/2017/000331].

REFERENCES

Alves, D., & Skole, D. (1996). Characterizing land cover dynamics using multi-temporal imagery. *International Journal of Remote Sensing*, *17*(4), 835–839. doi:10.1080/01431169608949049

Amirtham, L. R., Devadas, M. D., & Perumal, M. (2009). Mapping of micro-urban heat islands and land cover changes: A case in Chennai City, India. *The International Journal Climate Change: Impacts and Responses*, *1*(2), 71–84. doi:10.18848/1835-7156/CGP/v01i02/37258

Artis, D. A., & Carnahan, W. H. (1982). Survey of emissivity variability in thermography of urban areas. *Remote Sensing of Environment*, *12*(4), 313–329. doi:10.1016/0034-4257(82)90043-8

Asmala, A., & Abdullah, M. (2007). Mapping urban heat island phenomenon: Remote sensing approach. *Journal-The Institution of Engineers, Malaysia*, *68*(3), 25–30.

Avdan, U., & Jovanovska, G. (2016). Algorithm for automated mapping of land surface temperature using LANDSAT 8 satellite data. *Journal of Sensors*, *2016*, 2016. doi:10.1155/2016/1480307

Badarinath, K. V. S., Chand, T. K., Madhavilatha, K., & Raghavaswamy, V. (2005). Studies on urban heat islands using ENVISAT AATSR data. *Photonirvachak (Dehra Dun)*, *33*(4), 495–501. doi:10.1007/BF02990734

Carlson, T. N., & Ripley, D. A. (1997). On the relation between NDVI, fractional vegetation cover, and leaf area index. *Remote Sensing of Environment*, *62*(3), 241–252. doi:10.1016/S0034-4257(97)00104-1

Census of India. (2011). *Haryana, Series-07 Part XII-B, District Census Handbook Faridabad Village and Town Wise Primary Census Abstract (PCA)*. Directorate of Census Operations Haryana.

Chadchan, J., & Shankar, R. (2009). Emerging urban development issues in the context of globalisation. *Journal of ITPI*, *6*(2), 78–85.

Chadchan, J., & Shankar, R. (2012). An analysis of urban growth trends in the post-economic reforms period in India. *International Journal of Sustainable Built Environment*, *1*(1), 36–49. doi:10.1016/j.ijsbe.2012.05.001

Chen, L., Li, M., Huang, F., & Xu, S. (2013). Relationships of LST to NDBI and NDVI in Wuhan City based on Landsat ETM+ image. In *2013 6th International Congress on Image and Signal Processing (CISP)* (Vol. 2, pp. 840-845). IEEE.

Cristóbal, J., Jiménez-Muñoz, J. C., Sobrino, J. A., Ninyerola, M., & Pons, X. (2009). Improvements in land surface temperature retrieval from the Landsat series thermal band using water vapor and air temperature. *Journal of Geophysical Research, D, Atmospheres*, *114*(D8), D08103. doi:10.1029/2008JD010616

Dai, A., Trenberth, K. E., & Karl, T. R. (1999). Effects of clouds, soil moisture, precipitation, and water vapor on diurnal temperature range. *Journal of Climate*, *12*(8), 2451–2473. doi:10.1175/1520-0442(1999)012<2451:EOCSMP>2.0.CO;2

Ding, H., & Shi, W. (2013). Land-use/land-cover change and its influence on surface temperature: A case study in Beijing City. *International Journal of Remote Sensing*, *34*(15), 5503–5517. doi:10.1080/01431161.2013.792966

Du, C., Ren, H., Qin, Q., Meng, J., & Zhao, S. (2015). A practical split-window algorithm for estimating land surface temperature from Landsat 8 data. *Remote Sensing*, *7*(1), 647–665. doi:10.3390/rs70100647

Duan, S.-B., Li, Z.-L., Wang, C., Zhang, S., Tang, B.-H., Leng, P., & Gao, M.-F. (2019). Land-surface temperature retrieval from Landsat 8 single- channel thermal infrared data in combination with NCEP reanalysis data and ASTER GED product. *International Journal of Remote Sensing*, *40*(5-6), 1763–1778. doi:10.1080/01431161.2018.1460513

Estoque, R. C., Murayama, Y., & Myint, S. W. (2017). Effects of landscape composition and pattern on land surface temperature: An urban heat island study in the megacities of Southeast Asia. *The Science of the Total Environment*, *577*, 349–359. doi:10.1016/j.scitotenv.2016.10.195 PMID:27832866

Fabrizi, R., Bonafoni, S., & Biondi, R. (2010). Satellite and ground-based sensors for the urban heat island analysis in the city of Rome. *Remote Sensing*, *2*(5), 1400–1415. doi:10.3390/rs2051400

Fan, C., Myint, S., Kaplan, S., Middel, A., Zheng, B., Rahman, A., Huang, H. P., Brazel, A., & Blumberg, D. (2017). Understanding the impact of urbanization on surface urban heat islands—A longitudinal analysis of the oasis effect in subtropical desert cities. *Remote Sensing*, *9*(7), 672. doi:10.3390/rs9070672

Gallo, K., Hale, R., Tarpley, D., & Yu, Y. (2011). Evaluation of the relationship between air and land surface temperature under clear-and cloudy-sky conditions. *Journal of Applied Meteorology and Climatology*, *50*(3), 767–775. doi:10.1175/2010JAMC2460.1

Ghosh, S., & Ghosh, S. (2017). Evaluating Patterns of Urban Growth in Faridabad, Sub-Region of NCR, India Using Remote Sensing, GIS and Entropy Approach. *Indian Cartographer*, *37*, 258.

Govind, N. R., & Ramesh, H. (2019). The impact of spatiotemporal patterns of land use land cover and land surface temperature on an urban cool island: A case study of Bengaluru. *Environmental Monitoring and Assessment*, *191*(5), 283. doi:10.100710661-019-7440-1 PMID:30997556

Gupta, S., & Roy, M. (2012). Land Use /Land Cover classification of an urban area- A case study of Burdwan Municipality, India. *International Journal of Geomatics and Geosciences*, *2*(4), 1014–1026.

Hansen, J., Ruedy, R., Sato, M., & Lo, K. (2010). Global surface temperature change. *Reviews of Geophysics*, *48*(4), RG4004. Advance online publication. doi:10.1029/2010RG000345

Hua, A. K., & Ping, O. W. (2018). The influence of land-use/land-cover changes on land surface temperature: A case study of Kuala Lumpur metropolitan city. *European Journal ofRemote Sensing*, *51*(1), 1049–1069. doi:10.1080/22797254.2018.1542976

Intergovernmental Panel on Climate Change (IPCC). (2014). In R. K. Pachauri & L. A. Meyer (Eds.), *Climate Change2014; Contribution of Working Groups I, II and III to the Fifth Assessment Report of the Intergovernmental Panel on Climate Change; Core Writing Team* (p. 151). IPCC.

Jiménez-Muñoz, J. C., Cristóbal, J., Sobrino, J. A., Sòria, G., Ninyerola, M., Pons, X., & Pons, X. (2008a). Revision of the single-channel algorithm for land surface temperature retrieval from Landsat thermal-infrared data. *IEEE Transactions on Geoscience and Remote Sensing*, *47*(1), 339–349. doi:10.1109/TGRS.2008.2007125

Jiménez-Muñoz, J. C., & Sobrino, J. A. (2008b). Split-window coefficients for land surface temperature retrieval from low-resolution thermal infrared sensors. *IEEE Geoscience and Remote Sensing Letters*, *5*(4), 806–809. doi:10.1109/LGRS.2008.2001636

Jiménez-Muñoz, J. C., Sobrino, J. A., Skoković, D., Mattar, C., & Cristóbal, J. (2014). Land surface temperature retrieval methods from Landsat-8 thermal infrared sensor data. *IEEE Geoscience and Remote Sensing Letters*, *11*(10), 1840–1843. doi:10.1109/LGRS.2014.2312032

Kayet, N., Pathak, K., Chakrabarty, A., & Sahoo, S. (2016). Spatial impact of land use/ land cover change on surface temperature distribution in Saranda Forest, Jharkhand. *Modeling Earth Systems and Environment*, *2*(3), 127. doi:10.100740808-016-0159-x

Keramitsoglou, I., Kiranoudis, C. T., Ceriola, G., Weng, Q., & Rajasekar, U. (2011). Identification and analysis of urban surface temperature patterns in Greater Athens, Greece, using MODIS imagery. *Remote Sensing of Environment*, *115*(12), 3080–3090. doi:10.1016/j.rse.2011.06.014

Kumar, S., Ghosh, S., Hooda, R. S., & Singh, S. (2019). Monitoring and Prediction of Land Use Land Cover Changes and its Impact on Land Surface Temperature in the Central Part of Hisar District, Haryana Under Semi-Arid Zone of India. *Journal of Landscape Ecology*, *12*(3), 117–140. doi:10.2478/jlecol-2019-0020

Kumari, B., Tayyab, M., Mallick, J., Khan, M. F., & Rahman, A. (2018). Satellite-Driven Land Surface Temperature (LST) Using Landsat 5, 7 (TM/ETM+ SLC) and Landsat 8 (OLI/TIRS) Data and Its Association with Built-Up and Green Cover Over Urban Delhi, India. *Remote Sensing in Earth Systems Sciences*, *1*(3-4), 63–78. doi:10.100741976-018-0004-2

Lazzarini, M., Marpu, P. R., & Ghedira, H. (2013). Temperature-land cover interactions: The inversion of urban heat island phenomenon in desert city areas. *Remote Sensing of Environment*, *130*, 136–152. doi:10.1016/j.rse.2012.11.007

Li, H., & Liu, Q. (2008). Comparison of NDBI and NDVI as indicators of surface urban heat island effect in MODIS imagery. In *International conference on earth observation data processing and analysis (ICEODPA)* (Vol. 7285, p. 728503). International Society for Optics and Photonics. 10.1117/12.815679

Li, Z. L., Tang, B. H., Wu, H., Huazhong, R., Guangjian, Y., Wan, Z., Trigo, I. F., & Sobrino, J. A. (2013). Satellite-derived land surface temperature: Current status and perspectives. *Remote Sensing of Environment*, *131*, 14–37. doi:10.1016/j.rse.2012.12.008

Liang, S. (Ed.). (2018). *Comprehensive Remote Sensing*. Elsevier.

Liu, H., & Weng, Q. (2008). Seasonal variations in the relationship between landscape pattern and land surface temperature in Indianapolis, USA. *Environmental Monitoring and Assessment*, *144*(1-3), 199–219. doi:10.100710661-007-9979-5 PMID:17899413

Liu, J., & Niyogi, D. (2019). Meta-analysis of urbanization impact on rainfall modification. *Scientific Reports*, *9*(1), 1–14. doi:10.103841598-019-42494-2 PMID:31086196

Liu, L., & Zhang, Y. (2011). Urban heat island analysis using the Landsat TM data and ASTER data: A case study in Hong Kong. *Remote Sensing*, *3*(7), 1535–1552. doi:10.3390/rs3071535

Malik, M. S., Shukla, J. P., & Mishra, S. (2019). Relationship of LST, NDBI and NDVI using Landsat-8 data in Kandaihimmat Watershed, Hoshangabad, India. *Indian Journal of Geo-Marine Sciences*, *48*(1), 25–31.

Mallick, J., Kant, Y., & Bharath, B. (2008). Estimation of land surface temperature over Delhi using Landsat-7 ETM+. *The Journal of Indian Geophysical Union*, *12*, 131–140.

Mallick, J., Rahman, A., & Singh, C. K. (2013). Modeling urban heat islands in heterogeneous land surface and its correlation with impervious surface area by using night-time ASTER satellite data in highly urbanizing city, Delhi-India. *Advances in Space Research*, *52*(4), 639–655. doi:10.1016/j.asr.2013.04.025

Markham, B. L., & Barker, J. L. (1986). Landsat MSS and TM post-calibration dynamic ranges, exoatmospheric reflectances and at-satellite temperatures. *EOSAT Landsat Technical Notes, 1*, 3-8. http://landsathandbook.gsfc.nasa.gov/pdfs/L5_cal_document.pdf

Mathew, A., Sreekumar, S., Khandelwal, S., & Kumar, R. (2019). Prediction of land surface temperatures for surface urban heat island assessment over Chandigarh city using support vector regression model. *Solar Energy*, *186*, 404–415. doi:10.1016/j.solener.2019.04.001

Mukherjee, S., Joshi, P. K., & Garg, R. D. (2017). Analysis of urban built-up areas and surface urban heat island using downscaled MODIS derived land surface temperature data. *Geocarto International*, *32*(8), 900–918. doi:10.1080/10106049.2016.1222634

Nathalia, D., Kumar, K. M., Kishore, N., & Krishnan, G. (2017). Environmental change detection using Geo-Spatial techniques in Aravalli hills and Environs (Faridabad District, Haryana). *International Journal of Applied Environmental Sciences*, *12*(5), 865–875.

Newbold, K. B., & Scott, D. (2013). Migration, commuting distance, and urban sustainability in Ontario's Greater Golden Horseshoe: Implications of the Greenbelt and Places to Grow legislation. *The Canadian Geographer. Geographe Canadien*, *57*(4), 474–487. doi:10.1111/j.1541-0064.2013.12044.x

Oke, T. R. (1982). The energetic basis of the urban heat island. *Quarterly Journal of the Royal Meteorological Society*, *108*(455), 1–24. doi:10.1002/qj.49710845502

Pal, S., & Ziaul, S. K. (2017). Detection of land use and land cover change and land surface temperature in English Bazar urban centre. *The Egyptian Journal of Remote Sensing and Space Sciences*, *20*(1), 125–145. doi:10.1016/j.ejrs.2016.11.003

Pandey, B., & Seto, K. C. (2015). Urbanization and agricultural land loss in India: Comparing satellite estimates with census data. *Journal of Environmental Management*, *148*, 53–66. doi:10.1016/j.jenvman.2014.05.014 PMID:24958549

Polydoros, A., & Cartalis, C. (2014). Assessing thermal risk in urban areas–an application for the urban agglomeration of Athens. *Advances in Building Energy Research*, *8*(1), 74–83. doi:10.1080/17512549.2014.890536

Prata, A. J., Caselles, V., Coll, C., Sobrino, J. A., & Ottlé, C. (1995). Thermal remote sensing of land surface temperature from satellites: Current status and future prospects. *Remote Sensing Reviews*, *12*(3-4), 175–224. doi:10.1080/02757259509532285

Qaid, A., Lamit, H. B., Ossen, D. R., & Shahminan, R. N. R. (2016). Urban heat island and thermal comfort conditions at micro-climate scale in a tropical planned city. *Energy and Building*, *133*, 577–595. doi:10.1016/j.enbuild.2016.10.006

Qin, Z., Karnieli, A., & Berliner, P. (2001). A mono-window algorithm for retrieving land surface temperature from Landsat TM data and its application to the Israel-Egypt border region. *International Journal of Remote Sensing*, *22*(18), 3719–3746. doi:10.1080/01431160010006971

Rai, B., & Nair, S. S. (2013). Change detection of Barkhal Lake in Faridabad District of Haryana using geo-informatic techniques. *International Journal of Remote Sensing & Geoscience*, *2*(2), 38–41.

Ramachandra, T. V., Aithal, B., & Durgappa, S. (2012). Insights to urban dynamics through landscape spatial pattern analysis. *International Journal of Applied Earth Observation and Geoinformation*, *18*, 329–343. doi:10.1016/j.jag.2012.03.005

Ranagalage, M., Estoque, R., Zhang, X., & Murayama, Y. (2018). Spatial changes of urban heat island formation in the Colombo District, Sri Lanka: Implications for sustainability planning. *Sustainability*, *10*(5), 1367. doi:10.3390u10051367

Rasul, A., Balzter, H., & Smith, C. (2015). Spatial variation of the daytime Surface Urban Cool Island during the dry season in Erbil, Iraqi Kurdistan, from Landsat 8. *Urban Climate*, *14*, 176–186. doi:10.1016/j.uclim.2015.09.001

Rousta, I., Sarif, M., Gupta, R., Olafsson, H., Ranagalage, M., Murayama, Y., Zhang, H., & Mushore, T. (2018). Spatiotemporal analysis of land use/land cover and its effects on surface urban heat island using Landsat data: A case study of Metropolitan City Tehran (1988–2018). *Sustainability*, *10*(12), 4433. doi:10.3390u10124433

Sahana, M., Dutta, S., & Sajjad, H. (2019). Assessing land transformation and its relation with land surface temperature in Mumbai city, India using geospatial techniques. *International Journal of Urban Sciences*, *23*(2), 205–225. doi:10.1080/12265934.2018.1488604

Saini, H. S., Kaura, L. P., Mujtaba, S. A. I., Verma, K. & Pant, N. C (2017). Assessment of the soils around Faridabad, NCR in the context of changing landuse. *Indian Journal of Geosciences, 70-71*(3&4), 219–226.

Senanayake, I. P., Welivitiya, W. D. D. P., & Nadeeka, P. M. (2013). Assessment of green space requirement and site analysis in Colombo, Sri Lanka–a remote sensing and GIS approach. *International Journal of Scientific and Engineering Research*, *4*(12), 29–34.

Sharifi, A. (2016). From Garden City to Eco-urbanism: The quest for sustainable neighborhood development. *Sustainable Cities and Society*, *20*, 1–16. doi:10.1016/j.scs.2015.09.002

Sharma, R., & Joshi, P. K. (2014). Identifying seasonal heat islands in urban settings of Delhi (India) using remotely sensed data–An anomaly based approach. *Urban Climate*, *9*, 19–34. doi:10.1016/j.uclim.2014.05.003

Sobrino, J. A., Jiménez-Muñoz, J. C., & Paolini, L. (2004). Land surface temperature retrieval from LANDSAT TM 5. *Remote Sensing of Environment*, *90*(4), 434–440. doi:10.1016/j.rse.2004.02.003

Sodoudi, S., Shahmohamadi, P., Vollack, K., Cubasch, U., & Che-Ani, A. I. (2014). Mitigating the urban heat island effect in megacity Tehran. *Advances in Meteorology*, *2014*, 1–19. doi:10.1155/2014/547974

Stathopoulou, M., & Cartalis, C. (2009). Downscaling AVHRR land surface temperatures for improved surface urban heat island intensity estimation. *Remote Sensing of Environment*, *113*(12), 2592–2605. doi:10.1016/j.rse.2009.07.017

Stewart, I. D. (2011). A systematic review and scientific critique of methodology in modern urban heat island literature. *International Journal of Climatology*, *31*(2), 200–217. doi:10.1002/joc.2141

Sultana, S., & Satyanarayana, A. N. V. (2018). Urban heat island intensity during winter over metropolitan cities of India using remote-sensing techniques: Impact of urbanization. *International Journal of Remote Sensing*, *39*(20), 6692–6730. doi:10.1080/01431161.2018.1466072

Tsou, J., Zhuang, J., Li, Y., & Zhang, Y. (2017). Urban heat island assessment using the Landsat 8 data: A case study in Shenzhen and Hong Kong. *Urban Science*, *1*(1), 10. doi:10.3390/urbansci1010010

United Nations. (2015). Transforming our world: The 2030 agenda for sustainable development. *General Assembley 70 session*. United States Geological Survey. Landsat8 Operational Land Imager and Thermal Infrared Sensor Calibration Notices. Available online: https://landsat.usgs.gov/calibration_notices.php

United Nations. (2018). *World Urbanization Prospects, The 2018 Revision*. United Nations. Available online: https://www.un.org/development/desa/publications/2018-revision-of-world-urbanization-prospects.html

United Nations General Assembly. (1987). Report of the world commission on environment and development: Our common future. Oslo, Norway: United Nations General Assembly, Development, and International Co-operation. *Environment*.

Valsson, S. & Bharat, A (2009). Urban Heat Island: Cause for Microclimate Variations. *Architecture-Time Space & People*, 21–25.

Van, T. T., & Bao, H. D. X. (2010). Study of the impact of urban development on surface temperature using remote sensing in Ho Chi Minh City, northern Vietnam. *Geographical Research*, *48*(1), 86–96. doi:10.1111/j.1745-5871.2009.00607.x

Verbesselt, J., Hyndman, R., Newnham, G., & Culvenor, D. (2010). Detecting trend and seasonal changes in satellite image time series. *Remote Sensing of Environment*, *114*(1), 106–115. doi:10.1016/j.rse.2009.08.014

Wadhawan, M., & Ahmad, S. (2010). *11th ESRI India User Conference 2010*. Academic Press.

Wang, L., Lu, Y., & Yao, Y. (2019). Comparison of three algorithms for the retrieval of land surface temperature from Landsat 8 images. *Sensors (Basel)*, *19*(22), 5049. doi:10.339019225049 PMID:31752412

Wang, Y. C., Hu, B. K. H., Myint, S. W., Feng, C.-C., Chow, W. T. L., & Passy, P. F. (2018). Patterns of land change and their potential impacts on land surface temperature change in Yangon, Myanmar. *The Science of the Total Environment*, *643*, 738–750. doi:10.1016/j.scitotenv.2018.06.209 PMID:29957438

Weng, Q. (2001). Modeling urban growth effects on surface runoff with the integration of remote sensing and GIS. *Environmental Management*, *28*(6), 737–748. doi:10.1007002670010258 PMID:11915963

Weng, Q. (2009). Thermal infrared remote sensing for urban climate and environmental studies: Methods, applications and trends. *ISPRS Journal of Photogrammetry and Remote Sensing*, *64*(4), 335–344. doi:10.1016/j.isprsjprs.2009.03.007

Yang, Y., Zhang, X., Lu, X., Hu, J., Pan, X., Zhu, Q., & Su, W. (2018). Effects of Building Design Elements on Residential Thermal Environment. *Sustainability*, *10*(2), 57. doi:10.3390u10010057

Yu, X., Guo, X., & Wu, Z. (2014). Land surface temperature retrieval from Landsat 8 TIRS—Comparison between radiative transfer equation-based method, split window algorithm and single channel method. *Remote Sensing*, *6*(10), 9829–9852. doi:10.3390/rs6109829

Yue, W., Xu, J., Tan, W., & Xu, L. (2007). The relationship between land surface temperature and NDVI with remote sensing: Application to Shanghai Landsat 7 ETM+ data. *International Journal of Remote Sensing*, *28*(15), 3205–3226. doi:10.1080/01431160500306906

Zhan, W., Zhou, J., Ju, W., Li, M., Sandholt, I., Voogt, J., & Yu, C. (2014). Remotely sensed soil temperatures beneath snow-free skin-surface using thermal observations from tandem polar-orbiting satellites: An analytical three-time-scale model. *Remote Sensing of Environment*, *143*, 1–14. doi:10.1016/j.rse.2013.12.004

KEY TERMS AND DEFINITIONS

Climate Change Adaptation: To enhance the capability environmental, social, and economic systems to adjust the climate change impact by reducing their vulnerability towards actual and expected climate.

Climate Change Mitigation: To reduce the rate of global warming effects by limits the factors responsible for it.

Land Surface Temperature: Degree of hotness of the surface of the Earth when one touches it on any location.

Land Use Land Cover Change: The changed situation of land from its natural condition after anthropogenic intrusion to occupy the land for human benefit.

Spectral Indices: Mathematical equations considering two or more wavelengths based on the spectral properties of different features of topography of earth surface to highlight the features of interest.

Sustainable Development: Sustainable development is the combination of principles and approaches for the growth of the present society by judiciously using resources required to meet human development goals without compromising the stability of the ecosystem and conserve the resources for the future generations too.

Sustainable Urbanism: It is the way to construct, plan and manage the city with the principles to maintain equity between urban growth and natural resources to prevent ecological imbalance and make the city climate resilient as well.

Urban Cool Island: Cooler urban area compared to the non-urban/rural surroundings.

Urban Heat Island: Small urban pockets significantly warmer compared to the rural surrounding.

Chapter 9
Remote Sensing and GIS for Modelling Green Roofs Potential at Different Urban Scales

Teresa Santos
https://orcid.org/0000-0001-7061-7346
Interdisciplinary Centre of Social Sciences (CICS.NOVA), NOVA FCSH, Universidade NOVA de Lisboa, Portugal

José António Tenedório
https://orcid.org/0000-0003-3047-0807
Interdisciplinary Centre of Social Sciences (CICS.NOVA), NOVA FCSH, Universidade NOVA de Lisboa, Portugal

Caio Silva
https://orcid.org/0000-0001-8910-1841
Universidade de Brasília, Brazil

Thiago Montenegro Góes
https://orcid.org/0000-0002-5745-229X
Universidade de Brasília, Brazil

ABSTRACT

Urban greening is an essential dimension of sustainable urban development. Through green planning, cities can develop strategies towards climate change and mitigation actions and reduce emissions. This chapter aims to evaluate the combined use of remote sensing data and GIS to produce different levels of sustainable indicators, starting from the city level to the neighborhood and the building levels. Green roofs can be part of the solution through the ecosystem services provided. Its benefits are evaluated for the different urban scales through scenario analysis. The impact in the outdoor thermal comfort is assessed at the neighborhood level, while the overall potential to improve energy efficiency is evaluated at the building level. The methodology is implemented in the city of Lisbon, Portugal through a pilot

DOI: 10.4018/978-1-7998-2249-3.ch009

case study. The results can provide rationales for the city when formulating new incentives to encourage the uptake of green roofs in Lisbon.

INTRODUCTION

The United Nations' 2030 Agenda for Sustainable Development (SD) defined 17 SD goals for the next fifteen years. The aim is to achieve a better and more sustainable future, by addressing global challenges like poverty, inequality, climate, environmental degradation, prosperity, peace, and justice (United Nations, 2016). Among these goals, Goal 11 is dedicated to Sustainable Cities and Communities, and its targets include, among others, increasing the number of cities and human settlements that adopt and implement integrated policies and plans for mitigation and adaptation to climate change.

Green infrastructures such as trees, plants, or forests are vital assets that through spatial planning can help cities adapt to climate change (Demuzere et al., 2014; Zölch et al., 2016). They provide services like managing high temperatures and flooding or reducing greenhouse gas emissions through direct carbon sequestration (Chen, 2015; Jayasooriya & Ng, 2014; Klemm, Heusinkveld, Lenzholzer, & van Hove, 2015; Muñoz-Vallés et al., 2013; Norton et al., 2015; Yang & Wang, 2017). Green infrastructures also have other social, economic and environmental benefits like improving health and well-being by filtering air pollutants and increase recreational opportunities, reducing buildings energy usage and heating and cooling costs by providing shading and wind-blocking, or fostering biodiversity through habitats provision (Camps-Calvet et al., 2016; Connop et al., 2016; Elmqvist et al., 2015; EPA - United States Environmental Protection Agency, 2014; Tzoulas et al., 2007).

Sustainable development requires green spaces but, in densely built city centers, unoccupied land is rarely available, and green spaces are few and far between (Haaland & van den Bosch, 2015; Jim, 2004). The loss of green urban areas during the last decades and the consequent soil sealing opened the way to novel mitigation strategies for restoring the environment and protect human health. Typically, open green spaces are the most desired interventions; nevertheless, available space is generally sparse for this purpose. However, roofs are valuable empty grey spaces that can be used to compensate for the lack of permeable land and help cities adapt to climate change (Bates et al., 2013; Francis & Lorimer, 2011; Haaland & van den Bosch, 2015; Santos et al., 2016). Green Roofs (GR) are strategies to promote urban greenspace, and that can be explored in a sustainable development context. GR constitute a solution to restore the environmental quality in densified neighborhoods, providing benefits for the buildings, and for the community as well (Getter & Rowe, 2006; Jim, 2013), in the context of energy transition in the built environment.

For property owners, the installation of a GR reduces the amount of energy needed to maintain the indoor temperature. In fact, GR improve the thermal performance of buildings by reducing the thermal load in the summer and promoting insulation in the winter (Saadatian et al., 2013). Lowering the buildings' heating and cooling cost with air-conditioning improves energy efficiency. This reduced energy consumption also lowers the building carbon footprint, thus helping to promote low-carbon communities (Ascione et al., 2013; Lehmann, 2014). Furthermore, the presence of a green cover increases the longevity of the waterproof membrane, decreasing the exposure to large temperature fluctuations and ultraviolet radiation (Bianchini & Hewage, 2012). Increasing membrane durability also reduces material waste. Another benefit is the improved acoustical insulation characteristics of the roof. The green coverage attenuates the noise inside the building by mitigating the sound waves over the roof (Saadatian et al., 2013).

Public benefits include Urban Heat Island (UHI) mitigation through shadowing and evapotranspiration. GR provides several ecosystem services: biodiversity and habitat, pollution control, improved air quality and carbon storage. Furthermore, they act as an effective stormwater retention management tool by reducing the amount of rainwater discharged into the sewers (Brandão et al., 2017); also, in peak flows the tempering effect allows designing drainage systems with reduced diameters. Other benefits include sound absorption leading to reducing noise exposure levels, aesthetical benefits and new green jobs in the local economy. Furthermore, GR can offer a novel space for leisure activities and can be used as educational laboratories or even can serve other functions like local food production.

There are two types of GR: intensive and extensive (Berardi et al., 2014). Extensive GR are covered with low growing plants like sedums, requiring little maintenance and no permanent irrigation. The soil depth is shallow (7-15 cm) and therefore, extensive green roofs have low weight, thus minimizing the installation costs and the structural load of the roof. On the contrary, intensive GR requires professional maintenance and complex irrigation systems. The soil depth is deeper (more than 15 cm), contributing to a heavier weight load on the building structure. However, while extensive roofs only have limited plant species options, intensive roofs can include a variety of vegetation, from herbaceous plants to small trees or shrubs (Vijayaraghavan, 2016). Plant selection may also include vegetables and fruits. These rooftop farms require a higher nutrient application and careful maintenance. In this paper, all roofs with a plant growing medium are under the category of "Green Roofs", and no distinction is made between intensive and extensive types.

Germany started back in the 1960s the conversations from conventional to green roofs and is now the European country with the highest rate of retrofittings. In Munich, for example, all suitable flat roofs with an area over 100 m2 must be covered with green. The same specifications apply to all new residential and commercial roofs in Linz, Austria. In Copenhagen, Denmark, this is mandatory for all new suitable roofs under 30º pitch, while in France, all roofs of new buildings in commercial areas must either be covered with vegetation or solar panels (Madre et al., 2014). In Toronto, Canada, it is stipulated that all new commercial, institutional or residential and industrial rooftops with a minimum gross floor area of 2000 m2 must have a green roof (TGRTAG, 2011).Recently, New York city, USA, approved the a policy that mandates that new and existing buildings under renovation located in specific groups must be all covered with green systems, and/or solar panels and/or wind turbines (Maya Stern et al., 2019).

Not every building can receive a green cover. Several factors that influence the decision-making of the potential adoption of rooftop greenery systems must be evaluated (Karteris et al., 2016). The first difficulty is the presence of tiles as cover material. Also pitched roofs with a slope higher than 20þ are not indicated to receive these green structures. Finally, many buildings would need structural upgrades to be able to hold the weight of the GR (Santos et al., 2016). So, planning for green development in cities should be based on information about local suitability for green retrofitting. Knowing the local capabilities would provide a solid foundation for decision-making. Sustainable development indicators are quantitative descriptors of urban environments, commonly used to assist the design of policies and incentives and to monitor its implementation. In this context, remote sensing data is one of the most efficient sources of information about the urban surface. The recent advances in sensor technologies and algorithms made it possible to acquire Very-High Resolution (VHR) images (5m to less than 1m), at global coverage, on a regular basis, at relatively low cost. Such detailed information about the surface enables multi-scale approaches, from cities to blocks, and blocks to buildings (Freire et al., 2014). Such VHR information includes radiometric and altimetric data. Combining different digital sensors, complex 3D models of the surface can be produced in Geographic Information Systems (GIS) environments and used to evaluate diverse factors that contribute to urban quality. Typically, derived data is presented as spatial indicators, that provide up-dated quantitative information about a specific object, from vegetation status to land surface temperature, water quality, or pollutant emissions.

Geographic-based indicators can be derived to support green planning, targeting incentives and schemes to those areas where the retrofitting is expected to be more successful (Santos et al., 2016). In a planning scenario, public buildings can be initial candidates for green retrofitting. Due to their higher visibility, promoting green roofs on suitable public buildings can create a positive impact in the city and encourage private owners to incorporate GR (Pevzner, 2013). In this context, public school buildings constitute a good case study for GR implementation. In one hand, the schools' unused roof area can be converted in a novel recreational amenity for the local community and be used as an educational resource and, in the other hand, the schools can also beneficiate from lower heating and cooling costs due to GR insulating effect.

The present research is a preliminary pilot-study that aims to demonstrate in Lisbon city, Portugal, how the combined use of Remote Sensing data and GIS can produce different levels of indicators to guide the implementation of GRs. Two scales of analysis are proposed: the city, the neighborhood and the building. There are three methodological steps in this research: the Lisbon GR potential map (Santos et al., 2016) is used to locate priority areas for promoting green roofing within the city according to geographic and social potential (educational use), then the benefits are analyzed 1) at the neighborhood level and, 2) at the school building level. The methodological results can provide rationales for the city when formulating new incentives to encourage the uptake of GR in Lisbon.

The chapter is structured as follows. A literature review was conducted to assess the current use of remote sensing data and Geographic Information Systems (GIS) to produce green indicators at the city and the local scale. Then, the methodology is described, including the data models used, followed by a detailed analysis and discussion of results. Key findings are then addressed in the conclusion section, as well as study limitations and future recommendations.

BACKGROUND

Remote Sensing For Land Cover Mapping

Remote Sensing is the observation of an object, area or phenomena from a distance using sensors onboard airborne or spaceborne platforms (Campbell & Wynne, 2011; Jensen, 2015; Lillesand et al., 2015). Areas of application include forestry, agriculture, oceanography, geomorphology, climatology, etc. Regarding the urban environment, aerial imagery has been the most common data source for mapping human activities (Ural et al., 2011; Vu et al., 2009) and only recently, images obtained by satellite or Unmanned Aerial Vehicle (UAV) have gained interest, as alternative data sources for mapping urban areas. Depending on the flight height, detailed information about land surface could be captured and used to study urban areas. Alternatively, sensors developed for remote detection capture, through a digital process, the energy reflected by the different elements on the surface of the Earth, and then converted into a Digital Number (DN) that constitutes the image. Most commercial sensors are confined to the region of the visible and near-infrared. In this sense, through the analysis of the DNs in the image is then possible to extract information about the elements on the surface and study their dynamics (Campbell & Wynne, 2011).

Landsat 1, launched in 1972, was the first generation of satellites designed for land observation, capturing large areas in several regions of the electromagnetic spectrum. However, its 80 meter-ground resolution did not allow for detail applications. The Landsat program evolved to lower orbit spacecraft and improved sensor instruments. Landsat 4 and 5, launched in 1982 and 1984 respectively, were able to capture images with 30-meter resolutions, and its sensor – Thematic Mapper (TM) – provided seven bands in several spectral regions, carefully chosen to provide information for vegetation and surface mapping (Fu et al., 2020). SPOT High-Resolution Visible with 20-m spatial resolution in its multispectral sensor and a 10-m resolution panchromatic sensor were also part of this second generation of sensors. The availability of free (Landsat), worldwide image data set with medium spatial and high spectral resolutions, boosted the development of digital image processing algorithms for extracting thematic information from satellite observation data. More recently, the Copernicus Open Access Hub, from the European Space Agency (ESA) also provides free and open access to all Sentinel data products, including multispectral imagery with 10, 20 and 60 m resolution. Nevertheless, and although second-generation sensors provided spatial resolution higher than first generation ones, it was still not significantly better for urban studies because it continued to be inferior to the spatial resolutions of urban elements (Le Bris & Chehata, 2019; Tran et al., 2011).

The declassification of military technology in the 1990s, that allowed developing the third generation of more accurate sensors (Aplin, 2003). The first Very-High Resolution (VHR) images were commercially available in 1999 when IKONOS-2 was launched (Fu et al., 2020). The VHR spatial resolution of 4 m resolution in the multispectral mode and 1 m in the panchromatic mode, compromised the spectral resolution of previous generations, with only four spectral bands (three in the visible and one in the near infrared region). QuickBird-2 satellite, launched in 2001, offered images with 2.44 m resolution in the multispectral mode and 0.61 m in the panchromatic mode, and with four spectral bands. Since then, other VHR satellites have been launched, with increased spatial and spectral resolutions (e.g., Pleiades-1, GeoEye-1, WorldView-3) (Satellite Imaging Corp, n.d.). Nowadays, there are different digital imaging devices, collecting remotely information. Examples include Charge Coupled Device (CCD) digital cameras, thermal cameras, LiDAR (Light Detection And Ranging), push broom scanners or by Synthetic Aperture Radar (SAR) (Fu et al., 2020).

The recent availability of digital cameras acquiring data representing elevation allowed improving the geometric correction (i.e., the process of assigning a coordinate system) of remotely sensed imagery. Furthermore, such accurate altimetric information also allows the generation of contour lines and terrain models, thus providing another source of information for analysis. Nowadays, Digital Surface

Models (DSM), collected through LiDAR sensors, can be used to generate slope and aspect models for integration into (land cover) thematic classification data sets or used as a stand-alone data layer.

Remote sensing data is widely used when geographic information is required. For mapping purposes, quantitative techniques are generally used to identify the objects in the images. The aim of a digital classification is, therefore, the analysis of multispectral data and application of statistics to assign each image-pixel to a thematic class (Jensen, 2015). When it comes to urban land cover mapping, approaches based on merging data from various sources, are generally used for improving classification's quality. To overcome the high heterogeneity of urban areas, additional data – also known as ancillary data – is commonly included in the classification process. Such data are often composed of map-based thematic data or terrain data like insolation, census data, building footprint, road network, rain intensity, land zoning, etc. In this context, joining spectral information from optical 2D images with 3D surface information has allowed more sophisticated studies of the urban environment. Tooke, Coops, Goodwin, and Voogt (2009) classified urban vegetation characteristics like species and condition, using an approach based on VHR QuickBird imagery, field-based land cover observations and shadow estimates derived from a LiDAR DSM. Raciti, Hutyra, & Newell (2014) explored the use of imagery and LiDAR-based tree-height estimates, coupled with field-based observations, to develop an urban biomass map of Boston, USA. The data was used to map canopy cover and aboveground tree carbon storage. Liu, Coops, Aven and Pang (2017) evaluated the potential of LiDAR and hyperspectral imagery to estimate the crown structure and canopy physiology of 15 urban tree species in a Canadian city. Results indicated an overall accuracy of 51.1%, 61.0%, and 70.0% using hyperspectral, LiDAR and the combined data respectively.

For regional and national scales, multispectral algorithms at the pixel level are generally used. However, for studying urban environments, the use of traditional pixel-based classifiers produces low accuracy maps (Tehrany et al., 2014; Weng, 2012). Only with the development of new integrated methodologies like GEographic Object Base Image Analysis – GEOBIA – was it possible to group image-pixels into objects based on information available from spectral, textural, contextual and spatial domains (Blaschke & Strobl, 2001).

According to Blaschke (2010), GEOBIA has emerged as an area of research due to: 1) the increasing availability of image data being produced in a broader range of spatial, spectral, radiometric and temporal resolutions; 2) the orchestrated supranational programs and systems for regular or on-demand surveys of the Earth's surface (e.g. Global Earth Observation System of Systems - GEOSS, or the Global Monitoring for Environment and Security - GMES), and 3) the availability of powerful, off-the-shelf software that bridges image processing and GIS functionalities in an object-based environment.

Urban Sustainability and Remote Sensing Data

Urban sustainability is one of the dominant paradigms in the design and management of cities (Heymans et al., 2019). On the one hand, because urban history has shown the irrationality of some urban planning solutions (or lack of planning, such as informal urban growth) such as vertical growth with very high densities or dominantly horizontal growth, dispersed and fragmented. Informal urban growth has revealed, especially in less developed countries, the inability of public policies to respond to the city's attraction to population and employment, as well as to economic activities and capital. On the other hand, because the political decision began to incorporate in its discourse the environmental concerns and the transition to the sustainability. The new discourses on the city translate these environmental concerns, and this influences how urban planners and managers project the city of the future.

The structure and form of urban settlements must be analyzed for planners to make intervention proposals. Such intervention proposals, after implementation, should be monitored. In fact, the city must be monitored so that policy makers can assess the adjustments and mismatches of urban interventions. It is to this process that remote sensing is useful, because it allows to observe in a multitemporal way the changes that occur in the cities resulting from the urban interventions.

Sustainable urban planning success is highly dependent on accurate and robust information. Remote sensing-based indicators are capable of producing spatial tools to inform and support the implementation of urban policies and initiatives (Behling, Bochow, Foerster, Roessner, & Kaufmann, 2015; Chen et al., 2018; Lakes & Kim, 2012; Zhang & Guindon, 2006).

Spatial urban indicators gather the information that allows planners to decide where and when should the policies be realized, what can be the realistic targets, and who is more apt to implement them. Chrysoulakis et al. (2014) provided an extensive list of indicators for urban planning based on Earth Observation data and concluded that these are appropriate to retrieve information suitable to describe air pollution and public health, energy efficiency, transportation accessibility, thermal comfort, urban green, and vulnerability to natural hazard.

When considering adaptation and mitigation measures to climate change, remote sensing data can provide relevant indicators to better estimate current needs and evaluate alternative planning scenarios. In this context, green infrastructures can produce climatic modifications at the micro-scale, but its efficiency depends on the implementation criteria (vegetation density, selection of species, etc). Ng et al. (2012) studied the cooling effects of greening in Hong Kong and conclueded that in high-density cities, the amount of tree planting needed to lower pedestrians level air temperature by around 1 °C is approximately 33% of the urban area. Lobaccaro & Acero (2015) studied the benefits of different green actions inside street canyons in Bilbao, Spain, and concluded that the maximum reduction of the Physiological Equivalent Temperature (PET) thermal index registered in this study was 10ºC in the compact low-rise and mid-rise areas and 8 ºC in the open-set high-rise area, in the scenario with grass and trees. Zölch et al. (2016) tested the impact of difference green scenarios in heat mitigation and concluded that planting trees had the strongest impact with an average PET reduction of 13% compared with existing vegetation and that another valuable adaptation option is green facades, which have mitigating effects of 5%–10%. In contrast, the effects of green roofs were negligible. Santos et al. (2019) tested the effect of promoting new green areas, at ground and at rooftop levels, in the thermal comfort of the surrounding urban area, using the requalification project of a neighborhood located in Lisbon, Portugal. The simulation results indicated that the presence of green infrastructures revealed an improvement in the comfort indexes for the morning and afternoon periods, and regarding the air temperature, a potential reduction up to 3ºC is expected.

Planning for green intervention in a city implies the existence of suitable areas on the ground, however, in high-density areas, available space is generally sparse. In this context, green roofs can constitute a valid alternative and knowing where the best locations in the city to receive such structures are located constitutes an essential tool when it comes to green planning (Santos et al., 2019). Mallinis et al. (2014) proposed a methodology to estimate green roof retrofitting areas using orthoimages and a Digital Surface Model (DSM), while Wong & Lau (2013) used a virtual 3D model of the buildings and Google View images, and Tian & Jim (2011) studied the factors that influence the spatial pattern of sky gardens in Hong Kong, using orthophoto maps, buildings' footprints, and zoning plans. When considering green roofs potential, high pitched, tiled and shadow roofs are not good candidates for retrofit. Some authors considered criteria such as roofing material, area and shadowing (Mallinis et al., 2014), sunlight and roof space (Wong & Lau, 2013) or roof slope (Silva et al., 2017). However, Santos et al. (2016) went further beyond and included all these criteria - sunlight hours, roofs' material, area and slope - and innovated by considering and quantifying the individual impact of each criterion in the selection of candidates for green roofs. Using a 3D DSM and VHR imagery, the current vegetation at the ground-level was mapped and the GR potential at the rooftop level was assessed, at the city scale. More recently, Mutani & Todeschi (2020) replicated Santos et al. (2016) methodology developed for Lisbon to identify potential green roofs in some areas of Turin, thus demonstrating its transferability.

Santos et al., 2016, using Lisbon as a case study, mapped GR potential, and identified local capabilities. Starting from this point, it is now pertinent to assess the role of public buildings to the sustainable development of the city, through a pilot-study.

STUDY AREA AND DATASET

Study Area

The selected study area is in Lisbon, located on Portugal, on the western coast of Europe. The city occupies an area of 84 km^2 and has approximately 500 000 inhabitants (census 2011). Lisbon has a Mediterranean climate (Köppen classification: Csa) ("Köppen-Geiger Climate Classification," 2010), and is subject to sudden, intense rain, overwhelming the catchment systems and triggering flash floods. The UHI has an average intensity of 3°C, with higher occurrence in more densely built-up areas near the Tagus river and along the central circulation axis (Alcoforado et al., 2009).

The built environment is very diverse, from single-family houses to multi-story apartments, mix-use buildings, and industrial pavilions. Buildings, in the studied neighborhood, are not very high, ranging from two-floor houses up to 8 to 9-floor mid-rise buildings. According to the 2011 census, 77% of residential buildings in Lisbon were built before 1970 (20% are before 1919, typically located in the historical area). In this sense, GR are not a common infrastructure in Lisbon. On one hand, many buildings have limited structural capacity and, on the other hand, the typical covering material – clay – is used in 80% of the city buildings (Santos et al., 2016). Furthermore, there are no financial incentives to encourage investments in GR installation.

Geographic Dataset

The dataset used in this study has in remote sensing its source of information about the surface, along with other auxiliary data:

- An orthophoto from 2012 (Direção-Geral do Território, n.d.), with 0,50 m^2 of spatial resolution was used to identify the land cover, from roofing materials to pavements, soil and green areas;
- A normalized Digital Surface Model (nDSM) from 2006, with a pixel of 1 m (Santos, 2011) was used to identify the height of all buildings in the study area;
- The buildings' footprints from 2006 were updated from the municipal cartography 1:1000 (Santos, 2011), and used to identify its location and area;

- The public schools' geographic information is available at the City Hall open data portal and it was used to identify which schools have green roof potential (Câmara Municipal de Lisboa, n.d.);
- The city's climatic characterization includes the seasonal information for the summer and weather periods (Silva et al., 2018);
- The starting point for the analysis is a map with the green roof potential of all rooftops in Lisbon (Santos et al., 2016) (Figure 1). This map includes the location of all rooftops across the city with geographic characteristics that enable its green coverage: being flat roofs (£ 11º), without red-tiles, with available area higher than 100 m^2 and with an average daily solar insolation of 3 hours. According to these criteria, 2890 buildings were classified with green roof potential.

Figure 1. The Green Roof Potential Map for the city of Lisbon, Portugal (Santos et al., 2016)

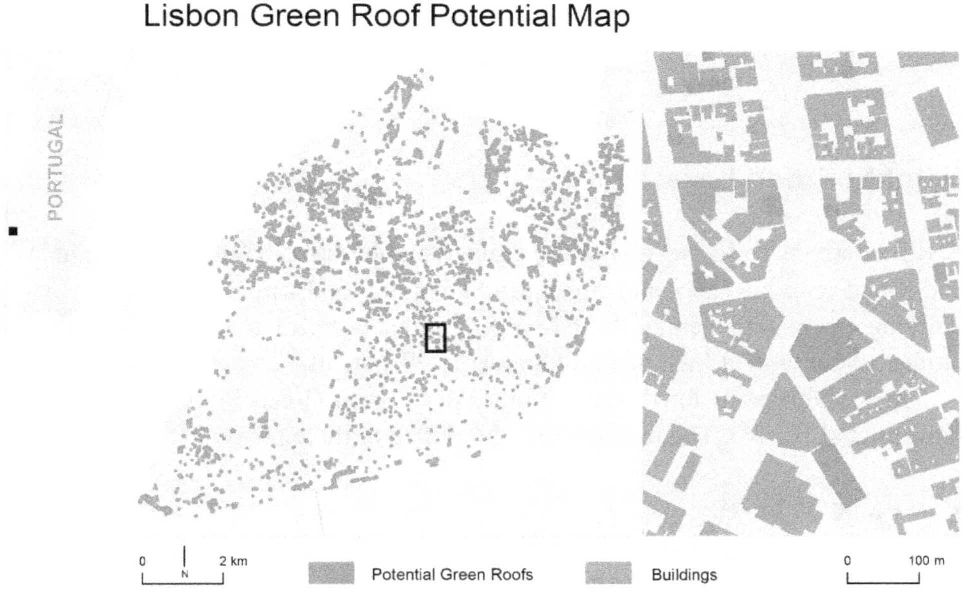

METHODS

The methodological framework comprises three broad steps to evaluate the GR benefits starting with the information available at the city scale to identify one suitable neighborhood and the public building in it. Then, the thermal comfort will be assessed at the neighborhood level and the thermo-energy will be applied at the building level (Figure 2). Following is a detailed description of each phase.

Figure 2. The methodological framework to assess the thermal comfort at the neighborhood and to evaluate the thermo-energetic condition at the building level

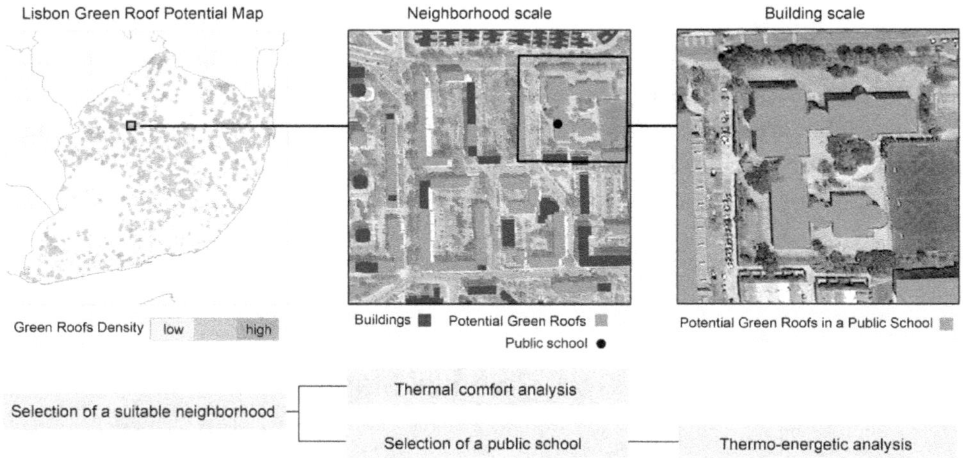

Neighborhood Level

The first step is to locate within the city, a neighborhood based on geographical and social criteria. The goal is to select a neighborhood with 1) a high density of buildings with GR potential (based on geographic criteria) and, 2) that includes a public school with at least one building with GR potential (social criteria).

The starting point for this selection is the Lisbon Green Roof Potential Map (Santos et al., 2016). The methodology is applied in three phases:

- Based on the city map, a point density analysis is applied, using the Spatial Analyst Tool available in ArcGIS® software (Esri). Using the green roofs' centroids as the input point feature, the output is a raster map with the magnitude-per-unit of the green roofs that fall within an area around each cell. The result is then a density map, where the buildings with of GR potential are concentrated within the city. This map allows selecting suitable neighborhoods according to a geographic criterion, i.e., the density/concentration;
- Then, the public schools that have one or more buildings with GR potential are identified. This is accomplished by overlaying the schools' layer with the Lisbon Green Roof Potential Map.

- Lastly, a neighborhood is selected for detailed evaluation by overlaying the information of the public schools' buildings with GR potential with the density map.

The second step, at the neighborhood level, assesses the contribution of the GR in the thermal comfort of the public areas. The aim is to evaluate thermal comfort based on the following four physical parameters:

- Air temperature – measures the energy of molecular motion (i.e., how quickly are the air molecules moving) and indicates how cold the air is. It is generally presented in Kelvin, Fahrenheit or Celsius scales (the latter being adopted in this publication). Values of 20 up to 22 °C indicate typical comfortable temperatures (Givoni, 1992);
- Relative humidity – measures the amount of water vapor present in the air. It is expressed as a percentage of the amount of air needed to achieve saturation at a given temperature. Low values indicate drier environments, where respiration becomes difficult, while high values indicate wetter environments, where sweating capability is decreased. Comfortable relative humidity values are within the range of 30 and 70% (Givoni, 1992);
- Wind Speed – measures the air velocity in m/s units. The higher the air velocity, the higher is the heat exchange between a human body in space and the air around it. Comfortable air velocity values are generally 1 m/s (Olgyay, 2015);
- Mean radiant temperature – it summarizes all radiant heat fluxes (long and short-wave components) of radiating surrounding surfaces that reach the human body. In sunny days, and in outdoor conditions, the mean radiant temperature can be more than 30 °C higher than air temperature (Kántor & Unger, 2011). Although even for shaded locations, mean radiant temperature can be higher than air temperature due to the diffuse and reflected solar radiation components (Ali-Toudert & Mayer, 2007).

The thermal comfort is associated with these variables: in hot weather, air flow helps evaporative cooling from sweating. If the relative humidity is high, then the sweating process is harder, but air velocity can help by promoting ventilation and removing saturated air near the skin with fresh lower-humidity air. However, in cold conditions, air velocity can increase the evaporative cooling beyond comfort levels. Wind speed also influences mean radiant temperature.

Nevertheless, thermal comfort is a subjective feeling that also depends upon other variables. In this sense, this research uses the Predicted Mean Vote (PMV) index, one of the most known models of thermal comfort, adopted in the ISO Standard

7730 (2005). This index was developed by Fanger (1972) and indicates the mean thermal sensation vote on a standard scale for a large group of persons for any given combination of the thermal environmental variables, activity (i.e., metabolic rate) and clothing levels (). The reference person for PMV is a 35-years-old male, with a height of 1.75 m, and a weight of 75 kg. The PMV scale is generally defined between -3 (cold) to +3 (hot) (ASHRAE comfort scale). Zero is an ideal value, meaning a neutral situation. Although PMV was initially developed for indoor applications, it can also be adapted to outdoor situations (e.g., in ENVI-met software it is adapted following German VDI 3787 standard).

All simulations are performed in ENVI-met® v4 software, developed by Bruse and Fleer (1998), which is a is a high-resolution numerical simulation model of the urban environments based on local thermodynamics, fluid simulation and vegetation physiology. Its scientific approach allows quantifying the effects of architectural and urban planning on-site thermal characteristics. ENVI-met operates in three steps:

- As ENVI-met adopts a numerical model, the study area must be a grid cell with the identification of all elements at the surface. So, the first step is to design a map with the location and the characterization of the materials of every building, vegetation, soil and other urban elements. Remote sensing data can be used in this stage to aid the classification process. Buildings are the first elements to be modelled. This is done using the orthophoto to locate the built-up areas and to identify the covering material, and the nDSM to inform about the respective height. Then, the next element to be modelled is the vegetation cover, using ENVI-met plant database. The orthophoto is used to match the local vegetation with the species available in the database. For the GR cover, the plant type used was Luzerne (*Medicago sativa*) with 18 cm, a perennial flowering plant. This the species available at Envi-met that most resembles those used in Portugal. The last elements to be modelled are the soil and the surface, also using the orthophoto image. This modelling stage can be complex and take much time, depending on the required detail;
- After map generation, local climatic data must be configurated, based on a typical day. Two climatic archives were built to take describe Lisbon's summer and winter conditions. The data included air temperature, relative humidity, wind velocity at 10 m height, wind direction and relative humidity at 2500 m height (Silva et al., 2018);
- The last step the microclimate of outdoor spaces is simulated based on local climate parameters. The simulations typically cover 24-48 h ahead and result in a raster file for each elevation in the 3D surface model, for each variable, for each hour, in a daily basis. The output is a set of 2D coloured maps

(or sections) representing the selected variables: air temperature, relative humidity, wind velocity, mean radiant temperature and PMV.

To test the impact of implementing GR in the most suitable buildings, an initial input file map is produced considering two situations: 1) the baseline scenario, with the current land cover; and 2) the green roof scenario, where the suitable buildings are identified. Then, the urban microclimate is simulated for two characteristic climatic periods: August and January, which represent the summer and winter, respectively. The same approach is applied for both scenarios to allow proper comparison. The impact of GR in the outdoor thermal comfort of the selected neighborhood is then evaluated through the physical parameters and the PMV. The methodological framework is presented in Figure 3.

Building Level

In the third step, at the building level, it is analyzed the GR impact on the energy consumption of a school building. The main analysis purpose is to evaluate the overall GR potential to improve a building energy efficiency, in a building stock standpoint, and not a retrofit analysis of the school itself. The focus is to enable a better understanding of the possible energy reduction of the GR implementation in a broader sense, to foster better policy decision-making.

For this thermo-energetic analysis of the school, it is carried a dynamic building simulation in EnergyPlus 8.5 with the aid of the graphical user interface DesignBuilder 6.0. Nine different scenarios are evaluated for the energy consumption, a baseline scenario with a traditional roof and eight different GR alternatives. Both heating and cooling load are analyzed for the whole school year.

The thermophysical characterization of the building envelope, as well as its systems, schedules and loads, are defined by the standard of the Portuguese Secondary School Building Modernization Program (www.parque-escolar.pt). The overall measurements were obtained from Google Earth.

The following parameters are considered in the energy consumption analysis:

- All activities and equipment schedules are set from 8:00 to 18:00 all weekdays. Also, school vacation (from July 15th to September 15th) is not considered in the calculations;
- Occupancy of 0.57 and the metabolic rate of 0.75, adequate to children;
- The plug load is set to 3.6 W/m^2 according to study from Srinivasan, Lakshmanan, Srivastav and Santosa (2011);

Figure 3. A conceptual framework for modelling the impact of green roofs at the neighborhood level using ENVI-met.

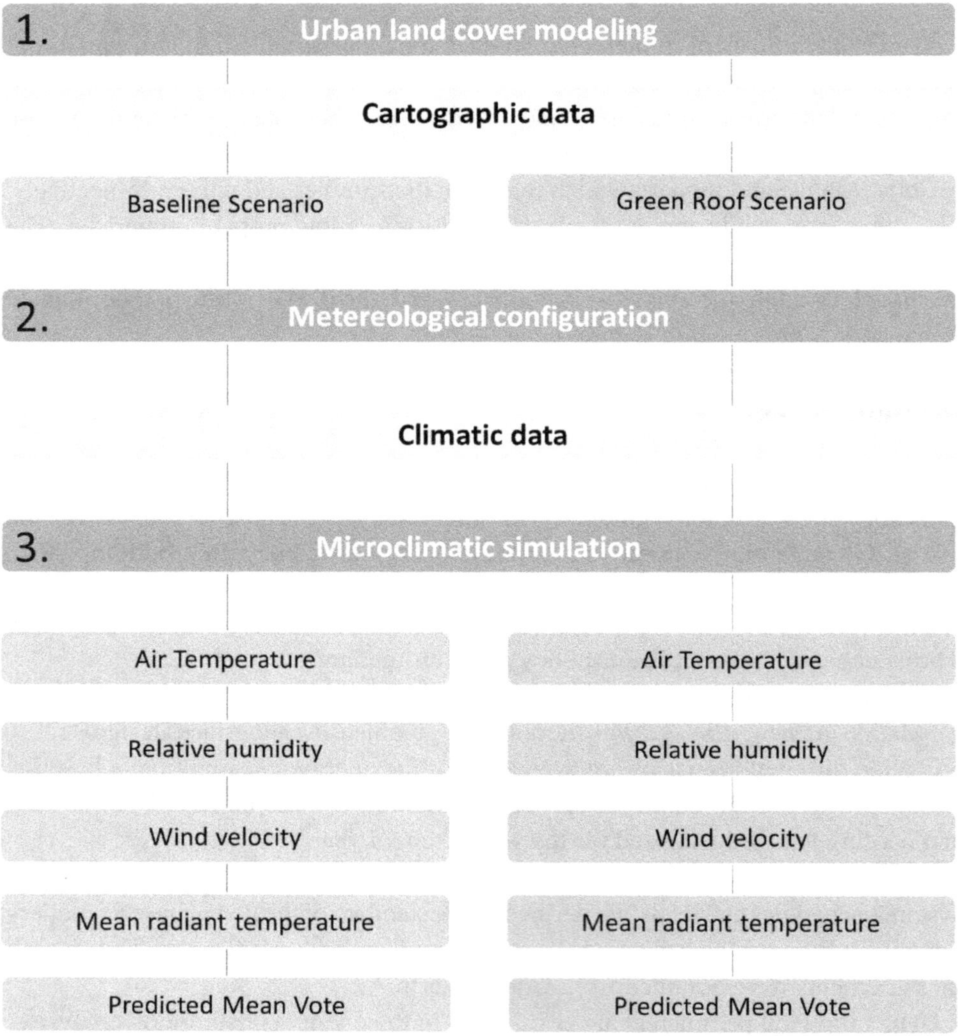

- The artificial light power density is 2.4 W/m^2 - 100 lux according to *"Regulamento de Desempenho Energético dos Edifícios de Comércio e Serviços"* (RECS) (Ministérios do Ambiente, Ordenamento do Território e Energia e da Solidariedade, Emprego e Segurança Social, 2013);
- Heating and cooling equipment follow the Portuguese energy efficient regulation *"Regulamento De Desempenho Energético Dos Edifícios De Habitação"* (REH) guidelines and have a Coefficient of Performance

(COP) of 2.8 for heating and 2.5 for cooling for air-exchanger rooftop units (Ministério da Economia e do Emprego, 2013);

- The weather file from Laboratório Nacional de Engenharia e Geologia (LNEG) for Lisbon is used in the simulations (www.lneg.pt);+8
- The envelope is characterized according to the Portuguese energy efficient regulation *"Regulamento das Características de Comportamento Térmico dos Edifícios"* (RCCTE) (Ministério das Obras Públicas, Transportes e Comunicações, 2006), as expressed in Table 1.
- For the school blueprint, it is employed the same value from the study of Bernardo, Antunes, Gaspar, Pereira and da Silva (2017) with a 21% of window to floor ratio. The thermal physical characterization of the window system is shown in Table 2

Table 1. Envelope characteristics according to the Portuguese regulation

System	Layers	U-value (W/ m².K)
Wall	External Render + Brickwork + Plaster	1.80
Partition	Plaster + Brickwork + Plaster	2.00
Roof	Asphalt + Fiberboard+ XPS Extruded Polystyrene + Cast Concrete	1.25
Flor & Slab	Timber Flooring + Floor/Roof Screed + Cast Concrete	1.65

Table 2. Thermal characteristics of the window system used in the simulation

Window system	SHGC (W/W)	Light transmission (W/W)	U-value (W/m².K)
Double glazed	0.56	0.56	4.50

Besides the baseline scenario, with a standard roof, it is also simulated eight different GR solutions, made up of three parameters with two levels each one. The GR alternatives vary according to substrate (soil) thickness, vegetation height, and Leaf Area Density (LAD). Substrate thickness and vegetation height area evaluated at 0.20 meters and 0.40 meters and LAD for levels are 2.0 and 4.0. These values are a generalization of two types of GR, one with lawn and another with plants, small bushes, according to Silva (2014). However, instead of evaluating only two possible solutions of GRs, the present research aims to evaluate how each parameter affects energy consumption, so a combination number of these two solutions are tested,

Table 3. Green roof scenarios characteristics regarding substrate thickness, vegetation height and LAD

	Substrate Thickness (m)	Vegetation Height (m)	LAD (m^2/m^3)
GR 1	0.20	0.20	2.00
GR 2	0.20	0.20	4.00
GR 3	0.20	0.40	2.00
GR 4	0.20	0.40	4.00
GR 5	0.40	0.20	2.00
GR 6	0.40	0.20	4.00
GR 7	0.40	0.40	2.00
GR 8	0.40	0.40	4.00

and therefore the eight different scenarios. Table 3 states all scenarios and their parameters combinations as well as their labels.

All GR scenarios use the same soil thermal-physical characterization, derived from earth material available in the DesignBuilder software library. With specific heat of 880 J/kg.K, conductivity of 1.28 W/m.K and density of 1460 kg/m^3. The leaf has a thermal emission of 0.9 W/W and an absorptance of 0.7 W/W. The irrigation applied in all models use the is the smart irrigation type that wets the soil to maintain it with more than 30% of humidity and takes into consideration the precipitation according to the weather file. The Green Roof model applied is the Ecoroof model which uses the Fast All-Season Soil STrength (FASST) developed by Frankenstein and Skahill (2004) as the energy balance calculations algorithm.

RESULTS AND DISCUSSION

To test the contribution of green roofs as a sustainable and comfort solution in the city of Lisbon, two levels of analysis were conducted, from the neighborhood to the building level. The results are presented following each level of analysis.

Neighborhood Level

Starting from the location of suitable rooftops in the city, available at the Lisbon Green Roof Potential Map, a density analysis was performed using the roofs' centroids as the input point features and a circle shaped area with a radius of 100 m. The result was a 25 m resolution raster file, with the places with higher and lower GR densities.

All public schools were overlaid with the city's potential map and from the initial set of 182 schools, 66 (37%) were identified as having at least one building suitable for receiving green roofing.

The next step was selecting a neighborhood that had a high density of buildings with GR potential and a public school, also with GR potential. By overlaying these two layers, Telheiras, a recent area, developed in the 1970s, was selected as the case study for the neighborhood level and the Telheiras EB, an elementary school identified for the building level analysis (Figure 4).

The Telheiras study area includes 159 buildings, mostly of residential use. Building heights vary from 1 to 2 floors in single-family houses, up to 8 and 9 floors in apartment buildings. The average footprint is 247 m^2. Among this set, there are 37 buildings with green roof potential.

To compare the impact of GR in Telheiras, the ENVI-met software was selected as the measuring tool.

Following the methodological framework, the same approach was applied in two land cover scenarios: the baseline scenario and the proposed GR scenario. The study area's morphology was typified for better computation (Figure 5). This is a generalization operation, that replaces a large number of objects by a smaller number of objects while trying to maintain the typical urban structure.

The model size is then an input area of 135 X 1135 X 30 grid, with a grid size of dx = 2.5 m, dy=2.5 m, and dz = 2 m. After modelling the two scenarios, using the available remote sensing data set, the area of GR is only 7% of the total area under analysis (Table 4).

After site modelling, the basic meteorological variables were introduced. The ENVI-met model simulated the climatic variables and comfort indicator for both scenarios, considering the summer and winter periods. From the hourly outputs, four were selected to describe the impact of GR at the neighborhood level: 9 am, 12 pm, 3 pm and 6 pm.

The first variable to be simulated was air temperature (Table 5 and Figure 6). In the summer, a more significant temperature drop is observed in the spaces located near the buildings where the green roof was tested, ranging between 1 and 2 °C (Table 3). In the winter, there is very little impact of GR buildings. This effect on temperature drop was already anticipated since green areas can reduce urban temperatures due to the thermal energy absorption occurring during evapotranspiration.

Figure 4. Selection of the neighborhood (C) and the school building (D) for detailed comfort and energy analysis, based on the density analysis at the buildings with green roof potential (B) available for Lisbon city (A)

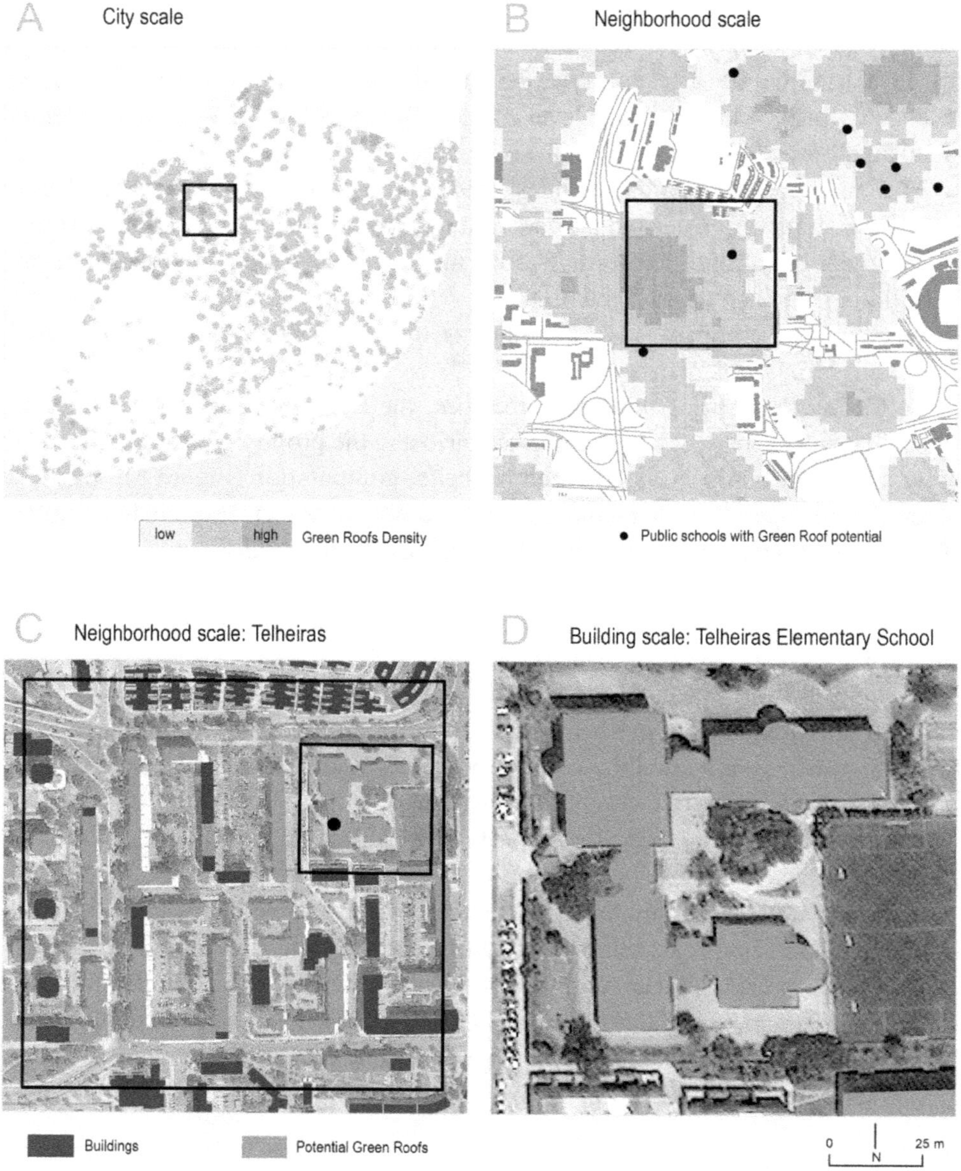

Figure 5. Baseline and Green Roof scenarios modelled in ENVI-met

Regarding the relative humidity, and for the summer period, the neighborhood mean values did not vary among scenarios (Table 6). However, a sharp improvement was detected near the GR buildings, with an increased value of 5%. In the winter simulation, the GR has a greater benefit in the whole neighborhood area, with a global increase of 5-10% in relative humidity, at 3 pm and 6 pm. These positive effects are a result of the evapotranspiration process, that contributes to an increased humidity, making the local environment more comfortable. In fact, the relative humidity recommended by the ISO 7730 considering thermal comfort parameter is 30–70%, and value of 50% is regarded as the most comfortable humidity level for a thermal environment (Givoni, 1992; Jing et al., 2012).

Table 4. Percentage of each land cover element in the two scenarios

Land Cover	Baseline Scenario	Green Roof Scenario
Green areas	20%	27%
Sidewalks	34%	34%
Buildings	16%	9%
Streets	30%	30%

Table 5. Simulation results for the air temperature, considering the baseline scenario and the green roof scenario, in Telheiras neighborhood, Portugal, during summer and winter

		Air Temperature (°C)		
		Baseline Scenario	Green Roof Scenario	
		Neighborhood	Neighborhood	Near Green Roof Buildings
Summer	9 am	24 °C	24 °C	- 1 °C
	12 pm	28 °C	28 °C	- 2 °C
	3 pm	28-32 °C	28-32 °C	- 1 °C
	6 pm	27-30 °C	27-30 °C	- 2 °C
Winter	9 am	19 °C	19 °C	No variation
	12 pm	19-22 °C	19-22 °C	- 0.2 °C
	3 pm	19-24 °C	19-24 °C	- 0.2 °C
	6 pm	19-22 °C	19-22 °C	- 0.2 °C

For the wind velocity, during summertime, the mean value for the neighborhood is 1-4 m/s, at all hours, and there is a consistent slight increase of 1 m/s near the GR buildings, making those areas more comfortable due to the induced ventilation. During winter there is no variation between scenarios.

The last climatic variable to be simulated is the mean radiant temperature (Table 7 and Figure 7). The simulation in ENVI-met shows high values in all scenarios, however, the green scenario shows some local improvements. In the summer, at 12 pm and 6 pm periods, the area with lower values of mean radiant temperature (less 10 °C) is higher near the buildings with GR. The same dynamics is observed in the winter. This cooling contribution of the GR is due to the vegetation's low emissivity.

Along with the four climatic variables, the PMV was also simulated to measure the effect of GR in the human thermal comfort (Table 8 and Figure 8). In the summer, only the morning and evening periods show variability between the two scenarios. In the green scenario, the PMV indicates less area with thermal discomfort. While the whole area is hot, with values higher than 2, the areas near the buildings are less uncomfortable, with PMV values ranging between +1 and +1.5 (warm). In the winter, it is also visible that near building with GR the environment is more comfortable. At 9 am, part of the cold areas near the buildings are rated as slightly cold in the green scenario. At noon and 3 pm, there is also an improvement near the buildings with GR, that are found to be more comfortable.

Figure 6. Air temperature simulated in ENVI-met, for the baseline scenario and the green roof scenario, at 9 am, 12 pm, 3 pm, and 6 pm, for the summer and winter time periods.

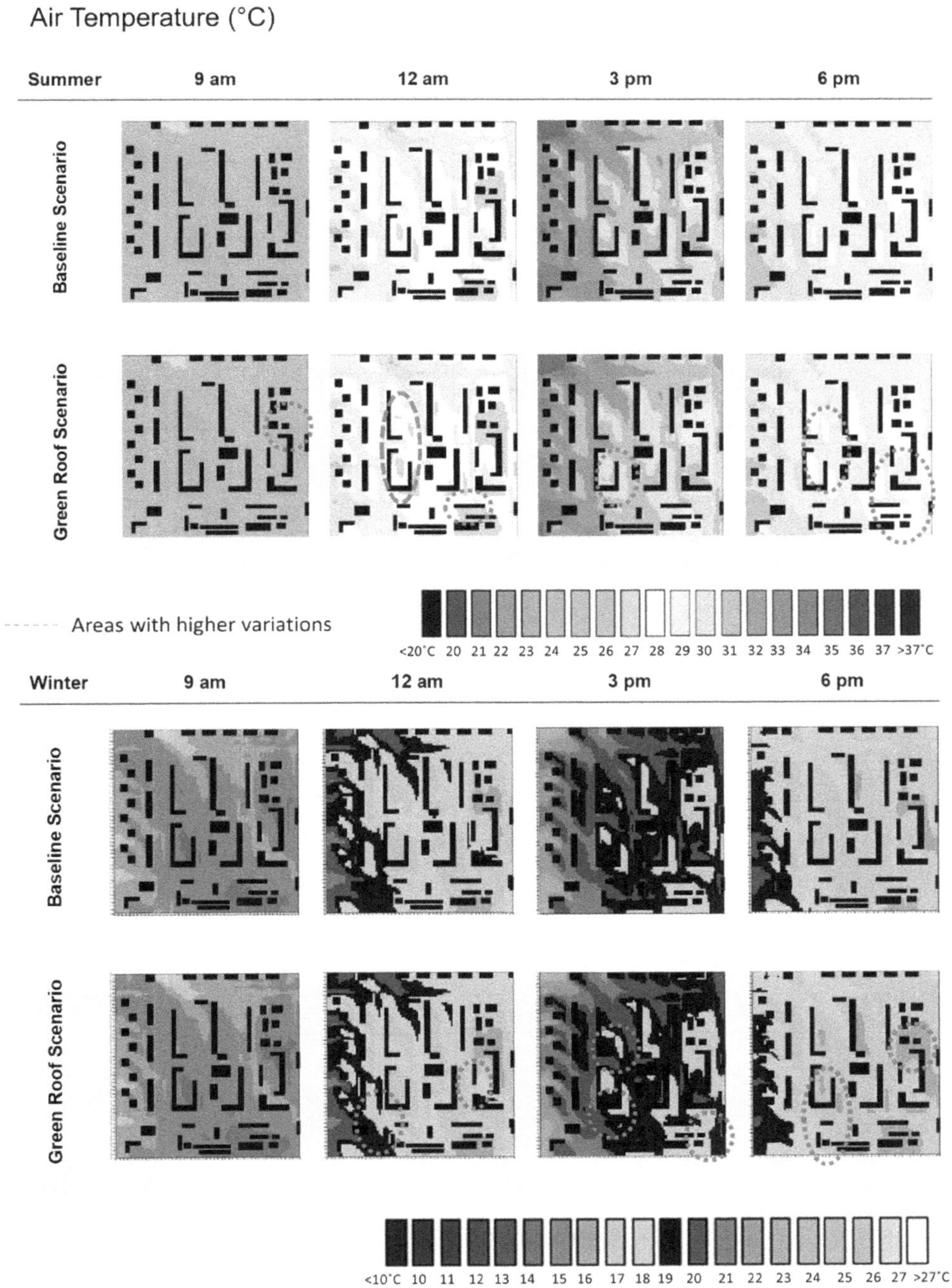

Table 6. Simulation results for the relative humidity, considering the baseline scenario and the green roof scenario, in Telheiras neighborhood, Portugal, during summer and winter

		Relative Humidity (%)		
		Baseline Scenario	Green Roof Scenario	
		Neighborhood	Neighborhood	Near Green Roof Buildings
Summer	9 am	50-70%	50-70%	+ 5%
	12 pm	30-60%	30-60%	+ 5%
	3 pm	25-35%	25-35%	+ 5%
	6 pm	25-35%	25-35%	No variation
Winter	9 am	30-65%	30-65%	+ 5-10%
	12 pm	30-65%	30-65%	+ 5-10%
	3 pm	20-40%	35-75%	+ 5-10%
	6 pm	15-20%	20-25%	+ 5-10%

Table 7. Simulation results for the mean radiant temperature, considering the baseline scenario and the green roof scenario, in Telheiras neighborhood, Portugal, during summer and winter

		Mean Radiant Temperature (°C)		
		Baseline Scenario	Green Roof Scenario	
		Neighborhood	Neighborhood	Near Green Roof Buildings
Summer	9 am	40 – 65 °C	40 – 65 °C	No variation
	12 pm	50 – 65 °C	50 – 65 °C	Reduction of 10 °C
	3 pm	55 – 65 °C	55 – 65 °C	Reduction of 5 °C
	6 pm	35 – 65 °C	35 – 65 °C	Reduction of 10 °C
Winter	9 am	30 – 65 °C	30 – 65 °C	No variation
	12 pm	30 – 65 °C	30 – 65 °C	Reduction of 10 °C
	3 pm	55 – 65 °C	55 – 65 °C	Reduction of 5 °C
	6 pm	25 – 50 °C	25 – 50 °C	Reduction of 20 °C

From the simulation process at the neighborhood level, two conclusions can be made: 1) season matters and, 2) GRs in multi-story buildings have little impact in the thermal comfort of the neighborhood area, and its benefits are most felt near the buildings:

Figure 7. Mean radiant temperature simulated in ENVI-met, for the baseline scenario and the green roof scenario, at 9 am, 12 pm, 3 pm, and 6 pm, for the summer and winter time periods.

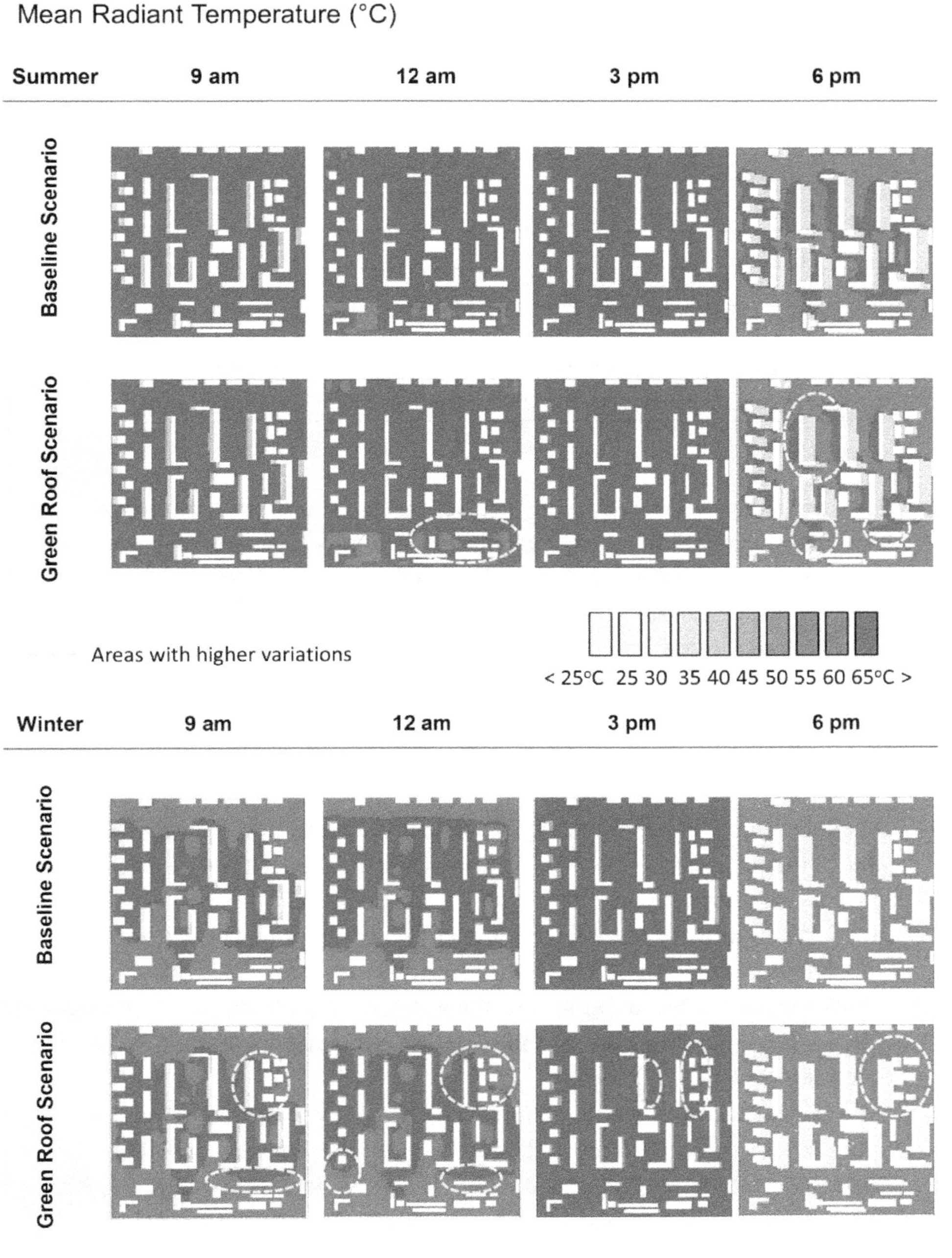

Table 8. Simulation results for the Predicted Mean Vote - PMV, considering the baseline scenario and the green roof scenario, in Telheiras neighborhood, Portugal, during summer and winter

		Predicted Mean Vote		
		Baseline Scenario	**Green Roof Scenario**	
		Neighborhood	**Neighborhood**	**Near Green Roof Buildings**
Summer	**9 am**	From +0,5 to +2,5	From +0,5 to +2,5	Reduction of 0,5
	12 pm	Around +2,5	Around +2,5	No variation
	3 pm	Around +2,5	Around +2,5	No variation
	6 pm	From +1,5 to +2,5	From +1,5 to +2,5	Reduction of 0,5
Winter	**9 am**	From -2,0 to +0,5	From -2,0 to +0,5	Reduction of 0,5
	12 pm	From -0,5 to +2,0	From -0,5 to +2,0	Reduction of 0,5
	3 pm	From +0,5 to +2,5	From +0,5 to +2,0	Reduction of 0,5
	6 pm	From -1,0 to +1,0	From -1,0 to +1,0	No variation

- In summertime, the presence of GRs promotes a more comfortable urban environment. Because of evapotranspiration – the natural process of transferring water from soil and plant to the air – the GRs tend to lower air temperature, promote higher relative humidity rate and ventilation. All these benefits are more evident near the buildings with GR. In winter time there is less impact towards comfort, with similar air temperature and wind velocity in booth scenarios. Nevertheless, the higher humidity rate in the afternoon induced by the presence of GRs indicates a more comfortable environment in this scenario.
- In evaluating all five parameters in the microclimatic simulation, it was found that at the neighborhood scale, there was no variation among scenarios. Other authors also reported similar results (e.g., Lobaccaro & Acero, 2015; Ng et al., 2012; Santos et al., 2019; Zölch et al., 2016). This can be explained by the fact that the area of GR in the study area is only 7% of the total area. However, and when looking into the areas near the building with GR, it is always more comfortable than in more distant areas, meaning that GR effect becomes lower by distance.
- The parameter that was less impacted by GR was the wind velocity. This situation was already expected, since the typical herbaceous plants used in extensive GR are less successful at lowering the wind speed than small trees.

Figure 8. PMV simulated in ENVI-met, for the baseline scenario and the green roof scenario, at 9 am, 12 pm, 3 pm, and 6 pm, for the summer and winter time periods.

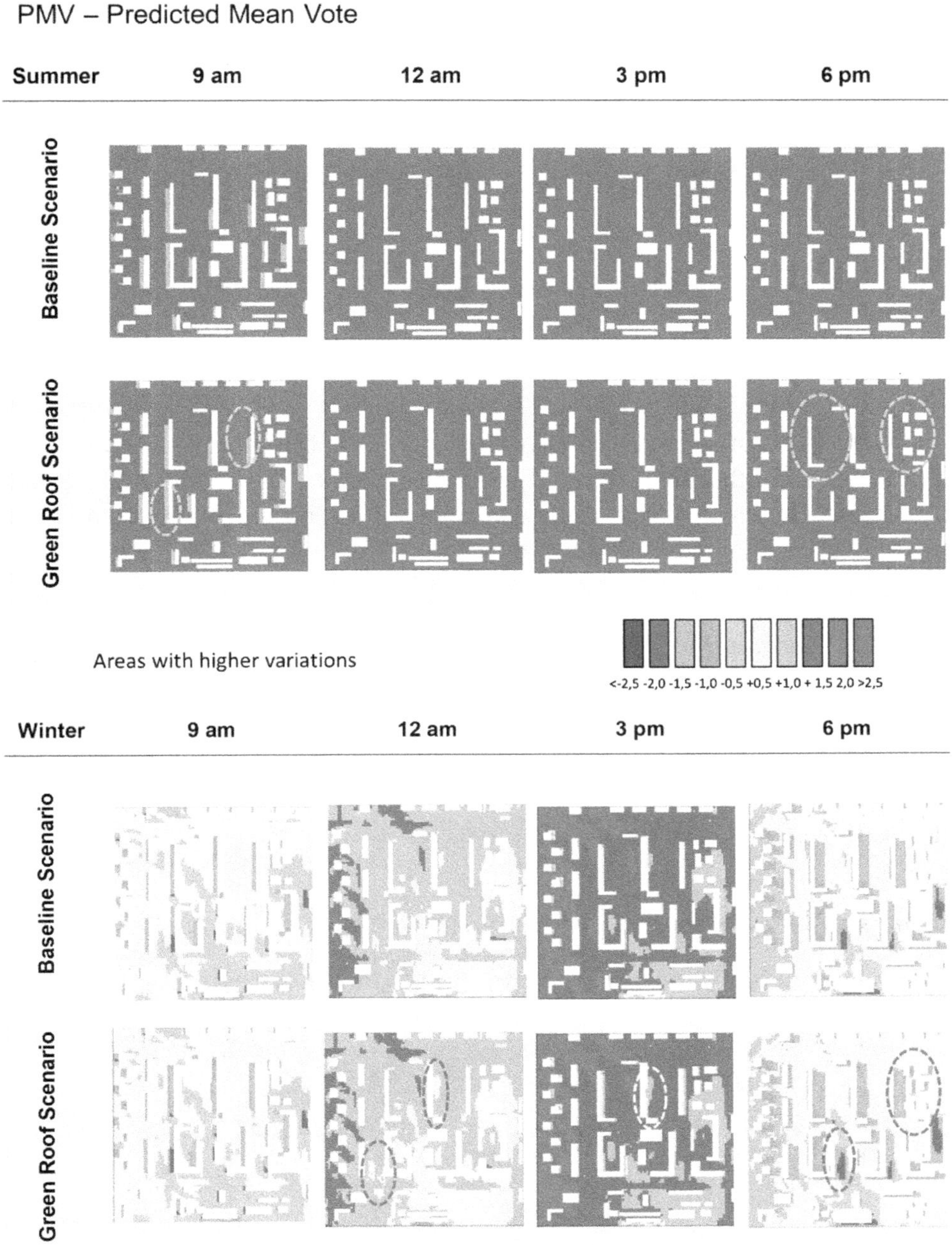

Figure 9. School blocks modelled for the energy consumption analysis

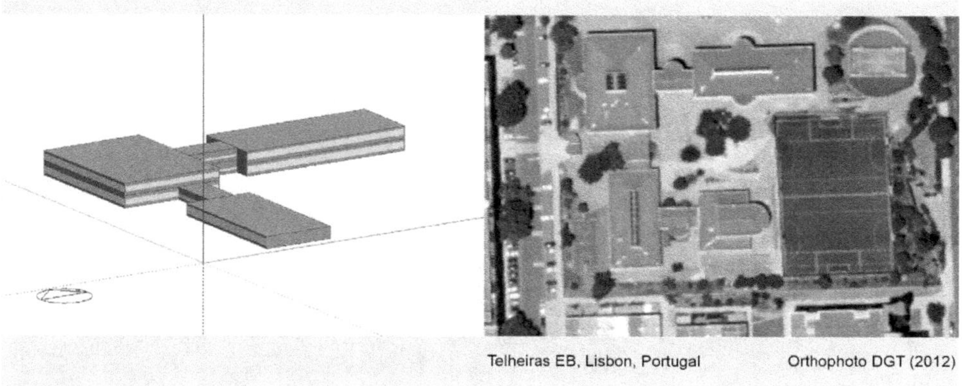

Assessing planning interventions with higher GR benefits for the neighborhood scale, requires further analysis. For future researches, testing variables like larger GR areas in combination with distinct urban densities (in area and in volume) and different levels of green areas is recommended.

Building Level

The impact of GRs in the energy consumption of a school was tested in a Telheiras EB building. Even though the school itself has not been refurbished under the Portuguese Secondary School Building Modernization Program, it is the one adopted since it is the national golden standard and the perspective of this analysis is more global, of the building stock.

Figure 10. Energy consumption breakdown for the baseline

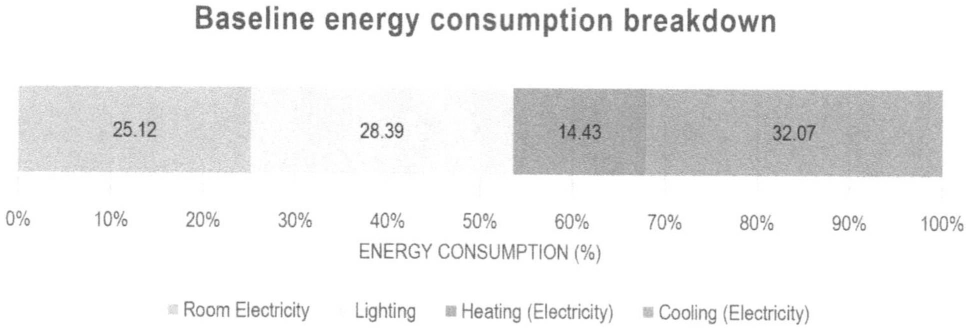

Figure 11. Energy reduction for each Green Roof scenario

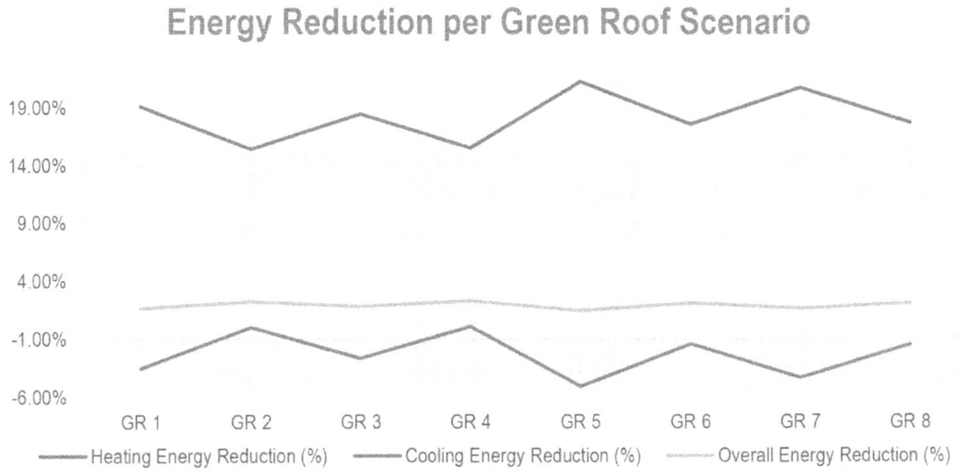

A thermo-energetic simulation is carried out to evaluate the school and define a baseline for energy consumption for further comparison. Only the main building blocks were modelled, as shown in the Figure 9. The surroundings and the gymnasium were not considered. All blocks have a simplified geometry since the EnergyPlus calculates its results basically from the volume and the surfaces with different exposure condition. The overall measurements were assessed using Google Earth.

The three-dimensional model follows school volumetry, with two blocks with two floors, and another one with only one floor with circulation blocks connecting each one of them. The main blocks are characterized as classrooms without any division as an individual thermal zone, and the circulation blocks are defined as circulation, each one with only one thermal zone as well. An adaptation of the default DesignBuilder template Teaching Areas – Education Non-Residential was used for the classrooms and Circulation Areas – Education Non-Residential is used in the thermal zones assigned as circulations.

Table 9. Energy reduction values for each scenario

	GR 1	GR 2	GR 3	GR 4	GR 5	GR 6	GR 7	GR 8
Heating Energy Reduction (%)	19.15%	15.42%	18.47%	15.52%	21.30%	17.57%	20.78%	17.71%
Cooling Energy Reduction (%)	-3.51%	0.03%	-2.69%	0.05%	-5.12%	-1.47%	-4.38%	-1.48%
Overall Energy Reduction (%)	1.64%	2.23%	1.80%	2.26%	1.43%	2.06%	1.59%	2.08%

The blueprint of the school employed is the same value from the study of Bernardo et al. (2017) with a 21% of window to floor ratio, which in the present building represent 33% of window to wall ratio. Figure 10 presents the energy consumption breakdown for the end use of the baseline. The heating and cooling represent almost half of all energy consumption.

The results in Figure 11 show that the GR implementation improves the heating energy efficiency, with a reduction of up to 21.30%. On the other hand, the cooling efficiency does not change as much, with a reduction of its consumption of 0.05% in the best scenario (Table 9). However, in most scenarios, the GR implementation led to an increase in cooling energy consumption, up to 5.12%. The results indicate that the implementation of GR should be studied specifically to be tailored to each situation. Also, the graph in Figure 8 indicates that in scenarios where there is a more considerable improvement in cooling, a milder improvement in heating happens, pointing out that different solutions of GR have a different impact in those end uses. Moreover, despite the heating and cooling energy consumption oscillation per GR scenario, the GR implementation enables a steady improvement of 1.43% to 2.26 of the overall energy consumption.

The different GR parameters analyzed have a different impact on energy consumption, especially if analyzed for cooling or heating. Figure 12 shows a matrix of distribution graphs for each parameter, for cooling, heating, and overall energy consumption, as well as the results of F-value and P-value from one-way variance analysis (ANOVA) between the different levels of each parameter.

The variation of vegetation height studied does not seem to impact energy consumption at any level. However, the soil thickness and the LAD appear to have a greater impact and have opposing influences. On one hand, the increase of the GR soil thickness reduces the heating energy consumption, but it increases cooling energy consumption. On the other hand, the increase of LAD reduces cooling energy consumption, however, increases heating energy consumption. It may be connected to basic energy conservation measures. The thermal mass increase may improve heating systems energy efficiency since it increases thermal inertia and makes more difficult heat transmission, and that is what increasing the soil thickness means to the building envelope. Yet, a greater LAD represents more leaf layers, which implies in more shading, one of the most employed strategies for decreasing thermal load, which reduces cooling energy.

Even though the differences found in the results, from a statistical standpoint, only the LAD presents any significant effect in all studied energy consumption levels: heating, cooling, and overall. However, it is worth pointing out that the analyzed sample is limited and must not be considered in generalization, especially for other climates and typologies. The main purpose of this experiment is to present an alternative to evaluate GR itself and its different solutions as well its readiness and adequateness to Lisbon climate. However, the results point out best scenarios are not the ones with greater improvement in heating energy efficiency, but the ones with greater cooling energy reduction, since cooling energy represents a greater end use. So, higher LAD and lower soil thickness means a better overall energy reduction, even though an opposite trend represents more improvement in heating energy efficiency.

Figure 12. Graphs for each parameter – thickness, height and LAD -, for cooling, heating, and overall energy consumption, as well as the results of F-value and P-value from one-way variance analysis (ANOVA)

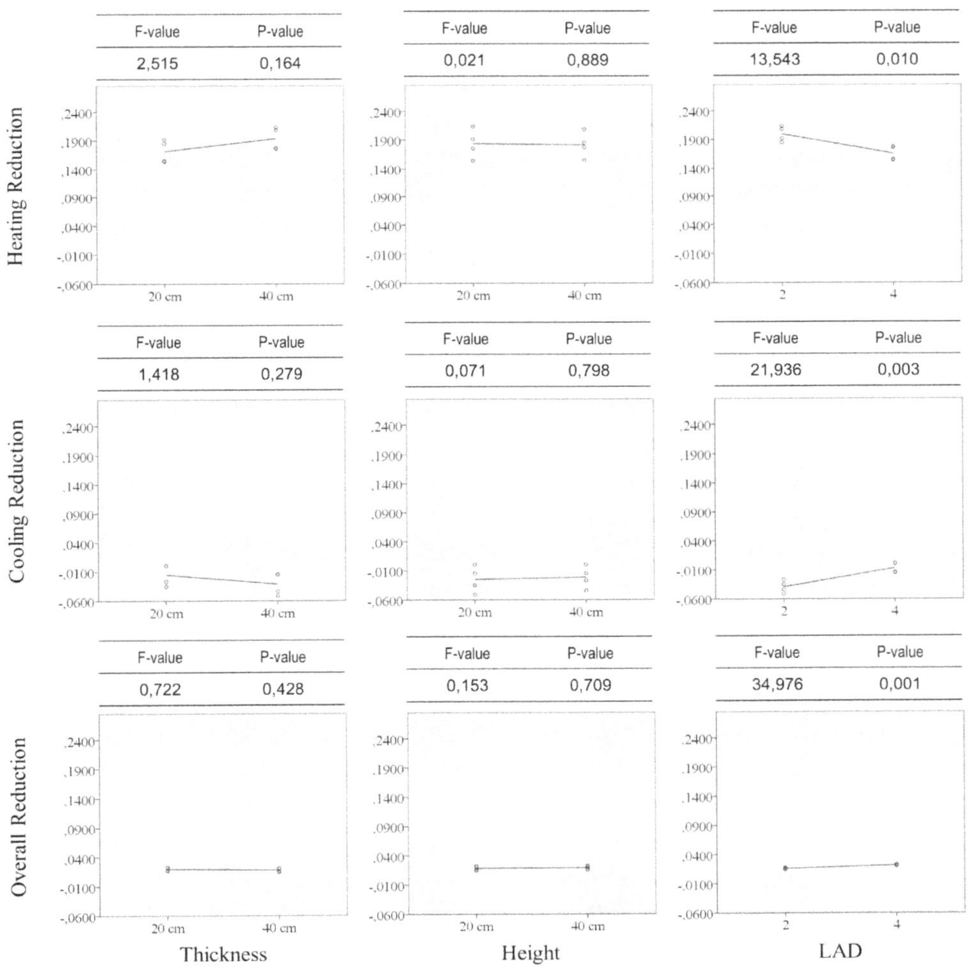

CONCLUSION

Increasing green areas in cities is a strategy to improve thermal outdoor comfort. A scenario-based analysis is proposed to evaluate the impact of future green plans. Such technique considers alternative outcomes and presents different options for future development strategies. Using scenario analysis within the planning process is a way of select the best strategies and practices to meet the cities' sustainable goals, and to determine priorities and investment plans. The developed framework provides also a useful tool for other professionals such as architects and landscape architects.

However, the analysis is as good as the quality of the data used to model the phenomena. Using the proposed methodology, remote sensing data, with the support of GIS, can be used to produce relevant indicators about the urban environment. Due to the high spatial resolution and acquisition capability over large areas, input data from remote sensing was successfully used to 1) evaluate all buildings within the city that meet the geographic criteria for receiving GR, 2) select the most suitable and, 3) assess the implications at the neighborhood and at the building scale.

This work shows the effect of GR at a range of scales: neighborhood and building. In all scales, there is an alignment of the results that suggests a GR implementation potential in more horizontal and low-rise building typologies, either for the citizen thermal comfort in public areas, either for the building energy consumption. Nevertheless, this research only addressed these two GR effects. In fact, GR benefits are more complete, and include storm water retention, carbon sequestration, enhanced biodiversity, decline of air and noise pollution, provision of additional leisure areas, and by extension improve public health and quality of life. This indicates the contribution of this multiscale and multidisciplinary approach to qualify urban planning and design.

This work shows thermal comfort performance distributed in a neighborhood. On one hand, the simulation results are presented as map files, thus facilitating its interpretation and analysis. On the other hand, a limitation of this study was the unique type of vegetation used (Luzerne). New studies could use native species or vegetables. Additionally, despite only 7% GR was implemented, the research showed spatial influences in different parts of the neighborhood area. The point where it is possible to observe higher differences is near the public school and other low-rise buildings. Although most results were pure, it was possible to emphasize that the most comfortable values were near the lower buildings. It means that the green roof of low-rise buildings has more impact on the thermal comfort of the pedestrian.

In all different scales, there is an alignment of the results that suggests a more significant GR implementation potential in more horizontal building typologies, either for the citizen thermal comfort in public areas, either for the building energy consumption. This indicates the contribution of this multiscale and multidisciplinary approach to qualify urban planning and design.

UHI effect is another matter that requires mitigation measures like green area increase. In this context, GR can be part of the solution through the services provided like temperature attenuation and relative humidity increasement. Furthermore, by greening its buildings, the municipality can contribute to a more sustainable and resilient city, while also encouraging private owners investing in green infrastructures.

ACKNOWLEDGMENT

This work is financed by national funds through FCT - Foundation for Science and Technology, I.P., within the scope of the project «UIDB / 04647/2020» of CICS. NOVA - Centro Interdisciplinar de Ciências Sociais da Universidade Nova de Lisboa.

The first author was financed by national funds through the FCT, under the Norma Transitória – DL 57/2016/CP1453/CT0004. The second author was financed by the University of Brasilia and the Public Research Foundation of Federal District of Brasília (FAP-DF).

REFERENCES

Alcoforado, M.-J., Andrade, H., Lopes, A., & Vasconcelos, J. (2009). Application of climatic guidelines to urban planning: The example of Lisbon (Portugal). *Landscape and Urban Planning*, *90*(1), 56–65. doi:10.1016/j.landurbplan.2008.10.006

Ali-Toudert, F., & Mayer, H. (2007). Effects of asymmetry, galleries, overhanging façades and vegetation on thermal comfort in urban street canyons. *Solar Energy*, *81*(6), 742–754. doi:10.1016/j.solener.2006.10.007

Aplin, P. (2003). Comparison of simulated IKONOS and SPOT HRV imagery for classifying urban areas. In Remotely-Sensed Cities. CRC Press.

Ascione, F., Bianco, N., de' Rossi, F., Turni, G., & Vanoli, G. P. (2013). Green roofs in European climates. Are effective solutions for the energy savings in air-conditioning? *Applied Energy*, *104*, 845–859. doi:10.1016/j.apenergy.2012.11.068

Bates, A. J., Sadler, J. P., & Mackay, R. (2013). Vegetation development over four years on two green roofs in the UK. *Urban Forestry & Urban Greening*, *12*(1), 98–108. doi:10.1016/j.ufug.2012.12.003

Behling, R., Bochow, M., Foerster, S., Roessner, S., & Kaufmann, H. (2015). Automated GIS-based derivation of urban ecological indicators using hyperspectral remote sensing and height information. *Ecological Indicators*, *48*, 218–234. doi:10.1016/j.ecolind.2014.08.003

Berardi, U., GhaffarianHoseini, A. H., & GhaffarianHoseini, A. (2014). State-of-the-art analysis of the environmental benefits of green roofs. *Applied Energy*, *115*, 411–428. doi:10.1016/j.apenergy.2013.10.047

Bernardo, H., Antunes, C. H., Gaspar, A., Pereira, L. D., & da Silva, M. G. (2017). An approach for energy performance and indoor climate assessment in a Portuguese school building. *Sustainable Cities and Society*, *30*, 184–194. doi:10.1016/j.scs.2016.12.014

Bianchini, F., & Hewage, K. (2012). Probabilistic social cost-benefit analysis for green roofs: A lifecycle approach. *Building and Environment*, *58*, 152–162. doi:10.1016/j.buildenv.2012.07.005

Blaschke, T. (2010). Object based image analysis for remote sensing. *ISPRS Journal of Photogrammetry and Remote Sensing*, *65*(1), 2–16. doi:10.1016/j.isprsjprs.2009.06.004

Blaschke, T., & Strobl, J. (2001). What's wrong with pixels? Some recent developments interfacing remote sensing and GIS. *GeoBIT/GIS*, *6*, 12–17.

Brandão, C., Cameira, M., Valente, F., Cruz de Carvalho, R., & Paço, T. A. (2017). Wet season hydrological performance of green roofs using native species under Mediterranean climate. *Ecological Engineering*, *102*, 596–611. doi:10.1016/j.ecoleng.2017.02.025

Bruse, M., & Fleer, H. (1998). Simulating surface–plant–air interactions inside urban environments with a three dimensional numerical model. *Environmental Modelling & Software*, *13*(3), 373–384. doi:10.1016/S1364-8152(98)00042-5

Câmara Municipal de Lisboa. (n.d.). *Escolas Públicas—Secundário*. Retrieved March 26, 2020, from http://geodados.cm-lisboa.pt/datasets/680ce63e07fd4c5eb165b9d47498bef5_4

Campbell, J. B., & Wynne, R. H. (2011). *Introduction to Remote Sensing* (5th ed.). The Guilford Press.

Camps-Calvet, M., Langemeyer, J., Calvet-Mir, L., & Gómez-Baggethun, E. (2016). Ecosystem services provided by urban gardens in Barcelona, Spain: Insights for policy and planning. *Environmental Science & Policy*, *62*, 14–23. doi:10.1016/j.envsci.2016.01.007

Chen, W., Huang, H., Dong, J., Zhang, Y., Tian, Y., & Yang, Z. (2018). Social functional mapping of urban green space using remote sensing and social sensing data. *ISPRS Journal of Photogrammetry and Remote Sensing*, *146*, 436–452. doi:10.1016/j.isprsjprs.2018.10.010

Chen, W. Y. (2015). The role of urban green infrastructure in offsetting carbon emissions in 35 major Chinese cities: A nationwide estimate. *Cities (London, England)*, *44*, 112–120. doi:10.1016/j.cities.2015.01.005

Chrysoulakis, N., Feigenwinter, C., Triantakonstantis, D., Penyevskiy, I., Tal, A., Parlow, E., Fleishman, G., Düzgün, S., Esch, T., & Marconcini, M. (2014). A Conceptual List of Indicators for Urban Planning and Management Based on Earth Observation. *ISPRS International Journal of Geo-Information*, *3*(3), 980–1002. doi:10.3390/ijgi3030980

Connop, S., Vandergert, P., Eisenberg, B., Collier, M. J., Nash, C., Clough, J., & Newport, D. (2016). Renaturing cities using a regionally-focused biodiversity-led multifunctional benefits approach to urban green infrastructure. *Environmental Science & Policy*, *62*, 99–111. doi:10.1016/j.envsci.2016.01.013

Demuzere, M., Orru, K., Heidrich, O., Olazabal, E., Geneletti, D., Orru, H., Bhave, A. G., Mittal, N., Feliu, E., & Faehnle, M. (2014). Mitigating and adapting to climate change: Multi-functional and multi-scale assessment of green urban infrastructure. *Journal of Environmental Management*, *146*, 107–115. doi:10.1016/j.jenvman.2014.07.025 PMID:25163601

Direção-Geral do Território. (n.d.). *DGTerritório—Ortofotos Digitais*. Retrieved March 26, 2020, from https://www.dgterritorio.pt/cartografia_e_geodesia/cartografia/cartografia_de_base___topografica_e_topografica_de_imagem/ortofotocartografia/ortofotocartografia_pos_2004/

Elmqvist, T., Setälä, H., Handel, S., van der Ploeg, S., Aronson, J., Blignaut, J., Gómez-Baggethun, E., Nowak, D., Kronenberg, J., & de Groot, R. (2015). Benefits of restoring ecosystem services in urban areas. *Current Opinion in Environmental Sustainability*, *14*, 101–108. doi:10.1016/j.cosust.2015.05.001

EPA - United States Environmental Protection Agency. (2014). *The Economic Benefits of Green Infrastructure: A Case Study of Lancaster, PA* (EPA 8 00-R-14-007). EPA.

Fanger, P. O. (1972). *Thermal comfort analysis and applications in environmental engineering*. McGraw-Hill.

Francis, R. A., & Lorimer, J. (2011). Urban reconciliation ecology: The potential of living roofs and walls. *Journal of Environmental Management*, *92*(6), 1429–1437. doi:10.1016/j.jenvman.2011.01.012 PMID:21306818

Frankenstein, S., & Skahill, B. (2004). *FASST Soil Moisture Prediction Error at Yuma, Arizona* (ERDC/CRREL SR-04-1). Engineer Research and Development Center. https://www.researchgate.net/publication/235096218_FASST_Soil_Moisture_Prediction_Error_at_Yuma_Arizona

Freire, S., Santos, T., Navarro, A., Soares, F., Silva, J. D., Afonso, N., Fonseca, A., & Tenedório, J. (2014). Introducing mapping standards in the quality assessment of buildings extracted from very high resolution satellite imagery. *ISPRS Journal of Photogrammetry and Remote Sensing*, *90*, 1–9. doi:10.1016/j.isprsjprs.2013.12.009

Fu, W., Ma, J., Chen, P., & Chen, F. (2020). Remote Sensing Satellites for Digital Earth. In H. Guo, M. F. Goodchild, & A. Annoni (Eds.), *Manual of Digital Earth* (pp. 55–123). Springer., doi:10.1007/978-981-32-9915-3_3

Getter, K. L., & Rowe, D. B. (2006). The Role of Extensive Green Roofs in Sustainable Development. *HortScience*, *41*(5), 1276–1285. doi:10.21273/HORTSCI.41.5.1276

Givoni, B. (1992). Comfort, climate analysis and building design guidelines. *Energy and Building*, *18*(1), 11–23. doi:10.1016/0378-7788(92)90047-K

Haaland, C., & van den Bosch, C. K. (2015). Challenges and strategies for urban green-space planning in cities undergoing densification: A review. *Urban Forestry & Urban Greening*, *14*(4), 760–771. doi:10.1016/j.ufug.2015.07.009

Heymans, A., Breadsell, J., Morrison, G. M., Byrne, J. J., & Eon, C. (2019). Ecological Urban Planning and Design: A Systematic Literature Review. *Sustainability*, *11*(13), 3723. doi:10.3390u11133723

Jayasooriya, V. M., & Ng, A. W. M. (2014). Tools for Modeling of Stormwater Management and Economics of Green Infrastructure Practices: A Review. *Water, Air, and Soil Pollution*, *225*(8), 2055. doi:10.100711270-014-2055-1

Jensen, J. R. (2015). *Introductory Digital Image Processing: A Remote Sensing Perspective* (4th ed.). Pearson.

Jim, C. Y. (2004). Green-space preservation and allocation for sustainable greening of compact cities. *Cities (London, England)*, *21*(4), 311–320. doi:10.1016/j.cities.2004.04.004

Jim, C. Y. (2013). Sustainable urban greening strategies for compact cities in developing and developed economies. *Urban Ecosystems*, *16*(4), 741–761. doi:10.100711252-012-0268-x

Jing, S., Li, B., Tan, M., & Liu, H. (2012). Impact of Relative Humidity on Thermal Comfort in a Warm Environment. *Indoor and Built Environment*. Advance online publication. doi:10.1177/1420326X12447614

Kántor, N., & Unger, J. (2011). The most problematic variable in the course of human-biometeorological comfort assessment—The mean radiant temperature. *Open Geosciences*, *3*(1). Advance online publication. doi:10.247813533-011-0010-x

Karteris, M., Theodoridou, I., Mallinis, G., Tsiros, E., & Karteris, A. (2016). Towards a green sustainable strategy for Mediterranean cities: Assessing the benefits of large-scale green roofs implementation in Thessaloniki, Northern Greece, using environmental modelling, GIS and very high spatial resolution remote sensing data. *Renewable & Sustainable Energy Reviews*, *58*, 510–525. doi:10.1016/j.rser.2015.11.098

Klemm, W., Heusinkveld, B. G., Lenzholzer, S., & van Hove, B. (2015). Street greenery and its physical and psychological impact on thermal comfort. *Landscape and Urban Planning*, *138*, 87–98. doi:10.1016/j.landurbplan.2015.02.009

Köppen-Geiger Climate Classification. (2010). *Encyclopedia of Geography* (B. Warf, Ed.). SAGE Publications, Inc., doi:10.4135/9781412939591.n671

Lakes, T., & Kim, H.-O. (2012). The urban environmental indicator "Biotope Area Ratio"—An enhanced approach to assess and manage the urban ecosystem services using high resolution remote-sensing. *Ecological Indicators*, *13*(1), 93–103. doi:10.1016/j.ecolind.2011.05.016

Le Bris, A., & Chehata, N. (2019). Urban morpho-types classification from SPOT-6/7 imagery and Sentinel-2 time series. *2019 Joint Urban Remote Sensing Event*, 1–4. doi:10.1109/JURSE.2019.8808988

Lehmann, S. (2014). Low carbon districts: Mitigating the urban heat island with green roof infrastructure. *City. Cultura e Scuola*, *5*(1), 1–8. doi:10.1016/j.ccs.2014.02.002

Lillesand, T., Kiefer, R. W., & Chipman, J. (2015). *Remote Sensing and Image Interpretation* (7th ed.). Wiley.

Liu, L., Coops, N. C., Aven, N. W., & Pang, Y. (2017). Mapping urban tree species using integrated airborne hyperspectral and LiDAR remote sensing data. *Remote Sensing of Environment*, *200*, 170–182. doi:10.1016/j.rse.2017.08.010

Lobaccaro, G., & Acero, J. A. (2015). Comparative analysis of green actions to improve outdoor thermal comfort inside typical urban street canyons. *Urban Climate*, *14*, 251–267. doi:10.1016/j.uclim.2015.10.002

Madre, F., Vergnes, A., Machon, N., & Clergeau, P. (2014). Green roofs as habitats for wild plant species in urban landscapes: First insights from a large-scale sampling. *Landscape and Urban Planning*, *122*, 100–107. doi:10.1016/j.landurbplan.2013.11.012

Mallinis, G., Karteris, M., Theodoridou, I., Tsioukas, V., & Karteris, M. (2014). Development of a nationwide approach for large scale estimation of green roof retrofitting areas and roof-top solar energy potential using VHR natural colour orthoimagery and DSM data over Thessaloniki, Greece. *Remote Sensing Letters*, *5*(6), 548–557. doi:10.1080/2150704X.2014.933276

Ministério da Economia e do Emprego. (2013). *Diário da República n.º 159/2013, Série I de 2013-08-20*. Diário da República Eletrónico. https://dre.pt/pesquisa/-/search/499237/details/maximized?jp=true/en/en

Ministério das Obras Públicas. Transportes e Comunicações. (2006). *Decreto-Lei 80/2006, 2006-04-04*. Diário da República Eletrónico. https://dre.pt/pesquisa/-/search/672456/details/maximized?jp=true

Ministérios do Ambiente, Ordenamento do Território e Energia e da Solidariedade, Emprego e Segurança Social. (2013). *Portaria 349-D/2013 de 2013-12-02*. Diário da República Eletrónico. https://dre.pt/web/guest/pesquisa/-/search/263693/details/maximized?print_preview=print-preview

Muñoz-Vallés, S., Cambrollé, J., Figueroa-Luque, E., Luque, T., Niell, F. X., & Figueroa, M. E. (2013). An approach to the evaluation and management of natural carbon sinks: From plant species to urban green systems. *Urban Forestry & Urban Greening*, *12*(4), 450–453. doi:10.1016/j.ufug.2013.06.007

Mutani, G., & Todeschi, V. (2020). The Effects of Green Roofs on Outdoor Thermal Comfort, Urban Heat Island Mitigation and Energy Savings. *Atmosphere*, *11*(2), 123. doi:10.3390/atmos11020123

Ng, E., Chen, L., Wang, Y., & Yuan, C. (2012). A study on the cooling effects of greening in a high-density city: An experience from Hong Kong. *Building and Environment*, *47*, 256–271. doi:10.1016/j.buildenv.2011.07.014

Norton, B. A., Coutts, A. M., Livesley, S. J., Harris, R. J., Hunter, A. M., & Williams, N. S. G. (2015). Planning for cooler cities: A framework to prioritise green infrastructure to mitigate high temperatures in urban landscapes. *Landscape and Urban Planning*, *134*, 127–138. doi:10.1016/j.landurbplan.2014.10.018

Olgyay, V. (2015). *Design with Climate: Bioclimatic Approach to Architectural Regionalism - New and expanded Edition*. Princeton University Press. doi:10.2307/j.ctvc77kqb

Pevzner, A. (2013). *Living Architecture: Green Roofs for Public Buildings*. General Services Administration: Energy & Sustainability / Safety & Environmental Branch Facilities Management & Services Programs Division PBS, Region 9. https://sftool.gov/Content/attachments/Living_Architecture_Green_Roofs_for_Public_Buildings.pdf

Raciti, S. M., Hutyra, L. R., & Newell, J. D. (2014). Mapping carbon storage in urban trees with multi-source remote sensing data: Relationships between biomass, land use, and demographics in Boston neighborhoods. *The Science of the Total Environment*, *500–501*, 72–83. doi:10.1016/j.scitotenv.2014.08.070 PMID:25217746

Saadatian, O., Sopian, K., Salleh, E., Lim, C. H., Riffat, S., Saadatian, E., Toudeshki, A., & Sulaiman, M. Y. (2013). A review of energy aspects of green roofs. *Renewable & Sustainable Energy Reviews*, *23*, 155–168. doi:10.1016/j.rser.2013.02.022

Santos, T., Silva, C., & Tenedório, J. A. (2019). Promoting Citizens' Quality of Life Through Green Urban Planning. In L. Ragia, R. Laurini, & J. G. Rocha (Eds.), *Geographical Information Systems Theory, Applications and Management* (pp. 153–175). Springer International Publishing. doi:10.1007/978-3-030-06010-7_10

Santos, T., Tenedório, J. A., & Gonçalves, J. A. (2016). Quantifying the City's Green Area Potential Gain Using Remote Sensing Data. *Sustainability*, *8*(12), 1247. doi:10.3390u8121247

Satellite Imaging Corp. (n.d.). *Satellite Sensors and Specifications*. Retrieved March 23, 2020, from https://www.satimagingcorp.com/satellite-sensors/

Silva, B., Silva, C., & Santos, T. (2018). Protocolo de simulação computacional do microclima urbano para a cidade de Lisboa, Portugal. *Península Ibérica No Mundo: Problemas e Desafios Para Uma Intervenção Ativa Da Geografia*, 1066–1073.

Silva, C. M., Flores-Colen, I., & Antunes, M. (2017). Step-by-step approach to ranking green roof retrofit potential in urban areas: A case study of Lisbon, Portugal. *Urban Forestry & Urban Greening*, *25*, 120–129. doi:10.1016/j.ufug.2017.04.018

Silva, M. F. da R. (2014). *Simulação energética de coberturas verdes*. https://fenix.tecnico.ulisboa.pt/cursos/mec/dissertacao/2353642480248

Srinivasan, R. S., Lakshmanan, J., Srivastav, D., & Santosa, E. (2011). Benchmarking plug-load densities for k-12 schools. *Proceedings of Building Simulation, 2011*, 2746–2752.

Stern, M., Peck, S. W., & Joslin, J. (2019). *Green roof and wall policy in North America Regulations, Incentives, and Best Practices*. Green Roof for Healthy Cities. https://static1.squarespace.com/static/58e3eecf2994ca997dd56381/t/5d84dfc371cf0822bdf7dc29/1568989140101/Green_Roof_and_Wall_Policy_in_North_America.pdf

Storch, H., & Downes, N. K. (2011). A scenario-based approach to assess Ho Chi Minh City's urban development strategies against the impact of climate change. *Cities (London, England)*, *28*(6), 517–526. doi:10.1016/j.cities.2011.07.002

Tehrany, M. S., Pradhan, B., & Jebuv, M. N. (2014). A comparative assessment between object and pixel-based classification approaches for land use/land cover mapping using SPOT 5 imagery. *Geocarto International*, *29*(4), 351–369. doi:10.1080/10106049.2013.768300

TGRTAG -Toronto Green Roof Technical Advisory Group. (2011). *Toronto Green Roof Construction Standard (TGRCS)*. Office of the Chief Building Official, Toronto Building, City of Toronto. Canada. http://www.toronto.ca/greenroofs/pdf/GreenRoof-supGuidelines.pdf

Tian, Y., & Jim, C. Y. (2011). Factors influencing the spatial pattern of sky gardens in the compact city of Hong Kong. *Landscape and Urban Planning*, *101*(4), 299–309. doi:10.1016/j.landurbplan.2011.02.035

Tooke, T. R., Coops, N. C., Goodwin, N. R., & Voogt, J. A. (2009). Extracting urban vegetation characteristics using spectral mixture analysis and decision tree classifications. *Remote Sensing of Environment*, *113*(2), 398–407. doi:10.1016/j.rse.2008.10.005

Tran, T. D.-B., Puissant, A., Badariotti, D., & Weber, C. (2011). Optimizing Spatial Resolution of Imagery for Urban Form Detection—The Cases of France and Vietnam. *Remote Sensing*, *3*(10), 2128–2147. doi:10.3390/rs3102128

Tzoulas, K., Korpela, K., Venn, S., Yli-Pelkonen, V., Kaźmierczak, A., Niemela, J., & James, P. (2007). Promoting ecosystem and human health in urban areas using Green Infrastructure: A literature review. *Landscape and Urban Planning*, *81*(3), 167–178. doi:10.1016/j.landurbplan.2007.02.001

United Nations. (2016). *Sustainable Development Goals Report 2016*. UN.

Ural, S., Hussain, E., & Shan, J. (2011). Building population mapping with aerial imagery and GIS data. *International Journal of Applied Earth Observation and Geoinformation*, *13*(6), 841–852. doi:10.1016/j.jag.2011.06.004

Vijayaraghavan, K. (2016). Green roofs: A critical review on the role of components, benefits, limitations and trends. *Renewable & Sustainable Energy Reviews*, *57*, 740–752. doi:10.1016/j.rser.2015.12.119

Vu, T. T., Yamazaki, F., & Matsuoka, M. (2009). Multi-scale solution for building extraction from LiDAR and image data. *International Journal of Applied Earth Observation and Geoinformation*, *11*(4), 281–289. doi:10.1016/j.jag.2009.03.005

Weng, Q. (2012). Remote sensing of impervious surfaces in the urban areas: Requirements, methods, and trends. *Remote Sensing of Environment*, *117*, 34–49. doi:10.1016/j.rse.2011.02.030

Wong, J. K. W., & Lau, L. S.-K. (2013). From the 'urban heat island' to the 'green island'? A preliminary investigation into the potential of retrofitting green roofs in Mongkok district of Hong Kong. *Habitat International*, *39*, 25–35. doi:10.1016/j.habitatint.2012.10.005

Wulder, M. A., Loveland, T. R., Roy, D. P., Crawford, C. J., Masek, J. G., Woodcock, C. E., Allen, R. G., Anderson, M. C., Belward, A. S., Cohen, W. B., Dwyer, J., Erb, A., Gao, F., Griffiths, P., Helder, D., Hermosilla, T., Hipple, J. D., Hostert, P., Hughes, M. J., ... Zhu, Z. (2019). Current status of Landsat program, science, and applications. *Remote Sensing of Environment*, *225*, 127–147. doi:10.1016/j.rse.2019.02.015

Yang, J., & Wang, Z.-H. (2017). Planning for a sustainable desert city: The potential water buffering capacity of urban green infrastructure. *Landscape and Urban Planning*, *167*, 339–347. doi:10.1016/j.landurbplan.2017.07.014

Zhang, Y., & Guindon, B. (2006). Using satellite remote sensing to survey transport-related urban sustainability: Part 1: Methodologies for indicator quantification. *International Journal of Applied Earth Observation and Geoinformation*, *8*(3), 149–164. doi:10.1016/j.jag.2005.08.005

Zölch, T., Maderspacher, J., Wamsler, C., & Pauleit, S. (2016). Using green infrastructure for urban climate-proofing: An evaluation of heat mitigation measures at the micro-scale. *Urban Forestry & Urban Greening*, *20*, 305–316. doi:10.1016/j.ufug.2016.09.011

ADDITIONAL READING

Congalton, R. G. (2010). Remote Sensing: An Overview. *GIScience & Remote Sensing*, *47*(4), 443–459. doi:10.2747/1548-1603.47.4.443

de la Barrera, F., Reyes-Paecke, S., & Banzhaf, E. (2016). Indicators for green spaces in contrasting urban settings. *Ecological Indicators*, *62*, 212–219. doi:10.1016/j.ecolind.2015.10.027

European Commission. (2015). *Towards an EU research and innovation policy agenda for nature-based solutions & re-naturing cities: Final report of the Horizon 2020 expert group on 'Nature-based solutions and re-naturing cities': (full version).* Retrieved from https://publications.europa.eu/en/publication-detail/-/publication/fb117980-d5aa-46df-8edc-af367cddc202

Gupta, K., Kumar, P., Pathan, S. K., & Sharma, K. P. (2012). Urban Neighborhood Green Index – A measure of green spaces in urban areas. *Landscape and Urban Planning*, *105*(3), 325–335. doi:10.1016/j.landurbplan.2012.01.003

Karteris, M., Theodoridou, I., Mallinis, G., Tsiros, E., & Karteris, A. (2016). Towards a green sustainable strategy for Mediterranean cities: Assessing the benefits of large-scale green roofs implementation in Thessaloniki, Northern Greece, using environmental modelling, GIS and very high spatial resolution remote sensing data. *Renewable & Sustainable Energy Reviews*, *58*, 510–525. doi:10.1016/j.rser.2015.11.098

Lang, S., Schöpfer, E., Hölbling, D., Blaschke, T., Moeller, M., Jekel, T., & Kloyber, E. (2008). Quantifying and Qualifying Urban Green by Integrating Remote Sensing, GIS, and Social Science Method. In I. Petrosillo, F. Müller, K. B. Jones, G. Zurlini, K. Krauze, S. Victorov, ... W. G. Kepner (Eds.), *Use of Landscape Sciences for the Assessment of Environmental Security* (pp. 93–105). Springer Netherlands.

Mauree, D., Naboni, E., Coccolo, S., Perera, A. T. D., Nik, V. M., & Scartezzini, J.-L. (2019). A review of assessment methods for the urban environment and its energy sustainability to guarantee climate adaptation of future cities. *Renewable & Sustainable Energy Reviews*, *112*, 733–746. doi:10.1016/j.rser.2019.06.005

Perini, K., & Magliocco, A. (2014). Effects of vegetation, urban density, building height, and atmospheric conditions on local temperatures and thermal comfort. *Urban Forestry & Urban Greening*, *13*(3), 495–506. doi:10.1016/j.ufug.2014.03.003

Pinto, N. N., Tenedório, J. A., Antunes, A. P., & Cladera, J. R. (2001). *Technologies for Urban and Spatial Planning: Virtual Cities and Territories*. Retrieved from https://www.igi-global.com/gateway/book/75471

Tenedório, J. A., & Rocha, J. (2018). *Introductory Chapter: Spatial Analysis, Modelling, and Planning*. Spatial Analysis, Modelling and Planning. doi:10.5772/intechopen.81049

KEY TERMS AND DEFINITIONS

3D Urban Models: Mathematical representation of a urban objects based on its location (x, y) and height (z).

Geographic Information: Data with a known location on the earth's surface.

Green Planning: Constitutes a set of environmental strategies to ensure the protection of the natural environment.

Impervious Land: Soils that become sealed due to the urbanization process, and no longer contribute to the rainwater infiltration, biodiversity, plantation or carbon storage.

Land Planning: The technical and political process that evaluates, organizes, controls, and prepares the actual and future use of land and its resources.

Land Use Land Cover: Land use describes the social and economic activity occurring in the site (housing, transports, parks), while land cover indicates the material at the site surface (e.g. water, trees, pavement).

Microclimate: A climatic condition in a relatively small area, localized within a few meters or less above and below the Earth's surface.

Spatial Analysis: Methods and techniques that produce knowledge from data with geographic representation.

Sustainable Urban Development: Promoting urban growth in line with the present needs without compromising the needs of future generations.

Urban Green Spaces: Vegetated areas within the urban fabric, including parks, lawns, street trees, residential gardens, cemeteries, sport fields, green walls, or green roofs.

Chapter 10
Green Walls as an Environmental Strategy:
A Case Study in Brasília

Pedro Dias Boa Sorte
Universidade de Brasília, Brazil

Abner L. Calixter
Universidade de Brasília, Brazil

Caio Silva
https://orcid.org/0000-0001-8910-1841
Universidade de Brasília, Brazil

Marta A. B. Romero
https://orcid.org/0000-0002-4222-8463
Universidade de Brasília, Brazil

ABSTRACT

Vegetation is an excellent strategy to tackle global warming and climate change. In this context, the green wall is a kind of green infrastructure that is used as a bioclimatic strategy. Also, it can be used as an approach to insert green areas in an urban context, contributing to urban areas to become more sustainable. This chapter aims to evaluate the thermal performance of green walls using three different native climbing vegetation from the Cerrado biome. Thus, the authors used a computational tool named ENVI-met (v.3) as a method to analyze and evaluate which plants would bring the best performance in terms of improving thermal comfort. The database the authors obtained from the simulations will serve as a comparison of the thermal performance between walls without vegetation cover, as well as clarify the morphological and physiological characteristics that influence its thermal performance. The results demonstrate that Arrabidaea pulchra species enabled green walls with lower temperatures and higher air humidity rate in most situations.

DOI: 10.4018/978-1-7998-2249-3.ch010

INTRODUCTION

Mitigating actions to tackle global warming and climate change are urgent. The plans to combat the consequences of global warming and climate change have two distinct scales, global and local, and each one has its specific measures and efficiency. The built environment can promote a better condition of comfort for all occupants. Before, we need to understand the thermal comfort better and promote a more comfortable condition in our cities.

Thermal comfort is an individual perception which implies a neutral sense to the body within a particular environment (Lin, 2009), not being necessary to cool down or heat up the air temperature, humidity, or wind speed. In its norm 55, the American Society of Heating, Refrigerating, and Air Conditioning Engineers (ASHRAE, 2010) asserts that thermal comfort is a "condition of mind that expresses satisfaction with the thermal environment and is assessed by subjective evaluation" (p. 3). In this sense, thermal energy exchange results in the thermic conditions in the environment (Costa, 1974).

Surface materials influence thermic conditions in urban areas substantially. For example, we can observe this process in the urban heat island (UHI) phenomenon, first described by the meteorologist Luck Howard in 1833, who measured the weather variation in London (Howard, 1833). Heat is absorbed by the urban surfaces with heat capacity (Taleghani, 2018) and released to the environment, raising the air temperature.

Urban areas can also contribute to extreme heat events (EHE) and increase mortality and morbidity in urban population (Gabriel & Enddlicher, 2011), as seen in Paris, France (2003) (Dhainaut, Claessens & Riou, 2003), Melbourne, Australia (2009) (Department of Human Services, 2009) and Moscow, Russia (2010) (Revich, 2011). Replacing artificial surface materials is one strategy to reduce UHI and EHE (Norton et al, 2014).

For this study, the authors analyzed exchanges in thermal energy among the natural environment, the built environment, and anthropogenic activities. Costa (1974) argues that these exchanges are influenced by several processes, such as human activity, clothes, and climatic conditions, or the intrinsic characteristics of the materials. Indeed, surface materials have a direct impact on thermal comfort in cities.

In this sense, it is possible to control thermal energy exchanges in the built environment by choosing and incorporates adequate materials that would favor or decrease such exchanges. This observation enables cities with less energy consumption, more thermal comfort and consequently more secure. Also, it is possible to make achieve those goals as we make the cities more morphologically similar to natural environments. All this contributes to more sustainable cities. In this research, we consider vegetation as a composite material for the second skin of the building facade. More specifically, climbing plants from Brazilian Cerrado.

As many of the strategies for thermal comfort can be evaluated through computational simulations, the authors decided to use this approach to evaluate the impact of vegetation in the built environment.

Most of the studies in this field, especially in hot climates, have focused on the performance of the building and its cooling systems (Olivieri, Olivieri, & Neila, 2014). However, Reinhart and Davila (2016) developed a proposal for climate compression to launch mitigation strategies, and Panchen et al. (2014) highlighted the role of vegetation in climate seasonality studies, and this is the discussion promoted in this research.

Willis et al. (2010) state that many exotic species are less vulnerable to climate change. In this sense, exotic landscaping may prevail over the use of native species, which might be considered unsustainable (Primack & Rodrigues, 2001).

NATURAL VS. BUILT LANDSCAPE

The landscape represents a spatial part on earth's surface, which is composed of complex systems and formed by rock dynamics, water, salt, vegetation, men, and other animals. It is a unity (Zonneveld, 1995), or as a heterogeneous portion of land. This distinct and measurable unity which is defined by a group and repeated ecosystems, such as geomorphological elements and disturbance (Forman & Godron, 1986).

Zonneveld's (1995) definition is better suited to evaluate the natural landscape of urban surroundings, because, importantly, a portion of land contains the result of natural elements acting for billions of years. As a result, this landscape is considered the standard for life to proceed in this area. If the goal is to allow nature to thrive on the urban landscape, it is necessary to consider that natural record.

A natural landscape is part of an ecological network, which is a complex system that contemplates the biodiversity in all ecological levels. The physical connection between these portions of the natural landscape is an important topic when conserving ecological networks is an issue. The connection allows the transit of animals, vegetation, and information, and the development of all species in that portion.

On the other hand, portions of the natural landscape can be isolated promoted by a barrier effect. The barrier effect represents any physical barriers, such as roads, buildings, farms, industry, and urban environment. In comparison, anthropogenic events can act more severely than natural barriers (i.e., sea, ocean, and desert), once it can cause effects on a much bigger spatial and temporal scales (Battisti, 2004).

The human ability to transform natural spaces has led to several impacts on the ecosystem. The consequences of these impacts are still unexplored, especially regarding the interactions between human processes and the ecological dynamics in this area (Alberti, 2008).

The richness of some species with the same ecological function has distinct performances in urban areas. The characteristics of the reminiscent vegetal surface is changed such as the size, complexity, and vegetation structure. Consequently, the colonizing population suffers for the loss of a niche and opportunities in ecological interactions (Aleixo, 2001; Bennett, Hinsley, Bellamy, Swetna, & Mac Nally 2004; Lord & Norton, 1990; McIntyre, 1995; Savard, Clergeau, & Mennechez, 2000).

As the urban population grows, the landscape is modified. The dominant landscape is replaced and causes environmental restrictions on this area (Bocquier, 2014; Paul & Meyer, 2001). Also, these new restrictions act as a filter: Some species respond negatively and disappear, others respond positively and become dominant in the area. This causes a homogenization of the species in urban areas (Seto, Güneralp, & Hutyra, 2012).

The isolation of natural landscapes in smaller areas than the original is a phenomenon which is called "habitat fragmentation" and is an essential treat for biodiversity (Collinge, 1996). Changes in the earth's soil are a human activity that affects biodiversity on fragmentation, habitat opportunity, flow modifications of resources, disturbance regime, and species composition (Kinzig & Grove, 2001).

In addition to landscape modification and fragments, urban fragments often receive new vegetation. As many times this vegetation does not occur on the natural surroundings, it is also called "exotic vegetation."

This modification is often led by architects and urbanists, who see the vegetation primarily as an esthetic element. Unfortunately, electing vegetation by architects and landscape planners is too far from scientific questions or guidance, sometimes acting as pure art, which is one of the reasons for the major quantity of exotic vegetation in the urban environment. Thus, the authors concur this can be a technical instrument to reflect on concepts and propose a creative solution to the issues of contemporary landscape (Silva, 2007).

GREEN WALLS

The vegetation of a green wall provides a lower transfer of thermal energy, compared to the bare wall (i.e., without vegetation cover). The vegetation layer decreases the solar incidence in the wall that makes the outer fence of the building. In this sense, the shading of the facade, which is located behind the vegetation, is the most efficient strategy to avoid thermal energy gain (Romero, 2013). It can reduce the solar absorption coefficient of a wall by 66% (Eumorfopoulou & Kontoleon, 2009). Also, vegetation has a lower solar absorption coefficient (α) and emissivity coefficient (ε), lower thermal conductivity (λ), higher thermal resistance (R), and higher specific heat (c) than most traditional materials for coating, such as apparent concrete. Green walls can be categorized as an urban green infrastructure (UGI), which are a network of planned and unplanned green spaces, spanning both the public and private realms, and managed as an integrated system to provide a range of benefits (Lovell &Taylor, 2013; Tzoulas et al, 2007)

As a result of these characteristics, most of the researches involving green walls, such as Olivieri et al.'s (2014), Pan and Chu's (2016), and Wong and Baldwin's (2016) studies, among others, demonstrate that their use in hot weathers leads to lower temperatures inside the building when compared to a bare wall. Nevertheless, differences in the methodology may result in numerical differences. For example, Olivieri et al. (2014) verified that the presence of vegetation on the facades provided a maximum temperature of 5 °C lower than a situation without vegetation.

Djedjig et al (2016) evaluate the hygrothermal performance of a model of green walls, and a model of mass flows in street canyons at the warm climate. The model was created to assess the green wall's hygrothermal impact and to evaluate the developed numerical tools. This research emphasizes the outcomes of both urban microclimate and buildings energy loads, especially in the summer period. The experimental and numerical results obtained with green facades underline the advantage of this modeling approach for the design of passive cooling for buildings and mitigation of extreme thermal conditions within street canyons in dense cities, especially in warm climates.

Liang et at (2014) evaluate how vertical greenery can influence outdoor thermal comfort. This research analyzes the radiant temperature (tmrt) of two green walls (Green Wall A and Green Wall B) using customized globe thermometers. The diurnal tmrt profile was altered significantly due to the installation of vertical greenery. When both green walls were present, the peak tmrt occurred at 15:00 h, but with more concrete exposed when one wall was removed, the peak occurred at 17:00 h and was 10.9–12.9° higher at 0.5 m away from the wall. The authors concluded that vertical greenery helped to reduce tmrt both during the day and at night, to varying extents.

Galagoda et al (2018) present the concept of living wall as technological solutions to replace the declined greenery at the urbanized environment and also reliable applications for thermal regulation in buildings through insulation effect and escalates the energy use efficiency. They evaluated three types of green infrastructures, such as (T1) living walls, (T2) indirect green facades and (T3) direct green facades located in the Colombo metropolitan in Sri Lanka, using a satisfaction survey as well. The study revealed that vertical greenery systems were highly effective on outside wall surface temperature reductions at 1100 h–1500 h time zones. T1 and T2 accounted for superior temperature reduction in the range of 1.61 °C–1.72 °C through the facade relative to the distance than T3. Maximum temperature reduction compared to the bare wall control was obtained for the T1 (0.28 °C–8.0 °C) followed by T2 (1.34 °C–7.86 °C) and T3(1.34 °C–6.64 °C). Likewise, parameters such as relative humidity and CO2 emissions are calculated. This research highlighted how green-facades works as a sustainable approach to microclimatic changes and human thermal comfort.

Consequently, by providing temperatures closer to thermal comfort, green walls reduce the need for active cooling and, thus, save energy. Pan and Chu (2016) compared the energy expenditure for cooling in buildings that are equipped with green-walled vs. those with bare walls. In all weather conditions, energy expenditure in green wall buildings was lower, and in August and September, the hottest summer months, the reduction reached 16%. Wong and Baldwin (2016) argue that a green wall with niches, and separate from the facade can reduce up to 20% of the energy for air conditioning, or produce an annual saving of 8%. In a study that took into account the specificity of the vegetation, Dunnet and Kingsbury (2008) stated that the appropriate choice of the species could bring different indices of economy, which can reach up to 64% of savings.

Also, if compared to other UGI, such as green roofs or urban parks, green walls have the advantage of becoming part of the entire urban fabric. Urban surfaces which are not used, such as the building envelope, could easily be covered with green walls and alter the microclimate of the built environment, as well as the local climate of the city (Alexandri & Jones, 2008), contributing to sustainable urbanism.

This way, there is essential potential of lowering the urban temperature when the building envelope is covered with vegetation (Alexandri & Jones, 2008). Ecological benefits can be added up when specific vegetation species are taken into account.

VEGETATION

The vegetation of climbing habits called "trepadeiras" in Portuguese (i.e.,"climbing plants" or "vines" literally in English) belongs to the same botanical group of lianas and vines, which also have specializations for ascension and vertical development (Pujals, 2011). Pteridophytes, gymnosperms, and angiosperms have species and several botanical families (Pujals, 2011), and they are an important component of forest communities (Engel, Fonseca, & Oliveira,1998). Although 90% of the species of this kind of plants are located in the tropics (Walter, 1971), these plants can succeed in any climate and plant community that offers enough physical support. Vines tend to be intolerant to shade, but this condition may stimulate the sudden elongation of the stem, as in etiolated plants (Walter, 1971). The growth rate can vary between 5cm/day and 130cm/day (Peñalosa, 1985), reaching up to 15cm of stem diameter and up to 70 meter in length (Jacobs, 1988).

Such ranges in taxonomy, morphology, and ecological characteristics of the vines reinforce that the vegetation is not an inert element, but a living being which establishes relations with its surroundings, to which it can bring even undesirable results. Therefore, the origin of the species should not be ignored. When choosing the vegetation that will compose a green wall, the authors recommend native species from the local biome.

This recommendation relies on the fight to preserve biodiversity. Exotic species are one of the main treats of biodiversity conservation, as they can move as nascent species in a competition for resources, change their habitat, and lead themselves to extinction, which indirectly means reducing resources (e.g., medicines, water, and other raw materials) which are appealing for a human species (Primack & Rodrigues, 2001).

Importantly, the native species have social ethos, since they constitute an important cultural and economic patrimony for the local populations. Their use then represents the possibility of reconciling the objectives of anthropic intervention with those of biodiversity conservation and allows the integration of the built environment into the natural environment (Nascimento & Oliveira, 2005).

SIMULATION AND COMPUTATIONAL TOOLS

Architects and engineers regularly use different kinds of software as tools to inform decision-making in their projects. Nowadays, it is possible to simulate the impact that a green wall will pose on the climatic parameters of its surroundings, and compare it with a bare wall. In this research, the authors selected the ENVI-met software (v.3), which understands the interaction between the climate, the natural environment, and the built environment (Bruse, 2004). As Simon et al. (2018) demonstrated, ENVI-met can model the transpiration process and leaf temperature, thus informing the real impact over the urban microclimate. This software allows choosing the available species from its vegetation database for the development of the simulations.

In this sense, the possibility of altering the parameters of the vegetation to characterize a species or a specific taxonomic group is strategic. However, its default database includes only generic vegetation, such as grass, corn, and soybeans, which are not suitable for the reality of thermal simulation of green walls, nor do they represent local floras. Obtaining parameters that represent specific vegetation, demands field and laboratory research, thus little new information is usually added. Therefore, this research aims to provide such parameters (i.e., primary data), as they can characterize native species of the Cerrado biome, which makes possible its use in hygrothermal simulations with ENVI-met and promotes its application to green walls. This use of green technologies allows a greater understanding of thermal comfort in the building, through the evaluation of the impacts of the reduction of energy expenditure with active strategies of cooling, also contributing to the conservation of biodiversity.

STATE OF THE ART

Many types of research dedicate their goals to understand how urban development and a lack of natural vegetation contribute to increased temperatures in cities. This part of the chapter presents the relevant state of the art through a literature review. Some studies discuss the relationship between vegetation as a means to promote thermal comfort in an environmental context, and also the green walls and green roofs as strategies of adaptation or mitigation. Most of these studies used ENVI-met as an evaluation model. . The authors selected ten different kinds of research in different climate and scale contexts.

The scale of urban heat island (UHI) offers relevant research in two different climate contexts. Maleki and Mahdavi (2016) discuss mitigation strategies using a three-dimensional urban microclimate model. In their study, they investigated the effects of the variation of physical and geometrical properties of the urban fabric on the urban microclimate and outdoor thermal comfort using ENVI-met. The authors concluded that results that increase the amount of vegetation and permeable pavements can cool the air temperature down by up to 3K.

In another climate, Razzaghmanesh, Beecham, and Salemi (2016) present an experimental investigation at the several climate levels. They use simulation at the micro and macroscale of a typical urban environment in South Africa, evaluating the potential for mitigating the UHI effect in the city of Adelaide. They discuss changing an urban environment and replacing vegetated surfaces with low albedo materials is one of the reasons for increasing temperatures in an urban environment, and consequently, also one of the essential causes of UHI effects. The results also showed that they have significant cooling effects in the summertime and could behave as an insulation layer to keep buildings warmer in the winter. Finally, the authors affirm that the use of green walls and street trees together with the adoption of high albedo materials is recommended for achieving the optimum efficiency in terms of reducing urban temperatures and mitigating UHI effects.

Pastore, Corrao, and Kvols Heiselberg (2017) show a new methodology using a microurban to a building scale assessment by linking the ENVI-met software to the building energy simulation program EnergyPlus. The authors developed a method to analyze the influence of plants on urban microclimates to enhance outdoor and indoor temperatures. Additionally, this method allows to assess the benefits of different technical solutions which are applied to buildings. These authors' findings indicated a significant reduction in both areas, in outdoor and indoor temperatures. Their work highlighted the importance of multi and interscale approaches for high-accuracy investigations into the thermal performance of buildings.

Morakinyo et al (2019) aim the passive techniques as green facades could be a great strategy to improve the outdoor thermal environment and reduce indoor energy use. Specifically, in high-density cities like Hong Kong, buildings' surface area is much higher than the roof and ground surface areas combined, offering a vertical surface platform for greening. The authors used ENVI-met model simulation to understand the potential air cooling and thermal comfort improvement of a neighborhood of varying densities. They found that 30–50% of facades in the high-density urban setting of Hong Kong must be greened to potentially cause about1 °C reduction in both daytime and nighttime air temperature. At the same time, it helps to improve daytime pedestrian thermal comfort by at least one thermal class.

Fahmy et al (2017) presented the application of green cover and its effect on domestic energy efficiency in the present (2020) and in the prescriptions for the future (2050 and 2080) in Egyptian urban communities. The authors used two different software, such as ENVI-metV4.0 simulations (accounting for outdoor conditions) with indoor simulations using DesignBuilderV4.2. Results showed cool spots at the center of communities witch improvements decrease by 2080 due to the effects of climate change. Whole site averages of (Ta) showed increased records for the adapted cases owed to the suggested green facade coverage, which draws attention to the sensitive plantation of building walls as well as the coverage percentage of urban trees that might traps heat. In comparison to their un-adapted cases, the least energy efficiency result for the whole site was 10.0%, corresponding to 23.8% cost saving at 2080 in case two. In contrast, the maximum was 21.3%, matching to 35.7% cost savings today. Summing energy savings until the end of the century, case one payback period was 20 years (in 2037), and case two was 15 years (in 2032).

Ahmad, Babak, Davoud, and Suhardi (2019) discuss how local climates negatively affect UHIs and outdoor thermal comfort. These authors present recent studies that focus on the effects of green spaces and plants on the microclimate and thermal comfort of recent years. As a methodology, this research accosts different parameters, such as physical properties of plants, location and vegetation cover, planting densities and crown density, plant element, leaf type, planting patterns and arrangement, and albedo. The results highlighted that several factors with impact on the cooling effect of plants had not received adequate attention in previous studies, yet.

Zölch, Maderspacher, Wamsler, and Pauleit's (2016) research focuses on understanding the performance of urban green infrastructure (UGI). These authors discuss how comparable information concerning the performance of different UGI types to moderate such impacts is mostly lacking. This research investigates how different types and quantities of UGI (i.e., trees, green roofs, and green facades) affect pedestrian thermal comfort. The study shows the physiological equivalent temperature (PET) values by 2.4 K; however, this could be vastly reduced by different UGI scenarios. The article proved that planting trees had the most positive impact, with an average PET reduction of 13%, compared with existing vegetation. Specifically, in green facades, it has mitigating effects of 5-10%. Other results indicate that increasing the share of green cover does not directly correspond to the magnitude of the PET reduction as well.

Herath, Halwatura, and Jayasinghe (2018) evaluated the mitigation and resilience strategy of UGI in tropical Colombo, the commercial Sri Lanka capital, with a computer-based simulation for a hot and humid end-summer sunny climate conditions in August. The method uses seven different UGI scenarios which these authors defined under real scale modeling by ENVI–met, with green roofs considering vegetation percentage and existing geometry. The results highlighted that different green infrastructures significantly reduced the temperature with the vegetative fraction, and the researchers recorded the highest reduction of outdoor air temperature at 15:00 on a typical day. The best scenarios show a reduction of about 2°C in comparison with the outdoor temperature, and present a temperature reduction of 1.64°C for green roof 100%, compared to real UHI current scenario. This research highlighted the power of green to combat extreme temperatures.

Salata, Golasi, Vollaro, and Vollaro (2015) analyze the effect of the vegetation and high albedo materials characterizing horizontal and vertical boundaries of the site and the cloister by Giuliano da Sangallo, a historical site in Rome. As a methodology, the authors developed five different scenarios, with different vertical and horizontal materials of the present buildings. They simulated the scenarios and analyzed the variations of physical quantities (i.e., air temperature, mean radiant temperature, relative humidity, and wind speed) affecting the perception of environmental comfort through the predicted mean vote index. Their results showed the vegetation can be a significant benefit to the environment, and high albedo materials can ease the thermal load of the buildings with higher thermal stress for the pedestrians.

In relation to building scale, Laureti and Battisti (2017) investigate the influence of building materials (e.g., cool colored materials) and vegetated urban surfaces (e.g., green roofs, green walls, and trees) on the urban microclimate and on pedestrian outdoor thermal comfort in Rome. The study highlights the multiple effects of cool colored materials on human comfort, and investigates which could be the best situation in terms of mitigation of ambient temperatures and thermal discomfort. Finally, it shows the negative effect of cool colored materials on human thermal comfort when they are applied isolated to surfaces, such as pavements and urban facades, but it also underlines the benefit which is associated with a mixed combination of cool materials and vegetation, especially trees, setting the way for further research.

CASE STUDY

Definition of Green Walls

A green wall might be defined as a surface which is partially covered by vegetation or a kind of vertical elements of vegetation in buildings, such as walls, partitions or facades. This research presents three types of the design concept of green wall, in terms of several layers, modular logic, and habits of the vegetation This research analyzes green wall which uses a climbing plant with support to hold on the surface.

Climber plants have more advantages than a simple herbal plant, in terms of infrastructure costs and environmental complexity. They can grow faster than other species and have a harder structure. Therefore, their behavior reduces the demand for artificial features and creates a natural appearance.

This research is organized in two steps. The first one is dedicated to collect and use primary data. It initiates the investigation on which species would be more suitable in terms of climate adaptation, due to its availability in Cerrado biome and capacity of providing shade rapidly.

The second step is dedicated to inserting and processing the data of the plants in relation to leaves density in the ENVI-met software. Previously, the authors organized the database in tables, to fulfill the configuration file in ENVI-met program and run the analysis.

Selection of the Species

The authors selected the species in terms of three characteristics, in relation to research goals:

1. The species must be native in Cerrado biome.
2. The species must be resistant to solar exposition along the day.
3. The species must have beautiful flowers or leaves to adorn buildings.

Then, the authors searched the literature for climbing species with these characteristics. The book Trepadeiras Ornamentais do Cerrado (Ramalho & Proença, 2004), was useful as a database for this study.

After that, the researchers selected five different species to study:

1. Banisteriopsis pubipetala.
2. Banisteriopsis anisandra.
3. Psiguria ternate.
4. **Cissus erosa.**

5. Arrabidaea pulchra.

All of them are native species; therefore, it was not easy to find them in commercial stores. Thus, the authors searched in areas of native vegetation in public parks in Brasilia.

Then, they consulted the Herbarium database of the Botany Institute of the University of Brasilia (http://www.florescer.unb.br/bol/UB/Home/Index). This database has about 200.000 species already registered.

The available information is extremely relevant, such as the specific geographic coordinates of each collected species. Each species with the restriction was located inside the Federal District territory.

The authors checked the geographic coordinates of all species and carried out 10 field trips between June and August 2014 in the following locations:

1. Ecological Reserve of Brazilian Institute of Geography and Statistics. Geographic coordinates: 15°57'02.70"S - 47°51'55.24"O.
2. Água Limpa Farm (FAL – UnB). Geographic coordinates: 15°56'55.91"S/47°56'08.45"W.
3. Olhos d'Água Park, Brasília - DF. Geographic coordinates: 15°44'33.16"S/47°53'10.11"W.
4. University Campus Darcy Ribeiro. Geographic coordinates: 15°45'48.29"S/47°52'11.08"W.
5. Olympic Center of the University of Brasilia. Geographic coordinates: 15°45'51.44"S/47°51'28.20"W.

Based on the information the authors obtained in the field, they chose three species. The selection criteria were ease of access and possibility to be studied. Figures 1, 2, and 3 represent the species.

Choice of Parameters

The ENVI-met software has several parameters to characterize the vegetation. The researchers selected these parameters according to their relevance for this research, such as resistance, nativeness, and appearance.

In this way, the parameters the authors extracted for this chapter are: Carbon fixation (C?), minimum stomatal resistance (rs_min), shortwave albedo (a_f), and leaf area density (LAD). The methodology to obtain these data was direct and destructive to reach significant differentiation among the species. The authors used standard values for all the other parameters the software required.

Figure 1. Cissus erosa

Species: *Cissus erosa* (CE)

Family: Vitaceae

Popular Name: cipó de fogo

Geographic Coordinate: 15°57'02.70"S - 47°51'55.24"O. Ecological Reserve of Brazilian Institute of Geography and Statistics (RECOR).

Figure 2. Arrabidaea pulchra

Species: *Arrabidaea pulchra* (AP)

Family: Bignoniaceae

Popular Name: *Not avaliable*

Geographic Coordinate 15°44`33.16"S /

47°53`10.11"W. Olhos d'Água Park– Brasília - DF.

Figure 3. Banisteriopsis anisandra

Species: *Banisteriopsis anisandra* (BA)

Family: Malpighiaceae

Popular Name: Cipó prata

Geographic Coordinate 15°56'55.91"S / 47°56'08.45"W.

Primary Data Collection

The researchers used the infrared gas analyzer (IRGA), model LCi, ADC BioScientific (UK) to determine the values of maximum assimilation of CO2 (Amax), transpiration rate (E), and stomatal conductance (gs).

The equipment does not directly reveal stomatal resistance values, but stomatal conductance. Hence, it is necessary to perform a mathematical conversion. Besides, the equipment does not perform measurements of albedo directly, which the authors obtained from an indirect method.

Thus, the researchers extracted leaves of each species and placed them on a regular surface until they formed a homogeneous layer. After that, they measured the incident light between the leaves with the chamber near them and with the IRGA LCi light sensor facing the sky.

Then, they measured the light the leaves reflected with the camera facing down near the surface of the leaves, avoiding shading. They performed 10 measurements of each situation and established an average. The researchers found the albedo index by calculating the ratio of the mean of the reflected light to the average of the incident light.

In stable conditions of temperature and ambient light, the authors calculated the measurements of each species. They checked the data twice for equipment or user errors. If in the data confirmed any discrepancy, the researchers performed a new measurement.

LAD is a parameter composed of two variables: Leaf area and volume expressed per m2/m3. The authors summarized each by collecting a 1-m-long stem, measured from the younger portion to the more mature portion. For the determination of LAD, they used a direct and destructive method.

In order to obtain the leaf volume of the stem, the authors considered the leaves and the space between them. They compared this volume with a cone trunk volume, which is a three-dimensional geometric figure that represents the real model, where the base in formed by more adult leaves (larger radius) and the smaller base is composed of the top, formed by the younger leaves (smaller radius), corresponding to a height of 1m. The researchers removed all the leaves of this stem and immediately took them to the laboratory for measurement on the leaf area scanner. The scanner was a model CI-202 of the brand CID Inc. The value of LAD is given for a ratio of the total leaf area on the volume of the pyramid trunk.

Hygrothermic Simulation

After the data collection of the studied plant species, the researchers carried out the first round of simulation with the software ENVI-met. 3.1. The objective of the simulation was to verify how the primary data they obtained for each species would influence the climatic parameters. The primary data reflected the morphophysiological characteristics of each species, and they differed between them. Besides, the authors expected a significant correlation between these characteristics and the thermal simulation results.

Before starting the simulation process in ENVI-met, it is necessary to configure the data that will be used: Climate, vegetation, research area template, date, and simulation duration. Thus, the authors configured the climate data according to the city of Brasilia and the vegetation data according to the collected data. The researchers developed the modeling of the virtual elements of the research area so that no interference occurred between each building and that they could analyze the facades in all the orientations independently.

Lastly, it is necessary to select the output data. For this research, the authors selected climatic parameters that could be influenced by the vegetation and that could modify the thermal comfort sensation.

Weather Data

The climate data the researchers used in the simulation are from the "brasilia_seca" file, corresponding to the dry period (May to September), elaborated by Castelo Branco (2009).

Vegetation Data

In order to enter the data of each species of vegetation, the authors used the command "Edit PLANTS.DAT ...", inside the "Database" menu, which provides the standard database of the program. They entered these values in the archive database, where the three vegetation species were available to be selected and used in the simulation.

Area Template

For this study, the authors designed a mesh of 40m x 40m). This mesh was composed of smaller units of square shape and with 1m in each dimension.

The building the authors modeled for the study has the shape of a cube, with dimensions of 3m x 3m x 3m. They modeled the green wall using the insertion of vegetation in the closest squares to one side of the cube (green facade). In this way, each green wall was dimensioned in 1m of length, 1meter of width, and 3m of height. They modeled four buildings in each 40m by 40m grid, each with green walls in a distinct orientation (i.e., north, south, east and west), equidistant from each other by 20m in X and Y, so that no interference occurred between them; the other points were loamy soil.

Data and Time

The simulation lasted 24 hours, starting on September 8, 2008 (date which corresponding to the climatic data used), at 6:00, for each climbing species and a control scenario, with bare walls. The authors used data from the hours 9:00, 12:00, 15:00, 18:00, and 21:00.

Data Output

The researchers marked the following parameters as "Output Data:"

1. Pot_Temperature
2. Relative_Humidity

The authors generated the maps of the results of the simulations with the program LEONARDO 3.75. The resulting values for each parameter are very relevant for this research. The authors considered the value that suffered less interference from other factors and that was more reliable to the vegetation present in the green wall. Therefore, from the three squares of 1m x 1m x 3m that corresponded to the green wall, the researchers chose the value which was in the center square. They discarded the values they obtained in the lateral squares.

RESULTS

Primary Data

Table 1 shows the data bank for each vegetation species the authors obtained from the analysis of primary data from their field and laboratory studies.

Table 1. Primary data for each vegetation species

Plant Parameters	Plant Species		
	Arrabidaea Pulchra	**Banisteriopsis Anisandra**	**Cissus Erosa**
ID	AP	BG	CE
C?	C3	C3	C3
TY	3	3	3
rs_min	47	247	214
a_f	0.07	0.05	0.06
HH.HH	03.00	03.00	03.00
TT.TT	00.50	00.50	00.50
LAD 1 to LAD 10	1.682	0.75	97
RAD 1 to RAD 10	0.1	0.1	0.1
NAME	Arrabidaea pulchra	Banisteriopsis anisandra	Cissus Erosa

Control Scenario

For a more appropriate understanding of the simulated data, the researchers made a comparative analysis with the control scenario. They established a comparison value for air temperature and air humidity. They calculated this value through an arithmetic mean they obtained by the numeric value of bare walls on all geographical orientations and hours. As a result, they were able to obtain a value that represented the global thermal behavior on a single hour. This value allowed for comparisons between geographical orientations, hours, and species. The researchers could also evaluate the thermal behavior during the day, as well as determinate the hours and geographical orientations that presented the highest and lowest values.

Table 2 and 3 report air humidity and the air temperature the authors obtained in every hour and geographical orientation, and the calculated arithmetic mean, which is the value for the control scenario.

Control Scenario for Species Scenario

The authors compared the values of air temperature and air humidity they obtained from the control scenario to the values of each species. The numeric difference is showed in ΔT°C and percentile points (pp). Figures 4 to 8 illustrate the differences (°C), in every hour, of air temperature of each species compared to the control scenario. Figures 9 to 13 illustrate the differences (pp), in every hour, of air humidity of each species compared to the control scenario

Table 2. Air humidity for control scenario

Façade Orientation	Time (h)					
	9	12	15	18	21	Average Air Relative Humidity (%) on Orientation Analysis
North	66,34	68,27	62,8	66,84	65,72	65,99
South	69,33	70,28	65,65	70,33	68,86	68,89
East	66,4	69,11	63,14	67,27	65,92	66,368
West	61,45	64,19	60,13	65,16	62,85	62,76
Average Air Humidity (%) on Time Analysis	65,88	67,96	62,93	67,4	65,84	

Statistical Analysis

The statistical analysis of this research consists of the principal components analysis (PCA) . The authors compared the raw values of air temperature and air humidity, and not the numeric variation related to the mean, with the primary data: rs_min, LAD, and a_f.

The values of air temperature and air humidity relative did not present normal distribution, so the authors had to normalize them through Equation 1. As there were distinct units, they standardized the values they had obtained from Equation 1 by Equation 2. There was no need to manipulate the values the researchers obtained from the primary data, as they presented numeric differentiation amongst themselves.

Table 3. Air temperature for control scenario

Façade Orientation	Time (h)					
	9	12	15	18	21	Average Temperature (°C) on Orientation Analysis
North	24,3	32,51	34,01	30,82	26,39	29,61
South	24,07	32,23	33,29	30,43	25,96	29,2
East	24,43	32,71	33,73	30,86	26,47	29,64
West	25,54	32,16	33,77	31,1	26,86	29,89
Average Temperature (°C) on Time Analysis	24,59	32,4	33,7	30,8	26,42	

Figure 4. Air temperature variation at 9:00

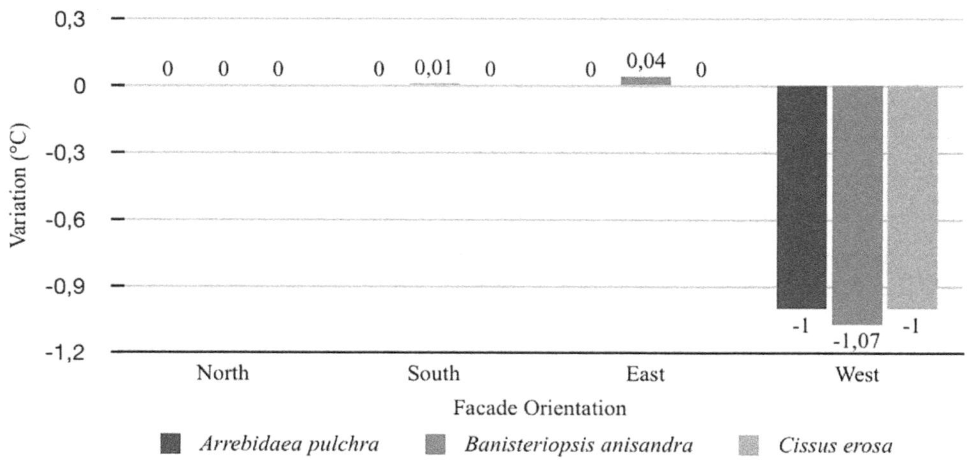

$$Yi' = log10(Yi + 1) \tag{1}$$

where Yi' is the standardized variable.

$$Zi = (Yi' - Yii) / sy' \tag{2}$$

Figure 5. Air temperature at 12:00

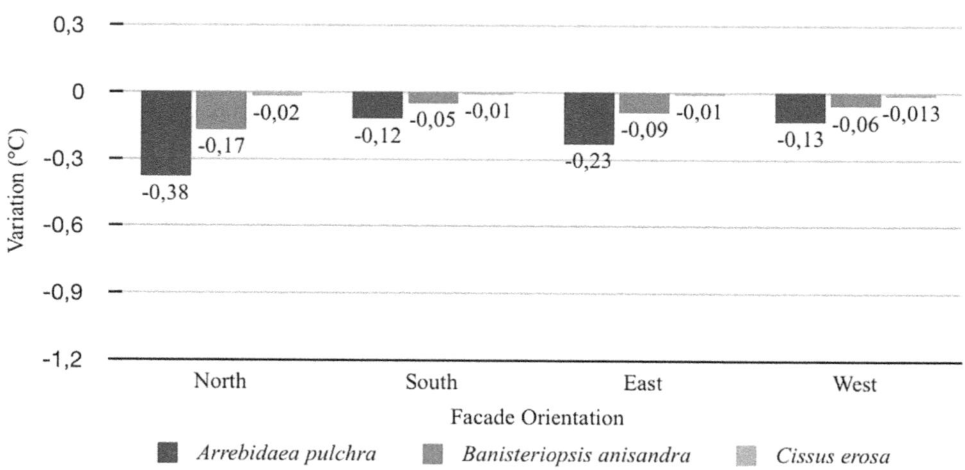

where Zi is the standardized variable, Yi' is the normalized variable, Yii is the normalized variable mean, and sy' is the normalized variable mean standard deviation.

Figure 6. Air temperature at 15:00

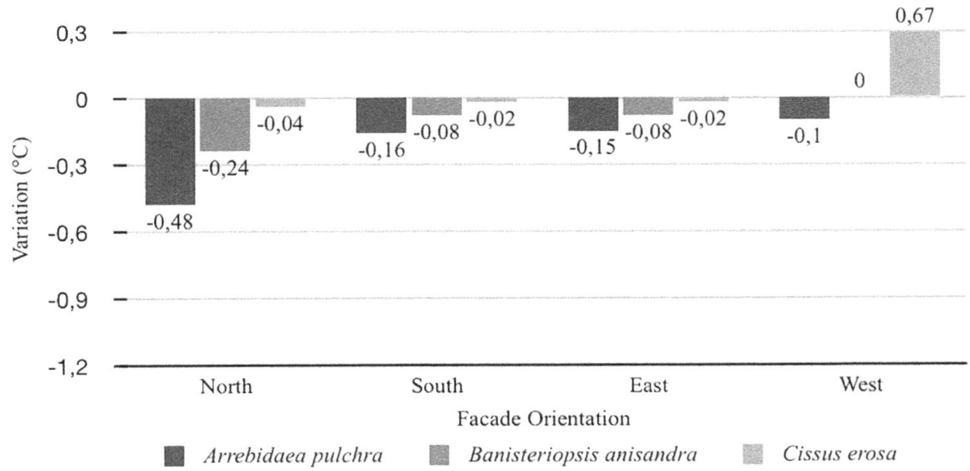

Figure 7. Air temperature at 18:00

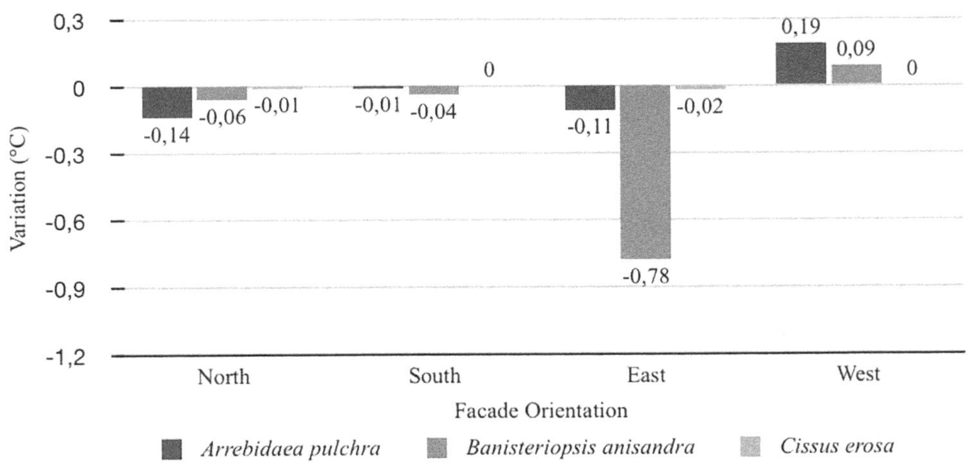

Figure 8. Air temperature at 21:00

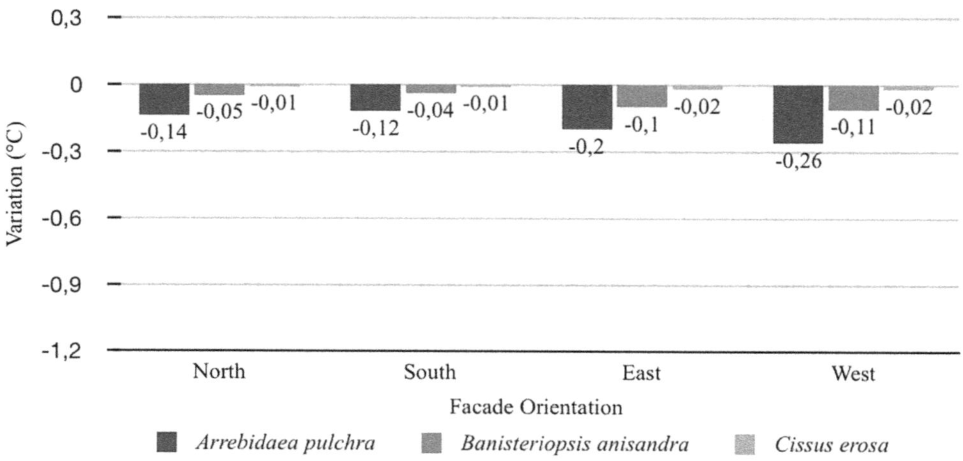

Figure 9. Air humidity variation at 9:00

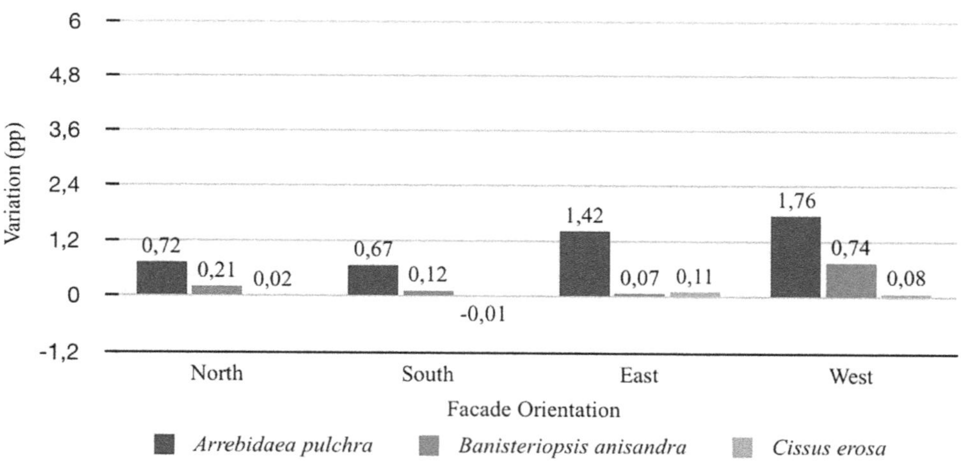

Species Differentiation through Primary Data

The PCA using the primary data (rs_min, LAD, and a_f) showed that Arrabidaea pulchra has a significant differentiation from other species (e.g., Banisteriopsis anisandra and Cissus erosa) that have morphological similarities (Figure 14).

Figure 10. Air humidity variation at 12:00

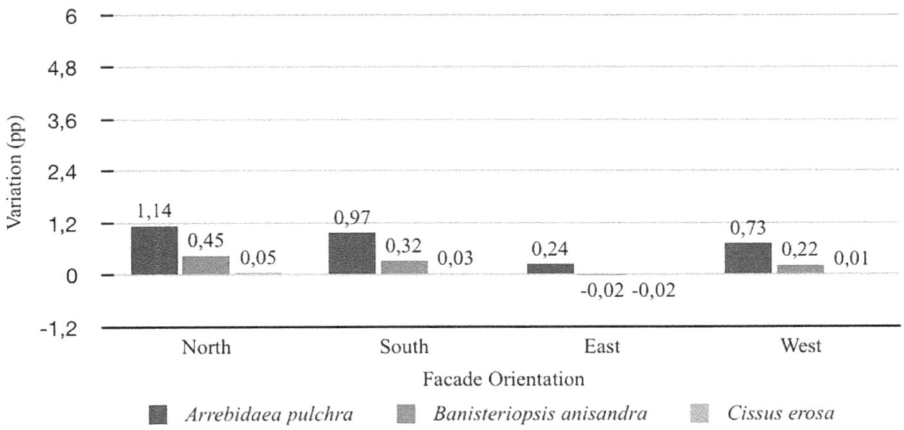

Figure 11. Air humidity variation at 15:00

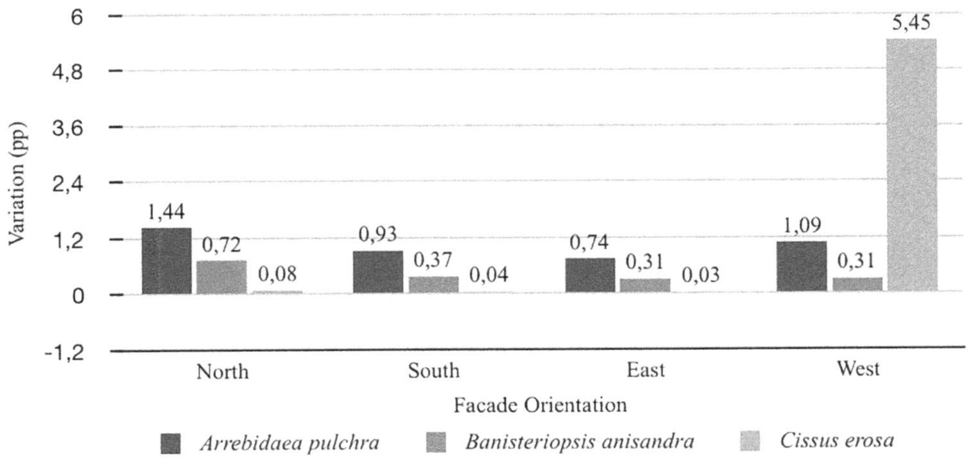

Primary Data and Their Climate Influence

The analysis of the data the authors obtained by the simulation showed that the hour 15:00, in the north facade, presented the highest value for air temperature and the lowest air humidity. The statistical analysis aimed to understand the relationship between primary data (morphological characteristics of the species) and the values for air temperature and air humidity at the 15:00 hour on the north facade. The researchers used a linear regression; it showed a significant result for the LAD, which has the strongest influence on the values of air temperature and air humidity. The LAD showed to be the most significant parameter to modify the climate conditions on the control scenario (Figures 15 and 16).

Figure 12. Air humidity variation at 18:00

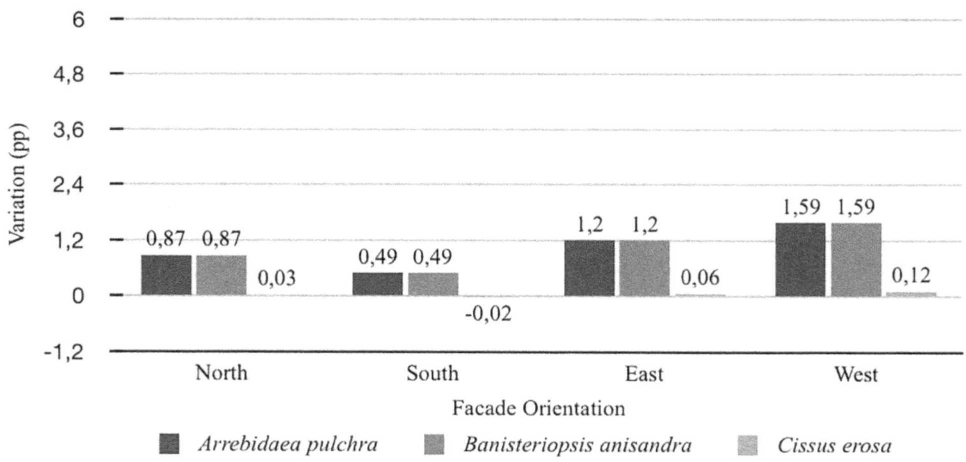

Figure 13. Air humidity variation at 21:00

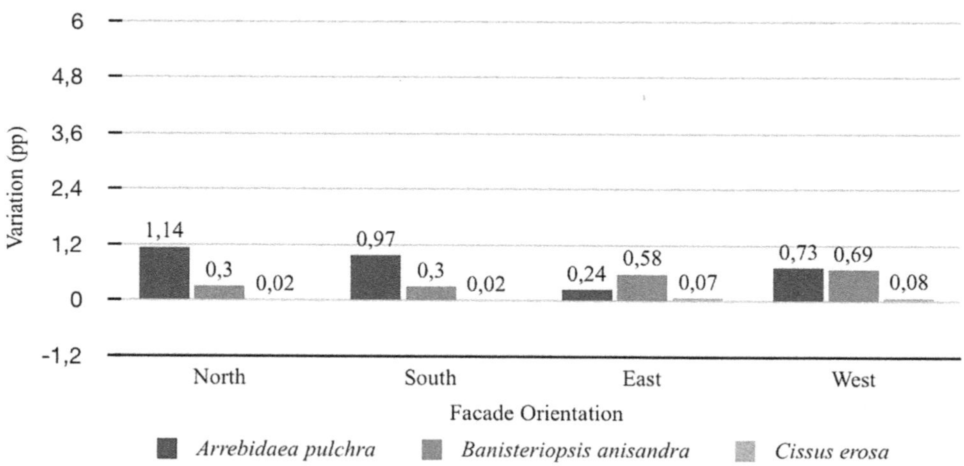

Figure 14. Principal components analysis related to species

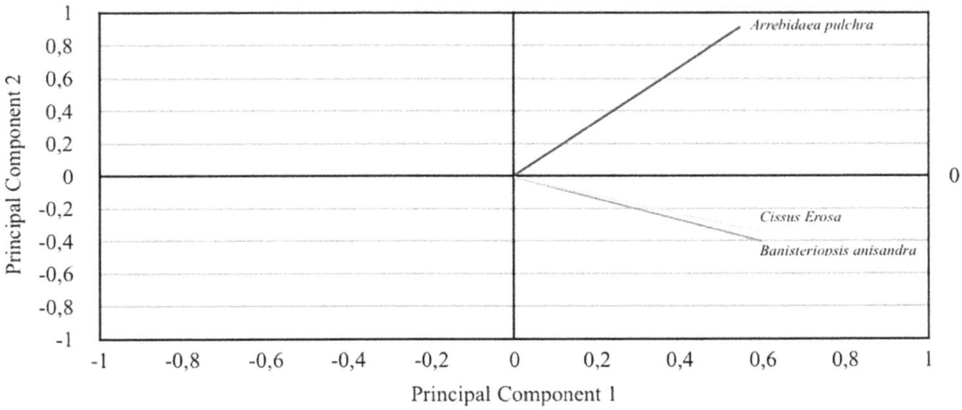

However, when the authors analyzed how each species influences the climate, they observed that all have similar impacts, as well as they all differentiate from the control scenario (Figure 17).

As to air humidity, all species are distinct from the control scenario. However, Arrabidaea pulchra stands out, in comparison with Banisteriopsis anisandra and Cissus erosa (Figure 18).

Figure 15. Leaf area density and air temperature

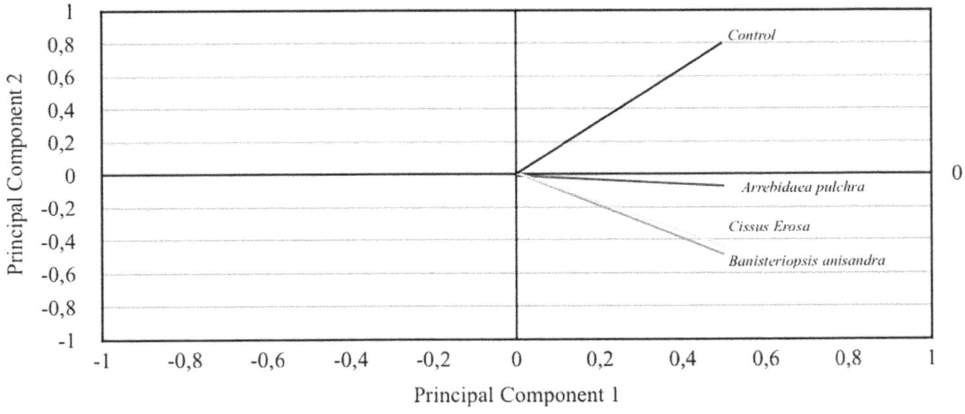

Figure 16. Leaf area density and air humidity

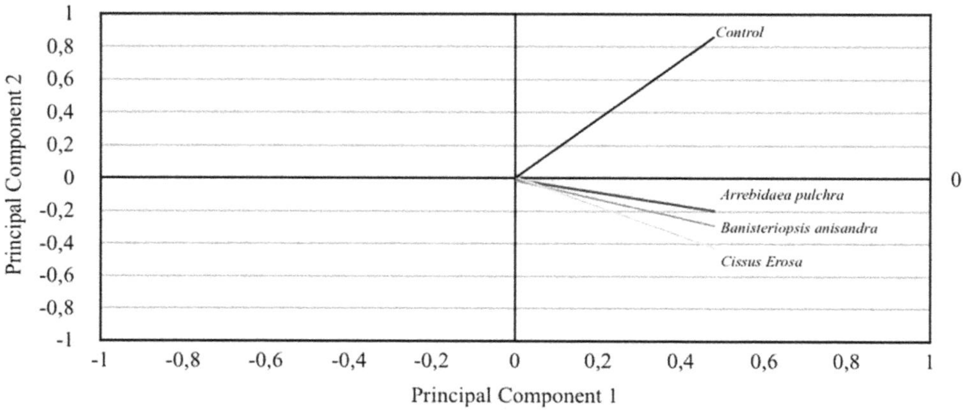

Arrabidaea pulchra (AP) obtained the highest air temperature decrease and the highest air humidity increments. *Banisteriopsis anisandra* (BA) also obtained significant results related to the air temperature, but less than *Arrabidaea pulchra* when compared to air humidity increments. *Cissus erosa* (CE) showed results that are close to the control scenario parameters. The analysis of Figures 13 to 22 and of the PCA allows to verify that *Arrabidaea pulchra* stands out for air humidity increments. All three have a similar statistical impact on air temperature. The LAD showed to be the most significant parameter to influence the climatic conditions on a control scenario.

Figure 17. Principal components analysis related to air temperature

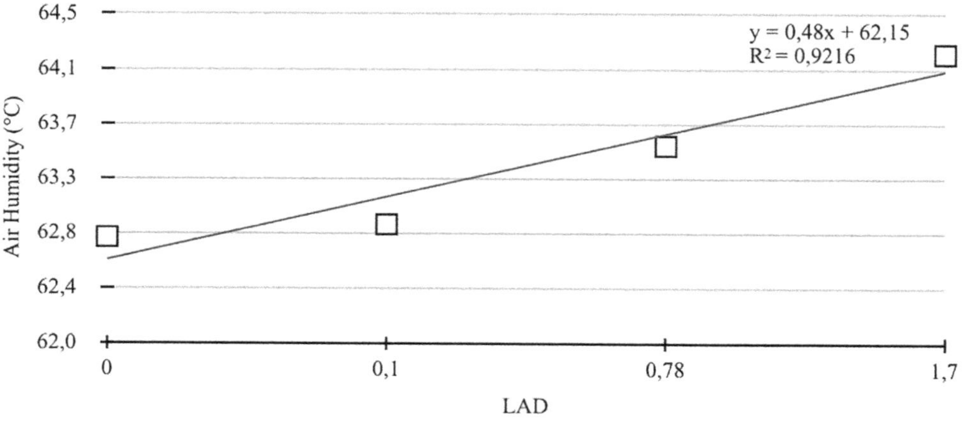

Figure 18. Principal components analysis related to air humidity

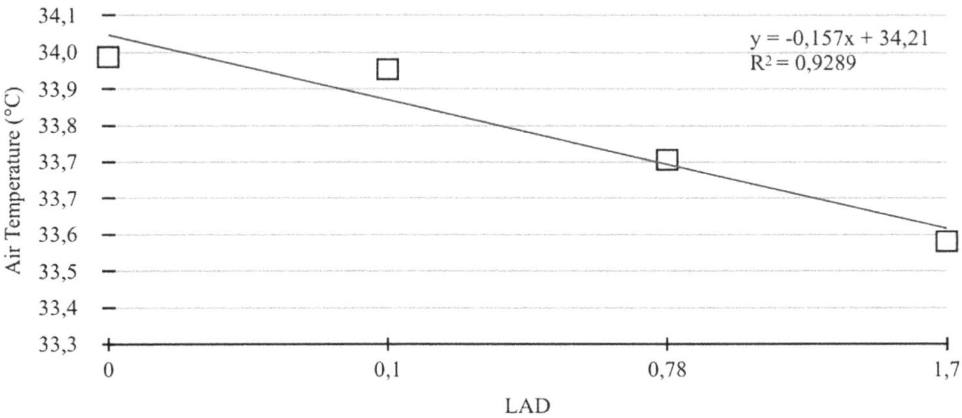

On the west and east facades, which receive more solar radiation, simulation showed that the use of green walls provided a more comfortable condition, in comparison with the control scenario. The advantage of the use of green wall is also perceptible on the south and north facades.

This thermal study presented data to point out that each species has a distinct impact on climatic conditions. Air temperature differences between the species were not higher than 2°C, while the difference of air humidity was not superior to 2 pp. However, it is important to consider that the wall dimension was 9m2 and that, if this area was larger, results could be more representative. The same can happen in a neighborhood, once the bigger is the number of green walls, more control it has on air temperature and humidity. Therefore, it is important to consider the results of this research, because they indicate that green walls have better climatic behavior than a control scenario, and, thus, indicate that there are differences between different species.

This research did not describe the growth rate of each species, while it would be very useful to determine the viability of their use in a green wall. Thus, a future research should analyze the viability, germination, and growth rates of the seeds.

CONCLUSION

The hygrothermal evaluation process showed that green walls reduce air temperature and increase air humidity near the building facades. This issue is evidenced by other studies. Wong et al (2010) investigated the surface temperature of eight different green wall systems and demonstrated that bare walls were 4,36 °C hotter. Taleghani et al (2014) also did a similar experiment and observed that bare walls were 4,7 °C hotter than green walls.

However, the behavior of vegetation on green walls changes according to the position of the sun. This study highlighted that each part of a wall receives a specific amount of solar radiation. The vegetation, as well, as a living being, responds physiologically to its environmental stimulus, such as radiation, wind, and humidity, thus resulting in different shapes of green.

The species *Arrabidaea pulchra*, therefore, has a higher performance than the others, in terms of blocking solar radiation. Although the species *Cissus erosa* presented timid results, it performed better than the control scenario. As a result, the *Cissus erosa* might also be indicated for green walls.

The statistical analysis indicated a strong correlation between the LAD in decreasing air temperature and increasing air humidity. This correlation evidences that the LAD is an important parameter for choosing species in green wall projects. Species with a higher density of leaves improve the evapotranspiration process and deliver a larger shading area.

The statistical analysis showed that morphological and physiological aspects have a positive correlation with improving thermal performance. Among the analyzed species, the *Arrabidaea pulchra* had the best results. In terms of methodological tools, the ENVI-met software accepts specific morphological and physiological characteristics of these plants.

The results from ENVI-met software went through a statistical analysis and confirmed the correlation between vegetation parameters and climate impact. However, the absence of real green walls for in-situ measurements does not make ENVI-met totally efficient for this kind of research. It would be necessary to use a real green wall to compare results from an in-situ scenario and a controlled one.

However, it was not possible to build a real model. Thus, the computational simulation process was an easy and affordable methodology, although using a computer was not always possible due to the lack of numeric data about the study subject. In this case, there were three species with no morphological and physiological data. As a result, the collection of these data was an important methodological step, in this research. Therefore, this chapter is a relevant scientific contribution in terms of implementing more vegetation data and creating computational data in the ENVI-met platform, specially the native ones.

The authors suggest future research to be focused on the cultivation and monitoring of the growth of species in controlled situations. This kind of test could improve relevant data, and confirm and validate the computational simulations. In addition, the authors recommend the study of more species of the Cerrado biome. These studies could discuss the viability analysis, species germination, and different ways of green applications in buildings and architectural projects. Besides, this study would contribute to diffuse the seed of these species.

Reducing excessive air temperature was obtained with the use of green walls. Further, the presence of native vegetation in an urban area can prove advantages in terms of fauna and flora conservation. Additionally, the authors argue that mitigating solutions or strategies for energy efficiency through the use of vegetation should seek the use of native species and incorporate them into the design process.

Ultimately, the authors conclude that green infrastructure promotes thermal comfort and increases the environmental quality for the human being.

REFERENCES

Ahmad, H., Babak, A., Davoud, Z., & Suhardi, B. M. (2019). Assessing the thermal comfort effects of green spaces: A systematic review of methods, parameters, and plants' attributes. *Sustainable Cities and Society*, 49.

Alberti, M. (2008). *Advances in urban ecology: Integrating humans and ecological processes in urban ecosystems*. Springer. doi:10.1007/978-0-387-75510-6

Aleixo, A. (2001). Conservação da avifauna da Floresta Atântica: Efeitos da fragmentação e a importância de florestas secundárias. In J. L. B. Albuquerque J.F. Cândido Junior, F.C. Straube & A.L. Roos (Eds.), Ornitologia e conservação: Da ciência às estratégias (pp. 199-206). Tubarão, SC: Unisul.

Alexandri, E., & Jones, P. (2008). Temperature decreases in an urban canyon due to green walls and green roofs in diverse climates. *Building and Environment*, *43*(4), 4280–4493. doi:10.1016/j.buildenv.2006.10.055

American Society of Heating, Refrigerating, and Air-Conditioning Engineers. (2010). *Standard no 55: Thermal environmental conditions for human occupancy*. Retrieved from http://arco-hvac.ir/wp-content/uploads/2015/11/ASHRAE-55-2010.pdf

Battisti, C. (2004). *Frammentazione ambientale, connettività, reti ecologiche: Un contributo teorico e metodologico con particolare riferimento alla fauna selvatica*. Roma: Provincia di Roma, Assessorato alle politiche ambientali, Agricoltura, e Protezione civile.

Bennett, A. F., Hinsley, S. A., Bellamy, P. E., Swetnam, R. D., & Mac Nally, R. (2004). Do regional gradients in land-use influence richness, composition and turnover of bird assemblages in small woods? *Biological Conservation*, *119*(2), 191–206. doi:10.1016/j.biocon.2003.11.003

Bocquier, P. (2014). World urbanization prospects. The 2014 review. New York: United Nations.

Bruse, M. (2004, March). *ENVI-met 3.0: Updated model overview*. Retrieved from http://www.envi-met.net/documents/papers/overview30.pdf

Castelo Branco, L. M. B. (2009). *Análise da arborização em Brasília. O caso da superquadra 308 sul* (Unpublished master's thesis). Universidade de Brasília, Brazil.

Costa, E. C. (1974). Física aplicada à construção: Conforto térmico (2nd ed.). São Paulo: Blücher.

Department of Human Services. (2009). *January 2009 heatwave in Vicoria: an assessment of health imapcts*. Victorian Government Department of Human Services, Melbourne, Victoria. Retrieved from: https://www2.health.vic.gov.au/about/publications/researchandreports/January-2009-Heatwave-in-Victoria-an-Assessment-of-Health-Impacts

Dhainaut, J., Claessens, Y., Gisnburg, C., & Riou, B. (2004). Unprecedent heat-related death during the 2003 heat wave in Paris: Consequences on emergency departments. *Critical Care (London, England)*, *8*(1), 1–2. doi:10.1186/cc2404 PMID:14975035

Dunnet, N., & Kingsbury, N. (2008). *Planting green roofs and living walls*. Timber Press.

Engel, V. L., Fonseca, R. C. B., & Oliveira, R. E. (1998). Ecologia de lianas e o manejo de fragmentos florestais. *Série Técnica IPEF*, *12*(32), 43–64.

Eumorfopoulou, E. A., & Kontoleon, K. J. (2009). Experimental approach to the contribution of plant-covered walls to the thermal behaviour of building envelopes. *Building and Environment*, *44*(5), 1024–1038. doi:10.1016/j.buildenv.2008.07.004

Forman, R. T. T., & Godron, M. (1986). *Landscape ecology*. John Wiley.

Gabriel, K. M. A., & Endlicher, W. R. (2011). Urban and rural mortality rates during heat waves in Berlin and Brandenburg, Germany. *Environment and Pollution*, *159*(8-9), 2044–2050. doi:10.1016/j.envpol.2011.01.016 PMID:21295389

Herath, H. M. P. I. K., Halwatura, R. U., & Jayasinghe, G. Y. (2018). Modeling a tropical urban context with green walls and green roofs as an urban heat island adaptation strategy. *Procedia Engineering*, *212*, 691–698. doi:10.1016/j.proeng.2018.01.089

Howard, L. (1833). *The climate of london. deduced from meteorological observations made in the metropolis and various places around it*. Harvey and Darton.

Kinzig, A. P., & Grove, J. M. (2001). Urban-suburban ecology. In S. A. Levin (Ed.), Encyclopedia of biodiversity (pp. 733-745.). Amsterdam: Elsevier Science & Technology. doi:10.1016/B0-12-226865-2/00278-9

Laureti, F., & Battisti, A. (2017). *A study of vegetation and cool materials effects on pedestrian comfort in Rome*. Paper presented at International Conference on Urban Confort and Environmental Quality: URBAN – CEQ, University of Genova, Italy.

Lin, P. (2009). Thermal perception, adaptation, and attendance in a public square in hot and humid regions. *Building and Environment*, *44*(10), 2017–2026. doi:10.1016/j.buildenv.2009.02.004

Lord, J. M., & Norton, D. A. (1990). Scale and the spatial concept of fragmentation. *Conservation Biology*, *4*(2), 197–202. doi:10.1111/j.1523-1739.1990.tb00109.x

Lovell, S. T., & Taylor, J. R. (2013). Supplying urban ecosystem services through multifunctional green infrastructure in the United States. *Landscape Ecology*, *28*(8), 1447–1463. doi:10.100710980-013-9912-y

Maleki, A., & Mahdavi, A. (2016). Evaluation of urban heat islands mitigation strategies using 3dimentional urban micro-climate model envi-met. *Asian Journal of Civil Engineering*, *17*(3), 357–371.

McIntyre, N. E. (1995). Effects of forest patch size on avian diversity. *Landscape Ecology*, *10*(2), 85–99. doi:10.1007/BF00153826

Nascimento, M. S. B., & Oliveira, M. E. (2005). *Diversidade e uso de plantas nativas*. Retrieved from https://www.embrapa.br/imprensa/artigos/2005/artigo.2005.05.8820647706/

Olivieri, F., Olivieri, L., & Neila, J. (2014). Experimental study of the thermal-energy performance of an insulated vegetal facade under summer conditions. *Building and Environment*, *77*, 61–76. doi:10.1016/j.buildenv.2014.03.019

Pan, L., & Chu, L. M. (2016). Energy saving potential and life cycle environmental impacts of a vertical greenery system in Hong Kong: A case study. *Building and Environment*, *96*, 293–300. doi:10.1016/j.buildenv.2015.06.033

Panchen, Z. A., Primack, R. B., Nordt, B., Ellwood, E., Stevens, A., Renner, S. S., Willis, C. G., Fahey, R., Whittemore, A., Du, Y., & Davis, C. C. (2014). Leaf out times of temperate woody plants are related to phylogeny, deciduousness, growth habit, and woody anatomy. *The New Phytologist*, *203*(4), 1208–1219. doi:10.1111/nph.12892 PMID:24942252

Pastore, L., Corrao, R., & Kvols Heiselberg, P. (2017). The effects of vegetation on indoor thermal comfort: The application of a multi-scale simulation methodology on a residential neighborhood renovation case study. *Energy and Building*, *146*, 1–11. doi:10.1016/j.enbuild.2017.04.022

Paul, M. J., & Meyer, J. L. (2001). Streams in the urban landscape. *Annual Review of Ecology, Evolution, and Systematics*, *32*(1), 333–365. doi:10.1146/annurev.ecolsys.32.081501.114040

Peñalosa, J. (1985). Dinâmica de crescimento de Lianas. In A. Gomezpompa & R. S. Del Amo (Eds.), Investigaciones sobre la regeneración de selvas altas en Veracruz, México (pp. 147-169). México: Alhambra Mexicana.

Primack, R. B., & Rodrigues, E. (2001). *Biologia da conservação* (8th ed.). Planta Editora.

Pujals, A. (2011). *Ecologia de lianas e cipós em fragmentos florestais: Dualidade e propostas de manejo* (Unpublished master thesis). Universidade de Maringá.

Ramalho, C. L., & Proença, C. E. B. (2004). Trepadeiras ornamentais do cerrado. Brasília: Universidade de Brasília e Embrapa Cerrados.

Razzaghmanesh, M., Beecham, S., & Salemi, T. (2016). The role of green roofs in mitigating urban heat island effects in the metropolitan area of Adelaide, South Australia. *Urban Forestry & Urban Greening*, *15*, 89–102. doi:10.1016/j.ufug.2015.11.013

Reinhart, C. F., & Davila, C. C. (2016). Urban building energy modeling – A review of a nascent field. *Building and Environment*, *97*, 196–202. doi:10.1016/j.buildenv.2015.12.001

Revich, B. (2011). Heat-wave, air quality and mortality in European Russia in summer 2010: Preliminary assessment. *Yekologiya Cheloveka. Human Ecology*, *7*, 3–9.

Romero, A. B. (2013). Princípios bioclimáticos para o desenho urbano (2nd ed.). Brasília: Editora UnB. doi:10.7476/9788523013486

Salata, F., Golasi, I., Vollaro, A. D., & Vollaro, R. D. (2015). How high albedo and traditional buildings' materials and vegetation affect the quality of urban microclimate. A case study. *Energy and Building*, *99*, 32–49. doi:10.1016/j.enbuild.2015.04.010

Savard, J. P. L., Clergeau, P., & Mennechez, G. (2000). Biodiversity concepts and urban ecosystems. *Landscape and Urban Planning*, *48*(3-4), 131–142. doi:10.1016/S0169-2046(00)00037-2

Seto, K. C., Güneralp, B., & Hutyra, L. R. (2012). Global forecasts of urban expansion to 2030 and direct impacts on biodiversity and carbon pools. *Proceedings of the National Academy of Sciences of the United States of America*, *109*(40), 16083–16088. doi:10.1073/pnas.1211658109 PMID:22988086

Silva, F. D. S., & Blamires, D. (2007). Avifauna urbana no Lago Pôr do Sol, Iporá, Goiás, Brasil. *Lundiana: International Journal of Biodiversity*, *8*, 17–26.

Simon, H., Lindén, J., Hoffmann, D., Braun, P., Bruse, M., & Esper, J. (2018). Modeling transpiration and leaf temperature of urban trees - A case study evaluating the microclimate model ENVI-met against measurement data. *Landscape and Urban Planning*, *174*, 33–44. doi:10.1016/j.landurbplan.2018.03.003

Taleghani, M. (2018). Outdoor thermal comfort by different heat mitigantion strategies – A review. *Renewable & Sustainable Energy Reviews*, *81*, 2011–2018. doi:10.1016/j.rser.2017.06.010

Taleghani, M., Sailor, D. J., Tenpierik, M., & van den Dobblesteen, A. (2014). Thermal assessment of heat mitigation strategies: The case of Portland state university, Oregon, USA. *Building and Environment*, *73*, 138–150. doi:10.1016/j.buildenv.2013.12.006

Tzoulas, K., Korpela, K., Venn, S., Yli-Pelkonen, V., Kazmierczak, A., Niemela, J., & James, P. (2007). Promoting ecosystem and human health in urban areas using green infrastructure: A literature review. *Landscape and Urban Planning*, *81*(3), 167–178. doi:10.1016/j.landurbplan.2007.02.001

Walter, H. (1971). *Ecology of tropical and subtropical vegetation*. Van Nostrand Reinhold.

Willis, C. G., Ruhfel, B. R., Primack, R. B., Miller-Rushing, A. J., Losos, J. B., & Davis, C. C. (2010). Favorable climate change response explains non-native success in Thoreau's woods. *PLoS One*, *5*(1), e8878. doi:10.1371/journal.pone.0008878 PMID:20126652

Wong, I., & Baldwin, A. N. (2016). Investigating the potential of applying vertical green walls to high-rise residential buildings for energy-saving in sub-tropical region. *Building and Environment*, *97*, 34–39. doi:10.1016/j.buildenv.2015.11.028

Wong, N., Kwang, T., Chen, Y., Sekar, K., Tan, P., Chan, D., & Chiang, K. (2010). Thermal evaluation of vertical greenery systemns for building walls. *Building and Environment*, *45*(3), 663–672. doi:10.1016/j.buildenv.2009.08.005

Wong, N., Tay, S., Wong, R., Ong, C., & Sia, A. (2014). Life cycle cost analysis of rooftop gardens in Singapore. *Building and Environment*, *38*(3), 499–509. doi:10.1016/S0360-1323(02)00131-2

Zölch, T., Maderspacher, J., Wamsler, C., & Pauleit, S. (2016). Using green infrastructure for urban climate-proofing: An evaluation of heat mitigation measures at the micro-scale. *Urban Forestry & Urban Greening*, *20*, 305–316. doi:10.1016/j.ufug.2016.09.011

Zonneveld, I. S. (1995). *Land ecology: An introduction to landscape ecology as a base for land evaluation. Land management and conservation*. SPB, Academic Publishing.

KEY TERMS AND DEFINITIONS

Biophilic Design: Biophilic design is a natural principle to design with nature. This sort of design incorporates natural elements like light, vegetation, water, and other natural aspects.

Double Skin Facade: Exterior wall composed of two constructive layers.

Environmental Strategy: Urban or architectural strategy that has the main goal to make a human intervention more friendly to the natural environment.

Green Walls: Exterior or interior wall that has vegetation adhered to its surface.

Nature Conservation: Science or principle that has the main goal to protect nature.

Thermal Comfort: Thermal Comfort is a personal stage of mind that expresses satisfaction with the thermal environment and is assessed by subjective evaluation (ASHRAE Standart 55).

Urban Resilience: Urban resilience means the capacity of individuals, communities, and cities of support, grow, and adaptation in a sustainable development way.

Chapter 11
An Overview of the Potential of UAV Applications to the Built Environment:
A Role in Sustainable Urbanisation

David R. Green
UCEMM, University of Aberdeen, UK

Billy J. Gregory
DroneLite, UK & UCEMM, University of Aberdeen, UK

Jason J. Hagon
GeoDrone Survey Ltd., UK

Alex R. Karachok
UCEMM, University of Aberdeen, UK

Jakob Larsen
University of Aalborg, Denmark

Alastair Skitmore
UCEMM, University of Aberdeen, UK

ABSTRACT

This chapter presents an overview of the potential role of UAVs for monitoring, mapping, surveying, modelling, and visualising the 'built environment' and their role in sustainable urbanisation. The 'built environment' includes rural, urban, and underwater environments. Together with low-cost image processing and softcopy photogrammetry, fixed-wing and multi-rotor UAVs can collect a wide range of imagery for generating 3D models of individual buildings, and record and analyse architecture and infrastructure and terrain models. Consideration is given to non-imaging sensors carried on UAVs. Input to GIS provides a basis to create visually realistic models of the 'built-environment' for urban and rural planning and decision-making for sustainable urbanisation. 3D visualisation software, virtual, and augmented reality will allow public engagement with the spatial planning process. Safety and operational considerations are needed for UAV flights. The chapter concludes by examining how this technology will develop in the future to play a role in sustainable urbanisation.

DOI: 10.4018/978-1-7998-2249-3.ch011

INTRODUCTION

Currently there is a wide range of commercial small-scale airborne platforms available for acquisition of low-cost, high-resolution aerial remote sensing data. These include kites, model aircraft, balloons or blimps, helicopters, Unmanned Aerial Vehicles (UAVs), Unmanned Aerial Systems (UASs), and Drones (Green et al., 2020a). They have recently been complemented by waterborne drones operating both on the water surface and underwater carrying hydrographic sensors to monitor underwater structures and the impact of floating structures on the environment (De Lima et al., 2015).

In the past ten years, drones, UAVs and UASs have evolved from, relatively simple 'toys' with the capability to carry miniaturised RGB cameras for aerial photography and video footage, into mature aerial photographic systems for survey. The drone industry has quickly become professional and highly respected with major developments in aerial platforms and sensors including autonomous flight software, advanced flight controllers and First Person View (FPV) systems, accompanied by flight regulations, requirements for pilot training, and competency tests. The total market value of battery powered UAVs is expected to reach over 1000 million dollars by 2023 (Harrop and Harrop, 2014).

The rapid development of microelectronics and microprocessors, GPS and navigation systems, together with reduced costs over the last five years, have all helped to trigger an unprecedented demand for, and growth in, the use of UAV platforms and sensors for civilian applications (Watts et al., 2012; Colomina and Molina, 2014). This has been aided by advances in battery technology to extend duration of flight, the range of cameras and sensors available, and both softcopy photogrammetry and digital image processing software to generate mosaics, orthophotos, and DSMs and DTMs. In a very short time, drones have been applied to many new areas of work, including applications using vertical and oblique photography, precision agriculture and viticulture, river monitoring, vegetation mapping, forest and pasture surveys, coastal surveys, archaeological sites, and other examples including the 'built-environment'. Besides traditional aerial photography, videography, and other imagery, the acquisition of stereo-photography to build surface models with the aid of softcopy photogrammetry has led to many examples where houses and other buildings have been accurately surveyed and modelled and the results viewed in an interactive environment for visual communication via the Internet.

This contribution seeks to present a contextual overview of the potential role of UAVs and the related technologies for monitoring, mapping. surveying, modelling, and visualisation of the 'built-environment' as the basis to explore their role in sustainable urbanisation.

Definition of the 'built-environment' includes applications of UAVs to both rural, city and underwater environments with emphasis placed on the collection of geospatial data and imagery at the building and community level. Both fixed-wing and multi-rotor UAVs have the capability to collect a wide-range of RGB, multi-spectral, hyperspectral, thermal and Lidar imagery. Together with low-cost digital image processing and softcopy photogrammetric software these provide the basis for generating 3D terrain and surface models of individual buildings, architecture, infrastructure, and communities, as well as asset monitoring, maintenance, and management. In the underwater environment, sonar, bathymetric Lidar, video, and multibeam sensors, add another dimension to the potential of these robotic platforms for data and image acquisition.

Consideration will be given to the use of other types of miniaturised non-imaging sensors (e.g. air quality) that can be carried on UAVs to capture environmental data. Input to Geographical Information Systems (GIS) and modelling/simulation software provides a basis for more visually realistic models of the 'built-environment' for utilisation in planning and decision-making. 3D visualisation software, Virtual Reality (VR), and Augmented Reality (AR) also offer new opportunities to engage the public in the planning process through participatory GIS and landscape fly-throughs with the potential to contribute to sustainable urbanisation.

Successful flight operation in built-up areas requires discussion about regulations and safety requirements. Illustrated with a few examples, the chapter concludes by examining ways that this technology will develop and evolve in the future.

UAVS AND SENSORS

UAVs

Today there are a wide range of drone platforms that include multi-rotors (quadcopter (4), hexacopter (6), and octacopter (8)), fixed-wing, and VTOLs (Vertical Takeoff and Landing). In addition, some small-scale helicopters have been used for remote sensing.

When drones first became available, few included cameras. Many had very low-quality miniature cameras only suitable for capturing low-resolution images and video. Within a short period of time, however, drones e.g. DJI's Phantom 1 carried small recreational cameras such as the GoPro (https://gopro.com/) popular for activity-based photography (Figure 1). Although not specifically designed for aerial photography their small physical size and relatively light weight made them ideal for mounting on small 'hobby drones', initially on a fixed bracket, and later DJI's Zenmuse series of 2D and 3D gimbals. High-quality gimbals now support a wide range of DSLR cameras of varying sizes, and additional camera-based sensors e.g. NIR and NDVI cameras. Larger cameras and sensors can be carried by drones with suitable payload capacity and lift (Figure 2).

Figure 1. GoPro camera on a DJI Phantom 1 Quadcopter (Source: David R Green – UCEMM – University of Aberdeen)

Figure 2. Panasonic GH4 DSLR camera carried on an Align 800 Helicopter (Source: David R Green – UCEMM – University of Aberdeen)

Drones with RGB cameras soon began to demonstrate the potential for aerial work, typically oblique and vertical photographs, and video. Although early cameras were not ideal for aerial work the development of low-cost softcopy photogrammetric software (e.g. AgiSoft (https://www.agisoft.com/)) soon facilitated correction of the imagery, generation of orthophotos and photomosaics, DTMs, and DSMs. Other cameras extended the visual spectral range to include multispectral wavelengths e.g. the Near-Infrared (NIR), and hyperspectral sensors with more spectral bandwidths to gather additional environmental information.

Figure 3. MapIR camera (Source: David R Green – UCEMM – University of Aberdeen)

Sensors

Although RGB cameras are still in widespread use, there is a demand for other sensors, both non-imaging and imaging, and low-cost, higher-resolution systems. These range from modified GoPro cameras to similar types of miniaturised cameras with filters to detect the Near Infrared (NIR) and other parts of the Electromagnetic Spectrum (EMS). Examples include the MapIR range (https://www.mapir.camera/) (Figure 3) and IR-Pro (now discontinued) from the USA. There have been developments in small multispectral sensors aimed at agricultural, horticultural crop, forest canopy, and vegetation monitoring. The Parrot Sequoia 5-band multispectral camera is one example (https://www.parrot.com/business-solutions-us/parrot-professional/parrot-sequoia). A larger version of this sensor is the MicaSense RedEdge (https://micasense.com/rededge-mx/). More costly than the MapIR sensors these provide five (5) bands of imagery to generate NDVI images. Other companies such as Sentera (www.sentera.com) provide dedicated NDVI cameras.

More advanced sensors are also available including thermal, hyperspectral, and Lidar. Whilst these have huge potential, their cost and need for specialised aerial platforms also means that they have not yet found quite such widespread use. Although low-cost thermal imagers only show relative surface temperatures of surfaces (cold to hot) providing an image suitable for manual on-screen interpretation, more expensive options yield absolute temperature data. Several manufacturers offer cameras where the temperature at a location is an average of the pixels surrounding the one of interest, whilst others give the temperature value the of the pixel (Flir (https://www.flir.co.uk/suas/) and Optris (https://www.optris.co.uk/infrared-cameras)). Many drone manufacturers sell thermal camera-ready platforms either with their own thermal cameras or one from an established thermal camera manufacturer e.g. Yuneec (https://www.yuneec.com/) and DJI (https://www.dji.com/). The popularity of thermal cameras lies with their potential for monitoring heat loss from buildings (Figure 4), and locating people and animals in environmental, police, fire and search and rescue operations (https://www.citylab.com/life/2016/12/the-urban-drone-invasion-is-nigh/511496/).

Figure 4. Heat loss in buildings (FLIR thermal camera) (Source: David R Green – UCEMM – University of Aberdeen)

Hyperspectral cameras for drones are also available from different manufacturers e.g. Headwall (https://www.headwallphotonics.com/uav-integration). These have use in many applications because of the number of different wavelengths that can be sensed (up to 255 channels) increasing the dimensionality of the spectral dataset and the potential to acquire more information. However, added costs, need for specialist aerial platforms, and requirement for training to extract the information makes them less popular.

Another sensor, Lidar (both terrestrial and bathymetric) can be mounted on several specialist UAV platforms e.g. Riegl (http://www.riegl.com/). One benefit of Lidar is the generation of DTMs and DSMs, and the potential to see through vegetation. However, a cheaper DTM source uses softcopy photogrammetry software (e.g. AgiSoft (https://www.agisoft.com/), Pix4D (https://www.pix4d.com/), and Blue Marble's Global Mapper (https://www.bluemarblegeo.com/products/global-mapper.php)) and stereo-photographs. This lower-cost alternative is easier to utilise and has been widely demonstrated for numerous model generation applications.

Bathymetric Lidar sensors can be mounted on small aerial platforms with the capability to generate depth information in water up to 20-40m deep, providing the water is clear and not too turbulent e.g. Riegl's Bathycopter (http://www.riegl.com/products/unmanned-scanning/bathycopter/). Of less obvious use than terrestrial Lidar in the context of the 'built-environment', they do have the capability to generate 3D information of surfaces in shallow water of interest to studying the 'built-environment' of archaeology (Benjamin et al., 2019).

Radar systems are gradually being integrated into large UAV systems (e.g. military) for navigation and image acquisition. Recently, studies have revealed success with small, low-cost, high-resolution radar systems specifically designed for operation on small unmanned aerial vehicles (Li and Ling, 2015; Scannapieco et al., 2017; Ball, 2019).

TECHNOLOGY

In a very short time, drone technology has moved away from being basic to very advanced and increasingly customised for specific applications. Several key developments in the technology have facilitated this transformation and their potential for a wider range of applications.

Batteries and Flight Duration

One of the major constraints of early platforms was battery technology. Batteries were a limited power source often giving just 6-8 minutes of flight time. This was not a problem for small-area coverage but accounting for cold air, wind strength, and take-off, return to home, and landing functions, it reduced the actual flight time considerably. Within a few years, batteries evolved into smart batteries with longer flight times e.g. 20-25 minutes and supplying operational information to the operator (battery status, remaining flight time, and flight characteristics). These batteries have become more compact and average flight time has increased to >30 mins. Fixed-wing aircraft, by contrast, have longer flight times, ranging from 45-55 minutes, as they are lighter and offer less resistance in flight, making them more suitable for large area coverage.

Controllers

Initially handset controllers used ones from Radio-Controlled Model Aircraft (RCMA). More expensive drones had specialised controllers to aid in flying the aircraft, camera, and gimbal operation. Standardisation of drone controllers has been slow. Some manufacturers have customised the array and positioning of the controls to facilitate ease of operation, and most now have a smaller footprint, are simpler to use, and come with significantly improved visual and audible information to aid the pilot/operator before, during, and after flight. Some drones rely on use of a mobile phone/tablet for the FPV (First Person View) screen, others providing a built-in view screen. With growing concerns about safe operation of UAVs, companies have begun to build-in onscreen tutorials, including reference information, into the controller for educational purposes. The outcome is RTF (Read-To-Fly) platforms reducing the learning curve for the operator.

Autonomous Flight

Early UAVs were usually flown manually by the pilot, often without an FPV (First Person View), and stereo-imagery was acquired through skilled flying by the pilot where overlapping flight-lines were visually assessed to provide the required stereo-imagery for photo-mosaics and 3D models. Sophisticated phone and tablet apps for Android and iPhone - provided by drone companies (DJI Go4) or through third parties - allow customisation of the overflight pattern e.g. straight line or grid, the camera tilt, speed, and the flying height of the aircraft. The data/image capture task has now become a one button push for take-off and one for landing. The technology has made it much easier to conduct repeat flights of the same area to gather multi-temporal imagery of a site. Most drones now have autonomous flight capability allowing them to be used for more advanced aerial survey tasks.

UAV Swarms

The development of smaller drone platforms, miniaturisation of sensor technology, and autonomous flight have led to simultaneous flying of multiple drone platforms, known as swarms. Swarm technology offers possibilities for aerial monitoring including additional flight duration and aerial coverage (Hildmann et al., 2019; Wu et al., 2017). Greater aerial coverage and Beyond the Line of Visual Sight (BLOVS) can now be achieved using mobile telecommunications networks.

SAFE OPERATION

The practicality of overflights in the 'built-environment' unfortunately presents considerable risk in relation to obstacles encountered when flying; in order to overfly the tallest buildings means that the altitude of the platform may be above the legal altitude of 400 feet. Urban areas are often close to airfields and airports, and other buildings including industrial sites with factories, towers, and chimneys are part of the infrastructure. Legislation in many parts of the world currently restricts the flying of urban areas to only those pilots who have special permissions to fly because of the high risk involved.

With the increasing frequency of media coverage about drones, many documenting near misses between UAVs and civilian aircraft, as well as daily reports of drone incidents emerging around the globe, and the rapidly growing use of drones ranging from aerial imagery acquisition to the proposed delivery of parcels, there is a growing call worldwide for tightening-up of the current rules and regulations concerning who, what, where, and when UAV platforms can actually be flown.

In many countries (e.g. the UK, USA, Australia, and Canada), there are moves to control and regulate UAV flights, to require education and training for operators and pilots (including the provision of best-practice guidelines), and a requirement for operator insurance. Many countries are moving to deal with the rapid proliferation of a wide range of UAV technology to cover everyone from the hobbyist to the commercial service provider. More effort is being focused on ensuring safe operation of these small to large aerial platforms and their payloads, and the potential risks associated with the likely increase in air traffic that will result if drone-based data and image collection continues to grow (https://www.questuav.com/media/case-study/safety-standards-in-urban-environment/). Other steps are also being taken to minimise operational risk associated with urban flights. For example, several drone capture methods have been developed to physically capture drones in a net or by 'drone snatching' using another drone when necessary (Skitmore, 2018; 2020).

Whilst not all reports of drone incidents have been substantiated, growing awareness of drones has raised public concern, whether it be related to their use, safe operation, the invasion of privacy, or operations where risk to other activities is perceived to be high. The CAA (Civil Aviation Authority - https://www.caa.co.uk/home/) in the UK (and in other countries (Stocker et al., 2017)) has begun to recommend best-practice guidelines, and actioned the need to register drones on a national database following purchase. The CAA has provided strict regulations on both leisure and commercial operators, preventing platforms from being operated unless the operator has received appropriate training, is registered, and flying outside of a drone free zone, or in the case of commercial operators has special permissions to operate a drone in a built up area. In the UK, drones may be operated in urban areas either recreationally in wide-open spaces such as parks or by commercial operators with special permissions and exemptions from the Civil Aviation Authority (CAA). In the UK, drone use is governed by the Air Navigation Order (ANO) 2016 (https://www.dpreview.com/articles/5101939047/how-to-safely-operate-a-drone-in-urban-areas). This has recently been updated given the unusual situation surrounding the outbreak of COVID-19 in the UK with the publication of CAP1915 UAS guidance for COVID-19 BVLOS Operations to help UAS operators *'to apply for UAS BVLOS authorisations aimed at supporting the COVID-19 response more effectively and efficiently'* (https://www.caa.co.uk/Consumers/Unmanned-aircraft/Our-role/Updates-about-drones/). A similar approach is being taken by the FAA (Federal Aviation Authority) in the USA and in many other countries.

More advanced flight apps are now also beginning to become available allowing operators to: refresh their flying knowledge and skills prior to any flight; be aware of who else is flying around them; be visible to others e.g. civilian and military aircraft; and be aware of the designation of no-fly zones for drones. The DJI Mavic Air 2 now comes equipped with new AirSense technology utilising ADS-B aviation technology for receiving signals from nearby airplanes and helicopters and displaying their locations on the on-screen map of the DJI Fly App (Abbott, 2020; https://www.techradar.com/uk/reviews/dji-mavic-air-2-review).

It is likely that such regulations will be tightened in the future to ensure some of the incidents recorded to date are not repeated, particularly with an increase in lower-level airborne traffic. An ongoing evaluation of technologies to detect and track small unmanned aircraft systems (sUAS) in dense urban environments for military, and potentially homeland security, applications is also making progress as the program enters its next phase (). In the USA, as of 2019, an air traffic control system for drones is already well-advanced.

THE POTENTIAL OF DRONES IN THE BUILT ENVIRONMENT

As indicated earlier in this chapter, coverage of the 'built-environment' will be broad ranging and will include both the terrestrial and underwater environment, rural and urban, buildings and infrastructure, and current and historical aspects.

Potential Role

The potential role for drones in urban environments is increasingly a topic of interest to both commercial businesses and the public and goes far beyond just data acquisition to include the generation of 3D models and geo-visualisation of the urban environment (see: https://www.smartcitiesworld.net/news/news/what-is-a-drones-role-in-a-city-3159).

The future impact of drones in all aspects of the 'built-environment' is widely viewed now as being very significant (https://www.alphr.com/life-culture/1005190/how-will-drones-reshape-our-cities). For example, already drones are being trialled for urban deliveries (https://newcities.org/blog-urban-deliveries-with-drones/) in many countries, including India, drone delivery systems to high rise and apartment buildings have already been trialled (newcities.org) (https://www.orfonline.org/expert-speak/drones-in-smart-cities-55717/). The recent COVID-19 pandemic has also rather unexpectedly triggered a growing interest in the current and future role of drones to assist with the remote delivery of packages, food and medicine within urban areas, as well as for their use in monitoring e.g. social distancing and disinfecting sites (Green et al., 2020b).

With progress in the development of remote sensing technology – including UAV sensors and platforms – together with advanced digital image processing and softcopy photogrammetry software – new opportunities have also emerged for the development of aerial monitoring, mapping and modelling, with the possibility for high frequency multi-temporal data collection that can provide a comparatively inexpensive and spatially precise means to capture data when compared to traditional sources of aerial photography and satellite imagery.

The high-resolution data and imagery acquired by drones is seen as complementary to other sources of remote sensing. Drones provide an important option with several advantages over the other more traditional aerial platforms such as light aircraft and helicopters, as well as satellite data and imagery. These advantages include improved mission safety, flight repeatability, the potential for reduction in operational costs, fewer weather-related flying limitations (e.g. a UAV can fly below the clouds), as well as higher spatial and temporal resolution imagery. Many drone platforms are now also waterproof and can fly in both damp and wet conditions.

The current drone and sensor technology available, much of it low-cost (between £500 and £1500), allows virtually anyone to fly a drone autonomously and to collect vertical and/or oblique aerial stereo-photography. Mobile phone apps permit the operator to specify the exact photography required for direct input to softcopy photogrammetric software to generate orthophotos, mosaics, and 3D terrain/surface/models. The ease with which the required data can now be acquired and processed means that it is relatively quick to generate a 3D model of a building or area and to view, interact and navigate with the aid of 3D viewers e.g. embedded on a web page (Figure 5) to provide public access. There are also possibilities to use Virtual Reality (VR) and Augmented Reality (AR).

Figure 5. 3D building in virtual online viewer (Source: Jason J. Hagon - GeoDrone Ltd.)

In the next section of this chapter, a brief look is taken at some of the different ways that UAVs/drones and the associated technologies are currently being used to monitor, map, and model the 'built-environment'. In the following examples, the scope of the work referred to will showcase some of the existing uses and future potential that can now be achieved using such data, imagery and information to record and document the 'built environment'. These examples will provide the basis to consider the role of drone technology in sustainable urbanisation.

UAVs In The 'Built-Environment'

As mentioned at the beginning of this chapter, rather than restricting the 'built-environment' to just 'urban', a much wider-range of examples will be used to illustrate the potential of UAVs that includes: buildings in both the urban and rural environment such as houses and farm buildings, historic buildings, villages and communities, commercial and industrial buildings, urban and rural infrastructure, as well as some coastal and marine examples where aerial and surface platforms and sensors are used to document underwater archaeology and industrial infrastructure e.g. harbour walls.

Many of the examples relate to monitoring, mapping, and modelling the outside fabric, architecture, and visual appearance of buildings and infrastructure as the basis to create photographic and sensor records.

Urban Areas and Infrastructure

Urban areas present a much more diverse built 'landscape' than a small rural community, with many more buildings that occupy larger footprints and more than one or two storeys.

Rakha and Gorodetsky (2020) mention the opportunity to inspect and analyse building stocks, where specialists with different backgrounds can document building performance, visualise heat transfer using infrared imaging, and create digital models using 3D photogrammetry. Other sensors provide the basis for measurements such as dimensions and volume using laser-scanning technology. Small drones – such as the DJI Mavic Mini – that can fly inside without GPS using instead small camera-based object/collision avoidance systems to permit safe flying – offer the possibility to acquire imagery inside a building or with modification (such as the addition of a protective cage) from within infrastructure e.g. pipelines and tanks. The potential role of drone-based imagery also extends to the rapid evaluation of the status of the architectural infrastructure following a serious seismic event such as an earthquake as part of an emergency response (Achille et al., 2015).

Applications with potential in urban areas include those that can help to analyse green spaces using NDVI images, heat and energy efficiency using thermal sensors and visible light outputs (http://www.geogeoglobal.com/case-studies/2015/towards-a-smart-campus-aimee-maps-university-of-glasgow). There are many good examples also of the commercial use of drones to monitor railway infrastructure, onshore, nearshore, and offshore wind turbines, and other towers e.g. radio/TV towers. 3D models of infrastructure, including renewable energy constructions such as wind turbines, provide a quick way to generate a detailed record of building structures and together with the DTMs (Digital Terrain Models) and DSMs (Digital Surface Models) can form the basis to complete 3D models of the structure and its maintenance (Figure 6). Monitoring and surveying of onshore, inshore and offshore oil and gas structures have been well-developed by a number of companies including, for example, CyberHawk (https://thecyberhawk.com/). The role of UAV imagery can also be extended to the creation of 3D models showing the impact of a planning development such as wind parks on land or in the coastal environment. Bridges have also successfully been monitored with drones (Reagan et al., 2017; Zink and Lovelace, 2015) and as the basis not only for examining the condition of the structure, but also to record the 3D structure using a 360-degree mosaic (Figure 7). Other features that can be photographed can include roads, walls, and fences where aerial imagery and mosaics can be used as the basis for constructing visual montages in planning applications (Figure 8). Drones with video cameras can also be used to monitor activities on construction sites when connected to desk computers in the office and are useful for safety management on construction sites. (Chen et al., 2016; Li et al., 2016) (Figure 9).

Figure 6. Wind turbine (DJI Phantom) (Source: Jason J. Hagon – GeoDrone Ltd.)

There are many possibilities for drones to be used in the architectural and urban fields (Labib, 2017), some being: 3D models of buildings; thermal images of buildings; 3D site scanning; urban air quality, acoustic and microclimate studies (e.g. temperature, humidity); environmental management (Gallacher, 2015); the movement of people in public spaces; island heat mapping; urban simulations; and urban design. Drones can be used to autonomously map building construction (https://www.fosterandpartners.com/news/archive/2017/10/foster-partners-and-betomorrow-exploring-use-of-drones-to-autonomously-map-buildingconstruction /?lipi=urn%3Ali%3Apage%3Ad_flagship3_feed%3BBkl7Km2tSvi1aVSusiF5SA %3D%3D; Bognot et al., 2018; Gheisari and Esmaeili, 2016.).

As noted by João Pernão at the University of Lisbon *'With drones, it is possible to view and understand places from above, go to hidden parts of the 'urban fabric' and to survey details in buildings that otherwise would be practically and cost-wise impossible as well as allowing visual interaction with built structures*

Figure 7. 360-degree mosaic with bridge (DJI Mavic 2 Pro) (Source: Billy J. Gregory – DroneLite)

(https://www.researchgate.net/post/What_are_the_applications_of_drone_technology_in_architecture_and_the_built_environment_field).

Figure 8. Visual Montage (DJI Mavic 2 Pro) (Source: Billy J. Gregory – DroneLite)

Figure 9. Construction site in Forres, Moray, Scotland, UK (DJI Mavic) (Source: Billy J. Gregory – DroneLite)

Drones are therefore an important and innovative resource to help understand and design better urban spaces and better architecture, and as discussed by Green (2017) drones have the potential to allow for the conscious re-moulding of the large-scale physical environmental imageability and to make completely new landscapes in a short time.

Rural

In much the same way as 'urban' environments, buildings and infrastructure of rural areas e.g. houses (both old and new), villages and farms (houses and farm buildings) can easily be imaged and monitored with using drones and sensors through the capture of aerial photographs and stereo-photographs as the basis for 3D models of the various different types of buildings, and the creation of ortho-mosaics (Figure 10 a, b and Figure 11 a,b,c,d). Aerial photographs can also be used to monitor the condition of building roofs needing repair (Figure 12). Indeed, there are now quite a number of software solutions using AI (Artificial Intelligence) and Deep Learning that can automatically process the drone imagery and present a detailed information resource for the client about the roof condition (for example: EZ3D (https://ez3d.io/) and IMGING (https://www.lovelandinnovations.com/drone-inspection/)).

Figure 10. (a,b): Farm building/model (DJI Phantom) (Source: Jason J. Hagon – GeoDrone Ltd.); (c) Farm buildings (Mavic 2 Pro) (Source: Billy J. Gregory – DroneLite); and (d) Farm buildings and infrastructure (DJI Mini) (Source: Andrew Smith – University of Aberdeen)

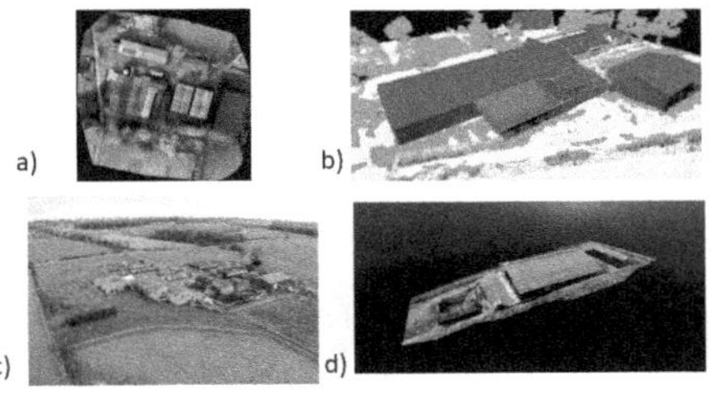

Commercial, Industrial and Harbours

Likewise, images of commercial and industrial buildings can also be captured using drones and visually documented, not only providing a record of the assets, but also the basis for a detailed estate management system, allowing for the exploration of both the external fabric and the inside of a building.

Figure 11. Different buildings and structures (DJI Spark) (Source: (a,b,c) Alex R. Karachok – UCEMM, University of Aberdeen; (d) (Yuneec Typhoon H) David R. Green – UCEMM – University of Aberdeen)

Figure 12. House roof (DJI Spark) (Source: David R. Green – UCEMM – University of Aberdeen)

Figure 13. Underwater cable (Source: Matthew J. Brannan – EIVA)

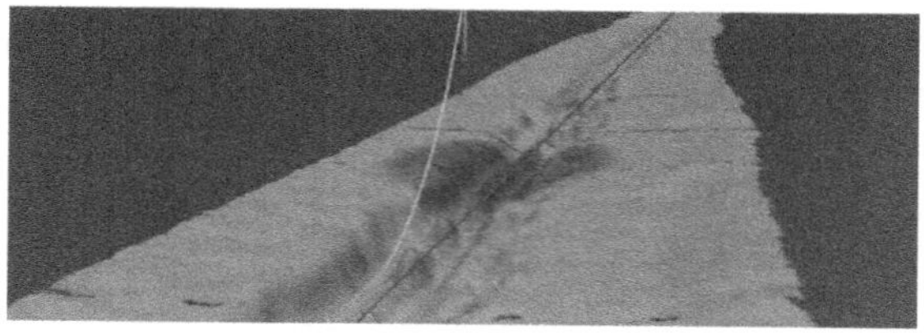

The capture of imagery of any marine structure e.g. pipeline (Gómez and Green, 2015) and harbour wall, can also be documented, and even edited to accommodate temporal changes to the building. A good example is that of an Australian harbour/ jetty that was flown at low-tide with stereo-imagery allowing the condition of the jetty to be examined and monitored over time with the aid of UAV images taken at low water. The high-resolution imagery provided a detailed record of the structure as an alternative to more costly sonar/multibeam imagery acquired with an ASV (Autonomous Surface Vehicle) (see e.g. https://youtu.be/Tpe0vHeLpB8). Other similar examples, captured by waterborne survey and processed in EIVA software (https://www.eiva.com/) provide an opportunity to visualise the underwater built-environment including pipelines, and oil and gas platforms (Figures 13, 14, 15) (courtesy of Matthew Brannan, EIVA).

Figure 14. Underwater pipeline (Source: Matthew J. Brannan – EIVA)

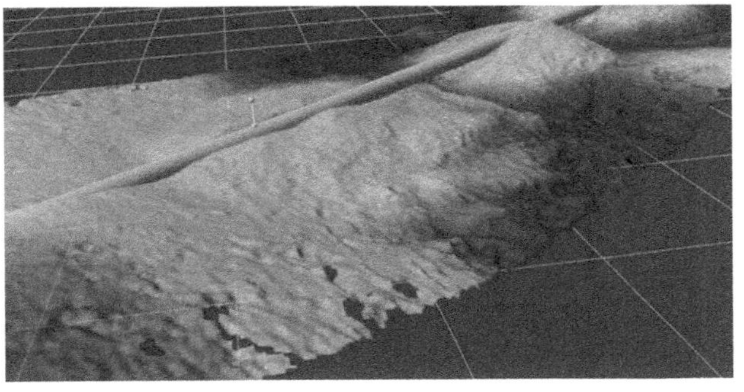

Figure 15. Oil and gas platform (Source: Billy J. Gregory – DroneLite)

Real Estate

Another role for drones that is essentially very similar to those applications described earlier, is that of taking aerial photographs of properties that are to be placed on the market by Estate Agents (Realtors) for clients (see e.g. http://www.geogeoglobal.

com/case-studies/2015/10/26/analysing-modelling-real-estate-in-high-res-3d). These often require just basic oblique and vertical aerial photographs of houses and gardens, but with software to generate 3D interactive models for electronic display, communication, and interaction, the visuals offered to prospective buyers are considerably more advanced than in the past (Figure 16).

Figure 16. Real estate (Source: David R. Green – UCEMM – University of Aberdeen)

Community Projects

The use of stereo-imagery to generate 3D models of buildings offers the possibility to create a semi-realistic model of a community environment for spatial planning. In the past, computer-based graphics and software products such as Google Earth (https://www.google.co.uk/intl/en_uk/earth/) and Google's Sketchup (https://www.sketchup.com/) have been used to create buildings and to simulate a range of rural, urban and other built environments, including for example windfarms. Photographs and graphic textures are often used to make the buildings look more visually realistic, and with the aid of Google Earth and other software tools semi-realistic built landscapes can be *'built'*, navigated, and flown through. More sophisticated and custom solutions are to be found in the form of CommunityViz (https://communityviz.city-explained.com/index.html) that has evolved into an add-on to the ESRI (Environmental Systems Research Institute) ArcGIS software suite (https://www.esri.com/en-us/arcgis/about-arcgis/overview).

Figure 17. Underwater archaeology: a 3D visualisation of the remains of part of the submerged town of Dunwich (Source: Professor David Sear Department of Geography and Environment, University of Southampton (https://www.southampton.ac.uk/news/2013/05/secret-streets-of-britains-atlantis.page))

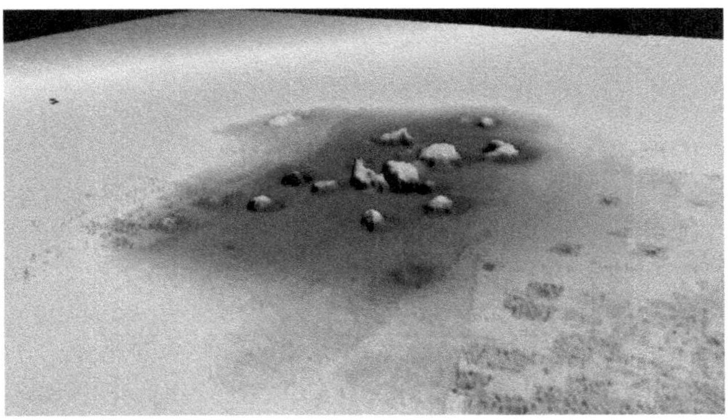

Today the use of softcopy photogrammetry software can be used to transform stereo-aerial photography into 3D models of buildings and landscapes which can then be integrated into fly-throughs. With the evolving role of Virtual Reality (VR) headsets and Augmented Reality (AR), it is now possible to create more realistic community models that can then be visually and interactively experienced by the individual. Combinations of drone imagery, image processing software, gaming software and VR (Virtual Reality) have already been used to develop and to construct immersive experiences to provide realistic 'cityscapes' with potential for urban planning (http://www.briteliteimmersive.com/blog/drone-imagery).

Historical and Archaeological

Extending the definition of 'built-environment' to include the historical 'built-environment' makes it possible to include the use of drones to map and model archaeological sites and, when integrated with other software, to generate reconstructions of historical sites and buildings (see for example: https://www.wessexarch.co.uk/archaeological-services/unmanned-aerial-vehicle-uav; http://

www.geogeoglobal.com/case-studies/dumbartonrock3d). Autonomous Underwater Vehicles (AUVs) can also make use of hydrographic sensors to capture underwater historical and archaeological structures (Mindell and Bingham, 2001; Roman and Mather, 2016; Allotta et al., 2018; Benjamin, 2019) (Figure 17).

Drones For Constructing a Building

A PBS (Public Broadcasting Service) TV programme – 'The Age of the Drone' – reported on the potential use of drone swarms to build a structure 'brick' by 'brick'. Although only a demonstration, it reveals the current payload capacity of some drones and the capability of autonomous flight to undertake small-scale building construction activities (PBS – https://pbsinternational.org/programs/the-age-of-the-drone/).

Other Environmental Data Collection

Whilst many people associate UAVs solely with photographic sensors many miniaturised non-imaging sensors are now available making it possible to carry them onboard small aerial platforms. These include environmental sensors to monitor airborne pollutants (e.g. Rutkauskas et al., 2019; Gómez and Green, 2017), as well as chemical and noise pollution. Spatial recording of sensor data provides the basis for mapping at a resolution not otherwise achievable (e.g. environmental contaminants and noise) providing a layer of information that may be input to a Geographical Information System (GIS) for integration with other environmental data layers, geo-visualisation, and spatial analysis.

SUSTAINABLE URBANISATION

So far this chapter has considered the status of the drone and sensor technology currently available, examples of some of the different ways that the technology is already being used to monitor, record, detect change, and visualise the 'built-environment', to integrate the data and information collected with other geospatial technologies, and to highlight some of the constraints for use within the 'built-environment'. In this section, some of the different ways drone technology is already being used to help address the vision of sustainable urban development will be explored.

Urban Living

The growing number of people both moving to and residing in urban areas around the world is now considered to be one of the most significant trends of the 21st century and urban areas are increasingly becoming the focus of global development. Indeed, cities and urban areas are now the main abode of human beings (currently around 50% and expected to rise to 66% withing a generation) in both developed and developing countries. Furthermore, rather than just being somewhere to live and work, cities have now become the centres of economic growth, power, and political influence serving to 'shape' society at every level. They are also considered to be a major factor in environmental trends and sustainability processes

(https://www.un.org/millenniumgoals/pdf/Think%20Pieces/18_urbanization.pdf).

Increasingly rapid urbanisation comes at a price, however, with major impacts on the physical environment and the ecological footprint likely. Growing urban populations require land, space, homes, jobs, services, and a transport network as well as a water supply and waste disposal facilities. To achieve economic and socially equitable growth therefore requires cities to be planned, governed and to provide services to the population. Sustainable urbanisation (developing current cities that are sustainable into the future) is considered key to future urban development and will require careful management.

Many challenges now face future urban development and growth such as implementing green transport, reducing pollution, generating renewable energy, building construction, industry, improving air quality, providing adequate water supplies for consumption, increasing biodiversity, planning for disaster preparedness and adapting to climate change, all of which need to be addressed together as part of sustainable urbanisation; the future prioritisation of which is becoming increasingly important. In some parts of the world, planning for climate change is also becoming even more crucial as many large cities located at the coast are clearly becoming vulnerable to climate change manifested through sea-level rise and the increased frequency of riverine flooding that could soon displace millions of people from their homes, placing jobs and homes, vital services, such as power supplies and transport, at risk. (https://www.worldbank.org/en/news/immersive-story/2018/01/31/3-big-ideas-to-achieve-sustainable-cities-and-communities). To fully understand how urban centres both contribute and need to adapt to climate change requires a better grasp of how transport, buildings, heating and cooling systems, industries and other urban activities currently act as emitters and directly cause climate change.

The speed and scale of urbanization now being faced will clearly bring tremendous challenges to the current and future urban environment, especially in developing countries, where cities are now facing very rapid and major changes in response to the growing population and its needs alongside the uncertainty of the likely impacts of climate change. Our current lifestyle and consumption choices are changing the size, density and structure of cities and urban areas, with many cities in the developing world experiencing rapid and largely uncontrolled population growth and a lack of land-use planning. Demand for housing, basic urban services, and consumer goods are growing, and preference for suburban living is starting to have a negative impact on both rural and urban environments. Widening income gaps, worsening pollution, and aging buildings and bridges are all signs that today's cities are struggling to keep up with city dwellers' aspirations for a sustainable and prosperous future. Note, however, that this change post COVID-19 where cities have become the focal hot-spots of the infections and spread of the coronavirus, with predictions that the future may bring an exodus of people from urban living to the rural environment.

Without planning and the development of appropriate infrastructure many cities will increasingly begin to experience serious problems associated with these unprecedented changes and the consequences of the unprecedented negative impacts arising from urbanisation, not only because of climate change but also because of the likely reduction in economic growth, quality of life, and increased social instability that will result without careful and sustainable urban planning. Climate friendly urban planning and design are therefore not only considered necessary for the security of a city's inhabitants but also to safeguard livelihoods and protect the urban economy. Protecting infrastructure and services will and enable economic activity to flourish. Compact cities, with well-designed services and infrastructure will help to reduce the cost of energy provision, transport, and other services that businesses need, and will increase productivity and efficiency. To this end there is now a growing global commitment to address sustainable urbanization in the future.

Drones and The Sustainable Urban Environment

As discussed earlier, drone technology has advanced quickly in a very short period and can clearly provide many opportunities in relation to both its use as a remote sensing platform and in other ways. The question remains, however, as to the role that drone technology might play in helping society to develop sustainable urban environments in the future. The following section provides some insights into how their use can play a role using some examples to illustrate.

Drone-based analytics, in combination with GIS approaches, have the unique advantage of cost-effective and accurate data. In addition to accessible visualisation and spatial analysis this will form an important basis for urban infrastructure planning (https://medium.com/dronamaps/the-future-of-urban-planning-drone-based-3d-maps-and-pipelines-in-lucknow-26d73f6e4cd6).

Urban development requires data and information as the basis for planning and decision-making. With the rapid growth of cities there is a growing need for more data and information as data-driven decision-making will become ever more essential. Traditional ground-based tools and techniques, however, are very time consuming and costly to operate. Although the use of drones has grown very quickly, drone technology is only at the beginning of having an impact on urban planners and landscape architects. But, as the earlier examples discussed have already shown, this is rapidly changing as the technology becomes more commonplace and more developed. It is now possible to gather a lot of data from drones which can quickly be viewed, transformed, and analysed with the assistance of artificial intelligence (AI) and machine-based learning.

With the increase in population in urban areas, resource planning and allocation will be crucial. Drones can provide a quick and efficient way to collect the large amounts of 'big data' needed through automated mapping, cadastral surveying, corridor surveying, and volumetric calculations. Together with AI (Artificial Intelligence)-based processing, or machine-based learning, data can be gathered, processed, and analysed very quickly. The ease with which data can be collected using drones also provides scope for the individual (the citizen) to become involved, by flying a drone, collecting photographs, and sharing the results with organisations such as local authorities. As a form of crowd-sourced data and information there is potential for public participation in planning and decision-making to increase.

Although traditional aerial photographs and satellite imagery have often been used in the past to understand the changing image and spatial dimensionality of cities, UAV technology is now offering a new and exciting way to collect higher-resolution data and imagery (both aerial photography and videography), and in addition helping to visualise the urban environment quickly, on-demand, and without the need for much expertise. The flexibility offered to planners and decision-makers from real-time acquisition of high-resolution imagery will lead to new ways to visualise the fine detail of the urban environment, to view a city, and for visual communication, the provision of and access to information, with the potential to stimulate and promote greater public participation in the spatial planning process. Using UAV technology can help landscape architects and planners to examine the existing social and environmental conditions of city sites, and to document the movement of people, cars, buses, and green transport, such as bicycles, through transit corridors, to understand changing urban and demographic patterns, to see topographical and hydrologic changes, and to monitor environmental degradation.

Combining drone imagery with more traditional sources of visual information, such as aerial and ground photography, will also provide a valuable way to help reveal both empirical and experiential observations, as a basis for community engagement, to explore the diversity of housing types, the voids of under-utilized open space, the movement of traffic patterns, and the flow of natural systems, the nodes of community activity, and to understand the role of illumination of the urban landscape. Drone data and imagery will also help to make complex problems more understandable to the public and may even lead to greater involvement and improved dialogue.

Planners and engineers have already begun to use drones to acquire data from previously difficult and inaccessible areas to help fill information gaps, faster, more cheaply, and more safely than before. Examples include building inspections that are usually invasive and time-consuming, and the use of thermal cameras to quickly find air leaks to help improve the energy efficiency of buildings (https://www.simulyze.com/blog/drones-and-the-future-of-urban-planning). Developers are also experimenting with integrating UAV technology to improve things like energy efficiency, project progress monitoring, sewage design, utility establishment and detect illegal constructions, and encroachments.

Because of the ease with which drone imagery can now be captured and the manoeuvrability of the platform and camera it is possible not only to capture many unique photographic shots but also to monitor the city in real-time. This can provide city planners with a deeper understanding of city contexts by giving them access to street-level imagery. Rather than using personnel to inspect transport infrastructure, drones provide a more cost-effective and safer solution for law enforcement and city engineers. Furthermore, drones can play an important role in improving urban resilience. Using drones to detect danger zones after a natural disaster help the delivery of aid whereby drones can spot trees or electric wiring that have fallen on roads after a storm, helping to improves safety in high-risk areas https://thecityfix.com/blog/Friday-fun-drones-mapping-cities-road-safety-maoqi-sun-erin-staver-erin-kenney/.

Water supply management is crucial in cities, and planning is needed to allow for a robust provision. The expansion of cities and asphalt roads, along with the industrialisation of agriculture and forests disrupts local water cycles by diverting water from its natural routes leading to frequent droughts and dehydration. Urban structures can also cause water to rapidly pool in a localised area, causing flooding after torrential rain, leading to instability, and having a knock-on effect on local economies (https://news.microsoft.com/europe/features/this-slovak-startup-is-using-ai-and-drones-to-help-preserve-natural-water-cycles/). Once again, drones can provide the means to provide rapid data acquisition, a surveying and mapping capability, and access to crucial up-to-date information (https://xponentialworks.com/why-use-drones-for-surveying-and-mapping/).

Environmental sensors attached to drones will also help to influence urban design where they can help to model the airflow between buildings, influence placement of vegetation, and even direct the way pedestrians access and walk through the city (https://www.simulyze.com/blog/drones-and-the-future-of-urban-planning). Using drone technology will help urban planners to create places to live for growing communities and ensure that a city has the proper flexibility to adapt to the future, and to provide better-informed decision-making to benefit urban planning (https://www.simulyze.com/blog/drones-and-the-future-of-urban-planning).

Spatial mapping of air quality and pollution monitoring is an emerging field in the use of drones providing the means to monitor the urban environment and to provide the basis to provide a temporal database and understanding of environmental status (Gallacher, 2015). This information could be used to identify sources of pollution, their spread, and even information that could be used to regulate and prosecute. Models of wind movement may be useful for assessing the impact of a new construction on surrounding areas or to optimize the placement of a wind farm (Jensen et al., 2009).

Drones will also undoubtedly play a crucial role in helping to manage land-use problems such as land shortage and the availability of affordable housing in urban areas as they grow (Wihbey, 2017; https://www.lincolninst.edu/publications/articles/drone-revolution). Drones are being used in many countries to enrich and improve land policy and management, to modernise land registries for contemporary uses, and to help address other large-scale challenges, including the implementation of property tax. Drone technology is already helping to facilitate a movement toward multi-stakeholder Multi-Purpose Cadastres (MPCs) that can facilitate better urban planning, more equitable taxation, increase revenues, and provide a broader tax base. The use of drones is proving crucial in the initial and provisional identification of physical property limits in cities and jurisdictions where there is still no formal land administration system and the land structure is unknown (Wihbey, 2017).

Drones may also help the urban environment to become more sustainable and part of a major rethink about certain aspects of the urban environment. One example is the development of a re-configurable architectural roof canopy system deployed by drones that can be adjusted to follow the direction of the sun, with the canopy providing shade resulting in less reliance on ventilation systems. https://www.weforum.org/agenda/2020/07/how-drones-and-aerial-vehicles-could-change-cities?hss_channel=tw-15579320

In China urban green space (UGS) has been studied with the help of landscape metrics generated using drones. These can have an important part in reducing the problems associated with urbanization by helping to evaluate the extent and the environmental and climatic benefits of UGS (https://www.sciencedirect.com/science/article/pii/S0169204617300877?via%3Dihub).

Another aspect of sustainable urbanisation is the development of Smart cities. Drones can be used on a regular basis in various ways to, for example, monitor traffic and crowds, key infrastructures, and urban development. They can also be used to provide services and opportunities such as smart policing and disaster management. Future cities will likely need to improve sustainable urban transport and drones can be used to 'map sites for a metro project, bus transit or even a bicycle path' (https://www.geospatialworld.net/blogs/how-drones-are-crucial-for-smart-cities/). Drones offer flexibility allowing surveyors to map long corridors efficiently at the start of projects and to collect in-depth data to aid decision-making at an earlier stage. Not only can they monitor congestion, but they can help to remove it by monitoring from the sky (https://www.geospatialworld.net/blogs/how-drones-are-crucial-for-smart-cities//https://www.researchgate.net/publication/324028974_The_evolution_of_UAVs_applications_in_urban_planning/https://www.int-arch-photogramm-remote-sens-spatial-inf-sci.net/XLI-B6/131/2016/isprs-archives-XLI-B6-131-2016.pdf). The potential for utilizing drones to improve city traffic information and management has already been explored by a multitude of law enforcement agencies in various countries (https://thecityfix.com/blog/Friday-fun-drones-mapping-cities-road-safety-maoqi-sun-erin-staver-erin-kenney/). Other uses in this context and highlighted by Geo.Geo (http://www.geogeoglobal.com/), include the potential role of high-resolution imagery to support the concepts of the Smart city (https://www.orfonline.org/expert-speak/drones-in-smart-cities-55717/) and urban sustainability, with the means to generate everything from detailed 3D urban models and realistic visualisations, to optimising the siting of solar panels within urban areas where building layout and height are important considerations (http://www.geogeoglobal.com/case-studies/edinburgh3d). Some of the advances include exploring how our new 'view' from drones on the urban landscape can help to change the urban building exteriors to facilitate drone traffic navigation in cities (https://medium.com/predict/the-three-dimensional-city-how-drones-will-impact-the-future-urban-landscape-5103af61af72).

As an extreme vision of the city of the future, drones may not only be used as monitoring and data collection platforms, but also as an aerial form of green urban transport. To achieve this will necessitate a transformation in the architecture and infrastructure of future cities including adaptations such as building drone landing pads, solar photovoltaic panels for energy efficiency, charging points for delivery drones, and environmental landscaping to mitigate noise emissions (https://www.weforum.org/agenda/2020/07/how-drones-and-aerial-vehicles-could-change-cities?hss_channel=tw-15579320).

SUMMARY AND CONCLUSION

In a very short space of time, drone platforms and sensors have evolved from toys and customised solutions to mainstream remote sensing technology with the potential to capture a wide range of data, that is spatially referenced, and can be captured autonomously. The data can be rapidly processed into ortho-mosaics and 3D models using softcopy photogrammetric software and integrated into Geographical Information Systems for geo-visualisation, communication, and participatory activities.

Drone technology undoubtedly has immense potential for monitoring, mapping, and modelling the 'built-environment', old or new, and the related infrastructure on both land and in the water. Today there are many applications of UAVs to the 'built-environment' where drones have the potential to provide general aerial photography (vertical and oblique), videography, monitoring capability, thermal mapping, map updates, derivation of terrain and surface models, and the generation of 3D models of buildings and infrastructure. Already these platforms and sensors are in routine day to day use and it is rapidly becoming clear that they will continue to contribute towards addressing the issues facing sustainable urbanisation.

However, in a recent online document entitled: *Big City Life Awaits Drones in Final Year of NASA Research* (https://www.nasa.gov/feature/ames/big-city-life-awaits-drones-in-final-year-of-nasa-research), it is noted that the landscape of the 'built-environment' hold unique challenges for flying drones, and NASA (National Aeronautics and Space Administration) are already actively engaged in undertaking research into much needed drone traffic management.

To release the potential to contribute to the development of sustainable urban areas including the development of new urban transportation networks, reduced air pollution in cities, and efficient routing of emergency services in built-up areas amongst other things (https://www.iceniprojects.com/the-drones-are-coming/) will require both national policy makers and local planning authorities to respond quickly to the needs of this very rapidly evolving technology by introducing regulations that cover safety, noise (https://arc.aiaa.org/doi/pdf/10.2514/6.2019-2685), and security issues (https://www.defensedaily.com/darpa-showing-progress-small-drone-detection-urban-environments/homeland-security//https://www.westerncity.com/article/preparing-drone-age-your-city-potential-benefits-and-issues and https://www.cnbc.com/2019/02/20/nasa-to-test-urban-drone-traffic-management-in-texas-and-nevada.html).

Increasing use of drones in the future will have a major impact on the use of low-level airspace and affect many other proposed activities. Whilst future regulations may continue to limit who can fly drones, especially in urban areas, adaptation, development and exploitation of new 4G and 5G mobile telecommunications will be needed to facilitate parcel delivery, monitoring of roads and traffic jams, as well as crowd monitoring and policing. The use of drone technology in any 'built-environment', and particularly urban areas, where people are present/resident also necessitates consideration of the constraints that may need to be placed on drone overflights for reasons of safety, as well as considerations of invasion of privacy as a result of a sensor (e.g. a camera) being mounted on an aerial platform.

ACKNOWLEDGMENT

To my co-authors for their support and for making this chapter possible.

REFERENCES

Abbott, J. (2020). *DJI Mavic Air 2 Review – The Best Drone Around for Beginners and Hobbyists*. https://www.techradar.com/uk/reviews/dji-mavic-air-2-review

Achille, C., Adami, A., Chiarini, S., Cremonesi, S., Fassi, F., Fregonese, L., & Taffurelli, L. (2015). UAV-Based Photogrammetry and Integrated Technologies for Architectural Applications/Methodological Strategies for the After-Quake Survey of Vertical Structures in Mantua (Italy). *Sensors (Basel)*, *15*(7), 15520–15539. doi:10.3390150715520 PMID:26134108

Allotta, B., Costanzi, R., Ridolfi, A., Salvetti, O., Reggiannini, M., Kruusmaa, M., Salumae, T., Lane, D.M., Frost, G., Tsiogkas, N., Cocco, M., Gualdesi, L., Lacava, G., Roig, D., Gundogdu, H.T., Dede, M.I.C., Baines, S., Tusa, S, Latti, P., & Scaradozzi, D. (2018). The ARROWS Project: robotic technologies for underwater archaeology. *IOP Conf. Ser.: Mater. Sci. Eng.*, *364*. 10.1088/1757-899X/364/1/012088

Altuntas, C. (2019). Urban area change visualization and analysis using high density spatial data from time series aerial images. *Reports on Geodesy and Geoinformatics.*, *107*(1), 1–12. doi:10.2478/rgg-2019-0001

Ball, M. (2019). *High-Resolution SAR Imagery Captured with Small UAS. Unmanned Systems Technology*. https://www.unmannedsystemstechnology.com/2019/02/high-resolution-sar-imagery-captured-with-small-uas/

Benjamin, J., McCarthy, J., Wiseman, C., Bevin, S., Kowlessar, J., Astrup, P. M., Naumann, J., & Hacker, J. (2019). Integrating Aerial and Underwater Data for Archaeology: Digital Maritime Landscapes in 3D. In 3D Recording and Interpretation for Maritime Archaeology. Coastal Research Library.

Bognot, J. R., Candidoa, C. G., Blancoa, A. C., & Montelibanoc, J. R. Y. (2018). Building Construction Progress Monitoring Using Unmanned Aerial System (UAS), Low-Cost Photogrammetry, and Geographic Information System (GIS). *ISPRS Annals of the Photogrammetry, Remote Sensing and Spatial Information Sciences*, *4*(2), 41-47.

Chen, B., Chen, Z., Deng, L., Duan, Y., & Zhou, J. (2016). *Building Change Detection with RGB-D Map Generated from UAV Images*. Academic Press.

Colomina, I., & Molina, P. (2014). Unmanned Aerial Systems for Photogrammetry and Remote Sensing: A Review. *ISPRS Journal of Photogrammetry and Remote Sensing*.

De Lima, R. L. P., Boogaard, F. C., De Graaf, R., Dionisio Pires, M., & Sazonov, V. (2015). Monitoring the Impacts of Floating Structures on the Water Quality and Ecology Using an Underwater Drone. *E-Proceedings of the 36th IAHR World Congress*.

Djimantoro, M. I., & Suhardjanto, G. (2017). The Advantage by Using Low-Altitude UAV for Sustainable Urban Development Control. In *Proceedings of The International Conference on Eco Engineering Development 2017 (ICEED 2017)*. IOP Publishing. 10.1088/1755-1315/109/1/012014

Gallacher, D. (2015). Drone Applications for Environmental Management in Urban Spaces: A Review. *International Journal of Sustainable Land Use and Urban Planning*, *3*(4), 1-14.

Gheisari, M., & Esmaeili, B. (2016). Unmanned Aerial Systems (UAS) for Construction Safety Applications. *Proceedings of the Construction Research Congress 2016*, 2642-2650. 10.1061/9780784479827.263

Gómez, C., & Green, D. R. (2015). *Small-Scale Airborne Platforms for Oil and Gas Pipeline Monitoring and Mapping*. Unpublished UCEMM Report for Redwing Ltd.

Gómez, C., & Green, D. R. (2017). Small Unmanned Airborne Systems to Support Oil and Gas Pipeline Monitoring and Mapping. Arabian Journal of Geosciences, 10(9). doi:10.100712517-017-2989-x

Green, D. R., Gregory, B. J., & Karachok, A. R. (Eds.). (2020a). Unmanned Aerial Remote Sensing UAS for Environmental Applications. CRC Press.

Green, D. R., Karachok, A. R., & Gregory, B. J. (2020b). The Potential of Drone Technology in Pandemics. In COVID-19: Geospatial Information and Community Resilience. Taylor and Francis.

Green, J. L. (2017). *Drones Will Elevate Urban Design*. https://www.smartcitiesdive.com/ex/sustainablecitiescollective/drones-will-elevate-urbandesign/1053491/

Harrop, P., & Harrop, J. (2014). Electric Drones: Unmanned Aerial Vehicles (UAVs) 2015-2025. IDTechEx Report.

Hildmann, H., Kovacs, E., Saffre, F., & Isakovic, A. F. (2019). Nature-Inspired Drone Swarming for Real-Time Aerial Data-Collection Under Dynamic Operational Constraints. *Drones (Basel)*, *2019*(3), 71. doi:10.3390/drones3030071

Jensen, A. M., Morgan, D., Chen, Y., Clemens, S., & Hardy, T. (2009). Using Multiple Open-Source Low-Cost Unmanned Aerial Vehicles (UAV) for 3D Photogrammetry and Distributed Wind Measurement. In *ASME/IEEE 2009 International Conference on Mechatronic and Embedded Systems and Applications*; 20th Reliability, Stress Analysis, and Failure Prevention Conference (Vol. 3, pp. 629–634). Retrieved from http://ebooks.asmedigitalcollection.asme.org/data/Conferences/IDETC/CIE 2009/70856/629_1.pdf

Labib, R. (2017). *What are the applications of drone technology in architecture and the built environment field?* [Commentary]. https://www.researchgate.net/post/What_are_the_applications_of_drone_technology_in_architectur_and_the_built_environment_field

Li, C. J., & Ling, H. (2015). Synthetic Aperture Radar Imaging Using a Small Consumer Drone [Unpublished Powerpoint Presentation]. 10.1109/APS.2015.7304729

Li, M., Nan, L. L., Smith, N., & Wonka, P. (2016). Reconstructing Building Mass Models from UAV Images. *Computers & Graphics*, *54*(February), 84–93. doi:10.1016/j.cag.2015.07.004

Mindell, D. A., & Bingham, B. (2001). New Archaeological Uses of Autonomous Undersea Vehicles. *MTS/IEEE Oceans 2001. An Ocean Odyssey. Conference Proceedings*, 555-558. 10.1109/OCEANS.2001.968781

Rakha, T., & Gorodetsky, A. (2020). Unmanned Aerial System (UAS) Applications in the Built Environment: Towards Automated Building Inspection Procedures Using Drones. In Unmanned Aerial Remote Sensing UAS for Environmental Applications. CRC Press.

Reagan, D., Sabato, A., & Niezrecki, C. (2017). Unmanned Aerial Vehicle Acquisition of Three-Dimensional Digital Image Correlation Measurements for Structural Health Monitoring of Bridges. Conference: SPIE Smart Structures and Materials + Nondestructive Evaluation and Health Monitoring.

Roggema, R. (2016). The Future of Sustainable Urbanism: A Redefinition. *City, Territory and Architecture*, *3*(22).

Roman, C., & Mather, R. (2016). Autonomous Underwater Vehicles as Tools for Deep-Submergence Archaeology. *Proceedings of the Institution of Mechanical Engineers, Part M: Journal of Engineering for the Maritime Environment*, *224*(4), 327-340.

Roman, C., & Mather, R. (2016). Autonomous Underwater Vehicles as Tools for Deep-Submergence Archaeology. *Journal of Engineering for the Maritime Environment, 224*(4), 327-340.

Rutkauskas, M., Asenov, M., Ramamoorthy, S., & Reid, D. T. (2019). Autonomous multi-species environmental gas sensing using drone-based Fourier-transform infrared spectroscopy. Optics Express, 27(7).

Scannapieco, A. F., Renga, A., Fasano, G., & Moccia, A. (2017). Ultralight Radar for Small and Micro-UAV Navigation. *The International Archives of the Photogrammetry, Remote Sensing and Spatial Information Sciences, Volume XLII-2/W6, 2017. International Conference on Unmanned Aerial Vehicles in Geomatics*, 333-338. 10.5194/isprs-archives-XLII-2-W6-333-2017

Skitmore, A. (2018). *Launch and Recovery System for Improved Fixed-Wing UAV Deployment in Complex Environments* [Unpublished MSc Thesis]. Department of Maritime and Mechanical Engineering. Liverpool John Moores University.

Skitmore, A. (2020). Launch and Recovery System for Improved Fixed-Wing UAV Deployment in Complex Environments. In Unmanned Aerial Remote Sensing: UAS for Environmental Applications. CRC Press.

Stocker, C., Bennett, R., Nex, F., Gerke, M., & Zevenbergen, J. (2017). Review of the Current State of UAV. *Remote Sensing*, *9*(5), 459–486. doi:10.3390/rs9050459

Watts, A. C., Ambrosia, V. G., & Hinkley, E. A. (2012). Unmanned Aircraft Systems in Remote Sensing and Scientific Research: Classification and Considerations of Use. *Remote Sensing*, *4*(6), 1671–1692. doi:10.3390/rs4061671

Wihbey, J. (2017). The Drone Revolution: UAV-Generated Geodata Drives Policy Innovation. In Land Lines. Lincoln Institute of Land Policy.

Wu, D., Arkhipov, D. I., Kim, M., Talcott, C. L., Regan, A. C., McCann, J. A., & Venkatasubramanian, N. (2017). ADDSEN: Adaptive Data Processing and Dissemination for Drone Swarms in Urban Sensing. *IEEE Transactions on Computers*, *66*(2), 183–198.

Zink, J., & Lovelace, B. (2015). *Unmanned Aerial Vehicle Bridge Inspection Demonstration Project*. Research Project Final Report 2015-40. Minnesota Department of Transportation.

ADDITIONAL READING

Albeaino, G., Gheisari, M., & Franz, B. W. (2019). A Systematic Review of Unmanned Aerial Vehicle Application Areas and Technologies in the AEC Domain. *Journal of Information Technology in Construction*, *24*, 381–405.

Gevaert, C. M., Sliuzas, R., Persello, C., & Vosselman, G. (2018). Evaluating the Societal Impact of Using Drones to Support Urban Upgrading Projects. *ISPRS International Journal of Geo-Information*, *2018*(7), 91. doi:10.3390/ijgi7030091

Koeva, M., Muneza, M., Gevaert, C., Gerke, M., & Nex, F. (2018). Using UAVs for map creation and updating. A case study in Rwanda. *Survey Review*, *50*(361), 312–325. doi:10.1080/00396265.2016.1268756

Li, X., Lv, Z., Hijazi, I., & Hongzan, J. (2016). Assessment of Urban Fabric for Smart Cities. *IEEE Access: Practical Innovations, Open Solutions*, *4*, 373–382. doi:10.1109/ACCESS.2016.2517072

Nex, F., & Remondino, F. (2019). *Preface: Latest Developments, Methodologies, and Applications Based on UAV Platforms*. Drones.

Primatesta, S., Guglieri, G., & Rizzo, A. (2019). A Risk-Aware Path Planning Strategy for UAVs in Urban Environments. *Journal of Intelligent & Robotic Systems*, *95*(2), 629–643. doi:10.100710846-018-0924-3

KEY TERMS AND DEFINITIONS

DSM: Digital surface model is a three-dimensional model of the surface of a landscape.

DTM: Digital terrain model is a three-dimensional model of the topographic variation in the landscape.

FPV: First person view is a device that provides the drone operator with the view seen by the drone camera.

GIS: Geographical information system is desktop or mobile software to collect, store, edit, analyse, visualise, and communicate spatial data and information.

GPS: Global positioning system is a device used to find out the location or co-ordinates of a point on the Earth's surface.

Mosaic: A collection of overlapping digital aerial photographs that have been stitched together into a single photograph.

Orthophoto: A digital photograph that has been corrected to remove distortions arising from yaw, pitch and roll of the aerial platform.

Photogrammetry: A manual or digital technique used to analyse stereo photographs, to create 3D models, and to extract quantitative information.

UAS: Unmanned aerial system comprising the both the airborne vehicle and the camera.

UAV: Unmanned aerial or airborne vehicle.

Chapter 12
Application of Conventional UAVs for the Identification and Classification of Dense Green Spaces

Josep Roca
https://orcid.org/0000-0003-3970-6505
Technical University of Catalonia (CPSV), Spain

Blanca Arellano
Technical University of Catalonia (CPSV), Spain

ABSTRACT

The objective of this chapter is to show the usefulness of conventional UAVs for the identification, inventory, and classification of trees in the context of dense green spaces. The aim is to demonstrate the potential of low-cost drones (with traditional red, green, blue [RGB] sensors) to identify and classify trees in public parks. A case study is discussed on Turó Parc in Barcelona, in which a 3D model was developed and an exercise to identify and classify the vegetation was carried out using the information provided by a UAV. The example confirms that conventional drones could be useful for studying green urban spaces characterized by a high density of plant species. Non-professional UAVs have a potential that should not be undervalued, as they enable three-dimensional point clouds to be obtained of high spatial density.

DOI: 10.4018/978-1-7998-2249-3.ch012

INTRODUCTION

The special significance of green spaces in cities has been fully demonstrated. Recent studies have supported an increase in green spaces and a greater commitment to health promotion policies (Nieuwenhuijsen et al., 2017; van den Bosch and Nieuwenhuijsen, 2017). The inhabitants of cities are exposed to environmental pollution, which may critically affect their health (Shrestha et al., 2016). Increasing the number of green spaces is a measure that mitigates the loss of natural environments within cities due to densification processes and improves human health and the well-being of a large section of urbanized societies (Dallimer et al., 2011; Pauleit et al., 2005). Green spaces are key elements to improve the quality of urban life. They contribute to human well-being by providing ecosystem services such as climate regulation, the capture of pollutants or flood control. They also promote contact between residents and community integration, and offer a favourable place for health, relaxation and contemplation of nature (Chiesura, 2004; Lee and Maheswaran, 2011; Dobbs et al., 2014; Larondelle et al., 2014; Carrus et al., 2015; Marselle et al., 2015). In addition, greener environments have lower crime rates (Kuo and Sullivan, 2001). They tend to have a positive effect on people and induce mental vitality (Takayama et al., 2014). Green spaces can help to reduce urban heat islands and generate true cold islands in urbanized environments (Norton et al., 2015; Arellano and Roca, 2018; García and Arellano, 2018). Finally, urban parks provide economic value for cities, including an increase in the value of properties in their proximity (Roca, 1988; Harnik and Crompton, 2014).

The study of urban green space has developed exponentially with the use of satellite and airborne sensors. In particular, the analysis of the Normalized Difference Vegetation Index (NDVI), Leaf Area Index (LAI), Fraction of Photosynthetically Absorbed Aadiation (FAPAR) and Fractional Vegetation Cover (fcover) have revealed the quality of urban vegetation with a high level of detail. At the same time, aerial and satellite images with high radiometric and spatial resolution have been used to draw up inventories of trees in cities, and even to determine the trees' biological health status (http://iarbol.starlab.es). However, the use of high-resolution satellites and conventional aerial flights to obtain data is usually economically inaccessible. In addition, it can be difficult to identify and classify trees from the data that is collected: it is hard to interpret the results when the vegetation is very dense, as is often the case in urban parks. For this reason, using remote sensing to catalogue trees in urban parks is not straightforward. For example, Barcelona has a magnificent inventory of trees on public roads (https://opendata-ajuntament.barcelona.cat/data/en/dataset/arbrat-viari) but does not have a similar inventory for the dense vegetation of urban parks.

Drones are already popular in society. They can fly at low speed and low height, and have the advantages of high security and high reliability at low cost. The great flexibility of the unmanned aerial vehicle (UAV) means that high-resolution images can be obtained in a very short time, which is not the case for satellite and airborne images. For these reasons, drones provide effective support to complement information obtained from remote sensing satellites and from conventional photography. This text analyses the structure of the conventional, low-profile non-professional drone photography system and highlights its advantages: agility, mobility, economy and speed in obtaining information in comparison with traditional remote sensing technology. The use of drones is an efficient way to obtain high-resolution photographs in a short time. UAVs have been used in a wide range of studies. From obtaining detailed digital elevation models (Cruz et al., 2017) to classifying urban and rural land (Pan et al., 2020). Drones provide a good platform for quickly and cheaply obtaining the parameters of vegetation canopies (Ma, 2018), including the tree height and canopy radius (Chang et al., 2020). However, they have been little used in the identification of vegetation / species in small green areas located in urban context. In this field, drones can provide additional knowledge to that provided by conventional satellite and aerial remote sensing systems. The higher spatial and temporal resolution of the images obtained by UAVs is decisive for an adequate recognition of the existing plant species in urban parks.

In this context, the objective of this chapter is to show the usefulness of conventional UAVs for the identification, inventory and classification of trees in the context of dense green spaces. The aim is to demonstrate the potential of low-cost drones (with traditional red, green, blue [RGB] sensors) to identify and classify trees in public parks. A case study is discussed on Turó Parc in Barcelona, in which a 3D model was developed and an exercise to identify and classify the vegetation was carried out using the information provided by a UAV (Ma, 2018).

USE OF REMOTE SENSING TECHNOLOGIES IN THE STUDY OF URBAN GREEN AREAS: ADVANTAGES AND DISADVANTAGES OF THE USE OF CONVENTIONAL UAVS

Access to remote sensing (RS) has enabled the study of vegetation in urban systems to an extent that has never been reached before. Vegetation indices (VI) have been developed over the past decades to measure the general health status of the vegetation. Healthy vegetation absorbs the energy provided by red wavelengths (chlorophyll absorbs photons that are then stored as energy through photosynthesis), and efficiently reflects the energy in the near infrared (NIR) region of the spectrum. Consequently, a set of vegetation indices have been created, including:

- The Ratio Vegetation Index (Birth and McVey, 1968), calculated by dividing the reflectance values of the NIR band by those of the red band.
- The Normalized Difference Vegetation Index (NDVI, Rouse et al., 1974), obtained as NIR-RED/NIR+RED, provides a measure of the amount of green vegetation present in a pixel, so that green vegetation can be separated from its background soil brightness. This is the most commonly used VI as it minimizes topographic effects and produces a statistically normal distribution.
- Leaf area index (LAI, Chen and Black, 1992, Asner, Braswell, et al., 1998) quantifies the amount of leaf area in an ecosystem and is a critical variable in processes such as photosynthesis, respiration, and precipitation interception. It represents the amount of leaf material in an ecosystem and is geometrically defined as the total one-sided area of photosynthetic tissue per unit ground surface area. This ratio can be related to gas-vegetation exchange processes such as photosynthesis, evaporation and transpiration, rainfall interception, and carbon flux (Zheng and Moskal, 2009).
- The fraction of photosynthetically absorbed radiation (FAPAR) is defined as the fraction of the photosynthetically radiation (PAR) absorbed by the green leaves. It is a weighted sum of the direct FAPAR and diffuse FAPAR, depending on the source of the incoming radiation (Tao et al., 2016). LAI and FAPAR have been recognized as Essential Climate Variables (ECV) by GCOS (2011), for their key roles in energy, mass and momentum exchanges between the land surface and the atmosphere.
- Fcover, the fraction of the green vegetation in the nadir direction, is used to separate vegetation and soil in energy balance processes, including temperature and evapotranspiration. Both fcover and PV (Carlson and Ripley, 1997; Valor and Caselles, 1996) have been used as indicators of the proportion of land occupied by vegetation.
- The Enhanced Vegetation Index (EVI) is a modified NDVI which improves the quality of the index by optimizing the vegetation signal in the Leaf Area Index (LAI). Whereas the Normalized Difference Vegetation Index (NDVI) is chlorophyll sensitive, the EVI is more responsive to canopy structural variations, including Leaf Area Index (LAI), canopy type, plant physiognomy and canopy architecture. The two vegetation indices complement each other in global vegetation studies and improve the detection of vegetation changes and the extraction of canopy biophysical parameters (Huete, et al., 1997; Weier, et al., 2000).
- The Atmospherically Resistant Vegetation Index (ARVI; Kaufman and Tanré, 1992), an enhancement of the NDVI, is designed to minimize atmospheric-induced variations in the VI on a pixel by pixel basis.

- The Soil Adjusted Vegetation Index (SAVI; Huete, 1988), a modification of the NDVI, is one of the main VIs developed to minimize the influence of soil brightness in areas where little vegetation cover is present.

However, despite their great scientific interest, these VIs are not designed to quantify the concentration of any given vegetation component (Colaninno, 2016). Instead, they are generally used to measure the quality of photosynthetic material in vegetation, and to map the overall amounts of vegetation components. Remote sensing via satellites provides information on landforms and the general distribution of vegetation types over large or medium areas. However, it is unlikely to provide adequate spatial or temporal resolution to determine the distributions of individual species or fine-scale differentiation among landscape features and vegetation types. Moreover, available satellite images may not represent optimal phenological stages for the identification of species and vegetation types (Cruzan et al., 2016).

Therefore, the use of remote sensing can make it difficult to obtain significant information for urban and territorial planning. Specifically, the low spatial resolution of most satellite images prevents accurate quantification of the trees that exist in the urban public space and the detailed identification of species. Even when images captured by airborne sensors of greater spatial resolution are available, it is almost impossible to obtain detailed indicators that are useful for urban management, given the lack of temporal resolution in the acquisition of these images. Figure 1 presents for the same area an example of the NDVI obtained using Landsat8 (30 metres/pixel) and Sentinel2 (10 metres/pixel), and an aerial image supplied by the Cartographic and Geological Institute of Catalonia (ICGC) (0.9 m/pixel).

Figure 1. NDVI by Lansat8, Sentinel2 and ICGC
Source: compiled by the authors

A range of new satellite sensors have become operational over the past decades, offering data at finer than 5 metres of spatial resolution. Examples are Quickbird, IKONOS, GeoEye-1, OrbView-3, WorldView-2 or Pleiades. Such data are useful for urban studies, but three operational constraints remain: (1) a high cost per scene; (2) suitable repeat times are often only possible if oblique view angles are used, which distort geometric and radiometric pixel properties; and (3) cloud contamination, which can obscure features of interest (Loarie et al. 2007).

One of the projects carried out using remote sensing that has achieved greatest detail in the analysis of urban green space is i-arbol, developed by the company Star Lab (see figure 2) (https://www.esmartcity.es/comunicaciones/iarbol-gestion-arboles-espacios-verdes-ciudad) for Barcelona City Council. This project was designed to implement on the city's inventory of trees (https://opendata-ajuntament.barcelona.cat/data/ca/dataset/arbrat-zona) a methodology aimed at assessing the health of urban vegetation. The spatial resolution of the satellite images that were used (Pleiades, with 0.7 m/panchromatic pixel and 2.8 for the rest of the spectral bands, including infrared) enabled a zone analysis of the health and vigour of the urban green space by means of the NDVI and other vegetation indices. However, it did not enable accurate identification of trees, or the determination of existing plant species in the city[1].

Figure 2. I-arbol Barcelona
Source: Starlab

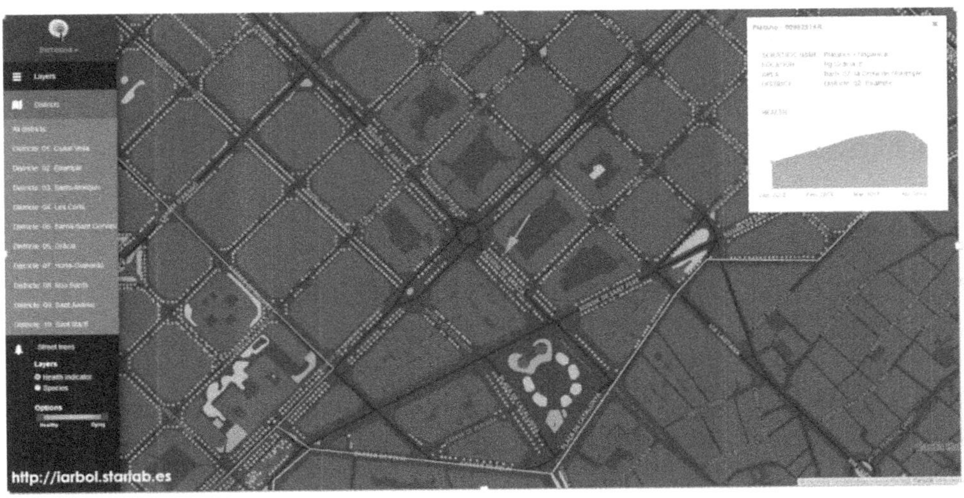

More recently, the Barcelona City Council has developed a website on biodiversity in the city (https://ajuntament.barcelona.cat/atlesbiodiversitat/ca/). This website provides for the first time information on the existing vegetation not only in the streets but also in urban parks. This atlas (figure 3) is a series of maps that show some of the most important data on Barcelona's biodiversity. It includes the following layers: a) the city's trees; b) the parks, gardens, squares and green plots of land that Barcelona City Council manages, with the species of plants that live there; c) the birds that nest in the city; d) The butterflies that are found in the city; e) vertebrate animals (small mammals, amphibians, reptiles); f) the fauna (vertebrates and invertebrates) that live in naturalised ponds; g) all the plant communities that can be found in Barcelona, whether natural or semi-natural; and h) the extent of plant cover the city has, seen from the air.

However, this Atlas has been possible with the help of the biologists who manage the green areas of the city, through manual and individualized work. Not through remote sensing techniques.

Figure 3. Barcelona Biodiversity Atlas
Source: Barcelona City Council

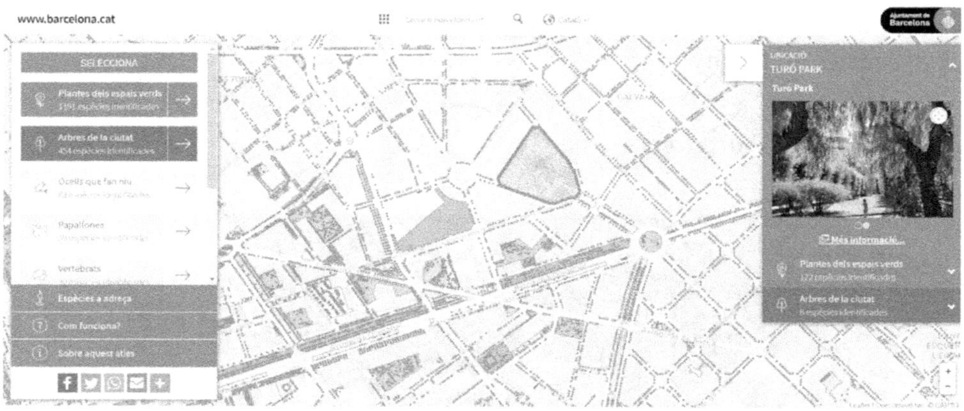

3D MODELLING OF URBAN PARKS THROUGH INFORMATION OBTAINED WITH UAVS

With the continuing development of UAV systems in recent years, aerial photography systems via drones have become a strong complement to traditional remote sensing methods. New methods for fine-scale remote sensing are emerging and may revolutionize urban studies, spatial ecology and environmental science. UAVs are lightweight, low-cost aircraft platforms operated from the ground that can adapt with great flexibility to specific data collection needs. UAVs offer researchers and planners a promising path to timely, cost-effective monitoring of urban environmental phenomena at spatial and temporal resolutions that are appropriate in many urban studies.

With the diversification of studies using UAVs, the possibility of remote sensing research on urban green spaces is increasing. The use of micro-unmanned aerial vehicles (UAVs, unmanned aerial systems [UAS], small aerial drones) may provide adequate levels of image detail to estimate the distribution of individual plant species or vegetation types over several hectares at a relatively low cost (Anderson and Gaston, 2013).

UAS have been successfully used to monitor small areas of natural spaces (Cruzan et al., 2016), provide scale-appropriate data for describing canopy gap metrics relating to the floristic biodiversity of a forest understory (Getzin et al., 2012), and, in general, for the analysis of rural landscapes (with a focus on precision agriculture). However, they have still not been widely used in studies on urban and territorial planning.

The potential of UAVs in urban analysis has two main limitations:

- First, the low spectral resolution of the sensors used in drones, which are usually limited to visible RGB bands.
- The limited capacity of image processing software developed for traditional remote sensing methods, which cannot resolve problems derived from UAV new technology, such as the high number of images or the characteristics of the photographs that are obtained[2].

Therefore, it is clear that new software needs to be developed that is suitable for processing photographs obtained by UAVs. Recently, powerful data processing software has emerged, which can combine images taken in different modes. Some examples are Virtuo Zo NT, Agisoft PhotoScan, Inpho Photogrammetry or Drone2Map for ArcGIS.

Specifically, Drone2Map has revolutionized the ArcGIS platform by integrating the Pix4D algorithm; a system that can process a large number of images and automatically extract orientation points through image pre-processing. It is generated using original data from UAVs and provides digital surface model (DSM) and 3D grid texture products. It has a simple user interface, automated workflow and perfect integration with ArcGis or ArcGIs Online.

This software has revolutionized spatial analysis through the use of UAVs, even allowing the obtaining of important spatial information by simply obtaining RGB images (Chadwick et al., 2020).

CASE STUDY: IDENTIFICATION AND CLASSIFICATION OF DENSE GREEN URBAN AREAS (PARKS) BY UAV IN THE TURÓ PARC OF BARCELONA

The objective of the research is to show the usefulness of conventional UAVs for the identification, inventory and classification of trees in the context of dense green spaces. The aim is to demonstrate the potential of low-cost drones (with traditional RGB sensors) to identify and classify trees in public parks. A case study was carried out for the Turó Parc in Barcelona. A 3D model was developed, and the vegetation was identified and classified according to the information provided by UAV.

The research had the following specific objectives:

- Assess the procedures and technical characteristics for obtaining detailed images of urban spaces characterized by dense vegetation using conventional drones and low-altitude flights.
- Use lateral photographs obtained by drones to construct digital elevation models (DEM), obtain slopes and generate an orthomosaic.
- Identify the main typology and number of trees in the urban green space under study.
- Identify and classify the main tree species in the research area.

The methodology consisted of the steps below:

- Analyse "in situ" the zone under study and verify with municipal specialists the characteristics of the main species of tree that are found in it.
- Programme and carry out field research to obtain images of the study zone from an altitude of 60 metres, using a high-resolution RGB camera installed in a conventional drone[3].
- Use Drone2map software to georeference the images and draw up an orthoreferenced mosaic of them.
- Use Drone2map for ArcGis to construct a high-resolution digital elevation model (DEM).
- Use the Multi-Resolution Segmentation function from the eCognition programme to analyse the situation in detail and obtain a model of trees in the zone using the nearest neighbor configuration classification method.
- Compare the model with the field information obtained "in situ" so that the main species of trees can be distinguished and classified, and the trees in the zone can be identified and inventoried.

In this research, the Turó Parc was selected as the study area. The park was constructed in 1934, has a surface area of 2.88 Ha and is in the Sarriá-San Gervasio neighbourhood of Barcelona, Spain (figure 4). It is located at 41° 23' 40.5" N latitude and 2° 08' 27.6" E longitude. The park contains a small copse of holm oaks, with an understory of Mediterranean shrubs and climbers, such as laurustinus. In addition, there are magnolias, false pepper trees, Himalayan cedar, lindens, plane trees and judas trees. There are large specimens of date palm, Canary island date palm and Mexican fan palm. In summary, 122 plant species, 32 species of nesting birds and 7 species of vertebrates.

Figure 4. Turó Parc (Barcelona)
Source: Compiled by the authors

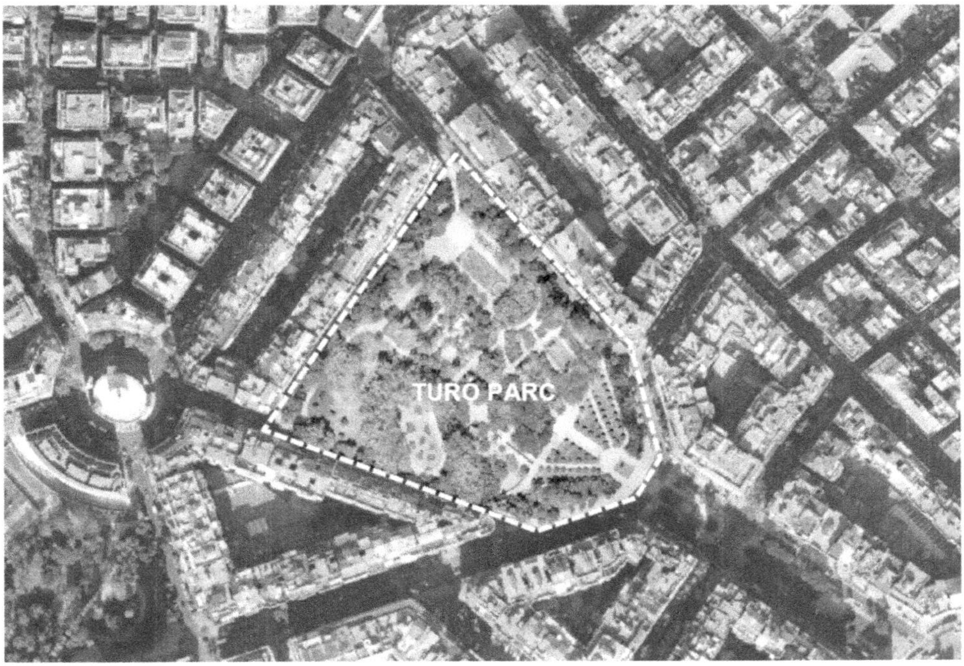

In this research, we used a Mavic Pro drone with 4 rotors for non-professional use, which was launched in 2016 by the company DJJ[4].

Figure 5. Ground control system
Source: Compiled by the authors with Pix4Dcapture (Ma, 2018)

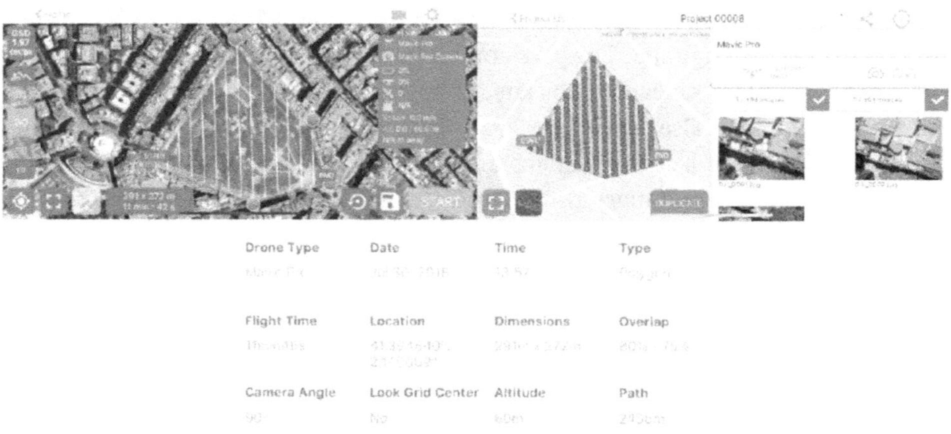

Table 1. Main species of trees in Turó Parc

	Nombre del arbol	Cantidad
1	Schinus molle var. areira	18
2	Quercus ilex	95
3	Casuarina cunninghamiana	7
4	Robinia pseudoacacia	14
5	Phoenix canariensis	7
6	Cupressus sempervirens	22
7	Olea europaea var. europaea	6
8	Phytolacca dioica	3
9	Platanus hispanica	33
10	Washingtonia filifera	17
11	Acer campestre	2
12	Tipuana tipu	94
13	Tilia tomentosa	28
14	Jacaranda mimosifolia	3
15	Pinus pinea	13
16	Magnolia grandiflora / Magnolia grandiflora	13
17	Populus alba 'Pyramidalis'	5
18	Morus alba	1
19	Ginkgo biloba	2
20	Euonymus japonicus	2

Source: Esperanza Huerta

Figure 6. Orthomosaic image processed by Drone2map
Source: Compiled by the authors (Ma, 2018)

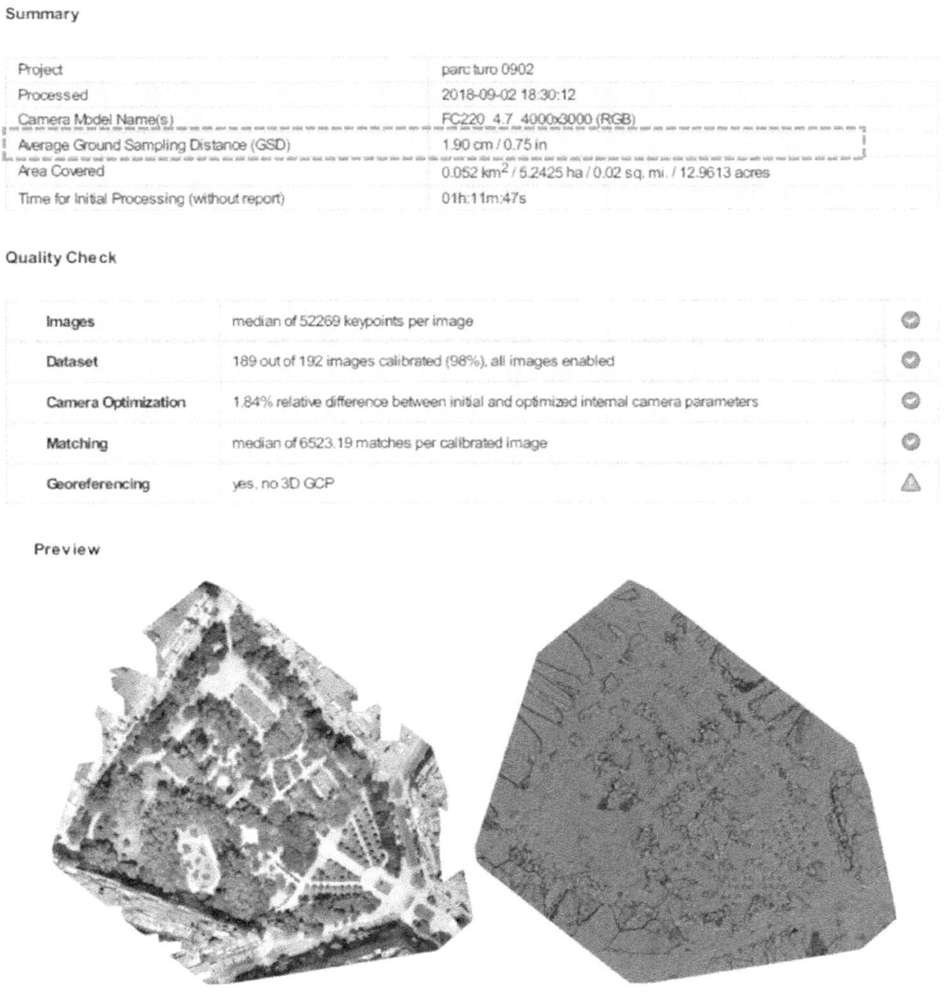

Summary

Project	parc turo 0902
Processed	2018-09-02 18:30:12
Camera Model Name(s)	FC220_4.7_4000x3000 (RGB)
Average Ground Sampling Distance (GSD)	1.90 cm / 0.75 in
Area Covered	0.052 km^2 / 5.2425 ha / 0.02 sq. mi. / 12.9613 acres
Time for Initial Processing (without report)	01h:11m:47s

Quality Check

Images	median of 52269 keypoints per image
Dataset	189 out of 192 images calibrated (98%), all images enabled
Camera Optimization	1.84% relative difference between initial and optimized internal camera parameters
Matching	median of 6523.19 matches per calibrated image
Georeferencing	yes, no 3D GCP

Preview

Pix4Dcapture software was used to develop the flight plan[5] (figure 5). The flight was carried out in summer 2018 in cloudy/sunny weather, with high visibility and wind below level 3, which is suitable for aerial photographs. The aerial photography process with the drone was carried out with manual take-off, according to the previously planned routes, automatic return, etc. The flight lasted 14 minutes, and a total of 194 images of the study area were obtained with an approximate resolution of 2.6 cm/pixel. The image quality was good and there was no obvious blurring.

With the help of park staff[6] and using the park's official website, information was obtained on the main tree species in the park in the study area: 20 main types of trees, of which the most common are Quercus ilex, Platanus hispánica and Tipuana tipu (see Table 1).

Once the field information had been analysed and the images obtained as described above, Drone2map software for Arcgis and SfM[7] technology were used to obtain the orthorectified mosaic image and the DEM of the study area. Figure 6 shows the results of processing in Drone2map. As can be seen, the dataset of 189 images was used to construct a mosaic with a ground sampling distance (GSD) of 1.9 cm per pixel, which means that the resolution of the image obtained was remarkably high.

Through ArcGis, the DEM was processed (Figure 7a) to obtain a slope model (Fig. 7b) and the corresponding orthomosaic (Fig. 7c), which were exported to eCognition for analysis and classification.

Figure 7. Models generated in ArcGis
Source: Compiled by the authors (Ma, 2018)

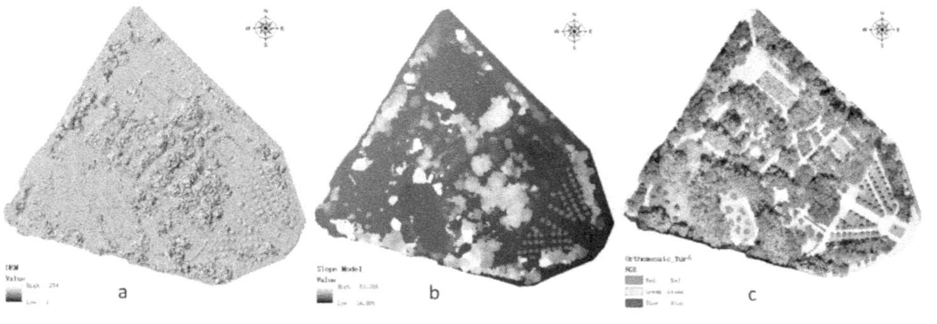

By means of the multiresolution segmentation algorithm in eCognition and using a process of adjustment of the parameters of scale, shape and compactness, the percentage of difference in the altitude of the elevation models could be highlighted to emphasise the limits of the trees. When these parameters were adjusted, the segmentation result shown in figure 8 was generated.

Once this initial segmentation had been obtained, a training sample was identified for the tree classification, based on information obtained in the "in situ" analysis. Figure 9 shows the training sample for the main species of trees in the study area: Schinus molle var. Areira, Quercus ilex, Robinia pseudoacacia, Cupressus sempervirens, Platanus hispánica, Tipuana tipu, Tilia tomentosa, Pinus pinea and Magnolia grandiflora.

Figure 8. Result of the segmentation
Source: compiled by the authors using eCognition

The final classification, once the results had been checked, enabled identification of the species of trees in the case study. Figure 10 shows this final classification.

Table 2 shows the good performance of the classification. The accuracy rate for the set of nine main species was 78%, and for six species it was above 85%. Due to the clear characteristics of the shape of the crown, Quercus ilex and Tilia were classified with high accuracy rates of 97% and 96%, respectively. In contrast, considerable errors were found for some species, such as Platanus hispanica. The main reason is that the image that was generated did not show sufficient information on the crown of this type of trees. The results for Magnolia grandiflora were not as good as expected. This is a small type of tree, with many trunks of high density, so dense that they are difficult to count.

Figure 9. Training sample
Source: Compiled by the authors

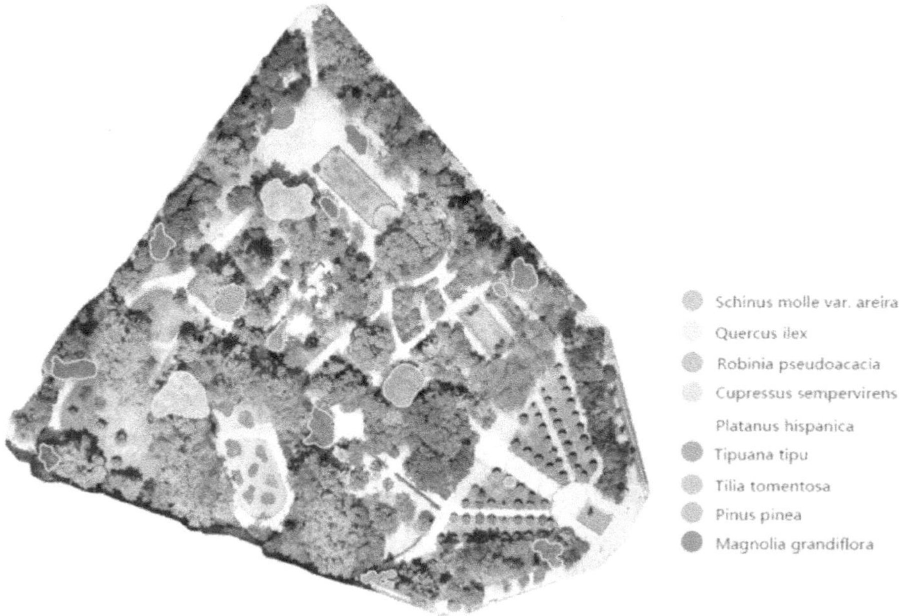

Figure 10. Final classification
Source: compiled by the authors

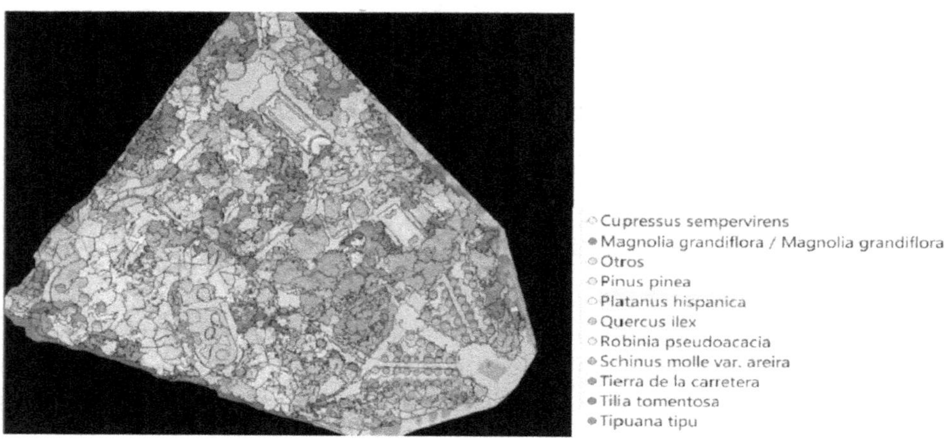

Table 2. Results of the classification

	Species	Real no.	Classification	Accuracy
1	Schinus molle var. areira	18	16	89%
2	Quercus ilex	95	92	97%
4	Robinia pseudoacacia	14	16	86%
6	Cupressus sempervirens	22	19	86%
9	Platanus hispanica	33	55	33%
14	Tipuana tipu	94	67	71%
15	Tilia tomentosa	28	27	96%
19	Pinus pinea	20	23	81%
20	Magnolia grandiflora	13	4	31%
	Total	**337**	**319**	**78%**

Source: compiled by the authors

CONCLUSION

The above results confirm that tree species can be identified and classified using high-definition aerial images obtained from a UAV. The study shows that a rigorous analysis of urban spaces with a high density of plant species can be undertaken after simply obtaining digital RGB bands. Obviously, the use of multispectral images would improve the accuracy of the analyses.

In the research method that was developed, there are two reasons why the results could be considered acceptable. The first is the use of an object-based classification methodology. The combination of visible spectral bands with the elevations and the appearance of the objects (the crowns of the trees) reveals regularities that this type of classification can identify as different tree species. Previous studies demonstrated that object-based classification is more accurate than pixel-based classification (Tarabalka et al., 2010). However, this methodology alone does not enable us to perfectly identify and classify each tree. Therefore, the segmentation method needs to be improved by suitably adjusting the segmentation parameters so that more accurate identification of the number and type of trees can be attained. The second reason for the good performance of the implemented model is based on the quality of the information obtained "in situ". The quality of this expert information is crucial to be able to select a training sample that can lead to suitable learning that culminates in the most accurate classification possible of the tree structure in the study area.

There are many uncertainties in the exercise that was undertaken, such as the adjustment of the multi-resolution segmentation parameters (parameters that are estimated subjectively and critically affect the classification results). However, the example confirms that conventional drones could be useful for studying green urban spaces characterized by a high density of plant species. Non-professional UAVs have a potential that should not be undervalued (Chadwick et al., 2020), as they enable three-dimensional point clouds to be obtained of high spatial density. Further study of 3D models obtained from oblique images of the ground and elevations, in different positions and angles, should enable through a combination of 2D and 3D analysis even more accurate results to be obtained than those in this study. A work combining both UAV LiDAR and photogrammetry should be considered, as LiDAR can provide a more accurate 3D canopy structure. In this case, the combination of techniques, such as obtaining multiple oblique images together with the use of LIDAR, will allow the construction of more detailed 3D models, which will help to better delimit the tree canopy, its LAI, as well as the identification of plant species.

As we said, the use of multispectral images would improve the precision of the analyzes. The analysis of the spectral bands of the near and middle infrared, together with the conventional RGB bands, can serve to better understand the health of the trees, by obtaining the traditional indices such as NDVI, LAI and fapar.

Furthermore, given that the form of the tree canopy varies over the year and the colour of the leaves varies continuously, the exercise of identifying and classifying trees would undoubtedly improve if the data collection via UAV could be repeated at different times of year. The introduction of a seasonal variable would doubtless help to increase the efficiency of the classification.

It should be noted the crucial contribution that drones can make for the implementation of urban microclimatic studies. To further enhance the ability to simulate surface heat budgets of land surface models and improve the land surface parameterization scheme of climate and meteorological models, work combining (through the use of UAVs) LiDAR and multispectral photogrammetry should be considered. This combination of technologies can provide a more accurate knowledge of 3D tree canopy structure, which would allow more adequate modeling of land temperature (LST), a greater knowledge of the leaf surface of plants, and therefore of their function of photosynthesis and of sensible and latent heat transfer to air temperature (LSAT). Combined data from LIDAR and spectral bands can help to model wind circulation, ground shading, plant photosynthesis activity, evapotranspiration, sensible and latent heat flux, as well as the temperature regulation function made by plants and trees.

In this study, we have presented a workflow for the automatic identification of individual tree crowns by the use of true colour orthomosaics obtained thanks UAVs conventional RGB images. Conventional drones allow a fast and detailed obtaining of accurate 3D models of vegetation, through the construction of digital elevation models (DEMs). The segmentation of the 3D information obtained by drones serves to individually identify existing trees in urban areas with dense vegetation.

In addition, the semi-automatic classification (through training samples obtained thanks to the collaboration of biology experts) allows to identify the arboreal species with a high degree of accuracy.

Therefore, the methodology designed in this paper helps to improve stakeholder information regarding the existing vegetation in urban parks with dense vegetation. Collaboration with the public managers of the parks, as well as with experts in plant biology, allows the construction of intelligent models for the development of inventories of urban biodiversity. Experiences such as the Biodiversity Atlas developed by the Barcelona City Council can be replicated in a generalized way through the information provided by drones in other localities with fewer technical resources.

The information obtained through conventional UAVs can complement, quickly and cheaply, that obtained through free satellite images (Landsat, Sentinel, ...), without the need to spend money and time in obtaining images from private satellites or in programming and conducting air flights. The agility, versatility and economy of UAVs allow them to compete with an advantage over other means of remote sensing. Its higher spatial and temporal resolution makes it possible to obtain more detailed information on the vegetation than that resulting from satellite images. This more detailed information is crucial for the development of biodiversity inventories in dense urban areas. And this is possible with the use of conventional drones!!

Therefore, we can conclude that the low cost and ease of use of commercial drones enable an in-depth analysis of green urban spaces of high tree density and a regular review of the health of the species that form urban parks.

ACKNOWLEDGMENT

The authors would like to thank Weizhao Ma for his collaboration in this study. Ma was a student of ours on the master's degree in Advanced Studies in Architecture, in the specialization in Urban and Architectural Management and Valuation. His master's thesis was on the capacity of drones to monitor green urban spaces (Ma, 2018).

REFERENCES

Anderson, K., & Gaston, K. J. (2013). Lightweight unmanned aerial vehicles will revolutionize spatial ecology. *Frontiers in Ecology and the Environment*, *11*(3), 138–146. doi:10.1890/120150

Arellano, B., Roca, J., & Batlle, E. (2018, September 18). Green areas and urban heat island: Combining remote sensed data with ground observations. *Proc. SPIE 10767. Remote Sensing and Modeling of Ecosystems for Sustainability, 15*, 1076705.

Asner, G. P., Braswell, B. H., Schimel, D. S., & Wessman, C. A. (1998). Ecological research needs from multiangle remote sensing data. *Remote Sensing of Environment, 63*(2), 155–165. doi:10.1016/S0034-4257(97)00139-9

Birth, G., & McVey, G. (1968). Measuring the color of growing turf with a reflectance spectroradiometer. *Agronomy Journal, 60*(6), 640–643. doi:10.2134/agronj1968.00021962006000060016x

Carlson, T. N., & Ripley, D. A. (1997). On the Relation between NDVI, Fractional Vegetation Cover, and Leaf Area Index. *Remote Sensing of Environment, 62*(3), 241–252. doi:10.1016/S0034-4257(97)00104-1

Carrus, G., Scopelliti, M., Lafortezza, R., Colangelo, G., Ferrini, F., Salbitano, F., & Sanesi, G. (2015). Go greener, feel better? The positive effects of biodiversity on the well-being of individuals visiting urban and peri-urban green areas. *Landscape and Urban Planning, 134*, 221–228. doi:10.1016/j.landurbplan.2014.10.022

Chadwick, A. J., Goodbody, T. R. H., Coops, N. C., Hervieux, A., Bater, C. W., Martens, L. A., White, B., & Röeser, D. (2020). Automatic Delineation and Height Measurement of Regenerating Conifer Crowns under Leaf-O Conditions Using UAV Imagery. *Remote Sensing, 12*(24), 4104. doi:10.3390/rs12244104

Chang, M., Zhu, S., Cao, J., Chen, B., Zhang, Q., Chen, W., Jia, S., Krishnan, P., & Wang, X. (2020). Improvement and Impacts of Forest Canopy Parameters on Noah-MP Land Surface Model from UAV-Based Photogrammetry. *Remote Sensing, 12*(24), 4120. doi:10.3390/rs12244120

Chen, J. M., & Black, T. A. (1992). Defining leaf area index for non-flat leaves. *Plant, Cell & Environment, 45*(4), 421–429. doi:10.1111/j.1365-3040.1992.tb00992.x

Chiesura, A. (2004). The role of urban parks for the sustainable city. *Landscape and Urban Planning, 68*(1), 129–138. doi:10.1016/j.landurbplan.2003.08.003

Colaninno, N. (2016). *Semi-automatic land cover classification and urban modelling based on morphological features. Remote sensing, geographical information systems, and urban morphology: defining models of land occupation along the Mediterranean side of Spain* (doctoral thesis). Technical University of Catalonia. Barcelona. Upccommons. https://upcommons.upc.edu/handle/2117/96372

Cruz Josa, D., Sacristán, M. A., & Mier Gutiérrez, S. (2017). Sistemas LIDAR embarcados en Drones - TerraSystem LidarPod. In L. A. Ruiz, J. Estornell, & M. Erena (Eds.), *Teledetección. Nuevas plataformas y sensores aplicados a la gestión del agua, la agricultura y el medio ambiente*. UPV.

Cruzan, M. B., Weinstein, B. G., Grasty, M. R., Kohrn, B. F., Hendrickson, E. C., Arredondo, T. M., & Thompson, P. G. (2016). Small Unmanned Aerial Vehicles (Micro-Uavs, Drones) in Plant Ecology. *Applications in Plant Sciences*, *4*(9), 1600041. doi:10.3732/apps.1600041 PMID:27672518

Dallimer, M., Tang, Z., Bibby, P. R., Brindley, P., Gaston, K. J., & Davies, Z. G. (2011). Temporal changes in greenspace in a highly urbanized region. *Biology Letters*, *7*(5), 763–766. doi:10.1098/rsbl.2011.0025 PMID:21429910

Dobbs, C., Kendal, D., & Nitschke, C. R. (2014). Multiple ecosystem services and disservices of the urban forest establishing their connections with landscape structure and sociodemographics. *Ecological Indicators*, *43*, 44–55. doi:10.1016/j.ecolind.2014.02.007

Garcia-Haro, A., & Arellano, B. (2018). Isla de frío de los parques urbanos de Barcelona. Estudio de caso del Turó Parc y el Parc del centre del Poblenou. *Proceedings, CTV 2018. XII Congreso Internacional Ciudad y Territorio Virtual. CPSV, 2018.*

GCOS. (2011). *Systematic Observation Requirements for Satellite-Based Products for Climate, 2011 Update, Supplemental Details to the Satellite-Based Com-ponent of the Implementation Plan for the Global Observing System for Climate in Support of the UNFCCC (2010 Update).* World Meteorological Organization (WMO).

Getzin, S., Wiegand, K., & Schöning, I. (2012). Assessing biodiversity in forests using very high-resolution images and unmanned aerial vehicles. *Methods in Ecology and Evolution*, *3*(2), 397–404. doi:10.1111/j.2041-210X.2011.00158.x

Harnik, P., & Crompton, J. L. (2014). Measuring the total economic value of a park system to a community. *Managing Leisure*, *19*(3), 188–211. doi:10.1080/13606719.2014.885713

Huete, A. (1988). A soil-adjusted vegetation index (SAVI). *Remote Sensing of Environment*, *25*(3), 295–309. doi:10.1016/0034-4257(88)90106-X

Huete, A., Liu, H., Batchily, K., & Leeuwen, W. (1997). A Comparison of Vegetation Indices Over a Global Set of TM Images for EOS-MODIS. *Remote Sensing of Environment*, *59*(3), 440–451. doi:10.1016/S0034-4257(96)00112-5

Kaufman, Y. J., & Tanré, D. (1992). Atmospherically resistant vegetation index (ARVI) for EOS-MODIS. *IEEE Transactions on Geoscience and Remote Sensing*, *30*(2), 261–270. doi:10.1109/36.134076

Kuo, F. E., & Sullivan, W. C. (2001). Environment and crime in the inner city: Does vegetation reduce crime? *Environment and Behavior*, *33*(3), 343–367. doi:10.1177/0013916501333002

Larondelle, N., Haase, D., & Kabisch, N. (2014). Mapping the diversity of regulating ecosystem services in European cities. *Global Environmental Change*, *26*, 119–129. doi:10.1016/j.gloenvcha.2014.04.008

Lee, A. C., & Maheswaran, R. (2011). The health benefits of urban green spaces: A review of the evidence. *Journal of Public Health*, *33*(2), 212–222. doi:10.1093/pubmed/fdq068 PMID:20833671

Loarie, S. R., Joppa, L. N., & Pimm, S. L. (2007). Satellites miss environmental priorities. *Trends in Ecology & Evolution*, *22*(12), 630–632. doi:10.1016/j.tree.2007.08.018 PMID:17996978

Ma, W. (2018). *Aplicación de los UAV convencionales para la identificación y clasificación de los espacios verdes densos. Estudio de caso de la Parc Turó-Barcelona* (Master's thesis). Technical University of Catalonia. Upccommons. https://upcommons.upc.edu/handle/2117/330117

Marselle, M., Irvine, K., Lorenzo-Arribas, A., & Warber, S. (2015). Moving beyond green: Exploring the relationship of environment type and indicators of perceived environmental quality on emotional well-being following group walks. *International Journal of Environmental Research and Public Health*, *12*(1), 106–130. doi:10.3390/ijerph120100106 PMID:25546275

Nieuwenhuijsen, M. J., Khreis, H., Triguero-Mas, M., Gascon, M., & Dadvand, P. (2017). Fifty shades of green. *Epidemiology (Cambridge, Mass.)*, *28*(1), 63–71. doi:10.1097/EDE.0000000000000549 PMID:27525811

Norton, B. A., Coutts, A. M., Livesley, S. J., Harris, R. J., Hunter, A. M., & Williams, N. S. (2015). Planning for cooler cities: A framework to prioritise green infrastructure to mitigate high temperatures in urban landscapes. *Landscape and Urban Planning*, *134*, 127–138. doi:10.1016/j.landurbplan.2014.10.018

Pan, L., Gu, L., Ren, R., & Yang, S. (2020, August 20). Land cover classification based on machine learning using UAV multispectral images. *Proc. SPIE 11501. Earth Observing Systems*, *25*, 115011F.

Pauleit, S., Ennos, R., & Golding, Y. (2005). Modeling the environmental impacts of urban land use and land cover change—A study in Merseyside, UK. *Landscape and Urban Planning*, *71*(2-4), 295–310. doi:10.1016/S0169-2046(04)00083-0

Roca, J. (1988). *La estructura de los valores urbanos: análisis teórico empírico*. Instituto de Estudios de Administración Local.

Rouse, J. J., Haas, R., Deering, D., Schell, J., & Harlan, J. (1974). Monitoring the Vernal Advancement and Retrogradation (Green Wave Effect) of Natural Vegetation. NASA/GSFC Type III Final Report, Greenbelt, MD.

Shrestha, R., Flacke, J., Martinez, J., & Van Maarseveen, M. (2016). Environmental health related socio-spatial inequalities: Identifying "hotspots" of environmental burdens and social vulnerability. *International Journal of Environmental Research and Public Health*, *13*(7), 691. doi:10.3390/ijerph13070691 PMID:27409625

Takayama, N., Korpela, K., Lee, J., Morikawa, T., Tsunetsugu, Y., Park, B. J., & Kagawa, T. (2014). Emotional, restorative and vitalizing effects of forest and urban environments at four sites in Japan. *International Journal of Environmental Research and Public Health*, *11*(7), 7207–7230. doi:10.3390/ijerph110707207 PMID:25029496

Tao, X., Liang, S., Hea, T., & Jina, H. (2016). Estimation of fraction of absorbed photosynthetically active radiation from multiple satellite data: Model development and validation. *Remote Sensing of Environment*, *184*, 539–557. doi:10.1016/j.rse.2016.07.036

Tarabalka, Y., Chanussot, J., & Benediktsson, J. A. (2010). Segmentation and classification of hyperspectral images using watershed transformation. *Pattern Recognition*, *43*(7), 2367–2379. doi:10.1016/j.patcog.2010.01.016

Valor, E., & Caselles, V. (1996, September). Mapping land surface emissivity from NDVI: Application to European, African, and South American areas. *Remote Sensing of Environment*, *57*(3), 167–184. doi:10.1016/0034-4257(96)00039-9

van den Bosch, M., & Nieuwenhuijsen, M. (2017). No time to lose-Green the cities now. *Environment International*, *99*, 343–350. doi:10.1016/j.envint.2016.11.025 PMID:27923587

Weier, J., & Herring, D. (2000, August 30). *Measuring Vegetation (NDVI and EVI)*. Retrieved 2013, from Earth Observatory: https://earthobservatory.nasa.gov/Features/MeasuringVegetation/

Zheng, G., & Moskal, L. M. (2007). Retrieving Leaf Area Index (LAI) Using Remote Sensing: Theories, Methods and Sensors. *Sensors (Basel)*, *9*(4), 2719–2745. doi:10.339090402719 PMID:22574042

KEY TERMS AND DEFINITIONS

Accuracy: The degree to which a measured value conforms to true or accepted values. Accuracy is a measure of correctness. It is distinguished from precision, which measures exactness.

DEM: Acronym for digital elevation model. The representation of continuous elevation values over a topographic surface by a regular array of z-values, referenced to a common datum. DEMs are typically used to represent terrain relief.

Drone2Map: Drone2Map for ArcGIS is a desktop application that turns raw still imagery from drones into information products in ArcGIS.

Image Segmentation: Segmentation is the process of portioning a remote sensing mage into non-overlapping regions or segments.

Land Cover: The classification of land according to the vegetation or material that covers most of its surface; for example, pine forest, grassland, ice, water, or sand.

NDVI: Acronym for Normalized Difference Vegetation Index. Introduced by Rouse et al. (1974) is obtained from NIR-RED/NIR+RED and provides a measurement of the amount of green vegetation present in a pixel.

Object-Based Classification: While pixel-based classification is based solely on the spectral information in each pixel, object-based classification uses both spectral and spatial information for classification. The process involves categorization of pixels based on their spectral characteristics, shape, texture, and spatial relationship with the surrounding pixels.

Spatial Resolution: Refers to the size of the smallest feature that can be detected by a satellite sensor or displayed in a satellite image (pixel size). The dimensions on the ground of a single cell in a raster, measured in map units.

Spectral Bands: Refers to the bands across the electromagnetic spectrum. A conventional camera captures light across three wavelength bands in the visible spectrum: red, green, and blue (RGB). Spectral imaging encompasses a wide variety of wavelength bands that go beyond RGB. Spectral imaging may use the infrared, the visible spectrum, the ultraviolet, x-rays, etc.

Turó Parc: Park located in Barcelona. It was constructed in 1934 and has a surface area of 2.88 Ha.

ENDNOTES

1 The inventory of trees in the city quantifies the number of trees, available in GIS format in open data, as well as the species that can be found in Barcelona. However, this information is limited to trees on public roads. There is no general information base that includes trees in urban parks, where the greater density of vegetation makes the work of inventorying species difficult. In 2018 (https://www.bcn.cat/estadistica/catala/dades/anuari/cap01/C0101100.htm), in the municipality of Barcelona, there were 156,477 trees on the streets, and 84,779 in urban parks (this figure does not include the trees in the protected natural space of Collserola, which would increase the city's tree "capital" to 1.4 million.

2 Raw data obtained by UAV usually consist of hundreds of digital images acquired in different lighting conditions and at varying flight altitudes. The preprocessing of UAV imagery aims to orthorectify each image and mosaic them into a single one. This is a difficult task because raw images suffer from lighting variation, platform sway and sensor displacement.

3 The measuring system is based on a UAV flight platform, with the help of a camera sensor as a device to obtain images of the ground. It incorporates high range technologies such as automated flight, remote sensing science and GPS navigation. The aerial photography system of conventional drones is mainly comprised of a flight platform, a flight control system, a ground monitoring system, mission equipment (sensor), a data transmission system and a transmission and recovery system.

4 The drone is comprised of a flight system with remote sensing control, cameras with a cloud platform and mobile devices that support this use (for example, an iPad). The drone weighs 743 grams. It is equipped with an image stabilization system that can compensate for the intense movement of the aircraft and the control accuracy is ±0.03 degrees. The flight system is integrated into the body of the drone, with dual inertial measurement units (IMU) and compass systems incorporated in the front part of the drone. It has a joystick and the operating frequency is 2.400 ~ 2.483 GHz. The drone can be interconnected with the mobile device to control the camera and the aircraft. The flight can also be planned. DJI Lightbridge, a high-definition image transmission system, is incorporated; the image transmission and remote sensing control can reach distances of 7 km. The UAV has a system to detect obstacles in front of it and can take the initiative to avoid obstacles during flight. Mavic Pro can receive GPS and GLONASS satellite positioning signals simultaneously to obtain image position information with an accuracy of 5 m. The integrated camera in Mavic Pro is a low-cost camera. The focal length is 26 mm (35 mm in equivalent

format), with a CMOS image sensor and a resolution of 12 million pixels. It can capture visible spectral information (RGB). To adapt better to the antenna, the manufacturer has optimised the camera lens has with a 78.8° angle of view, f/2.2 aperture of the lens with ultra-low distortion, and a clear image. Mavic Pro has a high-energy-density battery and a highly efficient smart flight power system. The maximum flight speed is 20 m/s. The maximum battery duration, for a single battery, is approximately 30 minutes.

5 Pix4Dcapture is excellent software for planning UAV routes. It can be used with a wide range of UAV models, including the DJI series. It is only necessary to determine the flight range on the base map provided by the software before the flight. When parameters such as flight altitude, maximum flight speed and image overlay are configured, the software will automatically generate the route, the estimated flight time, and the number of aerial images. During the flight, the software can show the position, altitude, speed, camera parameters and information such as the wireless signal strength, which are useful for users to control the aerial photography process. The design has an altitude of 60 m, the vertical overlay of the image is 80% and the lateral overlay is 75% (Figure 5).

6 The authors would like to thank Esperanza Huerta for her assistance.

7 SfM technology, based on multi-view stereo photography, can effectively reconstruct the 3D ground from images with many angles of view and simultaneously calculate the camera parameters of images (Smith et al., 2015). It provides an efficient solution for processing images from drones given that it does not require the precise position of the photograph, the scale of the image and the focal length of the shot. Hence, it has become the main method for processing images obtained using UAV.

Chapter 13
Counter-Mapping Practices in Land Regularization Through Geoinformation Technologies:
The ÉPURA Group Experience in Cuiabá, Brazil

Andrea Figueiredo Arruda
Faculdade de Arquitetura, Universidade Federal de Mato Grosso, Brazil

Cristina Delgado Henriques
https://orcid.org/0000-0003-0100-5586
CIAUD, Faculdade de Arquitetura, Universidade de Lisboa, Portugal

Claudio Santos de Miranda
Faculdade de Arquitetura, Universidade Federal de Mato Grosso, Brazil

ABSTRACT

This chapter outlines the experience of the Épura Research and Extension Group of the Federal University of Mato Grosso for the identification, characterization, and demarcation of informal settlements in Cuiabá (Brazil). This involved cooperating with the local residents' associations and public authorities in recognizing local demands and the need to open up urban land regularization processes. The cartographic issue is central to the recognition and legitimization of these territories and the use of geoinformation technologies, namely geographic information systems and unmanned aerial vehicles, can increase the speed and accuracy of the mapping of the neighborhoods. Additionally, the demands for urban and housing improvements and provision of space for everyday activities can be highlighted through the visualization and measuring tools that are embodied in these technologies. Furthermore, the experience requires the university to play a role in seeking more participatory and proactive practices within urban planning processes.

DOI: 10.4018/978-1-7998-2249-3.ch013

INTRODUCTION

The right to an adequate standard of living, including housing was recognized by The Universal Declaration of Human Rights as early as 1948. The new global agenda of the millennium on Sustainable Development Goals reaffirms the need to ensure access for all to adequate, safe and affordable housing (United Nations, 2015).

The concept of adequate housing according to the United Nations Human Settlements Programme (2014) is based on the principle that it must provide "more than four walls and a roof", and give an assurance that it can meet the following criteria: legal security of tenure, affordability, habitability, availability of essential services, appropriate building materials, basic facilities and infrastructure, accessibility, a suitable location, and cultural identity. Although the right to housing is recognized as a constitutional right in many countries, there are still millions of families living in informal settlements, slums, inadequate and precarious forms of habitat. These self-produced territories are consolidated as the possible place to live, in light of the global demographic growth.

With regard to effective measures taken for land regularization, there is a consensus on the need for continuous efforts to identify, measure, characterize, and demarcate informal settlements so that they can be incorporated in an ¨intervention¨ agenda as a part of a broader public policy for housing, although there are still many challenges at the administrative, technical and political levels that need to be overcome. With more or less success, some experiences have already been taken to tackle this situation, mainly through the use of new technologies for surveying and land registration processes, to determine the priority areas.

This chapter presents the experience and practice of the Research Group on Urban and Regional Planning (Épura), at the Department of Architecture and Urbanism, Federal University of Mato Grosso. It combines a set of initiatives with the aim of giving visibility to informal settlements, by identifying and characterizing these territories, and comparing the local demarcations with those of the official cartography of the municipality. The purpose of this is to put on the agenda their recognition and address people´s urban demands.

Attention is drawn here to attempts to produce participatory mapping in three informal settlements within the municipality of Cuiabá (capital of the state of Mato Grosso, Brazil), by making use of geoinformation technologies. The methodology carried out to achieve this goal consists of identifying informal settlements and confronting them with data and demarcations of official records (municipal and federal) through the analysis and manipulation of georeferenced data in Geographic Information System (GIS). In a second stage, the boundaries of the urban demarcation and the households of the settlements are ascertained from Unmanned Aerial Vehicles (UAVs) images.

The experience described in this chapter is aimed at filling the huge institutional and administrative gap in policymaking, especially with regard to the practice of identification and demarcation of settlements or informal urban areas in the municipality of Cuiabá. The results of its application in several case studies (Getúlio Vargas, Salim Felício, and Vila Verde), as well as making an evaluation of its feasibility in other contexts of delimitation of informal settlements, are also presented here.

Background

Informal settlements provide shelter for a large number of people in the world and, although they are built of different materials and have different spatial conditions, they embody an extensive allocation of household resources and host a thriving social life and activities. Our understanding of informal settlements follows Dovey et al. (in press) as being "Self-organized modes of production through which the urban poor produce affordable housing and urban infrastructures". Lutzoni (2016) adds that "A very wide range of situations may be included, like spontaneous processes of occupation of the territory, absence of property titles, self-building of houses, illegal inhabiting in contexts with rapid urbanization, temporary uses of space, forms of self-organization and development of urban areas at city edges, etc.". In Brazil there are many expressions that designate this type of popular settlements in a city, both at the academic and legislative level (federal and municipal) which introduces a serious obstacle to endowing these territorial areas with legitimacy and granting them recognition, both in fact and in law.

Informal settlements encompass a vast morphological diversity of land, and housing dimensions, which are revealed in wide-ranging population densities, urbanization patterns, degrees of infrastructure consolidation, construction features, and land ownership issues.

Land tenure can be understood as "the mode by which land is held or owned, or the set of relationships among people concerning land or its product" (Payne, 2001: 416). The problem of land tenure irregularity results from the lack of property rights, which leads to insecurity of tenure for the occupants of the informal settlements and hence the constraints of social exclusion and a lack of urban rights. In fact, deprivation of property is a key issue to understand the urbanization pattern of cities in developing countries and, above all, the exclusion and the socio-spatial segregation process that accompanies it. The classification of irregular concerning land tenure is given because it is in conflict with the legal framework. When there are failings in the regularization process, owing to the absence of property deeds, the insecurity of ownership means the residents suffer from the stigma of being "illegal" and "clandestine".

A sustainable approach to overcome problems concerning the irregularity of land tenure and urban and housing inadequacies of informal settlements involves, in the first place, determining where they exist and then putting them into official documents and maps. Being included in a map means being part of a certain city, municipality, or locality, and is a first step to demanding basic infrastructural facilities and services and thus a means of instigating the process of obtaining recognition of land tenure. Several frameworks for achieving this goal have been established either by national and international agencies, or by non-governmental organizations and research groups at universities. Map Kibera, in Nairobi, Kenya, is probably the most widely cited example of where "the residents could use localized information to influence policy and development" (Hagen, 2011:70), but there are also some "counter-mapping" experiences that seek to "power dynamics in the necessary processes of collaboration" (Radjawali et al., 2017) and empower "invisible" communities.

Although the United Nations has defined the criteria for identifying situations of inadequacy or urban informality (United Nations Human Settlements Programme, 2008), there is still a need to measure and demarcate these settlements within urban areas and as a result, find out which social groups require more attention in the formulation of urban and social public policies.

Raposo (2017) refers to the emergence of professional practices and academic training based on alternative, collaborative, and sustainable processes and underlines the role played by academia in the "emancipation of the present," through the production of another space that, in the wake of Lefebvre (1974), results from the active intervention of the participants, by means of a "counter-project" or "counter-plan.".

In this context, and in line with the social demands arising from the uneven process of urbanization in Brazil in the 1980s/90s, movements and practices are emerging, which regard urbanization and the regularization of informal settlements as giving prerogative powers for intervention. By having a closer look at these territories, and thus being able to face the problem *in loco*, these movements seek to guarantee the social and urban rights of the residents. Having as precursors Turner (1977) and Santos (1981), among others, these practices took place throughout the 1990s and increased in the 2000s, by establishing a more interactive relationship with popular movements, in a participatory and counter-hegemonic way (Martins, 1997; Lopes & Rizek, 2006; Lopes, 2011; Raposo, 2017). The Pró-Favela movement in Belo Horizonte (1983), PREZEIS (Regularization Plan for Special Areas of Social Interest) in Recife (1987), Favela Bairro Program in Rio de Janeiro (1994), and Bairro Legal Program in São Paulo (2001) are some examples of local initiatives that either ran slum urbanization programs, projects, and actions or sought to improve the informal settlements.

From an urban perspective, the irregularity of these settlements originates from the modern matrices of urban planning and the legislation that governs them, especially when they do not meet the legal requirements regarding the subdivision, use, and occupation of urban land, such as compliance with the urban land parceling rules, to standards established on the road network, the size of the plots, or the requirements of urban infrastructure. These are very common situations in settlements with a high density of residents. Maricato (2003:153) highlights the fact that "detailed modernist patterns of construction and land occupation, present in zoning laws, construction codes, land parceling laws, among others, live together with the huge illegal city, where the contravention is the rule.". Being illegal, these settlements largely end up by not being included in the official data or not being represented on official maps.

However, some significant experiments have been designed, that seek to overcome the shortcomings of conventional cartography, by incorporating new ways of thinking about, representing and intervening in the informal settlements, in order to recognize the rights of the inhabitants to remain in the occupied territory. Counter-mapping practices have proved to be an alternative to the official mapping process (Radjawali et al., 2017); these allow the territory to can be identified by the social group itself, and can be an important claim instrument for its residents. Acselrad (2010) refers to several networks and communities that have combined to design a new method of cartographic representation that can act as a participatory system that challenges conventional cartography. By including the local community in the practices of social cartography, the experiences "show that new issues are brought into play in cartographic disputes that contribute to territorialize spatial descriptions" (Acselrad, 2010: 12). Panek & Netek (2019:1) advocate that "Collaborative mapping and community engagement became new trends in the development discourse that allow local communities to become part of the power structures and influence what is mapped and what/who is on the map".

These practices can also be enhanced by means of new technologies to survey spatial information, whether this information is of a social, ethnic, political, morphological, or land-based nature. With regard to useful information about informal settlements, the use of new technologies has proved to be effective in the speed of production of planimetry and altimetry, and also for the analysis of environmentally- sensitive areas and risk situations, such as natural hazards or landslides. Interesting work has also been carried out that provides details of the pathologies and opportunities for building adaptations of units in slums, as shown by Sliuzas et al. (2018) and Gomes &Pedrassoli (2018).

Innovative technological advances within the field of land regularization, namely those that resort to the use of Geographic Information Systems (GIS), Remote Sensing, and Global Positioning System (GPS), are becoming more prevalent in Brazil. Several projects that are involved in the conflicts issues concerning the registry of land, make use of these technologies either by mapping unmapped areas or by gathering information in a georeferenced database (Freitas & Lima, 2012; Perez et al., 2017). With regard to the information that Remote Sensing can provide, satellite or manned aerial imagery are the most common products used in land regularization processes. Google Earth is very popular as it provides free access to high resolution images. However, when the community is engaged in a land regularization process, as well as an aerial image of the site, 3D modelling and visualization are a powerful means for achieving success. Low-cost technology and data quickly obtained by unmanned aerial vehicles (UAVs) offer many more advantages than those obtained by traditional means. As Stöcker et al. (2018) points out, "UAVs are able to bridge the gap between time-consuming but high accuracy field surveys and the timeless fashion of classical aerial surveys." In addition, very high-resolution aerial and expressive videos can assist by ensuring that everyone in community meetings and discussions (Radjawali et al., 2017) can enhance the quality of the field demarcations for land regularization actions.

MAPPING INFORMAL SETTLEMENTS: FROM "INVISIBILITY" TO THE OPPORTUNITY FOR THE RECOGNITION OF URBAN AND HOUSING RIGHTS IN CUIABÁ

Concepts and Legal Framework

In Brazil, informal settlements have been the object of attention in urban policymaking since the Federal Constitution of 1988. Law No. 1.0257/2001, also called the "Statute of the City", regulates the Articles 182 and 183 of the Constitution, and sets out the general guidelines for urban policies, including the guarantee of "the right to sustainable cities", and incorporating the right to urban land, housing and "land tenure regularization and urbanization of areas occupied by low-income populations" (Brasil, 2001). Articles 182 and 183 defined the legal basis for the social functions of the city and the recognition of informal settlements as special areas, or special zones of social interest (ZEIS). This was considered to be a priority for public funding and involved policies resulting from urban and land regularization.

Since then, there has been a considerable effort to identify precarious urban housing and the demarcation of ZEIS in Brazilian municipalities, especially in more progressive municipal administrations, such as those shown in the studies by Denaldi et al. (2013) and Rolnik & Santoro (2013). The demarcation of ZEIS in the master plans and the municipal land use laws has been a key legal and urban instrument for the management of urban land. It also suggests that within these areas, special patterns of urbanization, land use and occupation of land can be adopted, by recognizing the "existing occupation and foreseeing the elaboration of plans with direct participation of the residents to define their future configuration, planning, and management." (Rolnik & Santoro, 2013, p. 1).

However, in general terms, it is recognized that there are still significant institutional and administrative obstacles with regard to the measuring, characterization and mapping of the real requirements of these territories, and a need to formulate effective policies, plans, programs and actions.

In alignment with the general principles of urban policymaking, established by the Federal Constitution and the Statute of the City, the National Housing Policy (PNH) provides a deeper understanding of the social function of property and framing the right to housing, by citing urban integration as one of its components, as well as urbanization and the need to regularize the land tenure of precarious settlements. These last were defined as "sectors of urban territory of varying sizes and typologies, which have the following in common: predominantly residential areas inhabited by low-income families; the precariousness of housing conditions, characterized by numerous shortages and inadequacies; [...] the same historical origins, based on strategies used by low-income groups to find an autonomous solution for their housing needs." (Brasil, 2010: 9).

Following the publication of the PNH, Law No. 11.977/2009 was a more forward-looking Act in Brazilian urban legislation. As well as devoting a whole chapter to the subject, it interprets the potential for Land Regularization on the basis of a set of legal, urban, environmental and social measures, by linking them to constitutional principles and to the notion of a citizen´s right to housing and to the amenities of the city (Brasil, 2009). The new legislation introduced key legal instruments such as urban demarcation, ownership legitimacy, and the requirements of the land regularization project, including the introduction of basic infrastructure and social facilities. In July 2017, Law No. 13.465/2017 (known as the *Reurb* Law) was enacted; this, repealed the previous law and provided a new definition of urban land regularization policies. It also integrated a set of measures that were designed to incorporate informal urban settlements in urban planning and giving its occupants title deeds (Brasil, 2017).

The *Reurb* Law disregards the resident´s income, origin, and nature of the occupation and values the aspect of land tenure irregularity, reinforcing the stigma of clandestine and marginalization that traditionally accompanies the formation of these settlements, affecting, therefor, its territorial recognition. Endorsing De Soto (2001), this Law has a strong focus on the title deeds to the detriment of its articulation with the urban, social, and environment dimensions. It deconstructs the understanding of the concept of full regularization, defined in the previous wording of the Law, which combined legal land tenure with urban regularization, and involved the implementation of urban infrastructure, basic facilities and the provision of public services. On the other hand, the Law lists the administrative procedures for the demarcation of an "informal urban nucleus" ("That was clandestine and irregular or in which it was not possible to grant title deeds to its occupants" (BRASIL, 2017, art. 11) and leads to new conditions for carrying out regularization processes, as well as requiring technical measures to be taken by the municipal administration.

In statistical terms, the Brazilian Institute of Statistical Geography (Instituto Brasileiro de Geografia e Estatística - IBGE) has estimated the number of households within the urban perimeter based on the concept of "subnormal agglomerate." This is defined as: "a set of at least 51 housing units (shacks, dwellings, etc.) occupying or having occupied, until a recent period, land owned by others (public or private) generally arranged in a disorderly and dense manner. Most of them lack essential public services" (Instituto Brasileiro de Geografia e Estatística, 2010).

At the municipal level, the Local Plan for Housing of Social Interest (PLHIS) was drawn up in 2013, on the basis of the National Housing Policy. During this period, and according to the last census, 51,057 people (14,000 households) were living in 9 subnormal agglomerates in Cuiabá. Furthermore, the João Pinheiro Foundation for the Housing Deficit found that there were 18,089 urban households in an inadequate condition in the municipality, occupied by families with a low income (Fundação João Pinheiro, 2013). However, the PLHIS of Cuiabá laid emphasis on the Quantitative Deficit, that is, the need to construct new housing units. It only identified seven settlements in the capital as being precarious, and these were defined as "partially urbanized settlements that need additional infrastructural features and simple urbanization, without the need for removal" (Município de Cuiabá, 2013b), although, in practice, it is evident that this demand is significantly higher. In the same year, an official document was issued, which compiled a list of neighbourhoods in the capital (Composition of Neighbourhoods). In this, the terms "informal settlement", ¨invasion¨ and "grilo [illegal possession]" were defined, and referred to the "occupation of an area of public or private property, predominantly for housing purposes, without the authorization of the owner, with or without access to public roads" (Município de Cuiabá, 2013a).

In addition, the Law of Land Use and Occupation (LUOS), enacted in 2015, characterized the "special zone of social interest" as a place of residence for low-income groups, living in precarious conditions, which required the special attention of the municipality. The Special Zone of Social Interest 1, which corresponds to a more global understanding of informal settlements, is defined as representing "irregular installations, irregular public or private housing estates, occupied by low-income groups that, owing to the way they have been established, are in need of partial or complete regularization" (Município de Cuiabá, 2015, art. 35 and 36).

Table 1 shows the contradictions in the legislation with regard to understanding and assessing informal settlements, both at a federal and municipal level.

Table 1. Summary of definitions of informal settlements at federal and municipal level

Institutions and devices		Term	Definition	Quant.
Federal Level – BRASIL	PNH/2004	Precarious settlement	Sections of urban land of various sizes and typologies which have the following in common: predominantly residential areas inhabited by low-income families; very poor living conditions; [...] historical origins, related to strategies adopted by the people in the low-income groups to find the means of adapting to their residential needs, in an autonomous manner	-
	Law 11.977/2009	Informal settlement	Settlements divided into informal or irregular lots, located in public or private urban areas and mainly used for residential purposes	-
	Law 13.465/2012	Informal urban nucleus	Clandestine or irregular dwellings or those where it was impossible to trace the title deeds of the occupants	-
Federal Level – Dados Estatísticos (para Cuiabá)	IBGE/Dados FJP (2010)	Subnormal Settlement	a set of at least 51 housing units (shacks, dwellings, etc.) occupying or having occupied, until a recent period, land owned by others (public or private) generally arranged in a disorderly and dense manner.	9 (in the case of Cuiabá)
	FJP (2013)	Housing inadequacy	Inadequacy due to the lack of infrastructure services (electricity, water supply, sewage and garbage collection); lack of exclusive toilet; excessive densification.	18,089 houses
Municipal Level – Cuiabá	Composição de Bairros	Informal settlement "*invasão*" or "*grilo*"	Occupation of a property in a public or private area predominantly for residential purposes without the authorization of the legal owner, with or without access to public roads	30
	PLHIS	Precarious settlement	Partly urbanized settlements that need additional infrastructural facilities and simple urbanization, without the need for removal	7
	LUOS	ZEIS	Irregular lots public or private irregular housing complexes, occupied by people from low-income groups	59

Being slums, informal occupations or irregular settlement, in all cases, there are situations where housing inadequacy, lack of a basic infrastructure, public services, and essential facilities coexist, to a greater or lesser degree. Apart from the definition, which undoubtedly is essential for public policymaking, little progress has been made in identifying and demarcating these settlements in urban territory. The formulation of policies, plans, programs, and projects of full land regularization to face of enormous demands, has met with institutional and administrative obstacles for effectively achieving their aims, especially at the municipal level.

Counter-Mapping in Getúlio Vargas, Salim Felício and Vila Verde

Informal settlements in Cuiabá are occupations in public areas reserved in formal allotments for institutional purposes, green areas, and, very often, in areas of permanent conservation, which is, evidence that they are environmentally-threatened communities, and sometimes the residents are at risk. Identifying and characterizing the inadequacies and irregular issues of these territories is essential to recognize the procedures required for their regularization.

In light of this, the Épura Research and Extension Group of the Federal University of Mato Grosso seeks, first, to understand how to establish a framework for understanding the "purpose" of the land regularization policy. This means it is s necessary to assess the definitions already formulated, by the federal authorities' normative tools (and approved in the last two decades), and also take them as a reference-point for the definitions and approaches at the municipal level, where possible measures in the area of policymaking can be territorialized.

Secondly, the identification and characterization of the informal settlements through an analysis of the official data on precariousness and irregularity in the municipality of Cuiabá was carried out, through an interpretation of the PLHIS, the register of Neighborhood Composition, and the LUOS of Cuiabá, respectively with 7, 30, and 52 demarcated locations. The definitions and demarcations of the subnormal agglomerates of IBGE were also taken into account, with 9 demarcated locations. Although the LUOS has a wider dimension of the ZEIS 1, comprising 52 settlements, the official demarcation does not correspond to the demarcation of the settlements recognized by the residents' associations, which means there is a need to redefine their perimeters.

The settlements show significant differences in relation to the perimeters presented in the official devices, either when partially or fully demarcated (Figure 1).

In addition to divergences over defining what the "object" of the land tenure regularization policy should be, both at the federal and municipal levels, another point that arises is the demarcation of these settlements within the urban area. This raises questions regarding their characterization, the measurement of households and families, socio-urban indices, and above all, their physical and territorial boundary-lines. Before the regularization procedures can move forward, the new law requires, for example, a report about the area needing regularization, which must include a list of residents, measurements of the perimeters and area, a representation of adjoining areas, coordinates, and a map overlay of the situation in the property registry (Brasil, 2017, art. 19). Later, when drawing up the land regularization project, there is also the need for a georeferenced planimetric and cadastral survey and to establish the perimeters of the informal urban area.

Visits were made to informal settlements of the localities of Getúlio Vargas, Salim Felício, and Vila Verde, which assisted in characterizing the land and building registers, their respective uses, and typological patterns, as well as the possible demands of the settlement, such as measures for improving housing or increasing the facilities. Interviews were also carried out with local leaders and families, to understand the history of the locality and the origins of the families to form a preliminary register of the profile of the residents (Figure 2).

The demarcation of the settlements was carried out by mapping the perimeters in accordance with the local conditions, as well using surveying geotechnologies, namely UAV and Real Time Kinematic (RTK). This task makes it possible to define the actual boundaries of the settlement and existing buildings, as well as any significant physical and environmental features in the area.

Figure 1. Official demarcation of informal settlements in Cuiabá
Source: Epura, 2020

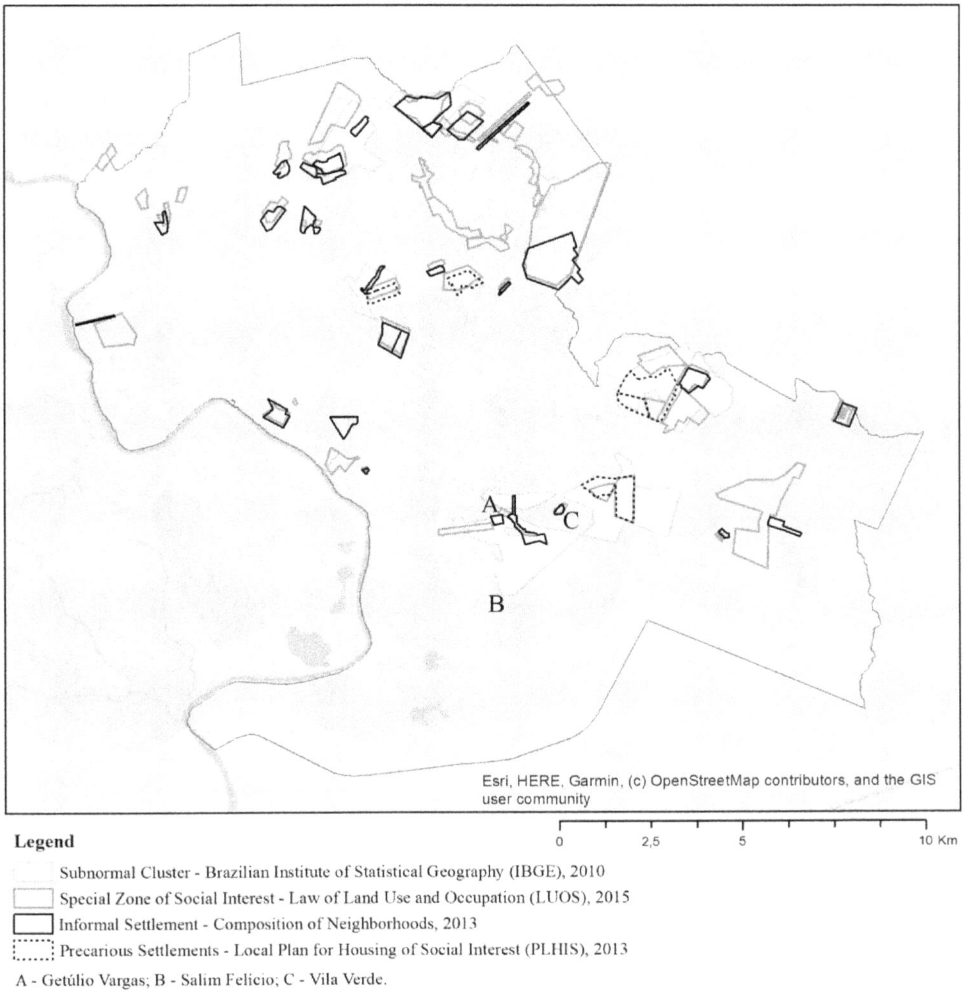

In terms of official demarcations, in Figure 1 we highlighted the demarcations of a "subnormal agglomerate" (IBGE, 2010), a "Special Zone of Social Interest" (Cuiabá, 2015), and "informal settlements" (Cuiabá, 2013a). The "precarious settlements" were not demarcated, because, although they are consolidated settlements, (especially in the case of Getúlio Vargas and Vila Verde), the Local Social Housing Plan (Cuiabá, 2013b) did not recognize them, which shows the limited scope of the official document in terms of identifying the requirements of all the neighborhoods in the capital. In addition, this was, the demarcation recognized by the Residents'

Association, which corresponded precisely with the demarcation of households that require land tenure regularization.

Figure 2. Field work activities with students and the Residents' Association – data collection with families (2a), identification of territorial boundaries and urban requirements (2b), a survey conducted with UAVs - Unmanned Aerial Vehicles, RPAs (Drone) and the setting of points with RTK (2c), identification of lots in the aerial image with students and residents, meeting with the Residents' Association (2d)
Source: Epura, 2020

The Getulio Vargas informal settlement comprises three different typologies called: Getúlio Vargas (the same name of the settlement), Presidente III, and Jardim Santa Fé. These represent three situations that differ, depending on factors related to locality, environment, time, and origin of occupation, but which are recognized collectively within a set (Figure 3). The Getúlio Vargas settlement has been occupied for a longer period, (approximately thirty years), which increases the complexity of reading the demarcations, and reveals how slow the municipal authority has been in effecting its regularization. Its first residents came from other locations in the city, such as Dom Aquino, Praeirinho, Canjica, and Novo Horizonte.

Getúlio Vargas, the oldest locality, is an informal occupation that is mainly located in an area of permanent conservation (Córregos do Machado and São Gonçalo). It also occupies the green area of the Residential Coxipó and the institutional subdivision of Jardim Presidente II, alongside a stream. Presidente III is also an informal occupation. However, in this case, there has been a splitting and irregular sale of the green area of the formal Jardim Presidente II subdivision. Jardim Santa Fé on the other hand, is an irregular subdivision in an area belonging to the Federal Government; this was previously subdivided adhered to standards that approximate to the requirements of the law governing the subdivision of urban land, although without prior registration by the competent bodies of the municipality, or notaries. As in the case of President III, there was an irregular sale of lots. In morphological terms, the subdivision or irregular subdivision presents greater regularity of the lots and blocks, nearing the standards of urban planning. This differed from the informal occupation, where there was a more organic layout and, in this case, was situated alongside a stream.

Despite the differences in origin and territorial demarcations, the three localities of the Getúlio Vargas informal settlement present similar issues of precariousness with regard to their households, environment, basic infrastructure, and insecurity of land tenure. An analysis of satellite images, identified 700 sites, and according to information from the Residents' Association, there are 1300 families residing there.

Salim Felício can be regarded as an irregular occupation in the municipality of Cuiabá, with an area designated for institutional purposes and for the green area of the formal settlement called "Parque Real" (Figure 4). It has been occupied for approximately seven years, mainly from sales of property or irregular donations. The green area and the area for social facilities of Cuiabá were parceled into lots and divided into four blocks. This occupation also extends to the natural reserve area of the Figueirinha Stream.

Both Salim Felicio and Favela do Quiabo are worth examining because they are not demarcated in any official document, and, although they have only been occupied since 2013, the Land Use Law of 2015 did not demarcate them as ZEIS. In this sense, the action is important for giving visibility to these areas so that they are incorporated into the planning actions of the municipality.

Vila Verde, like Getúlio Vargas, is an older occupation, that dates back to the late 1980s, when the process of urbanization intensified in the capital of Mato Grosso. Although they are still in a situation of insecure tenure, there are currently an average of 150 families in 110 houses with a considerable degree of consolidation. The settlement is officially located in an "institutional" area and green area of the formal "Jardim Presidente I" subdivision (Figure 5). As it is a more traditional and consolidated settlement, its boundaries were delimitated as an "informal settlement" by the official document that formulated the Neighborhood Composition (Município de Cuiabá, 2013a) which was, more recently, updated as "ZEIS 1" by the Land Use Law (Município de Cuiabá, 2015).

Figure 3. Juxtaposition of formal demarcations, official demarcation, and local demarcation: Getúlio Vargas, Presidente III and Jardim Santa Fé
Source: Epura, 2020

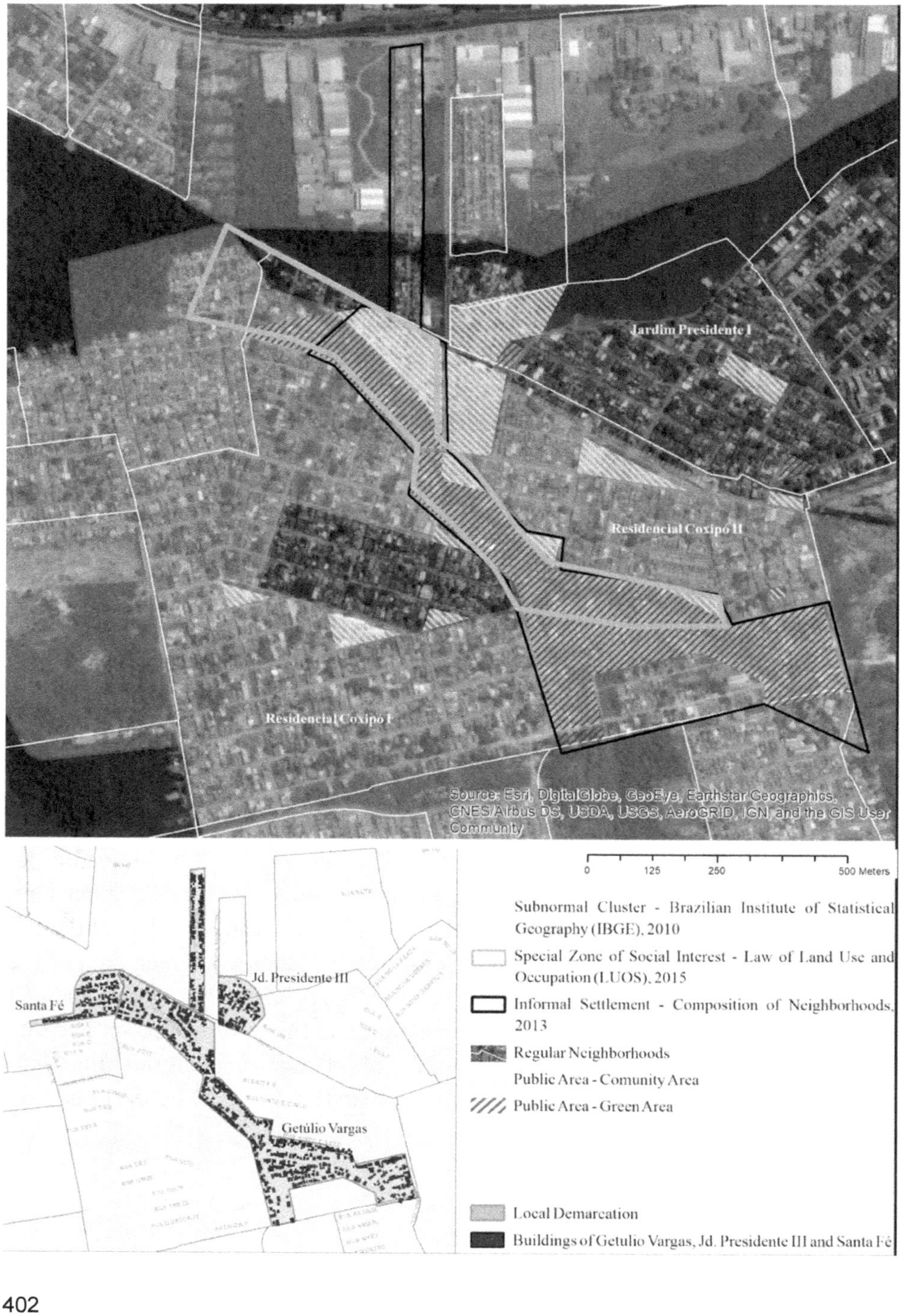

Figure 4. Juxtaposition of formal demarcation, official demarcation and local demarcation: Salim Felicio and Favela do Quiabo
Source: Epura, 2020

Use Of UAVs (drones) For The Demarcation of Informal Settlements

The Reurb Law enacts a land regularization project in one of its stages (Brasil, 2013, Art. 28), and must contain, among other requirements, a planimetric, altimetric, and cadastral survey, with a georeferenced map containing the perimeters of the informal urban nucleus and the non-conformities of the legal, urban and environmental situations, as well as the technical studies and the solutions to the issues involved.

The use of the term "georeferenced" for the urban environment, is an innovation for the land registration in the country, and its need can be explained by the accumulated irregular and inaccurate registries and maps present throughout the national territory, that resulted in an endless process of legal-land conflicts since the 19th century.

The georeferencing was sanctioned by official legislation (Brasil, 2016) in Decree 8.764/2016, which led to the setting up of the National Territorial Information Management System (SINTER). The purpose of SINTER is to integrate, (in a spatial database), the cadastral and geospatial data of urban and rural properties created by the Federal Government, the States, the Federal District, and the Municipalities.

This georeferencing achieves a very high degree of accuracy, for the scale of urban plots, and has a spherical position error with a radius less than, or equal to, 8cm. In other words, in addition to the need for planimetric coordinates to meet the requirements of the legislation in force, mapping the altitude is also necessary. Until now, the legal requirements of georeferencing have been restricted to the Instituto Nacional de Colonização e Reforma Agrária (National Institute of Colonization and Agrarian Reform) which has the competence to adopt rural land regularization procedures. Its norms can accept a planimetric error with a radius less than, or equal to 50cm.

However, the practical experiences of Épura and its theoretical research found that the topographic survey was a hindrance to the actions required. This was owing to the following factors: i) the lack of adequate topographic equipment, logistical, and funding for fieldwork, as well as logistical difficulties: ii) when outsourcing this activity, there is a need to delegate decision-making for matters such as the selection of notable milestones and references for the identification and demarcation of the existing blocks and lots. The reason for this is that, as the streets are unpaved and there are no surface drainage channels, the boundaries between the public areas of the roads and the private areas of the lots, do not have visible and easily identifiable landmarks.

Figure 5. Juxtaposition of formal demarcations, official demarcation and local demarcation: Vila Verde
Source: Epura, 2020

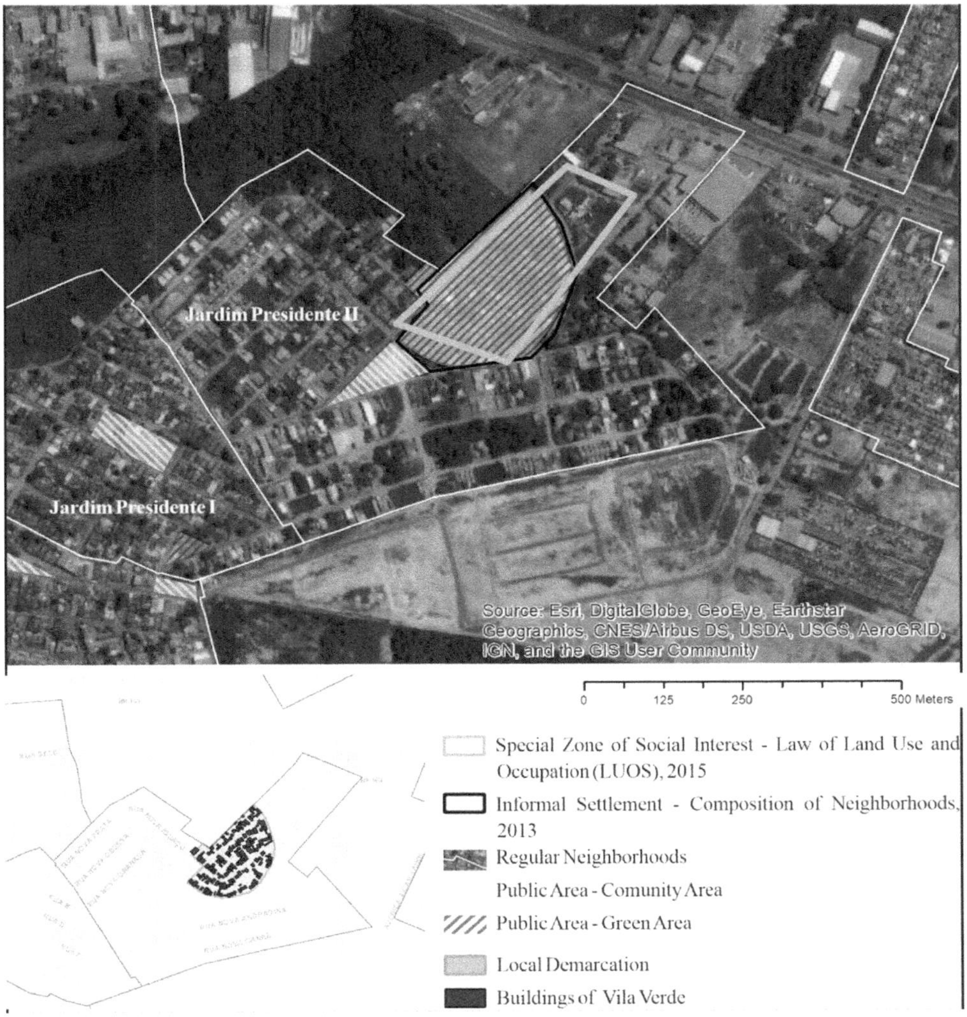

These concerns led the ÉPURA Group to conduct the process of defining the land features inside its office by means of a photogrammetric critical evaluation. As a result, it was decided to obtain the physical registration of the settlement "indirectly", by replacing the topographic process (which makes use of a total station) with a system of mapping that relies on the use of aerial images obtained by UAVs, and follows photogrammetry procedures to check the centimeter-level accuracy required.

This process reduced the time needed for the fieldwork to survey the studied settlements to half a working day. In addition, the resolution of the image with a materialized pixel (ground sample distance - GSD) of 7.5cm, enabled a search to be conducted of details for decision-making in the process of the demarcation of the field objects, with complete coverage regarding the constructions, their walls, artifacts, and the existing vegetation.

In the case studies, an UAV was used, fitted with a digital camera that had a focal distance lens of 8.8mm, combined with a gimbal to neutralize flight inclinations, and a 12 mega pixel sensor. An inertial measurement unit with a gyroscope, accelerometer and a location system integrated with the GPS and GLONASS constellations, are embedded in the UAV. The aerophotogrammetric survey was carried out in an automated way and controlled by a flight control application for a) mapping at an altitude of 250m and b) acquiring images with 80% vertical and lateral overlap (Table 2).

Table 2. UAV settlement surveys

Settlement	Area (ha)	Fight duration (min.)	Photos (nr.)
Vila Verde	2	8	45
Salim Felicio (Real Parque)	45.4	43	211
Getúlio Vargas	35.32	35	241

To achieve the desired degree of accuracy, 11 support points were placed on the ground, (7 control points, and 4 points for error checking in Salim Felicio). 12 control points and 12 check points were placed in Getúlio Vargas and 3 control points and 3 check points were placed in Vila Verde. In addition, reference points were also provided to ensure greater accuracy in determining the vertices of the existing lots and walls.

These support points were demarcated using dual -frequency GNSS receiver pairs configured to ensure millimeter-level accuracy of the reference-points. By means of the georeferencing of the tie points of the point cloud with the control points, through post-processing, it was possible to ensure a suitable degree of precision which could be confirmed by checking the control points.

In the post-processing, and after the images had been aligned, the camera calibration process was carried out by applying the correction parameters for the decentralized radially symmetric distortion of the lenses, of the georeferencing system of the images using centroids, and of the support points system.

This photogrammetric process enabled the densified point cloud to be constructed and surfaces to be formed that allowed the production of the digital elevation model, digital terrain model, contour lines extraction, and the orthophoto mosaic which in turn allowed the vectorization according to the Figure 6.

Figure 6. Digital surface and terrain model, point cloud, with perspective center of images obtained with drone (a), identification of Salim Felício permanent conservation areas (b) and Vila Verde orthophoto (c)
Source: Epura, 2020

In the case of Vila Verde and Salim Felício, the definition of streets and lots is more regular than that of Getúlio Vargas, where there is an intense occupation in the areas of permanent conservation. Thus, given the lower complexity of parceling and lower number of households, it was possible to detect the households from the orthophoto mosaic and compare them with information about the numbering of the blocks and lots at the time of the on-site visits (Figure 7).

Figure 7. Formal demarcations in Residencial Coxipó and Jardim Presidente I (7a), Real Parque (7b) and Jardim Presidente II (7c), with an emphasis on informal occupied public areas, (Getúlio Vargas, Salim Felício and Vila Verde respectively). Example of vectorization from orthophoto mosaic of Vila Verde (7d): and real demarcation of households and community spaces
Source: Epura, 2020

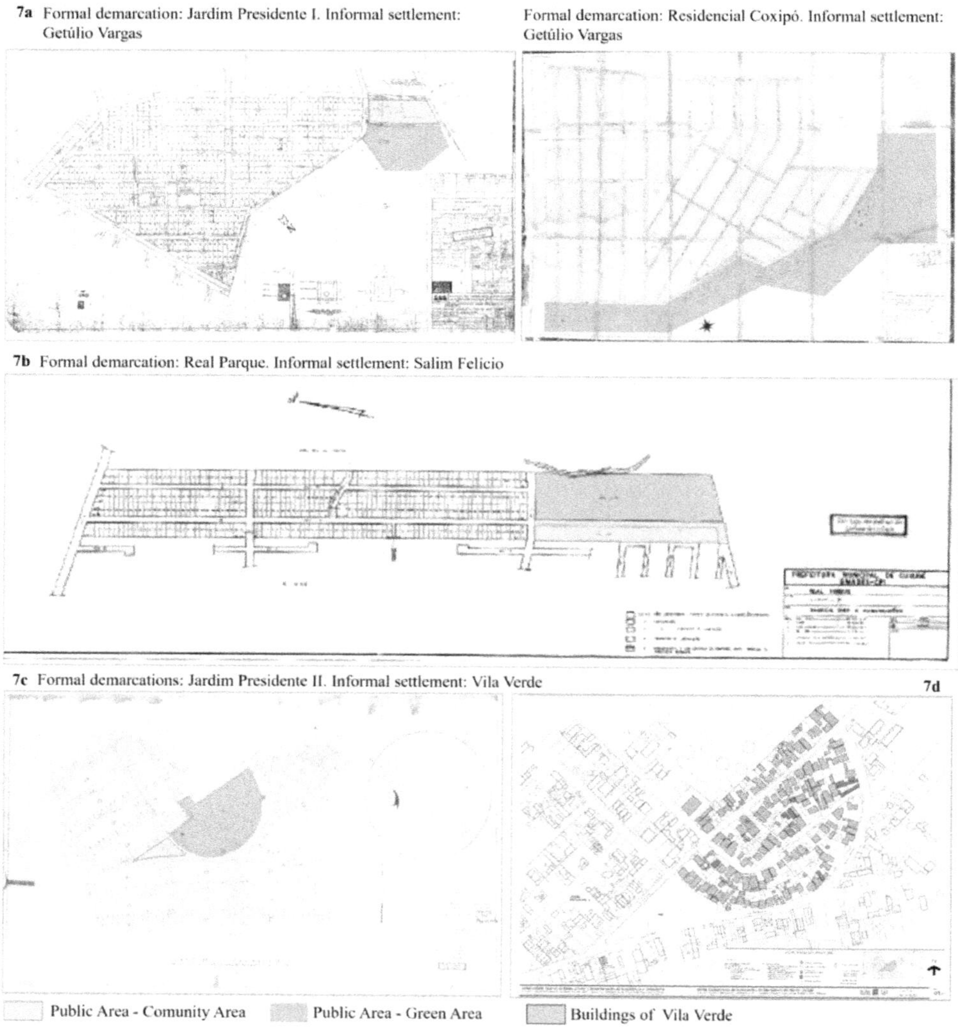

This exercise made it possible a) to prepare the planimetric and altimetric maps more quickly and in a more precise way, b) to list the requirements for infrastructural facilities and urban improvements in the settlements, and c) to identify the physical features of the settlements (number of lots, number of houses per lot, empty lots, and construction patterns of the homes). In addition, it meant that the registration database of the families could be updated for the Residents' Association, an important stage for compiling the list of families.

FUTURE RESEARCH DIRECTIONS

The experience presented here was a first approach of the use of new technologies to identify informal settlements that are not demarcated or that contradict the official demarcations.

On the basis of the objectives of the New Law of Reurb, it is becoming possible to support the views of the residents' associations concerning the respective informal settlements under study and with regard to the identification of "informal urban centers" that must be regularized and provided with essential public services (Brasil, 2013, art. 10). It was possible to compile the official information available and the information of the associations, which included the following: a) the on-site identification of the real perimeters, b) the number of resident families, c) an assessment of the time and history of the occupation, and d) the most urgent demands of the localities. This procedure is important, because through the "land legitimation" tool, it allows the regularization to take place in an original way, that is, the demarcations adopted may fail to take account of previous demarcations, and the referred instrument consolidates the real demarcation shown in the act of opening of the Reurb process (Brasil, 2013, artº 23).

The Reurb Law stipulates that residents' associations, civil society organizations, and the Defensoria Pública (Public Defender's Office) are legitimate agencies that can order the regularization process. After the data obtained from the settlements has been collated and updated, the residents' associations, with the support of the Defensoria Pública and the University, are in a position to request the recognition of ownership by setting in motion the land regularization process through a public entity, in this case the Housing and Land Regularization Secretariat of the City of Cuiabá (Secretaria Municipal de Habitação e Regularização Fundiária).

The idea of counter-mapping through the use of new technologies for a more reliable demarcation of the lots and the settlement as a whole, has proved to be a feasible way of recognizing and legitimizing a place in the city. There is a growing number of experiences of participatory counter-mapping that supplement hegemonic cartography, with varying degrees of participation and autonomy.

Even though the university played a major role in this strategy, particularly in image vectorization and map production, the internal evaluations made by the Épura group might assist in increasing the involvement of the residents in the technological handling process, which will allow the residents' associations to have more autonomy to update the cadastral base and demands of the neighborhood.

Another challenge which has also been identified is how to extend the activities to other neighborhoods in the capital, beginning with settlements in the southern region, where there is a higher rate of land tenure irregularity and social demands.

Like other countries, Brazil is establishing an improved legal framework for drone flights. This process of drawing up rules for the use of UAVs by different agents is quite recent and follows a number of experiments. In light of this new framework, the Épura group is preparing to incorporate the conditions required for the use of these tools, and to register the UAVs in the Civil Aircraft National Agency. This means a license will have to be requested, at the Aerial Space Control Department (DECEA), for all the missions undertaken by the Épura group and a certificate obtained by our technical group in the Defense Ministry for processing aerial surveys. This is important to legitimize the efforts and work of the Épura Group and the Residents' Association to obtain recognition of the informal settlements in Cuiabá.

Far from putting ourselves at the forefront of a public service that, we know, is the duty of the State, we are simply taking part in urban planning, by fostering greater activism and participation of the residents in civil society. As well as monitoring and inspecting, this should put pressure on the public policymakers to meet local demands and regularize their land rights.

CONCLUSION

Although advances have been made in urban legislation in Brazil, serious gaps remain, especially with regard to the inclusion of the informal settlements in urban areas, which makes it difficult to implement effective public policies.

In the case of Cuiabá, it is estimated that 60% of the neighborhoods are in a situation of irregular land tenure. The lack of codified laws for title deeds to the property nullifies any request to the public authorities to make urban improvements. Generally, the families themselves are responsible for not only constructing their houses, but also for the infrastructure of the neighborhoods, such as the opening up of roads, and extension of the water supply, drainage and, electricity systems, even if this means making irregular connections.

This situation reveals the dominant forces involved in urban planning, and in the process of producing official maps, of what is considered to be both a city and not a city. As a result, there are territories that are entitled to receive public funding and others that remain "invisible", "informal" and "irregular".

Establishing the conditions for the recognition of informal settlements as an integral part of the city, is more than a "counter-project," it is an ethical commitment and a social responsibility which entails a recognition of the right of these territories to exist.

The practical work of the Épura Research and Extension Group is based on the university's teaching, research, and extension triad. It involves a search for theoretical references that can be used to identify and characterize the informal settlements. This can make it easier to take measures for urban and land regularization in the municipality, by encouraging the use of new technologies. The actions are also linked to teaching activities, within the scope of the Urbanism Project III discipline (2013-2019), offered in the official grade of the course, which are aimed at making an intervention in the informal settlements.

The study of the Getúlio Vargas, Salim Felício, and Vila Verde settlements led to a better understanding of their urban, legal, and land tenure dimensions, since it meant they could be characterized as informal occupations in environmental conservation and green areas, and ¨irregular subdivisions¨ of public and private property.

In legal terms, it was important to recognize the contradictions that exist at the legislative level, because there is an excessive number of definitions and overlapping of the demarcations, as well as the invisibility of part of the settlements in official records, reveal the need to adapt and update them at both a federal and municipal level. It was also important to understand the constraints of the official data and the institutional fragilities especially with regard to the inability of the technical and administrative system to meet urban demands.

In the situations presented, the use of new technologies has proved to be efficient. The procedure involved has allowed a more rapid processing of geographical data, and made it possible to demarcate the locality in the municipal territory, as well as to identify and count the households.

Moreover, it draws attention urban and land inadequacies, and underlines the urgency of the demands for essential services, such as paving, construction of a sewage system, improvement of public and community spaces, and finally, the regularization of land ownership. By mapping and collating data and information about the locality, the residents' association is in a position to request this regularization from the municipality to secure land tenure rights.

Addressing the mapping issue is essential for the recognition and legitimation of informal settlements, especially if the purpose of urban land is regarded as a social right and means of access to the city. It is necessary to change the traditional methods of conventional mapping since these are based on an understanding of land tenure as representing an exclusive right to private property.

Counter-mapping is committed to the recognition of the legitimacy of occupation, for those who have the right to live in these settlements. It will always be a process of discovery and learning. On the one hand, it heightens the political awareness of families that are beginning to recognize their situation during the process of urban planning and management, and are thus in a stronger position to fight for their social and territorial rights. On the other hand, from the standpoint of the university, recognizing the limitations of the official data available, means urban challenges can be discussed. Hence, the university can assist in the procedure both by a remote reading of the place and in situ, by listening to its residents, and taking note of their real demands and needs.

Above all, this can lead to a fruitful exchange of knowledge, between the academic community and the residents, and thus clarify the nature of the existing land conflicts and the procedures needed for their regularization. Linking research to practice, puts the university at the service of the community.

Moreover, a reflective process of listening to people and their demands is a means of stimulating students and, enhancing teaching-learning process. It is hoped that it can somehow shape their professional careers through a transformative process which can help to mitigate the country's social and territorial inequalities.

This chapter was framed by the extension activities of Épura Group. This activities were supported by "Pró-reitoria de Cultura, Extensão e Vivência" of the Federal University of Mato Grosso (Edital PBEXT nº 001/2018).

ACKNOWLEDGMENT

This research is part of the Research Project *Africa Habitat, From the sustainability of the habitat to quality of the inhabit in the urban margins of Luanda and Maputo* (Ref. 333121392), coordinated by Isabel Raposo, Faculty of Architecture in the University of Lisbon, funded by Program "Knowledge for Development Initiative" from FCT and AKDN (Portuguese national Funding agency for science and Technology and AgaKhan Development Network).

REFERENCES

Acselrad, H. (2010). *Cartografia social e dinâmicas territoriais: marcos para o debate*. Universidade Federal do Rio de Janeiro, Instituto de Pesquisa e Planejamento Urbano e Regional.

Brasil. (1988). *Constituição da República Federativa do Brasil*. Diário Oficial da União.

Brasil. (2001). *Lei nº 10.257- Estatuto da Cidade*. Diário Oficial da União.

Brasil. (2009). *Lei nº 11.977*. Diário Oficial da União.

Brasil. (2016). *Decreto nº 8.764*. Diário Oficial da União.

Brasil. (2017). *Lei nº 13.465*. Diário Oficial da União.

Canavarros, A. F. A. (2016). *A Consolidação de um Tipo Urbano e Arquitetônico de Moradia para os Pobres: Velho Modelo, Novas Periferias no Espaço Urbano Mato-Grossense* [Unpublished doctoral dissertation]. Faculdade de Arquitetura, Universidade de Lisboa, Lisbon, Portugal.

Cardoso, A. L. (2008). Contextualização/caracterização. In J. Santa Rosa (Ed.), Política habitacional e a integração urbana de assentamentos precários: parâmetros conceituais, técnicos e metodológicos (pp. 13-45). Secretaria Nacional de Habitação. Ministério das Cidades.

Corrêa, R. L. (1999). *O espaço urbano*. Ática.

De Soto, H. (2001). O Mistério do Capital. *Record*.

Denaldi, R. (Ed.). (2013). *Planejamento habitacional: notas sobre precariedade e terra nos planos locais de habitação*. Annablume.

Dovey, K., Shafique, T., van Oostrum, M., & Chatterjee, I. (in press). Informal Settlements is not a Euphemism for 'Slum'. *International Development Planning Review*.

Fernandes, E. (2011). *Regularização de assentamentos informais na América Latina*. Lincoln Institute of Land Policy.

Freitas, C., & Lima, L. (2012). *Cadastro Territorial i SIG no Processo de Regularização fundiária: o caso da ZEIS do Planalto Pici em Fortaleza*. IV Simpósio Brasileiro de Ciências Geodésicas e Tecnologias da Geoinformação. Universidade Federal de Pernambuco – UFPE.

Gomes, J., & Pedrassoli, J. (2018). *Levantamento de características de moradias em favelas com o uso de drone: um experimento na Comunidade do Alto das Pombas, Salvador/Ba* [Paper presentation]. III Seminário Nacional Sobre Urbanização de Favelas – Urbfavelas.

Hagen, E. (2011). Mapping Change. Community Information Empowerment in Kibera. *Innovations*, *6*(1), 69–94. doi:10.1162/INOV_a_00059

Instituto Brasileiro de Geografia e Estatística. (2010). *Base de informações do censo demográfico 2010: resultados do universo por setor censitário*. https://goo.gl/wLOIwd

Lefebvre, H. (1974). La production de l'espace. *Anthropos.*

Lopes, J. M. de A. (2011). *Sobre arquitetos e sem-tetos: técnicas e arquitetura como prática política*. Rev. Brasileira Estudos Urbanos Regional. Universidade de São Paulo.

Lopes, J. M. de A., & Rizek, C. S. (2006). O mutirão autogerido como procedimento inovador na produção da moradia para os pobres: uma abordagem crítica. In A. L. Cardoso & A. K. Abiko (Eds.), *Procedimentos de gestão habitacional para população de baixa renda*. ANTAC.

Lutzoni, L. (2016). In-formalised urban space design. Rethinking the relationship between formal and informal. *City. Territory and Architecture*, *3*(20), 1–14.

Maricato, E. (2000). As idéias fora do lugar e o lugar fora das idéias: planejamento urbano no Brasil. In O. Arantes, C. Vainer, & E. Maricato (Eds.), *A cidade do pensamento único - desmanchando consensos* (pp. 121–192). Vozes.

Maricato, E. (2003). *Metrópole, legislação e desigualdade*. Estudos Avançados.

Martins, M. L. R. (1997). *São Paulo: movimentos populares urbanos e o direito à cidade*. FASE.

Ministério das Cidades. (2004). *Política Nacional de Habitação*. Cadernos MCidades, Habitação 4. Espalhafato Comunicação.

Município de Cuiabá. (2013a). *Composição dos bairros de Cuiabá*. https://www.cuiaba.mt.gov.br/upload/arquivo/composicao_bairros.pdf

Município de Cuiabá. (2013b). *Plano Local de Habitação de Interesse Social*. Unpublished.

Município de Cuiabá. (2015). *Lei Complementar nº 389 - Lei de Uso e Ocupação do Solo (LUOS)*. https://www.legisweb.com.br/legislacao/?id=320141

Panek, J., & Netek, R. (2019). Collaborative Mapping and Digital Participation: A Tool for Local Empowerment in Developing Countries. *Information (Basel)*, *10*(255), 1–14. doi:10.3390/info10080255

Payne, G. (2001). Urban land tenure policy options: Titles or rights? *Habitat International*, *25*(3), 415–429. doi:10.1016/S0197-3975(01)00014-5

Perez, L. P., Bischof dos Santos, R., Almeida, G., & Carvalho, G. C. (2017). Spatial data in the Global South: A case study of alternative land management tools for cities with limited resources. In *2017 IEEE Global Humanitarian Technology Conference (GHTC)* (pp. 1-10). 10.1109/GHTC.2017.8239224

Pinheiro, F. J. (2013). *Déficit habitacional municipal no Brasil*. Centro de Estatística e Informações., https://goo.gl/euvcpK

Radjawali, I., Pye, O., & Flitner, M. (2017). Recognition through reconnaissance? Using drones for counter-mapping in Indonesia. *The Journal of Peasant Studies*, *44*(4), 817–833. doi:10.1080/03066150.2016.1264937

Raposo, I. (2017). Intervir nas margens urbanas: o papel da academia. *Revista Espaços vividos e espaços construídos. Estudos sobre a cidade, 5*, 29-39.

Rolnik, R., & Santoro, P. F. (2013). *Zonas Especiais de Interesse Social (ZEIS) em Cidades Brasileiras: Trajetória Recente de Implementação de um Instrumento de Política Fundiária*. Lincoln Institute of Land Policy. https://www.lincolninst.edu/sites/default/files/pubfiles/rolnik-wp14rr1po-full_0.pdf

Santos, C. (1981). *Movimentos urbanos no Rio de Janeiro*. Zahar Editores.

Sliuzas, R., Gevaert, C. M., Persello, C., & Vosselman, G. (2018). Evaluating the Societal Impact of Using Drones to Support Urban Upgrading Projects. *ISPRS International Journal of Geo-Information*, *7*(3). https://www.mdpi.com/2220-9964/7/3/91

Stöcker, C., Ho, S., Koeva, M., Nkerabigwi, P., Schmidt, C., Zevenbergen, J., & Bennett, R. (2018). Towards UAV-based Land Tenure Data Acquisition in Rwanda: Needs Assessment and Technology Response. *FIG Congress 2018: Embracing our smart world where the continents connect - Enhancing geospatial maturity of societies*. http://www.fig.net/resources/proceedings/fig_proceedings/fig2018/papers/ts08c/TS08C_stocker_koeva_et_al_9428.pdf

Turner, J. (1977). *Vivienda: todo el poder para los usuarios*. H. Blume.

United Nations. (2015). *Transforming our world: the 2030 Agenda for Sustainable Development*. https://sdgs.un.org/2030agenda

United Nations Human Settlements Programme. (2008). *Secure Land Rights for All*. UN-Habitat/Global Land Tool Network.

United Nations Human Settlements Programme. (2014). *The Right to Adequate Housing*. Fact Sheet No.21, United Nations https://www.ohchr.org/Documents/Publications/FS21_rev_1_Housing_en.pdf

Valladares, L. P. (1980). *Passa-se uma casa: análise do programa de remoção de favelas do Rio de Janeiro*. Zahar Editores.

ADDITIONAL READING

Canavarros, A. F. A. (2017). Entre o lugar de origem e o lugar de destino: pela consolidação do direito à habitação na estrutura social urbana de Cuiabá. *Espaços vividos, construídos, 6*(1), 42-52.

Casagrande, G., Sik, A., & Szabo, G. (Eds.). (2018). *Small Flying Drones. Applications for Geographic Observation*. Springer. doi:10.1007/978-3-319-66577-1

Denaldi, R. (Ed.). (2013). *Planejamento habitacional: notas sobre precariedade e terra nos planos locais de habitação*. Annablume Editora.

Fernandes, E. (2011). *Regularização de assentamentos informais na América Latina*. Lincoln Institute of Land Policy. https://www.lincolninst.edu/sites/default/files/pubfiles/regularizacao-assentamentos-informais-full_1.pdf

Henriques, C., Domingues, A., & Pereira, M. (2020). What is Urban After All? A Critical Review of Measuring and Mapping Urban Typologies in Portugal. *ISPRS International Journal of Geo-Information, 9*(11), 630. Advance online publication. doi:10.3390/ijgi9110630

Ralphs, M. P., & Wyatt, P. (2003). *GIS in Land and Property Management*. Taylor & Francis. doi:10.4324/9780203470374

Tenedório, J. A., Rebelo, C., Estanqueiro, R., Henriques, C., Marques, L., & Gonçalves, J. A. (2014). New developments in Geographical Information Technology for Urban and Spatial Planning. In N. N. Pinto, J. A. Tenedório, A. P. Antunes, & J. R. Cladera (Eds.), Technologies for Urban and Spatial Planning: Virtual Cities and Territories (pp.196-227). IGI Global. doi:10.4018/978-1-4666-4349-9.ch010

KEY TERMS AND DEFINITIONS

Counter-Mapping: Cartographic representation that challenges (or opposes) conventional, technical and official cartography. It is characterized by the real spatial demarcation legitimizing the historical, symbolic and territorial occupation dimensions by a social group.

Informal Settlements: Urban territories, mostly self-produced, where, to a greater or lesser degree, coexist insecurity of tenure, needs for housing and urban improvements namely basic infrastructures, implementation of services and essential public amenities.

Insecurity of Tenure: Includes different situations concerning land occupation: illegal occupation, tolerated occupation, occupation legitimized by a social group or customary practices but not considered as legal by government or local authorities.

Official Demarcation: Land tenure delimitation officially recognized by the municipal public power.

Precarious Settlement: Official designation used by the National Housing Policy to define, at a national level, urban housing inadequacies where low-income families live.

Subnormal Agglomerate: Official designation used by the Brazilian Institute of Geography and Statistics to define, identify, and characterize urban informal settlements, based on information from census data.

Urban Land Regularization: Set of legal, urban, environmental and social measures that aim to comply with the constitutional principle of the right to housing and access to urban land.

Chapter 14
Three-Dimensional Modelling for Cultural Heritage

Luis Marques
https://orcid.org/0000-0002-6198-8417
CICS.NOVA, Universidade NOVA de Lisboa, Portugal & CPSV, Technical University of Catalonia, Spain

Josep Roca
https://orcid.org/0000-0003-3970-6505
Technical University of Catalonia (CPSV), Spain, Spain

ABSTRACT

The creation of 3D models of urban elements is extremely relevant for urbanists constituting digital archives and being especially useful for enriching maps and databases or reconstructing and analyzing objects/areas through time, building/recreating scenarios and implementing intuitive methods of interaction. The widespread data available online offer new opportunities to generate realistic 3D models without the need to go physically to the place. This chapter aims to demonstrate the potential 3D modeling and visualization/interaction of urban elements in the city for multiple purposes, and it is organized in four main topics: The first deals with the theoretical framework regarding the bases of the human perception of the spatial environment and the importance of 3D modelling. The second and third deal with technical procedures on terrestrial/aerial data acquisition and demonstrate alternatively data gathered online to generate 3D models for the visualization of urban elements of the city, and the fourth introduces 3D model visualization within an augmented reality environment.

DOI: 10.4018/978-1-7998-2249-3.ch014

INTRODUCTION

Recent technical-scientific research has tried do analyze, replicate, and develop technological structures and aptitudes analogous to the human brain. Several technological innovations and concepts have been introducing significant words, such as smart or intelligence, tied with the human brain and its capabilities. Computers have been promoting digital data acquisition, processing, and dissemination; however, with the emergence of mobile devices a revolution in Human-Machine or Real-Virtual environment interfaces was initiated, regarding multiple functions and applications and their special features, such as portability and mobility (fitting in the pocket and taken almost anywhere by its user). The use of smartphones had a huge impact on society, changing human behavior, individually and between each other, with information and communication technologies, and their perception of the virtual and real spatial environment. Large amounts of information in digital format have become more accessible to the user through the Internet at fast growing speed and broadband access progression (increasing amount and size of transmitted content). Social networks and other tools provide instruments to communicate and share information within a synthetized virtual environment (parallel and associated with the real environment), where people interact with each other in nearly real time. The digital revolution has now penetrated our culture so deeply that the many new forms of communication that have been transforming spatial environments are now revealing previously unknown aspects in the form of large databases (Big Data), providing us with opportunities for analysis and modelling, quite different from the methods applied earlier (Batty, 2013). This Volunteered Geographic Information (Goodchild, 2007) is particularly relevant in the field of cultural heritage, comprehending the value that each patrimonial object has for each individual, providing tools to a continuous valuation, evaluation and modelling.

The concept of Cultural Heritage assumes the notion of value and cultural inherency, often related with unicity, rareness, importance, memory, or symbolic representation of one or more specific objects, events, milestones, achievements, or historical significance. Value has always been the reason underlying heritage conservation. However, the notion of valuation might be identified or increased, when supported individually or collectively by external and/or internal representations as a process of increasing the intrinsic value of the cultural heritage. Digital representations of cultural heritage are highly valuable for scientific study, conservation, educational or tourism purposes and more recently also relevant for virtual users in times of confinement (i.e. as due to a virus outbreak). Technology permits to extend the real world to interdependent mixed virtual environments, allowing a global access to unlimited information, virtually, without barriers, physically anywhere. The public space is no longer limited to the physical space to become virtually perpetual, unlimited and reachable by any user (Innerarrity, 2010). Individuals can visualize, evaluate, discuss, and intervene virtually and collectively through the past, present, future and beyond human capabilities, simulating re-presentations of cultural heritage and its increased value. However, user expectations embrace accurate and rigorous representations of the real world as authentic (Barrado-Timón, 2019).

THEORETICAL FRAMEWORK: REPRESENTATIONS, SPATIAL ENVIRONMENT AND CULTURAL HERITAGE

Representations play a central role in a cognitive system, supporting perceptions, which are at the basis of how individuals recognize the physical world. The concept of space has been considered an innate organizing principle of the mind, through which the world is, and must be, perceived. Thus, representations are fundamental in the brain, being at the source of all empirical knowledge and conception of the world. People seem to have quite accurate mental representations of their spatial environment, coherently capturing the categorical spatial relations between elements that support perception, reorientation, and spatial inferences (Golledge, 1999). However, mental (internal) representations of the spatial environment diverge from the real world, eliminating and/or adding entities, frequently introducing distortions. Reconstruction generates systematic errors; therefore, humans rely on external representations to serve as cognitive aids for memory and information processing. Visuospatial representations have many similarities to those that reside in the mind. External cognitive tools function to extend the powers of the mind by exploring the human skills at spatial reasoning (Tversky, 2005). These external affordances enhance human cognition and help to understand the spatial environment.

Technology has been allowing humans to extend (represent) the physical world. Technological advances, especially those regarding the ability to represent the physical world and modelling capabilities, have led to the creation of an increasing number of digital information, objects, and environments. However, Technology itself does not increase the value of Cultural Heritage. The value of Cultural Heritage is a quality attributed by users where technology plays an important role (Marques et al, 2017).

The use of Digital Representations (often analogous to internal representations) and geo-science applications can provide many useful functions, such as data acquisition, navigation, object operation, modelling, visualization, dissemination, among others. The use of Digital Representations (DR) in research offers numerous advantages, especially the ability to create detailed models of possible pasts and futures. By modelling several different variables, we can test and change them to explore different hypothetical situations. This is not possible in the real world, as circumstances on site are unpredictable and constantly changing. Spatial modelling techniques can be used to develop visual simulations, including diverse attribute data through time (as the world is not static).

Often two-dimension (2D) representations are considered insufficient to address real 3D planning problems, especially at the scale (in area and height) of the modern city. 3D representations generate more intuitive and accurate methods of interaction between the user and the represented data. Considering that most of the public does not possess skills to comprehend entirely Geographical Information Systems (GIS), 3D urban environments can increase the engagement of the user, making it much easier to understand, interact, and participate by visualizing interactively, for example, volumes, textures, shadows, sights and their relation. This principally occurs because of its high level of similarity to reality, congregating detailed information and volume property, stimulating more enthusiasm and simplifying creative solutions for detected problems after a walk-drive-fly-through re-presentation (embodying the user in that environment or presence), improving understanding and communication. Thus, 3D representations and analysis tools are more adequate for enriching spatial representations. 3D settings can support positional cognition by offering a realistic/ meaningful environment where the learning process can effectively be located giving the user the possibility of interacting with objects in their actual context (Rebelo *et al.*, 2015).

In the technical-scientific field, there is a growing interest among technicians for the usage of 3D virtual models, taking advantage of their enormous potential for pedagogical, documentary, and exploratory activities, as well as of their predictive power. This is especially relevant for urban studies since it facilitates interventions in historical sites, rehabilitation and planning of buildings, neighborhoods and cities, providing opportunities to simulate past, present, or future interventions, and their respective evolution and management, where people can move through and interact with the space within a virtual/mixed environment (Marques *et al.*, 2017). There have been major advances in the ability to characterize and monitor the world in real-time, using networks of sensors and information through the Internet. Thus, the third spatial dimension is becoming more important in an extensive range of applications and it seems likely that in the future geospatial technologies will operate in the full four dimensions (three spatial dimensions, plus time – 4D) of the geographic environment. 4D digital representations can be built to reflect the past, existing and possible future conditions of cultural heritage, showing the current or proposed appearance, façade, style, material and size, applied to buildings, urban settlements, sites and landscapes, being able to alter and improve, on demand, while integrating all the agents in this process. Thus, virtual environments can offer enhanced opportunities and provide additional tools for people to communicate with others and interact with the urban models. Modelling aids planners and other agents to understand the past, objects evolution and positional relations with the environment, understanding patterns and experiment scenarios to envision the impact of various plans and policies. Modelling several different variables allows testing and adjusting features to explore different hypothetical situations. Synthetized environments enabled to conduct experiments, introducing or modifying variables and adjusting conditions, such as transforming the topography, structures and their extension or height, planning, visualizing sights, colors, and lighting conditions. This is especially valuable in the field of Cultural Heritage considering the high level of technical and historical understanding, preservation, adaptation, and dissemination. When creating these models, there are endless possibilities to run experiments with empirical results that the community can measure, compare and recreate, allowing to understand, analyze, intervene, and distribute Cultural Heritage information about an object or site.

The acquisition of, dissemination of, and search for Digital Representations has had an extraordinary growth, in part due to the emergence of mobile devices, especially those with broadband Internet access, which started a revolution in the Human-Machine and Real-Virtual environments. Mobility has increased the influence of (social) networks in which individuals are able to react to or share almost instantaneously posts, events, or photos, allowing the fast and global dissemination of information, replicating most human social relations on a universal scale (Pires, 2016). Often, this volunteered information is associated with a place or coordinate position. Society has unprecedented access to large amounts of information, sharing knowledge and communicating, while at the same time it is metaphorically developing a synthetized environment, albeit linked to the real world. These developments have brought main digital changes that revolutionized the concept and usage of spatial representations. This vast provided sensed human data, when shared can provide valuable information and tools to recreate and recognize the value attributed by each user, considering individual and collective experiences, with the advantage of being available digitally and extensively.

There is a Smart Public Space (Figure 1) when there is a combination of HS + Tech. HS are in physical, tangible, sensitive interaction with real space. Hence an internal representation of this real space. Using Tech (mobile devices, Apps, AR, VR, Wi-Fi signals) the HS enrich these internal representations of public space, itself enhanced by enrichment with virtual data. In this context, we can admit that there is a Smart Public Space when there is convergence between the physical public space and the virtual public space resulting from the HS + Tech interface.

Buildings with historical value, as well as all other structures, are part of the architectural composition of the city and of the places in general. The built heritage is a space of memory, overlapped, as if structured in layers (due to the passage of time) and in fusion (for example, of architectural styles, forms, methods and construction materials). All built heritage has a background, which synthesizes collectively shared representations. Many of these representations remain through time, they are part of the cities and the public space that compose them; are historically appropriate identity and identification references. The city is understood as an open landscape constituted by three-dimensional pieces in an organized environment of which the open spaces are part. Many of these open spaces are public: squares, farmer's markets, downtown shopping malls, public parks, playgrounds, greenways, waterfronts, etc. This public space is the place of sociability manifestations such as walking, talking, studying, shopping, interacting with others. The built heritage is a fundamental element of the public space.

Figure 1. Smart Public Space: Human Sensors (HS) + Technology (Tech)

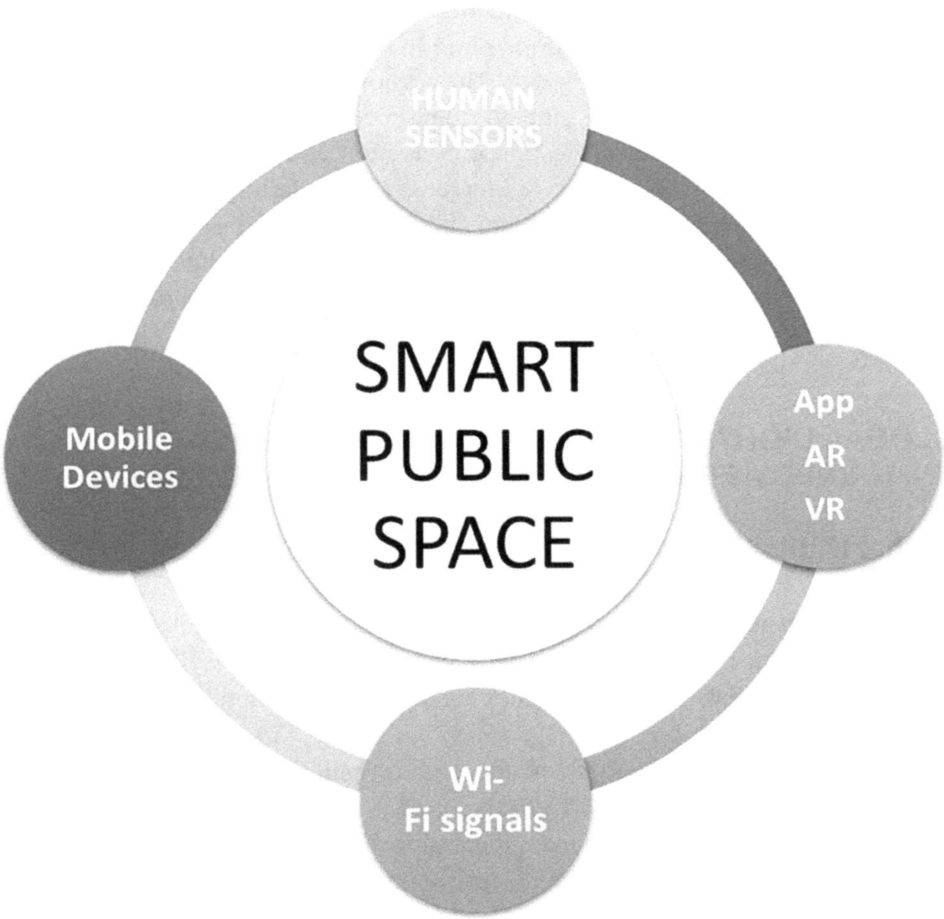

The decade of nineties of the 20th century was marked by the emergence of Virtual Reality (VR). The second decade of this century lives the Augmented Reality (AR). The overlay in "synchronous time" of virtual images with elements of tangible reality is becoming an unprecedented social, cultural and tourist practice. Today, the diversity of mobile platforms providing access to real-time geographic information (synchronization) and the ability to support Apps (understood as data and technology that supports space-time extension) built on AR enable what we can call of expansion of the public space with patrimonial or other value; ease the synchronization between physical public space and virtual public space. This is also relevant not only for the fusion of both real and virtual environments, but also when it verifies the impossibility to go to place as in times of confinement (i.e. virus outbreak). Apps can be understood as the dematerialized extension of the public space with patrimonial value. The recreation of the public space with patrimonial value (physical and limited) also gives place to the virtual public space and leads to what we can call smart public space. The acknowledgment of Cultural Heritage value and valuation techniques is strategic to promote substantiated, integrative and oriented procedures for the identification, preservation, and dissemination of tangible and intangible heritage (Tenedório & Marques, 2017). Heritage industry has been supported economically through tourism and in particular urban cultural tourism. Historical heritage and cities have the capacity to attract tourist's attention. Through technology, the experience of heritage can be enhanced beyond the physical object and directing greater benefits for the heritage conservation (Barrado-Timón, 2019).

GLOBAL DIGITAL HERITAGE AND TERRESTRIAL/ UNMANNED AERIAL VEHICLES 3D POINT CLOUD DATA

Global Digital Heritage

Heritage preservation includes mainly research, field exploration, archive management, conservation, data visualization, and use. Technology has been playing an important role in a transversal way, especially in terms of data acquisition, management, and visualization.

The use of photographs, besides drawings or paintings of the most valuable constructions could then be gathered, stored, and visualized. According to Andrés & Pozuelo (2009), catalogues of monuments and their surroundings was one of the first applications of image. These archives supported the reconstruction of several monuments after the World Wars. Since World Wars, policies, techniques and science of the modern conservation movement have evolved in an international context, increasingly involving all sectors of society, with the establishment of international treaties and documents such as: the Athens Charter – 1931; the Venice Charter – 1964; the World Heritage Convention – 1972; the European Charter of the Architectural Heritage – 1975; the Convention for the Protection of the Architectural Heritage of Europe – 1985; or the Horizon 2020 in the Reflective Societies: Cultural Heritage and European Identities. These figures leaded to the creation of international organizations such as: United Nations Educational, Scientific and Cultural Organization (UNESCO, 1946); the International Council of Museums (ICOM, 1946); the International Centre for the Study of the Preservation and Restoration of Cultural Property (ICCROM, 1956); and the International Council on Monuments and Sites (ICOMOS). These documents and organizations often resort to the use of existing resources, techniques and accessible technology to acquire, record, and promote the heritage (ICOMOS; UNESCO, 2019).

Technological developments have facilitated faster and more efficient register, inventory, research, access and visualization of historical data (Templin, 2019). A good example of this idea is the simple process of digitalization that allows the expansion of the storing capacity, in this case applied to heritage information (in terms of physical space in the archives) prolonging the lifecycle (higher security), easy duplication, research and access to the documents (through internet virtually from everywhere). Digital use can synthetically preserve the historical, cultural and scientific values of cultural heritage items by applying modern science and technology to heritage preservation, and the research has further extended to heritage information acquisition, digitally aided research, digital conservation, and digital exhibition and utilization.

Heritage management information practices are currently still associated with two-dimensional imaging, mapping and analysis. More recently as the increasing use of portable devices such as smartphones and with the positional and navigational capabilities, increased the use of Geographic Information Systems (GIS) to associate databases with texts, drawings and pictures to a spatial reference. This capability has great advantages when compared with the also relevant heritage inventories, based on entries and registers that are exclusively descriptive, albeit less dynamic and interactive. Thus, the use of heritage inventories with spatial references offers interesting forms of supporting technical assistance and for tourism as they can support the user to the place or even attract him to an unknow (or unseen, as for example underground) object by proximity using positional features (such as GPS). The possibility to visualize three-dimensional (3D) data within an AR environment, using the device's camera, allows new possibilities to enrich the real space with virtual data and even virtually time traveling - 4D. The addition of 3D visualization and modelling capabilities (through time) became extremely valuable for planners to communicate simultaneously simplifying the study of complex urban environments in their full spatial extent including shape, size, volume, and the spatial configuration in all x, y and z dimensions (adapted from Yao *et al.*, 2006 in Marques, 2017).

In archaeological sites, mapping, registration, documentation and visualization of cultural heritage is very important for heritage conservation, but also understanding of the historical importance and to transmit to public. Traditionally, mapping and documentations studies are carried out through terrestrial measurements performed during excavation. Technical improvements in laser scanning and/or image acquisition, terrestrial and aerial (using UAV) have become an alternative to register and document archaeological areas, historic sites and cultural heritage buildings/structures with accurately data and efficiently (Ilci *et al.*, 2019). Digital recording, documentation and protection ensure that the heritage is protected against attrition, war, natural disaster, weather, urbanization damage, climate change and human neglect through-out history. Therefore, detailed documentation and mapping studies should be performed with great care, accurately and as quickly as possible in all parts of heritage sites. In the case of archaeology, digitally documentation and evaluation are extremely important for the archaeological findings. The architectural survey and restitution studies are documentation methods and basic preliminary stages for the preparation of a restoration. The information recorded in the archaeological campaigns can be materialized in a virtual reconstruction and shared with the scientific community (Carvajal-Ramírez et al., 2019). The work that has been developed in partnership between Geodrone[1] and Cromeleque companies (Portugal) is a good example of the potential to collect actual real data on site (using UAV) and later virtually reconstruct the building and/or structure (e.g. Virtual reconstruction of the Nossa Senhora da Ajuda Bridge; Santa Maria do Bispo Chuch in Montemor-o-Novo; Paço dos Alcaides). These combination between the real and computer-generated model allows to "virtual time travel" and visualize structures as they were in past (or future), giving a clear perception of the relevance of the element, adding knowledge and value to the cultural heritage.

When a natural disaster occurs, there is often significant damage to vitally important infrastructure. Technicians must quickly locate the damage structures that need immediate attention. These technical groups need to determine how to allocate their resources most efficiently to save time and money without having to assess each area individually. Therefore it is extremely relevant to first acquire 3D data of the heritage in different situations and moments: before the disaster occurs surveys must be conducted as a prevention dataset of information (allowing to faithful reconstitution of the original object); after the disaster occurring it is relevant to collect data to support technically the affected areas (Eckert *et al.*, 2019); and many relevant and valuable heritage sites are usually interest landmarks, being often photograph by the public from which an increasing number of relevant data has been shared online, which can be used to generate 3D models. This information can be used to model an object with very interesting results.

3D models archive of global heritage start to be developed to support solutions providing digital data in x, y, z axes to governments, regional institutions, museums, local scholars, and the public. Example of that is the Global Digital Heritage, a not-for-profit, private research and education organization dedicated to documenting, monitoring, and preserving the global cultural and natural heritage, using digital visualization, 3D virtualization, geospatial informatics, and open access solutions to provide digital data[2]. These datasets of 3D data might be extremely relevant for the future as might be needed to restauration or recover of a building / structure.

Terrestrial/Aerial (UAV) and Passive/Active Sensors For 3D Point Cloud Data

A 3D model can be generated totally within a virtual environment (e.g. using computer software such as 3D Studio Max or Sketchup) and ideal for futuristic planned objects, however modelling can be acquired from scanning the object from different angles (from the ground: terrestrial or from the air: aerial) using passive or active sensors.

Active sensors supply their own light energy, which can be controlled. Some advantages that active systems have over passive sensors are that they do not require solar illumination of surfaces or perfect weather conditions to collect useful data. Consequently, they can be employed at night or in conditions of haze, clouds, or light rain (depending on the wavelength of the system). There are two main types of active sensors: Radio Detection And Ranging (RaDAR) and Light Detecting And Ranging (LiDAR) systems. RaDAR uses microwave wavelengths (ranging from 1 millimetre to 1 meter). Microwave pulses are transmitted to a target or surface, and the timing and intensity of the return signal is recorded. Information about the structure and composition of objects and surfaces can be detected. LiDAR uses laser light as an illumination source. A short pulse of light is emitted from a laser and a detector receives the light energy (photons) after it has been reflected, or absorbed and remitted by an object or surface. The possible wavelengths range from about 0.3 to 1.5 micrometres, which covers the ultraviolet through near-infrared spectral range. The principle advantages of laser scanning are the provision of precise x, y, z and the high level of automation (Heritage et al., 2009 *in* Marques, 2017), increasing data acquisition in a short period of time, and avoiding the common mistakes generated by traditional surveying (Castillo & Almirall, 2006 *in* Marques, 2017). In this sense, LiDAR systems have established a new era in terms of the three-dimensional (Gambino et al., 2005). Laser scanning become a wide commonly used technology to register, measure, and model Cultural Heritage elements (widely used and discussed in the scientific community), even if the application of this technology in architectural survey does not replace, however complements the widely accepted methods to register architectural heritage (Castillo & Almirall, 2006 in Marques, 2017).

Passive sensors measure light reflected or emitted naturally from surfaces and objects. Such instruments merely observe and depend primarily on solar energy as the ultimate radiation source, illuminating surfaces and objects. They do not emit their own radiation but receive natural light and thermal radiation from the Earth's surface. The most common sensor system is the photographic camera and its basic element systems, consisting fundamentally in optics, film /CCD, and filters . Cameras can be used on nearly every platform (Space, Aerial and Ground-based) and consist in one of the most common practices of (re)presenting the real world (Marques, 2017).

Photogrammetry is the art, science, and technology of obtaining information about objects and the environment through processes of recording, interpreting and measuring photographic images and patterns of electromagnetic energy (Wolf and Dewitt, 2000; McGlone, 2004 *in* Aber *et al.*, 2010). Photogrammetry is primarily concerned with making precise measurements of three-dimensional objects and terrain features from two-dimensional photographs. Applications include the measuring of coordinates; the quantification of distances, heights, areas, and volumes (Aber *et al.*, 2010). Photogrammetry can be generally divided in two main types, according with the position of the sensor. Terrestrial photogrammetry, with the camera handheld or on a tripod or platform, usually dealing with near and smaller object and distances. Aerial photogrammetry uses aircrafts to collect the imagery from the air. Combining terrestrial and aerial imagery has the advantage to associate the capacity to cover higher and bigger objects with high image detail.

Unmanned Aerial Vehicle (UAV) refers to a class of aircrafts that can fly without the presence of a pilot (on-board). UAV-systems are quickly increasing in photogrammetric applications (Lin *et al.*, 2011), considering the reduction of costs, especially due to the development of navigation systems and other geographic information technologies, when compared with traditional methods of aerial survey. The usage of a UAV greatly simplifies the assignment of collecting aerial data and according with the following processing phase. Although, these data are authentic with an impressive level of detail (usually using high resolution cameras/sensors). While terrestrial acquired data usually doesn't require permits to collect the data, the UAV survey usually requires a certain number of preparation tasks:

1. Locate and verify the level of authorization needed (e.g. using the site or app "voanaboa" – Portugal – Figure 2);
2. Authorization permits to use the aerial space (request to civil/military authorities);

3. Verify the weather and safety circumstances, light conditions, radio frequency and signal interference tests;
4. Onsite GPS and compass calibration;
5. Flight path determination (considering flight height and surrounding objects);
6. Imagery acquisition according with the processing phase and object: height, overlapping (lateral/frontal), time, format, focal distance, camera data, spatial resolution, image coordinates.

Figure 2. Areas of air exclusion, voanaboa mobile APP and military (Autoridade Aeronáutica Nacional) or civilian (Autoridade Nacional da Aviação Civil) authorization forms / online request

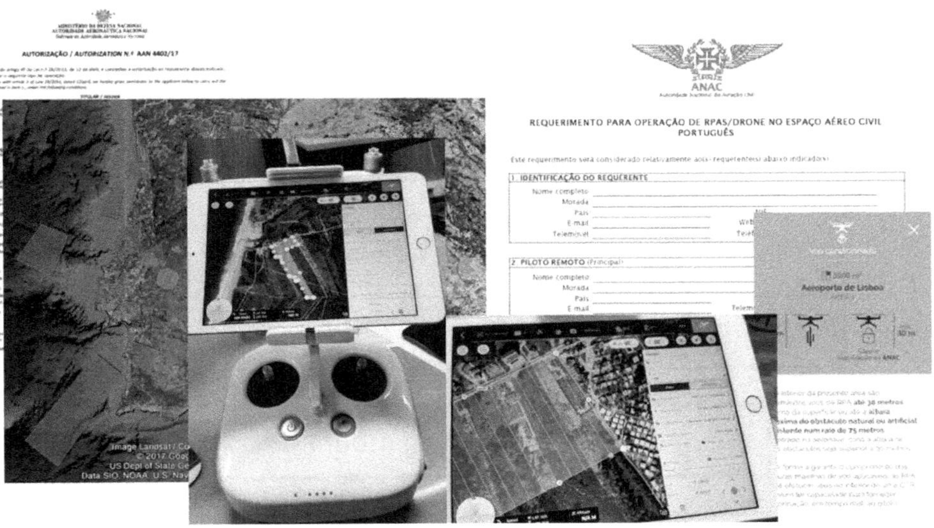

Figure 3 synthesizes schematically the methodology, procedures and fluxes of data concerning the visualization of 3D models, gathered (internet catalogues) or produced, using 3D modelling software or 3D scanning (terrestrial or aerial - UAV).

Based on the information collected by the UAV, the next stage involves transferring the data to the computer and preparing the processing phase. The 3D geometry is obtained by creating images of the same object from different positions. This makes a single point on the object visible as a pixel in multiple images (figure 4 and 5). These lines will intersect at one point, which is the 3D location of the object point. Position and orientation of each image requires to be known. Tie points are usually used to link the images together, being a recognizable point identified in images where it occurs. A group of tie points will allow for the reconstruction of the relative position of all images. This stage allows to begin the modeling process, generating the point cloud of the object and the following stages of creating a mesh (surfaces) and the texture applied for a realistic 3D model.

Figure 3. Schematic approach to modelling, represent and visualize 3d objects

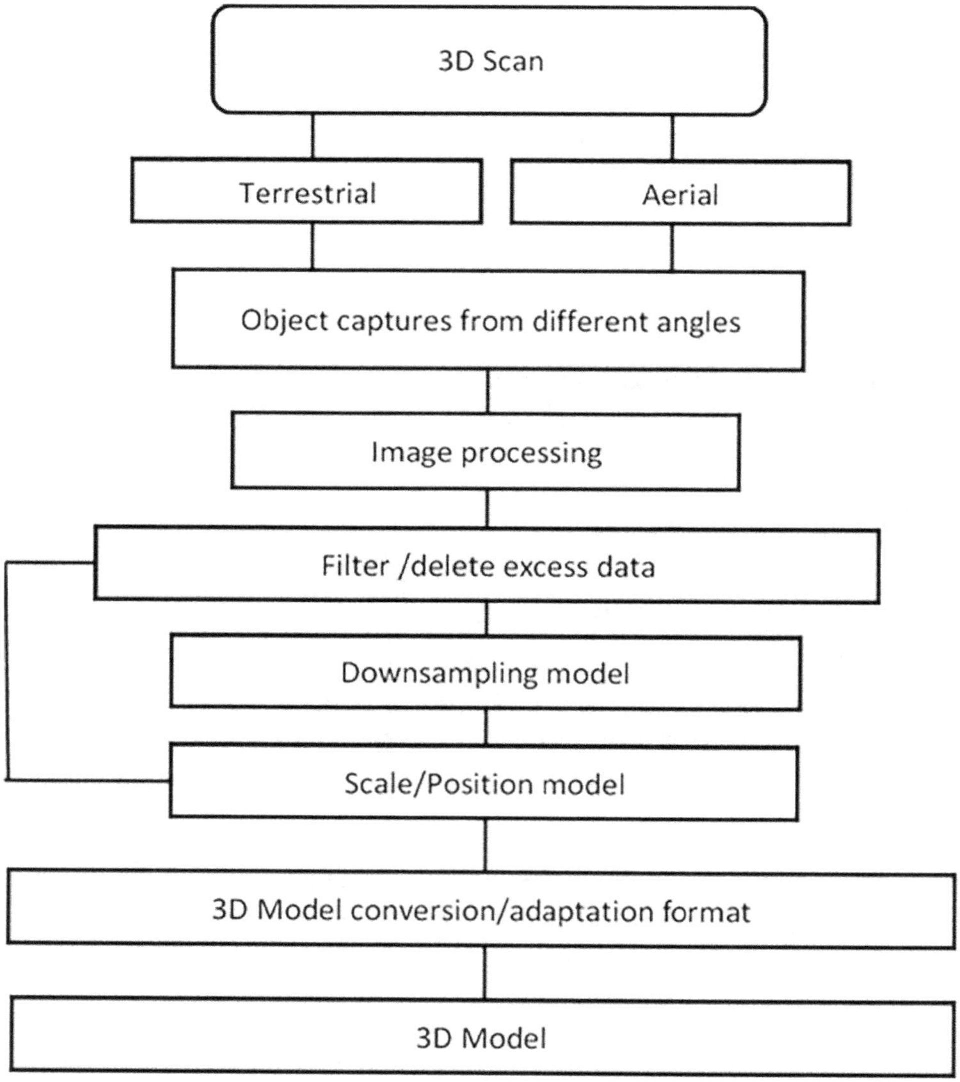

There are in the market several software to process UAV data such as: Agisoft Metashape; Bentley context capture; ESRI Drone2map; Pix4D; as the most commonly used by the 3D modelling community. Some of these tools have a higher level of atomization and others requires the insertion of parametrization in different stages of the imagery processing by the user. However, the workflow implemented in different software packages reveal many similarities (Figure 6). Agisoft PhotoScan solution is normally referred as highly user-friendly concerning modelling 3D data of terrestrial or aerial photographic surveys.

Figure 4. Dense Image Matching: terrestrial 3D point cloud through 2D images, Lisbon, Portugal

Considering the Image Processing stage from the 3D Scan procedures referred previously, the following sequence of technical procedures details and supports the fluxes of tasks to process and modelling 3D objects and explains how a 3D model is generated (Figure 6).

Figure 5. Dense Image Matching using UAV in Almada, Portugal.

Figure 6. Schematic approaches to image and 3D modelling processes

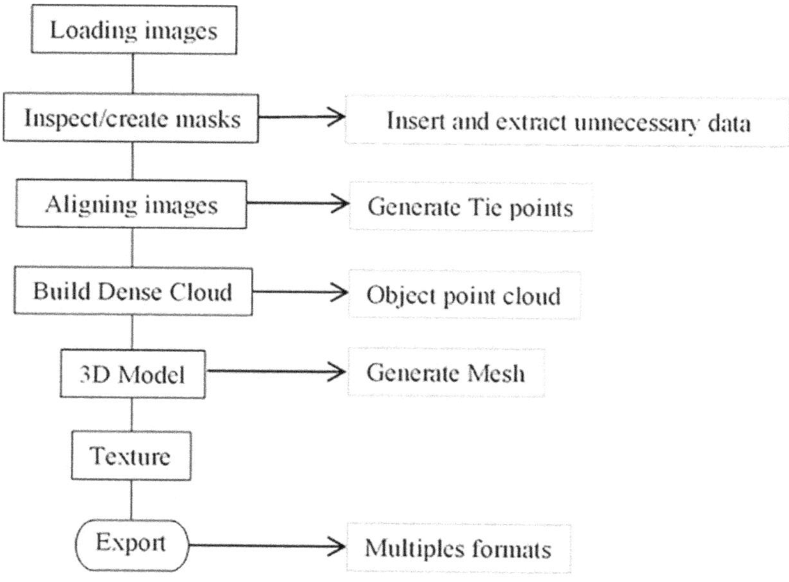

The results of the UAV survey consist in several outputs and visualization in different perspectives, such as 3D Point Cloud; Mesh (solid, wireframe); Texturized 3D model (anaglyph stereo visualization); DEM; DTM; and Ortho Image.

3D MODELLING AND PUBLIC INFORMATION IN SOCIAL NETWORKS

The advantages of using 3D point clouds and meshes (with or without textures) consist of the ability to rapidly gather higher temporal and spatial resolution information, practically on demand (high frequency). The source data can be acquired from 2D multimedia, digital cartography, 3D built model or 3D scan from different angles (terrestrial or aerial). However, considering the vast resources available in the internet, valuable data can be retrieved allowing to process this voluntary geographic information without the necessity to be physically present in the place and time (e.g. lost or inexistence object). This is especially more relevant after the state of emergency and confinement declared by numerous countries during the 2020 SARS-CoV-2 outbreak. Technology have been allowing to faster access to human sensed data through virtually environments. In this sense the real space can be represented creating virtual environments using only virtually distributed volunteered geographical data. The aim is to collect open source data and recreate objects and spaces from the actual place. These methods are more valuable when we apply remotely the methodology to places that we cannot access physically; and/or for some reason were disappearing; and/or from the past. For example, colleting images from one object that no longer exists allow us to comprehend the historical site and evolution of a determinate place in time, knowing the evolution of the surrounding space and modelling what an identity and part of the city was once. Examples of collecting data from the past were built with successful results, using data repositories of collections of ancient images. Using internet resources allowed to create remotely 3D models of distant objects and visualise them as good as data collected on site. After verifying the results of the remotely develop models and their coherency, allow us to confirm the robustness of the algorithms and methods. The application of these technics to ancient cultural heritage in the past recurring to old data archives will offer new possibilities to virtual time travel as well as register and inventory of past elements with relevant identity of the place.

Example of modelling 3D models from data available through internet were developed considering official inventories of the national and international repositories and using social networks as resources of public shared data online.

Modeling The Iconic Lisbon - Tower of Belém

Built between 1514 and 1520 strategically in the margins of the Tagus river, the Belém Tower in Lisbon is classified as world heritage by UNESCO since 1983 (DGPC and UNESCO, 2019). This example aimed to use municipal and national imagery archives and other online imagery resources data bases, such as Instagram, Google images or the former geotagged photo sharing mashup database: Panoramio. Most of the imagery available online are from the land view of the tower, being difficult to collect enough imagery (with similar color, texture, format and angles of capture) from the river side to establish key points. 43 images with better resolution, similar characteristics and without spatial reference were loaded in the software, however only 33 were align and used for the tie point stage (see Figure 6). Beside the lack of position on the river side, there were identified a conflict in the software likely associated to older pictures, probably due to its analog bases, although scanned. However, the result of the 3D model of the Tower is quite impressive, considering the number of images and the low time consuming for data acquisition and preparation (Figure 7).

Figure 7. Lisbon Tower of Belém 3D model using images from municipal and national archives and other online resources such as Google, Instagram and Panoramio

Modeling The Barcelona Arc De Triomf

Another example of remotely 3D modelling was searched for the city of Barcelona. Using only data retrieved from the Google Earth street view, for this trial was searched one object with captures of all sides. The Arc de Triomf was designed by the architect Josep Vilaseca i Casanovas and was built in 1888 as the main entry to the Barcelona International Exposition.

The Arc de Triomf is located near a round abound with a good coverage of images from the northern side. As the source of data is only one and the images tend to have the similar characteristics (e.g. color, texture, resolution, format), there is a higher probability of the software recognizing key points. From 44 images loaded, only 5 were not align (see Figure 6). The result was very accurate with the measurements of the 3D object very close to the real (Figure 8).

Figure 8. Barcelona Arc de Triomf 3D model using images from Google Earth Street View

Modeling The Paris Cathedral of Notre Dame

The cathedral is one of the most widely recognized symbols of the city of Paris and the French nation. A quintessential Gothic cathedral, building started in 1163 and lasted over 170 years (Centre des Monuments Nationaux, 2019). Included on the Paris, Banks of the Seine as a UNESCO World Heritage site, the cathedral is associated to Victor Hugo's novel, The Hunchback of Notre-Dame, published in 1831. On 15 April 2019 the cathedral caught fire, destroying the spire and the oak frame and lead roof.

After the fire of 2019, a growing number of information about the monument became available online. The objective of developing a 3D model of the monument was not to replace the existing models, then demonstrate that it is feasible to create a virtual reconstruction using simply open source data available online, the robustness of the algorithm and to modelling through time. The importance and relevance of the cathedral has the support and enough archive data to perform a desirable restoration. It became highly recommendable to gather relevant information over the city elements and monuments to preserve digitally enough information as might be indispensable for the future.

After some research over available data pre-fire online, it was used two videos available on YouTube with a circular aerial footage of the cathedral, to collect the frames from different angles. These sources of data had some differences in the time of acquisition and dissimilarities in the building structure and imagining with problems in the existing scaffoldings in the center and back of the building. The option was to include only the areas with the main structure of the building. The high resolution of videos, characteristics of the image and coverage allowed to align the total number of images loaded, however was needed to create masks for all the frames collected aiming to reduce the noise of the city and nearest buildings (see Figure 6). The result is very interesting with some highly detailed areas as visible in the figures 9 and 10.

Figure 9. Paris Notre Dame Cathedral model using images two YouTube videos

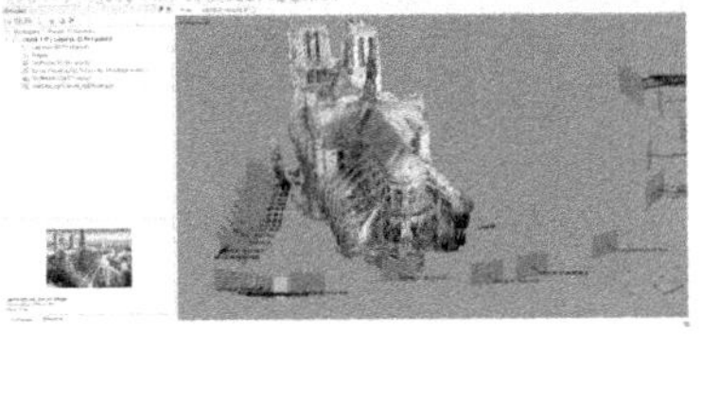

Figure 10. Paris Notre Dame Cathedral model from different perspectives and details

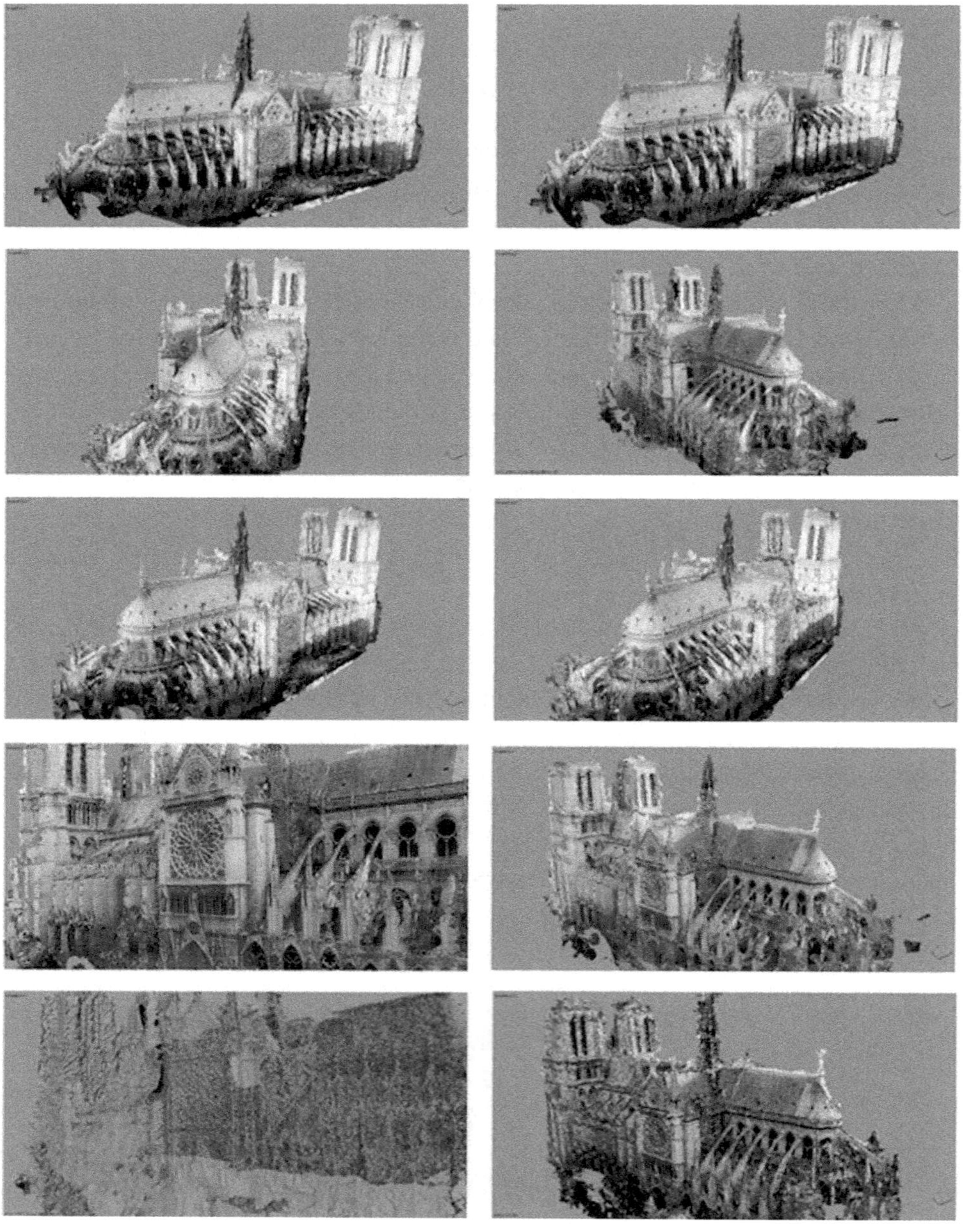

The aim of this work was not to create a high detailed 3D model of the cathedral, once there are existing acquired detailed models prior fire (please note that lase scan surveys were already done and highly detailed models are existing for this iconic monument), but to demonstrate the potential and robustness of the technology at low or no cost. Therefore, the use of shared data though social networks or technical archives allow generating models extremely realistic and certainly valuable for objects that were lost or are for some reason inexistent.

AUGMENTED REALITY VISUALIZATION

The mobile market has been expanding, exceeding by far computer sales. Smartphones combine in a small device, features and functions with multiple advantages, such as portability (size and weight) and associated capabilities; computational and graphical capabilities; wide range of connectivity spanning form 3G, 4G, GPRS[3], HSPA(HSPA+)[4] to Wi-Fi; Bluetooth and Infrared interfaces, built in GPS/GLONASS receivers (plus Assisted capability); electronic compass; gyroscope; accelerometers; touch screen capability; light and proximity sensors; front and rear cameras; battery life (longer than laptops); microphone; speaker; and a varied offer of applications and functions with or without cost for the end user. (adapted from Magliochheti *et al.*, 2012 *in* Marques, 2017). Using mobile devices for exploring and interacting with the user's physical environment is among the most promising improvements for commercially successful future telecommunication services and applications. The geographical positioning features are essential for mobile devices that need to identify their accurate position systematically and resorting to geographical information offering multiple solutions to operate interactively answering several requests, such as viewing the position in a map; multiple navigation types (vehicles or on foot); or showing nearby objects in different representational ways, including their evolution in time. This capability is especially relevant in mixed environments where the device needs to (re)present virtual information in the real place (or vice versa) and thus it may be considered as the fundamental base connection feature between the synthetized and the real worlds.

Two-dimensional digital representations of a 3D environment require cognitive resources and topological reasoning to relate the mobile representation to reality. The development of new techniques that bring together location-aware portable devices and visualization of interactive 3D representations is an exciting pervasive environment for users. High-quality geographic data sources are becoming increasingly more consistent as a resource that contributes to territorial management and the creation of detailed 3D models of the real environment. Prototypical representations for mobile devices can be developed to experiment different input types depending on the device technical features. Examples of this are the 3D representations on devices, which allow to be rotated using the numerical or cursor keys; or if the device is equipped with acceleration sensors, then the view can be adapted to the measured angle; or if a digital compass enables an intuitive orientation it can rotate the view mimicking the user's movement.

Figure 11. Paris Notre Dame Cathedral model within an AR environment

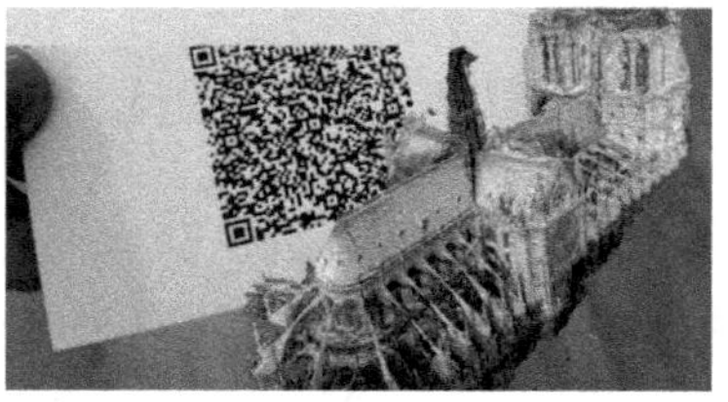

The availability of extensive and detailed 3D models offers completely new visualization possibilities on mobile devices, e.g. in navigation or guidance (video or audio) and representation of geo-referenced content. Representations of virtual environments were until now limited to desktop computers. Nowadays, they can take full advantage of the unique attributes provided by mobile computing (as the examples referred before, such as portability, connectivity, context-awareness, and multimodal interfaces).

While VR require features that are more associated with impressiveness, usually demanding more static background and higher level of realism in order to embrace a larger amount of human senses to activate the user's absence from that place and feel existing in the virtual environment; by contrast, AR requires a more dynamic system that is portable and in continuous interaction with the real world in order to view the physical environment augmented with digital representations (Pinto *et al.*, 2013).

AR enhances the real world instead of replacing it. The users can view the real world enriched with additional 3D graphics superimposed to their field of view. The possibility of combining real and virtual objects will allow a huge amount of applications. In the field of Cultural Heritage, AR is of the most outstanding technologies, considering the possibilities to recreate 3D representations, to "virtual time travel" (fundamental for cultural elements that are often historically relevant), and combine the virtual with the real world (e.g. recreate some object from the past or viewing several phases of construction and using the spatial environment as a marker). This is more valuable when using mobile devices since these are portable and can be used on site, knowing the user's position through navigation systems. Thus, AR made it possible to enhance the real world with data and synthetized 3D detailed representations, changing the way we perceived the world and having spatial aids to navigate and alert us to our interests' proximity in the field (e.g. emitting a sound or vibrating while navigating near some relevant object), and recreate / foresee objects in evolution or presently inexistent (therefore: re-presented).

AR and VR as tools can enable an increase interest and access to information and thus learning capabilities (Longo, 2018). The application of this technology to promote cultural heritage are attracting new visitor (tourist) or technical profiles by improving their experience. In addition, AR and VR capabilities have been converting gradually into mechanisms for the protection and conservation of cultural heritage, either through digital recovery of lost cultural assets or by recreating virtual worlds that are technically inaccessible for diverse reasons or in a higher level, to support rehabilitation (Barrado-Timón & Hidalgo-Giralt, 2019).

The figure 11 shows the result tiled 3D model generated using online virtual data within an AR environment. It was used a QR code card as marker and when the user rotates the card, can visualize the 3D model from all the angles and interact with.

CONCLUSION

Technological developments in hardware and software have been allowing gradually the use of higher level of detailed and realistic 3D models of the real world. On the side of acquisition, 3D modelling techniques (such as photogrammetry) are supporting the generation of more realistic and truthfully representations of the cultural heritage. The technical information, which can acquire through the processes of 3D modelling, are becoming increasing relevant in digital inventories and/or archives. Thus, datasets of 3D models are essential to be built to support the study, preservation and intervention in heritage at a global level. However, the volume of objects and the need of constant actualization might create a problem on the side of the acquisition. As these objects are often a place of interest, the use of volunteered data such as pictures of videos available through the internet offer new possibilities to generate such models. In addition, it is possible to integrate historical data or future scenarios. On the side of the user, the widespread smartphones and other devices with a progressively capacity of computation allows to the extensive in the use of these information. The visualization of 3/4D data within an VR environment allow to immerse in a virtual world parallel to the real being able to navigate and visualise the object from different perspectives and adding or subtracting elements or data which enhances the user experience, supporting potentially past configurations or evolution of the place and/or simulations of future interventions. Allows also to virtual travel and time travel over valuable heritage virtually from anyplace (especially relevant during confinement). Instead of a total immersion in the virtual world, AR enhances the real place with synthetic information, allowing to fuse the virtual data with the real. These capabilities have the possibility to increase the level of information available and the interest of this heritage. The demonstrated capabilities of these technologies are not only interesting for visualization and archive as they increase the intrinsic value of cultural heritage for the users: public, decision makers, technicians and promoters. Thus, also supports social-economic, technical-scientific and political-administrative symbolic-religious development.

ACKNOWLEDGMENT

This work is financed by national funds through FCT - Foundation for Science and Technology, I.P., within the scope of the project «UIDB / 04647/2020» of CICS. NOVA - Centro Interdisciplinar de Ciências Sociais da Universidade Nova de Lisboa.

REFERENCES

Aber, J., Marzolff, I., & Ries, J. (2010). *Small-Format Aerial Photography, Photogrammetry*. Elsevier.

Andrés, M., & Pozuelo, F. (2009). Evolution of the architectural and heritage representation. *Landscape and Urban Planning*, *91*(2), 105–112. doi:10.1016/j.landurbplan.2008.12.006

Barrado-Timón, D., & Hidalgo-Giralt, C. (2019). The Historic City, Its Transmission and Perception via Augmented Reality and Virtual Reality and the Use of the Past as a Resource for the Present: A New Era for Urban Cultural Heritage and Tourism? *Sustainability*, *11*(10), 2835. doi:10.3390u11102835

Batty, M. (2013). *The New Science of Cities*. MIT Press. doi:10.7551/mitpress/9399.001.0001

Carvajal-Ramírez, F., Navarro-Ortega, A., Agüera-Vega, F., Martínez-Carricondo, P., & Mancini, F. (2019). Virtual reconstruction of damaged archaeological sites based on Unmanned Aerial Vehicle Photogrammetry and 3D modelling. Study case of a southeastern Iberia production area in the Bronze Age. *Measurement*, *136*, 225–236. doi:10.1016/j.measurement.2018.12.092

Eckert, G., Cassidy, S., Tian, N., & Shabana, M. E. (2019). Using Aerial Drone Photography to Construct 3D Models of Real World Objects in an Effort to Decrease Response Time and Repair Costs Following Natural Disasters. In K. Arai & S. Kapoor (Eds.), *Advances in Computer Vision* (pp. 317–325). Springer International Publishing.

Golledge, R. (Ed.). (1999). *Wayfinding Behavior: Cognitive Mapping and Other Spatial Processes*. Johns Hopkins University Press.

Goodchild, M. (2007). Citizens as sensors: The world of volunteered geography. *GeoJournal*, *69*(4), 211–221. doi:10.100710708-007-9111-y

Ilci, V., Ozulu, I., Bilgi S., & Alkan, R. (2019). The Usage on Unmanned Aerial Vehices for 3D Mapping of Archaeological Sites. Fresenius Environmental Bulletin, 968–974.

Innerarrity, D. (2010). *O Novo Espaço Público* (M. Ruas, Trans.). Teorema. (Original work published 2006)

Lin, Y., Hyyppa, J., & Jaakkola, A. (2011). Mini-UAV-Borne LIDAR for Fine-Scale Mapping. *IEEE Geoscience and Remote Sensing Letters*, *8*(3), 426–430. doi:10.1109/LGRS.2010.2079913

Longo, F., Nicoletti, L., & Padovano, A. (2018). An interactive, interoperable and ubiquitous mixed reality application for a smart learning experience. *International Journal of Simulation and Process Modelling*, *13*(6), 589–603. doi:10.1504/IJSPM.2018.095864

Marques, L. (2017). *Augmented Valuation of Cultural Heritage through Digital Representation based upon Geographic Information Technologies (GIT)* (Unpublished doctoral dissertation). Universitat Politècnica de Catalunya. Departament de Tecnologia de l'Arquitectura, PhD thesis, 341. Retrieved from: http://luisfilipemarques.com/site/wp-content/uploads/2017/10/LM_PhDthesis_UPC_ETSAB_2017.pdf

Marques, L., Tenedório, J., Burns, M., Romão, T., Birra, F., Marques, J., & Pires, A. (2017). Cultural heritage 3D modelling and visualisation within an augmented reality environment, based on geographic information technologies and mobile platforms. *Architecture. City and Environment*, *11*(33), 117–136. doi:10.5821/ace.11.33.4686

Pinto, N., Tenedório, J., Antunes, A., & Cladera, J. (2013). *Technologies for Urban and Spatial Planning: Virtual Cities and Territories*. IGI Global. doi:10.4018/978-1-4666-4349-9

Pires, A. (2016). *Combining paper maps and smartphones in the exploration of cultural heritage* (Unpublished master dissertation). Mestrado em engenharia informática, Faculdade de Ciências e Tecnologia da Universidade Nova de Lisboa. Retrieved from https://run.unl.pt/handle/10362/19962

Rebelo, C., Rodrigues, A., Tenedório, J., Goncalves, J., & Marnoto, J. (2015). Building 3D City Models: Testing and Comparing Laser Scanning and Low-Cost UAV Data Using FOSS Technologies. In Computational Science and Its Applications -- ICCSA 2015 (pp. 367–379). Springer International Publishing.

Templin, T., Brzezinski, G., & Rawa, M. (2019). Visualization of Spatio-Temporal Building Changes Using 3D Web GIS. *IOP Conference Series. Earth and Environmental Science*, *221*, 012084. doi:10.1088/1755-1315/221/1/012084

Tenedório, J., & Marques, L. (2017). How can 3D models and augmented reality visualization based on mobile platforms enhance the value of urban heritage? *International Federation of Surveyors*. Retrieved from https://www.fig.net/resources/monthly_articles/2017/tenedorie_etal_december_2017.asp

Tversky, B. (2005). Functional significance of visuospatial representations. In Handbook of Higher-Level Visuospatial Thinking, (pp. 1–34). Cambridge, UK: Cambridge University Press.

KEY TERMS AND DEFINITIONS

3D Models: Representation of objects based on its position x, y (2D) and z (height).

Augmented Reality: Enrichment of the real world with a complementary virtual world. The digital information or representations are merged with the real view via a device that combines both real and virtual. An AR system expands the real world scene allowing the user to maintain the sense of presence in that world.

Cultural Heritage: Assumes the notion of value and cultural inherency, often related with unicity, rareness, importance, memory, or symbolic representation of one or more specific objects, events, milestones, achievements, or historical significance. Value has always been the reason underlying heritage conservation. However, the notion of valuation might be identified or increased, when supported individually or collectively by processes of extend their intrinsic value.

Geographic Information: Data about places on the Earth's surface. Representations (analog or digital) of the external environment in space and time.

Geographic Information Technologies: The role of technics and tools to acquire, process, store, integrate and visualize geographic information (e.g. Global Navigation Satellite Systems, Geographic Information Systems, Remote Sensing, Unmanned Aerial Vehicles, mobile, apps, 3D, Virtual and Augmented Reality).

Mixed Reality: Complements reality/virtuality rather than completely replacing it, appearing to the user that the virtual and real coexist in the same environment.

Virtual Reality: Computer generated, interactive, three-dimensional environment in which a person is immersed. Alternative reality (illusionary and represented) that seems to possess all the attributes of reality, truly and objective.

Volunteered Geographic Information: Phenomenon known as user-generated content. A widespread engagement of large numbers of citizens, often with little in the way of formal qualifications, in the creation of geographic information. They are largely untrained and their actions are almost always voluntary, and the results may or may not be accurate. But, collectively, they represent a valuable set of data about the spatial environment in time.

ENDNOTES

1 URL: www.geodrone.pt (09 Jun 2019).

2 URL: https://globaldigitalheritage.org (09 Jun 2019).

3 System for mobile communication network (data service).

4 Mobile protocols.

Chapter 15
Unmanned Aerial Vehicles for Smart Cities:
Estimations of Urban Locality for Optimization Flights

Tetiana Shmelova
https://orcid.org/0000-0002-9737-6906
National Aviation University, Ukraine

Vitalii Lazorenko
National Aviation University, Ukraine

Oleksandr Burlaka
https://orcid.org/0000-0002-8883-6130
National Aviation University, Ukraine

ABSTRACT

In this chapter, the authors are presenting opportunities for the use of unmanned aerial vehicles (UAV) in town. Methods for the optimization of flight routes of UAVs in the dependence of target tasks in the city are presented, for example, area monitoring; search and rescue operations; retransmission of communication (in places, where the antenna coverage cannot be set due to terrain specifications); organization of logistics as the safe, cheap, and fast transportation method of goods; for aerial photography, for controlling traffic; for the provision of the first aid to people in emergencies; unmanned taxi. It is done using air navigation information and mathematical methods. Authors suggest dynamic programming methods, GRID analyses, expert judgment method, and fuzzy-logic methods for estimation of risk/safety of flights in the city. Optimization of flows and flexible redistribution of UAV routes in multilevel airspace is provided according to air navigation requirements and standards.

DOI: 10.4018/978-1-7998-2249-3.ch015

BACKGROUND

In recent years, Unmanned Aerial Vehicles (UAVs) became very popular in developed countries as one of the effective solutions to some tasks in city and agriculture areas (Austin, 2010; Chekunov, 2010; Gulevich, 2012; Sładkowski, 2019) in logistic as the safe, cheap and fast method of transportation of cargo; for aerial photography; for controlling road traffic. Nowadays UAVs are used to perform many tasks that were previously difficult to solve such as observation and monitoring missions in hard-to-reach places (forest, mountains, sea, rivers, lakes, big parks); monitoring forest fires; search and rescue operations; alternative performance of a difficult agricultural activity (aviation chemical work); relaying of communication signals in places where antenna coverage cannot be set because of terrain; for first aid to people in various life situations. Earlier since 1961 UAVs were widely used in some military operations (Austin, 2010; Chekunov, 2010). Now UAVs are effective both in military and civil aviation (Kreps, Zenko, 2014; Ignatiev, 2010; Sładkowski, 2019; Bondarev, Jafarzadeh, Kozub, 2014).

UAVs have several advantages, namely low operating cost, simplicity, availability. UAVs may be used there, where the usage of manned aircraft is impractical, expensive or dangerous. The main advantage of using UAVs is areas with extra high risks to humans or large and inaccessible areas with the necessity in control single or group UAVs flights in cities or agriculture terrain (Bondarev, Jafarzadeh, Kozub, 2014).

The use of a single-operating or a group of UAVs is opening a big variety and a principally new level of complexity of target tasks and missions lowering the presence of a human itself while task execution of smart governance in the future cities. The diverse application of UAVs is ensuring the utmost safety, economic efficiency, precision and quality, minimal environmental impact and complying with other essential requirements and parameters in future smart cities. Meeting these expectations and requirements would be possible with the introduction of operation management, decision support systems and control systems capable of supporting versatile and autonomous teams of UAV performing a wide range of complex tasks and activities with separation of roles in the team and with minimal human control while supporting the highest standards of safety (Austin, 2010; International Civil Aviation Organization (ICAO), 2015; Gulevich, 2012). In some cases, the use of a single UAV becomes ineffective. It is advisable to group (collective) use of UAV (Austin, 2010; Bondarev, Jafarzadeh, Kozub, 2014; Bondarev Shmelova, Kucherov, 2015; Shmelova, Bondarev, Kucherov, 2016; Shmelova, Bondarev, Lazorenko, Burlaka, 2019; Shmelova, Sterenharz, Burlaka, 2019; Shmelova, Kovalyov, Dolgikh, Burlaka, 2019):

- relay communications in places where it is impossible to install coverage antennas through difficult terrain;
- in agriculture (group field spraying);
- in aerial photography (group shooting of large territories);
- patrolling of large territories;
- monitoring of forest fires (group flights);
- monitoring of social behavior, noise and landscape pollution;
- movement of big parties of the cargo.

Additional useful properties of the group of UAVs comparing to one UAV are faster coverage of area fragment and consequently more efficient at photo/video monitoring, retransmission of communication and agricultural operations (Austin, 2010; Gulevich, 2012; Sładkowski, 2019).

For now, UAVs are going to be used in so called "Smart cities" for dynamic management processes and urban planning. A *Smart city* is a city in which new technologies are used to increase the life comfort of people, such as smart homes, traffic town control, planning town, flexible lighting control (González, 2016). The idea of "smart city" is characterized, first of all, by using the new achievements such as Artificial Intelligence (AI), Internet technologies in order to monitor the state of urban infrastructure facilities, their control, and based on the data obtained as a result of monitoring, optimal allocation of resources and ensuring the safety of citizens (Vyrelkin, Kucheryavy 2014; Gulevich, 2012). In "Inventing Future Cities", Michael Batty explores what needs to understand about cities in order to invent their future (Batty, 2019): "We cannot predict future cities, but we can invent them", so will be designed demands for a change of paradigm in Governance and Urban Planning. Some problems will persist (social cohesion, inaccessible territory, traffic jam, health care accessibility), but new technologies appear for optimal decisions such as privacy, safety, and security; data access, OpenGovData, Centre of Operation Intelligence for cities and data corporation; 5G and IoT; the using UAVs for difficult problems (Kitchin, 2014; Dolgikh, 2018).

Interesting are the shows in which groups of UAVs dance to famous music in the night sky. For example, controlled UAV with LED elements installed on them simultaneously depict various colorful figures in the air for use multi-colored lights. They were synchronously running and lighting in the night sky over the airport of the German city of Tornesch to the accompaniment of the orchestra performing classical music. Controlled on the ground by a crew using personal computers, the mass of UAV lit up the night sky in sync to a live performance of Beethoven's Fifth Symphony and executed a stunning light show resembling a fireworks display. It was a new Guinness World Record for group flight of UAVs (Guinness, 2017). The astonishing footage was shown for the first time during Intel CEO Brian Krzani's keynote on the opening day of CES technology tradeshow in Las Vegas, USA. A spectacular display of UAV technology by Intel Corporation involving 100 small aircraft being launched skywards information has earned a new world record title for the Most Unmanned Aerial Vehicles (UAVs) airborne simultaneously. The record of the show was set in collaboration with Airs Electronica Futurelab to push the limits of the UAV industry and to show what UAVs can be used for. The Forum "Urban Air Mobility" in November 2018 at Amsterdam discussed the future of UAVs in cities. Looked at from the perspective of cities and citizens, urban air mobility and the idea of Mobility as a Service (MaaS) provide a fascinating view of a possible future where a daily commute could seamlessly include a bicycle, train, and UAV service all as part of an integrated public transportation system (Drone life, 2018). In the UAV industry, UAVs as a mode of transportation are viewed primarily as a technical or regulatory problem, but Forum "Urban Air Mobility" view UAVs as one part of a city solution (Figure 1).

Figure 1. Forum "Urban Air Mobility" (Drone life, 2018)

Many of these tasks are for an urban locality and wherein effectively use single and group flight of UAVs (Kucherov, Kozub, 2015; 2016, 2017; Amelin, Antal, Vasiliev, 2013; Kharchenko, Shmelova, Bondarev, & Stratij, 2017; Shmelova & Bondarev, 2016; Shmelova, Bondarev, Lazorenko. Burlaka, 2019; Shmelova, Sterenharz, Burlaka, 2019; Shmelova, Kovalyov, Dolgikh, Burlaka, 2019). In this sense, the use of UAVs group flights is more appropriate, for example, for photo/video monitoring; group survey of large areas and patrol areas; delivery of cargo and the use of unmanned taxi to carry passengers. Noted additional useful properties such as faster coverage of big area fragment of urban and minimal risk in the movement of UAVs in so-called Smart City. Therefore, some disadvantages of UAVs connected to the limited capacity because of the small size of UAV can be satisfied by the use of group flight (Shmelova & Bondarev, 2015; 2016).

When planning UAV flights, it is important to comply with regulatory air navigation requirements and effective methods for flight operations (ICAO, 2009; 2011; 2015). The documents of ICAO are containing requirements and rules for the certification of UAVs and operators; UAV registration; rules for UAV operations; communication with the UAV; training of personnel for the operation of the UAV; emergency situations with UAV and flight safety; legal issues to ensure the possibility of performing safe, coordinated and effectively integrated flights UAVs (ICAO, 2005; 2011; 2015; 2017a, 2017b; 2019).

Emergencies may occur when changing modes of UAV operating. For operations carried out "manually", the human factor plays a very important role and a significant part of emergencies arise due to erroneous actions of the operator. Using a constant two-way radio leads to continuous Manual control of device parameters, which result into certain restrictions and inconveniences. The operator cannot be distracted from the controlling and takes full responsibility for the UAV and all other safety issues. Decision Support Systems (DSS) are creating for operators of UAVs which are including Data Bases and Data Models for UAVs' optimal management in a Smart city. The eGov technology may be used for DSS operators as the software, database and web design of target models (Shmelova, Bondarev, & Stratij, 2017). The using eGov methods provide comprehensive, innovative software and web design local governments and help for different problems in the city. The effectively use DSSs in emergency situations. Emergency situations and unforeseen circumstances include emergency landing/emergency landing sites; loss of the C2 line; interception; complicated meteorological conditions, technical problems with UAV, human factor problems, electromagnetic interference, disasters and the like. Flight planning of the UAV should be provided for the possibility of performing an emergency landing of the UAV in places where the risk to the safety of people or property on the ground will be minimal. As part of the pre-flight planning, alternate aerodromes/sites should be identified, respectively, to be returned in case of an emergency or an emergency situation due to meteorological conditions.

But despite a number of advantages, there are some drawbacks, namely the main problem associated with the use of airspace allocation of the frequency range for UAVs management and transmission of information from the board to the ground; lack of recommendation action algorithm of UAV operator in case of emergency situations (Kharchenko, Shmelova, Bondarev & Stratij, 2017; Shmelova & Bondarev, 2015; 2016). In (Kharchenko, Shmelova, Sikirda, 2012; 2016) an emergency engine stop, electric problems, and displaying altitude incompatibility have been investigated. In these situations, the parachute system is activating automatically, with transferring the coordinates of the forced landing to the operator's monitor. The use of a parachute landing system will not only provide reliable survival craft in an emergency but also simplify its operation. When a loss of communication with the UAV occurs, it reports to the control unit immediately. The report states the time and place of loss of communication, altitude of the UAV, the estimated remaining time of flight and the latest indication of heading (Kharchenko, Shmelova, Bondarev & Stratij, 2017; Shmelova & Stratij, 2015).

For the effective use of UAVs in a smart city, it is necessary to perform the following tasks:

- analysis of target tasks that can be solved using UAVs in a smart city;
- selection of the optimal flight (single / group) UAV for a specific target task using Expert Judgment method (EJM);
- determination of the group flight configuration and UAV - repeater using graph theory;
- risks assessment of UAVs movement on a certain territory of the city using EJM, fuzzy logic, Decision Making (DM) in Risk;
- determination of ways the minimum cost and levels flight using EJM, dynamic programming method;
- the building of DSS for UAV operators to determine the optimal actions in an emergency using DM in Risk and Uncertainty methods.

UAVs FOR SMART CITIES: ESTIMATIONS OF URBAN LOCALITY FOR FLIGHT OPTIMIZATION

The idea of smart-city became very popular in all countries. It is an urban area, which uses many electronic sensors to collect data and use it to manage assets and resources efficiently. The concept of "smart city" is characterized by using the new achievements such as using Artificial Intelligence (AI) and Internet of Things (IoT) in order to monitor the state of urban infrastructure facilities and control them, based on the data obtained as a result of monitoring, optimal allocation of resources and ensuring the safety of citizens (González, 2016; GSMA, 2016; Vyrelkin & Kucheryavy, 2014; Kirichek & Makolkina, 2016; Montgomery, Bekey, 1998). Nowadays, the using of UAVs may help to perform many tasks that were previously difficult to solve.

The effectiveness of presenting UAVs used for a "smart city" has some problems:

- the representation of buildings, roads, construction, recreation areas, and natural areas - flight risks should be considered;
- availability of specific orders of UAV flight, such as target use of UAV - optimal UAV flight choice should be considered;
- air navigation requirements for flight of manned and unmanned aircraft - rules of UAV flights should be considered (ICAO, 2011, 2015).

The "smart city" is an aggregate of several information and communication technologies, mathematical methods and AI (ICAO, 2017). The usage of UAVs in the smart city concept will help to solve some tasks as: traffic jams monitoring; search and rescue missions; photo/video observations; Wi-Fi hotspot or repeater; the transportation of goods; taxi service; ambulance. To effectively solve these problems the DSS operators of unmanned aircraft have been developed. For the control of the UAV, the system is using the next scheme of UAV control (Figure 2):

- Operator - single UAV;
- Group of operators – group of UAVs;
- Operator – Central Drone Repeater (CDR) – group of UAVs.

Figure 2. Control system by one or a group of the UAV

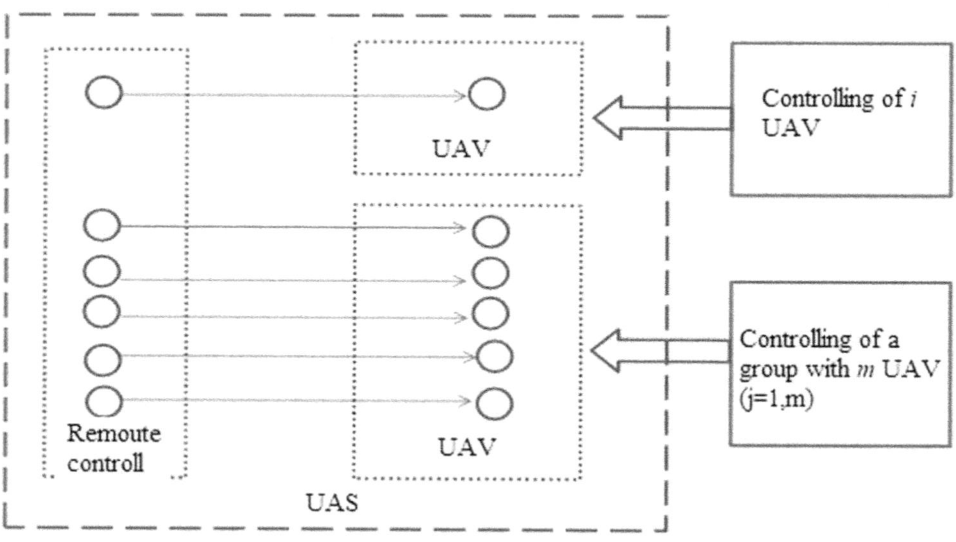

For the group flights of UAVs, it is advisable to apply specific group reliability criteria (personal computers or aircraft which are managed from center): connectivity, structural redundancy, uneven distribution of communications, structural compactness, degree of centralization in the system, survivability (Olifer, 2006)

Using graph theory, we can determine the effectiveness of different structures (topologies) in UAVs group formation. To control a group of UAVs from Remote Piloted Aircraft System (RPAS) suggested choosing and using a CDR to connect to the operator on the ground and control other UAVs (Figure 3) using the method of server selection in local computer networks (Olifer, 2006; Kuzin, 2011; Shmelova & Bondarev, 2015; 2016; Kucherov, Kozub, 2015; 2016, 2017).

Figure 3. Control system of UAVs group from RPAS using CDR-drone

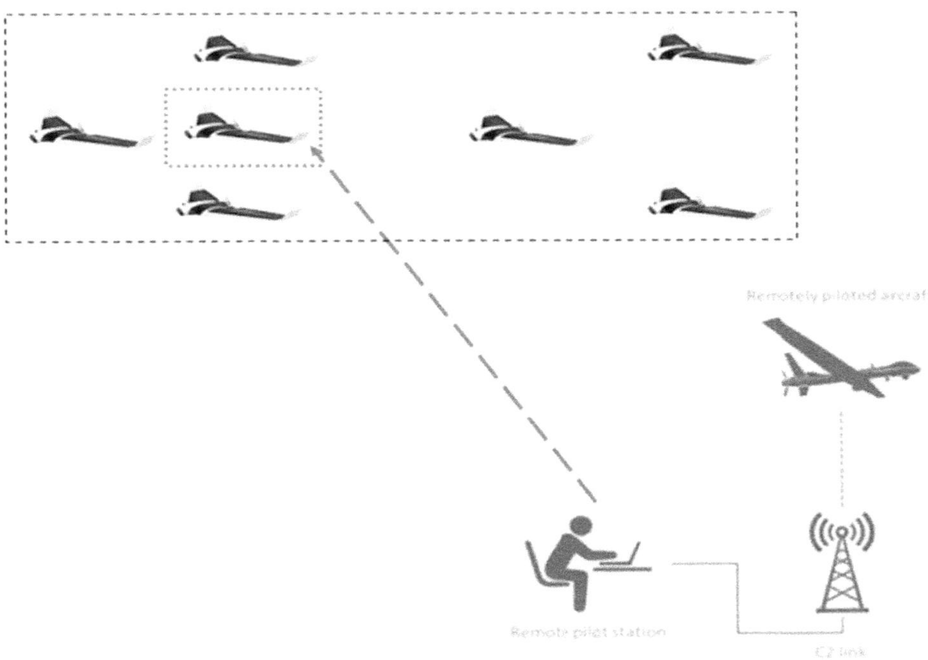

Distributed Decision Support System (DDSS) has been developed for the purposes of planning and flight control of UAV. DDSS represents a complex system with multiple interactions geographically distributed local remotely piloted aircraft (RPA). During the flight, UAVs may be controlled by remote piloting station (RPS). At any given time t_i k-UAV to be piloted by only one j-th RPS, if necessary, at time t_i+1 to be transmitted to the control (j + 1)-th RPS. This transfer flight control of the *j*-th RPS to ($j + 1$)-th RPS to be safe and effective, which is provided through the local UAV operators. To coordinate interaction and exchange of information between RPA pilots a local RPS NoSQL database has been developed (Tarasov & Gerasimov, 2007; Shmelova & Stratij, 2015). The computer program for DSS "Remote Expert Air Traffic Management System / Decision making in a common environment Flight & Flow Information for a Collaborative Environment (FF-ICE)" of the unmanned aircraft pilot (operator) has been developed and may be used for control of UAVs in a smart city (Khaletskiy, 2019). This tool uses 'blockchain' technology for the provision of interaction between RPS and RPA. By the point, the 'blockchain technology is a highly secured way of sharing the information, and it verifies learning achievements and Collaborative Decision Making (CDM) too.

Figure 4. Decentralized-distributed UAVs control system with blockchain connection between RPA

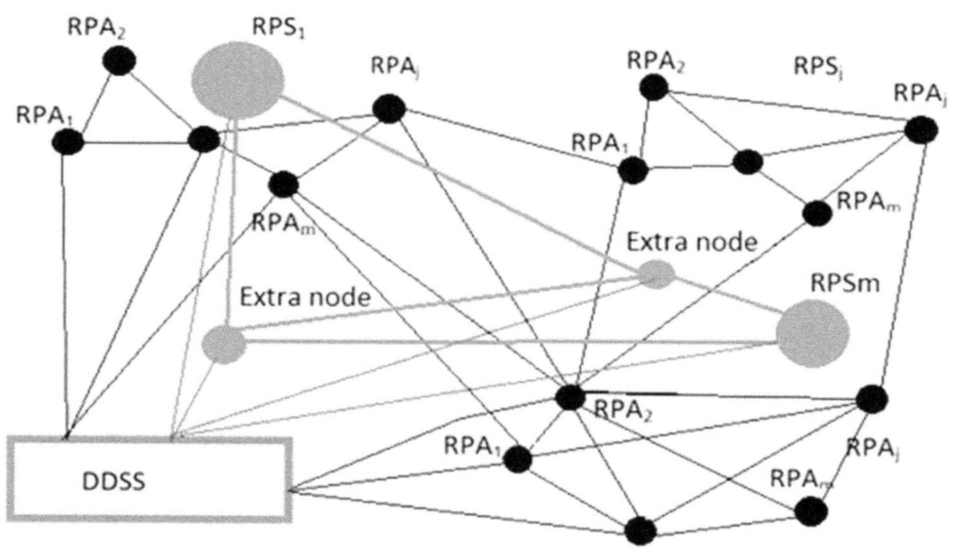

For today, the key to ensuring the safety of flights is the problem of the organization of CDM by all the operational partners based on general information on the flight process and ground handling of the manned and unmanned aircraft (Shmelova & Bondarev, 2015; 2016). There are many advantages of this approach such as enhanced security, big data analysis, and record keeping, real-time constant data exchange, (ICAO, 2018; Kitchin, 2014).

For the effective performance of work in a smart city, a classification of UAVs has been developed according to classic target tasks, tactical and technical characteristics of UAVs such as cargo capacity, flight range, construction, ways of take-off and landing; the number of UAVs in the flight. By the purpose, the UAV classified as: agricultural, forest and town works; surveillance and monitoring for situations such as natural, technical and social ordinary/emergency situations; organization of search and rescue; organization and controlling of traffic road; organization and improve performance logistics, and communications, urban planning; dynamic management and urban planning (Table 1).

For the management of the UAV by the task appropriate management system is proposed. Classification is taking into account all limitations and features of UAVs and theirs correspondences to target tasks. It is proposing to represent group of UAVs configuration as a graph with *n* tops (UAVs). They connected by *m* - arcs between each other - information data channels for UAVs control, the most common of which are: fully connected star, circle, tree, with a common tire, tire-star, star-circle, hybrid-pantry, (Olifer, 2006; Sizova O.A, 2012).

The effectiveness of UAVs group configurations is presented in Table 2. Evaluation criteria of UAVs group reliability are structured:

- Structural redundancy – *R;*
- The uneven distribution of connections – ε^2;
- Structural compactness – *D;*
- Relative distance – D_{rel};
- Level of system centralizing – *d;*
- Centrality - δ*;*
- Periphery – *P;*
- Survivability and moment of the network.

The Algorithm of finding Central Drone Repeater (CDR) in group of the UAVs in flight

1. To build graph *G (n; m)* for UAVs group
2. To build "adjacency matrix" $A = \|a_{ii}\|$ of graph *G (n; m) of* UAVs group:

$$a_{ij} = \begin{cases} 1, i \leftrightarrow i \\ 0, i \not\subset i \end{cases}$$

Table 1. Classification of UAV types

Nº	Class	Classification	Subclass	Code
1	A	UAV classification by purposes	Surveillance	A_1
			City works	A_2
			Other works (agricultural, forest, sea, river, steppe, etc.)	A_3
			Dynamics management and urban planning	A_4
			Relays communications	A_5
			Logistic	A_6
			Search and rescue operations;	A_7
			Organization of traffic, controlling traffic road, traffic problems	A_8
			Aerial photography	A_9
			First aid to people	A_{10}
			Noise and landscape pollution	A_{11}
			Monitoring nature emergency situation (forest fires, floods, earthquakes)	A_{12}
			Monitoring the social situations in the town (social behavior, criminal situations, social emergency situations, demonstrations, show on holidays, etc.)	A_{13}
			Monitoring the technical emergency situation	A_{14}
2	B	UAV classification by duration of the flight	UAV of a short flight (1 hour)	B_1
			Medium-flight UAV's (from 1 to 6 hours)	B_2
			Long flight UAV's (more 6 hours)	B_3
			Distant flight UAV's (more 30 hours)	B_4
3	C	UAV classification by weight.	Micro UAVs (to 1 kg).	C_1
			Small 1 - 100 kg.	C_2
			Lightweight 100 - 500 kg.	C_3
			Medium 500 - 5000kg.	C_4
			Heavy 5000 - 15000 kg.	C_5
			Super heavy 15000 kg or more	C_6
4	D	UAV classification by the type of aircraft	UAVs airplane (fixed-wing)	D_1
			UAVs helicopters (rotary-wing)	D_2
			UAVs with flapping wings	D_3
5	E	UAV classification by way of take-off	Airfield take off UAV	E_1
			Non-airfield UAV taking off from a catapult;	E_2
			Non-airfield UAV taking off from hands	E_3
7	F	UAV classification by landing way	Airfield landing UAV	F_1
			Non-airfield UAV landing with the help of parachute;	F_2
			Non-airfield UAV landing with the help of snares;	F_3
8	G	UAVs by the number of applications	UAV of single applications	G_1
			UAV of repeated applications	G_2
9	H	Number of UAVs in flight	UAV of single flight	H_1
			UAV of group flight	H_2

Table 2. Analysis the effectiveness of network topologies able type styles

Types of topologies	Criteria of reliability						
Network	R	$\varepsilon 2$	D	D_{rel}	d	δ	P
Tire	1,5	0	30	0,5	2	0	0
Circle	0,4	47,3	68	1,26	3	0,23	4,63
Star	2,5	105	50	1,5	4	0	0
Fully connected	0	13,3	50	0,6	2	1	22,5
Tire-star	0	5,42	96	1,28	4	0,0325	4,38
Star-circle	2,5	-23,8	80	3	6	0,33	3,3
Hybrid-pantry	1,26	-9,14	755	3,14	9	0,67	101,98

3. To build "matrix of introspection" $\rho = \|\rho_{ij}\|$ of UAVs group graph *G (n; m)*: $a_{ij} = \begin{cases} 1, i \leftrightarrow j \\ 0, i \not\subset j \end{cases}$
4. To build "distance matrix" $D = \|d_{ij}\|$ of UAVs group graph *G (n; m)* as the minimum distance d_{ij} between the nodes *i* and *j*
5. Determine the connectivity of UAVs group graph *G (n; m)*

$$L = \frac{1}{2}\sum\sum a_{ij} \geq n-1, i^{1} j$$

$$L_{\min} = n - 1$$

6. Determine the structural redundancy of *R* of UAVs group graph *G (n; m)*

$$R = \frac{1}{2}\left[\sum\sum a_{ij}\right]\frac{1}{n-1} - 1$$

7. Uneven distribution of relationships of UAVs group graph *G (n; m)*

$$\varepsilon^2 = \sum_{i=1}^{n}\left(\rho_i - \bar{\rho}\right)^2 = \sum \rho_i^2 - 4\frac{m^2}{n},$$

where $\rho_i = \sum_{i=1}^{n} \rho_{ij}$

8. Structural compactness D of UAVs group graph $G\ (n;\ m)$ shows the proximity of the parameters (D_i) to each other through the minimum chain length d_{ij} - the smallest distance between i and j. The common proximity of the elements in the network. Structural compactness is characterized by the indicator - the diameter of the structure: $d = max\ d_{ij}$. The values of D_{rel} and D integrally characterize the inertia of the processes in the system, with equal values of R and ε^2 their increase reflects the growth of the number of bonds that disconnect. This situation helps to reduce the reliability of the system as a whole.

$$D = \sum_i \sum_j d_{ij}$$

The relative distance:

$$D_{\text{rel}} = \frac{D}{D_{\min}} - 1;$$

where $D_{min} = n(n\text{-}1)$.

9. Degree of centralization in the system δ is determined using the index:

$$\delta = (n-1)(2Z_{\max} - n)\frac{1}{Z_{\max}(n-2)},$$

where $Z_{\max}$ maximum value of the indicator Z:

$$Z_i = \frac{D}{2}\left(\sum d_{ij}\right)^{-1}, i\text{=}1,n,\ i^1 j$$

10. Determine the centrality of the nodes C_i of the graph $G\ (n;\ m)$. The CDR in group of the UAVs in flight has $C_i = C_{\max}$:

$$C_i = \frac{\sum\sum d_{ij}}{\sum d_{ij}}$$

Relative periphery of the node

11. Determine the periphery P_i of the graph *G (n; m)*. The CDR in group of the UAVs in flight has *Pi* = 0. Relative periphery of the node

$$P_i = C_{max} - C_i$$

12. Determine the survivability of the network S, CRD in group of the UAVs in flight has S > 0:

$$S = \frac{\sum a_i - 2(n-1)}{2(n-1)}$$

13. Determine the moment of the network, *M* - characterizes the controllability of the CRD:

$$M = \frac{\sum_{i=1}^{n} a_i}{a_l^2} \sum_{i=1}^{n} (a_l - a_i)$$

14. Data table of parameters CRD in group of UAVs.

Depending on the "target task" the scheme of control has to be chosen as well as the topology of UAV group. For example: analysis and estimation of effectiveness of network topologies for performance the target task "monitoring" by UAVs group, single UAV and group of single UAV is presented in Table 3, where *R* – is a ranking scale of priorities of network topologies, effectiveness / weight coefficient *w* - the corresponding weight coefficients of the efficiency of the topology and type of flight.

A system of preferences of network topologies for the target task "monitoring":

$$S_m = R_1; R_2; R_7; R_8; R_9 \succ R_3 \succ R_4 \succ R_6 \succ R_5$$

Where effectiveness - weight coefficient w_j equal:

$$w_j = \frac{C_j}{\sum_{j=1}^{n} C_j};$$

$$C_j = 1 - \frac{R-1}{n}$$

Table 3. Estimation of effectiveness of network topologies for performance the target task "monitoring"

N	Network	*R*	*C*	*w*
1	Tire	3	0,78	0,16
2	Circle	3	0,78	0,16
3	Star	6	0,44	0,09
4	Fully connected	7	0,33	0,07
5	Tire-star	9	0,11	0,02
6	Star-circle	8	0,22	0,04
7	Hybrid-pantry	3	0,78	0,16
8	Single flights	3	0,78	0,16
9	Group of single flights	3	0,78	0,16

The Algorithm of building an Expert system (ES) for estimation of the performance of UAVs flights (group and single) for the decision of different target tasks in an urban locality:

1. *Estimation* of effective performance of the target task using the next systems: the group of separate UAVs with controls from separate operators; the UAVs group with control from CDR-UAV; single UAV with control single operator. If there is a UAV group with a need of control from CDR:
    a. Decomposition of the complex system on subsystems "network topologies - the target tasks", description of the characteristics of subsystems, and estimation of effectiveness of network topologies for performance the specific target task.
    b. The effectiveness of network topologies for performance the target task and definition of criteria estimation (definition the corresponding weight coefficients of the efficiency of the topology).

c. Estimation of network topologies of the UAV group for the specific target task using Expert Judgment Method (EJM) (definition of system preferences and coordination of experts' opinions too).

2. *Estimation* of UAVs' urban locality application and methods (GRID analyses of sector UAV flight, fuzzy-logic and EJM for estimation "risk").
3. *Aggregation* of subsystems to the new system (additive or multiplicative aggregation depend on the type of "target task").
4. *Graphical presentation of results* for Expert System (group UAV, single UAV or group of single UAVs), for example, estimation of effectiveness of network topologies for performance the target task "monitoring" by UAVs group of single UAVs.

The example of the performance of the target task "monitoring" by UAVs present in Figure 5.

Figure 5. Graphical presentation of results of Expert System, of the effectiveness of network topologies for performance the target task "monitoring" by UAVs

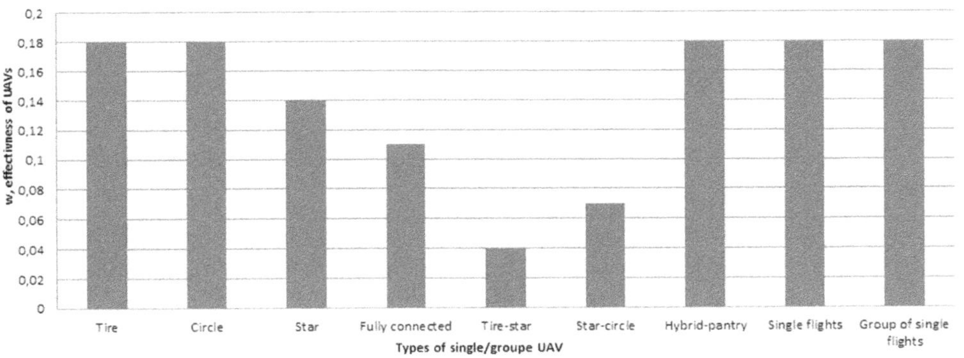

To evaluate the safety of UAVs flights in town, it is needed to obtain quantitative values of risks of flights in different territories of the town. Methods for evaluating "Risk" are the EJM and Fuzzy logic.

The air navigation rules to classify obstructions in town been chosen as "Restricted" and "Dangerous" areas, but they have nothing in common with ICAO's official definitions, this is an estimation of risks of UAV movement in smart-city. The "Restricted areas" in our case are such areas, where the risk of harming people is high, the "Dangerous areas" - the risk of harming people is very high.

Initial data for estimation risk:

a). *Buildings:* these are objects where people live and work (offices, factories, markets) and public places. Potential risk after these area penetrations:
– for UAVs ® very high;
– for people ® moderate to high.

b). *Columns and wired communication*: these objects are columns with its wires, masts, pipes antennas, which may endanger life and health of people nearby in case of breakdown. Potential risk after area penetration:
– for UAVs ® moderate;
– for people ® low to moderate.

c). *Trees and natural obstructions:* These objects are trees, hills, mountains. Potential risk after area penetration:
– for UAVs ® high to very high;
– for people ® very low.

d). *"Dangerous areas"* are attributed to the "Restricted area". "Dangerous areas" themselves are not hazardous, but permanent residence increases the risk directly proportional to the residence time. Potential risk at the moment of penetration:
– for UAVs ® very low;
– for people ® very low.
The potential risk when any UAV is staying for any period of time is a very complex task and depends on many factors, such as time, enclosing object, the previous trajectory of flight, maneuverability of UAV, aerodynamic aspects, environmental conditions.

f). *Track area:* it is a part of the planned flight path after UAV flight in which 99.99% UAV is or will be located according to "Flight plan" data:
– for UAVs ® high to very high;
– for people ® high to very high.

e). *Track conflict area:* It is unplanned part of space around "Track area":
– for UAVs ® high to very high;
– for people ® high to very high.

To evaluate the safety of UAV's flight in town it is recommended to use the EJM for building Expert System as AI. Nowadays ICAO documents define new approaches for achieving the main goal of ICAO -enhancing the effectiveness of global aviation security, and improving the practical and sustainable implementation of preventive aviation security measure. The Global Aviation Security Plan (GASP) identifies five key outcomes for improving effectiveness, such as (ICAO, 2017):

- enhancing awareness and response of risk;
- development of security culture and human capability;
- improving technological resources and foster innovation;
- improving oversight and quality assurance;
- increasing cooperation and support between states.

Thus, the quality of decisions depends on the using of developed and innovative technologies in aviation, such as AI (ICAO, 2018). Developing of AI in Air Navigation System (ANS) such as Expert Systems (ES), DSSs are considering new concepts in aviation need with using modern information technologies and modern courses: Data Science, Big Data, Data Mining, Multi-Criteria Decision Analysis, CDM, Blockchain, (Shmelova, 2015; 2016; 2017; IATA (International Air Transport Association), 2016; Kitchin, 2014).

EUROCONTROL is a great believer that new technology needed to ensure successful uptake of AI. For the future, aviation is moving towards a more interactive approach with aircraft trajectories being updated in real-time to adapt to changes. It is closer than one might think and it will certainly mean a step forward in the capabilities of our systems to cope with the flood of data and to make intelligent decisions. This will be essential if are to handle the levels of traffic predicted, as well as to cope with the new types of traffic, such as UAVs, that are on the horizon (EUROCONTROL, 2019).

Expert Judgment Method For Estimation Risk of UAVs Flights

The differentiating factor of any AI system from a standard software system is the ability to learn, to be able to improve, and to predict. Through training, AI system becomes able to generate knowledge and apply them to novel situations not encountered before. In AI, an ES is a computer system that simulates the decision making ability of a human. The ES is designed to solve complex problems by reasoning through bodies of knowledge, represented mainly as if-then rules rather than through conventional procedural code. The first ESs were created in the 1970s and then proliferated in the 1980s. Expert systems are one of forms of AI system. The AI document "White paper" presents the results of IATA research and development activities on AI in collaboration with airlines. The technologies of AI can be clustered in the following capabilities, such as: Expert Systems, Machine learning (ML), Natural Language Processing (NLP), Vision, Speech, Planning, Robotics (IATA, 2016). The ICAO documents recommend developing Intelligent Expert Systems in aviation to support decision making (DM) of operators (ICAO, 2016; 2018; 2019). For the education of aviation personnel in AI developed training, such as:

- Expert Judgment Method (EJM): Multi-criteria decision problems.
- Deterministic models: Network planning. Decision making in an emergency.
- Stochastic models: Decision making in Risk. Decision making in an emergency.
- Game Theory: DM in uncertainty. Optimal aerodrome of landing.
- Dynamic programming (DP) and GRID analyzes of the problem: The DP method to solve the problem of minimal cost, climbing an aircraft.

The quantitative estimation of the parameter's system using EJM uses to develop this system. This is one of the methods for solving multi-criteria decision or determines the quantitative values of quality indicators (more - smaller, harder - easier, more difficult - easy).

Algorithm of estimation of an area in fragment of the territory using EJM:

1) Questioners for experts, m – number of experts
2) Matrix of individual preferences - determine opinion of the experts and their systems of individual preferences - R_i
3) Matrix of group preferences R_{ij}
4) The experts' group opinion R_{grj}: $R_{grj} = \frac{\sum_{i=1}^{m} R_i}{m}$
5) The coordination of experts' opinion:
   a. Dispersion for each factors (obstructions): $Dj = \frac{\sum_{i=1}^{m} \left(R_{grj} - R_i\right)^2}{m-1}$
   b. Square average deviation: $\sigma_j = \sqrt{D_j}$
   c. Coefficient of the variation for each factors (obstructions): $v_j = \frac{\sigma_j}{R_{grj}} \bullet 100\%$

   If coefficient of a variation is $\nu_j < 33\%$ - opinion of the experts coordinated. If coefficient of a variation is $\nu_j > 33\%$ - opinion of the experts don't coordinated - to calculate Kendal's coefficient of concordance for evaluation of coordination of experts on all factors or to provide interrogation of the experts again.
6) Kendal's coefficient of concordance: $W = \frac{12S}{m^2(n^3 - n) - m\sum_{j=1}^{m} T_j}$,

If coefficient of concordance is $W > 0{,}7$ - opinion of the experts coordinated. If coefficient of concordance is W < 0,5…0,7- opinion of the experts don't coordinated, need provide interrogation of the experts again.

7) Compare system of preference R_{gr} and R_i, with the help of Rating correlation coefficient R_s (Spearman's coefficient): $R_{si} = 1 - \frac{6 \sum_{j=1}^{n} (x_{ij} - y_{ij})^2}{n(n^2 - 1)}$.

8) Significance of the calculations.

   a. The significance of the calculations W, criterion - χ^2:

$$Ç_f^2 = \frac{S}{\frac{1}{2} m(n+1) - \frac{1}{12(n-1)} \sum_{j=1}^{m} T_j} > Ç_t^2,$$

where

$Ç_f^2$ - factual value of variable;

$Ç_t^2$ - table value of variable.

   b. The significance of the calculations R_s for using Student's t - criterion:

$$t_{critical} = r_s \sqrt{\frac{n-2}{1-r_s^2}} > t_{st}.$$

9) Weight coefficient w_j *of* effectiveness – evaluation of obstructions in city:

$$w_j = \frac{C_j}{\sum_{j=1}^{n} C_j};$$

where $C_j = 1 - \frac{R-1}{n}$

10) Graphical or analytical presentation of weight coefficients.

Results of analytical estimation of an area in fragment of the territory before UAVs flights using EJM in Table 4.

Table 4. Estimation of an area in fragment of the territory

Name of obstruction	Parameter of obstruction	Identification	Results of estimation
Restricted areas			
Building	Restricted area	B-RA	10
Columns and wired communication	Restricted area	C-RA	9
Trees and natural obstructions	Restricted area	N-RA	8
Dangerous areas			
Vertical buffering area	Dangerous area	VBA	5
Horizontal buffering area	Dangerous area	HBA	7
Tracks			
Track area	Track	TA	50
Track Conflict area	Track	TCA	25
Flight area	Track	FA	1

Authors offer Methods for evaluating "Risk" using Fuzzy logic too.

Fuzzy Logic For Estimation Risk of UAVs Flights

Fuzzy logic methods have been applied to assess risk levels and is based on the logical rules "IF (condition) - TO (conclusion)". In this case, the corresponding probabilities of events and the size of possible outcomes are considered as Fuzzy sets P_j and L_{ij}, membership functions $\mu(P_j)$, $\mu(L_{ij})$. Risk R is determined as:

$$R = \mu(P_j) \times \mu(L_{ij})$$

Decision-making is the process of identifying and choosing one or more options among multiple alternatives based on the values and preferences. The right or the best choice directly proportional to the experience of the specialist and based on using as complete as possible information about each action.

In addition, it is necessary to isolate the information from the entire volume of the collected information. Depending on the *initial data*, two opposite cases could be distinguished: decision-making under conditions of certainty and decision-making under uncertainty. The decision is made in certainty if the person who makes the decision can precisely determine the outcome of each alternative solution possible in this situation. Decisions, where there are impossible to assess the probability of potential results in any way are considered to be accepted under uncertainty. Between these two cases, there are many solutions, the results of which are not certain, but the probability of each possible result can be counted. This is a decision that taken at risk. The sum of the probabilities of all alternative solutions equals 1.

For each abnormal, non-standard or emergency situation in aviation an expert may subjectively determine the level of risk using a qualitative scale. However, for formalizing the decision-making algorithm quantitative values of the variables are necessary. That is why, it is proposed to use the method of fuzzy sets to assess the qualitative risk level indicator. To be specific: we will place on the scale of the value of the linguistic variable "risk level":

1. "Very low risk" corresponds to the flight of UAV.
2. "Low risk" corresponds to restricted areas such as columns and wired communication;
3. "Average risk" corresponds to restricted areas such as a building;
4. "High risk" corresponds to dangerous areas;
5. "Very high risk" corresponds to the tracks area by busy of UAV.

The membership functions for the estimation of risk were obtained based on experimental data. Let's consider that the minimum risk level is zero units and the maximum is 100 units respectively. The fuzzy-logic functions of estimation in risk moving UAVs in flight, track conflict area, track area, restricted area, and dangerous area in Figure 6 (after the first round of the poll).

Figure 6. Fragment of the territory for estimation minimal cost and safety of UAVs movement

Let the minimum level of risk correspond to 0 units, and the maximum, respectively, 100 units. The resulting interval will be broken down into 10 segments. The degree of belonging of a certain value will be determined as the ratio of the number of responses in which the value of the linguistic variable occurs in a certain interval, to the maximum value of this number in all intervals.

Experts will be interviewed by the Delphi method in two rounds. There are 35 experts attend the survey. The results of the survey are listed in Table 5.

Table 5. Matrix of experts' opinion

Value	Interval, units									
	0-0,1	0,1-0,2	0,2-0,3	0,3-0,4	0,4-0,5	0,5-0,6	0,6-0,7	0,7-0,8	0,8-0,9	0,9-1
1	18	16	5	1	0	0	0	0	0	0
2	0	8	20	11	1	0	0	0	0	0
3	0	0	0	7	17	12	4	0	0	0
4	0	0	0	0	0	0	2	23	15	0
5	0	0	0	0	0	0	0	7	9	24
kj	18	24	25	19	18	12	6	30	24	24

To process the data are using a "*matrix of prompts*" which is a string with the elements:

$$k_j = \sum_{i=1}^{5} b_{ij},\ \ j = \overline{1,10}.$$

The matrix of prompts has the form:

$$M = \begin{Vmatrix} 18 & 24 & 25 & 19 & 18 & 12 & 6 & 30 & 24 & 24 \end{Vmatrix}$$

Choose from the matrix of prompts the maximum element and convert the elements:

$$k_{\max} = \max_j k_j = \max\{18; 24; 25; 19; 18; 12; 6; 30; 24; 24\} = 30;$$

$$c_{ij} = \frac{b_{ij} k_{\max}}{k_j}; i = \overline{1,5}$$

The results of calculations are included in the Table 6, based on which the functions of membership will be built.

Table 6. Matrix of experts' opinion

Value	Interval, units									
	0-0,1	**0,1-0,2**	**0,2-0,3**	**0,3-0,4**	**0,4-0,5**	**0,5-0,6**	**0,6-0,7**	**0,7-0,8**	**0,8-0,9**	**0,9-1**
1	30,0	20,0	6,0	1,6	0,0	0,0	0,0	0,0	0,0	0,0
2	0,0	10,0	24,0	17,4	1,7	0,0	0,0	0,0	0,0	0,0
3	0,0	0,0	0,0	11,1	28,3	30,0	20,0	0,0	0,0	0,0
4	0,0	0,0	0,0	0,0	0,0	0,0	10,0	23,0	18,8	0,0
5	0,0	0,0	0,0	0,0	0,0	0,0	0,0	7,0	11,3	30,0

To do this, we find the maximum elements in each line:

$$c_{i\max} = \max_j c_{ij}, \quad i = 1,2,\ldots,m, \quad j = 1,2,\ldots,n$$

$c_{1max} = 25{,}0$; $c_{2max} = 21{,}0$; $c_{3max} = 25{,}0$; $c_{4max} = 21{,}4$; $c_{5max} = 25{,}0$.

The value of the membership function is determined as:

Table 7. Matrix of experts' opinion

Areas	Interval, units									
	0–0,1	**0,1–0,2**	**0,2–0,3**	**0,3–0,4**	**0,4–0,5**	**0,5–0,6**	**0,6–0,7**	**0,7–0,8**	**0,8–0,9**	**0,9–1**
Flight UAV	1,00	0,67	0,20	0,05	0,00	0,00	0,00	0,00	0,00	0,00
Track conflict area	0,00	0,42	1,00	0,72	0,07	0,00	0,00	0,00	0,00	0,00
Track area	0,00	0,00	0,00	0,37	0,94	1,00	0,67	0,00	0,00	0,00
Restricted area	0,00	0,00	0,00	0,00	0,00	0,00	0,43	1,00	0,82	0,00
Dangerous area	0,00	0,00	0,00	0,00	0,00	0,00	0,00	0,23	0,38	1,00
	0,5	1,5	2,5	3,5	4,5	5,5	6,5	7,5	8,5	9,5

$$\mu = \frac{A_{ij}}{c_{i\max}}$$

The results of calculations are shown in the Table 7.

The membership functions for each value of the linguistic variable “Risk Level” are shown in Figure 7 according of the results of calculations from the Table 7.

Figure 7. Fuzzy-logic function of estimation of risk

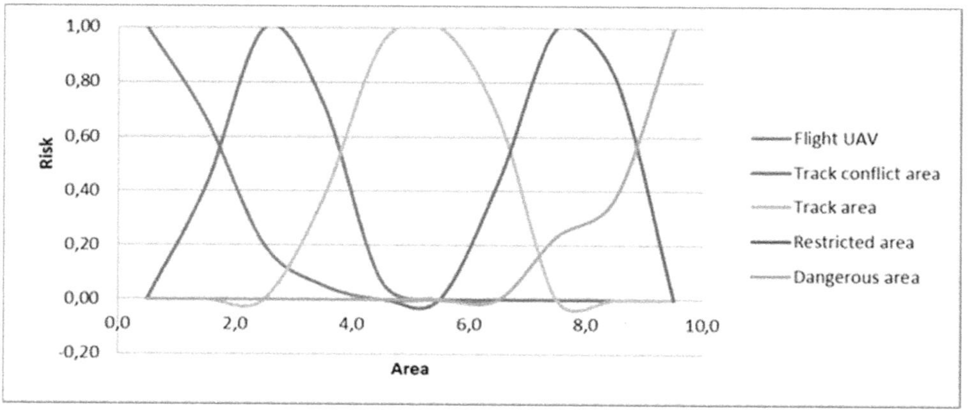

From the resulting diagrams, determined the quantitative indicators that correspond to the values of the linguistic variable “risk level”(after the second round of the poll):

“Very low risk” corresponds to the quantitative significance of the level of risk in 10.
“Low risk” corresponds to the quantitative significance of the level of risk in 35.
“Average risk” corresponds to the quantitative significance of the level of risk in 60.
“High risk” corresponds to the quantitative significance of the level of risk in 80.
“Very high risk” corresponds to the quantitative significance of the level of risk in 100.

The goals may be achieved by the execution of the following tasks: a comparative analysis of the effectiveness of the use of group flights of UAV for the intended purpose; categorization and coding of UAVs according to existing classifications; modelling UAV group flight for aerial photography using graph theory and determining the topology of a connected UAV system. To control a group of UAVs, the authors suggest using Drone-Repeater, which is connected to the operator on the ground and controls other UAVs.

Illustrative Example

The airspace over each region of the intended application of UAV is proposed to be classified as it was mentioned above and attributed by a system of minimal costs of moving which are reflecting estimated risks. For the estimation of risks, a method of Dynamic Programming (DP), EJM, and fuzzy logic shall be applied. Air navigation rules in each region of operation should also be taken into consideration for the question of legality of flight operations. Estimation of operating area for UAV as a fragment of the territory near some town is given in Figure 8.

Algorithm of definition of minimal cost and safety of UAVs movement ways in town is next:

1) Grid-analysis - cells are superimposing on a fragment of terrain (Figure 8).
2) Risk assessment of Grid cells depending on the type of area ("Restricted" or "Dangerous").
3) Finding the minimum cost path W1 for a UAV1 using the DP method for planning a flight in a level L1:

$$W_i(y_i) = y_{i-1}(RA;\ BA;\ TA;\ TCA;\ FA) + \min\left(y_i(RA;\ BA;\ TA;\ TCA;\ FA)\right)$$

Figure 8. Fragment of the territory for estimation minimal cost and safety of UAVs movement

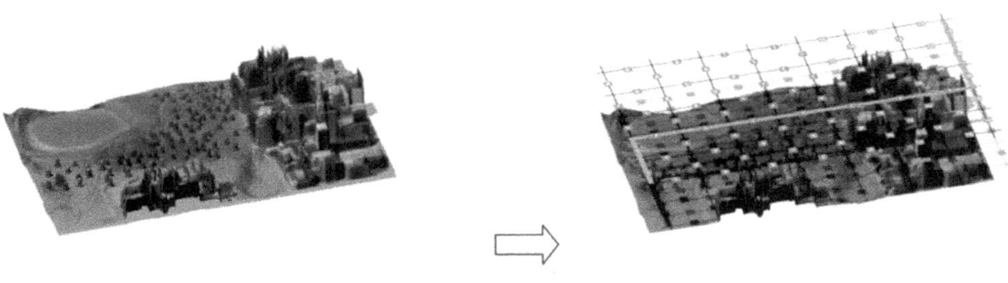

4) Assessing the path W_1 (level L1) of the UAV$_1$ as "Dangerous";

For example, estimation and finding the minimum cost path W1 for a UAV1 on Figure 8, W_1=39 for fragment of the territory from Figure 6. Risk assessment of Grid cells and minimum cost path W_1 for a UAV$_1$ present on Figure 9.

Figure 9. Risk assessment of Grid cells and minimum cost path W_1 for a UAV$_1$

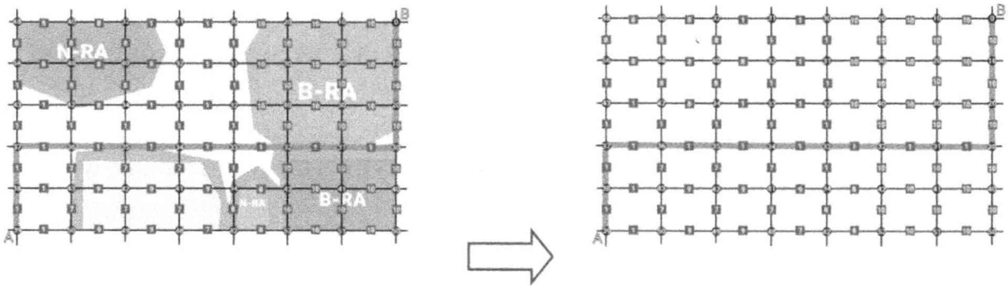

5) Finding the minimum cost path W2 for a UAV2 using the DP method for planning a flight in a level L_1, if necessary, the transition to the level L_2 (Figure 10).

Figure 10. Creating of root with flight levels and the transition to the level L_2

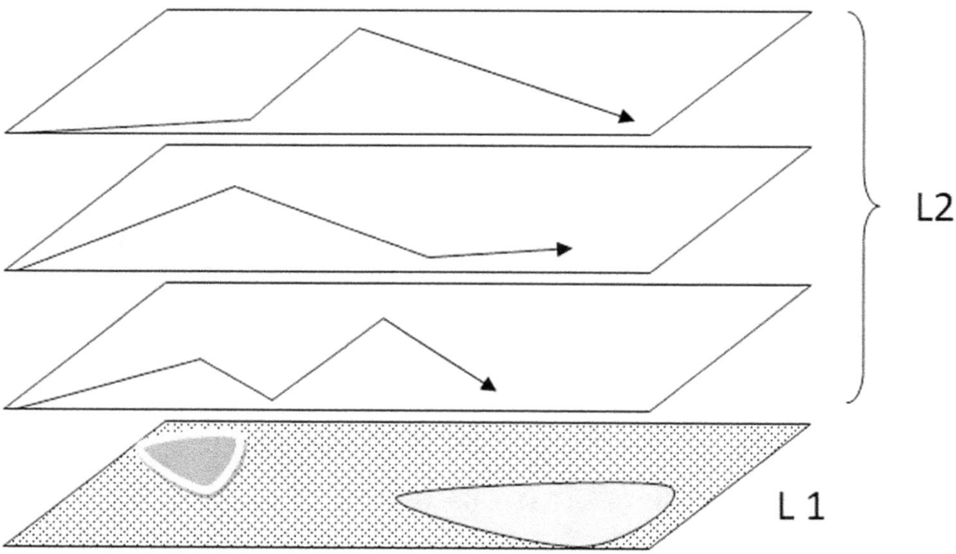

The Figure 11 shows an example of a simulated UAV flight in an urban area taking into account risks, aeronautical rules and the target assigned to UAV operator for levels L_1, L_2 and L_3.

Figure 11. Simulated UAV flight in an urban area

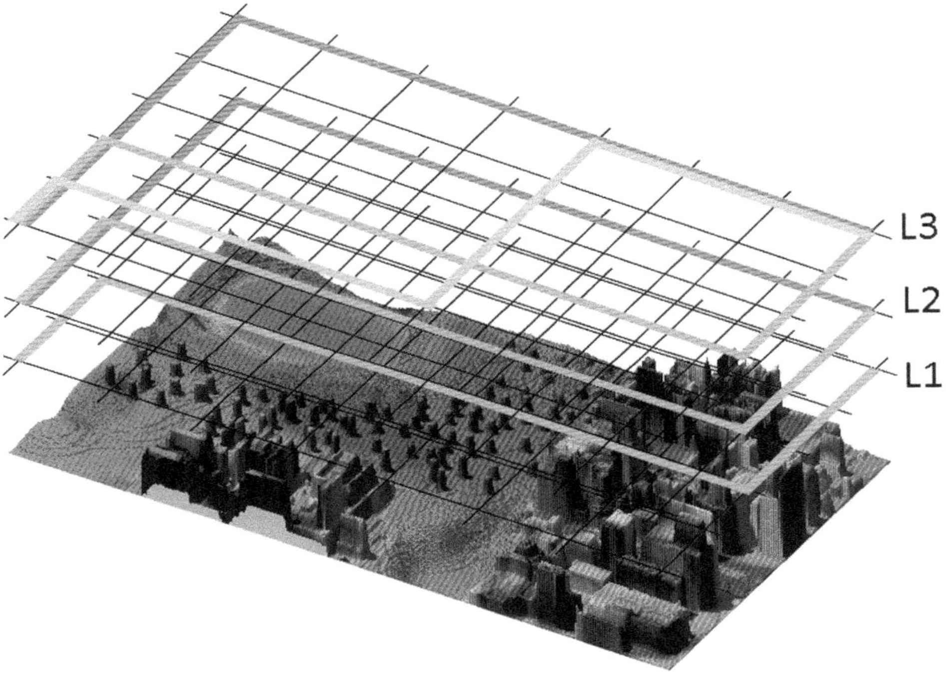

FUTURE RESEARCH DIRECTIONS

Further researches are referring to the findings of practical problems solution of UAV's operator activity in regular operations, in cases of emergency and on phases of the creation of software. Models of flight emergencies (FE) development and of DM in Risk and uncertainty by UAV's in FE will allow predicting the H-O's actions with the aid of the Informational-analytic and Diagnostics complex for research H-O behavior in extreme situation.

In the future, the authors will introduce the application of geometric approach to the determination of the optimal trajectory of UAV and UAV teams in a complex operational environment such as a modern city. Enhanced versatility in execution of complex tasks and missions is possible to achieve by separating management and control functions into a protocol stack model with conceptual levels encapsulating services and offering them to the higher levels of management. To this end, the authors will propose Control and Management of UAV Protocol Stack model (CMUPS) that has the capacity to significantly facilitate operations and management of UAV teams performing diverse tasks of ever-increasing complexity in the autonomous and collaborative organization while ensuring utmost control and safety of operation.

CONCLUSION

This chapter is revising some theoretical solutions for the implementation of UAV flights near/within a smart-town. Authors are predefining potential planning of group or single flights for the different potential activities of UAV (monitoring, data acquisition, transportations, urban survey, etc.) using methods of information technology, graph theory, and mathematics. The choice of the configuration is depending on the "target task" of UAVs flight and the effectiveness of flight in smart-city. The optimization of group flight routes for UAVs depends on the "target task" and results of estimation (cost/safety) of the flight area using methods of dynamic programming methods, GRID's analyses, Expert Judgment method, and fuzzy-logic methods. Algorithms and methods were presented such as of building an expert system for assessing the performance of UAVs flights (group or single); of finding central UAV (CDR) for effectiveness of connection in a group of the UAVs; of fuzzy-logic approach for estimation risk of UAVs flights; of determining ways of minimization of cost/safety of UAVs movement in the smart city. Optimization of flows and flexible redistribution of autonomous UAV routes in multilevel airspace was provided according to air navigation requirements and standards.

REFERENCES

Amelin, K. S., Antal, E., & Vasiliev, V. (2013). *Adaptive management of an autonomous group of unmanned aerial vehicles*. St. Petersburg State University.

Austin, R. (2010). *Unmanned Aircraft Systems: UAVs Design, Development and Deployment*. John Wiley & Sons Ltd. doi:10.1002/9780470664797

Batty, M. (2018). *Inventing Future Cities*. MIT Press. doi:10.7551/mitpress/11923.001.0001

Bondarev, D., Jafarzadeh, R., & Kozub, A. (2014). Effectiveness of Group Flight of Unmanned Aerial Vehicles. Systems of Information Processing, 6, 9-15.

Bondarev, D., Shmelova, T., & Kucherov, D. (2015). Models of group flights of unmanned aerial vehicles using graph theory. Science and Technology of the Air Forces of the Armed Forces of Ukraine - Scientific Periodicals, 3, 68-75.

Chekunov, E. (2010). Application of the UAV of US Armed Forces in Military Conflicts. *Foreign Military Review, 7*, 41-50.

Dolgikh, S. (2018). Spontaneous Concept Learning with Deep Autoencoder. International Journal of Computational Intelligence Systems, 12(1), 1-12.

Drone life. Urban Air Mobility. (2018). https://dronelife.com/2018/11/28/urban-air-mobility-the-first-uic2-forum-at-amsterdam-drone-week-shows-europes-commitment-to-smart-cities/

EUROCONTROL. (2019). *Artificial intelligence: a new era for aviation.* Skyway.

Fradkov, A., & Grigoriev, G. (2013). Decentral adaptive control of synchronization of networks of dynamical systems of dynamic systems with limited perturbations. *Automation and Telemechanics, 5*.

Ganin, S., & Karpenko, A. (1999). Unmanned aerial vehicles. Academic Press.

González, Á. (2016). Amazon lets users create their own smart home. *The Seattle Times for Government Technology*.

Guinness World Record. (2017). https://www.guinnessworldrecords.com/news-411677

Gulevich, S.P., Veselov, Y.G., Pryadkin, S.P., & Tirnov, S.D. (2012). *Analisis of factors affecting the safety of the flight of unmanned aerial vehicles*. Ru. Causes of accidents drones and methods of preventing them «Science and education».

IATA. (2016). https://www.iata.org/publications/Pages/AI-white-paper.aspxInternational

Ignatiev, O. (2010). UAV: The state and prospects of their use in the interests of NATO. *Foreign Military Review*, (12), 45–52.

International Civil Aviation Organization. (2005). *Global Air Traffic Management Operational Concept*. Doc. ICAO Doc 9854-AN/458.

International Civil Aviation Organization. (2009). Manual on Global Performance of the Air Navigation System. Doc. ICAO 9883.

International Civil Aviation Organization. (2011). *Unmanned Aircraft Systems (UAS)*. Circ. 328-AN/190.

International Civil Aviation Organization. (2015). Manual of Remoted Piloted Aircraft Systems (RPAS). Doc. ICAO 10019/AN 507.

International Civil Aviation Organization. (2017a). *Aviation Security Plan (GASP)*. Author.

International Civil Aviation Organization. (2017b). *Remotely piloted aircraft system (RPAS) Concept of operations (CONOPS) for international IFR operations*. ICAO RPAS Concept of Operations ICAO.

International Civil Aviation Organization. (2018) Potential of Artificial Intelligence (AI) in Air Traffic Management (ATM). *Thirteenth Air Navigation Conference ICAO*, 9-19.

International Civil Aviation Organization. (2018). *Model UAS Regulations: parts 149, 101, 102. In: Guidance material and tools*. ICAO. https://www.icao.int/safety/UA/Pages/default.aspx

International Civil Aviation Organization. (2019). *ICAO's Unmanned Aviation Bulletin*. ICAO. https://www.icao.int/safety/UA/Documents/ICAO%20UA%20Bulletin%202019-1.pdf

Kharchenko, V., Shmelova, T., Bondarev, D., & Stratiy, A. (2017). Modelling of Decision Making of Unmanned Aerial Vehicle's Operator in Emergency Situations. *Proceedings of the National Aviation University*, *1*(1), 20–28. doi:10.18372/2306-1472.70.11419

Kharchenko, V., Shmelova, T., & Sikirda, Y. (2012). *Decision-Making of Operator in Air Navigation System*. Kirovograd: KFA of NAU.

Kharchenko, V., Shmelova, T., & Sikirda, Y. (2016). *Decision-Making in Socio-Technical Systems: monograph*. NAU.

Kirichek, R., Makolkina, M., Sene, J., & Takhtuev, V. (2016). Estimation quality parameters of transferring image and voice data over zigbee in transparent model. *Communications in Computer and Information Science*, *601*, 260–267. doi:10.1007/978-3-319-30843-2_27

Kitchin, R. (2014). *The Data Revolution: Big Data, Open Data, Data Infrastructures and Their Consequences*. Sage Publications.

Kreps, S., & Zenko, M. (2014). *The Next Drone Wars*. Foreign Affairs. https://bmpd.livejournal.com/861105.html

Kucherov, D. P., & Kozub, A. N. (2015). Control System Objects with Multiple Stream of Information. *Proceedings IEEE 3rd International Conference Actual Problems of Unmanned Aerial Vehicles Developments (APUAVD)*, 290-293. 10.1109/APUAVD.2015.7346623

Kucherov, D. P., & Kozub, A. N. (2016). Group Behavior of UAVs in obstacle presence. *Proceeding IEEE 4th International Conference "Methods and Systems of Navigation and Motion Control (MSNMC)"*, 51-54. 10.1109/MSNMC.2016.7783104

Kucherov, D. P., Kozub, A. N., & Kostyna, A. N. (2017). Group of UAVs Moving on Smooth Control Law with Fixed Obstacles. *Advances in Science. Technology and Engineering Systems Journal*, *2*(3), 1034–1041. doi:10.25046/aj0203131

Kuzin, A. V. (2011). Computer networks. Moscow: FORUM: INFRA-M.

Nikiforov, A., & Munimaev, V. (2010). *Analysis of foreign unmanned aerial vehicles*. St. Petersburg Forestry Academy.

Olifer, V. G., & Olifer, N. A. (2006). Computer networks: principles, technologies, protocols. Academic Press.

Shmelova, T., & Bondarev, D. (2015). Theory Applying for Quantitative Estimation of UAV's Group Flight. *IEEE 3D International Conference Actual Problems of Unmanned Aerial Vehicles Developments (APPUAVD)*, 328-331.

Shmelova, T., & Bondarev, D. (2016). Graph Theory Applying for Quantitative Estimation of UAVs Group Flight into Aerial Photography. *Electronics and Control Systems*, *4*(46), 128–133.

Shmelova, T., Bondarev, D., & Znakovska, Y. (2016). Modeling of the Decision Making by UAVs Operator in Emergency Situations. *4th International Conference on Methods and Systems of Navigation and Motion Control (MSNMC)*, 31-34. 10.1109/MSNMC.2016.7783099

Shmelova, T., Kovalyov, Yu., Dolgikh, S., & Burlaka, O. (2019) Geometry-modeling based flight optimization for autonomous groups of UAVs *5D International Conference Actual Problems of Unmanned Aerial Vehicles Developments (APUAVD)*, 79-82. 10.1109/APUAVD47061.2019.8943856

Shmelova, T., Lazorenko, V., Bondarev, D., & Burlaka, O. (2019). Group flights of Unmanned Aviation Vehicles for Smart cities. *IEEE 9th International Conference on Advanced Computer Information Technologies (ACIT)*, 230-233. 10.1109/ACITT.2019.8780071

Shmelova, T., Sterenharz, O., & Burlaka, O. (2019) Optimization of flows and flexible redistribution of autonomous UAV routes in multilevel airspace. *IEEE Proceedings of the 15th International Conference on ICT in Education, Research and Industrial Applications. Integration, Harmonization and Knowledge Transfer; 5th International Workshop on Theory of Reliability and Markov Modelling for Information Technologies (TheRMIT), 4*, 703-714.

Shmelova, T., Stratij, A. (2015). *Computer Program "Aggregation of heterogeneous information flows": certificate of registration of copyright in the product N° 59466.* Academic Press.

Sładkowski, A., & Wojciech, K. (2019). Cases on Modern Computer Systems in Aviation. In Using Unmanned Aerial Vehicles to Solve Some Civil Problems. Academic Press.

Smart Cities Guide, G. S. M. A. Traffic Management, GSM Association. (2016) Available https://www.gsma.com/iot/gsma-smart-cities-guide-traffic-management/

Tarasov, V., & Gerasimov, B. (2007). Intelligent Decision Support Systems: Theory, Synthesis, Efficiency. Kyiv: International Academy of Computer Science and Systems.

Vyrelkin, A., & Kucheryavy, A. (2014). The usage of unmanned aircraft to solve the tasks of "a smart city". SPbGUT.

Yaroslav, K. (2019). *Remote expert air traffic management system "Decision making in a collaborative environment " FF-ICE" obese* [Unpublished Master's thesis]. National Aviation University, Ukraine.

ADDITIONAL READING

Ben-David, A. (1995). Monotonicity maintenance in information-theoretic machine learning algorithms. *Machine Learning, 19*(1), 29–43. doi:10.1007/BF00994659

Marsh, L., Calbert, G., & Gossink Tu, J. D. (2005). Multi-Agent UAV Path Planning. Defence Science and Technology Organization, 2188 – 2194.

Montgomery, J. F., & Bekey, G. A. (1998) Learning helicopter control through "teaching by showing". *IEEE Conference on decision and Control*. 10.1109/CDC.1998.761747

Sizova, O.A. (2012). *The application of graph theory in various types of scientific activity.* Scientific article Baku.

KEY TERMS AND DEFINITIONS

Air Navigation System (ANS): Is a complex of organizations, personnel, infrastructure, technical equipment, procedures, rules, and information that is used to provide airspace users of safe, regular, and efficient air navigation services.

Central Drone Repeater (CDR): Is remotely piloted aircraft in the UAVs group for relaying control signals from UAVs in a group and connect with an operator. The central drone has the best properties of the connection with other UAVs, its location has the centrality and minimal relative distance.

Collaborative Decision Making (CDM): Is a joint government/industry/peoples/ UAV operators initiative aimed at improving air traffic UAVs flow management through increased information exchange among urban community stakeholders.

Distributed Decision Support System (DDSS): Is a complex system with multiple interactions geographically distributed locally remotely piloted aircraft.

Expert System (ES): Is a computer system that emulates the decision-making ability of a human expert. The system is creating with the help of experts in the field of solving this problem.

Flight and Flow Information for a Collaborative Environment (FF-ICE): Is a new concept of the interaction of multiple participants in a collaborative environment s guided by the requirement to eliminate or reduce the limitations of the present Flight Plan and to accommodate the future environment, detailed in the Global Air Traffic Management Operational Concept (ICAO, 2005).

Unmanned Aerial Vehicle (UAV): Is aircraft without a human pilot onboard and maybe controlled with various degrees of autonomy such as either under remote control by an operator, autonomously by onboard computers or ground autonomous robot.

Chapter 16
Accuracy of UAV-Based Digital Elevation Models:
A Case Study in Colombia

Engelberth Soto-Estrada
EIA University, Colombia

Ann Wellens
Universidad Nacional Autónoma de México, Mexico

Andrés Pava-Restrepo
EIA University, Colombia

Camilo Hernández Mejía
École Polytechnique Fédérale de Lausanne (EPFL), Switzerland

ABSTRACT

UAV photogrammetry has simplified the measurement of the terrain's surface; however, the accuracy of the method is still a matter of research as it depends on multiple factors such as the UAV employed and the decisions made by the user during information acquisition and processing. This chapter analyses the accuracy of 27 digital elevation models (DEMs) built from the data acquired with a DJI Phantom 3 drone, which was flown at different heights and trajectories in a case study in the Colombian Department of Antioquia. The statistical analysis confirmed that the use of ground control points (GCPs) is fundamental to increase the accuracy of the map products. It also showed that accuracy, understood as the comparison between the coordinates of points measured in the field (CPs) with their corresponding values estimated in the orthophotomosaics, depends on the number of CPs and their location. An insufficient number of CPs or an inadequate distribution, for instance all or most of them located in a single land cover, will cause the accuracy to be overestimated.

DOI: 10.4018/978-1-7998-2249-3.ch016

INTRODUCTION

The reliability of a measurement depends on its proximity to a reference value, which is commonly referred to as accuracy. However, it is recognized that, although all the known or suspected sources of error have been evaluated and the appropriate corrections have been applied, a certain level of uncertainty about the correctness of the declared result still exists. Therefore, the Joint Committee for Guides in Metrology defined the uncertainty of measurement as the dispersion of the values that could reasonably be attributed to the measurand (Joint Committee for Guides in Metrology [JCGM], 2008).

The sources of error from measurements made with unmanned aerial vehicles (UAVs) have been widely examined in the scientific literature; among the best-known are flight altitude, image overlap, weather, and the number and distribution of ground control points (GCPs), which are marked targets on the ground of known location used to georeference UAV imagery (Rock et al., 2011; Leitao et al., 2016; Seifert et al., 2019; Barba et al., 2019). Different studies assess DEM quality for specific case studies; examples are viticulture in the Mediterranean (Pichon et al., 2016), surface mining in Slovakia (Kršák et al., 2016) and DEM generation for Sahitler hill in Turkey (Uysal et al., 2015).

The accuracy of map products, namely digital elevation models (DEMs) and orthomosaics, is determined by comparing the planimetric coordinates and elevations of well-defined points with information of the same points from an independent source of higher accuracy (Federal Geographic Data Committee of the United States of America [FGDC], 1998). The results are usually compared with standards developed for traditional procedures, such as ground-based total station surveys and manned aircraft photogrammetry, as most of the countries do not have consistent and appropriate accuracy standards that apply specifically to the new technologies for digital data (American Society for Photogrammetry and Remote Sensing [ASPRS], 2015).

The American Society for Photogrammetry and Remote Sensing proposed the Positional Accuracy Standards for Digital Geospatial Data (ASPRS, 2015), which are based on the United States of America standards for maps and spatial data, such as the National Map Accuracy Standard (NMAS) of 1947, the National Standard for Spatial Data Accuracy (NSSDA) of 1998, and the NDEP Guidelines for Digital Elevation Data of 2004. The ASPRS horizontal accuracy standard is divided into accuracy classes using root-mean-square-error (RMSE) statistics. The vertical accuracy standard uses RMSE statistics in non-vegetated terrain and 95th percentile statistics in vegetated land.

There are no quality standards for UAV-based measurements in Colombia. During 2018, as part of the research presented in this chapter, an online survey was conducted among companies that offer UAV photogrammetry in the country. They were asked about the accuracy of the information they commercialize and their quality assurance/quality control (QA/QC) procedures. About 30 companies were invited to participate, 12 of them responded. Only two indicated that they implement QA/QC practices. The companies that do not have quality controls mentioned that, usually, their clients do not request such verification, but that they (both the companies and the clients) trust in the precision reported by the photogrammetric software. It was also recognized that, commonly, GCPs are not used in UAV photogrammetry because most clients are not willing to pay for their measurement. In addition, two industry consortiums that regularly use UAV-based DEMs for mining extraction mentioned that the coordinates of the 3D models they receive are often out of phase by up to 15 m, which may suggest that the DEMs are reconstructed without geo-referenced GCPs. These findings show that there is still a lack of knowledge in the country about the accuracy of UAV map data, the influence of each process variable on the quality of the information - such as the use of GCPs or flight parameters - and the importance of having QA/QC procedures.

Considering the above, a research project was carried out to examine the accuracy of UAV-based DEMs. Different combinations of flight altitude, image overlap, and number and distribution of GCPs were analysed. Using a DJI Phantom 3, the most used drone in Colombia when the project began, 27 flights were carried out producing a DEM for each flight. It is expected that the results contribute to clarify the best practices for UAV photogrammetry in Colombia and in regions with similar characteristics: cloudy weather, dense vegetation, and steep relief.

The structure of the chapter is as follows. Section 2 presents a review of the state-of-the-art on the technological innovations that have popularized the use of UAVs for the acquisition of map data. This section contextualizes the reader about the opportunities offered by the latest UAV technology, which also justifies the need to continue updating the analysis on their performance, since the results of previous investigations may not apply to the new developments. Section 3 discusses error sources observed in UAV photogrammetry, and how they were considered in the research project presented in this chapter. The characteristics of the case study are described in section 4. This section also clarifies the analysed flight parameters, the GCPs combinations, and the data analysis procedure. Finally, the conclusions of the chapter are summarized in section 5.

UAV INNOVATIONS

Using UAVs for aerial mapping has become more and more popular because of recent technological advances in control and instrumentation, the development of more efficient and effective software for flight planning, mission execution, and data analysis, and the reduction in manufacturing costs (Colomina & Molina, 2014; Gerke, 2018; Goncalves et al., 2018, Giordan et al., 2020). This section presents a review of the state-of-the-art on these aspects.

Instrumentation and Control

Technological innovations, such as computer vision, embedded systems, miniaturization of low-cost cameras, inertial measurement, ultrasonic sensors, and laser rangefinders, have simplified the use of UAVs (Kanellakis & Nikolakopoulos, 2017; Kumar et al., 2017). Nowadays, it is possible to get accurate and useful information on board a UAV to carry out complex actions like obstacle detection and avoidance and real-time mapping (Da Jiang Innovations [DJI], 2019a; Flynt, 2019; Schatsky & Ream 2019). This data is combined with the information obtained by the commonly used sensors, i.e. GPS receiver, barometer, magnetometer, and accelerometer, allowing a better estimation of the UAVs' position and altitude (Kumar et al., 2017). These developments are leading to the automation of data collection (Floreano & Wood, 2015; Areias et al., 2018; Rakha & Gorodetsky, 2018), as already seen in automatic actions of consumer-grade drones, such as take-off and landing, and in UAVs developed for continuous tasks, such as monitoring of industrial facilities (Airobotics, 2020) or scheduled patrols (Flybase, 2020).

UAVs' design has evolved rapidly over the last decade (Rao et al., 2016). While the first drones were basic remote-controlled airplanes, the latest generation has enhanced intelligent piloting, full airspace awareness and automatic mission execution capabilities (Joshi, 2017). There are four main types of UAV platforms: single-rotor, fixed-wing, multi-rotor, and fixed-wing hybrid.

Single-rotor UAVs, commonly known as helicopters, are adequate to lift large payloads, such as a Lidar scanner, for longer flying times as they can be powered by a gas motor. These devices are more efficient than multi-rotors because they have longer rotor blades. However, helicopters are not as stable while flying and require good skills to keep it in the air. Single-rotor UAVs also require more maintenance because of their mechanical complexity (Villa et al., 2016). In contrast, fixed-wing drones are able to cover larger areas, as they are more efficient on the horizontal flight. The main drawback of fixed-wing UAVs is their inability to loiter over one point of interest (Australia UAV [AUAV], 2019). Launching and landing is also problematic, as a runway or a launcher is required.

New microcontrollers and propulsion systems enabled the development of multi-rotor UAVs (Al-Hajjaji et al., 2017). They are classified by the number of rotors, from tricopters to octocopters. The quadcopter or quad-rotor is the most used UAV because of its simpler dynamic structure (Cabreira et al., 2019), which makes it easy to operate in confined areas, such as urban or industrial environments. The major constraints of multi-rotors are the limited flying range, as most of the energy is used to maintain the aircraft in a stable flying position (Yacef et al., 2017), speed and carrying capacity. Currently, typical consumer-grade multi-rotors are capable of maximum flying times of 20 to 30 minutes with a minimal payload.

Hybrid UAVs integrate the advantages of vertical take-off and landing drones (VTOL) and fixed-wing UAVs. This UAV can operate in confined spaces, acting as multi-rotors, but they can also change the angle of their propellers to fly as a fixed-wing UAV and cover large areas. However, these devices are still under development. Smeur et al. (2018) indicate that three major challenges need to be addressed to control hybrid UAVs: altitude control, speed control, and guidance.

At present, the Chinese company Da-Jiang Innovations, DJI, is the dominant brand for UAV purchases, with a 74% global market share (Snow, 2019). DJI's dominance can be attributed to its aggressive product development, low prices, and industry-leading technology (Murison, 2018). The evolution of DJI UAVs is an example of the development of the drone industry in recent years. The Phantom series is one of the success products of the DJI company. The multi-rotor Phantom 1, released in January 2013, was the first ready-to-fly UAV that included an internal GPS system (Crane, 2018). This UAV had a flight time of 10 minutes and a flight range of about 300 m. In December of the same year, the Phantom 2 reached the market. It featured an improved battery that allowed up to 25 minutes of flight time and a 3-axis camera stabilizer, also known as gimbal. The Phantom 3 (April 2015) included a bottom-facing vision positioning system, which allowed the drone to maintain its position without a GPS signal, a full HD digital video downlink, and integrated the GLONASS signal, facilitating the aircraft to connect to more satellites for enhanced stabilization and position accuracy. In addition, the DJI Go App was introduced, allowing the drone pilots to easily access the camera settings and flight controls and be automatically alerted if any problems with the UAV were detected. Phantom 4, launched in March 2016, included an improved camera stabilizer system, a dual inertial measurement unit and a dual-compass set-up for added redundancy in flight and faster reaction time (DJI, 2019a). This UAV has an obstacle avoidance system through which the drone would intelligently stop or move around the objects detected in its path. The obstacle avoidance system led the way to more intelligent flight modes, such as the active track mode featured in UAVs like the DJI Mavic Pro, with which the drone automatically follows a subject and films a video at the same time (DJI, 2018), and the tripod mode that enables fine control of the UAV by limiting the maximum speed and reducing the braking distance. The tripod mode helps to perform flight missions in challenging conditions such as places with trees, bushes or other obstacles, or flying close to the ground or indoors (Flynt, 2019).

The growth of the global commercial drone market, which is estimated to exceed US$17 billion by 2020 (Goldman Sachs, 2016), suggests that the fast pace of technological innovations will continue, increasing the flying range, the quality of the information captured and the performance of onboard sensors, which will enhance UAV autonomy. In addition, Snow (2019) points out that the efficiency of the image processing software will also improve thanks to the advances in artificial intelligence, a subject that is addressed in the next section.

Mission Planning/Execution and Data Processing Software

UAVs' software has been developed at the same pace as their hardware. In just two years, from 2015 to 2016, venture investment in software-focused drone start-ups doubled, reaching an estimated US$335 million (Schatsky & Ream, 2016). Planning of flight missions, drone piloting and the processing of the captured information has been simplified by the new software, which is powered by cognitive technologies such as computer vision and machine learning (Schroth, 2018), that can automate drone operation on the one hand and data analysis on the other.

Applications such as DJI FlightPlanner and Mission Planner have facilitated flight mission planning. These applications enable drone pilots to carry out specific actions, such as draw the area of interest on an online satellite image, select the aircraft to be used, and introduce flight parameters. The software processes the information introduced by the drone pilot to optimize the flight plan, the updated version of which is transmitted to the UAV for flight execution.

Similarly, advances in navigation software have improved the drone's versatility. On-drone computer vision enables UAVs to identify and avoid obstacles, find safe landing spots, and track moving targets (Kanellakis & Nikolakopoulos, 2017; Marr, 2019). Schatsky & Ream (2016) illustrate the benefits of drones equipped with advanced navigation systems in a study case where wind turbines are inspected. They report that traditional methods with ground-based inspection allow the scrutiny of 2 to 5 turbines a day, while drones with manual flight control inspect from 10 to 12, and UAVs with advanced software with an autonomous flight from 15 to 20 turbines a day.

Innovations in software and hardware, i.e. high-performance graphic processing units (GPUs), are improving the analysis of the data collected by drones (Döllner, 2020). Photogrammetry software extracts geometric information from 2D images to create digital 3D models, using automatic feature identification and element matching throughout different images (Westoby et al., 2012). The extraction process, referred to as structure from motion (SfM - Micheletti et al., 2015), is based on the creation of 3D point clouds that are used to reconstruct digital surface models (DSM), which represent the ground surface and all objects on it like vegetation or buildings, and digital terrain models (DTM), which do not include the objects on the surface.

Artificial intelligence (AI), machine learning (ML) and deep learning (DL) have an enormous potential for disruptive innovations in geospatial data processing. In particular, ML-based techniques are being implemented in point classification, point cloud segmentation, object identification and shape reconstruction (Döllner, 2020). These innovations stimulated new applications for drones in several areas, such as the identification of intruders in restricted areas (Choudhary et al., 2018) or buried object detection (Schreiber et al., 2019), and reduced processing time. For example, the Pix4D software introduced a new image classification approach based on machine learning techniques. With the help of these methods, the software distinguishes between five different categories: buildings, high vegetation, human-made objects, roads, and ground. Previously, this classification required a lot of time and had to be done manually by isolating the data of interest, now the process requires minutes and ends with the automatic reconstruction of DTMs (van Rees, 2017).

Manufacturing Costs

There is a consensus in the scientific literature that the use of UAVs, for projects of a certain size, offers advantages over airplanes and satellites. These include low costs, high spatial resolution, and operational flexibility (Matese et al., 2015; Mikrut, 2016; Koci et al., 2017; Gruszczynski, 2017; Iizuka et al., 2018).

Matese et al. (2015) analysed the cost of acquisition, pre-processing and data processing using aerial, both manned and unmanned, and satellite platforms. Their findings show that for sites up to 5 ha UAVs are the best alternative, while for bigger areas the other two technologies should be preferred. The researchers point out that the costs depend, to a large extent, on the number of images to be processed. When using UAVs, the number of aerial images is proportional to the surveyed area, which is not the case of planes and satellites that cover larger areas with each photo.

The results of Matese et al. (2015) are valid under the conditions in which their research was carried out, i.e. the specific UAV and photogrammetric software used and the market prices available, but they may not correspond with the current technological context, as discussed in previous sections. For instance, UAV data analysis can be done now through cloud services, such as those offered by Hexagon Geospatial, Mosaicum, and Agribotix, thus avoiding the acquisition cost of photogrammetric software. There is a growing number of Free and Open Source Software (FOSS) for flight-mission planning, as for instance Px4 Open Source Autopilot, and for data processing. Drone Mapping Software is one of the latter. It features point cloud processing, 3D model reconstruction, multispectral analysis, and includes the possibility of cloud computing (OpenDroneMap, 2020). It is also possible to reduce data acquisition costs when manned aircraft are required. Companies like TerrAvion and GeoVantage offer advanced route planning web services to serve multiple consumers in a single flight. Technological innovations permanently change the advantages and disadvantages of aerial and satellite platforms, so absolute limits to choose one platform or the other cannot be determined. In general, the most cost-effective solution will depend on the information needs in terms of the area of analysis, the spatial resolution and the revisit period (DroneApps, 2019).

Despite the above, as the price of UAVs has fallen and their autonomy has been improved, the UAVs have emerged as major competition to airplanes and satellites. Since UAVs are made with smartphone parts, investments over the last 10 years in the development of these components have driven UAV prices down, increasing accessibility to consumers and business (Castellano, 2018). The price of the DJI Phantom 3 in the United States of America exemplifies this trend. The price of this UAV was expected to decrease from approximately US$1,000 in 2015 to 525 in 2020 (Reagan, 2015); however, the market exceeded expectations, as a Phantom 3 can now be purchased in Amazon for nearly US$400.

Experts agree that these trends will continue as a result of the increase in capital investments in the drone industry, which are attracted by the reduction in the prices of components such as batteries (Kapoor & MacDuffie, 2017), the massive commercial market potential (Meola, 2017), and the technological developments in artificial intelligence and analytics (Castellano, 2018).

ACCURACY OF UAV PHOTOGRAMMETRY

As mentioned in the introduction, the reliability of a measurement is determined by its level of uncertainty. A broader classification distinguishes uncertainty coming from uncontrollable errors and from imperfect correction of controllable errors, sometimes called systematic errors (JCGM, 2008). In UAV photogrammetry, uncontrollable errors are linked to weather, while controllable errors are related to how the technique is implemented. A third source of errors can be associated with the limitations of the procedure, for instance, the measurement of terrain under dense vegetation.

Conventional topographic survey methods have less uncertainty than the UAV-based techniques, since the former have fewer error components (Beretta et al., 2018). Extrapolating the error sources mentioned in the JCGM (2008) to the case of UAV photogrammetry, the following sources can be distinguished:

1. Imperfect data collection because of the use of inappropriate flight parameters, for instance, incorrect flight altitude or insufficient image overlap, or to obstacles on the surface, i.e. trees, vehicles or buildings.
2. Weather, such as cloudiness that may interfere with GPS reception (Leitao et al., 2016), wind affecting the UAV flight, sun elevation creating shadows (Ranjan-Martha et al., 2010), and poor light conditions hindering the identification of shapes and objects on the surface.
3. Insufficient quantity, inaccurate geo-referencing or inadequate distribution of GCPs.

These sources are not necessarily independent nor exhaustive, and it is possible that several of them contribute to the data uncertainty (Pérez, 2012). The main sources of error in UAV photogrammetry are discussed in this section.

Flight Altitude

Flight altitude is one of the main factors that determine the accuracy of point clouds (Kraus, 2012); it is directly related to the ground sampling distance (GSD), which is the distance between two consecutive pixel centres measured on the ground. Lower flight altitudes are required when a greater level of detail is needed. Incorrect selection of flight altitude may result in data with an insufficient spatial resolution to meet the user's requirements. In practice, a survey cannot be more accurate than the GSD, which is the reason why the quality standards for digital planimetric data, such as ASPRS (2015), are based on this parameter.

The formula used for GSD calculation is (Ippolito & Cigola, 2017):

$$\frac{f}{H} = \frac{ps}{GSD}$$

where:

f is the focal length of the UAV camera
H is the flight altitude
ps is the pixel size, i.e. the side distance of the CCD (charge-coupled device)/CMOS (complementary metal-oxide-semiconductor) matrix
GSD is the corresponding ground size of each pixel

The flight altitude should be low enough to keep the GSD smaller than the required accuracy and high enough to have the least number of images (Propeller

Aero, 2019). A GSD recommended by ASPRS (2015) for highly accurate geospatial information corresponds to 2.5 cm. To achieve this value, for example with a Phantom 3, the flight altitude can be determined from the technical specifications in table 1:

$$H = \frac{f * GSD}{ps} = \frac{0.357\text{cm} * 2.5\text{cm}}{0.0001529\text{cm}} = 5837\text{cm} = 58\text{m}$$

Some researchers have examined the influence of flight altitude on the quality of DEMs. Rock et al. (2011) carried out 19 flights at altitudes between 50 and 550 m, finding a significant relationship between flight altitude and vertical accuracy. At 50 m, the vertical error was estimated to be between 5 and 15 cm, while at 550 m it was between 30 and 120 cm. Using a LIDAR-equipped drone, Fuad et al. (2018) found similar results testing three flight altitudes: 20, 40 and 60 m; vertical errors of 32, 45 y 61 cm, respectively, were estimated.

Flight altitude also influences the spectral resolution of the data and the recognition of objects on the surface. Mesas-Carrascola et al. (2015) analysed three flight altitudes, 60, 80 and 100 m, for the creation of multispectral orthomosaics to support crop management. Using a spectralon panel, the researchers found a good geometric accuracy and spectral resolution at 60 m altitude, while at 100 m the transition between objects (i.e. crops, bare soil, weeds, and the spectralon panel) was poor, which limited the usefulness of the information as the row crops practically disappeared due to the low resolution.

Image Overlap

Image overlap, as shown in figure 1, is defined as the amount in which a photograph includes the area covered by another photograph. It is expressed as a percentage for both forward or frontal (between photos along the same flight line) and side or lateral (between photos on adjacent flight lines) directions; these overlaps are commonly referred to as respectively frontlap and sidelap in specialized software. Image overlap determines redundancy in the identification of tie points, which are common features visible in overlapping images. High image overlaps improve the accuracy of the point cloud and reduce distortions in the orthophoto or orthomosaic (Haala et al., 2013; Leitao et al., 2016). Pix4D (2019) recommends forward and side overlaps of respectively 75% and 65% for most terrain characteristics, and of respectively 85% and 70% for areas covered with dense vegetation.

Figure 1. Forward and side overlaps

Leitao et al. (2016) carried out 16 flights with overlaps between 60% and 80% to investigate the influence of this factor on DEM accuracy; their results did not evidence a significant relationship between both variables. The researchers point out that the variability in DEM accuracy may be introduced by other factors, such as weather, which were not investigated in their study. Dandois et al. (2015) analysed different overlaps in the reconstruction of DSM. They tested sidelaps between 20% and 80% and frontlaps between 60% and 96%. No significant differences in the estimated errors with different sidelaps were found; however, the best results, accuracies of 3.0 m, were obtained with frontlaps over 80%.

Weather and Terrain Characteristics

Weather influence image quality and UAV positioning, which affects the accuracy of the map data (DroneDeploy, 2017a). Wierzbicki et al. (2015) estimate that images acquired in bad weather reduce the accuracy of photogrammetric products by an average of 25%. The performance of onboard cameras depends largely on the lighting conditions (Opromolla et al., 2019), which also determine the contrast in the images and the ability of the processing software to discriminate details on the imagery (Thillainayagi & Kumar, 2019) and extract the final map products (Wierzbicki et al., 2015).

The angle of the sun over the horizon can be another limiting factor, as the shadows of the objects on the images affect the quality of orthomosaics and 3D models (Stensson, 2016). In some countries, this is a crucial aspect; for instance, the authority responsible for the real state division in Sweden has issued recommendations on the angle/height of the sun over the horizon (at least 30 degrees) and the shadow length ratio (1:1.7, i.e. shadows should be at most 1.7 times the object height) to obtain high-quality photogrammetric data (Landmäteriet, 2018). For the Stockholm area, for example, these recommendations are met from 6 in the morning to 4 in the afternoon, between days 80 and 270 of the year (namely, March to September, Swedish Meteorological and Hydrological Institute [SMHI], 2015).

The operation of UAV navigations systems, i.e. the global navigation satellite system (GNSS) receiver and the inertial measurement unit, can also be affected by climatological conditions. For instance, windy conditions can affect UAV flying capabilities and stability (Cimoli, 2015; Gerke & Przybilla 2016).

The accuracy of UAV photogrammetry is also conditioned by terrain characteristics such as dense vegetation cover, steep terrain (Manconi et al., 2019) and tall buildings (Zimmermann et al., 2017). Photogrammetry of densely vegetated terrain remains challenging due to the lack of exposed surfaces (Meng et al., 2017). Begashaw (2018) made a comparison between DTMs generated with UAV photogrammetry and airborne Lidar scanning (ALS) in open and vegetated areas. The author found no significant differences between UAV-DTMs and ALS-DTMs in open space, but differences between 3.4 and 4.1 m were estimated in densely vegetated areas. In general, vegetated zones are considered as low confidence areas in photogrammetry. The ASPRS (2015) standard states that the accuracy in these areas can be at least 1.5 times lower than in an open space with enough daylight.

Zimmermann et al. (2017) analysed the accuracy of UAVs' positioning in GPS challenging environments. They point out that the accuracy depends, essentially, on the local GPS measurement conditions. In urban areas these conditions can be challenging due to satellite shadowing, non-line-of sight receptions, signal diffraction or multipath effects. Therefore, the authors propose the construction of GPS geometry maps of the area of interest, which are maps containing the 3D environment of the planned flight and an estimation of the expected GPS measurement conditions, so that problematic zones can be avoided. Using these maps, the authors suggest identifying and eliminating satellite signals subject to errors. Initial tests carried out by Zimmermann et al. (2017) show that UAV positioning can be improved by 82% using such maps. Companies like Uber are working in the same direction to improve the accuracy of GPS receivers in urban canyons. Using a probabilistic analysis of the intensity of GPS signals, the company manages, through an Android application, to reduce the positioning error in challenging environments from 50 to 10 m (Iland et al., 2018).

GCPs

GCPs are marks on the ground, spaced strategically throughout the area of interest, that are used to establish the precise scale and orientation of the map data in relation to the ground data (Food and Agriculture Organization [FAO], 2020). GCP markers can be painted on the surface or on panels, large enough to be seen in the drone photos, with a colour or contrast that stands out. Since GCPs are used to improve the positioning of UAV images, their coordinates must be determined before the flight and with a higher accuracy than that provided by the drone's onboard positioning system. This can be achieved using a high accuracy GPS receiver, for instance, a centimetre-level real time kinematic (RTK) GPS, or by traditional surveying methods.

There is no consensus in the literature about the optimal number of GCPs, as it depends on the size and shape of the area to be mapped (FAO, 2020), the amount of image overlap (Pix4D, 2020), and the accuracy required by the user (Giordan et al., 2020). However, some researchers agree that at least 3 GCPs should be used in small areas (Prajwal et al., 2016; Oniga et al., 2018). Software developers have also issued recommendations on the minimum number of GCPs to guarantee the reconstruction of the 3D models; for instance, Pix4D (2019) and DroneDeploy (2017b) require a minimum of 5 GCP, whereas Agisoft (2020) indicates that 10-15 GCPs are needed to generate accurate products. Other sources, such as the Nevada Department of Transportation (Pix4D, 2018), have concluded that the use of over 5-10 GCPs had very little benefit in the point cloud optimization. Pix4D (2017) points out that the traditional use of GCPs in aerial mapping projects is evolving and that they may not be necessary in certain projects when using RTK-equipped drones.

In addition to the number of GCPs, their distribution has also been widely discussed in the literature. The following recommendations are suggested by the software developers DroneDeploy (2017c) and Pix4D (2019):

- A uniform distribution of GCPs is recommended, with enough separation to avoid confusion. If two or more GCPs appear in the same image, they are located too close to each other.
- Depending on the overlap of the flight, a buffer zone between the map's perimeter and any GCP should be considered, so that there is enough information for processing. DroneDeploy (2017c) suggests between 15 and 30 m.
- Terrain elevation changes, in combination with the orientation of the sensor, create image displacements (Jain et al., 2019). To reduce this distortion, it is recommended to locate a GCP at the main elevations of the study area.
- Obstructions like trees, buildings, shade or glare must be avoided.

At present, GCP-based georeferencing is carried out manually. It is based on the detection of the GCP markers on the UAV images and their labelling as control points. This must be done on all images that contain markers, which can be time-consuming, monotonous and prone to human errors (Jain et al., 2019). In addition, low-quality images, i.e. low spatial resolution, incorrect exposure level, motion blur, etc., can make it difficult to identify the centre of the markers. Figure 2 shows an aerial image acquired at 60 m altitude, GSD = 2.6 cm, with a GCP marked with a 20X20 cm checkerboard. The photo on the left (a) shows that the GCP's marker can be clearly identified; however, in the photo on the right (b), the identification of the centre of the marker is a more complicated task due to the scattering of the light on the white squares of the checkerboard. This suggests that the colours chosen for that marker are not suitable under the prevailing lighting conditions. The shape and size could also be improved. Jain et al. (2019) propose, for example, white "L" shaped markers of 65 X 65 cm.

Figure 2. Views of a ground control marker, scales 1:50 (a) and 1:10 (b) with the same GSD

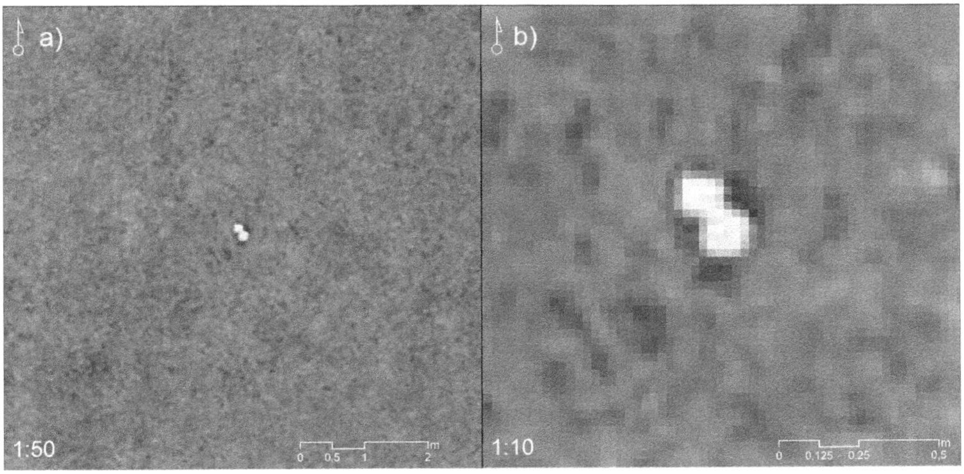

Besides GCPs, checkpoints (CPs) are required in UAV photogrammetry. Like GCPs, CPs must be marked on the surface to be identified in the aerial images, and measured with higher accuracy than that provided by the UAV's positioning system. CPs are not used for data processing; instead, they are used to validate the relative and absolute accuracy of the map products. This is done by comparing the known measured location of the checkpoints to the coordinates of the same points shown on the map (Dronedeploy, 2019).

Figure 3. Study area

CASE STUDY

Study Site

An area of approximately 3 ha in the province of Antioquia, northwest of Colombia, was examined. The site has nearly level areas and steep terrain close to the three streams that cross it. The land cover is composed of groups of isolated trees, shrubs, and areas covered with low vegetation. Figure 3 shows an orthomosaic of the study area.

MATERIALS AND METHODS

Mission Planning

A DJI Phantom 3 was used, which was the most widely used drone in the region at the time of the study. Table 1 summarizes its technical specifications. The flight missions were programmed with the DroneDeploy application.

Table 1. Technical specifications of the Phantom 3

Aircraft	Weight (battery and propellers included)	1216 g
	Flight speed	5 m/s ascent, 3 m/s descent
	Max flight speed	16 m/s
	Max flight time	25 min approx.
	Satellite positioning system	GPS
	Hover accuracy range	Vertical: ±0.5 m Horizontal: ±1.5 m
Gimbal	Stabilization	3-axis (pitch, roll, yaw)
	Controllable range	Pitch: -90° to +30°
	Max controllable angular speed	Pitch: 90°/s
	Angular vibration range	± 0.02°
Camera	Sensor	1/2.3" CMOS Effective pixels: 12M ps 0.0001529
	Lens	FOV 94° 20 mm (35 mm format equivalent) f/2.8 = 0.357 cm
	Image size	4000 X 3000

Source: DJI, 2019b

Three flight altitudes were tested: 60, 70 and 80 m. These flight altitudes, in combination with the optical characteristics of the UAV camera (see table 1) produced GSD values of approximately 2.5, 3.0 and 3.4 cm, respectively. These GSDs are recommended to produce high-quality maps for local accuracy determinations, scales 1:200 and 1:250, as stated in the ASPRS (2015) standard. Three frontlaps and sidelaps were examined: 70%, 75% and 80%, that are within the traditional range of values, 60% - 85%, suggested for photogrammetry procedures by authors such as Mikhail et al. (2001) and software developers like Pix4D (2019). To facilitate the analysis and presentation of results, the same front- and sidelap was used on each UAV-flight, actually generating a slightly larger number of photogrammetric products than required.

Figure 4. Distribution of GCPs and CPs

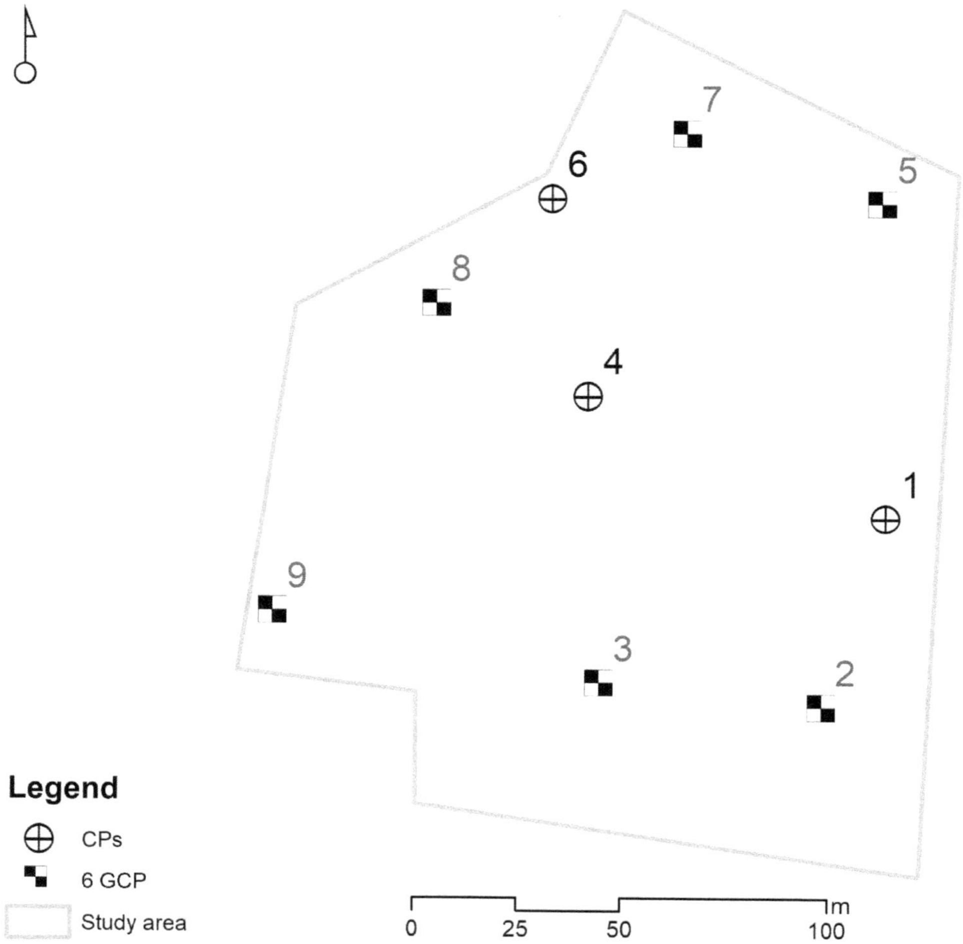

Data Acquisition

Six GCPs and three CPs were used in the study, as shown in figure 4. The GCPs were uniformly distributed, as suggested by Sanz-Ablanedo (2018), DroneDeploy (2019) and Pix4D (2019). The GCPs and CPs were measured with topographic equipment that provides millimetre-level accuracy. The local coordinates were transformed in UTM coordinates (UTM Zone 18N). An existing concrete survey monument was used for georeferencing both GCPs and CPs. The coordinates of each point were previously determined as part of an earlier project, using a high precision GPS and a post-processing method. Both GCPs and CPs were marked with 20 by 20 cm black and white checkerboards; their position is shown in figure 2.

Complying with the recommendations of Landmäteriet (2018), the flights were conducted on sunny days with some passing clouds and in the same week.

Data Processing

UAV imagery was processed with the software Agisoft PhotoScan, version 1.3. The parameters used during the processing are presented in table 2.

Table 2. Parameters used in Agisoft PhotoScan

Alignment parameters	Accuracy	High
	Pair preselection	Generic
	Key point limit	40.000
	Tie point limit	4.000
	Adaptive camera model fitting	Yes
	Optimized parameters	f (focal length) b_1 and b_2 (affinity and skew transformation coefficients) c_x and c_y (principal point coordinates) k_1 to k_4 (radial distortion coefficients) p_1 to p_4 (tangential distortion coefficients)
Dense point cloud	Quality	High
	Depth filtering	Moderate
DEM	Coordinate system	WGS 84 / UTM Zone 18N
	Surface type	Height field
	Source data	Dense
	Interpolation	Enabled
Orthomosaic	Blending mode	Mosaic
	Surface	Mesh

The DEMs and orthomosaics were reconstructed with 0, 3, and 6 GCPs, selected as shown in figure 5. Besides the number of GCPs, different flight altitudes and overlaps were analysed, as stated above. With these combinations, 27 DEMs and 27 corresponding orthomosaics were produced. The processing workflow is summarized in figure 6.

Figure 5. Scenarios with 0, 3 and 6 GCPs for DEM and orthomosaic generation

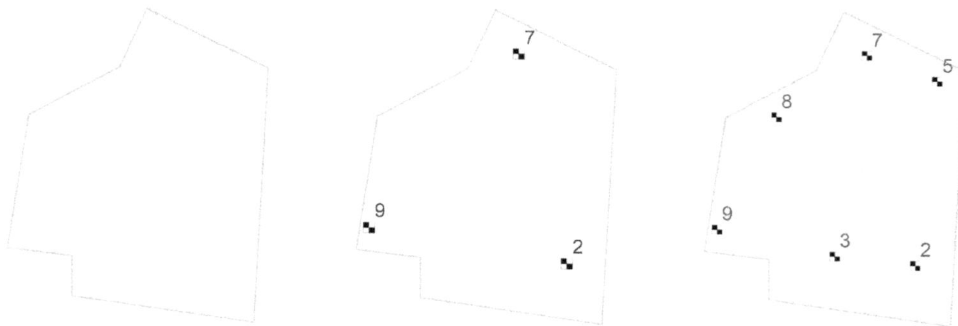

Figure 6. Data processing workflow

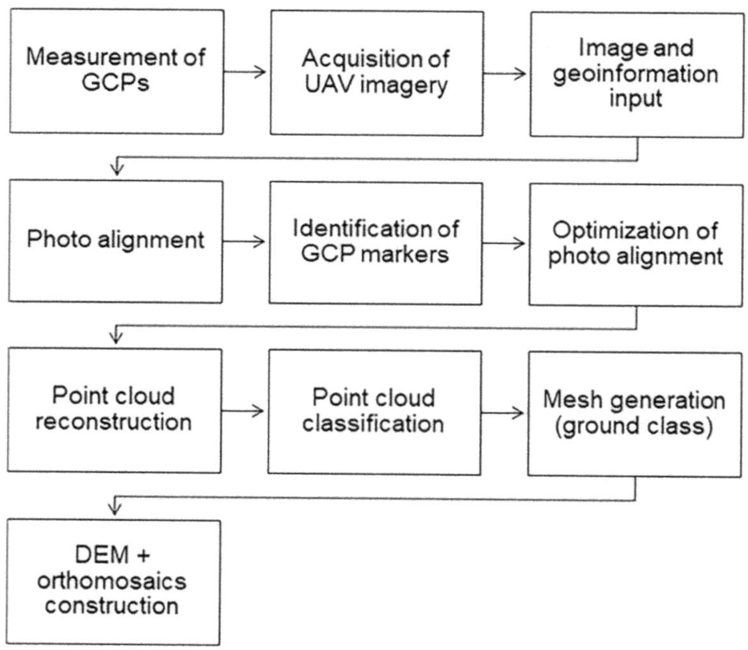

Accuracy Analysis

Horizontal and vertical accuracy were determined by comparing the measured CP coordinates (points 1, 4 and 6, see figure 4) with their corresponding values estimated by the processing software. The horizontal accuracy was estimated from orthomosaic x- and y-values, as follows (all formulas were taken from FGDC 1998):

$$RMSE_x = \sqrt{\frac{\sum_{I=1}^{n} \left(x\, data\, I - x\, check\, I\right)^2}{n}}$$

and

$$RMSE_y = \sqrt{\frac{\sum_{I=1}^{n} \left(y\, data\, I - y\, check\, I\right)^2}{n}}$$

where RMSE is the root-mean-square error, x and y data I are respectively the x- and y-coordinates of the Ith CP in the dataset, x and y check I are the coordinates of the Ith CP in the independent source of higher accuracy, n is the number of CPs tested and I is an integer ranging from 1 to n.

The horizontal RMSE is defined as:

$$RMSE_r = \sqrt{\left(RMSE_x^2 + RMSE_y^2\right)}$$

Finally, assuming that errors are normally distributed, the horizontal (radial) accuracy at the 95% confidence level was computed by the formula:

$$Accuracy_r = 1.73\ (RMSE_r)$$

In the same way, the vertical RSME and accuracy were estimated from the z-coordinates of the corresponding DEMs:

$$RMSE_z = \sqrt{\frac{\sum_{I=1}^{n} \left(z\, data\, I - z\, check\, I\right)^2}{n}}$$

where z check I is the vertical coordinate of the Ith CP in the dataset, z check I is the vertical coordinate of the Ith CP in the independent source of higher accuracy, n is the number of CPs tested and I is an integer ranging from 1 to n.

Assuming that errors are normally distributed, the vertical accuracy at the 95% confidence level is equal to:

$$Accuracy_z = 1.96\ (RMSE_z)$$

As can be noted from the previous expressions, both RMSE and accuracy correspond to dispersion measures and are expressed as distances. The RMSE indicates how much, on average, the measured and estimated coordinates differ, while the accuracy value exhibits the measurement error below which 95% of the estimated values are expected to be observed. In this sense, lower $Accuracy_r$ and $Accuracy_z$ stand for more precise procedures.

All statistical analyses were carried out with R (R Core Team, 2020), including packages *dplyr* (Wickham et al., 2020) and *basicTrendline* (Weiping & Guangchuang, 2018).

Influence of Mission Parameters on DEM Accuracy

After examining the accuracy of the map products with the procedure proposed by FGDC (1998), a factorial analysis of variance was conducted to assess the influence of flight altitude, overlap and number of used GCPs on horizontal and vertical DEM accuracy. Also, the effect of the interactions between these variables was analysed. Flight altitude and overlap (both front- and sidelap) considered the three previously mentioned levels, corresponding respectively to 60, 70 and 80 m for the first and 70, 75 and 80% for the second. For the number of GCPs, three levels were considered (0, 3 and 6).

When analysing the processing software reports, which include an error distribution map, the question arose whether the calculated accuracy was valid in the entire study area or only in the vicinity of the CPs. To answer this question, additional topographic surveys available in the study area were used for comparison. The information of 563 survey points, measured with total stations, was compared with the *z*-values estimated with each of the DEMs, according to the used GCPs. As their position was not marked with a checkerboard to be observed in the orthophotos, horizontal accuracy could not be determined with the help of these points.

The distribution of these additional 563 survey points is shown in figure 7. They correspond to 3 ground cover categories (1 = high vegetation or trees, 2 = grown vegetation or shrubbery, 3 = low vegetation or grass). A fourth category, artificial cover, corresponding to flat concrete floors, was not observed in the available data points. These categories were used to assess ground cover influence on DEM accuracy.

With the information from the topographic surveys, 5 levels of analysis were considered, corresponding to 2, 4, 8, 16 (figure 8) and all 563 survey points (figure 7). The selection of survey points used in the determination was done considering a uniform spread throughout the study area.

Figure 7. Additional survey points used in the error distribution analysis

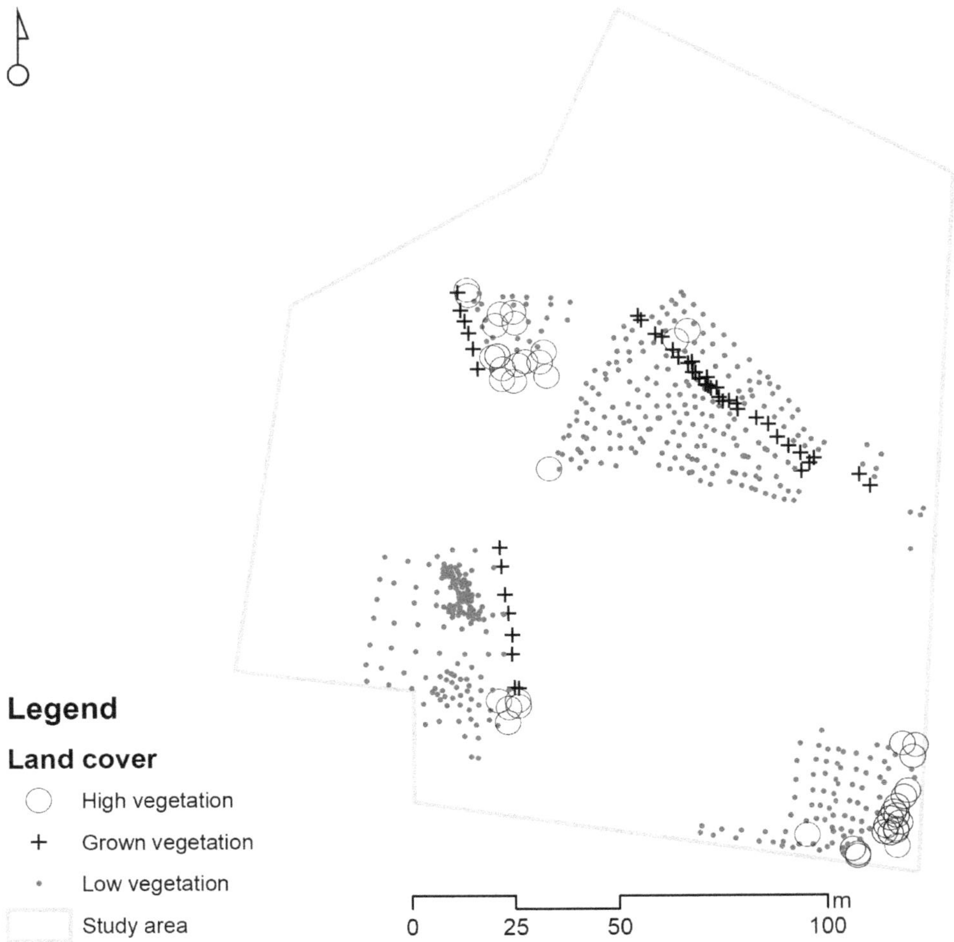

RESULTS AND ANALYSIS

Determination of Orthomosaic and DEM Accuracy

As mentioned above, horizontal errors were determined for the three fixed CPs (point numbers 1, 4 and 6 in figure 4) by subtracting the x- and y-values obtained for the UAV orthomosaics from the corresponding measurement, while vertical errors were calculated by subtracting the estimated DEM elevation from the corresponding CPs' z-values. The resulting RMSE and accuracy values were determined by the equations of section 4.2.4. Table 3 contains the combinations of flight altitude, overlap (both frontlap and sidelap) and number of GCPs used in the generation of map products. It also includes the horizontal and vertical RMSE and the corresponding accuracies; the latter are obtained by multiplying the horizontal and vertical RMSE by respectively 1.73 and 1.96.

Figure 8. Combinations of survey points (2, 4, 8, 16) used in the error distribution analysis

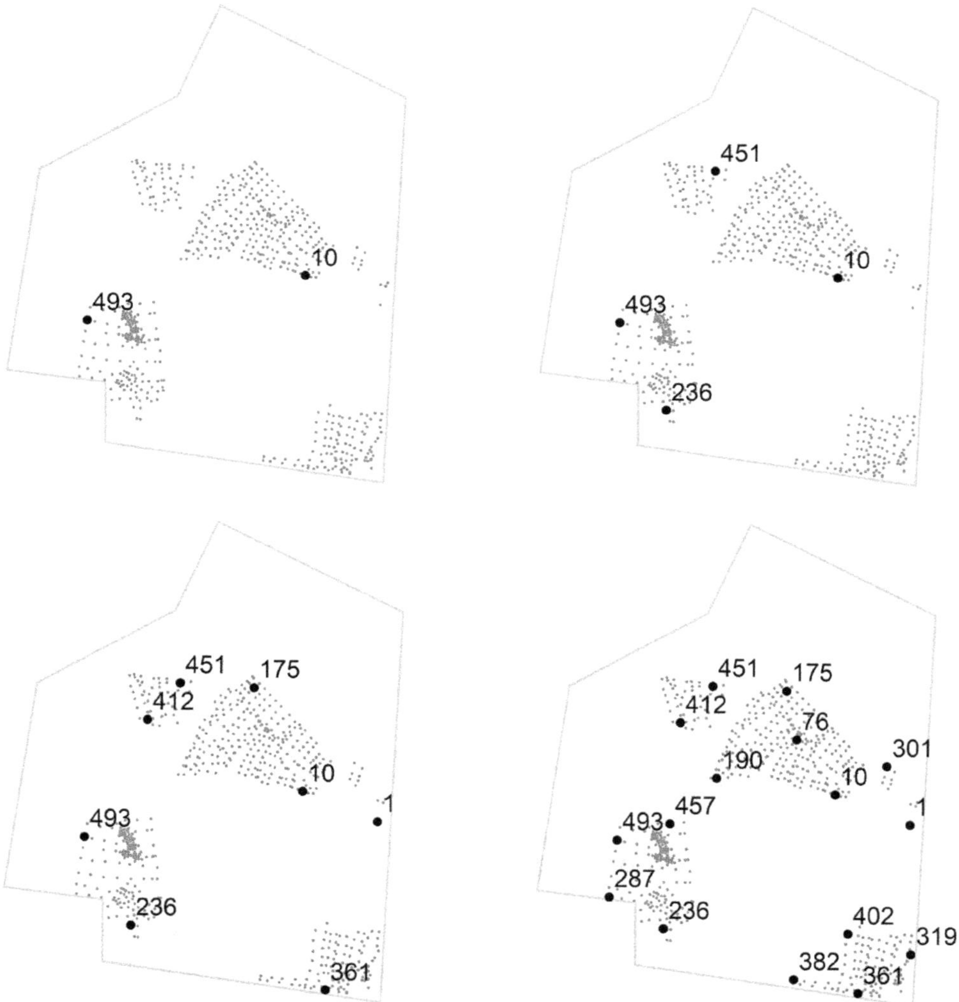

Table 3. Horizontal and vertical errors and accuracy

Flying altitude (m)	Overlap (%)	GCPs	RMSE$_r$ (m)	*Accuracy$_r$* (m)	RMSE$_z$ (m)	*Accuracy$_z$* (m)
60	70	0	6.793	**11.753**	3.099	**6.073**
		3	0.284	**0.492**	0.108	**0.211**
		6	0.086	**0.149**	0.196	**0.384**
	75	0	6.564	**11.355**	2.827	**5.541**
		3	0.293	**0.507**	0.186	**0.364**
		6	0.098	**0.170**	0.180	**0.353**
	80	0	6.929	**11.987**	0.616	**1.206**
		3	0.253	**0.438**	0.202	**0.397**
		6	0.144	**0.249**	0.113	**0.221**
70	70	0	6.685	**11.565**	1.003	**1.965**
		3	0.299	**0.518**	0.253	**0.495**
		6	0.097	**0.168**	0.076	**0.149**
	75	0	6.611	**11.438**	0.938	**1.839**
		3	0.286	**0.494**	0.051	**0.099**
		6	0.100	**0.174**	0.119	**0.233**
	80	0	6.778	**11.726**	0.652	**1.277**
		3	0.288	**0.498**	0.079	**0.156**
		6	0.096	**0.166**	0.106	**0.208**
80	70	0	6.281	**10.866**	2.290	**4.488**
		3	0.296	**0.513**	0.242	**0.474**
		6	0.103	**0.177**	0.087	**0.171**
	75	0	6.231	**10.780**	1.013	**1.985**
		3	0.288	**0.498**	0.056	**0.110**
		6	0.107	**0.185**	0.104	**0.204**
	80	0	6.401	**11.074**	1.356	**2.657**
		3	0.287	**0.496**	0.068	**0.132**
		6	0.101	**0.174**	0.055	**0.108**

The results in table 3 show that both horizontal and vertical precision improve with the number of GCPs. Horizontal accuracy (*Accuracy$_r$*), expressed as the 95% confidence interval limit for expected measurement errors, oscillated between 10.8 and 12.0 m when no GCPs were used. These values decreased rapidly to errors between 0.438 and 0.518 m when 3 GCPs were used, and between 0.149 and 0.249

m when 6 GCPs were used. For vertical accuracy (*Accuracy$_z$*), these values oscillated in the intervals [1.2 m; 6.1 m], [0.099 m; 0.495] and [0.108 m; 0.383 m] when using respectively 0, 3 and 6 GCPs. These results show that precision increased rapidly when GCPs are included in the process. On the other hand, the accuracy values found for the 0 GCP scenarios are equivalent to the accuracy of publicly available elevation datasets produced from some satellite-based sensors (ASPRS, 2015), making it unfit to use UAVs for photogrammetry when no GCPs are used (see table 3 and figure 9).

Figure 9. Improvement of horizontal (a) and vertical (b) accuracy with increased number of GCPs

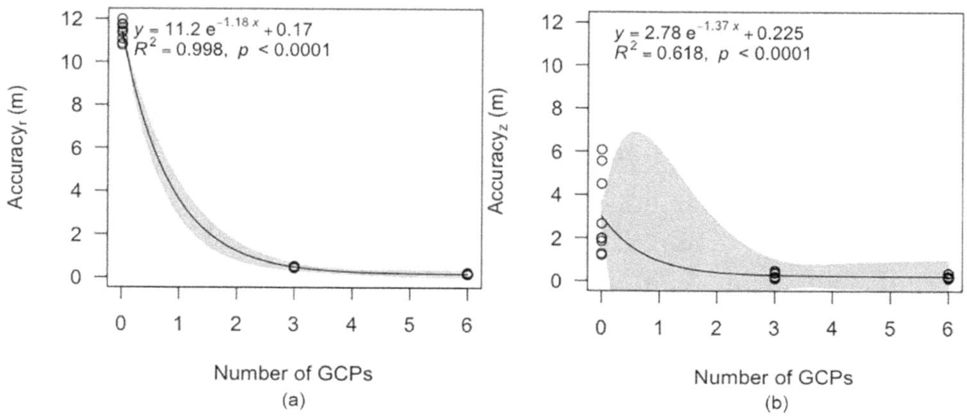

The horizontal precision improves in a more pronounced way with the number of GCPs than the vertical precision. For the scenarios where no GCPs were used for DEM and orthomosaic construction, the horizontal accuracy also presents less variation between the different flight configurations. However, vertical precision is on average twice as good as horizontal precision when 3 and 6 GCP are used (see figure 9). An exponential model could be adjusted in both cases; for the horizontal accuracy, the adjustment was good ($R^2 = 0.998$), while for vertical accuracy, the high variability of results when no GCPs are used caused a lower adjustment quality ($R^2 = 0.618$), mainly due to the important differences between flight scenarios for 0 GCPs.

The factorial analysis suggests that, at a 0.05 significance level, only the number of GCPs can be considered a decisive factor in DEM and orthomosaic precision. Nor flight altitude nor overlap (both frontlap and sidelap) could be proved to be statistically significant. In the determination of vertical accuracy, *Accuracy$_z$*, only the 0 GCPs scenario differs from the other two at the same significance level.

Figure 10. Orthomosaics constructed with 0 and 6 GCPs

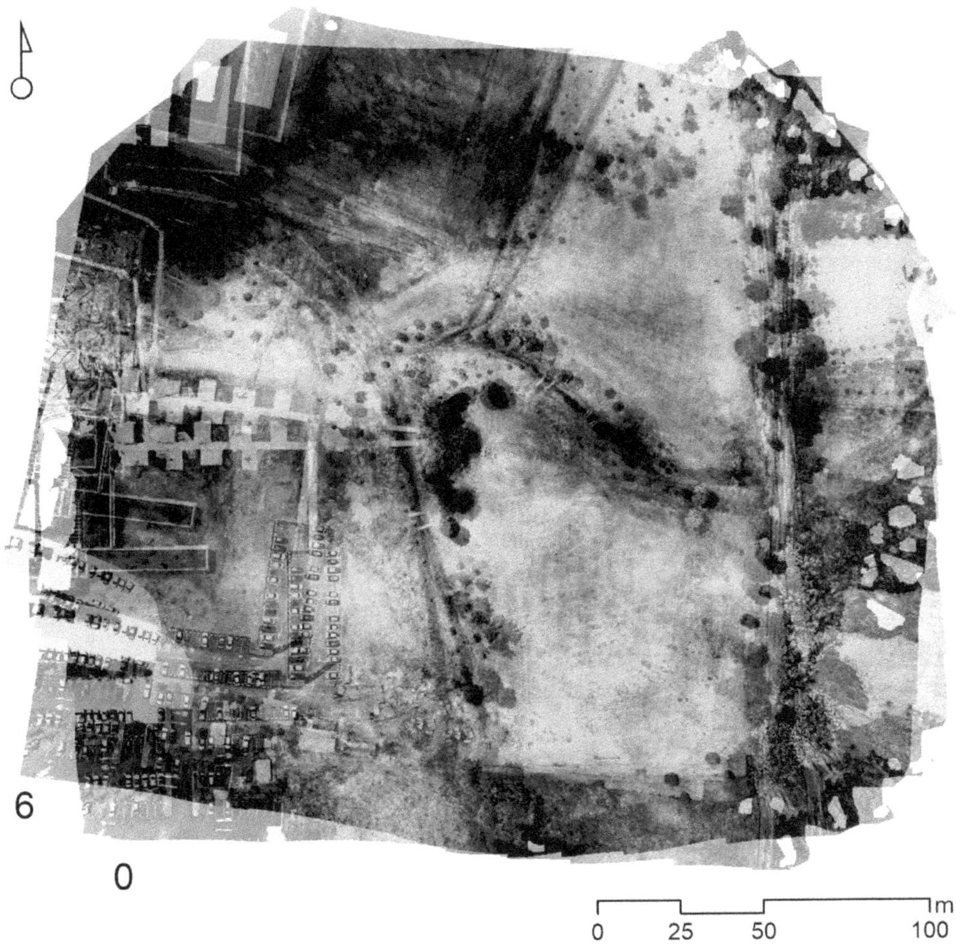

A detailed analysis of the estimated DEM values indicates that, in over 83% of the flight scenarios, the height of the CPs was underestimated. Underestimations ranged from 3.5 to 25 cm, averaging around 8.5. If the value was overestimated, the effect was only half the amount. The above might suggest a systematic error in the processing software, which was also reported by Alidoost & Arefi (2017). To reduce this error, a different processing software could be used, for instance Pix4D, as suggested by Costales-Acurio (2018).

Rotation Observed When No GCPs Are Used

As stated by other authors, the results of this research confirm that the absence of GCPs is not recommended to carry out UAV photogrammetry successfully. The software developer Pix4D (2019) points out that "if no GCPs are used, the final results have no scale, orientation, and absolute position information. Therefore, they cannot be used for measurements, overlay, and comparison with previous results. Additionally, the results may produce an inverted 3D model and the reconstruction may not preserve the shape of the surveyed area."

Figure 10 shows two overlapping orthomosaics of the study area, which were processed with 0 and 6 GCPs. Both orthomosaics coincide in the central region which corresponds to the UAV's take-off and landing zone, but they diverge at the edges. The orthomosaic constructed with 0 GCPs presents a counter clockwise rotation that creates a displacement of approximately 20 m at the edges of the image, in comparison with the orthomosaic georeferenced with 6 GCPs. This rotation may explain the magnitude of the horizontal errors observed in map products without GCPs (see table 3) and may also explain why this effect is more important in the horizontal than in the vertical dimension.

Statistical Analysis with Additional Survey Points

The accuracy analysis of section 4.3.1 corresponds to the analysis method proposed by FGDC (1998). It assumes that errors approximate a normal distribution and that controllable errors, such as failures in the measurement of CPs and GCPs or calibration errors in the processing software, were eliminated as best as possible. ASPRS (2015) notes that there may be areas with low confidence, such as areas beneath dense vegetation, or where the imagery is otherwise obstructed, in deep shadow, or with changing topography that is not adequately represented by the interpolation applied by the photogrammetric software. In these areas, the confidence level criterion is not recommended. Instead, the analysis shall be performed with the 95th percentile to ensure that 95% of errors fall at or below this value. On the other hand, the FGDC-method (1998) is rather undetailed on the location, number and characterization of checkpoints that should be used in DEM accuracy determination.

To complement the accuracy analysis, existing topographic survey information was used to make an additional comparison, considering more data points than the 3 available CPs. These data (563 survey points) were used to assess the normality of height estimation errors, as well as to determine the influence of both the number and characteristics (e.g. land cover) of comparison points on the vertical accuracy. The vertical errors were obtained by subtracting the survey point height from the corresponding DEM information obtained in the 27 flight missions with 0, 3 and 6

GCPs. This analysis only assesses vertical accuracy, because, to be able to determine horizontal accuracy, the survey points must be identified in the aerial images. As the topographic surveys were carried out as part of other projects and prior to the flight missions, this information was not available. In the evaluation of the influence of the mission parameters, the 0 GCPs scenario was not included because of the huge difference with the scenarios where GCPs were used to construct DEMs.

Assessment of Error Normality

A Shapiro-Wilk goodness-of-fit test was carried out on the distribution of estimation errors in each of the 563 survey points. None of the scenarios produced errors that could be fitted to the normal distribution, showing high values of kurtosis and probability functions biased to the right. In general, the scenarios based on 3 or 6 GCPs presented a similar error distribution, regardless of flight altitude and overlap. Nevertheless, a higher standard deviation for the probability distribution (and thus higher RMSEz) was observed for scenarios where no GCPs were used for DEM construction. Two examples of the vertical error distribution, together with the corresponding normal distribution, are presented in figure 11.

Figure 11. Error distributions and corresponding normal distributions for a flight altitude of 60 m and overlap of 70%, (a) 0 GCPs, (b) 6 GCPs

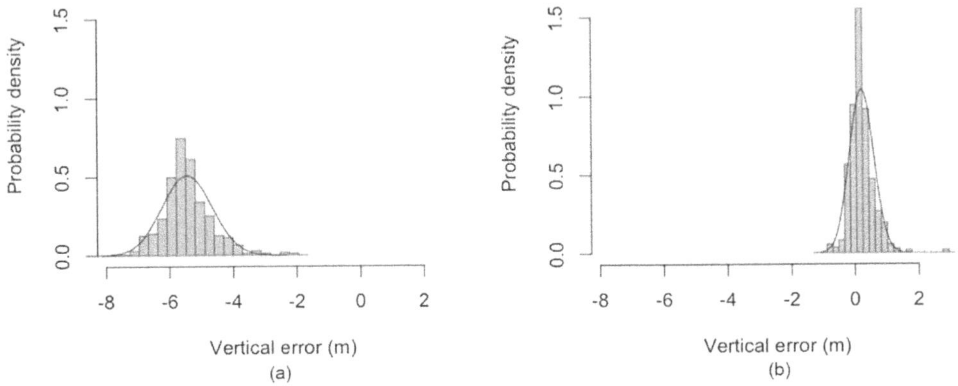

Influence of The Land Cover in The Accuracy Determination

The low but existing probability of quite high errors might indicate the superposition of different probability functions. To analyse this possibility in more depth, the survey points were classified by land cover (1 = high vegetation or trees, 2 = grown vegetation or shrubbery, 3 = low vegetation or grass, see section 4.2.5). The corresponding goodness-of-fit analysis showed that the error data corresponding to high vegetation or trees (8% of the survey points) cannot be considered normally distributed and present the largest errors, as well as bias to the right. The survey points located in grown vegetation or shrubbery (also 8% of the survey points) could be adjusted to the normal distribution for the DEMs built with 3 and 6 GCPs, while for the survey points located in low vegetation or grass (the remaining 84% of the survey points), normality was accepted only for some flight scenarios. These results show that the normality of error distribution is a condition that should not be taken for granted and should be assessed before applying the procedures proposed by FGDC (1998).

Figure 12. Increase of the vertical accuracy value with increased number of survey points

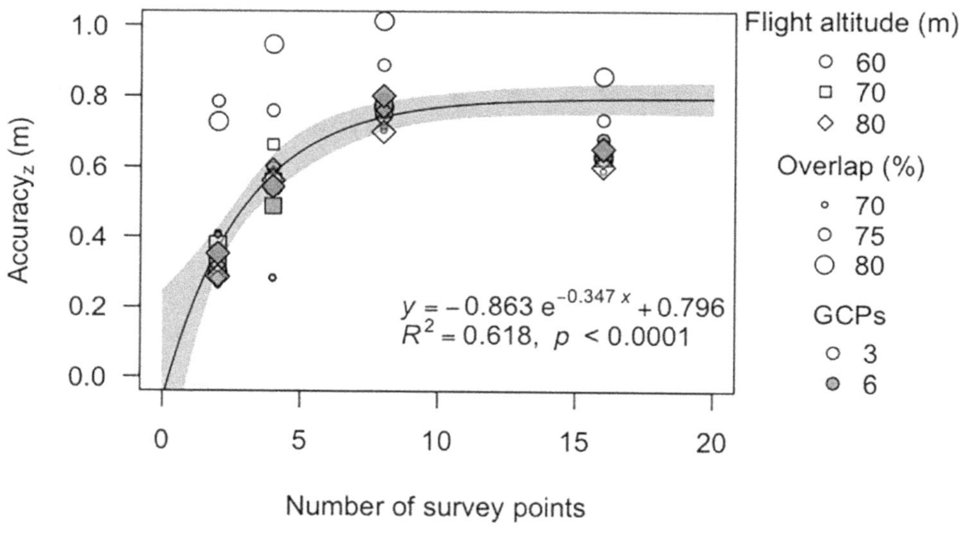

Influence of The Number of Data Points

As explained in the methodology, the complementary analysis of vertical accuracy was carried out using the survey point combinations of figures 7 and 8. Figure 12 presents the vertical accuracy, *Accuracy*$_z$, for different numbers of survey points and flight parameters. The flight altitude is depicted by three different symbols while the overlap is represented by the size of the used symbol. The distinction between the number of GCPs is done by shadowing the symbol. The results show a clear increase in the vertical accuracy value when an increasing number of survey points is used for the accuracy determination (figure 12).

Figure 12 shows differences between the different flight scenarios, but in all cases the accuracy value associated with vertical accuracy increased initially with the number of survey points and afterwards seems to stabilize at around 8 survey points. The scenarios based on 6 GCPs showed little difference between distinct flight scenarios. The flight scenarios were 3 GCPs where used presented more variability, especially the mission with a low flight altitude, high overlap and 3GCPs presented a more elevated error, regardless of the number of survey points used for the evaluation. As precision is expected to increase for lower flight altitudes and higher overlaps (Rock et al., 2011; Haala et al., 2013; Dandois et al., 2015; Mesas-Carrascola et al., 2015; Leitao et al., 2016; Fuad et al., 2018) this result seems to be against expectations; however, further analysis is required to be conclusive in this regard.

A factorial analysis to assess the significance of the mission parameters when using more survey points (or, by extension, CPs) indicates that, when GCPs are used to construct photogrammetric products, both the number of GCPs and survey points are statistically significant for DEM accuracy at a 0.05 significance level, with respective p-values of 0.0003 and $< 2.2\ e^{-16}$. The previous conclusion on flight altitude and overlap not being significant was also confirmed in this extended analysis.

The value ranges presented in figure 12 are comparable with the results discussed in section 4.3.1; as the specific location of used CPs and survey points do not coincide in both analysis, slight changes in value are observed. The previous result suggests that, if an insufficient number of CPs is selected to evaluate DEM quality, precision can be estimated to be better than it really is, because the selected CPs can be (a) not representative (the probability of non-representativeness diminishes when increasing the number of CPs), or (b) not represent the total variation of errors in the study area. A look at figure 12 also shows that the vertical accuracy estimate is around 30 cm when 2 survey points are used in accuracy determination, while this error increases to around 80 cm when using 8 survey points. Thus, the factor of 1.96 proposed by FGDC (1998) to account for the 95% of the error values might not be sufficient if CPs are not selected appropriately.

Influence of The Ground Cover

The set of 563 survey points was also used to assess the influence of ground cover on vertical accuracy. The corresponding factorial analysis showed that, although the number of GCPs used in the DEM construction is still a significant factor at a 0.05 significance level (p-value = 0.024), the most important factor is ground cover (p-value $< 2.2\ e^{-16}$). When filtering by ground cover, interactions with both flight altitude and overlap become slightly significant (not at a 0.05 level, but at a 0.1 level). Clearly, the taller the coverage at the location of survey points, the higher the accuracy value (see figure 13) or, in other words, the lower the precision of the constructed orthophotos and DEMs.

Figure 13. Decrease of vertical accuracy value with lower ground cover (1 = high vegetation or trees, 2 = grown vegetation or shrubbery, 3 = low vegetation or grass)

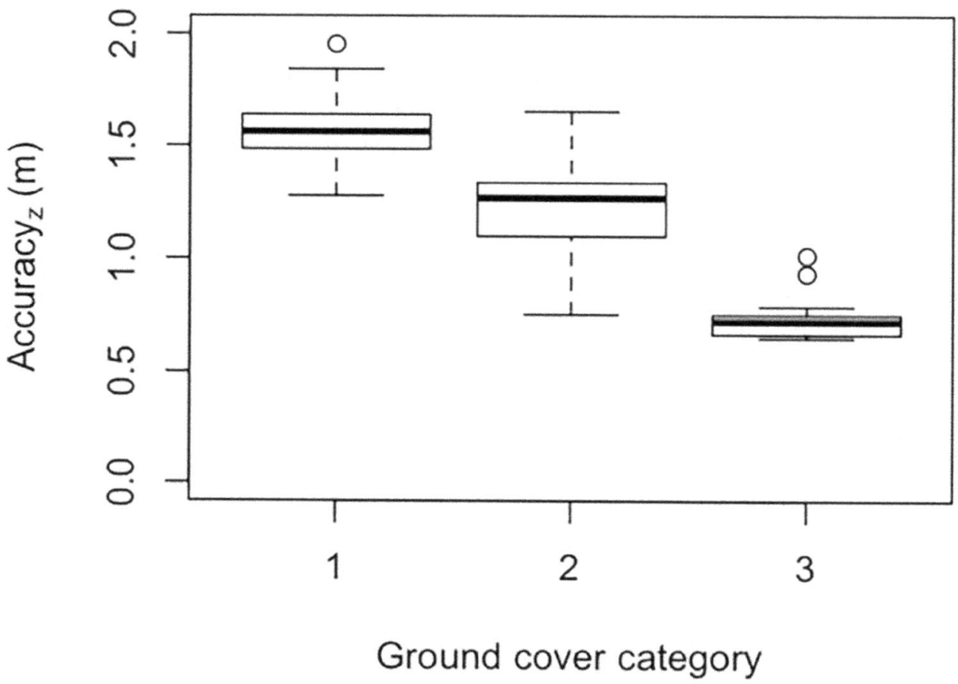

Errors seem to increase almost linearly with the ground cover height at the survey points, although more analysis is required to come to final conclusions. Most of the additional survey points (84%) were located in ground cover category 3, where the corresponding accuracy values were between 0.5 and 1 m. For areas covered with trees, this error can ascend to twice as much. Also, the lower the vegetation, the less difference in results between different flight missions, as can be observed by the variation for each of the land covers in figure 13.

The differences in accuracy found between distinct coverages suggest that CP selection for DEM accuracy determination should be carried out considering their ground cover. The presented results indicate that the accuracy of the DEM is preferably determined by CPs located in areas with low ground cover, or, that if all CPs present the same ground cover, that the estimated DEM accuracy is only valid for data points located in areas with the same ground cover.

Finally, it was also found that when the speed of the UAV exceeded the capacity of the sensor to capture clear images, the observed errors increased from 2 to 10 times, mainly in the case where no GCPs were used. The flights that generated this type of information were repeated to ensure an acceptable quality level of the input data.

CONCLUSION

After analysing 27 DEMs and orthomosaics, their precision can be concluded to improve with the number of GCPs. When these are not used, the quality of the data decreases enormously. Another finding is that the horizontal accuracy seems to present less variation than the vertical accuracy when using different flight configurations; however, the vertical accuracy was on average twice as good as horizontal accuracy. It was also found that, apparently, the vertical coordinate is more commonly underestimated than overestimated. This was observed in 83% of the flight scenarios, where the underestimated corresponded on average to 8.5 cm, indicating a possible systematic error introduced by the processing software.

The factorial analysis indicates that the number of GCPs is statistically significant for DEM accuracy at a 0.05 significance level, which is consistent with previous findings in the study field.

The number of control points used to determine the accuracy of map products is also statistically significant. Although in this study only 3 CPs were used, the additional analysis with 563 pre-existing survey points indicates that with 2 survey points, the vertical accuracy was estimated around 30 cm, while with 8 survey points the accuracy value increased to 80 cm. This decrease in precision suggests that the factor of 1.96 proposed by FGDC (1988) to account for the 95% of error values might not be sufficient if CPs are not selected appropriately. The use of an insufficient number of checkpoints can lead to unrepresentative results, such as a better estimate of accuracy than actually exists. The ground cover of the spots chosen to evaluate accuracy also influences the determination. Results indicate that, in order to achieve a representative result, points with different land cover must be considered and that if only one land cover is used, the results will be valid only in the areas with the same ground cover.

ACKNOWLEDGMENT

This research was supported by the Universidad EIA.

REFERENCES

Agisoft. (2020). *Orthophoto and DEM Generation (with GCPs) with Agisoft PhotoScan Professional.* https://www.agisoft.com/index.php?id=31

Airobotics. (2020). *Automated drone applications for industrial purposes.* https://www.airoboticsdrones.com/applications/

Al-Hajjaji, K., Ezzin, M., Khamdan, H., & El Hassani, A. (2017). *Design, Development and Evaluation of a UAV to Study Air Quality in Qatar.* https://arxiv.org/ftp/arxiv/papers/1709/1709.05628.pdf

Alidoost, F., & Arefi, H. (2017). Comparison of UAS-based photogrammetry software for 3D point cloud generation: A survey over a historical site. In *Proceedings of the 4th International GeoAdvances Workshop* (vol. 4, pp. 55-61). 10.5194/isprs-annals-IV-4-W4-55-2017

Areias, B., Humberto, N., Guardalben, L., & Fernandes, J. M. (2018). Towards an Automated Flying Drones Platform. In *Proceedings of the 4th International Conference on Vehicle Technology and Intelligent Transport System* (vol. 1, pp. 529-537). Setúbal, Portugal: Science and Technology Publications. 10.5220/0006792405290536

ASPRS. (2015). Positional Accuracy Standards for Digital Geospatial Data. *Photogrammetric Engineering and Remote Sensing*, *81*(3), 1–26. doi:10.14358/PERS.81.3.A1-A26

AUAV. (2019). *Types of Drones: Multi - Rotor vs Fixed - Wing vs Single Rotor vs Hybrid VTOL.* https://www.auav.com.au/articles/drone-types/

Barba, S., Barbarella, M., Di Benedetto, A., Fiani, M., Gujski, L., & Limongiello, M. (2019). Accuracy Assessment of 3D Photogrammetric Models from an Unmanned Aerial Vehicle. *Drones (Basel)*, *3*(4), 79. doi:10.3390/drones3040079

Begashaw, S. (2018). *Accuracy of DTM derived from UAV imagery and its effect on canopy height model compared to airborne Lidar in part of tropical rainforest of Berklah, Malaysia* [Unpublished master thesis]. University of Twente, Enschede, The Netherlands.

Beretta, F., Shibata, H., Cordova, R., de Lemos Peroni, R., Azambuja, J., & Coimbra-Leite, J. F. (2018). Topographic modelling using UAVs compared with traditional survey methods in mining. *REM - International Engineering Journal*, *71*(3), 463–470. doi:10.1590/0370-44672017710074

Cabreira, T. M., Brisolara, L. B., & Ferreira, P. R. Jr. (2019). Survey on Coverage Path Planning with Unmanned Aerial Vehicles. *Drones (Basel)*, *3*(1), 4. doi:10.3390/drones3010004

Castellano, F. (2018). *Commercial Drones Are Revolutionizing Business Operations.* https://www.toptal.com/finance/market-research-analysts/drone-market

Choudhary, G., Sharna, V., You, I., Yim, K., Chen, I. R., & Cho, J. H. (2018). Intrusion Detection Systems for Networked Unmanned Aerial Vehicles: A Survey. In *Proceedings of 2018 14th International Wireless Communications & Mobile Computing Conference (IWCMC)* (vol. 1, pp. 560-565). 10.1109/IWCMC.2018.8450305

Cimoli, E. (2015). *Determining Snow Depth Distribution from Unmanned Aerial Vehicles and Digital Photogrammetry* [Unpublished master thesis]. Technical University of Denmark. Lyngby, Denmark.

Colomina, I., & Molina, P. (2014). Unmanned aerial systems for photogrammetry and remote sensing: A review. *ISPRS Journal of Photogrammetry and Remote Sensing*, *92*, 79–97. doi:10.1016/j.isprsjprs.2014.02.013

Costales-Acurio, G. A. (2018). *Análisis comparativo entre los software de prueba Agisoft Photoscan y Pix4D para el procesamiento de datos obtenidos con fotogrametría de vehículo aéreo no tripulado (UAV) de bajo costo aplicado a proyectos de medio ambiente* [Unpublished undergraduate thesis]. Escuela Politécnica Nacional de Ecuador. Quito, Ecuador.

Crane, A. (2018). *From Start-up to Empire: The Evolution of DJI Drones.* https://dronelife.com/2018/05/14/from-startup-to-empire-the-evolution-of-dji-drones/

Dandois, J. P., Olano, M., & Ellis, E. C. (2015). Optimal Altitude, Overlap, and Weather Conditions for Computer Vision UAV Estimates of Forest Structure. *Remote Sensing*, *7*(10), 13895–13920. doi:10.3390/rs71013895

DJI. (2018). *You Can Film Like A Pro with DJI Drone Active Track.* https://store.dji.com/guides/film-like-a-pro-with-activetrack/

DJI. (2019a). *Phantom 4 Specs.* https://www.dji.com/phantom-4/info

DJI. (2019b). *Phantom 3 Standard.* https://www.dji.com/phantom-3-standard/info

Döllner, J. (2020). Geospatial Artificial Intelligence: Potentials of Machine Learning for 3D Point Clouds and Geospatial Digital Twins. *Journal of Photogrammetry, Remote Sensing and Geoinformation Science*, *2020*(1), 15–24. Advance online publication. doi:10.100741064-020-00102-3

DroneApps. (2019). *Price wars: counting the cost of drones, planes and satellites.* https://droneapps.co/price-wars-the-cost-of-drones-planes-and-satellites/

DroneDeploy. (2017a). *Seven Ways to Improve the Accuracy of Your Drone Maps.* https://blog.dronedeploy.com/seven-ways-to-improve-the-accuracy-of-your-drone-maps-9d3161f05956

DroneDeploy. (2017b). *What Are Ground Control Points (GCPs) and How Do I Use Them.* https://blog.dronedeploy.com/what-are-ground-control-points-gcps-and-how-do-i-use-them-4f4c3771fd0b

DroneDeploy. (2017c). *How Do I Use Ground Control Points? A guide to using ground control points with drone mapping software.* https://www.dronedeploy.com/blog/what-are-ground-control-points-gcps/

DroneDeploy. (2019). *GCP Checkpoints.* https://support.dronedeploy.com/docs/gcp-checkpoints-1

FAO. (2020). *Air surveys and cadastral maps.* Food and Agriculture Organization of the United Nations. http://www.fao.org/3/V4860E/V4860E04.htm

FGDC. (1998). *Geospatial Positioning Accuracy Standards. Part 3: National Standard for Spatial Data Accuracy*. Federal Geographic Data Committee of the United States of America.

Floreano, D., & Wood, R. J. (2015). Science, technology and the future of small autonomous drones. *Nature*, *521*(7553), 460–466. doi:10.1038/nature14542 PMID:26017445

Flybase. (2020). *Drone Automation Solution for Security & Surveillance.* https://flytbase.com/drone-security-solution/

Flynt, J. (2019). *DJI Tripod Mode: How it works and what it is?* https://3dinsider.com/dji-tripod-mode/

Fuad, N. A., Ismail, Z., Majid, Z., Darwin, N., Ariff, M. F. M., Idris, K. M., & Yussof, A. R. (2013). Accuracy evaluation of digital terrain model based on different flying altitudes and conditional of terrain using UAV LiDAR technology. In *Proceedings of IOP Conference Series: Earth and Environmental Science* (vol. 169, pp. 1-8). 10.1088/1755-1315/169/1/012100

Gerke, M. (2018). Developments in UAV-Photogrammetry. *Journal of Digital Landscape Architecture*, *3*, 262–272. doi:10.14627/537642028

Gerke, M., & Przybilla, H. J. (2016). Accuracy Analysis of Photogrammetric UAV Image Blocks: Influence of Onboard RTK-GNSS and Cross Flight Patterns. *Photogrammetrie, Fernerkundung, Geoinformation*, *2016*(1), 17–30. doi:10.1127/pfg/2016/0284

Giordan, D., Adams, M. S., Aicardi I., Alicandro, M., Allasia, P., Baldo, M., De Berardinis, P., Dominici, D., Godone, D., Hobbs, P., Lechner, V., Niedzielski, T., Piras, M., Rotilio, M., Salvini, R., Segor, V., Sotier, B. & Troilo, F. (2020). The use of unmanned aerial vehicles (UAVs) for engineering geology applications. *Bulletin of Engineering Geology and the Environment*. . doi:10.100710064-020-01766-2

Goldman Sachs. (2019). *Drones. Reporting for Work*. https://www.goldmansachs.com/insights/technology-driving-innovation/drones/

Goncalves, G. R., Perez, J. A., & Duarte, J. (2018). Accuracy and effectiveness of low cost UASs and open source photogrammetric software for foredunes mapping. *International Journal of Remote Sensing*, *39*(15-16), 5059–5077. doi:10.1080/01431161.2018.1446568

Gruszczynski, W., Matwij, W., & Cwiakala, P. (2017). Comparison of low-altitude UAV photogrammetry with terrestrial laser scanning as data-source methods for terrain covered in low vegetation. *ISPRS Journal of Photogrammetry and Remote Sensing*, *2017*(126), 168–179. doi:10.1016/j.isprsjprs.2017.02.015

Haala, N., Crammer, M., & Rothermel, M. (2013). Quality of 3D point clouds from highly overlapping UAV imagery. In Proceedings of International Archives of the Photogrammetry, Remote Sensing and Spatial Information Sciences (vol. XL-1/W2, pp. 183-188). doi:10.5194/isprsarchives-XL-1-W2-183-2013

Iizuka, K., Itoh, M., Shiodera, S., Matsubara, T., Dohar, M., & Watanabe, K. (2018). Advantages of unmanned aerial vehicle (UAV) photogrammetry for landscape analysis compared with satellite data: A case study in postmining sites in Indonesia. *Cogent Geoscience*, *4*(1), 1. doi:10.1080/23312041.2018.1498180

Iland, D., Irish, A., Madhow, U., & Sandler, B. (2018). *Rethinking GPS: Engineering Next-Gen Location at Uber*. https://eng.uber.com/rethinking-gps/

Ippolito, A., & Cigola, M. (2017). *Handbook of Research on Emerging Technologies for Digital Preservation and Information Modelling*. IGI Global. doi:10.4018/978-1-5225-0680-5

Jain, A., Mahajan, M., & Saraf, R. (2019). Standardization of the Shape of Ground Control Point (GCP) and the Methodology for Its Detection in Images for UAV-Based Mapping Applications. In *Advances in Computer Vision. Proceedings of the 2019 Computer Vision Conference (CVC)* (*vol. 1*, pp. 459-476). 10.1007/978-3-030-17795-9_34

JCGM. (2008). *Evaluation of measurement data - Guide to the expression of uncertainty in measurement*. Joint Committee for Guides in Metrology.

Joshi, D. (2017). *Exploring the latest drone technology for commercial, industrial and military drone uses.* https://www.businessinsider.com/drone-technology-uses-2017-7

Kanellakis, C., & Nikolakopoulus, G. (2017). Survey on Computer Vision for UAVs: Current Developments and Trends. *Journal of Intelligent & Robotic Systems*, *81*(1), 141–168. doi:10.100710846-017-0483-z

Kapoor, R., & MacDuffie, J. P. (2017). *Analysis Shows Continued Industry-Wide Decline in Electric Vehicle Battery Costs.* https://mackinstitute.wharton.upenn.edu/2018/electric-vehicle-battery-costs-decline/

Koci, J., Jariahani, B., Leon, J. X., Sidle, R. C., Wilkinson, S. N., & Bartley, R. (2017). Assessment of UAV and Ground-Based Structure from Motion with Multi-View Stereo Photogrammetry in a Gullied Savanna Catchment. *ISPRS International Journal of Geo-Information*, *6*(11), 328. doi:10.3390/ijgi6110328

Kraus, K. (2012). *Photogrammetrie. Geometrische Informationen aus Photographien und Laserscanneraufnahmen*. De Gruyter Lehrbuch.

Kršák, B., Blistan, P., Pauliková, A., Puskarova, P., Kovanič, Ľ., Palková, J., & Zelizňaková, V. (2016). Use of low-cost UAV photogrammetry to analyse the accuracy of a digital elevation model in a case study. *Measurement*, *91*, 276–287. Advance online publication. doi:10.1016/j.measurement.2016.05.028

Kumar, G. A., Kumar, A., Patil, R., Sill, S., & Chai, Y. (2017). A LiDAR and IMU Integrated Indoor Navigation System for UAVs and Its Application in Real-Time Pipeline Classification. *Sensors (Basel)*, *17*(6), 1268. doi:10.339017061268 PMID:28574474

Lantmäteriet. (2018). *Handbok i mät - och kartfagor Bilddata.* https://www.lantmateriet.se/sv/Om-Lantmateriet/Samverkan-med-andra/Handbok-i-mat--och-kartfragor-HMK/

Leitao, J. P., Moy de Vitry, M., Scheideffer, A., & Rieckermann, J. (2016). Assessing the quality of digital elevation models obtained from mini unmanned aerial vehicles for overland flow modelling in urban areas. *Hydrology and Earth System Sciences*, *20*(4), 1637–1653. doi:10.5194/hess-20-1637-2016

Manconi, A., Ziegler, M., & Blöchliger, T. (2019). Technical Note: Optimization of UAV flight planning in steep terrains. *International Journal of Remote Sensing*, *40*(7), 2483–2492. doi:10.1080/01431161.2019.1573334

Marr, B. (2019). *Artificial Intelligence in Practice: How 50 Successful Companies Used AI and Machine Learning to Solve Problems*. John Wiley & Sons.

Matese, A., Toscano, P., Di Gennaro, S. F., Genesio, L., Vaccari, F. P., Primicerio, J., Belli, C., Zaldei, A., Bianconi, R., & Gioli, B. (2015). Intercomparison of UAV, Aircraft and Satellite Remote Sensing Platforms for Precision Viticulture. *Remote Sensing*, *7*(3), 2971–2990. doi:10.3390/rs70302971

Meng, X., Shang, N., Zhang, X., Li, C., Zhao, K., Qiu, X., & Weeks, E. (2017). Photogrammetric UAV Mapping of Terrain under Dense Coastal Vegetation: An Object-Oriented Classification Ensemble Algorithm for Classification and Terrain Correction. *Remote Sensing*, *9*(11), 1187. doi:10.3390/rs9111187

Meola, A. (2017). *Drone market shows positive outlook with strong industry growth and trends*. https://www.businessinsider.com/drone-industry-analysis-market-trends-growth-forecasts-2017-7

Mesas-Carrascosa, F. J., Torres-Sánchez, J., Clavero-Rumbao, I., García-Ferrer, A., Peña, J. M., Borra-Serrano, I., & López-Granados, F. (2015). Assessing Optimal Flight Parameters for Generating Accurate Multispectral Orthomosaics by UAV to Support Site-Specific Crop Management. *Remote Sensing*, *7*(10), 12793–12814. doi:10.3390/rs71012793

Micheletti, N., Chandler, J. H., & Lane, S. N. (2015). Structure from Motion (SfM) Photogrammetry. In Geomorphological Techniques. British Society for Geomorphology.

Mikhail, E., Bethel, J., & McGlone, J. (2001). *Introduction to modern photogrammetry*. John Wiley & Sons.

Mikrut, S. (2016). Classical Photogrammetry and UAV - Selected aspects. In *Proceedings of XXIII ISPRS Congress* (vol. XLI-B1, pp. 947-952). 10.5194/isprsarchives-XLI-B1-947-2016

Murison, M. (2018). *DJI cements drone industry dominance with new enterprise products.* https://internetofbusiness.com/dji-drone-industry-domination-enterprise-hardware-and-software-launches/

Oniga, V. E., Breaban, A. I., & Statescu, F. (2018). Determining the Optimum Number of Ground Control Points for Obtaining High Precision Results Based on UAS Images. *Proceedings*, *2*(7), 352. doi:10.3390/ecrs-2-05165

OpenDroneMap. (2020). *Drone Mapping Software, Features.* https://www.opendronemap.org/webodm/

Opromolla, R., Inchingolo, G., & Fasano, G. (2019). Airborne Visual Detection and Tracking of Cooperative UAVs Exploiting Deep Learning. *Sensors (Basel)*, *19*(9), 4322. doi:10.339019194332 PMID:31591368

Pérez, M. (2012). Estimación de incertidumbres Guía GUM. *Revista Española de Metrología,* 113-130.

Pix4D. (2017). *Do RTK/PPK drones give you better results than GCPs?* https://www.pix4d.com/blog/rtk-ppk-drones-gcp-comparison

Pix4D. (2018). *Do more GCPs equal more accurate drone maps?* https://www.pix4d.com/blog/GCP-accuracy-drone-maps

Pix4D. (2019). *How to verify that there is enough overlap between the images.* https://support.pix4d.com/hc/en-us/articles/203756125-How-to-verify-that-there-is-enough-overlap-between-the-images

Pix4D. (2020). *Number and distribution of ground control points (GCPs) in corridor mapping.* https://support.pix4d.com/hc/en-us/articles/202559299-Number-and-distribution-of-ground-control-points-GCPs-in-corridor-mapping

Pichon, L., Ducanchez, A., Fonta, H., & Tisseyre, B. (2016). Quality of Digital Elevation Models obtained from Unmanned Aerial Vehicles for Precision Viticulture. *OENO One*, *50*(3). Advance online publication. doi:10.20870/oeno-one.2016.50.3.1177

Prawjal, M., Jain, R., Srinivasa, V., & Karthik, K. S. (2016). Optimal Number of Ground Control Points for a UAV based Corridor Mapping. *International Journal of Innovative Research in Science, Engineering and Technology*, *5*(9), 28–32. doi:10.15680/IJIRSET.2016.0505505

Propeller Aero. (2019). *What is Ground Sample Distance (GSD) and How Does it Affect Your Drone Data?* https://www.propelleraero.com/blog/ground-sample-distance-gsd-calculate-drone-data/

R Core Team. (2020). *R: A language and environment for statistical computing. R Foundation for Statistical Computing, Vienna, Austria.* https://www.R-project.org/

Rakha, T., & Gorodetsky, A. (2018). Review of Unmanned Aerial System (UAS) applications in the built environment: Towards automated building inspection procedures using drones. *Automation in Construction*, *93*, 252–264. doi:10.1016/j.autcon.2018.05.002

Ranjan-Martha, T., Kerle, N., van Westen, C. J., & Jetten, V. G. (2010). Effect of Sun Elevation Angle on DSMs Derived from Cartosat-1 Data. *Photogrammetric Engineering and Remote Sensing*, *76*(4), 429–438. doi:10.14358/PERS.76.4.429

Rao, B., Gopi, A., & Maione, R. (2016). The societal impact of commercial drones. *Technology in Society*, *45*, 83–90. doi:10.1016/j.techsoc.2016.02.009

Reagan, J. (2015). *How Low Will Drone Prices Go?* https://dronelife.com/2015/07/09/how-low-will-drone-prices-go/

Rock, G., Ries, J. B., & Udelhoven, T. (2011). Sensitivity analysis of UAV-Photogrammetry for creating Digital Elevation Models (DEM). In *Proceedings of Conference on Unmanned Aerial Vehicle in Geomatics* (vol. XXXVIII-1/C22, pp. 69-74). Academic Press.

Sanz-Ablanedo, E., Chandler, J. H., Rodríguez-Pérez, J. R., & Ordóñez, C. (2018). Accuracy of Unmanned Aerial Vehicle (UAV) and SfM Photogrammetry Survey as a Function of the Number and Location of Ground Control Points Used. *Remote Sensing*, *10*(10), 1606. doi:10.3390/rs10101606

Schatsky, D., & Ream, J. (2016). *Advanced software applications are driving commercial drone adoption.* https://www2.deloitte.com/insights/us/en/focus/signals-for-strategists/advanced-drone-software-applications.html

Schreiber, E., Heinzel, A., Peichl, M., Engel, M., & Wiesbeck, W. (2019). Advanced Buried Object Detection by Multichannel, UAV/Drone Carried Synthetic Aperture Radar. In *2019 13th European Conference on Antennas and Propagation (EuCAP)* (vol. 1, pp. 1-15). Academic Press.

Schroth, L. (2018). *Drones and Artificial Intelligence.* https://www.droneii.com/drones-and-artificial-intelligence

Seifert, E., Seifert, S., Vogt, H., Drew, D., van Aardt, J., Kunneke, A., & Seifert, T. (2019). Influence of Drone Altitude, Image Overlap, and Optical Sensor Resolution on Multi-View Reconstruction of Forest Images. *Remote Sensing*, *11*(10), 1252. doi:10.3390/rs11101252

Smeur, E. J. J., Bronz, M., & de Croon, G. C. H. E. (2018). *Incremental control and guidance of hybrid aircraft applied to the Cyclone tailsitter UAV.* https://arxiv.org/pdf/1802.00714.pdf

SMHI. (2015). *Vad är ett solbanediagram?* http://www.smhi.se/polopoly_fs/1.32097.1398236573!/image/stockholm_kontur.png_gen/derivatives/Original/stockholm_kontur.png

Snow, C. (2019). *Seven Trends That Will Shape the Commercial Drone Industry In 2019.* https://www.forbes.com/sites/colinsnow/2019/01/07/seven-trends-that-will-shape-the-commercial-drone-industry-in-2019/#31914dda7494

Stensson, L. N. (2016). *Spatial Accuracy in Orthophoto produced using UAV Photographic Images* [Unpublished master thesis]. Karlstads Universitet, Karlstad, Sweden.

Thillainayagi, R., & Kumar, K. S. (2019). Bi-dimensional Empirical Mode Decomposition Based Contrast Enhancement Technique for UAV Thermal Images. *Journal of the Institution of Electronics and Telecommunication Engineers*, 1–8. Advance online publication. doi:10.1080/03772063.2019.1619487

Uysal, M., Toprak, A. S., & Polat, N. (2015). DEM generation with UAV Photogrammetry and accuracy analysis in Sahitler hill. *Measurement*, *73*, 539–543. doi:10.1016/j.measurement.2015.06.010

van Rees, E. (2017). *Pix4D Extends Drone-based Imagery with Machine Learning Techniques.* https://www.commercialuavnews.com/energy/pix4d-extends-drone-based-imagery-machine-learning-techniques

Villa, T. F., Gonzalez, F., Miljievi, B., Ristovsky, Z. D., & Morawska, L. (2016). An Overview of Small Unmanned Aerial Vehicles for Air Quality Measurements: Present Applications and Future Prospectives. *Sensors (Basel)*, *16*(7), 1072. doi:10.339016071072 PMID:27420065

Weiping, M., & Guangchuang, Y. (2018). *BasicTrendline: Add Trendline and Confidence Interval of Basic Regression Models to Plot. R package version 2.0.3.* https://CRAN.R-project.org/package=basicTrendline

Westoby, M. J., Brasington, J., Glasser, N. F., Hambre, M. J., & Reynolds, J. M. (2012). "Structure-from-Motion" photogrammetry: A low-cost, effective tool for geoscience applications. *Geomorphology*, *179*, 300–314. doi:10.1016/j.geomorph.2012.08.021

Wickham, H., François, R., Henry, L., & Müller, K. (2020). *dplyr: A Grammar of Data Manipulation. R package version 0.8.5.* https://CRAN.R-project.org/package=dplyr

Wierzbicki, D., Kedzierski, M., & Fryskowska, A. (2015). Assessment of the influence of UAV image quality on the orthophoto production. In *Proceedings of International Conference on Unmanned Aerial Vehicles in Geomatics* (vol. XL-1/W4, pp. 1-8). 10.5194/isprsarchives-XL-1-W4-1-2015

Yacef, F., Rizoug, O., & Hamerlain, M. (2017). Optimization of Energy Consumption for Quadrotor UAV. In *Proceedings of 9th International Micro Air Vehicle Conference and Flight Competition (IMAV) 2017* (*vol. 1*, pp. 215-222). Academic Press.

Zimmermann, F., Eling, C., Klingbeil, L., & Kuhlmann, H. (2017). Precise positioning of UAVs - Dealing with challenging RTK-GPS measurement conditions during automated UAV flights. In *Proceedings of International Conference on Unmanned Aerial Vehicles in Geomatics* (vol. IV-2/W3, pp. 95-103). 10.5194/isprs-annals-IV-2-W3-95-2017

KEY TERMS AND DEFINITIONS

Checkpoints (CPs): Marks on the ground, measured with higher accuracy than that provided by the UAV's positioning system, that are used to validate the relative and absolute accuracy of the map products.

Digital Elevation Model (DEM): Three-dimensional representation of the surface without objects on it.

Digital Surface Model (DSM): Three-dimensional representation of the surface including all objects on it like vegetation or buildings.

Ground Control Points (GCPs): Marks on the ground that are used to determine the scale and orientation of the map data in relation to the ground data.

Ground Sampling Distance (GSD): Distance between two consecutive pixel centres measured on the ground.

Image Overlap: Amount in which a photograph includes the area covered by another photograph.

Map Accuracy: Difference between the position of known points and their estimated position in the map products.

Chapter 17
Impact-Resistant Flying Platform for Use in the Urban Construction Monitoring

Konstantin Dergachov
https://orcid.org/0000-0002-6939-3100
National Aerospace University KhAI, Kharkov, Ukraine

Anatolii Kulik
https://orcid.org/0000-0001-8253-8784
National Aerospace University KhAI, Kharkov, Ukraine

ABSTRACT

A case study drone that constitutes a shock-resistant aerial vehicle is discussed in the chapter. The aerial motor platform is placed in gimbal joints of the exclusive framework (shell). The platform is a helicopter type aerial vehicle powered with two coaxial rotors of contra rotation. Mathematical model of the platform spatial dynamics bases Lagrange's equations to bring reliable solutions so that advanced model-based control law design techniques can be used. Though the case study implies utilizing an automatic flight mode of the aerial vehicle, it can be piloted remotely on radio. The on-board video cameras and other sensors are used to bring about both navigational duties and surveillance missions such as building constructions monitoring.

DOI: 10.4018/978-1-7998-2249-3.ch017

BACKGROUND

This chapter presents urban grow concerns and trends and how unmanned aerial vehicle can help solve these difficulties.

Unmanned aerial vehicle (UAV) is a small-size air flight vehicle controlled remotely or flying automatically. A number of advantages of this vehicle compared to other traditional ones gave rise to their intensive expansion into practical civilian applications. They are successfully used in geodesy and mapping, monitoring the state of urban infrastructure, transport communications, engineering networks for various purposes, also monitoring environmental changes and warning emergency in engineering structures, impending natural hazards, and in a number of other areas.

Sociological studies on urbanization indicate that by now 55% of the world's population lives in cities (Kucharczyk & Hugenholtz, 2018) and by 2050 it may increase to 68% (Mohammed et al., 2014), the unprecedented urban population growth. So, in 1800, only 2% of people lived in cities. In 1950, only 30% of the world population was considered a city dweller. Every year almost 180 thousand people move to the cities. In developing countries, the urban population is replenished by 60 million people. This rate of the urban population growth, according to experts, will continue during the next 30 years (Jensen, 2016).

Intensive rate of the urban population growth has led to the emergence of a number of applied and scientific problems of stable urban development (Gallacher, 2016). According to the Director of the Earth Institute at Columbia University, Jeffy Sachs, cities around the world should move in three strategic directions that ensure acceptable living conditions for all residents of the city (Kucharczyk & Hugenholtz, 2018):

1. Urban planning, which includes well-designed water supply, sewerage and recycling systems, as well as public transport and health care systems;
2. City development strategy, i.e. setting goals taking into account the conditions of the region;
3. Effective city management.

The rapid growth of the urban population leads to an increase in the relevance of the problem of effective urban management (Gevaert et al., 2018). For a productive solution to this problem, several concepts have been proposed. The most popular and requested concept seems to be the formation of smart cities using UAVs. (Krieger et al., 2018). Mobility, low cost and ease of operation of UAVs allow using them as carriers of different types of sensors (Germen, 2016) that supply information to the urban control center about the technical condition of the urban infrastructure (Jensen, 2016).

Unmanned aerial vehicles, or drones, are usually mentioned as purposeful UAVs for smart cities organization (Mohammed et al., 2014). Such drones as quadcopter or hexacopter are helicopter-like aerial vehicles (in terms) from the standpoint of rotary-wing framework. Particular frame design depends on drones tasks and mission conditions. Besides the carrier frame and rotors, the small aerial vehicle brings a dedicated controller, communications and a payload (Liu et al., 2014). The thrust on these platforms is produced by electrically driven propellers of a fixed-pitch that are governed with electronic speed controller (ESC); also, the avionics contains attitude control sensors (IMU), microprocessor controller unit, facility to manage the payload as video cameras for instance. The controller and ESC govern rotors speed to attain navigation aims. The telemetry block serves for communications including information on the drone flight conditions.

Drones, representing a novel class of aerial vehicles, have a number of pros and undisputed advantages when bringing various missions that assist sustainable urban development. However, there are certain drawbacks of these vehicles, which are commonly due to extra ordinaries arising in the course of preparation for flight and in flight itself.

As a rule, emergency situations occur because of technical malfunctions on avionics facilities. These may be failures of the power supply, controller, telemetry unit etc. During the flight and carrying out a mission in condition of urban fabric and settings as dense buildings and trees, numerous power lines and communications so forth, drones collision with obstacles are possible. Impacts cause damages of propellers, structure elements, avionics, yet other critical even catastrophic effects can occur.

In the existing unmanned complexes, diagnostic processes of technical and functional condition of the drones are carried out, as a rule, manually by the operating personnel or the user. This requires a certain qualification, appropriate equipment and repair tools. Diagnostics may not be usually performed while carrying out the mission.

For ordinary current unmanned complexes, the diagnostic processes of the technical and functional condition of the drones are carried out with servicing personnel or by the user, as a rule, by hands. This asks for certain qualification, appropriate equipment and repair tools.

Failures during the flight mission, in most events, lead to a catastrophic emergency situation – such as the fall and destruction of the drone, the fall and damage of ground objects, the fall on flammable objects, traumas of living beings even their deaths, i.e. causing more hard consequences.

Newest materials and design tools allow you to create such shells for both propellers and devices that provide not only protection, but also shock resistance. That would dramatically reduce emergency occurrence risks and hazard environmental consequences.

Progress of on-board control systems that would implement artificial intelligence can significantly lower the impact of failures on operational quality of flying platforms during both the pre-flight control examination stages and operational phases in difficult conditions of piloted or automatic flight.

IMPACT RESISTANT AERIAL VEHICLE AS THE OBJECT OF CONTROL

Working Principle of Coaxial Rotors Aerial Vehicle

The impact resistant flying platform under the study is the aerial vehicle of vertical take-off and landing (VTOL) that is powered by coaxial rotors – two contra rotating propellers driven by electric motors. The rotor platform is suspended through gimbals in the special framework that, in addition, serves a shock-resistant shell of the drone (Figure 1).

The aerial vehicle (AV) has two pairs of control surfaces (actuators) aimed at longitudinal and lateral motion AV control. The surfaces deflection is carried with the associated servo drive that receives dedicated signals from the controller.

The protective shell is a reinforced cell type spatial construction made of rod elements. Inside the shell construction, there are gimbal rings fixed through the hinges to accommodate the platform. The hinges are all free; due to that the rotor platform may fluently rotate relative to the framework.

In terms of mechanics, the investigated drone is a material system, which structure may vary because of containing two independent movable rigid bodies – the flight rotor platform and the protecting framework; the latter is designed in the form of truncated icosahedron with attached gimbals in which the platform is free to turn.

Hence, the examined object exhibits in own motion a free material system that has changeable parts and the following parameters:

1. Rotors rotational speed and corresponding angular momentums
2. Control surfaces deviation and corresponding center of mass of the platform relocation resulting in change of the moment of inertia
3. The platform angular attitude changes in gimbals and resulting relocation of the platform center of mass with related change of the moment of inertia
4. The shell together the gimbal rings attitude in space changes that influence the aerial vehicle moment of inertia.

Figure 1. Exterior view of the aerial vehicle

Although, the two below parameters remain unchanged:

1. The aerial vehicle mass
2. The position of the center of mass of the housing (protective shell and rings assembly).

The center of mass of the protective shell collected together with gimbals rings is in the geometric center of their construction on the axis about which the motor unit (platform) rotates in the inner ring's hinges. The design supposes that axes of rotation of the control surfaces pass via their center of mass. Owing to that a variation of the center of mass position of the whole AV is negligible when the surfaces are shifting (about a few hundredth millimeter) and therefore can be neglected. As the result, AV's center of mass will always be on the straight that connects the centers of mass of the motor unit and the housing.

Working Principle of Coaxial Rotors Aerial Vehicle

Flying of the aerial vehicle is as below.

All the forces and moments needed for AV flying are formed as a result of air masses rejection produced and appropriately directed by the propellers and control surfaces

The lift force is produced by two bi-blade rotors of contra rotation and pointed in the rotors axis direction.

The propellers, which reject the air, perceive the air mass reaction in a form of the thrust produced and, besides, experience a reactive torque. The two are proportional to the rotors speed. The thrust force vectors of the both rotors coincide, hence they are added. At the same time, the reactive moments M_1 and M_2 have opposite signs because of the blades contra-rotation and, when being identical, they cancel each other on the axis of each rotor.

Thus, we can govern the traction force when synchronously changing the rotors speed. On the other hand, when increasing the speed of one rotor and reducing that of the other, the AV will get a rotation around the rotors co-axis due to the difference of reactive moments.

The steering control surfaces can deflect the rotors air flow in the longitudinal and lateral directions. A deflection of the air flow give rise to longitudinal and lateral control impacts generation. These will depend upon the total thrust, i.e. on the rotors speed, and the angle position of the control surfaces.

In the flight, the air flow meets AV's aerodynamic drag. The drag is pointed opposite the flight speed vector and proportional to the airspeed squared.

Consider features of the forces and moments. Prior to take-off, the platform should be positioned so that the spin axis of rotors is vertical and the control surfaces take a neutral position. On increasing rotor revs, the produced propeller thrust overcomes the weight and AV vertical lift begins. If the surfaces of longitudinal control have been deflected soon after the start, then the produced torque causes an angular acceleration of the housing in the direction of the acting moment and rotates the platform relative to the center of mass. The AV obtains a deviation from the vertical in the longitudinal plane. The thrust vector moves from the vertical as well. Its normal projection decreases although the horizontal one appears and produces the longitudinal thrust. The balance of forces acting on the vehicle was destroyed (the lift has dropped). To maintain flight altitude or to continue the climb an increase of the thrust is required. Under the total force of the force of longitudinal control and the horizontal projection of the thrust, the vertical path warps so that AV would acquire (in the vertical plane) the velocity, flightpath, other motion characteristics required to complete the flight assignment. Control of the lateral motion is carried out similarly.

When the vehicle faces an obstacle, the energy absorption occurs owing to elastic deformation of the protective shell, gimbals rings, and transverse beam. To decrease the shock failures, AV orientation should be implemented so that the velocity vector was pointed perpendicular to the transverse bar since a strain under bending is always bigger the contraction strain.

When AV perceives an angular velocity, the spinning rotors with propellers will get gyroscopic moments. If the rotors have the same angular momentum, then that gyroscopic moments cancel each the other. However, when there is a need to enforce a difference for the motors rpm to handle a maneuver, a difference of the angular momentums of spinning rotors takes place. That results in the difference of gyroscopic moments of the rotating parts and corresponding effects on the system flying conditions.

Deriving The Mathematical Model

To have clear understanding of features of the examined aerial vehicle in terms of controllability, a mathematical model of the control object has been derived. The approach implemented is based on Lagrange equations of the second kind.

The analysis requires certain steps to embody principles of theoretical mechanics, in particular methods of dynamics and analytical mechanics (Loitsyansky, 1983):

1. Finding the number of degrees of freedom of the vehicle as a material system and selection of independent generalized coordinates q_1 that fit the number of freedom degrees; a choice of the coordinate system and reference frames
2. Description of the active forces acting on the system (Dobronravov et al, 1974)
3. Forming a system of generalized forces $Q_1, Q_2, \ldots, Q_s$, in accordance with the chosen generalized coordinates
4. Calculating the kinetic energy T of the investigated material system. The energy has to be determined for the absolute motion and expressed via the generalized coordinates q_i and generalized velocities $\dot{q}_i$;

5. Finding kinetic energy partial derivatives on the generalized velocities $\dot{q}_1, \dot{q}_2, ..., \dot{q}_s$ namely $\frac{\partial T}{\partial \dot{q}_1}, \frac{\partial T}{\partial \dot{q}_2}, ..., \frac{\partial T}{\partial \dot{q}_s}$; then calculating time derivatives like $\frac{d}{dt}\left(\frac{\partial T}{\partial \dot{q}_1}\right), \frac{d}{dt}\left(\frac{\partial T}{\partial \dot{q}_2}\right), ..., \frac{d}{dt}\left(\frac{\partial T}{\partial \dot{q}_s}\right)$;
6. Finding kinetic energy T partial derivatives on the generalized coordinates q_1, q_2, ..., q_s such as $\frac{\partial T}{\partial q_1}, \frac{\partial T}{\partial q_2}, ..., \frac{\partial T}{\partial q_s}$;
7. 7. Substitution of the relations obtained into Lagrange equations:

$$\frac{d}{dt}\left(\frac{\partial T}{\partial \dot{q}_1}\right) - \frac{\partial T}{\partial q_1} = Q_1, \frac{d}{dt}\left(\frac{\partial T}{\partial \dot{q}_2}\right) - \frac{\partial T}{\partial q_2} = Q_2, ..., \frac{d}{dt}\left(\frac{\partial T}{\partial \dot{q}_s}\right) - \frac{\partial T}{\partial q_s} = Q_s .$$

The Degrees of Freedom of The Vehicle

The platform under the study has 11 degrees of freedom. We list them defining the generalized coordinate that corresponds to the freedom degree chosen (Figure 2):

1. The position vector of the center of mass of the system in the inertial frame – $r=q_1$;
2. The elevation angle of the center of mass – $\sigma=q_2$;
3. The azimuth of the center of mass – $\chi=q_3$;
4. Pitch angle – $\upsilon=q_4$;
5. Yaw angle – $\psi=q_5$;
6. Roll angle – $\gamma=q_6$;
7. The turning angle of the rotors platform relative to the gimbals inner ring – $\varphi_1=q_7$;
8. The turning angle of the outer ring relatively the protective shell – $\varphi_2=q_8$;
9. Rotation of the gimbals inner ring relatively outer ring – $\varphi_3=q_9$;
10. Rotation angle of the upper rotor – $\varepsilon_1=q_{10}$;
11. Rotation angle of the lower rotor – $\varepsilon_2=q_{11}$.

Figure 2. Generalized Coordinates of Investigated Material System

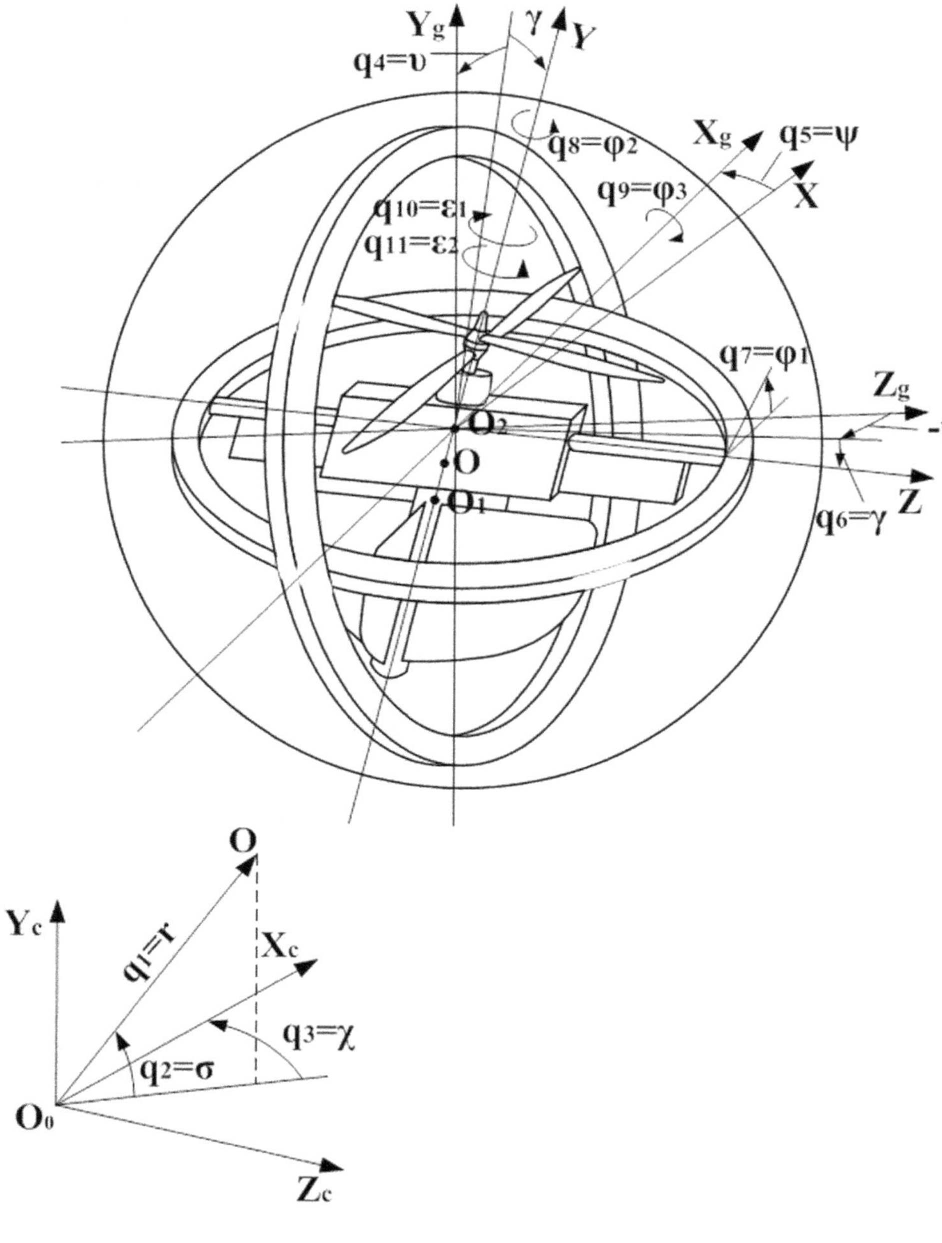

Choose of Coordinates and Reference Frames

In general, a motion of the material system of rigid bodies is specified by moving of the center of mass of the system and system bodies' motion relatively the center of mass. The motion of the center of mass of the material system is defined by the triad (independent coordinates) in the reference frame either by the corresponding position vector of the center of mass point.

Since the investigated AV assumes navigation in limited space, it is customary to examine its movement with respect to a starting point and in a restricted area like room. As a reference frame for the study of AV trajectories we accept the starting system $O_oX_cY_cZ_c$ with the origin O_o in the point of start. In accordance with the convention (Bat et al., 2013), axis O_oX_c lies in the horizontal plane and is pointed along the aiming line, axis O_oY_c points upwards, and axis O_oZ_c completes the right-hand reference triad. In the research framework, the stationary inertial frame is assumed to be the reference system.

When analyzing a motion of the material system's center of mass in the starting frame, one can use three Cartesian coordinates *x, y, z,* either polar coordinates, which are the position vector $\vec{r}$, and the azimuth χ and elevation σ angles. Next, we introduce two body coordinate systems (normal and body-fixed frames) that specify the position of bodies of the material system relative to the inertial reference frame. The axes of the starting frame along with object's body frames are shown in Figure 3. The origin of both body frames in the case study is attached to AV's center of mass.

The body frame of the aerial vehicle is defined by rotation about three dedicated angles relatively the inertial reference frame: pitch angle υ, roll angle γ, and yaw angle ψ.

When defining the AV body frame, we point the basis vectors of the body frame parallel to the principal axes of inertia of the vehicle. That simplifies writing of equations of motion since centrifugal moments of inertia relative to principal axes of inertia equal zero. Due to the fact that the investigated AV has a coaxial rotors scheme, the rotors spin axis is taken as the *OY* axis in the body frame; this is the vertical axis of AV physical symmetry, hence the principal central axis of inertia of the aerial vehicle.

Along with essential simplification of the mathematical model derivation, such a choice of the body frame for AV is expected to give rise to right algorithmic support on realizing a control strategy.

Next we need to take into consideration the fact that an obstruction can interfere the flying and an external impact on the protective shell is possible. There is a need to have a structural element at the flying platform that would absorb the impact energy together with the protective shell. Such an element in the case study design is the transverse beam hinged in the inner gimbals ring and carrying the rotors.

Figure 3. Starting(ground), normal, and body-fixed frames

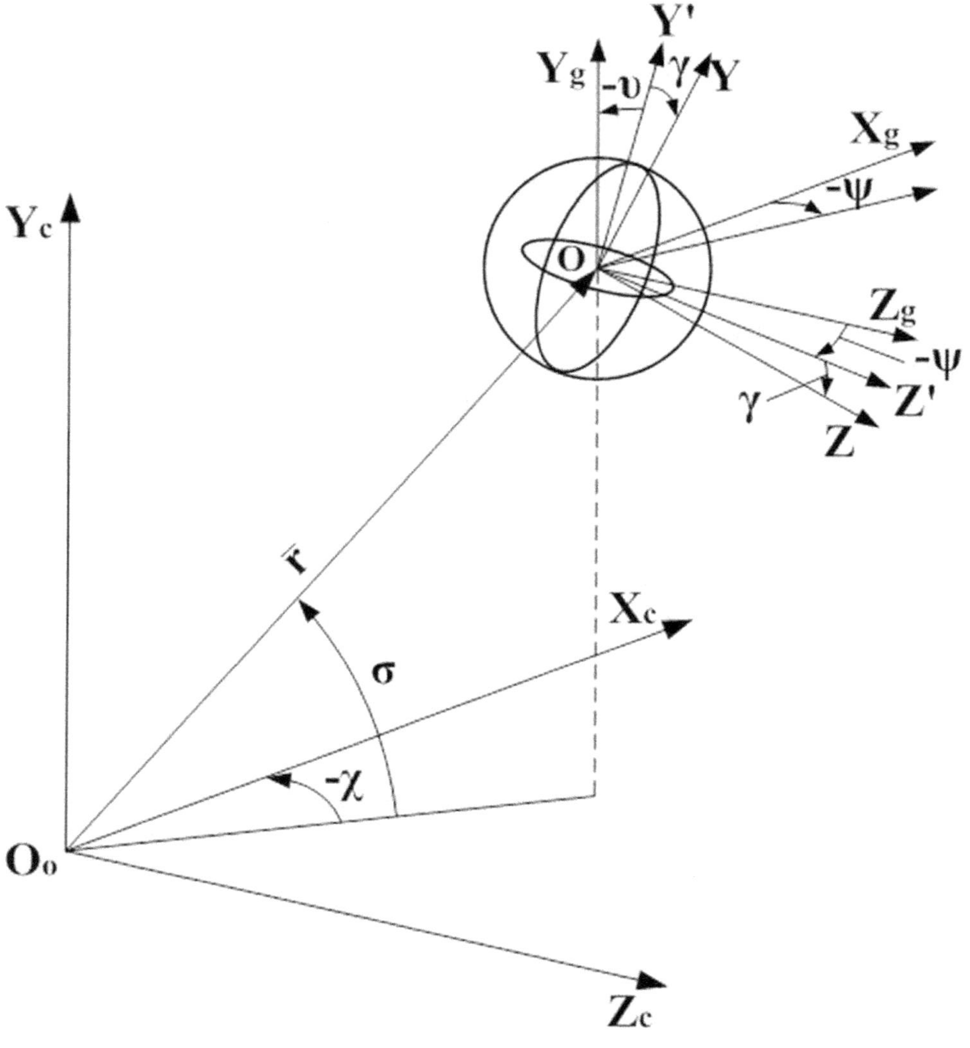

We accept the vector that is perpendicular to the beam with rotors and passing through the AV center of mass along the aiming line ahead, as the body frame axis OX. The axis perpendicular to the plane of material symmetry of the body, OY, will be the principal axis of inertia of the vehicle. The OZ axis that is pointing from the center of mass to the right and parallel to the beam completes the right-hand triad.

Gimbal rings are axisymmetric homogeneous bodies. Therefore, all the vectors both perpendicular to the rings plane and lying in those planes will be the principal axes of inertia.

Figure 4. Body frame transformations relative to normal frame

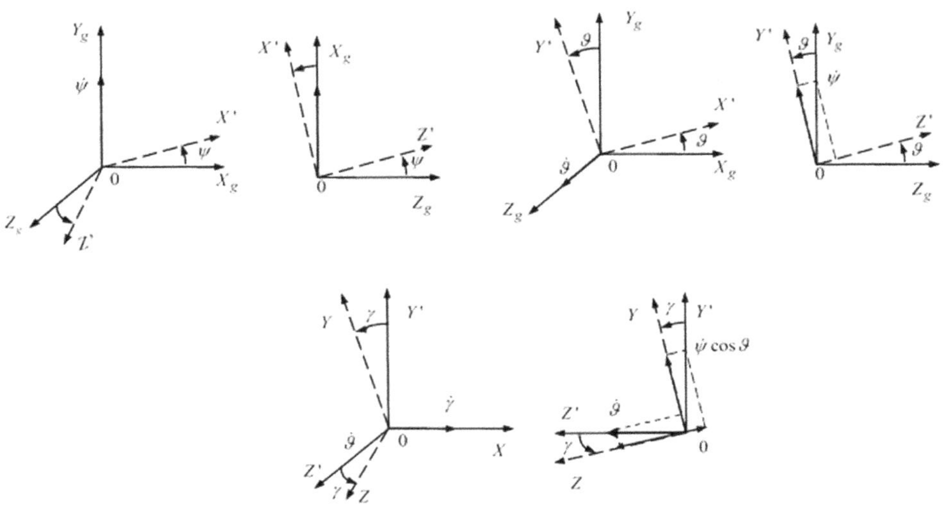

Rigid Body Kinematics

Soon after movement start, the examined material object will change the body frame axes position.

Let we analyze an arbitrary state of the material system with the goal of deriving equations of motion. To avoid misunderstanding in using signs of angles, angular velocities and vectors for the system arbitrary position we accept in the beginning the position that appropriates the convention

We need to know angular velocity of the body to determine the kinetic energy of rotational motion. To begin, we make use of the rule (Dobronravov et al, 1974) that defines relationships between the angles of pitch, yaw, and roll, on one hand, and the angular velocity vector $\bar{\omega}$ projections on the body frame axes, i.e. ω_x, ω_y, ω_z on the other.

Let axes of the body frame at an arbitrary moment of time are parallel to axes of the normal /inertial reference frame (Figure 4). We need to perform three successive rotations around the axis of intermediate frames so that the body frame takes up a new position. The turns should match the three specified above angles ψ, υ, and γ.

First rotate the connected system around the axis OYg at an arbitrary angle; this angle of rotation is, by definition, the yaw angle ψ (Figure 4).

Next, make an arbitrary rotation around the axis OZ'; this is pitch angle υ. Finally, turn the system around the axis OX to obtain roll angle γ.

As a result of these turns, we obtain three angular velocity vectors: when turning around the axis OYg – vector $\dot{\psi}$ around the axis OYg, during second turn – vector $\dot{\upsilon}$ around the axis OZ during third turn – vector $\dot{\gamma}$ around the axis OX.

The first rotation of the body frame is performed through the angle ψ about the OYg axis; by definition, this angle is a yaw angle. Next, we make a rotation about the OZ' axis; this is a pitch angle υ. Final rotation is provided for the roll angle γ (turn about the OX axis).

Resulted from these rotations, there will be specified three angular velocity vectors: vector $\dot{\psi}$ along OYg axis; vector $\dot{\upsilon}$ along OZ axis, vector $\dot{\gamma}$ along OX axis.

After projecting vectors $\dot{\psi}$, $\dot{\upsilon}$ and $\dot{\gamma}$ onto the axes OX, OY, OZ we get

$$\begin{cases} \omega_x = \dot{\gamma} + \dot{\psi}\sin\nu; \\ \omega_y = \omega_y' \cos\gamma + \dot{\upsilon}\sin\gamma = \dot{\psi}\cos\upsilon\cos\gamma + \dot{\upsilon}\sin\gamma; \\ \omega_z = \dot{\upsilon}\cos\gamma - \omega_y' \sin\gamma = \dot{\upsilon}\cos\gamma - \dot{\psi}\cos\upsilon\sin\gamma. \end{cases} \tag{1}$$

These equations govern the object angular velocity vector projections on the axes of the body-fixed frame as functions of the angles ψ, υ, γ and angle derivatives (angles rate of change).

Finding solution for $\dot{\psi}$, $\dot{\nu}$ and $\dot{\gamma}$ yields (Kulik, 2016):

$$\begin{cases} \dot{\upsilon} = \omega_y \sin\gamma + \omega_z \cos\gamma; \\ \dot{\gamma} = \omega_x - \mathrm{tg}\,\theta\left(\omega_y \cos\gamma - \omega_z \sin\gamma\right); \\ \dot{\psi} = \left(\omega_y \cos\gamma - \omega_z \sin\gamma\right) / \cos\upsilon. \end{cases} \tag{2}$$

Conventional limits for the angles look like: $-\pi<\psi\leq\pi$; $-\pi/2<\upsilon\leq\pi/2$; $-\pi<\gamma\leq\pi$.

The Forces and Moments on The System

During a flight, the following forces and moments act on the aerial vehicle (Figure 5):

1. The system parts gravity;
2. The rotors force $\overline{P}$ and reaction moments M_1, M_2;
3. Control forces $\overline{P}_P$ and $\overline{P}_B$;

4. Aerodynamic force R because of motion of the material system through the air ;
5. Moments of resistance to rotation in the gimbals hinges;
6. The collision force against the obstacle, $\bar{N}$.

Figure 5. Forces and moments on the system

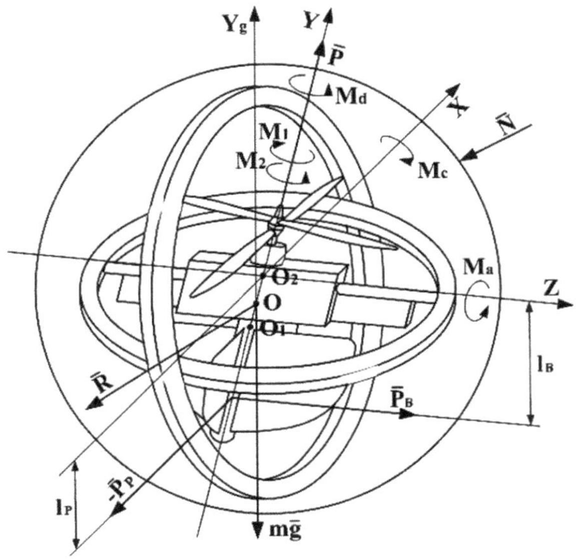

When you concern equations of motion of the center of mass of the system, all forces are reduced to the center of mass of the system by static zero.

Gravity forces. Taking into account the system structure and geometry we reduce gravity forces of all system parts to vehicle's total weight applied at the center of mass of the system O.

The total weight is made of the motor platform weight applied at the center of mass of the platform (point O1), and the protective housing weight, which includes those of the shell, gimbals and hinges and acts at the point O2 that denotes a geometric center of the structure (center of mass of the frame). The gravity, mg, is a central force pointing vertically down along the axis OYg.

Now, we determine gravity force projections on the position vector – $\bar{r}$, also on the directions σ and χ. The first projection constitutes $mg_r = -mg \sin \sigma$; the second and third components will be $mg_\sigma = -mg \cos \sigma$ and $mg_\chi = 0$, correspondingly.

Thrust motors. The forces and moments of the rotors are represented by the propellers thrust and blades' reaction moments on the frame. In the body frame, the thrust vector $\overline{P}$ is pointing vertically up in the axis OY direction. To determine the projection of the vector $\overline{P}$ on the position vector and directions σ and χ one needs to, first, perform the vector $\overline{P}$ projection on the axes of the body frame, then project the received results on the position vector and directions σ and χ.

The solutions for projections of the vector $\overline{P}$ on the axes of the body frame yield P_x=0, P_y=P, P_z=0, which resolve required components of the inertial reference frame:

$$\begin{aligned} P_{xg} &= P(\sin\psi \sin\gamma - \cos\psi \sin\upsilon \cos\gamma) \\ P_{yg} &= P\cos\upsilon\cos\gamma \qquad ; \\ P_{zg} &= P(\cos\psi \sin\gamma + \sin\psi \sin\upsilon \cos\gamma) \end{aligned} \tag{3}$$

The projections P_{xg} and P_{zg} are in the horizontal plane, hence may be readily consolidated on the horizontal component r_n of the position vector:

$$P_{rn} = P_{xg} \cos\chi + \mathrm{P}_{zg} \sin\chi \,. \tag{4}$$

The projection of the vector $\overline{P}$ on the position vector $\overline{r}$ is:

$$P_r = P_{rn} \cos\sigma + \mathrm{P}_{yg} \sin\sigma = (P_{xg} \cos\chi + P_{zg}\sin\chi)\cos\sigma + \mathrm{P}_{yg} \sin\sigma \,. \tag{5}$$

The projection of the vector $\overline{P}$ on elevation χ:

$$P_\chi = P_{xg} \sin\chi + P_{zg} \cos\chi \,. \tag{6}$$

The projection of the vector $\overline{P}$ on azimuth σ:

$$P_\sigma = P_{yg} \cos\sigma \,. \tag{7}$$

Drag force. The drag force R constitutes:

$$R = C_R\, \rho \frac{V^2}{2} S, \tag{8}$$

where C_R is the drag coefficient dependent on the Reynolds number and orientation of parts of the system relatively the stream speed;

ρ – air density;
V – the stream speed, that equals the object's speed if no wind impact;
S – AV's characteristic area, to which the aerodynamic drag factor C_R is being normalized.

The drag force R acts at the center of aerodynamic pressure and is pointing opposite the flight speed. We assume that the force R operates to a first approximation at the center of mass of the material system and produces no torques that would rotate the system. Let the aerodynamic drag be resolved into components on the direction of position vector and perpendicular to the latter one, namely: on σ and χ.

To provide for the vector convention $V^2 = \dot{r}^2 + r^2\dot{\sigma}^2 + r^2\dot{\chi}^2$, the drag force projections should be written on the directions of

- radius vector axis:

$$R_r = C_r\dot{r}^2, \tag{9}$$

where C_r is a coefficient that involves the projection C_R on the projection vector and the factor of $\rho S/2$;

- azimuth axis σ:

$$R_\sigma = C_\sigma r^2\dot{\sigma}^2, \tag{10}$$

where C_σ – a coefficient involving the projection C_R on the direction σ and the factor of $\rho S/2$;

- elevation axis χ:

$$R_\chi = C_\chi r^2\dot{\chi}^2, \tag{11}$$

where C_χ – a coefficient involving projection C_R on the direction χ and factor of $\rho S/2$.

All other generalized coordinates have no projections of the aerodynamic drag.

Obstructions impact force. During the object flying, collisions with obstacles are probable and the protective shell perceives an external kick force $\overline{N}$. A magnitude and direction of the force of interaction are uncertain and depend upon the smoothness and stiffness of both the obstacle and the protective shell. As the protective shell form is near a spherical, we adopt that the shell meets the obstacle from out the front hemisphere relative to velocity vector pointing.

In this case study, the force of interaction with an obstacle $\overline{N}$ is carried from the point of perceiving the kick to the center of mass of the system using a static zero rule, i.e. with the addition of the moment that equals a product of the force on the static lever arm; the latter supposes the shock point distance to the center of mass of the system. Hence, we conclude the outcomes as follows on the external interferences influence: the external impacts forces have been decomposed into appropriate components, σ and χ consequent torques that are acting on the rotating parts of the system through the movable housing and gimbals rings have been analyzed.

Forces of steering controls. We need to get the control force $\overline{P}_P$ projections on the position vector $\overline{r}$ and directions σ, χ, and υ. The input governing force $\overline{P}_P$ for l*ongitudinal control* is produced by the steering surface drive followed by a deflection of the air flow generated by the propellers. The resulting control force (as well as the thrust) is estimated at a first approximation as the sum of forces produced by each rotor individually. That force $\overline{P}_P$ is applied at the control surface's center of pressure, distant apart at l_p from the AV axis of rotation. The force can be calculated from

$$P_P = K_P P \sigma \mathrm{P}_{,} \tag{12}$$

where K_P specifies a thrust effect on the steering control; σ_P – deviation of the fin of steering.

The input force $\overline{P}_P$ lays in the plane perpendicular to $\overline{P}$, hence the force projections on the body frame axes are as follows:

$$P_{Px} = -P_P, P_{Py}=0, P_{Pz}=0. \tag{13}$$

The projections of $\overline{P}_P$ on the normal frame axes are

$$P_{Pxg} = -P_P \cos\psi \cos\upsilon;\ P_{Pyg} = -P_P \sin\upsilon;\ P_{Pzg} = -P_P \sin\psi \cos\upsilon\,. \tag{14}$$

Projecting of the horizontal components P_{Pxg} and P_{Pzg} onto the horizontal projection of position vector will yield

$$P_{PrP} = P_{Pxg} \cos\chi + P_{Pzg} \sin\chi\,. \tag{15}$$

The projection of the longitudinal control force $\overline{P}_P$ on the radius vector:

$$P_{Pr} = P_{PrP} \cos\sigma + P_{Pyg} \sin\sigma = \cos\sigma\ (P_{Pxg} \cos\chi + P_{Pzg} \sin\chi) + P_{Pyg} \sin\sigma\,. \tag{16}$$

The projection of the longitudinal control force $\overline{P}_P$ on the direction σ:

$$P_{P\sigma=}\ PP_{yg} \cos\sigma. \tag{17}$$

The projection of the longitudinal control force $\overline{P}_P$ on the direction χ:

$$P_{P\chi} = P_{Pxg} \sin\chi - P_{Pzg} \cos\chi\,. \tag{18}$$

The projection of the longitudinal control force $\overline{P}_P$ on the direction ν:

$$P_{P\nu} = P_P l_P. \tag{19}$$

Lateral control. Now we determine projections of the input force of lateral control, $\overline{P}_B$, on the position vector and directions σ, χ, υ. The lateral control is provided by the dedicated steering surface and operates in exactly the same way as the longitudinal governor works. The force $\overline{P}_B$ is applied at the control surface's center of pressure distant apart at l_B from the AV axis of roll:

$$P_B = K_B P\sigma B_{,} \tag{20}$$

where K_B specifies a thrust effect on the lateral control; σ_B – deviation of the fin of lateral control.

The input force $\overline{P}_B$ lays in the plane perpendicular to $\overline{P}$, hence the force projections on the body frame axes are as follows:

P_{Bx}=0, P_{By}=0, P_{Bz}= -P_B. (21)

The projections of $\overline{P}_B$ normal reference frame axes are

$$\begin{aligned} P_{Bxg} &= -P_B\left(\sin\psi\cos\gamma+\cos\psi\sin\upsilon\sin\gamma\right) \\ P_{Byg} &= P_B\cos\upsilon\sin\gamma \qquad ; \\ P_{Bzg} &= -P_B\left(\cos\psi\cos\gamma+\sin\psi\sin\upsilon\sin\gamma\right) \end{aligned} \tag{22}$$

Projecting horizontal components P_{Bxg} and P_{Bzg} onto the horizontal projection of position vector yields

$$P_{Br} = P_{Bxg}\cos\chi + P_{Bzg}\sin\chi\,. \tag{23}$$

Projection of the input lateral control force $\overline{P}_B$ on the position vector looks like

$$P_{Br} = P_{BrP}\cos\sigma + P_{Byg}\sin\sigma = \cos\sigma\ (P_{Bxg}\cos\chi + P_{Bzg}\sin\chi) + P_{Byg}\sin\sigma\,. \tag{24}$$

The projection of the lateral control force $\overline{P}_B$ on the direction σ:

$$P_{B\sigma} = P_{Byg}\cos\sigma\,. \tag{25}$$

The projection of the lateral control force $\overline{P}_B$ on the direction χ:

$$P_{B\chi} = P_{Bxg}\sin\chi - P_{Bzg}\cos\chi\,. \tag{26}$$

The projection of the lateral control force $\overline{P}_B$ on the direction ν:

$$P_{B\nu} = P_B l_B. \tag{27}$$

The resolved projections of active forces in the inertial reference frame make it possible to write the generalized forces and moments of the material system.

Definition of Generalized System Forces

For finding a generalized force Q_i that is associated with certain generalized coordinate q_i one needs to gather the sum of all works of active forces under certain generalized possible relocation δq_i, assuming all the rest generalized motions equal zero. The generalized force Q_i will then be a coefficient that scales δq_i. For the case study we have:

$$Q_1 = -mg\sin\sigma + \cos\sigma\left(P_{xg}\cos\chi + P_{zg}\sin\chi\right) + P_{yg}\sin\sigma - C_r\dot{r}^2 - N_z$$
$$+\left(P_{Pxg}\cos\chi + P_{Pzg}\sin\chi\right)\cos\sigma + P_{Pyg}\sin\sigma + \left(P_{Bxg}\cos\chi + P_{Bzg}\sin\chi\right)\cos\sigma + P_{Byg}\sin\sigma; \quad (28)$$

$$Q_2 = r\left(-mg\cos\sigma + P_{yg}\cos\sigma - C_\sigma r^2\dot{\sigma}^2 - N_\sigma + P_{Pyg}\cos\sigma + P_{Byg}\cos\sigma\right); \quad (29)$$

$$Q_3 = r\left(P_{xg}\sin\chi - P_{zg}\cos\chi - C_\chi r^2\dot{\chi}^2 - N_\chi + P_{Pxg}\sin\chi - P_{Pzg}\cos\chi + P_{Bxg}\sin\chi - P_{Bzg}\cos\chi\right); \quad (30)$$

$$Q_4 = P_P l_P;\ Q_5 = M_1 + M_2;\ Q_6 = P_B l_B;\ Q_7 = M_a;\ Q_8 = M_b; \quad (31)$$

$$Q_9 = M_d;\ Q_{10} = M_1;\ Q_{11} = M_2. \quad (32)$$

Relationships (28) – (32) for the generalized forces related to chosen generalized movements express the interrelation of the AV flight characteristics in a steady state. To obtain the equations of dynamics, it is necessary to determine the kinetic energy of the flying vehicle spatial motion.

The Kinetic Energy of The Vehicle

We determine the investigated aerial vehicle kinetic energy T in absolute motion. That energy includes kinetic energy of the moving center of mass T_{cm} and two energies of rotational motions about the center of mass, namely the base motor platform energy T_1 and energy of the protective shell with hinged gimbals, T_2:

$$T = T_{cm} + T_1 + T_2. \tag{33}$$

The kinetic energy of the center of mass motion constitutes

$$T_{cm} = \frac{1}{2} m \left(\dot{r}^2 + r^2 \dot{\sigma}^2 + r^2 \dot{\chi}^2 \right), \tag{34}$$

where: $m = m_1 + m_2$ is the system total mass including mass of the AV rotor platform m_1 and mass of the protective shell with gimbals m_2;

$\dot{r}$ – position vector length;
r – position vector rate of change;
$\dot{\sigma}$ – position vector angular velocity by elevation;
$\dot{\chi}$ – position vector angular velocity by azimuth.

Rotational kinetic energy of the motor platform. The studied subject is characterized by the kinetic energy T_1 of the platform center of mass motion relative to the center of mass of the entire system and the kinetic energy of the platform rotation about the own center of mass:

$$\begin{gathered} T_1 = \frac{1}{2} m_1 l_1^2 \left(\omega_x^2 + \omega_y^2 + \omega_z^2 \right) + \frac{1}{2} \Big[I_{1x} \omega_x^2 + I_{1y} \omega_y^2 + I_{1z} \omega_z^2 \\ -2 \left(I_{1xy} \omega_x \omega_y + I_{1xz} \omega_x \omega_z + I_{1yz} \omega_y \omega_z \right) \Big] + \frac{1}{2} I_{p1} \dot{\varepsilon}_1^2 + \frac{1}{2} I_{p2} \dot{\varepsilon}_2^2, \end{gathered} \tag{35}$$

where

l_1 is the distance between the centers of mass of the system and the platform;
ω_x, ω_y, ω_z – projections of motor platform angular velocities;
$I_{1x}, I_{1y}, I_{1z}, I_{1xy}, I_{1xz}, I_{1yz}$ – dedicated moments of inertia about body-fixed axes;
I_{p1}, I_{p2} – moments of inertia of spinning rotors;
$\dot{\varepsilon}_1, \dot{\varepsilon}_2$ – rotational speeds of the rotors.

Kinetic energy of the hinged shell. That is presented by the sum of kinetic energy of the framework center of mass motion relative to the center of mass of the entire system, T_2, and kinetic energy of the housing assembly rotation about own center of mass (recall that the mentioned assembly involves the protective shell and gimbal rings with joints):

$$T_1 = \frac{1}{2} m_2 l_2^2 \left(\omega_x^2 + \omega_y^2 + \omega_z^2\right) + \frac{1}{2}\left[I_{ob}(\dot{\varphi}_1^2 + \dot{\varphi}_2^2 + \dot{\varphi}_3^2) + I_{pn}\left(\dot{\varphi}_2^2 + \dot{\varphi}_3^2\right) + I_{pb}\left(\dot{\varphi}_1^2 + \dot{\varphi}_3^2\right) + I_{tn}\dot{\varphi}_1^2 + I_{tb}\dot{\varphi}_2^2\right], \quad (36)$$

where

l_2 - is the distance between the centers of mass of the system and the shell;

I_{ob} – moment of inertia of the protective shell (either of principal axes can be chosen for referencing since the shell shape is near the shape of a ball);

I_{pn}, I_{pb}, I_{tn}, I_{tb} – moments of inertia of the outer and inner gimbal rings relative to the axes $n-n$ and $m-m$;

$\dot{\varphi}_1, \dot{\varphi}_2, \dot{\varphi}_3$ – angular rates of each ring and protective shell, respectively.

In writing equations for the kinetic energy of rotation of the protective shell and gimbal rings it was assumed that terms of the second and higher orders of vanishing could be neglected (when the rings and the shell deviate very a few while AV is flying, less 5...7 degrees).

Hence, we can write the equation for kinetic energy of the material system as

$$\begin{aligned} T = {} & \frac{1}{2} m\left(\dot{r}^2 + r^2\dot{\sigma}^2 + r^2\dot{\chi}^2\right) + \frac{1}{2} m_1 l_1^2\left(\omega_x^2 + \omega_y^2 + \omega_z^2\right) + \\ & + \frac{1}{2}\left[I_{1x}\omega_x^2 + I_{1y}\omega_y^2 + I_{1z}\omega_z^2 - 2\left(I_{1xy}\omega_x\omega_y + I_{1xz}\omega_x\omega_z + I_{1yz}\omega_y\omega_z\right)\right] + \\ & + \frac{1}{2} m_2 l_2^2\left(\omega_x^2 + \omega_y^2 + \omega_z^2\right) + \frac{1}{2}\left[I_{ob}\left(\dot{\varphi}_1^2 + \dot{\varphi}_2^2 + \dot{\varphi}_3^2\right) + I_{pn}\left(\dot{\varphi}_2^2 + \dot{\varphi}_3^2\right) + \right. \\ & \left. + I_{pb}\left(\dot{\varphi}_1^2 + \dot{\varphi}_3^2\right) + I_{tn}\dot{\varphi}_1^2 + I_{tb}\dot{\varphi}_2^2\right] + \frac{1}{2} I_{p1}\dot{\varepsilon}_1^2 + \frac{1}{2} I_{p2}\dot{\varepsilon}_2^2. \end{aligned} \quad (37)$$

In the above relation, the angular velocity terms ω_x, ω_y, ω_z are defined via the generalized coordinates $\dot{\gamma}$, $\dot{\psi}$ and $\dot{\upsilon}$ as shown in (1). When you substitute that formulas, the dependences get cumbersome form; however, they collect numerous terms that have higher than the second order of smallness. Therefore, the approximations such as $\omega_x \approx \dot{\gamma}$, $\omega_y \approx \dot{\psi}$, $\omega_z \approx \dot{\upsilon}$ can be adopted if the angles γ, ψ and $\dot{\upsilon}$ are small.

If we take another assumption that a plane of material symmetry for the vehicle is the plane OXY, then I_{yz}=0 and I_{xz}=0.

With that above assumptions, the equation for the kinetic energy of the drone can be written in quite a simple form:

$$T = \frac{1}{2}m\left(\dot{r}^2 + r^2\dot{\sigma}^2 + r^2\dot{\chi}^2\right) + \frac{1}{2}\left(m_1 l_1^2 + m_2 l_2^2\right)\left(\dot{\gamma}^2 + \dot{\psi}^2 + \dot{\upsilon}^2\right) + \\ + \frac{1}{2}\left[I_{1x}\dot{\gamma}^2 + I_{1y}\dot{\psi}^2 + I_{1z}\dot{\upsilon}^2 - 2I_{1xy}\dot{\gamma}\dot{\psi}\right] + \frac{1}{2}\left[I_{ob}\left(\dot{\varphi}_1^2 + \dot{\varphi}_2^2 + \dot{\varphi}_3^2\right) + \\ + I_{pn}\left(\dot{\varphi}_2^2 + \dot{\varphi}_3^2\right) + I_{pb}\left(\dot{\varphi}_1^2 + \dot{\varphi}_3^2\right) + I_{tn}\dot{\varphi}_1^2 + I_{tb}\dot{\varphi}_2^2\right] + \frac{1}{2}I_{p1}\dot{\varepsilon}_1^2 + \frac{1}{2}I_{p2}\dot{\varepsilon}_2^2. \tag{38}$$

Writing the partial derivatives of kinetic energy by *generalized velocities* and their derivatives on time look like follows:

$$\frac{\partial T}{\partial \dot{q}_1} = \frac{\partial T}{\partial \dot{r}} = m\dot{r}; \quad \frac{d}{dt}\frac{\partial T}{\partial \dot{q}_1} = m\ddot{r}\,; \tag{39}$$

$$\frac{\partial T}{\partial \dot{q}_2} = \frac{\partial T}{\dot{\sigma}} = \frac{1}{2}mr^2 2\dot{\sigma} = mr^2\dot{\sigma}; \quad \frac{d}{dt}\frac{\partial T}{\partial \dot{q}_2} = mr^2\ddot{\sigma}\,; \tag{40}$$

$$\frac{\partial T}{\partial \dot{q}_3} = \frac{\partial T}{\partial \dot{\chi}} = \frac{1}{2}m2r^2 2\dot{\chi}; \quad \frac{d}{dt}\frac{\partial T}{\partial \dot{q}_3} = mr^2\ddot{\chi}\,; \tag{41}$$

$$\frac{\partial T}{\partial \dot{q}_4} = \frac{\partial T}{\partial \dot{\upsilon}} = \frac{1}{2}2\dot{\upsilon}\left(m_1 l_1^2 + m_2 l_2^2\right) + \frac{1}{2}2I_{1z}\dot{\upsilon} = \dot{\upsilon}\left(m_1 l_1^2 + m_2 l_2^2 + I_{1z}\right); \tag{42}$$

$$\frac{d}{dt}\frac{\partial T}{\partial \dot{q}_4} = \ddot{\upsilon}\left(m_1 l_1^2 + m_2 l_2^2 + I_{1z}\right); \tag{43}$$

$$\frac{\partial T}{\partial \dot{q}_5} = \frac{\partial T}{\partial \dot{\psi}} = \dot{\psi}\left(m_1 l_1^2 + m_2 l_2^2 + I_{1y}\right) - I_{1xy}\dot{\gamma}\,;$$

$$\frac{d}{dt}\frac{\partial T}{\partial \dot{\psi}} = \ddot{\psi}\left(m_1 l_1^2 + m_2 l_2^2 + I_{1y}\right) - I_{1xy}\ddot{\gamma}\,; \tag{44}$$

$$\frac{\partial T}{\partial \dot{q}_6} = \frac{\partial T}{\partial \dot{\gamma}} = \dot{\gamma}\left(m_1 l_1^2 + m_2 l_2^2 + I_{1x}\right) - I_{1xy}\dot{\psi}\,;$$

$$\frac{d}{dt}\frac{\partial T}{\partial \dot{\gamma}} = \ddot{\gamma}\left(m_1 l_1^2 + m_2 l_2^2 + I_{1x}\right) - I_{1xy}\ddot{\psi}\,; \tag{45}$$

$$\frac{\partial T}{\partial \dot{q}_7} = \frac{\partial T}{\partial \dot{\varphi}_1} = I_{ob}\dot{\varphi}_1 + I_{pb}\dot{\varphi}_1 + I_{tn}\dot{\varphi}_1 = \dot{\varphi}_1\left(I_{ob} + I_{pb} + I_{tn}\right); \tag{46}$$

$$\frac{d}{dt}\frac{\partial T}{\partial \dot{q}_7} = \ddot{\varphi}_1\left(I_{ob} + I_{pb} + I_{tn}\right); \tag{47}$$

$$\frac{\partial T}{\partial \dot{q}_8} = \frac{\partial T}{\partial \dot{\varphi}_2} = \dot{\varphi}_2\left(I_{ob} + I_{pn} + I_{tb}\right);$$

$$\frac{d}{dt}\frac{\partial T}{\partial \dot{\varphi}_2} = \ddot{\varphi}_2\left(I_{ob} + I_{pn} + I_{tb}\right); \tag{48}$$

$$\frac{\partial T}{\partial \dot{q}_9} = \frac{\partial T}{\partial \dot{\varphi}_3} = \dot{\varphi}_3\left(I_{ob} + I_{pn} + I_{pb}\right);$$

$$\frac{d}{dt}\frac{\partial T}{\partial \dot{\varphi}_3} = \ddot{\varphi}_3\left(I_{ob} + I_{pn} + I_{pb}\right); \tag{49}$$

$$\frac{\partial T}{\partial \dot{q}_{10}} = \frac{\partial T}{\partial \dot{\varepsilon}_1} = \dot{\varepsilon}_1 I_{\ 1};$$

$$\frac{d}{dt}\frac{\partial T}{\partial \dot{\varepsilon}_1} = \ddot{\varepsilon}_1 I_{\ 1}; \tag{50}$$

$$\frac{\partial T}{\partial \dot{q}_{11}} = \frac{\partial T}{\partial \dot{\varepsilon}_2} = \dot{\varepsilon}_2 I_{\ 2}; \ \frac{d}{dt}\frac{\partial T}{\partial \dot{\varepsilon}_2} = \ddot{\varepsilon}_2 I_{\ 2}. \tag{51}$$

Determine now the partial derivatives of kinetic energy by the generalized coordinates:

$$\frac{\partial T}{\partial q_1} = \frac{\partial T}{\partial r} = mr\left(\dot{\sigma}^2 + \dot{\chi}^2\right); \ \frac{\partial T}{\partial q_i} = 0, \ i = \overline{2,11}. \tag{52}$$

Relations (39) – (52) represent all system generalized forces and moments, for defining which the linear and angular accelerations of generalized coordinates have been used. It's possible to write equations of motion of the impact resistant drone in the form of Lagrange equations of the second kind now:

$$m\ddot{r} - mr\left(\dot{\sigma}^2 + \dot{\chi}^2\right) = -mg\sin\sigma + \cos\sigma\left(P_{xg}\cos\chi + P_{zg}\sin\chi\right) + P_{yg}\sin\sigma - C_r\dot{r}^2 - N_z$$
$$+\left(P_{Pxg}\cos\chi + P_{Pzg}\sin\chi\right)\cos\sigma + +P_{Pyg}\sin\sigma + \left(P_{Bxg}\cos\chi + P_{Bzg}\sin\chi\right)\cos\sigma + P_{Byg}\sin\sigma; \tag{53}$$

$$mr\ddot{\sigma} = -mg\cos\sigma + P_{yg}\cos\sigma - C_\sigma r^2\dot{\sigma}^2 - N_\sigma + P_{Pyg}\cos\sigma + P_{Byg}\cos\sigma; \tag{54}$$

$$mr\ddot{\chi} = P_{xg}\sin\chi - P_{zg}\cos\chi - C_\chi r^2\dot{\chi}^2 - N_\chi + P_{Pxg}\sin\chi - P_{Pzg}\cos\chi + P_{Bxg}\sin\chi - P_{Bzg}\cos\chi; \tag{55}$$

$$\ddot{\upsilon}\left(m_1 l_1^2 + m_2 l_2^2 + I_{1z}\right) = P_P l_P; \tag{56}$$

$$\ddot{\psi}\left(m_1 l_1^2 + m_2 l_2^2 + I_{1y}\right) - I_{1xy}\ddot{\gamma} = M_1 + M_2 ; \quad (57)$$

$$\ddot{\gamma}\left(m_1 l_1^2 + m_2 l_2^2 + I_{1x}\right) - I_{1xy}\ddot{\psi} = P_B l_B ; \quad (58)$$

$$\ddot{\varphi}_1\left(I_{ob} + I_{pb} + I_{tn}\right) = \mathrm{M}_a ; \quad (59)$$

$$\ddot{\varphi}_2\left(I_{ob} + I_{pn} + I_{tb}\right) = \mathrm{M}_b ; \quad (60)$$

$$\ddot{\varphi}_3\left(I_{ob} + I_{pn} + I_{pb}\right) = \mathrm{M}_d ; \quad (61)$$

$$\ddot{\varepsilon}_1 I_{\ 1} = M_1 ; \quad (62)$$

$$\ddot{\varepsilon}_2 I_{\ 2} = M_2 . \quad (63)$$

With the aim of studying transient processes of the vehicle and use the transfer functions technique to research control strategies, the above equations are customary linearized.

The Concept of Vision System For UAV

When developers choose a framework for the vision system (CVS) being set on the mobile transportation platform they need to take into account the following aspects (Dergachov et al., 2019):

- simplicity, reliability, accessibility and costs of the hardware;
- availability of effective yet common programming tools;
- accuracy, which can be realized over the computer vision system measurements;
- capability of CVS to hold the real time processing.

A solution for the vision system that bases on the drone carrier protected by the impact resistant shell is illustrated in Figure 6.

Figure 6. Probation flying of the aerial vehicle equipped with video sensors

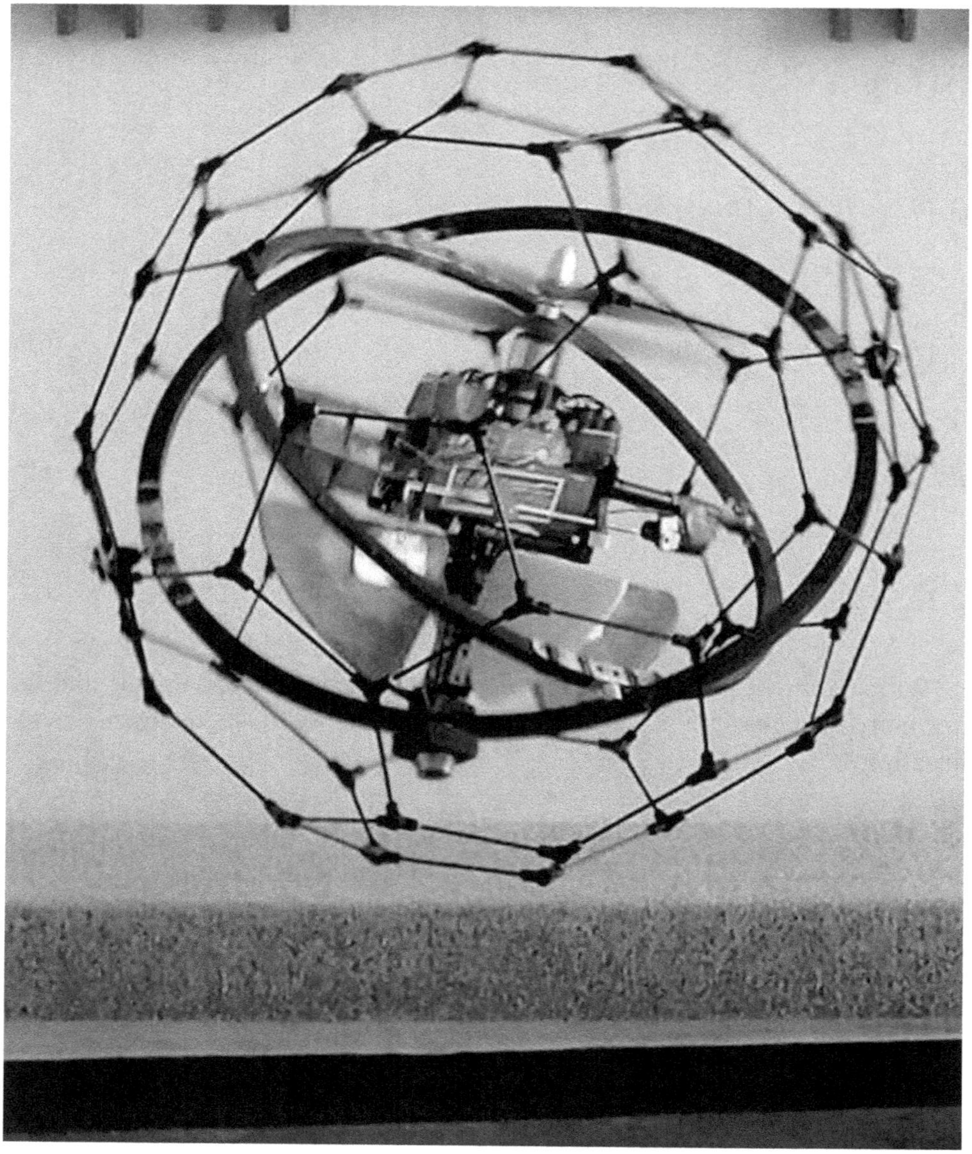

The use of such a platform allows using the built-in-shell gimbals for stabilizing the video sensor aiming while governing the drone travel and hovering. Currently, vision systems designers have a large scale of high-quality and inexpensive video sensors. As concerns the construction area subjects, computer vision complexes goaled to monitor constructions and buildings are proposed to set on board the aerial vehicle introduced above. The Raspberry Pi single-board microcomputer with the Raspbian operating system are chosen as the video system processor for handling video streams, and the Python is proposed as the basic programming language.

Efficient combination of the advanced software and hardware facilities is rational from the quality-price point of view.

Tools as 'Python Imaging Library' of Python – Pillow program suit can be used to carry up monitoring missions on construction sites using a computer vision system of the drone. The library contains a wide range of useful functions focused on fine image and video data processing. Though, lately a method of processing video data that is based on sharing resources of the Pillow library and Open CV library (Open Source Computer Vision Library) have become common (Shmelova et al., 2019). A combined use of these would allow the video data processing in real time if up-to-date algorithms were utilized.

One of topical task for constructions monitoring is the object detection followed by precise estimation of the scene progress in real time with the use of available hard- and software (Dergachov & Kulik, 2019). The issues (difficulties) are

- obtaining stable solutions over various lighting of the scene against the background of interference and
- making localization of the target in the frame.

That difficulties are solved typically using video streams analysis algorithms and basing on blocks matching techniques and correlation extremal algorithms. These methods possess a lack of accuracy, demand huge calculations, and therefore make them ineffective in mobile vision systems

As the task of recognition has to be solved and necessary algorithms realized, a distinct yet essential information feature of the stream image is pointed out to work with. It is proposed to rely on the most clear object's color that would distinguish the object from the scene background under various lighting conditions.

Then a procedure for converting the selected color component of the frame into a halftone image is fulfilled. The cycle is proceeded with the frame image binarization followed by the brightness clipping on the threshold level, which is matched appropriately for accurate spatial localization of the monitoring object. To finalize processing, it is necessary to calculate a set of invariant moments of zero and the first order, which determine the coordinates of the center of the captured object.

At the final stage of video data processing it is necessary to transform the image frame coordinates into the right central rectangular coordinate system; after, calculate the trajectory parameters of the monitoring object.

The fragmentary algorithm of monitoring construction sites with usage of the drone vision system will contain three blocks. Namely:

-provision of the video camera own verification, i.e. completion of camera internal calibration.

-specification of the reference points on the recorded scene images to be able to compute actual shooting scene geometry.

Calculation of the object displacement, which constitutes finding the rotation matrix coefficients and position vector coordinates in the given reference frame.

FUTURE RESEARCH DIRECTIONS

Some challenges call for onward development of the impact resistant UAV project.

First. Proper functioning of many perspective payloads asks for fine positioning of on-board cameras along with other sensors. That working depends much upon drone's attitude stability. This and other vehicle's operation characteristics and payload performance rely on ability of control system to counter probable impacts causing malfunctions. A disturbance may occur due to external effects such as atmospheric turbulence, strikes on obstructions, etc. and internal reasons (faults, break-down, malfunctioning, failures of elements or units). Flight controller, while using the on-board computer, has to make authentic and highly reliable on-line diagnosis on destabilization and ensure restoration of the platform angular position in real time. Mock-up research of the prospective design UAV indicates about expected gain from implementing up-to-date technology for such a diagnostics.

Secondly. A search for opportunity to use on-board video cameras without gimbals that brings to a simpler payload design and benefit weight-size and energy-saving characteristics. Along with that, conditions to obtain a video stream of proper quality, hence possibility to have fine estimates of the vehicle attitude and motion parameters are being complicated constantly, provision of monitoring both any other missions worsen. Therefore a task of constructing productive algorithms for operative and accurate processing of video streams of a camera that is housed on a traveling base is of substantial topicality.

Third. Providing for field experiments on construction sites with industrial designs of impact resistant UAVs will arise a broad set of problems to resolve - from algorithms and hardware approbation to vehicle control strategies verification and their final tuning to the conditions of technology items in urban construction; although, this requires substantial investments to provide large scale experimentations.

CONCLUSION

A unique UAV design bringing a flying platform in a shock stable protective shell is studied in the Chapter. The platform is set-up in gimbals fixed across the shell. Flight platform is a multi-rotor type drone powered by two Coaxial rotors with fix-pitch propellers of contra-rotation; the rotor constitutes a DC brushless motor.

A mathematical model of the drone that employs Lagrange equations of the second kind is studied. The presented model provides for a basis for designing a prospective automatic flight control system.

The aerial vehicle is designed to have both the full automatic control and the manual remote control by radio.

The drone is equipped with necessary flight control sensors and optional video cameras. The latter can serve two functions - besides navigation backing for the drone, they are aimed at monitoring construction processes and facilities. A conceptual feature of the vision system proposed for the project is the single-board microcomputer and open source software that is using special libraries for video data processing methods and functions.

REFERENCES

Bat, M. I., Dzhanelidze, G. Yu., & Kelzon, A. S. (2013). Theoretical Mechanics at Examples and Problems (10th ed.). Lan.

Dergachov, K., & Kulik, A. (2019). Ensuring the Safety of UAV Flights by Means of Intellectualization of Control Systems. In *Cases on Modern Computer Systems in Aviation* (pp. 287–310). IGI Global.

Dergachov, K., Kulik, A., & Zymovin, A. (2019). Environments Diagnosis by Means of Computer Vision System of Autonomous Flying Robots. In *Automated Systems in the Aviation and Aerospace Industries* (pp. 115–137). IGI Global.

Dobronravov, V. V., Nikitin, I. N., & Dvornikov, A. L. (1974). *Course in Theoretical Mechanics*. Vysshaya Shkola.

Gallacher, D. (2016). Drone applications for environmental management in urban spaces: A review. *International Journal of Sustainable Land Use and Urban Planning*, *3*(4).

Germen, M. (2016). *Alternative cityscape visualisation: drone shooting as a new dimension in urban photography*. Academic Press.

Gevaert, C., Sliuzas, R., Persello, C., & Vosselman, G. (2018). Evaluating the societal impact of using drones to support urban upgrading projects. *ISPRS International Journal of Geo-Information, 7*(3), 91.

GOST 20058-80. (1980). *Dynamics of Flying Vehicles in Atmosphere. Terms, Definitions, and Notations. Official edition*. Moscow: Goskomstandard.

Jensen, O. B. (2016). Drone city–power, design and aerial mobility in the age of “smart cities”. *Geographica Helvetica, 71*(2), 67–75. doi:10.5194/gh-71-67-2016

Krieger, Z., Wolf, H., & Reuter, T. (2018). *Advanced Drone Operations Toolkit: Accelerating the Drone Revolution*. Insight Report of World Economic Forum 91-93 route de la Capite CH-1223 Cologny.

Kucharczyk, M., & Hugenholtz, C. H. (2018). *Pre-disaster mapping with drones: an urban case study in Victoria*. Nat. Hazards Earth Syst. Sci. Discuss.

Kulik, A.S. (2016). *Elements of the theory of rational management of objects*. Kharkov: Nat. Aerospace. UN-T Them. N. Ye. Zhukovsky.

Laganière, R. (2014). *OpenCV Computer Vision Application Programming Cookbook* (2nd ed.). Packt Publishing Ltd.

Liu, P., Chen, A. Y., Huang, Y. N., Han, J. Y., Lai, J. S., Kang, S. C., ... Tsai, M. H. (2014). A review of rotorcraft unmanned aerial vehicle (UAV) developments and applications in civil engineering. *Smart Structures and Systems, 13*(6), 1065–1094.

Loitsyansky, L. (1983). Course of theoretical mechanics. Nauka.

Mohammed, F., Idries, A., Mohamed, N., Al-Jaroodi, J., & Jawhar, I. (2014). UAVs for smart cities: Opportunities and challenges. In *2014 International Conference on Unmanned Aircraft Systems (ICUAS)* (pp. 267-273). IEEE.

Shmelova, T., Sikirda, Y., Rizun, N., Kucherov, D., & Dergachov, K. (Eds.). (2019). *Automated Systems in the Aviation and Aerospace Industries*. Engineering Science Reference.

Targ, S. M. (1986). *Short course of theoretical mechanics*. Vyssh. Shkola Publ.

ADDITIONAL READING

Bradski, G., & Kaehler, A. (2008). *Learning OpenCV: Computer vision with the OpenCV library*. O'Reilly Media, Inc.

Dawson-Howe, K. (2014). *A practical introduction to computer vision with opencv*. John Wiley & Sons.

Hassaballah, M., & Hosny, K. M. (Eds.). (2018). *Recent Advances in Computer Vision: Theories and Applications* (Vol. 804). Springer.

KEY TERMS AND DEFINITIONS

AV Mathematical Model: A mapping of impact resistant drones functional performance with the use of Lagrangian formalism.

Computer Vision: Implies a integral process of extracting, identifying, and converting video information that includes six basic steps: 1) perceiving information, or sensing, 2) preconditioning or preprocessing, 3) segmentation, 4) description, 5) recognition, and 6) interpretation.

Computer Vision System: A sensory ensemble that provides shooting of work scenes and objects followed by the captured image conversion, processing and interpretation using the drone on-board computer; machined data go to payload management unit.

Destabilizing Effects: Are that objectively present in-flight disturbances which have to be identified and evaluated in course of carrying out respective missions.

Impact Resistant Aerial Vehicle: A UAV that is based on coaxial rotors platform and using bi-blade fixed-pitch propellers of contra-rotation; the platform is suspended in special protecting framework via gimbals.

Protective Framework: A power shell designed in the form of truncated icosahedron that carries function of the rotor platform defense from damage in case of the drone kick to obstruction.

Chapter 18
Future 3D Urbanism:
From UAV Data Modelling to Information Visualization

Rossana Estanqueiro
https://orcid.org/0000-0002-6087-408X
Interdisciplinary Centre of Social Sciences (CICS.NOVA), NOVA FCSH, Universidade NOVA de Lisboa, Portugal

José António Tenedório
https://orcid.org/0000-0003-3047-0807
Interdisciplinary Centre of Social Sciences (CICS.NOVA), NOVA FCSH, Universidade NOVA de Lisboa, Portugal

Carla Rebelo
Interdisciplinary Centre of Social Sciences (CICS.NOVA), NOVA FCSH, Universidade NOVA de Lisboa, Portugal

Joao Pedro Marques
GEODRONE, Portugal & Interdisciplinary Centre of Social Sciences (CICS.NOVA), NOVA FCSH, Universidade NOVA de Lisboa, Portugal

ABSTRACT

Urbanism has mainly used 2D data for both the urban analysis and diagnosis and the presentation of proposals for changes in the whole city or parts of the city. Even regarding the production of urban indicators, using, for example, the quantification of the existing green area in relation to the resident population, this practice is regularly based on the area and rarely on volume. This situation is mainly justified by the sluggishness and costs associated with obtaining 3D data. The recent development of data collection technology by unmanned aerial vehicles has triggered a change in this scenario. This chapter presents the UAV data acquisition and processing chain, analyses the positional accuracy of UAV data processing performed with GCP measurements obtained from GNSS, demonstrates how positional accuracy assessment and UAV workflow's quality control are relevant for ensuring the accuracy

DOI: 10.4018/978-1-7998-2249-3.ch018

of derived UAV geospatial products, and demonstrates the usability of 3D models in a theoretical 3D urbanism context.

INTRODUCTION

Cities and urban areas can be understood from a physical and material perspective as spaces formed by multiple strata or overlapping layers with various characteristics and attributes that, albeit different, confer on them distinctiveness and uniqueness as if they were living organisms. These overlapping layers or strata can be observed and analysed in various fields (social, demographic, cultural, political, economic, etc.) from the dynamics of interaction and very diverse flows. Nevertheless, the intrinsic physical element corresponds to the urban, built and material spatial expression, characterized by its typologies, shape geometries and land-use and occupation. The physical properties directly associated with these spaces are usually studied and analysed on two-dimensional (2D) surfaces.

"Horizontal" (2D) urban analysis has been explored in "classical" studies of the dialectic of compact versus dispersed ("urban sprawl") urban models of urban growth. Other approaches to the urban form, developed from a quantitative perspective, have made it possible to translate the fundamental characteristics or elements of urban form into quantifiable attributes by calculating indices or indicators – spatial metrics – of a more or less complex nature. These measures support the analytical study of the 2D urban form to support decision making in regional and urban planning, transport planning, environmental and energy planning, analysis and monitoring of urban-rural sustainability, public health and urban safety, socio-spatial exclusion/inclusion, among others. An example of this quantitative perspective is the metrics applied to the study of (sustainable) urban form based on geometric shape parameters (urban shape analysis) obtained in 2D (Estanqueiro, 2011). In this approach, the characteristics of the urban form are understood to be reduced to their fundamental elements or properties, which identify and come close to the phenomenon or reality to be represented or handled using metrics (quantifiable measures in the form of indices or indicators) with a vital spatial element. These can be addressed separately or combined to support urban analysis and monitor the evolution and change of the urban form considering urban sustainability.

Concurrently, more recent studies have explored vertical and volumetric urban analysis, supported by the potential of producing 3D (three-dimensional) urban models. 3D urban modelling is a high potential source of data for the extraction of parameters of interest for studies on urban sustainability in its different dimensions, from a more detailed analysis (intra-urban scale, block, neighbourhood, urban axial segment or set of buildings or urban elements). In fact, the way of thinking about cities and urban planning has resorted to geospatial data that go beyond the two-dimensional (2D). In this sense, the inherent two-dimensionality of classical urban studies is considered insufficient to understand and analyse the tangible domain of urban spaces. Huang, Liu & Lieske (2019) hold that "urban development is a multi-dimensional process that involves changes in urban form in both the horizontal and vertical dimensions" (2019:1). Thus, the use of three-dimensionality in analyses supported not only by horizontality but also by the vertical dimension and the volumetric arrangement of the set of urban elements and objects is additionally recommended. This idea is backed up by authors that admit the use of 3D data to support the production of relevant and useful geographical information for decision making in urban planning and urbanism (Shiode, 2001; Batty, 2000; Batty et al., 2000; Zhang, Zhu, & Wang, 2004; Yin & Shiode, 2014; Graham & Hewitt, 2012). Graham & Hewitt (2012) crucially argue that "(…) a fully volumetric urbanism is required which addresses the ways in which horizontal and vertical extensions, imaginaries, materialities and lived practices intersect and mutually construct each other within and between subterranean, surficial and suprasurface domains" (2012:74).

The progression and growth of urban areas at varying rates of transformation and complexity have posed a challenge to urban planning and town planning. Curbing the expansion of urban perimeters has benefited urban compression due to unprecedented vertical development. Skayannis (2010) warns that "(…) city planning may have to start foreseeing the third, Z, axis of cities, in order to avoid superficial and dangerous un-planned action in the future" (2010:2). Shiode (2001:263) emphasizes the incorporation of three-dimensionality in urban studies by pointing out that "(…) in addition to their designated applications, these models and data are potentially of great general importance to the understanding of urban structure as well as the mechanism of urban growth, spatial analysis and planning in the wider context". Graham & Hewitt (2012) stress the importance of attending to verticality in urbanism and urban planning, arguing that "this is especially so when contemporary urban processes involve: the radical horizontal extension of cities and urban regions and the engineering of unprecedented numbers of both super-high and super-deep structures; the unprecedented mass accessibility of the top-down visualities of computerized remote sensing and actual routine physical access to aerial and vertical spaces; and the rapid growth of horizontally distanciated networks of communication and growing exposure of every aspect of urban life to unblinking constellations of vertical imaging, tracking and surveillance" (2012:75). In line with the ideas set out by the aforementioned authors, Huang, Liu & Lieske (2019) justify the need to attend to three-dimensionality in urban studies given that "as the urban space is 3D space, 2D research and modelling cannot reflect the real urban morphology and the temporal and spatial change characteristics of a city. […] Research on the 3D space expansion of cities can detect and simulate 3D urban morphology changes more comprehensively and accurately. A full 3D model of a city will allow better understanding of the mechanisms of urban space expansion, and help predict 3D urban land use scenarios, which could provide critical information for urban planning." (2019:2). In this sense, the value and added value of 3D urban geographical information for the production of sustainable, inclusive and intelligible cities and urban spaces are admitted as a hypothesis but are still to be demonstrated.

Advances in the technology, means, processes and costs of obtaining data and producing geographical information have played a key role in supporting the exercise and practices of intervention, management and constant monitoring of urban evolution and cities. Nevertheless, Herbert & Chen (2015) have expressed their reluctance in the incorporation of 3D, stating that "(...) over time the theory, technology, and decision methodology have shifted from a 2D and expert system focus to a more flexible, decision support role incorporating 3D elements. Technology has developed better 3D data gathering and display techniques while the increasing use of 3D visualizations has driven research into 3D cartographic variables. However, there are concerns that these technological drivers may be limiting the potential of geovisualizations themselves (Sui 1998), and their effectiveness is still in question (Slocum et al. 2001). Despite the promise, there is no consensus on the effectiveness of 3D visualizations." (2015:23).

The latest evolution trends in technology related to the development of various types of sensors and new platforms, namely concerning spectral and geometric resolutions and their physical characteristics (size and weight), have brought about new work processes for the acquisition of geographical data. An example of this is the increasing use of unmanned aerial vehicles (UAVs) (or "drones", a widely used term). The relevance of using this type of platform is undeniable, as it has facilitated the speeding up of geographical data production processes due to speed, repeatability and cost-efficiency, offering new perspectives for research and the application of Geographic Information Technologies.

This chapter seeks to address the issue of 3D urbanism through the application of 3D UAV data acquisition to the Torcatas neighbourhood in the municipality of Almada, which is one of the fastest developing urban areas in the Metropolitan Area of Lisbon, Portugal. The chapter introduces the discussion and reflection on 3D models applied to urban areas in contrast to the classic view of urban analysis based on two-dimensional representations focusing on the contribution and potential of 3D UAV data to extract geographical information useful for urban planning and urbanism.

BACKGROUND

Three-Dimensionality In Urban Analysis

The term 3D is commonly used in a variety of technical and scientific contexts. From a classical perspective, the perception and visual representation of objects or observable elements in the real world are simultaneously associated with three dimensions (x, y, z) consisting of length, width and height, from which area and volume are extracted. Taking into account three-dimensionality allows us to integrate the vertical and volumetric aspects of a set of elements and objects (blocks of buildings, public spaces, green spaces, parks, vacant spaces, road infrastructures, etc.) and infer the relationships established in the urban structure and a city's shape (Holtier, Steadman, & Smith, 2000). McNeill (2020) examines the connection between spatial referents – verticality, surface, and underground – and volume, arguing that "rather than see these as a neat threefold division of urban space, [...] a critical stratigraphy might be deployed as a conceptual framework urban analysis, particularly within fine-grained examinations of how space is produced in major cities. [...] It is important to think about whether the key motif for understanding the 'landscapes of power' within cities is not that of the bluff façade, but rather the city in section, side-sliced to reveal the inner volumes that are economically programmed to drive and distribute value." (2020:15). Discussing the topic of "volumetric urbanism", McNeill (2020) lists four reasons for the interest and relevance of this research agenda, from which we would like to highlight the following ideas within the framework of urban theory (2020:13-14): (i) the lack of development and integration of vocabulary and a given mindset to conceive cities as being based around the organization of volume, and its measurement, although the design and construction, funding, and regulation of cities are currently much debated within urban theory; (ii) the contemporary urban theorist requires methodological innovation, with a working knowledge of specific technical practices and languages (for instance, a legal understanding of property rights relating to mass and volume and architectural and design interpretations of space, among others).

Julin et al. (2018) refer to the dissemination of the notion of 3D models but fail to provide a concrete definition of the term. Biljecki et al. (2015), relying on contributions from several authors, consider the 3D city model "(...) a representation of an urban environment with a three-dimensional geometry of common urban objects and structures, with buildings as the most prominent feature. A typical 3D city model is derived from various acquisition techniques, for instance, photogrammetry and laser scanning, extrusion from 2D footprints, synthetic aperture radar, architectural models and drawings, handheld devices, procedural modelling, and volunteered geoinformation." (2015:2843).

Three-Dimensionality and Urban Applications

Three-dimensionality as an element of the cartographic representation of space, of analysis and urban modelling has integrated technology-intensive approaches in different academic contexts. An example of this is the development and presentation of 3D models in the digital era in various technical and scientific contexts, as demonstrated in the literature on the subject.

Authors often resort to categorizing application approaches according to different perspectives of analysis or fields of use. Shiode (2001:267) defines four usage categories for 3D data and 3D models, namely: (1) Planning and design, including planning and detailed design reviews, problems of site location, community planning and public participation from information supported by 3D visualization; the visual representation of environmental impact from 3D models (hazards, impact of disasters and local pollutants at a fine scale. (2) Infrastructure and facility services, integrating water, sewerage, and electricity provision as well as road and rail network improvement and maintenance by detailed 2D and 3D data. The analysis of sightlines for mobile and fixed communications to ensure there is no interference to the reception of signals, especially due to tall buildings; and not least, the analysis and visualization of access routes to various locations used by the police, fire brigades, ambulances and other emergency services to maintain a safe environment. (3) Commercial sector and marketing, incorporating 2D and 3D models to support the visualization of the spatial distribution of customers and cater to market demands for specific economic activities, and the availability of space for development, also enabling "(...) the computation of detailed data concerning floorspace and land availability, land values and costs of development. Finally, virtual city models in 2D and 3D provide portals to virtual commerce through semi-realistic entries to remote trading and other commercial domains." (2001:267). And finally, (4) promoting and learning information about cities, focusing on the use of 3D visualization as a resource for the democratization of access to information about cities and the advancement of knowledge about cities by fostering, e.g., tourist attraction. The categorization proposed by Shiode (2001) reveals a cross-cutting, common field in the identified approaches, corresponding to harnessing the visualization potential enabled by 3D urban models. Regardless of the diversity of applicational approaches, Shiode (2001) emphasizes the sharing of "(...) one common feature in that GIS technology plays crucial role throughout the modeling process from data capturing to automated extrusion to use and maintenance of the product." (2001:268), enhancing the need and dependence on two fundamental factors in the development of 3D models: spatial information databases and remote sensing technology.

In reviewing the state of the art on 3D city model applications, Biljecki et al. (2015) demonstrate the existence of at least 29 use cases, which are only a sliver of the more than 100 application approaches. The categorization presented by the authors considers "use case" in relation to a specific field (an industry or sphere of activity and knowledge) to which it may provide a substantial benefit. The distinction of use cases is made based on the criterion of visualization and divided into two groups: "1. Non-visualisation use cases, which do not require visualising the 3D models and the results of the 3D spatial operations. [...] 2. Visualisation-based use cases. This includes: – Use cases that require running computations as in the group 1., but where visualisation is very important and the use cases would not make much sense without it (e.g., navigation, serious gaming, and urban planning). – Visualisation-only use cases such as communication of urban information and virtual reality, which do not necessarily rely on spatial operations, but where 3D city models have been found as an important component." (2015:2848). Group 1 use cases include five typical examples, namely: estimation of solar irradiation, estimation of energy demand, aid to positioning, determination of floor space, and classification of building types. Group 2, on the other hand, comprises the remaining 24 use cases, identified as geo-visualization and visualization enhancement, visibility analysis, estimation of shadows cast by urban features, estimation of the propagation of noise in an urban environment, 3D cadastre, visualization for navigation, urban planning, visualization for communication of urban information to citizens, reconstruction of sunlight direction, understanding SAR images, facility management, automatic scaffold assembly, emergency response, lighting simulations, radio-wave propagation, computational fluid dynamics, estimating the population in an area, routing, forecasting seismic damage, flooding, change detection, volumetric density studies, forest management, and finally, archaeology. The categorization proposed by Biljecki et al. (2015) and the respective systematization of use cases into two groups is indicative of the relevance and prevalence of visualization in 3D city model applications, proving to be, as they state, "(...) an inseparable part of the workflows involving 3D city models." (2015:2862). Biljecki et al. (2015) support Shiode's (2001) perspective by considering 3D visual representation as a dominant and highly explored aspect in application approaches developed since the genesis of 3D (digital) urban modelling (Batty et al., 2000).

Three-Dimensionality and Urban Communication (Visualization and Mapping)

This perspective of 3D exploration is highlighted by Batty et al. (2000) in a conceptual study on 3D city models and city visualization by establishing twelve categories, distributed over different application fields: emergency services, urban planning, telecommunications, architecture, facilities and utility management, marketing and economic development, property analysis, tourism and entertainment, e-commerce, environment, education and learning, and city portals. In this framework, 3D visual representation (or 3D visualization) and the use of 3D city models for visualization purposes and to support urban planning and urbanism have been explored since the advent of 3D and the intensive incorporation of technology along with digital evolution. The beginning of 3D model development is associated with representing a single object at a given moment in time using computer-aided architectural design (CAD). This type of models, albeit designed with volumetric methods and great geometric detail, was mainly intended for visual (aesthetic) appreciation, the illustration of plans from an empirical perspective, and/or the presentation of proposals for a building or small group of urban buildings. However, besides the high costs associated with these models, most of them had low analytical capabilities, i.e., they lacked attributes for spatial queries and advanced spatial analysis (Batty et al., 2000; Yin & Shiode, 2014). Conversely, GIS-based modelling has enabled the 2D visualization of different types of data (spatial, statistical, and quantitative data) associated with the study objects and offering the possibility of being used for querying and analytical purposes and 2.5D rendered image mapping. In fact, more recently, the advances in 3D GIS and the applicational developments using 3D data in a GIS environment, namely in urban planning and urban design, have enabled the construction of 3D GIS models by extruding 2D GIS data vertically to the respective height of each building (Yin & Shiode, 2014:154).

In this context, the contribution of 3D GIS to the growth and evolution of 3D-based applications is undeniable. However, there is still a fundamental technical question regarding the level of detail that should be implemented in these models, influenced by the objectives, purposes of use of the data and 3D geographical information, and the respective application field. This is the application of the notion of level of detail (LOD), which, according to Biljecki et al. (2014), corresponds to a measure that indicates grade and scale. The LOD can range from the most straightforward and most general (LOD0) to the most detailed and specific (LODn), depending on the characteristics of the part of the world that we wish to represent (object classes and urban elements). The lack of consensus and cross-cutting in the conceptual formalization of the term is highlighted by Biljecki et al. (2014) in a paper entitled "Formalisation of the level of detail in 3D city modelling". Although the authors consider LOD to be an essential property in 3D city models and widely used in GIS approaches, they warn that "(..) the main deficiency of the current LOD approaches is that it is not clear what the LOD is (it is merely clear that it is a conception of the design quality of the 3D city model and the amount of data) and what it comprises. If a series of multiple discrete LODs is available, it is not clear by what it is driven." (Biljecki et al., 2014:4). In this sense, they consider LOD "(…) the degree of correspondence between the model and the real-world object, being driven by the geometry, appearance, semantics and other related metrics which can be quantified, with separated exterior and interior concept. Such approach enables a consistent specification of the data, and facilitates the translation between the standards and the comparison of their LODs by decomposing them into quantified metrics since LODs are discretisations of continuous functions of the metrics." (Biljecki et al., 2014:25). Ross (2010) further adds that "the LOD concept and the appearance model can be used to encode 3D city objects with a very low level of geometric, semantic and visual detail as well as for the description of very complex objects which consist out of many object parts, include detailed semantic in-formation about geometry parts, and multiple appearance modes." (2010:72). Although the debate around LOD in the production of 3D city models persists, the benefits of 3D visualization in urban planning and urbanism are commonly agreed upon and unanimously acknowledged in different technical and scientific contexts. In this regard, Herbert & Chen (2015) emphasize two types of benefits considering, in brief, "(...) both intangible (in potentially improving site familiarity and the sense of context) and tangible (promising improved modeling of planning scenarios). Matching the visualization tools to the tasks required at each stage of the planning process promises to improve understanding of the site and issues involved, as well as enable better estimation of impacts. Ultimately, this should enable local authorities and the public to make better and more informed decisions." (Herbert & Chen, 2015:31-32).

Three-Dimensionality and Urban Analysis

Reviewing the state of the art has allowed us to assert that obtaining 3D data and producing 3D models is not only about the potential for visual representation and its use in and contribution to urbanism and urban planning. Urban planning and urbanism based on the third dimension consider four fundamental aspects systematized according to the (potential) interest and application and the objective(s) and goal(s) it aims to achieve, namely: (i) analysis (3D parameters and 3D urban data collecting, processing and analysing); (ii) modelling and simulation (3D urban geographical information production); (iii) monitoring and evaluating by integrating 2D and 3D urban geographical information to support decision making; and finally, (iv) communication using 3D urban visualization and mapping (Figure 1). These elements, taken individually or in combination with each other, are an integral part of and constitute 3D urbanism. 3D urbanism combines remote sensing, digital photogrammetry, and geographical information systems methods and techniques with the practice of city planning and urban design based on very high spatial and temporal resolution 3D data.

Figure 1. The main purposes of 3D urbanism

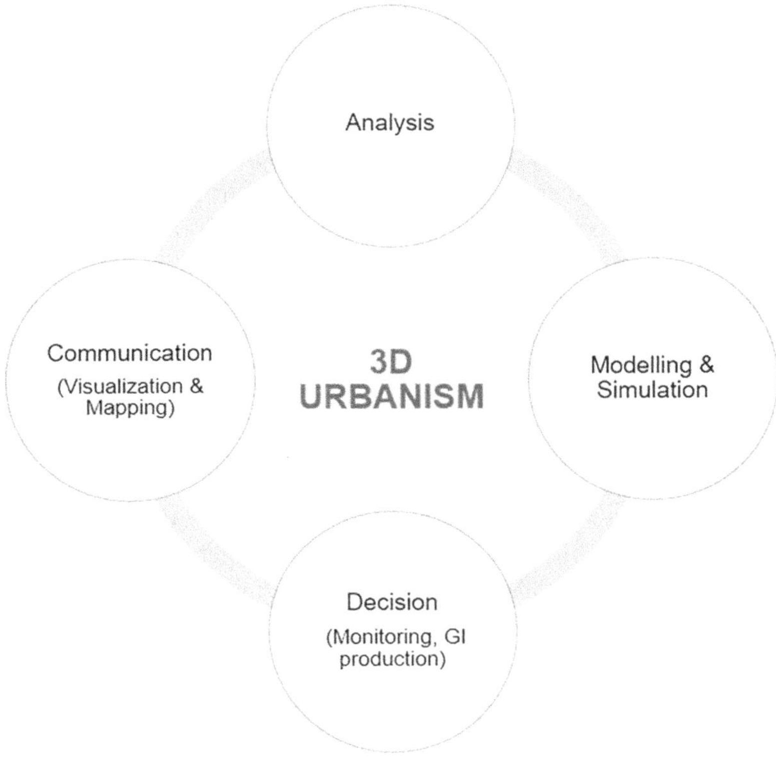

The difficulty lies in separating the four integrating parts of 3D urbanism. This idea is supported by the perspectives developed by Batty et al. (2000), Shiode (2001), and Biljecki et al. (2015) among others, for whom the visual representation and mapping component for communication and urban information purposes are inseparable from the production of 3D geographical information in urbanism and urban planning. The close association with and the advances made in GIS, remote sensing and digital photogrammetry have contributed to the development of 3D urbanism considering its analytical, modelling and simulation components as well as monitoring and evaluation to support decision making.

From 3D data, it is possible to analyse urban dynamics and the evolution of urban growth patterns in a GIS environment, as well as monitor land-use and land-cover change (2D) associated with verticality and volume parameters (3D) of built-up areas (Huang, Liu & Lieske, 2019) and urban green areas (Li et al., 2016). In this context, the application of spatial metrics to support the analysis of urban form incorporating the notions of compactness (Shirowzhan & Sepasgozar, 2019; Shirowzhan & Lim, 2014), density, and complexity (Shirowzhan & Lim, 2013) has been explored.

Holtier, Steadman, & Smith (2000) argue for the use of 3D data and 3D models based on the development of an object-oriented GIS (Smallworld) supported by a data model that ensures control of the geometric representation of entities and the topological interaction between these entities (called real-world objects, or RWOs). In the authors' understanding, "the special character of the GIS database for the four towns is that it combines a reasonably faithful representation of the gross 3-D form of the urban fabric, with data on the material and constructional characteristics of buildings, and data on occupants and their activities" (2000:71). This exploratory approach illustrates its applicability potential, "(…) where geometrical, fabric, and activity data need to be considered simultaneously (in contrast with purely geometrical models of the urban environment whose principal use is in visualisation); and to applications where many contiguous buildings need to be considered collectively (in contrast with CAD models of single buildings)" (2000:71).

3D modelling has benefited from the advances in sensors and platforms and remote sensing methods and techniques, considering that it uses data with a very high spatial resolution (< 0.5m). Some examples are the use of UAV data and the products derived from point cloud generation, which enable the creation of orthoimages, Digital Surface Models (DSM) and Digital Terrain Models (DTM) (Yao, Quin & Chen, 2019).

The Fourth-Dimension In Urban Analysis

Classical 2D geographical modelling approaches applied to urban phenomena, such as the evolution of spatial forms resulting from urban growth processes, directly or indirectly consider the integration of time, both in the production of static and dynamic models (Estanqueiro, 2011). In 3D urbanism and urban planning, the 4D in time urban analysis based on 3D models (3D plus time) gains relevance in tracking urban progress during a specific period (time series) or real-time analysis. The changes that occur in the urban space, whether by virtue of urban policies or due to interacting dynamics of very diverse and increasingly rapid nature and pace, have repercussions on the shape and spatial layout of cities. The observation, analysis and monitoring of urban permanence and change are based on successive time frames of varying duration (short, medium and long) and marked by different intensities and rhythms with impact on urban (re)configuration and land-use changes. Extreme phenomena and risk analysis with urban spatial incidence have also called for the incorporation of 4D in urban studies. In this sense, the notion of 4D has an intricate link to 3D urbanism and urban planning.

3D UAV Data and Urban Planning

The relevance of 3D modelling has been addressed and explored in different stages of local and city planning processes. It is commonly understood that the planning process consists of a set of dynamic and adaptive phases that begin with its development and end with the evaluation of any discrepancies between the provisions of the original document and the accomplished goals and objectives. Ahmed & Sekar (2015), in an integrative approach, refer to the possibility of incorporating 3D models in different stages of the planning process, namely "(...) data collection and checking, analysis, prediction, and presentations. [arguing that] They can lead to a superior understanding of real-world development." (Ahmed & Sekar, 2015:395). Corroborating this view, Huang, Liu & Lieske (2019), based on a review of the literature on extraction (definition and measurement), simulation and forecasting of vertical urban growth, emphasize the need to use 3D city models to support a better understanding of the mechanisms of urban space expansion and to support forecasting based on 3D urban land scenarios, admitting the possibility of providing critical information for urban planning.

Geographical information based on 3D UAV data is thus understood to enhance the procedural steps of urban planning and urban design (Figure 2). From a normative and formal perspective, its implementation is a result of the development of the plan. Allin (2010:13) suggests "(...) continuous improvements and, through compulsory plan monitoring and revision, for a more adaptive and responsive plan and policy development. This again has to be embedded in a consistent spatial strategy approach".

Figure 2. Geographical information based on 3D UAV data within the framework of a plan

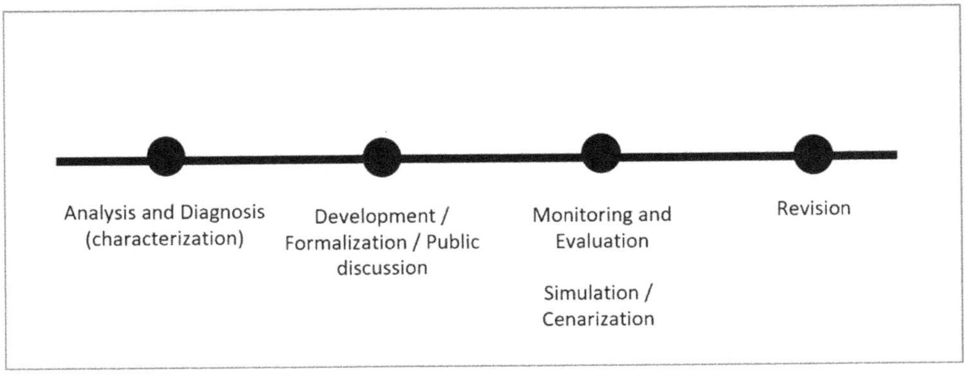

Rebelo (2016) argues that "the plan, the process, and the planning praxis require up-to-date geographical data at all times, both for monitoring actions and for the evaluation phases" (Rebelo, 2016:vii). We can state that 3D models of cities based on 3D UAV data encourage the participation of the society (citizens) and local stakeholders with very diversified interests and roles, resulting in added value to joint and sustained decision making in the collaborative and participatory planning of cities.

Estanqueiro (2011) suggests that current trends of urban growth, whether by filling intra-urban voids (urban infill) and urban compression (enhanced by verticalization or high-rise development) or by sprawl to increasingly distant "peripheries", pose challenges: on the one hand, addressing the quantitative measurement of the characteristics of urban physical and spatial growth, the evolution trend from a distant moment in time (t0) to a given more recent moment (t1), and the quantification of the intensity and nature of land-use and land-cover change; and on the other hand, considering the changing or evolving aspects of urban form filling (urban spatial qualification). From this interpretation, one can move on to a cross-analysis with other dimensions of urban form and structure, which considers the relationships between the tangible aspects of the whole and parts of the urban space and the intangible "contents". From this viewpoint, urban geographical information provided by the integration of 2D and 3D data is a highly significant resource for urban planning and urbanism, given the suitability of the acquisition scale(s), spatial resolution, and temporal resolution in order to ensure quality and accuracy across the chain of procedures to obtain the final product (from the data acquisition process to information extraction and visualization). Rebelo (2016) considers that "One of the crucial aspects of the plan is the quantification of the existing building volume. Another fundamental aspect is managing that volume: both regarding the existing volume and any additional volumes. Actually, the building volume in built areas has been one of the most sensitive topics on the densification of existing urban spaces or the design of new growing urban areas" (2016:vii) In this context, 3D UAV data acquisition and modelling reveal their potential contribution to the praxis of 3D urbanism.

UAV technology and data and the generated 3D geographical information play a relevant role by enabling urban analysis, modelling and simulation, and monitoring and evaluating the state and nature of urban phenomena in addition to supporting interdisciplinary communication between technicians and experts (urban planners and urban designers), different local stakeholders, and citizens.

From the moment UAVs became accessible for civilian purposes, they have provided a tremendous growth of new methodologies for geographical information extraction. Most of these new methodologies are based on previously existing knowledge and technologies, such as photogrammetry and LiDAR (Light Detection

And Ranging) (Remondino et al., 2011; Colomina & Molina, 2014). This technology offers an optimal relationship between the cost and the benefit of the survey, as well as the reduced time lapse between the moment the field survey takes place and the generation of the final product, along with the possibility of reducing the time interval between surveys (Colomina & Molina, 2014; Haala, Cramer, & Rothermel, 2013). In contrast, the use of LiDAR has been disregarded in favour of conventional cameras, since it is heavier and consumes more energy, which decreases the autonomy of the UAV and consequently the area surveyed by each battery (Leberl et al., 2010; Küng et al., 2011; Leitão, Moy de Vitry, Scheidegger, & Rieckermann, 2015; Rebelo et al., 2015; Yao, Quin & Chen, 2019).

Rebelo et al. (2015) in a comparative approach to test the performance of the LiDAR and UAV platforms with Free and Open Source Software (FOSS) technologies to extract building parameters (façade height and volume) demonstrate the potential of 3D UAV data as opposed to LiDAR data, arguing that "(…) the low-cost UAV imagery together with a robust methodology using FOSS tools can be very useful in the production of 3D buildings models for urban planning, unlike the LiDAR system. The accuracy of results shows that it can be enough for: i) a process of discussion and public participation in the planning process; ii) and for the monitorization of the built-area, such as the detection of illegal changes in the height of buildings" (2015:13).

The importance of three-dimensional data extracted from UAV imagery for urban analysis, and consequently for urban planning, has been highlighted by several authors (Shiode, 2001; Rebelo et al., 2015; Tenedório, Estanqueiro, Lima & Marques, 2016; Rebelo, 2016; Noor, Abdullah & Hashim, 2018; Yao, Quin & Chen, 2019; Rebelo, Rodrigues & Tenedório, 2019). This importance is supported by arguments based on exploratory studies with UAV data acquisition integration and methodological advances for 3D urban geographical information extraction. Noor, Abdullah & Hashim (2018) have objectively identified the applicational potential of UAV/drone remote sensing in urban areas, claiming that: "The unique aspects of spatial, temporal and spectral offered by drone platform considered an advantage for the users in urban study. The capability of highest spatial resolution up to 1 cm, fly on demand at critical stages and ability to integrate of multiple sensors. Furthermore, this growing availability of small, low cost components also provided urban stakeholder with opportunity to tailor a remote sensing platform to fit their own specific niche. […] although we acknowledged the integration of drones in remote sensing have proven to increase the efficiency of data acquisition, we should admit that their applications also are still at an experimental stage." (2018:1). Nevertheless, the progression and advances in the study and development of UAV-based remote sensing has been unprecedentedly explored in recent years in different scientific-technical contexts (Colomina & Molina, 2014). In turn, Yao, Quin & Chen (2019) point out that "As a means of RS data collection, UAV data and its derived products feature sounding promises to serve typical RS data analysis practices, of which two techniques/applications are mostly commonly used in the RS community: (1) land cover/image (LULC) classification or object detection; and (2) change detection." (2019:2). Yao, Quin & Chen (2019) further admit that "(…) with the available UAV data we are moving our observations from the static object level to dynamic object level (e.g., cars and individual pedestrians), where traditional techniques need to be upgraded to more advanced methods […] to deal with multi-modality data for accurate object interpretation. On the other hand, performing an RS analysis using UAV consists of joint efforts in both data acquisition and processing, while work analyzing the uncertainties associated with this two-step processes as well as work standardizing achievable interpretation results (classification and change detection) in a controlled way is still missing, which may greatly inform researchers in the community when adopting UAV-based RS methods for relevant applications." (2019:14). Wang & Li (2008) recommend the use of UAV data and techniques to support 3D information acquisition in a city, arguing that "(…) the general remote sensing techniques by satellite and aircraft can get the geographic information of large area, but there are some limitations in detail information acquisition of city. The satellite is restricted by the running cycle, height, cloud and some other factors. And the aircraft is also affected by many things such as airspace limit and weather." (2008:1). In their approach to the "classical" topic of change detection (or time-series data analysis) in Remote Sensing, Qin, Tian & Reinartz (2016) offer a summary of the main approaches and applications. In the comparative study between 2D and 3D data acquisition and between the UAV and LiDAR platforms, they highlight the main problems and (dis)advantages in 3D change detection, highlighting the applicational potential of UAV and future developments in urban, environmental, ecological, and civil applications.

Another emerging application field is UAV-based cadastral mapping (Manyoky et al., 2011; Crommelinck et al., 2016; Crommelinck et al., 2018). Crommelinck et al. (2016) suggest that "with UAVs being able to rapidly map small areas, the cadastral map could be kept up-to-date at low-cost in a sustainable way. These aspects together with the possibility of creating a transparent and participatory mapping process could contribute to another recent initiative, namely fit-for-purpose land administration published by the World Bank and the International Federation of Surveyors (FIG) (…)". (2016:17).

Rebelo, Rodrigues & Tenedório (2019) have shown the importance of 2D/3D control data, extracted from UAV data in dasymetric mapping to estimate a population using dasymetric mapping techniques within urban areas. A high-precision dasymetric mapping (using 3D information) was proposed to solve the issue of multi-temporal analysis of census datasets when there were changes in the boundaries of census tracts. This approach highlights the usage of UAV high-resolution data for the extraction of building block volume information under the development of a set of scripts based on a FOSS methodology.

Figure 3. UAV 3D data: from data acquisition to exploring information

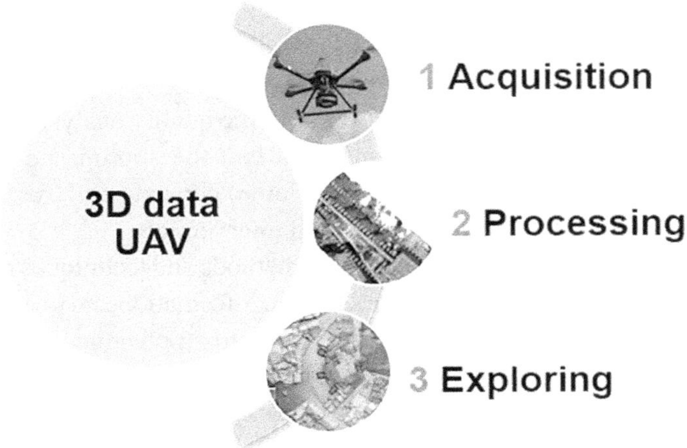

The revitalization of historic urban centres has been more recently explored using 3D UAV data (Tenedório, Estanqueiro, Lima, & Marques, 2016). The authors focus on the advantages of urban data collection by UAV to support the analysis and planning of urban revitalization areas, in itself a way to ensure efficient cities against the trend of urban sprawl. In this context, the authors consider that "Technologies used to acquire urban geographical information are an irreplaceable tool for the creation of information on the city. They also feed the computation of urban data as well as the 2D and 3D representations that are necessary to create urban plans." (2016:3). Theoretically, an urban plan is a formal planning instrument that promotes a city's efficiency and requires high-quality information based on computational processing. This may derive from two types of products: 2D ortho-images and 3D urban models. The potential for using these products in urban planning and urban design is exemplified by the authors, who highlight in particular the ability to "(…) measure and extract urban parameters, such as building façade height, building volume, solar exposure, public spaces, green areas, urban sustainability indicators, among other pieces of quantitative information that are useful for the plan. Besides the extraction of the abovementioned parameters, 3D visualization is fundamental for the promotion of the plan at different stages of the planning process, for example, at the time of public participation in the debate on urban revitalization proposals." (Tenedório, Estanqueiro, Lima, & Marques, 2016:6).

Data acquisition, processing and exploring are the three key steps (Figure 1) when using 3D UAV data in urban planning and urban design (Figure 3).

UAV surveyed data allow us to implement two chains of procedures to obtain products that support decision making in the context of urban planning and town planning (2D and 3D): on the one hand, the creation of point clouds (3D model), DSM, and orthoimage; and on the other hand, DTM, contour lines, profiles, and building volume.

The point cloud is a decisive step, along with the quality analysis of its outputs. Indeed, it is on the quality of the point cloud that the appropriate extraction of urbanistic parameters (perimeter, area and volume) depends on. These parameters are decisive both for urban analysis and urban intervention.

Starting with an orthoimage, we can apply methods and techniques that integrate digital image processing and GIS to extract urban information and produce thematic cartography relevant and useful for urbanism and urban planning. The most recent advances in automatic information extraction methods and techniques, namely by segmentation and object-based classification, associated with "traditional" visual image analysis techniques, have contributed to the production of urban information with great detail and thematic accuracy.

UAV DATA ACQUISITION, PROCESSING AND EXPLORING

UAV technology has undergone a significant evolution. These platforms, equipped with GNSS, RTK or PPK technology, can carry RGB, infrared, thermal, hyperspectral and multispectral sensors, as well as LiDAR systems. The new generations of UAVs have greater autonomy, considering flight duration, which translates into the possibility of surveying more extensive urban areas. We can say that currently, UAVs are the most cost-effective technology for urban information surveys.

Data acquisition starts with UAV survey planning, an operation that corresponds to the analysis of the required accuracy for topographic data; an accessibility analysis (topographic profile), taking into account the constraints imposed by legislation and regulations (closed or restricted airspace or prohibition of aerial imaging); and the identification of the type of targets in the area that require mapping.

UAV flight planning requires consideration of the following aspects: (i) data accuracy and reliability; (ii) number of flights and the timing of the flights; (iii) elevation angle of the sun to minimize the size of shadows in the images; (iv) weather conditions; (v) camera system and flight altitude; (vi) overlaps of adjacent images; (vii) topographic survey planning (location of GCPs and checkpoints);

(viii) location of the take-off site, which is very important for visually monitoring the operation and behaviour of the UAV; and (ix) rapid control of the quality of the images obtained (this can be important in deciding whether the flight needs to be repeated immediately).

Data processing uses a classic digital photogrammetry workflow: loading photos, aligning photos, building dense point clouds (Figure 4), modifying dense point clouds, building a mesh, texturing the 3D model.

Figure 4. Dense point cloud of the Torcatas neighbourhood in the city of Almada (Portugal)

Underlying the data processing software are multi-view stereophotogrammetry methods, Structure-from-Motion (SfM), and dense image matching. Optimizing parameter combination is very important at every stage of processing, and the accuracy of GCPs and checkpoints is decisive for the quality of the final results.

The workflow ends with calculating the DSM and the DTM, producing the orthoimage and exploring the geographical information of interest to urban planning (segmentation of the orthoimage, object-based classifications, contour lines, slopes, slope exposure, calculation of volumes and profiles). The quality of the outputs (maps, profiles, urban indicators) makes the urban analysis derived from these calculations all the more effective.

Accuracy assessment is essential for two reasons. First, because urban planning requires high accuracy when calculating indicators (perimeters, areas, volumes, etc.), i.e., the error in X, Y and the error in Z must be within the range accepted by scientific and international standards. The quality of urbanistic indicators depends on the accuracy assessment.

Figure 5. Stages of the application of quality control tasks

ACCURACY ASSESSMENT

The usage of UAVs for urban data production has become widely accessible to everyone. For this reason, the accuracy assessment of geospatial products obtained from UAV data should follow mapping accuracy standards (Popescu et al., 2016; Whitehead & Hugenholtz, 2015). It is also essential to introduce quality control (QC) procedures in the UAV photogrammetric workflow, where data quality is analysed, measured and compared with standards. Some authors (Amirebrahimi et al., 2018; Tmuši et al., 2020) have published guidelines and suggestions for the best practices on UAV usage for mapping applications, namely from data acquisition to the assessment of the derived geospatial products' quality.

The most frequently asked questions from users of urban data working with UAV technology are: *How do I know whether the geospatial product I've acquired has quality or if it meets the positional accuracy required for a particular application? How do I interpret the results obtained at the end of the UAV data processing?* In this subsection, we will clarify these issues using the case study of the Torcatas neighbourhood in the city of Almada (Portugal).

One of the most important concepts used for data quality assessment is positional accuracy. Positional accuracy represents the spatial proximity between the position of an element represented in a system and its position in the real world or in relation to a more accurate reference source (Drummond, 1995). According to Drummond (1995), the "true position" of a spatial entity, which will never be truly known, must be measured using the best and most accurate survey techniques available. The inconsistency between the estimated coordinate value and the "truth" or more accurate coordinate value represents the positional error for identical ground points (ASPRS, 2015). This type of errors can be seen at the end of the photogrammetric processing, particularly for each of the following: (i) GCPs used to improve the georeferencing accuracy of an image block; and (ii) independent check points (ICPs) to assess the accuracy of final geospatial products.

It is important to add that the positional error is frequently a result of the propagation of error as a combination of several random errors in measured quantities or the failure of some assumptions along with the sequential steps of a photogrammetric workflow (Wolf et al., 2014). Therefore, identifying the interconnectivity between each step and the potential variables involved in the UAV photogrammetric workflow contributes to a better understanding of the origin of positional errors while also assisting in defining quality control procedures to meet the accuracy required for geospatial data. Tmuši et al. (2020) summarize the diversity of factors and their interconnection that can influence the quality of DTM derived from UAV data.

Some quality control tasks can be applied across the UAV photogrammetric workflow before the geospatial product's accuracy assessment. These tasks that contribute to the geospatial product's accuracy can be divided into three main stages (Figure 5) of the UAV photogrammetric workflow.

Some of the tasks included in each of these stages are: (1) checking pre-flight conditions, which involves calibrating the sensor, inspecting UAV hardware, analysing environmental flight conditions (wind speed, cloudiness, sun's altitude, smog, dust), and ascertaining the need to place artificial photogrammetric targets on the ground; (2) UAV image analysis comprises checking the radiometric quality of the images (blurred or poorly exposed images must be excluded), verifying the presence of shadowed areas or cloudy shadows, looking for gaps in the overlap area and checking whether the frontal and lateral overlap meet the required percentage values; and (3) image block georeferencing implies the alignment of each UAV image in each strip, involving the quality assessment of the results obtained at the end of this process, namely the sparse point cloud, six exterior orientation (EO) parameters defining the location and orientation of each exposed image, and camera parameters. These results are obtained from the direct georeferencing system mounted on an aerial system and/or by a bundle block adjustment (BBA) within the Structure-from-Motion (SfM) algorithm (Iglhaut et al., 2019; He et al., 2018; Verykokou & Ioannidis, 2018). In practice, well-defined GCPs should be used to improve the accuracy of the BBA solution (Wolf et al., 2014). Besides, the number of GCPs should be reasonable with an optimal spatial distribution, including accurate measurements of those GCPs in the images and on the ground (Sanz-Ablanedo et al.; 2018; Tmuši et al., 2020). The traditional distribution for GCPs is near the block's borders, and some additional height control points in the centre of the block (Linder, 2016).

It should be noted that the image quality assessment, which is done before the BBA, has been fully automated in the processing software, which automatically identifies low-quality images.

In a UAV mapping project, the accuracy assessment is a mandatory task that should be taken into account for (i) the results of georeferencing or SfM processing and (ii) 2D/3D geospatial products (orthoimagery, 2D/3D vector data, DSM, DTM) obtained at the end of processing, whose evaluation is based on the usage of a reasonable number of well-defined ICPs (ground points not included in the processing) with an appropriated distribution (ASPRS, 2015).

The accuracy assessment includes error measurement (or positional error) and statistical value. The statistical value related to accuracy is the root mean square error (RMSE) as the degree of conformity to the "truth value" (Wolf et al., 2014). The RMSE for positional accuracy is computed from the positional errors obtained for identical points (ASPRS, 2015). High positional accuracy means that the RMSE value is near zero or below the RMSE threshold value.

Horizontal accuracy is assessed by using the linear RMSEx in the x-direction, RMSEy in the y-direction and planar $RMSE_{XY,}$ as shown in Equations (1), (2) and (3), respectively. Also, the RMSEz in the z-direction is used to evaluate the vertical accuracy of 3D geospatial data through Equation (4). The following RMSE equations for each direction were adapted for the accuracy assessment of the georeferencing step performed by SfM.

$$RMSE_X = \sqrt{\frac{\sum_{i=1}^{n}\left(X_{i(est)} - X_{i(GCP)}\right)^2}{n}} \tag{1}$$

$$RMSE_Y = \sqrt{\frac{\sum_{i=1}^{n}\left(Y_{i(est)} - Y_{i(GCP)}\right)^2}{n}} \tag{2}$$

$$RMSE_{XY} = \sqrt{RMSE_X^{\ 2} + RMSE_Y^{\ 2}} \tag{3}$$

$$RMSE_Z = \sqrt{\frac{\sum_{i=1}^{n}\left(Z_{i(est)} - Z_{i(GCP)}\right)^2}{n}} \tag{4}$$

Where:

$X_{i,est}, Y_{i,est}, Z_{i,est}$: ground coordinates estimated from SfM;

$X_{i,GCP}, Y_{i,GCP}, Z_{i,GCP}$: ground coordinates of control points from an independent source of higher accuracy;

n: total number of GCPs.

These equations are also used to assess the final geospatial products' accuracy, replacing GCPs with ICPs.

In the context of the accuracy assessment of photogrammetric products, some international and national organizations have addressed the accuracy or RMSE thresholds for general mapping or applications. In this case study, the results obtained at the end of the UAV data processing will be assessed by reference to the *Accuracy standards for digital geospatial data* developed by the American Society for Photogrammetry and Remote Sensing (ASPRS) (ASPRS, 2015). Some authors, such as Whitehead & Hugenholtz (2015) and Popescu et al. (2016), have addressed the usage of these accuracy standards for the assessment of UAV data.

The horizontal and vertical accuracy classes established in the ASPRS 1990 Accuracy Standards have been revised to the newer technologies related to the acquisition and processing of aerial images (ASPRS, 2015). In the last decades, the positional accuracy of mapping products has been tied to the quality of the aerial system (sensors, flight, and platform), image resolution (radiometric and spatial), and processing software (Popescu et al., 2016; ASPRS 2015).

In UAV mapping projects, where there are no specific accuracy requirements, the guidelines and thresholds set by the ASPRS 2015 standards help us to determine the level of positional accuracy that is expected from a geospatial product (orthoimage, planimetric or elevation data) or application. According to the ASPRS (2015), the assessment of the horizontal and vertical accuracies of a geospatial product must be performed with a minimum of 20 ICPs for areas less than 500 square kilometres.

In general, a centimetre-level accuracy obtained from UAV technology is enough for most cartographic and photogrammetric applications (Grubesic et al., 2020; ASPRS, 2015). However, this high level of accuracy is only achieved when there are several GCPs on the block during the SfM processing (Popescu et al., 2016; Sanz-Ablanedo et al., 2018) or using an onboard differential global navigation satellite system technique (Tmuši et al., 2020). Otherwise, the uncertainties still present in the direct georeferencing system of the UAV, which integrates a GNSS- Inertial Navigation System, will affect UAV mapping products' positional accuracy.

For this case study, only the accuracy assessment of georeferenced block and point cloud will be discussed in this subsection, based on the processing report delivered by the Agisoft Metashape software. In this report, a number of issues, parameters, and accuracy values were important to support the accuracy assessment, namely (a) overlap between flight lines and images, (b) self-camera calibration parameters and errors, and (c) root mean square (RMS) reprojection error averaged for all tie points or a sparse point cloud.

As illustrated in Figure 6, the UAV coverage of the study area shows a higher overlapping for the area of interest, without stereoscopic coverage failures or gaps, where the location camera of each image exposed appears at least in nine images.

Figure 6. Configuration of flight strips and the overlap area of UAV images

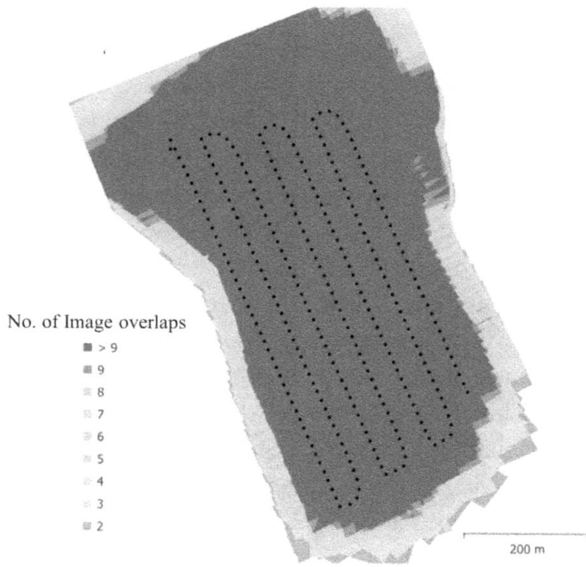

The camera's interior parameters (focal length, principal point displacement and lens distortion coefficients), refined by the self-calibration included in the BBA of SfM, were estimated with errors of a few microns. Capolupo et al. (2020) have demonstrated the influence of camera calibration quality on UAV data processing accuracy.

The reprojection error represents the discrepancy of a 3D point estimated from the SfM and its original projection detected on the image. Approximately 227 thousand tie points were acquired with an average reprojection error of less than one pixel (0.96 pixel), which means that the sparse point cloud's positional accuracy meets the value required by the standards. The magnitude of this error depends on the quality of the IO and EO parameters estimated for each exposed image.

For this case study, georeferencing the image block (or image alignment within the software) was performed to measure 11 GCPs on every image where they appear. Following the standard rules for the location of GCPs, these were initially evenly distributed close to the borders and in the centre of the area of interest. However, we could not measure GCPs in the northernmost area (Figure 7) because it is a private property to which we had no access permission.

At the end of the SfM processing, the GCP coordinates used in the BBA are estimated from the georeferenced block as well as their positional errors. It should be noted that the use of ICPs for image alignment could better measure the accuracy of the BBA solution (Wolf et al., 2014).

Figure 7. Horizontal and vertical errors and estimated GCP location.

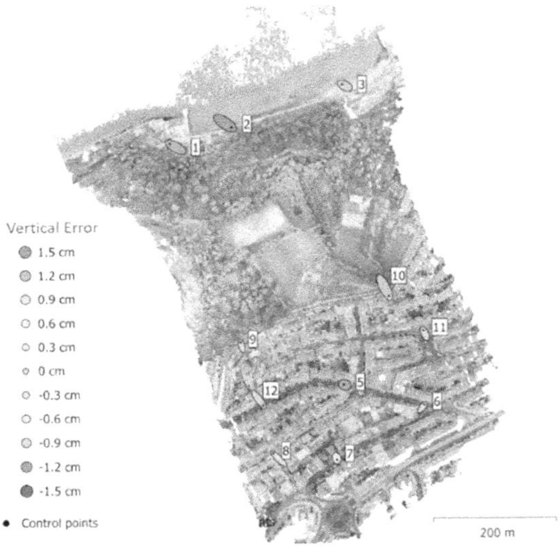

Table 1. Horizontal and vertical errors and mean absolute error (MAE) in centimetres

GCP	X	Y	XY	Z
1	-1.25	0.72	1.44	-1.04
2	1.39	-0.90	1.65	1.43
3	-0.62	0.35	0.71	-0.86
5	-0.21	0.07	**0.22**	-1.10
6	0.48	0.63	0.80	0.01
7	0.12	-0.37	**0.39**	0.61
8	-1.17	1.19	1.67	0.13
9	-0.23	0.88	0.91	0.03
10	0.90	-1.92	2.12	1.03
11	-0.33	0.80	0.86	-0.31
12	0.93	-1.33	1.62	-0.35
MAE	**0.69**	**0.83**	**1.13**	**0.63**

The location of the estimated GCPs is represented with a black dot in Figure 7. The positional error in the x-direction and y-direction is represented by the elliptical shape for each GCP (Figure 7). The smallest horizontal errors are identified for points 5 and 7 (Figure 7), where the shape of the ellipse is smaller and the planar error less than 5 mm (Table 1). About 55% of these GCPs have a planar error (XY error) of less than 1 cm.

The absolute vertical errors were less than 1.5 cm, and the maximum error was seen in GCP number 2 (Figure 7). The absolute vertical errors were less than or equal to 1 cm for 64% of the GCPs (Table 1).

The MAE shows that GCP measurements on images were performed within less than one pixel based on Ground Sample Distance (GSD) value (3.2 cm), ensuring georeferencing quality.

How do we know whether the georeferenced block has the positional accuracy required for a specific geospatial product or application using ASPRS (2015) accuracy standards? For aerial data processing, two different criteria of accuracy assessment can be established based on the type of geospatial product needed (ASPRS, 2015): (1) the horizontal $RMSE_{X/Y(BBA)}$ values should be less than or equal to half the $RMSE_{X/Y(2DMap)}$ values (or $RMSE_{X/Y(BBA)} \leq RMSE_{X/Y(2DMap)} / 2$) required for the 2D data (orthoimages and/or vector data), and vertical $RMSEz_{(BBA)}$ must be less than or equal to the $RMSE_{X/Y(2DMap)}$ values required; and (2) when the final geospatial product includes 3D data, the horizontal and vertical $RMSE_{X/Y/Z(BBA)}$ values should be less than or equal to half the $RMSE_{X/Y(2DMap)}$ or $RMSE_{Z(DSM)}$.

The RMSE values (Table 2) calculated from the errors indicated in Table 1 will allow us to assess whether the georeferenced block meets the positional accuracy required for the geospatial product or application.

Table 2. RMSE values (in centimetres) computed from the positional errors obtained from the BBA

$RMSE_X$	$RMSE_Y$	$RMSE_Z$	$RMSE_{XY}$
0.96	0.86	0.78	1.26

According to ASPRS (2015) accuracy standards, the $RMSE_{x/y(orthoimage)}$ thresholds for orthoimages are (1) less than 1 pixel (or better) for higher accuracy applications; (2) less than 2 pixels for standard mapping and GIS work; and (3) more than or equal to 3 pixels for visualization and less accurate work.

For this case study, the georeferenced block's positional accuracy is suited for the production of orthoimagery with a GSD higher than or equal to 5 cm. The horizontal $RMSE_{x/y(BBA)}$ values obtained from the BBA (Table 2) are less than half the recommended horizontal accuracy class $RMSE_{x/y(orthoimage)}$ (≤5cm). This means that the $RMSE_{x/y(BBA)}$ values of less than 2.5 cm can ensure the production of a very high-resolution orthoimage (GSD = 5 cm) with the level of quality required for higher accuracy applications. The vertical RMSE value ($RMSE_{Z(BBA)}$) is also relevant because of the DTM and the DSM used in orthoimage production. For this case, given a DSM with a vertical accuracy of a $RMSE_{z(DSM)}$ equal to 5 cm, the vertical accuracy obtained for the BBA meets ASPRS accuracy requirements.

Finally, the production of highly accurate 2D vector data acquired from source images with a GSD of about 3.5 cm should meet ASPRS Accuracy Standards of 5 cm regarding the horizontal accuracy class for RMSEx and RMSEy. Furthermore, the positional accuracy achieved for this exercise is also capable of producing accurate large-scale urban maps.

In addition to positional accuracy, there is another relevant element of spatial data quality that should be referred to in this chapter – *Lineage* (Morrison, 1995), which is related to identifying the history of a data set. In practice, it means that metadata production constitutes a relevant topic to understand data quality and preserve the valuable history of data sources for future UAV mapping projects. Recently, Tmuši *et al.* (2020) have suggested a metadata form for UAV mapping projects, which includes specific parameters and system configuration within each UAV survey step.

CONCLUSION AND FUTURE RESEARCH DIRECTIONS

Future 3D Urbanism will be based on 3D information (and 4D, 5D). Why? Because the volume of urban objects matters greatly to understand the role of buildings in conserving and dissipating heat and the vegetation's role in increasing urban thermal comfort, among other indicators. Scientific evidence has shown that the population's participation is more effective when the public discussion of urban planning proposals is based on 3D models. These aspects (urban heat, urban thermal comfort, participatory processes) are part of the 21st century's urbanistic discourse. This will continue to be so considering the current relevance of political agendas, the urban environmental goals of international organizations, and the academic environments, where one of the most innovative sciences is being developed: Sustainability Science.

The pace of change in urban environments has always been fast. But today, it is much more so due to the impacts of climate change at different scales. For this reason, there has never been so much talk about green urbanism, which is a concept that promises to implement concrete actions to change the ways we plan cities and develop urbanism.

Technology has become irreplaceable in the different stages of the urban planning process. The physical spaces of cities are virtualized, urban flows are increasing, and big data processing currently provides a city's representation for 24 hours repeatedly.

The processes of physical change in cities must be accompanied by monitoring. In other words, real-time data enable the permanent updating of urban environmental indicators. Therefore, today's rhythms are very different compared to those of past centuries. For this reason, data acquisition has to be fast and inexpensive.

UAV data can support urban databases with low-cost three-dimensional information, even if restricted to small geographical areas. Computing has made significant progress, and everything indicates that it will respond efficiently to the processing of large volumes of data.

UAV technology affordable for local administrations has two limitations: low autonomy of the UAVs to cover vast areas and heavy computational processing. But this does not preclude its increased application to the cities of the future.

Despite the limitations we have identified, we would like to highlight the advantages of modelling UAV data for urban analysis and for the creation of city plans: (i) low cost; (ii) not very demanding in terms of human resources; (iii) possibility to create 3D urban models to make measurements and extract urbanistic parameters; (iv) permanently updateable; (v) ability to monitor extreme weather events with urban impact (e.g., floods); (vi) create and maintain digital archives of urban evolution. Data acquisition and modelling technology for such data promote the implementation of the idea of smart cities. In fact, UAV for surveying and mapping can generate data and information that will serve the dimensions of city intelligence.

Future applied research should focus on optimizing the relationship between urbanism and technology. Currently, sustainable urbanism has a substantial technological basis (i.e., energy efficiency, waste management, water management, ecological efficiency, etc.). Urbanism in the future will be smart if it embeds technology to address urban challenges with 3D, 4D, and 5D representations. These future paths can design better cities.

ACKNOWLEDGMENT

This work is financed by national funds through FCT - Foundation for Science and Technology, I.P., within the scope of the project «UIDB/04647/2020» of CICS. NOVA - Centro Interdisciplinar de Ciências Sociais da Universidade Nova de Lisboa.

REFERENCES

Ahmed, F. C., & Sekar, S. P. (2015). *3D Volumetric Analysis in Everyday Urban Planning Processes. Applied Spatial Analysis, 8.* doi:10.100712061-014-9122-2

Allin, S. (2010). Trapped between Space and Time: Contemporary and Future Challenges to Land Use Planning in Germany. In *Proceedings of the 24th AESOP Annual Conference*. Aalto University School of Science and Technology.

Amirebrahimi, S., Quadros, N., Coppa, I., & Keysers, J. (2018). *UAV Data Acquisition in Australia and New Zealand*. FrontierSL.

ASPRS. (2015). ASPRS Positional Accuracy Standards for Digital Geospatial Data. *Photogrammetric Engineering and Remote Sensing*, *81*(3), 1–26. doi:10.14358/PERS.81.3.A1-A26

Batty, M. (2000). The New Urban Geography of the Third Dimension. *Environment and Planning. B, Planning & Design*, *27*(4), 483–484. doi:10.1068/b2704ed

Batty, M., Chapman, D., Evans, S., Haklay, M., Kueppers, S., Shiode, N., Smith, A., & Torrens, P. M. (2000). *Visualizing the city: communicating urban design to planners and decision-makers*. Working paper series, 26. CASA, UCL.

Biljecki, F., Ledoux, H., Stoter, J., & Zhao, J. (2014). Formalisation of the level of detail in 3D city modelling. *Computers, Environment and Urban Systems*, *48*, 1–15. doi:10.1016/j.compenvurbsys.2014.05.004

Biljecki, F., Stoter, J., Ledoux, H., Zlatanova, S., & Çöltekin, A. (2015). Applications of 3D City Models: State of the Art Review. *International Journal Geo-Information ISPRS*, *4*(4), 2842–2889. doi:10.3390/ijgi4042842

Capolupo, A., Saponaro, M., Borgogno Mondino, E., & Tarantino, E. (2020). Combining Interior Orientation Variables to Predict the Accuracy of RPAS–SfM 3D Models. *Remote Sensing*, *12*(17), 2674. doi:10.3390/rs12172674

Colomina, I., & Molina, P. (2014). Unmanned aerial systems for photogrammetry and remote sensing: A review. *Journal of Photogrammetry and Remote Sensing ISPRS*, *92*, 79–97. doi:10.1016/j.isprsjprs.2014.02.013

Crommelinck, S., Bennett, R., Gerke, M., Nex, F., Yang, M., & Vosselman, G. (2016). Review of Automatic Feature Extraction from High-Resolution Optical Sensor Data for UAV-Based Cadastral Mapping. *Remote Sensing*, *8*(8), 689. doi:10.3390/rs8080689

Crommelinck, S., Höfle, B., Koeva, M., Yang, M., & Vosselman, G. (2018). Interactive cadastral boundary delineation from UAV data. *ISPRS Annals of the Photogrammetry. Remote Sensing and Spatial Information Sciences*, *4*, 81–88.

Drummond, J. (1995). Positional Accuracy. In S. C. Guptill & J. L. Morrison (Eds.), *Elements of spatial data quality. International Cartographic Association* (pp. 31–58). Elsevier Science. doi:10.1016/B978-0-08-042432-3.50010-0

Eastman, C., Teicholz, P., Sacks, R., & Liston, K. (2011). BIM Handbook: A Guide to Building Information Modeling for Owners, Managers, Designers, Engineers and Contractors (2nd ed.). John Wiley & Sons Inc.

Estanqueiro, R. (2011). *Análise espacial da forma urbana em Sistemas de Informação Geográfica* (PhD Thesis). Universidade NOVA de Lisboa, Faculdade de Ciências Sociais e Humanas. http://hdl.handle.net/10362/7300

Graham, S., & Hewitt, L. (2012). Getting off the ground: On the politics of urban verticality. *Progress in Human Geography*, *37*(1), 72–92. doi:10.1177/0309132512443147

Grubesic, T. H., & Nelson, J. R. (2020). *UAVs and Urban Spatial Analysis*. Springer International Publishing. doi:10.1007/978-3-030-35865-5

Haala, N., Cramer, M., & Rothermel, M. (2013). Quality of 3D point clouds from highly overlapping UAV imagery. *The International Archives of the Photogrammetry, Remote Sensing and Spatial Information Sciences*, *40*(September), 4–6. doi:10.5194/isprsarchives-XL-1-W2-183-2013

He, F., Zhou, T., Xiong, W., Hasheminnasab, S. M., & Habib, A. (2018). Automated aerial triangulation for UAV-based mapping. *Remote Sensing*, *10*(12), 1952. Advance online publication. doi:10.3390/rs10121952

Herbert, G., & Chen, X. (2015). A comparison of usefulness of 2D and 3D representations of urban planning. *Cartography and Geographic Information Science*, *42*(1), 22–32. doi:10.1080/15230406.2014.987694

Holtier, S., Steadman, J. P., & Smith, M. G. (2000). Three-dimensional representation of urban built form in a GIS. *Environment and Planning. B, Planning & Design*, *27*(1), 51–72. doi:10.1068/bst5

Huang, Y., Liu, Y., & Lieske, S. (2019). Detecting, Modeling and Predicting Vertical Urban Growth: An exploratory review. In GeoComputation 2019. The University of Auckland. doi:10.17608/k6.auckland.9863141.v2

Iglhaut, J., Cabo, C., Puliti, S., Piermattei, L., O'Connor, J., & Rosette, J. (2019). Structure from Motion Photogrammetry in Forestry: A Review. *Current Forestry Reports*, *5*(3), 155–168. doi:10.100740725-019-00094-3

Julin, A., Jaalama, K., Virtanen, J., Pouke, M., Ylipulli, J., Vaaja, M., Hyyppä, J., & Hyyppä, H. (2018). Characterizing 3D City Modeling Projects: Towards a Harmonized Interoperable System. *ISPRS International Journal of Geo-Information*, *7*(55), 1–18. doi:10.3390/ijgi7020055

Kensek, K., & Noble, D. (2014). *Building Information Modeling: BIM in Current and Future Practice*. John Wiley & Sons. doi:10.4324/9781315797076

Küng, O., Strecha, C., Beyeler, A., Zufferey, J.-C., Floreano, D., Fua, P., & Gervaix, F. (2011). The Accuracy of Automatic Photogrammetric Techniques on Ultra-Light UAV Imagery. *ISPRS International Archives of the Photogrammetry, Remote Sensing and Spatial Information Sciences*, *38*(1), 125–130.

Leberl, F., Irschara, A., Pock, T., Meixner, P., Gruber, M., Scholz, S., & Wiechert, A. (2010). Point Clouds: LiDAR versus 3D Vision. *Photogrammetric Engineering and Remote Sensing*, *76*(10), 1123–1134. doi:10.14358/PERS.76.10.1123

Leitão, J. P., Moy de Vitry, M., Scheidegger, A., & Rieckermann, J. (2015). Assessing the quality of Digital Elevation Models obtained from mini-Unmanned Aerial Vehicles for overland flow modelling in urban areas. *Hydrology and Earth System Sciences Discussions*, *12*(6), 5629–5670.

Li, X., Li, W., Meng, Q., Zhang, C., Jancso, T., & Wu, K. (2016). Modelling building proximity to greenery in a three-dimensional perspective using multi-source remotely sensed data. *Journal of Spatial Science*, *61*(2), 389–403. doi:10.1080/14498596.2015.1132642

Linder, W. (2016). Digital photogrammetry: A practical course. In Digital Photogrammetry: A Practical Course. doi:10.1007/978-3-662-50463-5

Manyoky, M., Theiler, P., Steudler, D., & Eisenbeiss, H. (2011). Unmanned aerial vehicle in cadastral applications. In ISPRS. *The International Archives of the Photogrammetry, Remote Sensing and Spatial Information Sciences*, *38*(C22), 57–62.

Marques, L., Tenedório, J. A., Burns, M., Romão, T., Birra, F., Marques, J., & Pires, A. (2017). Cultural Heritage 3D Modelling and visualisation within an Augmented Reality Environment, based on Geographic Information Technologies and mobile platforms. *ACE: Architecture, City and Environment*, *11*(33), 117–136.

McNeill, D. (2020). The volumetric city. *Progress in Human Geography*, *44*(5), 815–831. doi:10.1177/0309132519863486

Morrison, J. L. (1995). Spatial data quality. In S. C. Guptill & J. L. Morrison (Eds.), *Elements of spatial data quality. International Cartographic Association* (pp. 1–12). Elsevier Science. doi:10.1016/B978-0-08-042432-3.50008-2

Popescu, G., Iordan, D., & Păunescu, V. (2016). The Resultant Positional Accuracy for the Orthophotos Obtained with Unmanned Aerial Vehicles (UAVs). *Agriculture and Agricultural Science Procedia*, *10*, 458–464. doi:10.1016/j.aaspro.2016.09.016

Qin, R., Tian, J., & Reinartz, P. (2016). 3D change detection - Approaches and applications. *ISPRS Journal of Photogrammetry and Remote Sensing*, *122*, 41–56. doi:10.1016/j.isprsjprs.2016.09.013

Rebelo, C. (2016). *3D Point Clouds in Urban Planning: Developing and Releasing High-end Methodologies based on LiDAR and UAV Data for the Extraction of Building Parameters* (PhD Thesis). Portugal, Universidade NOVA de Lisboa, Faculdade de Ciências Sociais e Humanas. http://hdl.handle.net/10362/19005

Rebelo, C., Rodrigues, A., & Tenedório, J. A. (2019). Dasymetric Mapping Using UAV High Resolution 3D Data Within Urban Areas. *Remote Sensing*, *11*(14), 1716. doi:10.3390/rs11141716

Rebelo, C., Rodrigues, A. M., Tenedório, J. A., Gonçalves, J. A., & Marnoto, J. (2015). Building 3D City Models: Testing and Comparing Laser Scanning and Low-Cost UAV Data Using FOSS Technologies. In Computational Science and Its Applications – ICCSA 2015. ICCSA 2015. Lecture Notes in Computer Science, (vol. 9157). Springer. doi:10.1007/978-3-319-21470-2_26

Ross, L. (2010). *Virtual 3D City Models in Urban Land Management—Technologies and Applications* (Ph.D. Thesis). Technische Universität Berlin, Berlin, Germany.

Sanz-Ablanedo, E., Chandler, J. H., Rodríguez-Pérez, J. R., & Ordóñez, C. (2018). Accuracy of Unmanned Aerial Vehicle (UAV) and SfM Photogrammetry Survey as a Function of the Number and Location of Ground Control Points Used. *Remote Sensing*, *10*(10), 1606. doi:10.3390/rs10101606

Shiode, N. (2001). 3D urban models: Recent developments in the digital modelling of urban environments in three-dimensions. *GeoJournal*, *52*(3), 263–269. doi:10.1023/A:1014276309416

Shirowzhan, S., & Lim, S. (2014). Three-Dimensional Spatial Metrics for Compactness Assessment of Urban Forms. *Proceedings of the 31st International Symposium on Automation and Robotics in Construction and Mining*. 10.22260/ISARC2014/0119

Shirowzhan, S., & Lim, S. (2013). Development of Urban Sustainability Index using 3-D spatial metrics. In *Proceedings of the ICSDEC 2012 International Conference on Sustainable Design, Engineering, and Construction* (pp. 238-245). 10.1061/9780784412688.028

Shirowzhan, S., & Sepasgozar, S. M. E. (2019). Spatial Analysis Using Temporal Point Clouds in Advanced GIS: Methods for Ground Elevation Extraction in Slant Areas and Building Classifications. ISPRS International Journal of Geo-Information, 8(3), 120.

Skayannis, P. (2010). Is urban underground space luxury? From infrastructures to subterranean structures. In *24th AESOP Annual Conference: Association of European Schools of Planning*. Aalto University.

Tenedório, J. A., Estanqueiro, R., Lima, A. M., & Marques, J. (2016). Remote sensing from unmanned aerial vehicles for 3D urban modelling: case study of Loulé, Portugal. In Virtual City and Territory "Back to the Sense of the City: International Monograph Book". Barcelona: Centre de Política de Sòl i Valoracions.

Tenedório, J. A., Rebelo, C., Estanqueiro, R., Henriques, C. D., Marques, L., & Gonçalves, J. A. (2016). New Developments in Geographical Information Technology for Urban and Spatial Planning. In Geospatial Research: Concepts, Methodologies, Tools, and Applications, (pp. 1965-1997). Hershey, PA: IGI Global. doi:10.4018/978-1-4666-9845-1.ch094

Tmuši, G. (2020). Current Practices in UAV-based Environmental Monitoring. *Remote Sensing*, *12*(6), 1001. doi:10.3390/rs12061001

Verykokou, S., & Ioannidis, C. (2018). A photogrammetry-based structure from motion algorithm using robust iterative bundle adjustment techniques. *ISPRS Annals of the Photogrammetry, Remote Sensing and Spatial Information Sciences, 4*(4/W6), 73–80. 10.5194/isprs-annals-IV-4-W6-73-2018

Whitehead, K., & Hugenholtz, C. H. (2015). Applying ASPRS accuracy standards to surveys from small unmanned aircraft systems (UAV). *Photogrammetric Engineering and Remote Sensing*, *81*(10), 787–793. doi:10.14358/PERS.81.10.787

Wolf, P. R., Dewitt, B. A., & Wilkinson, B. (2014). *Elements of Photogrammetry with Applications in GIS* (4th ed.). McGraw-Hill.

Yao, H., Quin, R., & Chen, X. (2019). Unmanned Aerial Vehicle for Remote Sensing Applications – A review. *Remote Sensing*, *11*(12), 1443. doi:10.3390/rs11121443

Yin, L., & Shiode, N. (2014). 3D spatial-temporal GIS modeling of urban environments to support design and planning processes. *Journal of Urbanism*, *7*(2), 152–169. doi:10.1080/17549175.2013.879452

Zhang, X., Zhu, Q., & Wang, J. (2004). 3D City Models Based Spatial Analysis to Urban Design. *Geographic Information Sciences: A Journal of the Association of Chinese Professionals in Geographic Information Systems*, *10*(1), 82–86. doi:10.1080/10824000409480658

Zhu, Z., Zhou, Y., Seto, K. C., Stokes, E. C., Deng, C., Pickett, S. T. A., & Taubenböck, H. (2019). Understanding an urbanizing planet: Strategic directions for remote sensing. *Remote Sensing of Environment*, *228*, 164–182. doi:10.1016/j.rse.2019.04.020

ADDITIONAL READING

Batty, M., Steadman, P., & Xie, Y. (2006). Visualization in Spatial Modeling. In J. Portugali (Ed.), *Complex Artificial Environments*. Springer., doi:10.1007/3-540-29710-3_4

Chen, J., Dowman, I., Li, S., Li, Z., Madden, M., Mills, J., Paparoditis, N., Rottensteiner, F., Sester, M., Toth, C., Trinder, J., & Heipke, C. (2016). Information from imagery: ISPRS scientific vision and research agenda. *ISPRS Journal of Photogrammetry and Remote Sensing*, *115*, 3–21. doi:10.1016/j.isprsjprs.2015.09.008

Pietsch, S. M. (2000). Computer Visualisation in the Design Control of Urban Environments: A Literature Review. *Environment and Planning. B, Planning & Design*, *27*(4), 521–536. doi:10.1068/b2634

KEY TERMS AND DEFINITIONS

3D Model: Is a digital representation of an object or scene in a three-dimensional reference system. This digital representation is made in two different formats: vector as geometric topological reconstruction using points, lines and areas; or raster in grid cells (voxel). The resolution of the 3D model will depend on the level of detail required in the representation of 3D information.

3D UAV Data: Are represented by a georeferenced dense point cloud, obtained from semi-automatic photogrammetric processing of UAV image pairs acquired by a small aerial platform system.

Positional Accuracy: Measures the spatial quality of geographical data, horizontally and vertically, by comparison with a more accurate geospatial technique, where the statistical root mean square error (horizontal and vertical) is calculated based on the mean of the squared positional errors at checkpoints.

UAV Data Processing: Corresponds to the sequential linking of a set of processes based on computer vision and traditional photogrammetry techniques. It consists of the processing of high-overlapping UAV image pairs to generate a 3D model represented by a dense point cloud.

Urban Big Data Processing: Are toolbox and programming models that allow the processing of large-scale data in permanent evolution to extract information and behaviours needed to support decision making in cities.

Urban Data: Are graphic and alphanumeric records with urban spatial reference, considering highly accurate scales of analysis and originating information necessary for understanding, knowledge and decision making in the fields of urbanism and urban planning.

Urban Planning: Is a process supported by politics, technology, techniques and methods, which originates or conditions the physical, environmental and social dynamics and the design of a city.

Compilation of References

Abbott, J. (2020). *DJI Mavic Air 2 Review – The Best Drone Around for Beginners and Hobbyists.* https://www.techradar.com/uk/reviews/dji-mavic-air-2-review

Aber, J., Marzolff, I., & Ries, J. (2010). *Small-Format Aerial Photography, Photogrammetry.* Elsevier.

Abhishek, N., Jenamani, M., & Mahanty, B. (2017). Urban growth in Indian cities: Are the driving forces really changing? *Habitat International, 69*, 48–57. doi:10.1016/j.habitatint.2017.08.002

Achille, C., Adami, A., Chiarini, S., Cremonesi, S., Fassi, F., Fregonese, L., & Taffurelli, L. (2015). UAV-Based Photogrammetry and Integrated Technologies for Architectural Applications/ Methodological Strategies for the After-Quake Survey of Vertical Structures in Mantua (Italy). *Sensors (Basel), 15*(7), 15520–15539. doi:10.3390150715520 PMID:26134108

Acselrad, H. (2010). *Cartografia social e dinâmicas territoriais: marcos para o debate.* Universidade Federal do Rio de Janeiro, Instituto de Pesquisa e Planejamento Urbano e Regional.

Agisoft. (2020). *Orthophoto and DEM Generation (with GCPs) with Agisoft PhotoScan Professional.* https://www.agisoft.com/index.php?id=31

Agyemang, F. S. K., & Silva, E. (2019). Simulating the urban growth of a predominantly informal Ghanaian city-region with a cellular automata model: Implications for urban planning and policy. *Applied Geography (Sevenoaks, England), 105*, 15–24. doi:10.1016/j.apgeog.2019.02.011

Ahmad, H., Babak, A., Davoud, Z., & Suhardi, B. M. (2019). Assessing the thermal comfort effects of green spaces: A systematic review of methods, parameters, and plants' attributes. *Sustainable Cities and Society*, 49.

Ahmed, F. C., & Sekar, S. P. (2015). *3D Volumetric Analysis in Everyday Urban Planning Processes. Applied Spatial Analysis, 8.* doi:10.100712061-014-9122-2

Airobotics. (2020). *Automated drone applications for industrial purposes.* https://www.airoboticsdrones.com/applications/

Aithal, B. H., & Ramachandra, T. V. (2013). Measuring urban sprawl in Tier II cities of Karnataka, India. *2013 IEEE Global Humanitarian Technology Conference: South Asia Satellite (GHTC-SAS)*, 321-329. 10.1109/GHTC-SAS.2013.6629939

Aithal, B. H., & Ramachandra, T. V. (2016). Visualization of urban growth pattern in Chennai using geoinformatics and spatial metrics. *Photonirvachak (Dehra Dun)*, *44*(4), 617–633. doi:10.100712524-015-0482-0

Alabi, M. O. (2020). Sustainable Urban form and challenges of open space utilization, Akure, Nigeria as a case study. *International Journal of Urban Sustainable Development*, *12*(3), 1–12. doi:10.1080/19463138.2020.1728691

Alberti, M. (2008). *Advances in urban ecology: Integrating humans and ecological processes in urban ecosystems*. Springer. doi:10.1007/978-0-387-75510-6

Albino, V.; Berardi, U.; Dangelico, R. M. (2015). Smart Cities: Definitions, Dimensions, Performance, and Initiatives. *Journal of Urban Technology, 22*(1), 3-21.

Alcoforado, M.-J., Andrade, H., Lopes, A., & Vasconcelos, J. (2009). Application of climatic guidelines to urban planning: The example of Lisbon (Portugal). *Landscape and Urban Planning*, *90*(1), 56–65. doi:10.1016/j.landurbplan.2008.10.006

Aleixo, A. (2001). Conservação da avifauna da Floresta Atântica: Efeitos da fragmentação e a importância de florestas secundárias. In J. L. B. Albuquerque J.F. Cândido Junior, F.C. Straube & A.L. Roos (Eds.), Ornitologia e conservação: Da ciência às estratégias (pp. 199-206). Tubarão, SC: Unisul.

Alexandri, E., & Jones, P. (2008). Temperature decreases in an urban canyon due to green walls and green roofs in diverse climates. *Building and Environment*, *43*(4), 4280–4493. doi:10.1016/j.buildenv.2006.10.055

Al-Hajjaji, K., Ezzin, M., Khamdan, H., & El Hassani, A. (2017). *Design, Development and Evaluation of a UAV to Study Air Quality in Qatar.* https://arxiv.org/ftp/arxiv/papers/1709/1709.05628.pdf

Alidoost, F., & Arefi, H. (2017). Comparison of UAS-based photogrammetry software for 3D point cloud generation: A survey over a historical site. In *Proceedings of the 4th International GeoAdvances Workshop* (vol. 4, pp. 55-61). 10.5194/isprs-annals-IV-4-W4-55-2017

Ali-Toudert, F., & Mayer, H. (2007). Effects of asymmetry, galleries, overhanging façades and vegetation on thermal comfort in urban street canyons. *Solar Energy*, *81*(6), 742–754. doi:10.1016/j.solener.2006.10.007

Allen, L., Lindberg, F., & Grimmond, C. S. B. (2011). Global to city scale urban anthropogenic heat flux: Model and variability. *International Journal of Climatology*, *31*(13), 1990–2005. doi:10.1002/joc.2210

Allin, S. (2010). Trapped between Space and Time: Contemporary and Future Challenges to Land Use Planning in Germany. In *Proceedings of the 24th AESOP Annual Conference*. Aalto University School of Science and Technology.

Allotta, B., Costanzi, R., Ridolfi, A., Salvetti, O., Reggiannini, M., Kruusmaa, M., Salumae, T., Lane, D.M., Frost, G., Tsiogkas, N., Cocco, M., Gualdesi, L., Lacava, G., Roig, D., Gundogdu, H.T., Dede, M.I.C., Baines, S., Tusa, S, Latti, P., & Scaradozzi, D. (2018). The ARROWS Project: robotic technologies for underwater archaeology. *IOP Conf. Ser.: Mater. Sci. Eng., 364*. 10.1088/1757-899X/364/1/012088

Al-sharif, A., & Pradhan, B. (2013). Monitoring and predicting land use change in Tripoli Metropolitan City using an integrated Markov chain and cellular automata models in GIS. *Arabian Journal of Geosciences*, *7*. doi:10.100712517-013-1119-7

Alsharif, A. A., & Pradhan, B. (2014). Urban sprawl analysis of Tripoli Metropolitan city (Libya) using remote sensing data and multivariate logistic regression model. *Photonirvachak (Dehra Dun)*, *42*(1), 149–163. doi:10.100712524-013-0299-7

Altuntas, C. (2019). Urban area change visualization and analysis using high density spatial data from time series aerial images. *Reports on Geodesy and Geoinformatics.*, *107*(1), 1–12. doi:10.2478/rgg-2019-0001

Alves, D., & Skole, D. (1996). Characterizing land cover dynamics using multi-temporal imagery. *International Journal of Remote Sensing*, *17*(4), 835–839. doi:10.1080/01431169608949049

Amelin, K. S., Antal, E., & Vasiliev, V. (2013). *Adaptive management of an autonomous group of unmanned aerial vehicles*. St. Petersburg State University.

American Society of Heating, Refrigerating, and Air-Conditioning Engineers. (2010). *Standard no 55: Thermal environmental conditions for human occupancy*. Retrieved from http://arco-hvac.ir/wp-content/uploads/2015/11/ASHRAE-55-2010.pdf

Amirebrahimi, S., Quadros, N., Coppa, I., & Keysers, J. (2018). *UAV Data Acquisition in Australia and New Zealand*. FrontierSL.

Amirtham, L. R. (2016). Urbanization and its impact on urban heat island intensity in Chennai Metropolitan Area, India. *Indian Journal of Science and Technology*, *9*(5), 1–8. doi:10.17485/ijst/2016/v9i5/87201

Amirtham, L. R., Devadas, M. D., & Perumal, M. (2009). Mapping of micro-urban heat islands and land cover changes: A case in Chennai City, India. *The International Journal Climate Change: Impacts and Responses*, *1*(2), 71–84. doi:10.18848/1835-7156/CGP/v01i02/37258

Amorim, A.; Pessoa, G. G.; Lemes, I. R.; Komazaki, J. M.; Jorge, L. C.; Galo, M.; Camargo, P. O. (2016). *Utilização de VANT para o georreferenciamento de imóveis rurais*. I Seminário Internacional UAV, Lisboa, Portugal.

Amorim, A., Pelegrina, M. A., & Julião, R. P. (2018). *Cadastro e gestão territorial. Uma visão luso-brasileira para a implementação de sistemas de informação cadastral nos municípios*. Editora UNESP Digital.

Anas, A., Arnott, R., & Small, K. A. (1998). Urban spatial structure. *Journal of Economic Literature*, *36*(3), 1426–1464.

Anderson, J. R., Hardy, E. E., Roach, J. T., & Witmer, R. E. (1976). *A land use and land cover classification system for use with remote sensor data*. Professional Paper., doi:10.3133/pp964

Anderson, K., & Gaston, K. J. (2013). Lightweight unmanned aerial vehicles will revolutionize spatial ecology. *Frontiers in Ecology and the Environment*, *11*(3), 138–146. doi:10.1890/120150

Anderson, T. W., & Goodman, L. A. (1957). Statistical Inference about Markov Chains. *Annals of Mathematical Statistics*, *28*(1), 89–110. doi:10.1214/aoms/1177707039

Andrés, M., & Pozuelo, F. (2009). Evolution of the architectural and heritage representation. *Landscape and Urban Planning*, *91*(2), 105–112. doi:10.1016/j.landurbplan.2008.12.006

Andriotti, J. L. (1988). Introdução à Geoestatística. *Estudos Tecnológicos – Acta Geológica Leopoldensia, São Leopoldo, 11*(27), 5-81. http://rigeo.cprm.gov.br/jspui/handle/doc/14835

Angel, S., Parent, J., & Civco, D. (2007). Urban sprawl metrics: an analysis of global urban expansion using GIS. *Proceedings of ASPRS 2007 Annual Conference*, *7*(11).

Angel, S., Sheppard, S., Civco, D. L., Buckley, R., Chabaeva, A., Gitlin, L., Kraley, A., Parent, J., & Perlin, M. (2005). *The dynamics of global urban expansion*. World Bank, Transport and Urban Development Department.

Ansari, T. A., Katpatal, Y. B., & Vasudeo, A. D. (2016). Spatial evaluation of impacts of increase in impervious surface area on SCS-CN and runoff in Nagpur urban watersheds, India. *Arabian Journal of Geosciences*, *9*(18), 702. doi:10.100712517-016-2702-5

Anselin, L. (1995). Local Indicators of Spatial Association-LISA. Geographical Analysis, 27(2), 93-115. doi:10.1111/j.1538-4632.1995.tb00338.x

Anselin, L., & Getis, A. (1992). Spatial Statistical Analysis and Geographic Information Systems. *The Annals of Regional Science*, *26*(1), 19–33. doi:10.1007/BF01581478

Aplin, P. (2003). Comparison of simulated IKONOS and SPOT HRV imagery for classifying urban areas. In Remotely-Sensed Cities. CRC Press.

Apostolos, L. (2008). Evaluating Urban Sprawl Patterns through Fractal Analysis: The Case of Greek Metropolitan Areas and Issues of Sustainable Development. *Scienze regionali*.

Ardalan, N. (2014). Towards Sustainable Urbanism in the Persian Gulf: Analysis of the Past. *International Journal of Islamic Architecture*, *3*(1), 171–186. doi:10.1386/ijia.3.1.171_1

Areias, B., Humberto, N., Guardalben, L., & Fernandes, J. M. (2018). Towards an Automated Flying Drones Platform. In *Proceedings of the 4th International Conference on Vehicle Technology and Intelligent Transport System* (vol. 1, pp. 529-537). Setúbal, Portugal: Science and Technology Publications. 10.5220/0006792405290536

Arellano, B., Roca, J., & Batlle, E. (2018, September 18). Green areas and urban heat island: Combining remote sensed data with ground observations. *Proc. SPIE 10767. Remote Sensing and Modeling of Ecosystems for Sustainability*, *15*, 1076705.

Arnfield, A. J. (2003). Two decades of urban climate research: A review of turbulence, exchanges of energy and water, and the urban heat island. *International Journal of Climatology*, *23*(1), 1–26. doi:10.1002/joc.859

Artis, D. A., & Carnahan, W. H. (1982). Survey of emissivity variability in thermography of urban areas. *Remote Sensing of Environment*, *12*(4), 313–329. doi:10.1016/0034-4257(82)90043-8

Artmann, M., Inostroza, L., & Fan, P. (2019). Urban sprawl, compact urban development and green cities. How much do we know, how much do we agree? *Ecological Indicators*, *96*(2), 3–9. doi:10.1016/j.ecolind.2018.10.059

Ascione, F., Bianco, N., de' Rossi, F., Turni, G., & Vanoli, G. P. (2013). Green roofs in European climates. Are effective solutions for the energy savings in air-conditioning? *Applied Energy*, *104*, 845–859. doi:10.1016/j.apenergy.2012.11.068

Ashley, W. S., Bentley, M. L., & Stallins, J. A. (2012). Urban-induced thunderstorm modification in the Southeast United States. *Climatic Change*, *113*(2), 481–498. doi:10.100710584-011-0324-1

Asmala, A., & Abdullah, M. (2007). Mapping urban heat island phenomenon: Remote sensing approach. *Journal-The Institution of Engineers, Malaysia*, *68*(3), 25–30.

Asner, G. P., Braswell, B. H., Schimel, D. S., & Wessman, C. A. (1998). Ecological research needs from multiangle remote sensing data. *Remote Sensing of Environment*, *63*(2), 155–165. doi:10.1016/S0034-4257(97)00139-9

ASPRS. (2015). Positional Accuracy Standards for Digital Geospatial Data. *Photogrammetric Engineering and Remote Sensing*, *81*(3), 1–26. doi:10.14358/PERS.81.3.A1-A26

Associação dos Municípios do Pontal do Paranapanema – UNIPONTAL. (2015). *PDEPP – Plano de Desenvolvimento Regional e Local dos Municípios do Pontal do Paranapanema*. Presidente Prudente.

Associação Nacional de Transportes Públicos - ANTP. (2011). *Sistema de Informações da Mobilidade Urbana - Relatório Geral 2011*. São Paulo: ANTP. Retrieved from http://www.antp.org.br/

Associação Nacional de Transportes Públicos - ANTP. (2019). *Environmental Impacts Simulator*. Retrieved from http://www.antp.org.br/simulador-de-impactos-ambientais/sobre-os-resultados.html

Atkinson, A. (2005). Urban development: Reviving and activating Utopian strategies. *City*, *9*(3), 279–295. doi:10.1080/13604810500392548

Atkinson, B. W. (1968). A preliminary examination of the possible effect of London's urban area on the distribution of thunder rainfall 1951-60. *Transactions of the Institute of British Geographers*, (44), 97–118. doi:10.2307/621751

AUAV. (2019). *Types of Drones: Multi - Rotor vs Fixed - Wing vs Single Rotor vs Hybrid VTOL*. https://www.auav.com.au/articles/drone-types/

Aurigi, A., & Odendaal, N. (2020). From "Smart in the Box" to "Smart in the City": Rethinking the Socially Sustainable Smart City in Context. *Journal of Urban Technology*, 1–16. doi:10.1080/10630732.2019.1704203

Austin, R. (2010). *Unmanned Aircraft Systems: UAVs Design, Development and Deployment*. John Wiley & Sons Ltd. doi:10.1002/9780470664797

Avdan, U., & Jovanovska, G. (2016). Algorithm for automated mapping of land surface temperature using LANDSAT 8 satellite data. *Journal of Sensors*, *2016*, 2016. doi:10.1155/2016/1480307

Azevedo, L. (2015). *Geolocalização, Geocolaboração e Redes Sociais. Estudo de caso aplicado ao turismo no Centro Histórico de Coimbra. tokusei* [Master Thesis]. Universidade de Coimbra, Portugal.

Azevedo, M. (1969). *O Plano Director de Urbanização de Lourenço Marques*. Separata do Boletim n. 7. Câmara Municipal de Lourenço Marques.

Badarinath, K. V. S., Chand, T. K., Madhavilatha, K., & Raghavaswamy, V. (2005). Studies on urban heat islands using ENVISAT AATSR data. *Photonirvachak (Dehra Dun)*, *33*(4), 495–501. doi:10.1007/BF02990734

Baik, J. J., Kim, Y. H., & Chun, H. Y. (2001). Dry and moist convection forced by an urban heat island. *Journal of Applied Meteorology*, *40*(8), 1462–1475. doi:10.1175/1520-0450(2001)040<1462:DAMCFB>2.0.CO;2

Bai, X., Dawson, R. J., Ürge-Vorsatz, D., Delgado, G. C., Barau, A. S., Dhakal, S., Dodman, D., Leonardsen, L., Masson-Delmotte, V., Roberts, D., & Schultz, S. (2018). Six research priorities for cities and climate change. *Nature*, *555*(7694), 23–25. doi:10.1038/d41586-018-02409-z

Ball, M. (2019). *High-Resolution SAR Imagery Captured with Small UAS. Unmanned Systems Technology.* https://www.unmannedsystemstechnology.com/2019/02/high-resolution-sar-imagery-captured-with-small-uas/

Balling, R. C. Jr, & Brazel, S. W. (1987). Recent changes in Phoenix, Arizona summertime diurnal precipitation patterns. *Theoretical and Applied Climatology*, *38*(1), 50–54. doi:10.1007/BF00866253

Banai, R., & DePriest, T. (2010). Urban sprawl: Definitions, data, methods of measurement, and environmental consequences. *Journal of Sustainability Education, 7.*

Barba, S., Barbarella, M., Di Benedetto, A., Fiani, M., Gujski, L., & Limongiello, M. (2019). Accuracy Assessment of 3D Photogrammetric Models from an Unmanned Aerial Vehicle. *Drones (Basel)*, *3*(4), 79. doi:10.3390/drones3040079

Barnes, G. (2005). The importance of trip destination in determining transit share. *Journal of Public Transportation Record*, *8*(2), 1–15. doi:10.5038/2375-0901.8.2.1

Barnes, K. B., Morgan, J. M. III, Roberge, M. C., & Lowe, S. (2001). *Sprawl development: Its patterns, consequences, and measurement*. Towson University.

Barow, I., Megenta, M., & Megento, T. (2019). Spatiotemporal analysis of urban expansion using GIS and remote sensing in Jigjiga town of Ethiopia. *Applied Geomatics*, *11*(2), 121–127. doi:10.100712518-018-0245-z

Barrado-Timón, D., & Hidalgo-Giralt, C. (2019). The Historic City, Its Transmission and Perception via Augmented Reality and Virtual Reality and the Use of the Past as a Resource for the Present: A New Era for Urban Cultural Heritage and Tourism? *Sustainability*, *11*(10), 2835. doi:10.3390u11102835

Barra, T. (1989). *Integrated Land Use and Transport Modelling: Decision Chains and Hierarchies (Cambridge Urban and Architectural Studies)*. Cambridge University Press. doi:10.1017/CBO9780511552359

Barros, A. P., & Hwu, W. (2002). A study of land-atmosphere interactions during summertime rainfall using a mesoscale model. *Journal of Geophysical Research, D, Atmospheres*, *107*(D14), 4227. doi:10.1029/2000JD000254

Bat, M. I., Dzhanelidze, G. Yu., & Kelzon, A. S. (2013). Theoretical Mechanics at Examples and Problems (10th ed.). Lan.

Bates, A. J., Sadler, J. P., & Mackay, R. (2013). Vegetation development over four years on two green roofs in the UK. *Urban Forestry & Urban Greening*, *12*(1), 98–108. doi:10.1016/j.ufug.2012.12.003

Battisti, C. (2004). *Frammentazione ambientale, connettività, reti ecologiche: Un contributo teorico e metodologico con particolare riferimento alla fauna selvatica*. Roma: Provincia di Roma, Assessorato alle politiche ambientali, Agricoltura, e Protezione civile.

Batty, M., Chapman, D., Evans, S., Haklay, M., Kueppers, S., Shiode, N., Smith, A., & Torrens, P. M. (2000). *Visualizing the city: communicating urban design to planners and decision-makers*. Working paper series, 26. CASA, UCL.

Batty, M. (2000). The New Urban Geography of the Third Dimension. *Environment and Planning. B, Planning & Design*, *27*(4), 483–484. doi:10.1068/b2704ed

Batty, M. (2005). *Cities and Complexity: Understanding Cities with Cellular Automata, Agent-Based Models, and Fractals*. The MIT Press.

Batty, M. (2013). *The New Science of Cities*. The MIT Press. doi:10.7551/mitpress/9399.001.0001

Batty, M. (2018). *Inventing Future Cities*. MIT Press. doi:10.7551/mitpress/11923.001.0001

Begashaw, S. (2018). *Accuracy of DTM derived from UAV imagery and its effect on canopy height model compared to airborne Lidar in part of tropical rainforest of Berklah, Malaysia* [Unpublished master thesis]. University of Twente, Enschede, The Netherlands.

Behling, R., Bochow, M., Foerster, S., Roessner, S., & Kaufmann, H. (2015). Automated GIS-based derivation of urban ecological indicators using hyperspectral remote sensing and height information. *Ecological Indicators*, *48*, 218–234. doi:10.1016/j.ecolind.2014.08.003

Benenson, I., & M Torrens, P. (2004). *Geosimulation: Automata-Based Modeling of Urban Phenomena*. doi:10.1002/0470020997

Benjamin, J., McCarthy, J., Wiseman, C., Bevin, S., Kowlessar, J., Astrup, P. M., Naumann, J., & Hacker, J. (2019). Integrating Aerial and Underwater Data for Archaeology: Digital Maritime Landscapes in 3D. In 3D Recording and Interpretation for Maritime Archaeology. Coastal Research Library.

Bennett, A. F., Hinsley, S. A., Bellamy, P. E., Swetnam, R. D., & Mac Nally, R. (2004). Do regional gradients in land-use influence richness, composition and turnover of bird assemblages in small woods? *Biological Conservation*, *119*(2), 191–206. doi:10.1016/j.biocon.2003.11.003

Benton-Short, L., Keeley, M., & Rowland, J. (2019). Green infrastructure, green space, and sustainable urbanism: Geography's important role. *Urban Geography*, *40*(3), 330–351. doi:10.1080/02723638.2017.1360105

Berardi, U., GhaffarianHoseini, A. H., & GhaffarianHoseini, A. (2014). State-of-the-art analysis of the environmental benefits of green roofs. *Applied Energy*, *115*, 411–428. doi:10.1016/j.apenergy.2013.10.047

Beretta, F., Shibata, H., Cordova, R., de Lemos Peroni, R., Azambuja, J., & Coimbra-Leite, J. F. (2018). Topographic modelling using UAVs compared with traditional survey methods in mining. *REM - International Engineering Journal*, *71*(3), 463–470. doi:10.1590/0370-44672017710074

Berling-Wolff, S., & Wu, J. (2004). Modeling urban landscape dynamics: A case study in Phoenix, USA. *Urban Ecosystems*, *7*(3), 215–240. doi:10.1023/B:UECO.0000044037.23965.45

Bernardo, H., Antunes, C. H., Gaspar, A., Pereira, L. D., & da Silva, M. G. (2017). An approach for energy performance and indoor climate assessment in a Portuguese school building. *Sustainable Cities and Society*, *30*, 184–194. doi:10.1016/j.scs.2016.12.014

Berry, W., & Feldman, S. (1985). *Multiple Regression in Practice*. doi:10.4135/9781412985208

Bhati, S., & Mohan, M. (2016). WRF model evaluation for the urban heat island assessment under varying land use/land cover and reference site conditions. *Theoretical and Applied Climatology*, *126*(1-2), 385–400. doi:10.100700704-015-1589-5

Bhatta, B. (2009). Analysis of urban growth pattern using remote sensing and GIS: A case study of Kolkata, India. *International Journal of Remote Sensing*, *30*(18), 4733–4746. doi:10.1080/01431160802651967

Bhatta, B., Saraswati, S., & Bandyopadhyay, D. (2010). Urban sprawl measurement from remote sensing data. *Applied Geography (Sevenoaks, England)*, *30*(4), 731–740. doi:10.1016/j.apgeog.2010.02.002

Bianchini, F., & Hewage, K. (2012). Probabilistic social cost-benefit analysis for green roofs: A lifecycle approach. *Building and Environment*, *58*, 152–162. doi:10.1016/j.buildenv.2012.07.005

Bian, T., & Ren, G. (2019). Reassessment of urbanization effect on surface air temperature trends at an urban station of North China. *Meteorology and Atmospheric Physics*, *131*(2), 237–250. doi:10.100700703-017-0568-z

Bibri, S. E. (2019). Data-driven smart sustainable urbanism: The intertwined societal factors underlying its materialization, success, expansion, and evolution. *GeoJournal*, 1–26. doi:10.100710708-019-10061-x

Biljecki, F., Ledoux, H., Stoter, J., & Zhao, J. (2014). Formalisation of the level of detail in 3D city modelling. *Computers, Environment and Urban Systems*, *48*, 1–15. doi:10.1016/j.compenvurbsys.2014.05.004

Biljecki, F., Stoter, J., Ledoux, H., Zlatanova, S., & Çöltekin, A. (2015). Applications of 3D City Models: State of the Art Review. *International Journal Geo-Information ISPRS*, *4*(4), 2842–2889. doi:10.3390/ijgi4042842

Birth, G., & McVey, G. (1968). Measuring the color of growing turf with a reflectance spectroradiometer. *Agronomy Journal*, *60*(6), 640–643. doi:10.2134/agronj1968.00021962006000060016x

Blaschke, T., & Strobl, J. (2001). What's wrong with pixels? Some recent developments interfacing remote sensing and GIS. *GeoBIT/GIS, 6*, 12–17.

Blaschke, T. (2010). Object based image analysis for remote sensing. *ISPRS Journal of Photogrammetry and Remote Sensing*, *65*(1), 2–16. doi:10.1016/j.isprsjprs.2009.06.004

Boavida-Portugal, I., Rocha, J., & Cardoso Ferreira, C. (2016). Exploring the impacts of future tourism development on land use/cover changes. *Applied Geography* (Vol. 77). doi:10.1016/j.apgeog.2016.10.009

Bocquier, P. (2014). World urbanization prospects. The 2014 review. New York: United Nations.

Bognot, J. R., Candidoa, C. G., Blancoa, A. C., & Montelibanoc, J. R. Y. (2018). Building Construction Progress Monitoring Using Unmanned Aerial System (UAS), Low-Cost Photogrammetry, and Geographic Information System (GIS). *ISPRS Annals of the Photogrammetry, Remote Sensing and Spatial Information Sciences*, *4*(2), 41-47.

Bondarev, D., Jafarzadeh, R., & Kozub, A. (2014). Effectiveness of Group Flight of Unmanned Aerial Vehicles. Systems of Information Processing, 6, 9-15.

Bondarev, D., Shmelova, T., & Kucherov, D. (2015). Models of group flights of unmanned aerial vehicles using graph theory. Science and Technology of the Air Forces of the Armed Forces of Ukraine - Scientific Periodicals, 3, 68-75.

Bone, C., Dragicevic, S., & White, R. (2011). Modeling-in-the-middle: Bridging the gap between agent-based modeling and multi-objective decision-making for land use change. *International Journal of Geographical Information Science*, *25*(5), 717–737. doi:10.1080/13658816.2010.495076

Borbora, J., & Das, A. K. (2014). Summertime urban heat island study for Guwahati city, India. *Sustainable Cities and Society*, *11*, 61–66. doi:10.1016/j.scs.2013.12.001

Bornstein, R. D., & Johnson, D. S. (1977). Urban-rural wind velocity differences. *Atmospheric Environment, 11*(7), 597-604.

Bornstein, R., & Lin, Q. (2000). Urban heat islands and summertime convective thunderstorms in Atlanta: Three case studies. *Atmospheric Environment*, *34*(3), 507–516. doi:10.1016/S1352-2310(99)00374-X

Bosch, M., Chenal, J., & Joost, S. (2019). Addressing urban sprawl from the complexity sciences. *Urban Science*, *3*(2), 60. doi:10.3390/urbansci3020060

Bouton, S., Knuper, S. M., Mihov, I., & Swartz, S. (2015). *Urban mobility at a tipping point. Insights on Sustainability*. McKinsey & Company. Retrieved from https://www.mckinsey.com/business-functions/sustainability/our-insights/urban-mobility-at-a-tipping-point

Brandão, C., Cameira, M., Valente, F., Cruz de Carvalho, R., & Paço, T. A. (2017). Wet season hydrological performance of green roofs using native species under Mediterranean climate. *Ecological Engineering*, *102*, 596–611. doi:10.1016/j.ecoleng.2017.02.025

Brasil. (1988). *Constituição da República Federativa do Brasil*. Diário Oficial da União.

Brasil. (2001). *Lei nº 10.257- Estatuto da Cidade*. Diário Oficial da União.

Brasil. (2009). *Lei nº 11.977*. Diário Oficial da União.

Brasil. (2016). *Decreto nº 8.764*. Diário Oficial da União.

Brasil. (2017). *Lei nº 13.465*. Diário Oficial da União.

Brubaker, K. L., & Entekhabi, D. (1996). Analysis of feedback mechanisms in land-atmosphere interaction. *Water Resources Research*, *32*(5), 1343–1357. doi:10.1029/96WR00005

Bruse, M. (2004, March). *ENVI-met 3.0: Updated model overview*. Retrieved from http://www.envi-met.net/documents/papers/overview30.pdf

Bruse, M., & Fleer, H. (1998). Simulating surface–plant–air interactions inside urban environments with a three dimensional numerical model. *Environmental Modelling & Software*, *13*(3), 373–384. doi:10.1016/S1364-8152(98)00042-5

Burian, S. J., & Shepherd, J. M. (2005). Effect of urbanization on the diurnal rainfall pattern in Houston. *Hydrological Processes: An International Journal*, *19*(5), 1089–1103. doi:10.1002/hyp.5647

Burke, L., Reytar, K., Spalding, M., & Perry, A. (2011). Reefs at Risk Revisited. Washington, DC: World Resources Institute (WRI).

Cabral, P., Augusto, G., Tewolde, M., & Araya, Y. (2013). Entropy in urban systems. *Entropy (Basel, Switzerland)*, *15*(12), 5223–5236. doi:10.3390/e15125223

Cabreira, T. M., Brisolara, L. B., & Ferreira, P. R. Jr. (2019). Survey on Coverage Path Planning with Unmanned Aerial Vehicles. *Drones (Basel)*, *3*(1), 4. doi:10.3390/drones3010004

Câmara Municipal de Lisboa. (n.d.). *Escolas Públicas—Secundário*. Retrieved March 26, 2020, from http://geodados.cm-lisboa.pt/datasets/680ce63e07fd4c5eb165b9d47498bef5_4

Campbell, J. B., & Wynne, R. H. (2011). *Introduction to Remote Sensing* (5th ed.). The Guilford Press.

Camps-Calvet, M., Langemeyer, J., Calvet-Mir, L., & Gómez-Baggethun, E. (2016). Ecosystem services provided by urban gardens in Barcelona, Spain: Insights for policy and planning. *Environmental Science & Policy*, *62*, 14–23. doi:10.1016/j.envsci.2016.01.007

Canavarros, A. F. A. (2016). *A Consolidação de um Tipo Urbano e Arquitetônico de Moradia para os Pobres: Velho Modelo, Novas Periferias no Espaço Urbano Mato-Grossense* [Unpublished doctoral dissertation]. Faculdade de Arquitetura, Universidade de Lisboa, Lisbon, Portugal.

Capolupo, A., Saponaro, M., Borgogno Mondino, E., & Tarantino, E. (2020). Combining Interior Orientation Variables to Predict the Accuracy of RPAS–SfM 3D Models. *Remote Sensing*, *12*(17), 2674. doi:10.3390/rs12172674

Cardoso, A. L. (2008). Contextualização/caracterização. In J. Santa Rosa (Ed.), Política habitacional e a integração urbana de assentamentos precários: parâmetros conceituais, técnicos e metodológicos (pp. 13-45). Secretaria Nacional de Habitação. Ministério das Cidades.

Carlson, T. N., & Ripley, D. A. (1997). On the relation between NDVI, fractional vegetation cover, and leaf area index. *Remote Sensing of Environment*, *62*(3), 241–252. doi:10.1016/S0034-4257(97)00104-1

Carmo, R. L., & Anazawa, T. M. (2014). Mortalidade por desastres no Brasil: O que mostram os dados. *Ciencia & Saude Coletiva*, *19*(9), 3669–3681. doi:10.1590/1413-81232014199.07432014 PMID:25184574

Carrus, G., Scopelliti, M., Lafortezza, R., Colangelo, G., Ferrini, F., Salbitano, F., & Sanesi, G. (2015). Go greener, feel better? The positive effects of biodiversity on the well-being of individuals visiting urban and peri-urban green areas. *Landscape and Urban Planning*, *134*, 221–228. doi:10.1016/j.landurbplan.2014.10.022

Caruso, G., Peeters, D., Cavailhes, J., & Rounsevell, M. (2009). Space–Time Patterns of Urban Sprawl, a 1D Cellular Automata and Microeconomic Approach. *Environment and Planning. B, Planning & Design*, *36*(6), 968–988. doi:10.1068/b34009

Carvajal-Ramírez, F., Navarro-Ortega, A., Agüera-Vega, F., Martínez-Carricondo, P., & Mancini, F. (2019). Virtual reconstruction of damaged archaeological sites based on Unmanned Aerial Vehicle Photogrammetry and 3D modelling. Study case of a southeastern Iberia production area in the Bronze Age. *Measurement*, *136*, 225–236. doi:10.1016/j.measurement.2018.12.092

Carvalho, T., Santos, L. B. L., Luz, E. F. P., Ishibashi, R., Jorge, A. A. S., & Londe, L. R. (2016). An Open Source Approach for Watershed's Delimitation to Support Flood Monitoring. *International Journal of Distributed Systems and Technologies, 7*(4), 77–88. doi:10.4018/IJDST.2016100105

Castellano, F. (2018). *Commercial Drones Are Revolutionizing Business Operations.* https://www.toptal.com/finance/market-research-analysts/drone-market

Castelo Branco, L. M. B. (2009). *Análise da arborização em Brasília. O caso da superquadra 308 sul* (Unpublished master's thesis). Universidade de Brasília, Brazil.

Cavion, R., & Lombardo, M. A. (2014). Floods and Urban Design Performance. In *Delft University of Technology, Faculty of Architecture and the Built Environment (Org.), New Urban Configurations - Delta Urbanism: Living with water in the urban deltas (pp. 1030-1035).* IOS Press.

Census of India. (2011). *Haryana, Series-07 Part XII-B, District Census Handbook Faridabad Village and Town Wise Primary Census Abstract (PCA).* Directorate of Census Operations Haryana.

Centro de Estudos e Pesquisas em Engenharia e Defesa Civil – CEPED. (2012). *Atlas Brasileiro de Desastres Naturais 1991 a 2010: volume Brasil. Centro Universitário de Estudos e Pesquisas sobre Desastres. Universidade Federal de Santa Catarina.* CEPED/UFSC.

Chadchan, J., & Shankar, R. (2009). Emerging urban development issues in the context of globalisation. *Journal of ITPI, 6*(2), 78–85.

Chadchan, J., & Shankar, R. (2012). An analysis of urban growth trends in the post-economic reforms period in India. *International Journal of Sustainable Built Environment, 1*(1), 36–49. doi:10.1016/j.ijsbe.2012.05.001

Chadwick, A. J., Goodbody, T. R. H., Coops, N. C., Hervieux, A., Bater, C. W., Martens, L. A., White, B., & Röeser, D. (2020). Automatic Delineation and Height Measurement of Regenerating Conifer Crowns under Leaf-O Conditions Using UAV Imagery. *Remote Sensing, 12*(24), 4104. doi:10.3390/rs12244104

Chang, C. R., Li, M. H., & Chang, S. D. (2007). A preliminary study on the local cool-island intensity of Taipei city parks. *Landscape and Urban Planning, 80*(4), 386–395. doi:10.1016/j.landurbplan.2006.09.005

Chang, H. I., Kumar, A., Niyogi, D., Mohanty, U. C., Chen, F., & Dudhia, J. (2009). The role of land surface processes on the mesoscale simulation of the July 26, 2005 heavy rain event over Mumbai, India. *Global and Planetary Change, 67*(1-2), 87–103. doi:10.1016/j.gloplacha.2008.12.005

Chang, M., Zhu, S., Cao, J., Chen, B., Zhang, Q., Chen, W., Jia, S., Krishnan, P., & Wang, X. (2020). Improvement and Impacts of Forest Canopy Parameters on Noah-MP Land Surface Model from UAV-Based Photogrammetry. *Remote Sensing, 12*(24), 4120. doi:10.3390/rs12244120

Changnon, S. A., Jr. (1981). METROMEX: A review and summary. Meteorological Monograph No. 40. American Meteorological Society.

Changnon, S. A. (2003). Urban modification of freezing-rain events. *Journal of Applied Meteorology*, *42*(6), 863–870. doi:10.1175/1520-0450(2003)042<0863:UMOFE>2.0.CO;2

Changnon, S. A., & Westcott, N. E. (2002). Heavy rainstorms in chicago: Increasing frequency, altered impacts, and future implications 1. *Journal of the American Water Resources Association*, *38*(5), 1467–1475. doi:10.1111/j.1752-1688.2002.tb04359.x

Chapman, S., Watson, J. E., Salazar, A., Thatcher, M., & McAlpine, C. A. (2017). The impact of urbanization and climate change on urban temperatures: A systematic review. *Landscape Ecology*, *32*(10), 1921–1935. doi:10.100710980-017-0561-4

Chauvin, J. P., Glaeser, E., Ma, Y., & Tobio, K. (2017). What is different about urbanization in rich and poor countries? Cities in Brazil, China, India and the United States. *Journal of Urban Economics*, *98*, 17–49. doi:10.1016/j.jue.2016.05.003

Chekunov, E. (2010). Application of the UAV of US Armed Forces in Military Conflicts. *Foreign Military Review, 7*, 41-50.

Chen, L., Li, M., Huang, F., & Xu, S. (2013). Relationships of LST to NDBI and NDVI in Wuhan City based on Landsat ETM+ image. In *2013 6th International Congress on Image and Signal Processing (CISP)* (Vol. 2, pp. 840-845). IEEE.

Chen, B., Chen, Z., Deng, L., Duan, Y., & Zhou, J. (2016). *Building Change Detection with RGB-D Map Generated from UAV Images*. Academic Press.

Chen, J. M., & Black, T. A. (1992). Defining leaf area index for non-flat leaves. *Plant, Cell & Environment*, *45*(4), 421–429. doi:10.1111/j.1365-3040.1992.tb00992.x

Chen, W. Y. (2015). The role of urban green infrastructure in offsetting carbon emissions in 35 major Chinese cities: A nationwide estimate. *Cities (London, England)*, *44*, 112–120. doi:10.1016/j.cities.2015.01.005

Chen, W., Huang, H., Dong, J., Zhang, Y., Tian, Y., & Yang, Z. (2018). Social functional mapping of urban green space using remote sensing and social sensing data. *ISPRS Journal of Photogrammetry and Remote Sensing*, *146*, 436–452. doi:10.1016/j.isprsjprs.2018.10.010

Chen, X., Li, B. L., & Allen, M. F. (2010). Characterizing urbanization, and agricultural and conservation land-use change in Riverside County, California, USA. *Annals of the New York Academy of Sciences*, *1195*(1), E164–E176. doi:10.1111/j.1749-6632.2009.05403.x PMID:20586769

Chen, Y., & Yu, S. (2016). Assessment of urban growth in Guangzhou using multi-temporal, multi-sensor Landsat data to quantify and map impervious surfaces. *International Journal of Remote Sensing*, *37*(24), 5936–5952. doi:10.1080/01431161.2016.1252473

Chiesura, A. (2004). The role of urban parks for the sustainable city. *Landscape and Urban Planning*, *68*(1), 129–138. doi:10.1016/j.landurbplan.2003.08.003

Ching, J., Mills, G., Bechtel, B., See, L., Feddema, J., Wang, X., ... Mouzourides, P. (2018). WUDAPT: An urban weather, climate, and environmental modeling infrastructure for the anthropocene. *Bulletin of the American Meteorological Society*, *99*(9), 1907–1924. doi:10.1175/BAMS-D-16-0236.1

Choudhary, G., Sharna, V., You, I., Yim, K., Chen, I. R., & Cho, J. H. (2018). Intrusion Detection Systems for Networked Unmanned Aerial Vehicles: A Survey. In *Proceedings of 2018 14th International Wireless Communications & Mobile Computing Conference (IWCMC)* (vol. 1, pp. 560-565). 10.1109/IWCMC.2018.8450305

Chouhan, H. A., Parthasarathy, D., & Pattanaik, S. (2018). Urban at the Edges: Mumbai's Coastline Urbanisms. Sustainable Urbanization in India, 279-293.

Christen, A., & Vogt, R. (2004). Energy and radiation balance of a central European city. *International Journal of Climatology: A Journal of the Royal Meteorological Society, 24*(11), 1395-1421.

Chrysoulakis, N., Feigenwinter, C., Triantakonstantis, D., Penyevskiy, I., Tal, A., Parlow, E., Fleishman, G., Düzgün, S., Esch, T., & Marconcini, M. (2014). A Conceptual List of Indicators for Urban Planning and Management Based on Earth Observation. *ISPRS International Journal of Geo-Information*, *3*(3), 980–1002. doi:10.3390/ijgi3030980

Chun, B., & Guldmann, J. M. (2014). Spatial statistical analysis and simulation of the urban heat island in high-density central cities. *Landscape and Urban Planning*, *125*, 76–88. doi:10.1016/j.landurbplan.2014.01.016

Cimoli, E. (2015). *Determining Snow Depth Distribution from Unmanned Aerial Vehicles and Digital Photogrammetry* [Unpublished master thesis]. Technical University of Denmark. Lyngby, Denmark.

Ciommi, M., Chelli, F. M., Carlucci, M., & Salvati, L. (2018). Urban Growth and Demographic Dynamics in Southern Europe: Toward a New Statistical Approach to Regional Science. *Sustainability*, *10*(8), 2765. doi:10.3390u10082765

Clarke, K., Parks, B., & Crane, M. (2002). *Geographic Information Systems and Environmental Modeling*. Prentice Hall.

Cobbinah, P. B., Erdiaw-Kwasie, M. O., & Amoateng, P. (2015). Rethinking sustainable development within the framework of poverty and urbanisation in developing countries. *Environmental Development*, *13*, 18–32. doi:10.1016/j.envdev.2014.11.001

Cohen, J. (1960). A Coefficient of Agreement for Nominal Scales. *Educational and Psychological Measurement*, *20*(1), 37–46. doi:10.1177/001316446002000104

Colaninno, N. (2016). *Semi-automatic land cover classification and urban modelling based on morphological features. Remote sensing, geographical information systems, and urban morphology: defining models of land occupation along the Mediterranean side of Spain* (doctoral thesis). Technical University of Catalonia. Barcelona. Upccommons. https://upcommons.upc.edu/handle/2117/96372

Colomina, I., & Molina, P. (2014). Unmanned Aerial Systems for Photogrammetry and Remote Sensing: A Review. *ISPRS Journal of Photogrammetry and Remote Sensing.*

Colomina, I., & Molina, P. (2014). Unmanned aerial systems for photogrammetry and remote sensing: A review. *ISPRS Journal of Photogrammetry and Remote Sensing*, *92*, 79–97. doi:10.1016/j.isprsjprs.2014.02.013

CONFEA, CREA, ABNT, MINISTÉRIO DAS CIDADES (2017). *Grupo Técnico Operacional. Norma de Engenharia para Georreferenciamento Urbano no Brasil.* Normas e Procedimentos de Engenharia para Cadastro Urbano no Brasil.

Connop, S., Vandergert, P., Eisenberg, B., Collier, M. J., Nash, C., Clough, J., & Newport, D. (2016). Renaturing cities using a regionally-focused biodiversity-led multifunctional benefits approach to urban green infrastructure. *Environmental Science & Policy*, *62*, 99–111. doi:10.1016/j.envsci.2016.01.013

Conolly, J., & Lake, M. (2006). Geographical Information Systems in Archaeology. *Cambridge Manuals in Archaeology*. doi:10.1017/CBO9780511807459

Conselho Municipal de Maputo. (2008). *Plano de Estrutura Urbana do Município de Maputo (PEUMM)*. Conselho Municipal de Maputo.

Corrêa, R. L. (1999). *O espaço urbano*. Ática.

Costa, E. C. (1974). Física aplicada à construção: Conforto térmico (2nd ed.). São Paulo: Blücher.

Costales-Acurio, G. A. (2018). *Análisis comparativo entre los software de prueba Agisoft Photoscan y Pix4D para el procesamiento de datos obtenidos con fotogrametría de vehículo aéreo no tripulado (UAV) de bajo costo aplicado a proyectos de medio ambiente* [Unpublished undergraduate thesis]. Escuela Politécnica Nacional de Ecuador. Quito, Ecuador.

Couclelis, H. (2005). "Where has the future gone?" Rethinking the role of integrated land-use models in spatial planning. *Environment & Planning A*, *37*(8), 1353–1371. doi:10.1068/a3785

Crane, A. (2018). *From Start-up to Empire: The Evolution of DJI Drones.* https://dronelife.com/2018/05/14/from-startup-to-empire-the-evolution-of-dji-drones/

Cristóbal, J., Jiménez-Muñoz, J. C., Sobrino, J. A., Ninyerola, M., & Pons, X. (2009). Improvements in land surface temperature retrieval from the Landsat series thermal band using water vapor and air temperature. *Journal of Geophysical Research, D, Atmospheres*, *114*(D8), D08103. doi:10.1029/2008JD010616

Crommelinck, S., Bennett, R., Gerke, M., Nex, F., Yang, M., & Vosselman, G. (2016). Review of Automatic Feature Extraction from High-Resolution Optical Sensor Data for UAV-Based Cadastral Mapping. *Remote Sensing*, *8*(8), 689. doi:10.3390/rs8080689

Crommelinck, S., Höfle, B., Koeva, M., Yang, M., & Vosselman, G. (2018). Interactive cadastral boundary delineation from UAV data. *ISPRS Annals of the Photogrammetry. Remote Sensing and Spatial Information Sciences*, *4*, 81–88.

Cruz Josa, D., Sacristán, M. A., & Mier Gutiérrez, S. (2017). Sistemas LIDAR embarcados en Drones - TerraSystem LidarPod. In L. A. Ruiz, J. Estornell, & M. Erena (Eds.), *Teledetección. Nuevas plataformas y sensores aplicados a la gestión del agua, la agricultura y el medio ambiente*. UPV.

Cruzan, M. B., Weinstein, B. G., Grasty, M. R., Kohrn, B. F., Hendrickson, E. C., Arredondo, T. M., & Thompson, P. G. (2016). Small Unmanned Aerial Vehicles (Micro-Uavs, Drones) in Plant Ecology. *Applications in Plant Sciences*, *4*(9), 1600041. doi:10.3732/apps.1600041 PMID:27672518

Cruz, M. F., & Fonseca, F. C. P. (2018). Vetores em contradição: Planejamento da mobilidade urbana, uso do solo e dinâmicas do capitalismo contemporâneo. *Cadernos Metrópole*, *20*(42), 553–576. doi:10.1590/2236-9996.2018-4212

Cutsinger, J., Galster, G., Wolman, H., Hanson, R., & Towns, D. (2005). Verifying the multi-dimensional nature of metropolitan land use: Advancing the understanding and measurement of sprawl. *Journal of Urban Affairs*, *27*(3), 235–259. doi:10.1111/j.0735-2166.2005.00235.x

Dadashpoor, H., Azizi, P., & Moghadasi, M. (2019). Land use change, urbanization, and change in landscape pattern in a metropolitan area. *The Science of the Total Environment*, *655*, 707–719. doi:10.1016/j.scitotenv.2018.11.267 PMID:30476851

Dadhich, P. N., & Hanaoka, S. (2011). Spatio-temporal Urban Growth Modeling of Jaipur, India. *Journal of Urban Technology*, *18*(3), 45–65. doi:10.1080/10630732.2011.615567

Dai, A., Trenberth, K. E., & Karl, T. R. (1999). Effects of clouds, soil moisture, precipitation, and water vapor on diurnal temperature range. *Journal of Climate*, *12*(8), 2451–2473. doi:10.1175/1520-0442(1999)012<2451:EOCSMP>2.0.CO;2

Dallimer, M., Tang, Z., Bibby, P. R., Brindley, P., Gaston, K. J., & Davies, Z. G. (2011). Temporal changes in greenspace in a highly urbanized region. *Biology Letters*, *7*(5), 763–766. doi:10.1098/rsbl.2011.0025 PMID:21429910

Dandois, J. P., Olano, M., & Ellis, E. C. (2015). Optimal Altitude, Overlap, and Weather Conditions for Computer Vision UAV Estimates of Forest Structure. *Remote Sensing*, *7*(10), 13895–13920. doi:10.3390/rs71013895

Dasgupta, A., Kumar, U., & Ramachandra, T. V. (2009). Urban Landscape analysis through spatial metrics. *Proceedings of International Conference on Infrastructure, Sustainable Transportation and Urban Planning,(CISTUP@ CiSTUP)*, 18-20.

De Lima, R. L. P., Boogaard, F. C., De Graaf, R., Dionisio Pires, M., & Sazonov, V. (2015). Monitoring the Impacts of Floating Structures on the Water Quality and Ecology Using an Underwater Drone. *E-Proceedings of the 36th IAHR World Congress.*

De Soto, H. (2001). O Mistério do Capital. *Record.*

Demuzere, M., Orru, K., Heidrich, O., Olazabal, E., Geneletti, D., Orru, H., Bhave, A. G., Mittal, N., Feliu, E., & Faehnle, M. (2014). Mitigating and adapting to climate change: Multi-functional and multi-scale assessment of green urban infrastructure. *Journal of Environmental Management, 146*, 107–115. doi:10.1016/j.jenvman.2014.07.025 PMID:25163601

Denaldi, R. (Ed.). (2013). *Planejamento habitacional: notas sobre precariedade e terra nos planos locais de habitação.* Annablume.

Departamento Nacional de Trânsito - DENATRAN. (2019). *Frota de veículos automotores licenciados por município.* Retrieved from http://www.denatran.gov.br/index.php/estatistica/237-frota-veiculos

Department of Human Services. (2009). *January 2009 heatwave in Vicoria: an assessment of health imapcts.* Victorian Government Department of Human Services, Melbourne, Victoria. Retrieved from: https://www2.health.vic.gov.au/about/publications/researchandreports/January-2009-Heatwave-in-Victoria-an-Assessment-of-Health-Impacts

Dergachov, K., & Kulik, A. (2019). Ensuring the Safety of UAV Flights by Means of Intellectualization of Control Systems. In *Cases on Modern Computer Systems in Aviation* (pp. 287–310). IGI Global.

Dergachov, K., Kulik, A., & Zymovin, A. (2019). Environments Diagnosis by Means of Computer Vision System of Autonomous Flying Robots. In *Automated Systems in the Aviation and Aerospace Industries* (pp. 115–137). IGI Global.

DESA. (2018). *Revision of world urbanization prospects.* Population Division of the UN Department of Economic and Social Affairs. https://population. un.org/wup

Deuskar, C., Baker, J. L., & Mason, D. (2015). *East Asia's changing urban landscape: Measuring a decade of spatial growth.* Academic Press.

Dewan, A. M., & Yamaguchi, Y. (2009). Using remote sensing and GIS to detect and monitor land use and land cover change in Dhaka Metropolitan of Bangladesh during 1960–2005. *Environmental Monitoring and Assessment, 150*(1-4), 237–249. doi:10.100710661-008-0226-5 PMID:18317939

DGT. (2018). *Especificações Técnicas da Carta de Uso e Ocupação do Solo de Portugal Continental para 1995, 2007, 2010 e 2015.* Retrieved from http://mapas.dgterritorio.pt/atom-dgt/pdf-cous/COS2015/ET-COS-1995-2007-2010-2015.pdf

Dhainaut, J., Claessens, Y., Gisnburg, C., & Riou, B. (2004). Unprecedent heat-related death during the 2003 heat wave in Paris: Consequences on emergency departments. *Critical Care (London, England), 8*(1), 1–2. doi:10.1186/cc2404 PMID:14975035

DiBari, J. N. (2007). Evaluation of five landscape-level metrics for measuring the effects of urbanization on landscape structure: The case of Tucson, Arizona, USA. *Landscape and Urban Planning*, *79*(3-4), 308–313. doi:10.1016/j.landurbplan.2006.04.004

Dickinson, R. E. (1995). Land-atmosphere interaction. *Reviews of Geophysics*, *33*(S2), 917–922. doi:10.1029/95RG00284

Dieleman, F., & Wegener, M. (2004). Compact city and urban sprawl. *Built Environment*, *30*(4), 308–323. doi:10.2148/benv.30.4.308.57151

Diem, J. E., & Brown, D. P. (2003). Anthropogenic impacts on summer precipitation in central Arizona, USA. *The Professional Geographer*, *55*(3), 343–355.

Dimopoulou, E. (2017). *GR_LADM*. The 6th Land Administration Domain Model (LADM) Workshop, Delft, The Netherlands.

Dimri, A. P. (2019). Comparison of regional and seasonal changes and trends in daily surface temperature extremes over India and its subregions. *Theoretical and Applied Climatology*, *136*(1-2), 265–286. doi:10.100700704-018-2486-5

Ding, H., & Shi, W. (2013). Land-use/land-cover change and its influence on surface temperature: A case study in Beijing City. *International Journal of Remote Sensing*, *34*(15), 5503–5517. doi:10.1080/01431161.2013.792966

Direção-Geral do Território. (n.d.). *DGTerritório—Ortofotos Digitais*. Retrieved March 26, 2020, from https://www.dgterritorio.pt/cartografia_e_geodesia/cartografia/cartografia_de_base___topografica_e_topografica_de_imagem/ortofotocartografia/ortofotocartografia_pos_2004/

Dixon, P. G., & Mote, T. L. (2003). Patterns and causes of Atlanta's urban heat island–initiated precipitation. *Journal of Applied Meteorology*, *42*(9), 1273–1284. doi:10.1175/1520-0450(2003)042<1273:PACOAU>2.0.CO;2

DJI. (2018). *You Can Film Like A Pro with DJI Drone Active Track*. https://store.dji.com/guides/film-like-a-pro-with-activetrack/

DJI. (2019a). *Phantom 4 Specs*. https://www.dji.com/phantom-4/info

DJI. (2019b). *Phantom 3 Standard*. https://www.dji.com/phantom-3-standard/info

Djimantoro, M. I., & Suhardjanto, G. (2017). The Advantage by Using Low-Altitude UAV for Sustainable Urban Development Control. In *Proceedings of The International Conference on Eco Engineering Development 2017 (ICEED 2017)*. IOP Publishing. 10.1088/1755-1315/109/1/012014

Doan, Q. V., Kusaka, H., & Ho, Q. B. (2016). Impact of future urbanization on temperature and thermal comfort index in a developing tropical city: Ho Chi Minh City. *Urban Climate*, *17*, 20–31. doi:10.1016/j.uclim.2016.04.003

Dobbs, C., Kendal, D., & Nitschke, C. R. (2014). Multiple ecosystem services and disservices of the urban forest establishing their connections with landscape structure and sociodemographics. *Ecological Indicators*, *43*, 44–55. doi:10.1016/j.ecolind.2014.02.007

Dobronravov, V. V., Nikitin, I. N., & Dvornikov, A. L. (1974). *Course in Theoretical Mechanics*. Vysshaya Shkola.

Doi, K., & Kii, M. (2012). Looking at sustainable urban mobility through a cross-assessment model within the framework of land-use and transport integration. *International Association of Traffic and Safety Sciences -. IATSS Research*, *35*(2), 62–70. doi:10.1016/j.iatssr.2012.02.004

Dolgikh, S. (2018). Spontaneous Concept Learning with Deep Autoencoder. International Journal of Computational Intelligence Systems, 12(1), 1-12.

Doll, C., Trinks, C., Sedlacek, N., Pelikan, V., Comes, T., & Schultmann, F. (2014). Adapting rail and road networks to weather extremes: Case studies for southern Germany and Austria. *Natural Hazards*, *72*(1), 63–85. doi:10.100711069-013-0969-3

Döllner, J. (2020). Geospatial Artificial Intelligence: Potentials of Machine Learning for 3D Point Clouds and Geospatial Digital Twins. *Journal of Photogrammetry, Remote Sensing and Geoinformation Science*, *2020*(1), 15–24. Advance online publication. doi:10.100741064-020-00102-3

Dong, T., Jiao, L., Xu, G., Yang, L., & Liu, J. (2019). Towards sustainability? Analyzing changing urban form patterns in the United States, Europe, and China. *The Science of the Total Environment*, *671*, 632–643. doi:10.1016/j.scitotenv.2019.03.269 PMID:30939316

Dovey, K., Shafique, T., van Oostrum, M., & Chatterjee, I. (in press). Informal Settlements is not a Euphemism for 'Slum'. *International Development Planning Review*.

Drone life. Urban Air Mobility. (2018). https://dronelife.com/2018/11/28/urban-air-mobility-the-first-uic2-forum-at-amsterdam-drone-week-shows-europes-commitment-to-smart-cities/

DroneApps. (2019). *Price wars: counting the cost of drones, planes and satellites.* https://droneapps.co/price-wars-the-cost-of-drones-planes-and-satellites/

DroneDeploy. (2017a). *Seven Ways to Improve the Accuracy of Your Drone Maps.* https://blog.dronedeploy.com/seven-ways-to-improve-the-accuracy-of-your-drone-maps-9d3161f05956

DroneDeploy. (2017b). *What Are Ground Control Points (GCPs) and How Do I Use Them.* https://blog.dronedeploy.com/what-are-ground-control-points-gcps-and-how-do-i-use-them-4f4c3771fd0b

DroneDeploy. (2017c). *How Do I Use Ground Control Points? A guide to using ground control points with drone mapping software.* https://www.dronedeploy.com/blog/what-are-ground-control-points-gcps/

DroneDeploy. (2019). *GCP Checkpoints.* https://support.dronedeploy.com/docs/gcp-checkpoints-1

Droste, A. M., Steeneveld, G. J., & Holtslag, A. A. (2018). Introducing the urban wind island effect. *Environmental Research Letters*, *13*(9), 094007. doi:10.1088/1748-9326/aad8ef

Drummond, J. (1995). Positional Accuracy. In S. C. Guptill & J. L. Morrison (Eds.), *Elements of spatial data quality. International Cartographic Association* (pp. 31–58). Elsevier Science. doi:10.1016/B978-0-08-042432-3.50010-0

Duan, S.-B., Li, Z.-L., Wang, C., Zhang, S., Tang, B.-H., Leng, P., & Gao, M.-F. (2019). Land-surface temperature retrieval from Landsat 8 single- channel thermal infrared data in combination with NCEP reanalysis data and ASTER GED product. *International Journal of Remote Sensing*, *40*(5-6), 1763–1778. doi:10.1080/01431161.2018.1460513

Du, C., Ren, H., Qin, Q., Meng, J., & Zhao, S. (2015). A practical split-window algorithm for estimating land surface temperature from Landsat 8 data. *Remote Sensing*, *7*(1), 647–665. doi:10.3390/rs70100647

Dunnet, N., & Kingsbury, N. (2008). *Planting green roofs and living walls*. Timber Press.

Dutta, V. (2012). Land use dynamics and peri-urban growth characteristics: Reflections on master plan and urban suitability from a sprawling north Indian city. *Environment & Urbanization Asia*, *3*(2), 277–301. doi:10.1177/0975425312473226

Eastman, C., Teicholz, P., Sacks, R., & Liston, K. (2011). BIM Handbook: A Guide to Building Information Modeling for Owners, Managers, Designers, Engineers and Contractors (2nd ed.). John Wiley & Sons Inc.

Eastman, J. (2009). *IDRISI Taiga Tutorial*. Graduate School of Geography, Clark University.

Eastman, J. R., Kyem, P., & Toledano, J. (1993). A procedure for multi-objective decision making in GIS under conditions of conflicting objectives. *European Conference on Geographical Information Systems*.

Eckert, G., Cassidy, S., Tian, N., & Shabana, M. E. (2019). Using Aerial Drone Photography to Construct 3D Models of Real World Objects in an Effort to Decrease Response Time and Repair Costs Following Natural Disasters. In K. Arai & S. Kapoor (Eds.), *Advances in Computer Vision* (pp. 317–325). Springer International Publishing.

EEA. (2017). *Landscapes in transition, An account of 25 years of land cover change in Europe. EEA Report No /10/2017*. Publications Office of the European Union.

Eidsvig, U. M. K., Kristensen, K., & Vangelsten, B. V. (2017). Assessing the risk posed by natural hazards to infrastructures. *Natural Hazards and Earth System Sciences*, *17*(3), 481–504. doi:10.5194/nhess-17-481-2017

Elmqvist, T., Setälä, H., Handel, S., van der Ploeg, S., Aronson, J., Blignaut, J., Gómez-Baggethun, E., Nowak, D., Kronenberg, J., & de Groot, R. (2015). Benefits of restoring ecosystem services in urban areas. *Current Opinion in Environmental Sustainability*, *14*, 101–108. doi:10.1016/j.cosust.2015.05.001

Engel, V. L., Fonseca, R. C. B., & Oliveira, R. E. (1998). Ecologia de lianas e o manejo de fragmentos florestais. *Série Técnica IPEF*, *12*(32), 43–64.

Enqvist, J. P., Tengö, M., & Bodin, Ö. (2019). Are bottom-up approaches good for promoting social–ecological fit in urban landscapes? *Ambio*, 1–13. PMID:30879271

EPA - United States Environmental Protection Agency. (2014). *The Economic Benefits of Green Infrastructure: A Case Study of Lancaster, PA* (EPA 8 00-R-14-007). EPA.

Epstein, J., Payne, K., & Kramer, E. (2002). Techniques for mapping suburban sprawl. *Photogrammetric Engineering and Remote Sensing*, *68*(9), 913–918.

Epstein, L. A., Bruner, E. M., Horowitz, M. M., Little, K. L., McCall, D. F., Mayer, P., Miner, H., Plotnicov, L., Schwab, W. B., & Schack, W. A. (1967). Urbanization and Social Change in Africa. *Current Anthropology*, *8*(4), 275–28410. doi:10.1086/200808

Esbah, H. (2007). Land use trends during rapid urbanization of the city of Aydin, Turkey. *Environmental Management*, *39*(4), 443–459. doi:10.100700267-005-0331-y PMID:17265112

Estanqueiro, R. (2011). *Análise espacial da forma urbana em Sistemas de Informação Geográfica* (PhD Thesis). Universidade NOVA de Lisboa, Faculdade de Ciências Sociais e Humanas. http://hdl.handle.net/10362/7300

Estoque, R. C., Murayama, Y., & Myint, S. W. (2017). Effects of landscape composition and pattern on land surface temperature: An urban heat island study in the megacities of Southeast Asia. *The Science of the Total Environment*, *577*, 349–359. doi:10.1016/j.scitotenv.2016.10.195 PMID:27832866

Eumorfopoulou, E. A., & Kontoleon, K. J. (2009). Experimental approach to the contribution of plant-covered walls to the thermal behaviour of building envelopes. *Building and Environment*, *44*(5), 1024–1038. doi:10.1016/j.buildenv.2008.07.004

EUROCONTROL. (2019). *Artificial intelligence: a new era for aviation*. Skyway.

Fabrizi, R., Bonafoni, S., & Biondi, R. (2010). Satellite and ground-based sensors for the urban heat island analysis in the city of Rome. *Remote Sensing*, *2*(5), 1400–1415. doi:10.3390/rs2051400

Fan, C., Myint, S., Kaplan, S., Middel, A., Zheng, B., Rahman, A., Huang, H. P., Brazel, A., & Blumberg, D. (2017). Understanding the impact of urbanization on surface urban heat islands—A longitudinal analysis of the oasis effect in subtropical desert cities. *Remote Sensing*, *9*(7), 672. doi:10.3390/rs9070672

Fan, F., Wang, Y., Qiu, M., & Wang, Z. (2009). Evaluating the temporal and spatial urban expansion patterns of Guangzhou from 1979 to 2003 by remote sensing and GIS methods. *International Journal of Geographical Information Science*, *23*(11), 1371–1388. doi:10.1080/13658810802443432

Fanger, P. O. (1972). *Thermal comfort analysis and applications in environmental engineering*. McGraw-Hill.

FAO. (2020). *Air surveys and cadastral maps*. Food and Agriculture Organization of the United Nations. http://www.fao.org/3/V4860E/V4860E04.htm

Fernandes, E. (2011). *Regularização de assentamentos informais na América Latina*. Lincoln Institute of Land Policy.

FGDC. (1998). *Geospatial Positioning Accuracy Standards. Part 3: National Standard for Spatial Data Accuracy*. Federal Geographic Data Committee of the United States of America.

Floreano, D., & Wood, R. J. (2015). Science, technology and the future of small autonomous drones. *Nature*, *521*(7553), 460–466. doi:10.1038/nature14542 PMID:26017445

Flybase. (2020). *Drone Automation Solution for Security & Surveillance*. https://flytbase.com/drone-security-solution/

Flynt, J. (2019). *DJI Tripod Mode: How it works and what it is?* https://3dinsider.com/dji-tripod-mode/

Forman, R. T. T., & Godron, M. (1986). *Landscape ecology*. John Wiley.

Fradkov, A., & Grigoriev, G. (2013). Decentral adaptive control of synchronization of networks of dynamical systems of dynamic systems with limited perturbations. *Automation and Telemechanics, 5*.

Francis, R. A., & Lorimer, J. (2011). Urban reconciliation ecology: The potential of living roofs and walls. *Journal of Environmental Management*, *92*(6), 1429–1437. doi:10.1016/j.jenvman.2011.01.012 PMID:21306818

Frankenstein, S., & Skahill, B. (2004). *FASST Soil Moisture Prediction Error at Yuma, Arizona* (ERDC/CRREL SR-04-1). Engineer Research and Development Center. https://www.researchgate.net/publication/235096218_FASST_Soil_Moisture_Prediction_Error_at_Yuma_Arizona

Freire, S., Santos, T., Navarro, A., Soares, F., Silva, J. D., Afonso, N., Fonseca, A., & Tenedório, J. (2014). Introducing mapping standards in the quality assessment of buildings extracted from very high resolution satellite imagery. *ISPRS Journal of Photogrammetry and Remote Sensing*, *90*, 1–9. doi:10.1016/j.isprsjprs.2013.12.009

Freire, S., Santos, T., & Tenedório, J. A. (2009). Recent urbanization and land use/land cover change in Portugal— The influence of coastline and coastal urban centers. *Journal of Coastal Research*, 1499–1503. http://www.jstor.org/stable/25738039

Freitas, C., & Lima, L. (2012). *Cadastro Territorial i SIG no Processo de Regularização fundiária: o caso da ZEIS do Planalto Pici em Fortaleza*. IV Simpósio Brasileiro de Ciências Geodésicas e Tecnologias da Geoinformação. Universidade Federal de Pernambuco – UFPE.

Friedmann, J., & Sorensen, A. (2019). City unbound: Emerging mega-conurbations in Asia. *International Planning Studies*, *24*(1), 1–12. doi:10.1080/13563475.2019.1555314

Fuad, N. A., Ismail, Z., Majid, Z., Darwin, N., Ariff, M. F. M., Idris, K. M., & Yussof, A. R. (2013). Accuracy evaluation of digital terrain model based on different flying altitudes and conditional of terrain using UAV LiDAR technology. In *Proceedings of IOP Conference Series: Earth and Environmental Science* (vol. 169, pp. 1-8). 10.1088/1755-1315/169/1/012100

Fujibe, F., Togawa, H., & Sakata, M. (2009). Long-term change and spatial anomaly of warm season afternoon precipitation in Tokyo. *Scientific Online Letters on the Atmosphere, 5*, 17–20. doi:10.2151ola.2009-005

Fundação Sistema Estadual de Análise de Dados – SEADE. (2016). *Gross Domestic Product (GDP) from municipalities of São Paulo, Brazil (XLS table)*. Retrieved from https://www.seade.gov.br/produtos/pib-municipal/

Fu, W., Ma, J., Chen, P., & Chen, F. (2020). Remote Sensing Satellites for Digital Earth. In H. Guo, M. F. Goodchild, & A. Annoni (Eds.), *Manual of Digital Earth* (pp. 55–123). Springer., doi:10.1007/978-981-32-9915-3_3

Gabriel, K. M. A., & Endlicher, W. R. (2011). Urban and rural mortality rates during heat waves in Berlin and Brandenburg, Germany. *Environment and Pollution, 159*(8-9), 2044–2050. doi:10.1016/j.envpol.2011.01.016 PMID:21295389

Gallacher, D. (2015). Drone Applications for Environmental Management in Urban Spaces: A Review. *International Journal of Sustainable Land Use and Urban Planning, 3*(4), 1-14.

Gallacher, D. (2016). Drone applications for environmental management in urban spaces: A review. *International Journal of Sustainable Land Use and Urban Planning, 3*(4).

Gallo, K., Hale, R., Tarpley, D., & Yu, Y. (2011). Evaluation of the relationship between air and land surface temperature under clear-and cloudy-sky conditions. *Journal of Applied Meteorology and Climatology, 50*(3), 767–775. doi:10.1175/2010JAMC2460.1

Galster, G., Hanson, R., Ratcliffe, M. R., Wolman, H., Coleman, S., & Freihage, J. (2001). Wrestling sprawl to the ground: Defining and measuring an elusive concept. *Housing Policy Debate, 12*(4), 681–717. doi:10.1080/10511482.2001.9521426

Ganeshan, M., Murtugudde, R., & Imhoff, M. L. (2013). A multi-city analysis of the UHI-influence on warm season rainfall. *Urban Climate, 6*, 1–23. doi:10.1016/j.uclim.2013.09.004

Ganin, S., & Karpenko, A. (1999). Unmanned aerial vehicles. Academic Press.

Garcia-Haro, A., & Arellano, B. (2018). Isla de frío de los parques urbanos de Barcelona. Estudio de caso del Turó Parc y el Parc del centre del Poblenou. *Proceedings, CTV 2018. XII Congreso Internacional Ciudad y Territorio Virtual. CPSV, 2018.*

García-Palomares, J. C., Gutierrez, J., & Mínguez, C. (2015). Identification of tourist hot spots based on social networks: A comparative analysis of European metropolises using photo-sharing services and GIS. *Applied Geography, Elsevier, 63*, 408–417. doi:10.1016/j.apgeog.2015.08.002

GCOS. (2011). *Systematic Observation Requirements for Satellite-Based Products for Climate, 2011 Update, Supplemental Details to the Satellite-Based Com-ponent of the Implementation Plan for the Global Observing System for Climate in Support of the UNFCCC (2010 Update).* World Meteorological Organization (WMO).

Gedzelman, S. D., Austin, S., Cermak, R., Stefano, N., Partridge, S., Quesenberry, S., & Robinson, D. A. (2003). Mesoscale aspects of the urban heat island around New York City. *Theoretical and Applied Climatology*, *75*(1-2), 29–42. doi:10.100700704-002-0724-2

Geoghegan, J., Pritchard, L., Jr., Ogneva-Himmelberger, Y., Chowdhury, R. R., Sanderson, S., & Turner, B. L. (1998). “Socializing the Pixel” and “Pixelizing the Social in Land-Use Land-Cover Change”. In D. Liverman, E. Moran, R. Rindfuss, & P. Stern (Eds.), People and Pixels. Linking Remote Sensing and Social Science (pp. 51-69). National Academic Press.

Gerke, M. (2018). Developments in UAV-Photogrammetry. *Journal of Digital Landscape Architecture*, *3*, 262–272. doi:10.14627/537642028

Gerke, M., & Przybilla, H. J. (2016). Accuracy Analysis of Photogrammetric UAV Image Blocks: Influence of Onboard RTK-GNSS and Cross Flight Patterns. *Photogrammetrie, Fernerkundung, Geoinformation*, *2016*(1), 17–30. doi:10.1127/pfg/2016/0284

Germen, M. (2016). *Alternative cityscape visualisation: drone shooting as a new dimension in urban photography*. Academic Press.

Getis, A., & Ord, K. (1992). The Analysis of Spatial Association by Use of Distance Statistics. Geographical Analysis, 24, 189-206. doi:10.1111/j.1538-4632.1992.tb00261.x

Getis, A. (1991). Spatial Interaction and Spatial Autocorrelation: A Cross-Product Approach. *Environment and Planning. SAGE Journals*, *23*(9), 1269–1277. doi:10.1068/a231269

Getter, K. L., & Rowe, D. B. (2006). The Role of Extensive Green Roofs in Sustainable Development. *HortScience*, *41*(5), 1276–1285. doi:10.21273/HORTSCI.41.5.1276

Getzin, S., Wiegand, K., & Schöning, I. (2012). Assessing biodiversity in forests using very high-resolution images and unmanned aerial vehicles. *Methods in Ecology and Evolution*, *3*(2), 397–404. doi:10.1111/j.2041-210X.2011.00158.x

Gevaert, C., Sliuzas, R., Persello, C., & Vosselman, G. (2018). Evaluating the societal impact of using drones to support urban upgrading projects. *ISPRS International Journal of Geo-Information*, *7*(3), 91.

Gheisari, M., & Esmaeili, B. (2016). Unmanned Aerial Systems (UAS) for Construction Safety Applications. *Proceedings of the Construction Research Congress 2016*, 2642-2650. 10.1061/9780784479827.263

Ghosh, S., & Ghosh, S. (2017). Evaluating Patterns of Urban Growth in Faridabad, Sub-Region of NCR, India Using Remote Sensing, GIS and Entropy Approach. *Indian Cartographer*, *37*, 258.

Giordan, D., Adams, M. S., Aicardi I., Alicandro, M., Allasia, P., Baldo, M., De Berardinis, P., Dominici, D., Godone, D., Hobbs, P., Lechner, V., Niedzielski, T., Piras, M., Rotilio, M., Salvini, R., Segor, V., Sotier, B. & Troilo, F. (2020). The use of unmanned aerial vehicles (UAVs) for engineering geology applications. *Bulletin of Engineering Geology and the Environment.* . doi:10.100710064-020-01766-2

Givoni, B. (1992). Comfort, climate analysis and building design guidelines. *Energy and Building, 18*(1), 11–23. doi:10.1016/0378-7788(92)90047-K

Goddard, I. L., & Tett, S. F. (2019). How much has urbanisation affected United Kingdom temperatures? *Atmospheric Science Letters, 20*(5), 896. doi:10.1002/asl.896

Gogoi, P. P., Vinoj, V., Swain, D., Roberts, G., Dash, J., & Tripathy, S. (2019). Land use and land cover change effect on surface temperature over Eastern India. *Scientific Reports, 9*(1), 8859. doi:10.103841598-019-45213-z PMID:31222135

Goldman Sachs. (2019). *Drones. Reporting for Work.* https://www.goldmansachs.com/insights/technology-driving-innovation/drones/

Golledge, R. (Ed.). (1999). *Wayfinding Behavior: Cognitive Mapping and Other Spatial Processes.* Johns Hopkins University Press.

Gomes, E., Abrantes, P., Banos, A., & Rocha, J. (2019). Modelling future land use scenarios based on farmers' intentions and a cellular automata approach. *Land Use Policy* (Vol. 85). doi:10.1016/j.landusepol.2019.03.027

Gomes, J., & Pedrassoli, J. (2018). *Levantamento de características de moradias em favelas com o uso de drone: um experimento na Comunidade do Alto das Pombas, Salvador/Ba* [Paper presentation]. III Seminário Nacional Sobre Urbanização de Favelas – Urbfavelas.

Gómez, C., & Green, D. R. (2017). Small Unmanned Airborne Systems to Support Oil and Gas Pipeline Monitoring and Mapping. Arabian Journal of Geosciences, 10(9). doi:10.100712517-017-2989-x

Gómez-Antonio, M., Hortas-Rico, M., & Li, L. (2016). The causes of urban sprawl in Spanish urban areas: A spatial approach. *Spatial Economic Analysis, 11*(2), 219–247. doi:10.1080/17421772.2016.1126674

Gómez, C., & Green, D. R. (2015). *Small-Scale Airborne Platforms for Oil and Gas Pipeline Monitoring and Mapping.* Unpublished UCEMM Report for Redwing Ltd.

Gonçalves, J. (2011). *DEM SRTM.* Retrieved July 2, 2019, from FCUP-University of Oporto website: https://www.fc.up.pt/pessoas/jagoncal/srtm

Goncalves, G. R., Perez, J. A., & Duarte, J. (2018). Accuracy and effectiveness of low cost UASs and open source photogrammetric software for foredunes mapping. *International Journal of Remote Sensing, 39*(15-16), 5059–5077. doi:10.1080/01431161.2018.1446568

Göndöcs, J., Breuer, H., Pongrácz, R., & Bartholy, J. (2017). Urban heat island mesoscale modelling study for the Budapest agglomeration area using the WRF model. *Urban Climate*, *21*, 66–86. doi:10.1016/j.uclim.2017.05.005

González, Á. (2016). Amazon lets users create their own smart home. *The Seattle Times for Government Technology*.

González, R. R., & Medina, J. S. (2004). Theories, models and urban realities. From New York to Kathmandu. *Dela*, (21), 69–81. doi:10.4312/dela.21.5.69-81

Goodchild, M. F. (2007). Citizens as sensors: The world of volunteered geography. *GeoJournal*, *69*(4), 211–221. doi:10.100710708-007-9111-y

GOST 20058-80. (1980). *Dynamics of Flying Vehicles in Atmosphere. Terms, Definitions, and Notations. Official edition*. Moscow: Goskomstandard.

Goswami, P., Shivappa, H., & Goud, B. S. (2010). Impact of urbanization on tropical mesoscale events: Investigation of three heavy rainfall events. *Meteorologische Zeitschrift (Berlin)*, *19*(4), 385–397. doi:10.1127/0941-2948/2010/0468

Govind, N. R., & Ramesh, H. (2019). The impact of spatiotemporal patterns of land use land cover and land surface temperature on an urban cool island: A case study of Bengaluru. *Environmental Monitoring and Assessment*, *191*(5), 283. doi:10.100710661-019-7440-1 PMID:30997556

Graham, S., & Hewitt, L. (2012). Getting off the ground: On the politics of urban verticality. *Progress in Human Geography*, *37*(1), 72–92. doi:10.1177/0309132512443147

Green, D. R., Gregory, B. J., & Karachok, A. R. (Eds.). (2020a). Unmanned Aerial Remote Sensing UAS for Environmental Applications. CRC Press.

Green, D. R., Karachok, A. R., & Gregory, B. J. (2020b). The Potential of Drone Technology in Pandemics. In COVID-19: Geospatial Information and Community Resilience. Taylor and Francis.

Green, J. L. (2017). *Drones Will Elevate Urban Design*. https://www.smartcitiesdive.com/ex/sustainablecitiescollective/drones-will-elevate-urbandesign/1053491/

Griffiths, P., Hostert, P., Gruebner, O., & van der Linden, S. (2010). Mapping megacity growth with multi-sensor data. *Remote Sensing of Environment*, *114*(2), 426–439. doi:10.1016/j.rse.2009.09.012

Grimmond, C. S. B., Roth, M., Oke, T. R., Au, Y. C., Best, M., Betts, R., ... Freitas, E. (2010). Climate and more sustainable cities: Climate information for improved planning and management of cities (producers/capabilities perspective). *Procedia Environmental Sciences*, *1*, 247–274. doi:10.1016/j.proenv.2010.09.016

Grubesic, T. H., & Nelson, J. R. (2020). *UAVs and Urban Spatial Analysis*. Springer International Publishing. doi:10.1007/978-3-030-35865-5

Gruszczynski, W., Matwij, W., & Cwiakala, P. (2017). Comparison of low-altitude UAV photogrammetry with terrestrial laser scanning as data-source methods for terrain covered in low vegetation. *ISPRS Journal of Photogrammetry and Remote Sensing*, *2017*(126), 168–179. doi:10.1016/j.isprsjprs.2017.02.015

GSDI. (2012). *Spatial Data Infrastructure Cookbook 2012 Update*. Global Spatial Data Infrastructure Association. Retrieved from http://gsdiassociation.org/images/publications/cookbooks/SDI_Cookbook_from_Wiki_2012_update.pdf

Guinness World Record. (2017). https://www.guinnessworldrecords.com/news-411677

Gulevich, S.P., Veselov, Y.G., Pryadkin, S.P., & Tirnov, S.D. (2012). *Analisis of factors affecting the safety of the flight of unmanned aerial vehicles*. Ru. Causes of accidents drones and methods of preventing them «Science and education».

Guo, W., Wang, X., Sun, J., Ding, A., & Zou, J. (2016). Comparison of land–atmosphere interaction at different surface types in the mid-to lower reaches of the Yangtze River valley. *Atmospheric Chemistry and Physics*, *16*(15), 9875–9890. doi:10.5194/acp-16-9875-2016

Gupta, R., Joshi, R., & Gupta, N. (2012). Urbanization and air environment in Jaipur. *Environment & Urbanization Asia*, *3*(2), 353–358. doi:10.1177/0975425312473231

Gupta, S., & Roy, M. (2012). Land Use /Land Cover classification of an urban area- A case study of Burdwan Municipality, India. *International Journal of Geomatics and Geosciences*, *2*(4), 1014–1026.

Haala, N., Crammer, M., & Rothermel, M. (2013). Quality of 3D point clouds from highly overlapping UAV imagery. In Proceedings of International Archives of the Photogrammetry, Remote Sensing and Spatial Information Sciences (vol. XL-1/W2, pp. 183-188). doi:10.5194/isprsarchives-XL-1-W2-183-2013

Haaland, C., & van den Bosch, C. K. (2015). Challenges and strategies for urban green-space planning in cities undergoing densification: A review. *Urban Forestry & Urban Greening*, *14*(4), 760–771. doi:10.1016/j.ufug.2015.07.009

Haberlie, A. M., Ashley, W. S., & Pingel, T. J. (2015). The effect of urbanisation on the climatology of thunderstorm initiation. *Quarterly Journal of the Royal Meteorological Society*, *141*(688), 663–675. doi:10.1002/qj.2499

Hackenberg, R. A. (1980). New patterns of urbanization in Southeast Asia: An assessment. *Population and Development Review*, *6*(3), 391–419. doi:10.2307/1972408

Hagen, E. (2011). Mapping Change. Community Information Empowerment in Kibera. *Innovations*, *6*(1), 69–94. doi:10.1162/INOV_a_00059

Hai, P. M., & Yamaguchi, Y. (2008). Characterizing the urban growth of Hanoi, Nagoya, and Shanghai city using remote sensing and spatial metrics. In *IGARSS 2008-2008 IEEE International Geoscience and Remote Sensing Symposium* (Vol. 5, pp. V-9). IEEE.

Hale, R. C., Gallo, K. P., & Loveland, T. R. (2008). Influences of specific land use/land cover conversions on climatological normals of near-surface temperature. *Journal of Geophysical Research, D, Atmospheres*, *113*(D14), D14113. doi:10.1029/2007JD009548

Hall, P. (2001). Sustainable cities or town cramming. *Planning for a Sustainable Future*, 101-114.

Han, S. S. (2010). Urban expansion in contemporary China: What can we learn from a small town? *Land Use Policy*, *27*(3), 780–787. doi:10.1016/j.landusepol.2009.10.010

Hansen, J., Ruedy, R., Sato, M., & Lo, K. (2010). Global surface temperature change. *Reviews of Geophysics*, *48*(4), RG4004. Advance online publication. doi:10.1029/2010RG000345

Harari, M. (2016). *Cities in bad shape: Urban geometry in India. Processed.* Wharton School of the University of Pennsylvania.

Harnik, P., & Crompton, J. L. (2014). Measuring the total economic value of a park system to a community. *Managing Leisure*, *19*(3), 188–211. doi:10.1080/13606719.2014.885713

Harrop, P., & Harrop, J. (2014). Electric Drones: Unmanned Aerial Vehicles (UAVs) 2015-2025. IDTechEx Report.

Hasse, J. E., & Lathrop, R. G. (2003). Land resource impact indicators of urban sprawl. *Applied Geography (Sevenoaks, England)*, *23*(2-3), 159–175. doi:10.1016/j.apgeog.2003.08.002

He, F., Zhou, T., Xiong, W., Hasheminnasab, S. M., & Habib, A. (2018). Automated aerial triangulation for UAV-based mapping. *Remote Sensing*, *10*(12), 1952. Advance online publication. doi:10.3390/rs10121952

Hengl, T. (2006). Finding the Right Pixel Size. *Computers & Geosciences*, *32*(9), 1283–1298. doi:10.1016/j.cageo.2005.11.008

Henriques, C. D. (2008). *Maputo. Cinco Décadas de Mudança Territorial. O uso do solo observado por tecnologias de informação geográfica.* Instituto Português de Apoio ao Desenvolvimento.

Henriques, C. D., & Tenedório, J. A. (2010). Mapping Urban Change to Plan the Future: Maputo City Change Model and the Municipal Structure Plan. In *Remote Sensing for Science, Education, and Natural and Cultural Heritage: Proceedings of the 30th EARSeL Symposium*. Paris: EARSeL.

Herath, H. M. P. I. K., Halwatura, R. U., & Jayasinghe, G. Y. (2018). Modeling a tropical urban context with green walls and green roofs as an urban heat island adaptation strategy. *Procedia Engineering*, *212*, 691–698. doi:10.1016/j.proeng.2018.01.089

Herbert, G., & Chen, X. (2015). A comparison of usefulness of 2D and 3D representations of urban planning. *Cartography and Geographic Information Science*, *42*(1), 22–32. doi:10.1080/15230406.2014.987694

Herold, M., Couclelis, H., & Clarke, K. C. (2005). The role of spatial metrics in the analysis and modeling of urban land use change. *Computers, Environment and Urban Systems*, *29*(4), 369–399. doi:10.1016/j.compenvurbsys.2003.12.001

Herold, M., Scepan, J., & Clarke, K. C. (2002). The use of remote sensing and landscape metrics to describe structures and changes in urban land uses. *Environment & Planning A*, *34*(8), 1443–1458. doi:10.1068/a3496

Hespanha, J. P. (2012). Development Methodology for an Integrated Legal Cadastre. ed. Delft, The Netherlands: TU Delft. Repository of Delft University.

Heymans, A., Breadsell, J., Morrison, G. M., Byrne, J. J., & Eon, C. (2019). Ecological Urban Planning and Design: A Systematic Literature Review. *Sustainability*, *11*(13), 3723. doi:10.3390u11133723

Hildmann, H., Kovacs, E., Saffre, F., & Isakovic, A. F. (2019). Nature-Inspired Drone Swarming for Real-Time Aerial Data-Collection Under Dynamic Operational Constraints. *Drones (Basel)*, *2019*(3), 71. doi:10.3390/drones3030071

Holtier, S., Steadman, J. P., & Smith, M. G. (2000). Three-dimensional representation of urban built form in a GIS. *Environment and Planning. B, Planning & Design*, *27*(1), 51–72. doi:10.1068/bst5

Hou, X., Change, B., & Yu, X. (2004). Land use change in Hexi corridor based on CA Markov methods. *Nongye Gongcheng Xuebao (Beijing)*, *20*(5), 286–291.

Howard, L. (1833). *The climate of london. deduced from meteorological observations made in the metropolis and various places around it*. Harvey and Darton.

Hua, A. K., & Ping, O. W. (2018). The influence of land-use/land-cover changes on land surface temperature: A case study of Kuala Lumpur metropolitan city. *European Journal of Remote Sensing*, *51*(1), 1049–1069. doi:10.1080/22797254.2018.1542976

Hua, L. J., Ma, Z. G., & Guo, W. D. (2008). The impact of urbanization on air temperature across China. *Theoretical and Applied Climatology*, *93*(3-4), 179–194. doi:10.100700704-007-0339-8

Huang, Y., Liu, Y., & Lieske, S. (2019). Detecting, Modeling and Predicting Vertical Urban Growth: An exploratory review. In GeoComputation 2019. The University of Auckland. doi:10.17608/k6.auckland.9863141.v2

Huang, J., Lu, X. X., & Sellers, J. M. (2007). A global comparative analysis of urban form: Applying spatial metrics and remote sensing. *Landscape and Urban Planning*, *82*(4), 184–197. doi:10.1016/j.landurbplan.2007.02.010

Huete, A. (1988). A soil-adjusted vegetation index (SAVI). *Remote Sensing of Environment*, *25*(3), 295–309. doi:10.1016/0034-4257(88)90106-X

Huete, A., Liu, H., Batchily, K., & Leeuwen, W. (1997). A Comparison of Vegetation Indices Over a Global Set of TM Images for EOS-MODIS. *Remote Sensing of Environment*, *59*(3), 440–451. doi:10.1016/S0034-4257(96)00112-5

Hurlimann, A. C., & March, A. P. (2012). The role of spatial planning in adapting to climate change. *Wiley Interdisciplinary Reviews: Climate Change*, *3*(5), 477–488. doi:10.1002/wcc.183

IATA. (2016). https://www.iata.org/publications/Pages/AI-white-paper.aspxInternational

IBGE – Instituto Brasileiro de Geografia e Estatística. (2010). *Demographic Census* [2010]. Retrieved from http://www.sidra.ibge.gov.br/cd/defaultcd2010.asp?0=4&i=P

IBGE – Instituto Brasileiro de Geografia e Estatística. (2018). *Population Estimate* [2018]. Retrieved from https://sidra.ibge.gov.br/tabela/6579

Iglhaut, J., Cabo, C., Puliti, S., Piermattei, L., O'Connor, J., & Rosette, J. (2019). Structure from Motion Photogrammetry in Forestry: A Review. *Current Forestry Reports*, *5*(3), 155–168. doi:10.100740725-019-00094-3

Ignatiev, O. (2010). UAV: The state and prospects of their use in the interests of NATO. *Foreign Military Review*, (12), 45–52.

Iizuka, K., Itoh, M., Shiodera, S., Matsubara, T., Dohar, M., & Watanabe, K. (2018). Advantages of unmanned aerial vehicle (UAV) photogrammetry for landscape analysis compared with satellite data: A case study in postmining sites in Indonesia. *Cogent Geoscience*, *4*(1), 1. doi:10.1080/23312041.2018.1498180

Iland, D., Irish, A., Madhow, U., & Sandler, B. (2018). *Rethinking GPS: Engineering Next-Gen Location at Uber.* https://eng.uber.com/rethinking-gps/

Ilci, V., Ozulu, I., Bilgi S., & Alkan, R. (2019). The Usage on Unmanned Aerial Vehices for 3D Mapping of Archaeological Sites. Fresenius Environmental Bulletin, 968–974.

INE. (2011). *Censos 2011 – Resultados Definitivos*. Retrieved from https://censos.ine.pt/xportal/xmain?xpid=CENSOS&xpgid=ine_censos_publicacao_det&contexto=pu&PUBLICACOESpub_boui=73212469&PUBLICACOESmodo=2&selTab=tab1&pcensos=61969554

Innerarrity, D. (2010). *O Novo Espaço Público* (M. Ruas, Trans.). Teorema. (Original work published 2006)

Inostroza, L., Baur, R., & Csaplovics, E. (2013). Urban sprawl and fragmentation in Latin America: A dynamic quantification and characterization of spatial patterns. *Journal of Environmental Management*, *115*, 87–97. doi:10.1016/j.jenvman.2012.11.007 PMID:23246769

Instituto Brasileiro de Geografia e Estatística. (2010). *Base de informações do censo demográfico 2010: resultados do universo por setor censitário*. https://goo.gl/wLOIwd

Intergovernmental Panel on Climate Change (IPCC). (2014). In R. K. Pachauri & L. A. Meyer (Eds.), *Climate Change2014; Contribution of Working Groups I, II and III to the Fifth Assessment Report of the Intergovernmental Panel on Climate Change; Core Writing Team* (p. 151). IPCC.

International Civil Aviation Organization. (2005). *Global Air Traffic Management Operational Concept*. Doc. ICAO Doc 9854-AN/458.

International Civil Aviation Organization. (2009). Manual on Global Performance of the Air Navigation System. Doc. ICAO 9883.

International Civil Aviation Organization. (2011). *Unmanned Aircraft Systems (UAS)*. Circ. 328-AN/190.

International Civil Aviation Organization. (2015). Manual of Remoted Piloted Aircraft Systems (RPAS). Doc. ICAO 10019/AN 507.

International Civil Aviation Organization. (2017a). *Aviation Security Plan (GASP)*. Author.

International Civil Aviation Organization. (2017b). *Remotely piloted aircraft system (RPAS) Concept of operations (CONOPS) for international IFR operations*. ICAO RPAS Concept of Operations ICAO.

International Civil Aviation Organization. (2018) Potential of Artificial Intelligence (AI) in Air Traffic Management (ATM). *Thirteenth Air Navigation Conference ICAO*, 9-19.

International Civil Aviation Organization. (2018). *Model UAS Regulations: parts 149, 101, 102. In: Guidance material and tools*. ICAO. https://www.icao.int/safety/UA/Pages/default.aspx

International Civil Aviation Organization. (2019). *ICAO's Unmanned Aviation Bulletin*. ICAO. https://www.icao.int/safety/UA/Documents/ICAO%20UA%20Bulletin%202019-1.pdf

Ippolito, A., & Cigola, M. (2017). *Handbook of Research on Emerging Technologies for Digital Preservation and Information Modelling*. IGI Global. doi:10.4018/978-1-5225-0680-5

Jaeger, J. A., Bertiller, R., Schwick, C., Cavens, D., & Kienast, F. (2010). Urban permeation of landscapes and sprawl per capita: New measures of urban sprawl. *Ecological Indicators*, *10*(2), 427–441. doi:10.1016/j.ecolind.2009.07.010

Jahangir, M. S., & Moghim, S. (2019). Assessment of the urban heat island in the city of Tehran using reliability methods. *Atmospheric Research*, *225*, 144–156. doi:10.1016/j.atmosres.2019.03.038

Jain, A., Mahajan, M., & Saraf, R. (2019). Standardization of the Shape of Ground Control Point (GCP) and the Methodology for Its Detection in Images for UAV-Based Mapping Applications. In *Advances in Computer Vision. Proceedings of the 2019 Computer Vision Conference (CVC)* (*vol. 1*, pp. 459-476). 10.1007/978-3-030-17795-9_34

Jain, M., Dimri, A. P., & Niyogi, D. (2016). Urban sprawl patterns and processes in Delhi from 1977 to 2014 based on remote sensing and spatial metrics approaches. *Earth Interactions*, *20*(14), 1–29. doi:10.1175/EI-D-15-0040.1

Jat, M. K., Garg, P. K., & Khare, D. (2008). Monitoring and modelling of urban sprawl using remote sensing and GIS techniques. *International Journal of Applied Earth Observation and Geoinformation*, *10*(1), 26–43. doi:10.1016/j.jag.2007.04.002

Jáuregui, E., & Romales, E. (1996). Urban effects on convective precipitation in Mexico City. *Atmospheric Environment*, *30*(20), 3383–3389. doi:10.1016/1352-2310(96)00041-6

Jayasooriya, V. M., & Ng, A. W. M. (2014). Tools for Modeling of Stormwater Management and Economics of Green Infrastructure Practices: A Review. *Water, Air, and Soil Pollution, 225*(8), 2055. doi:10.100711270-014-2055-1

Jaynes, E. T. (1957). Information theory and statistical mechanics. *Physical Review, 106*(4), 620–630. doi:10.1103/PhysRev.106.620

JCGM. (2008). *Evaluation of measurement data - Guide to the expression of uncertainty in measurement*. Joint Committee for Guides in Metrology.

Jenkins, P. (2000). Urban management, urban poverty and urban governance: Planning and land management in Maputo. *Environment and Urbanization, 12*(1), 137–152. doi:10.1177/095624780001200110

Jenkins, P. (2013). *Urbanization, Urbanism, and Urbanity in an African City. Home Spaces and House Cultures*. Palgrave Macmillan. doi:10.1057/9781137380173

Jenks, M., Burton, E., & Williams, K. (1996). Compact cities and sustainability: an introduction. *The Compact City: A Sustainable Urban Form*, 11-12.

Jensen, A. M., Morgan, D., Chen, Y., Clemens, S., & Hardy, T. (2009). Using Multiple Open-Source Low-Cost Unmanned Aerial Vehicles (UAV) for 3D Photogrammetry and Distributed Wind Measurement. In *ASME/IEEE 2009 International Conference on Mechatronic and Embedded Systems and Applications*; 20th Reliability, Stress Analysis, and Failure Prevention Conference (Vol. 3, pp. 629–634). Retrieved from http://ebooks.asmedigitalcollection.asme.org/data/Conferences/IDETC/CIE 2009/70856/629_1.pdf

Jensen, J. R. (2015). Introductory digital image processing: a remote sensing perspective (2nd ed.). Prentice-Hall Inc.

Jensen, J. R. (2015). *Introductory Digital Image Processing: A Remote Sensing Perspective* (4th ed.). Pearson.

Jensen, O. B. (2016). Drone city–power, design and aerial mobility in the age of "smart cities". *Geographica Helvetica, 71*(2), 67–75. doi:10.5194/gh-71-67-2016

Jha, R., Singh, V. P., & Vatsa, V. (2008). Analysis of urban development of Haridwar, India, using entropy approach. *KSCE Journal of Civil Engineering, 12*(4), 281–288. doi:10.100712205-008-0281-z

Jianping, L., Bai, Z., & Feng, G. (2005). RS-and-GIS-supported forecast of grassland degradation in southwest Songnen plain by Markov model. *Geo-Spatial Information Science, 8*(2), 104–109. doi:10.1007/BF02826848

Jim, C. Y. (2004). Green-space preservation and allocation for sustainable greening of compact cities. *Cities (London, England), 21*(4), 311–320. doi:10.1016/j.cities.2004.04.004

Jim, C. Y. (2013). Sustainable urban greening strategies for compact cities in developing and developed economies. *Urban Ecosystems, 16*(4), 741–761. doi:10.100711252-012-0268-x

Jiménez-Muñoz, J. C., Cristóbal, J., Sobrino, J. A., Sòria, G., Ninyerola, M., Pons, X., & Pons, X. (2008a). Revision of the single-channel algorithm for land surface temperature retrieval from Landsat thermal-infrared data. *IEEE Transactions on Geoscience and Remote Sensing*, *47*(1), 339–349. doi:10.1109/TGRS.2008.2007125

Jiménez-Muñoz, J. C., & Sobrino, J. A. (2008b). Split-window coefficients for land surface temperature retrieval from low-resolution thermal infrared sensors. *IEEE Geoscience and Remote Sensing Letters*, *5*(4), 806–809. doi:10.1109/LGRS.2008.2001636

Jiménez-Muñoz, J. C., Sobrino, J. A., Skoković, D., Mattar, C., & Cristóbal, J. (2014). Land surface temperature retrieval methods from Landsat-8 thermal infrared sensor data. *IEEE Geoscience and Remote Sensing Letters*, *11*(10), 1840–1843. doi:10.1109/LGRS.2014.2312032

Jing, S., Li, B., Tan, M., & Liu, H. (2012). Impact of Relative Humidity on Thermal Comfort in a Warm Environment. *Indoor and Built Environment*. Advance online publication. doi:10.1177/1420326X12447614

Joerin, J., Shaw, R., Takeuchi, Y., & Krishnamurthy, R. (2012). Assessing community resilience to climate-related disasters in Chennai, India. *International Journal of Disaster Risk Reduction*, *1*, 44–54. doi:10.1016/j.ijdrr.2012.05.006

Joshi, D. (2017). *Exploring the latest drone technology for commercial, industrial and military drone uses*. https://www.businessinsider.com/drone-technology-uses-2017-7

Joshi, R., Raval, H., Pathak, M., Prajapati, S., Patel, A., Singh, V., & Kalubarme, M. H. (2015). Urban heat island characterization and isotherm mapping using geo-informatics technology in Ahmedabad city, Gujarat state, India. *International Journal of Geosciences*, *6*(03), 274–285. doi:10.4236/ijg.2015.63021

JRC & ESA. (1994). *Modelling Deforestation Processes—A Review*. Luxembourg: ECSC-EC-EAEC.

Julião, R. P. (2010). Low-Cost SDI. The Portuguese example of building a SDI for small countries. *FIG Congress 2010*.

Julin, A., Jaalama, K., Virtanen, J., Pouke, M., Ylipulli, J., Vaaja, M., Hyyppä, J., & Hyyppä, H. (2018). Characterizing 3D City Modeling Projects: Towards a Harmonized Interoperable System. *ISPRS International Journal of Geo-Information*, *7*(55), 1–18. doi:10.3390/ijgi7020055

Kanellakis, C., & Nikolakopoulus, G. (2017). Survey on Computer Vision for UAVs: Current Developments and Trends. *Journal of Intelligent & Robotic Systems*, *81*(1), 141–168. doi:10.100710846-017-0483-z

Kantakumar, L. N., Kumar, S., & Schneider, K. (2016). Spatiotemporal urban expansion in Pune metropolis, India using remote sensing. *Habitat International*, *51*, 11–22. doi:10.1016/j.habitatint.2015.10.007

Kántor, N., & Unger, J. (2011). The most problematic variable in the course of human-biometeorological comfort assessment—The mean radiant temperature. *Open Geosciences*, *3*(1). Advance online publication. doi:10.247813533-011-0010-x

Kapoor, R., & MacDuffie, J. P. (2017). *Analysis Shows Continued Industry-Wide Decline in Electric Vehicle Battery Costs*. https://mackinstitute.wharton.upenn.edu/2018/electric-vehicle-battery-costs-decline/

Karteris, M., Theodoridou, I., Mallinis, G., Tsiros, E., & Karteris, A. (2016). Towards a green sustainable strategy for Mediterranean cities: Assessing the benefits of large-scale green roofs implementation in Thessaloniki, Northern Greece, using environmental modelling, GIS and very high spatial resolution remote sensing data. *Renewable & Sustainable Energy Reviews*, *58*, 510–525. doi:10.1016/j.rser.2015.11.098

Kaufmann, J. & Steudler, D. (1998). Cadastre 2014 – A vision for a future cadastral system. *FIG*.

Kaufmann, R. K., Seto, K. C., Schneider, A., Liu, Z., Zhou, L., & Wang, W. (2007). Climate response to rapid urban growth: Evidence of a human-induced precipitation deficit. *Journal of Climate*, *20*(10), 2299–2306. doi:10.1175/JCLI4109.1

Kaufman, Y. J., & Tanré, D. (1992). Atmospherically resistant vegetation index (ARVI) for EOS-MODIS. *IEEE Transactions on Geoscience and Remote Sensing*, *30*(2), 261–270. doi:10.1109/36.134076

Kayet, N., Pathak, K., Chakrabarty, A., & Sahoo, S. (2016). Spatial impact of land use/land cover change on surface temperature distribution in Saranda Forest, Jharkhand. *Modeling Earth Systems and Environment*, *2*(3), 127. doi:10.100740808-016-0159-x

Kensek, K., & Noble, D. (2014). *Building Information Modeling: BIM in Current and Future Practice*. John Wiley & Sons. doi:10.4324/9781315797076

Kenworthy, J., & Laube, F. (1999). An International Sourcebook of Automobile Dependence in Cities, 1960–1990. *Transportation Research Part A, Policy and Practice*, *33*, 691–723. doi:10.1016/S0965-8564(99)00006-3

Keramitsoglou, I., Kiranoudis, C. T., Ceriola, G., Weng, Q., & Rajasekar, U. (2011). Identification and analysis of urban surface temperature patterns in Greater Athens, Greece, using MODIS imagery. *Remote Sensing of Environment*, *115*(12), 3080–3090. doi:10.1016/j.rse.2011.06.014

Kharchenko, V., Shmelova, T., & Sikirda, Y. (2012). *Decision-Making of Operator in Air Navigation System*. Kirovograd: KFA of NAU.

Kharchenko, V., Shmelova, T., Bondarev, D., & Stratiy, A. (2017). Modelling of Decision Making of Unmanned Aerial Vehicle's Operator in Emergency Situations. *Proceedings of the National Aviation University*, *1*(1), 20–28. doi:10.18372/2306-1472.70.11419

Kharchenko, V., Shmelova, T., & Sikirda, Y. (2016). *Decision-Making in Socio-Technical Systems: monograph*. NAU.

Kii, M., Hirota, K., & Minato, K. (2005). A study on modal shift potential considering public transport operation. *Compendium of Papers of the 9th International conference on Computers in Urban Planning and Urban Management.*

Kii, M., & Hanaoka, S. (2003). Comparison of efficiency between private and public transport considering urban structure. *IATSS Research, 27*(2), 6–15. doi:10.1016/S0386-1112(14)60139-4

Kikon, N., Singh, P., Singh, S. K., & Vyas, A. (2016). Assessment of urban heat islands (UHI) of Noida City, India using multi-temporal satellite data. *Sustainable Cities and Society, 22*, 19–28. doi:10.1016/j.scs.2016.01.005

Kinzig, A. P., & Grove, J. M. (2001). Urban-suburban ecology. In S. A. Levin (Ed.), Encyclopedia of biodiversity (pp. 733-745.). Amsterdam: Elsevier Science & Technology. doi:10.1016/B0-12-226865-2/00278-9

Kiran, G., & Joshi, U. B. (2013). Estimation of variables explaining urbanization concomitant with land-use change: A spatial approach. *International Journal of Remote Sensing, 34*(3), 824–847. doi:10.1080/01431161.2012.720738

Kirichek, R., Makolkina, M., Sene, J., & Takhtuev, V. (2016). Estimation quality parameters of transferring image and voice data over zigbee in transparent model. *Communications in Computer and Information Science, 601*, 260–267. doi:10.1007/978-3-319-30843-2_27

Kishtawal, C. M., Niyogi, D., Tewari, M., Pielke, R. A. Sr, & Shepherd, J. M. (2010). Urbanization signature in the observed heavy rainfall climatology over India. *International Journal of Climatology, 30*(13), 1908–1916. doi:10.1002/joc.2044

Kitchin, R. (2014). *The Data Revolution: Big Data, Open Data, Data Infrastructures and Their Consequences.* Sage Publications.

Klemm, W., Heusinkveld, B. G., Lenzholzer, S., & van Hove, B. (2015). Street greenery and its physical and psychological impact on thermal comfort. *Landscape and Urban Planning, 138*, 87–98. doi:10.1016/j.landurbplan.2015.02.009

Knaap, G., & Talen, E. (2005). New urbanism and smart growth: A few words from the academy. *International Regional Science Review, 28*(2), 107–118. doi:10.1177/0160017604273621

Koci, J., Jariahani, B., Leon, J. X., Sidle, R. C., Wilkinson, S. N., & Bartley, R. (2017). Assessment of UAV and Ground-Based Structure from Motion with Multi-View Stereo Photogrammetry in a Gullied Savanna Catchment. *ISPRS International Journal of Geo-Information, 6*(11), 328. doi:10.3390/ijgi6110328

Kok, K., & Veldkamp, A. (2001). Evaluating Impact of Spatial Scales on Land Use Pattern Analysis in Central America. *Agriculture, Ecosystems & Environment* (Vol. 85). doi:10.1016/S0167-8809(01)00185-2

Konrad, C. P. (2003). *Effects of Urban Development on Floods.* USGS Fact Sheet FS-076-03. Retrieved from https://pubs.usgs.gov/fs/fs07603/pdf/fs07603.pdf

Köppen-Geiger Climate Classification. (2010). *Encyclopedia of Geography* (B. Warf, Ed.). SAGE Publications, Inc., doi:10.4135/9781412939591.n671

Kotharkar, R., & Bagade, A. (2018). Evaluating urban heat island in the critical local climate zones of an Indian city. *Landscape and Urban Planning*, *169*, 92–104. doi:10.1016/j.landurbplan.2017.08.009

Kotharkar, R., & Surawar, M. (2016). Land use, land cover, and population density impact on the formation of canopy urban heat islands through traverse survey in the Nagpur urban area, India. *Journal of Urban Planning and Development*, *142*(1), 04015003. doi:10.1061/(ASCE)UP.1943-5444.0000277

Kraus, K. (2012). *Photogrammetrie. Geometrische Informationen aus Photographien und Laserscanneraufnahmen*. De Gruyter Lehrbuch.

Kreps, S., & Zenko, M. (2014). *The Next Drone Wars*. Foreign Affairs. https://bmpd.livejournal.com/861105.html

Krieger, Z., Wolf, H., & Reuter, T. (2018). *Advanced Drone Operations Toolkit: Accelerating the Drone Revolution*. Insight Report of World Economic Forum 91-93 route de la Capite CH-1223 Cologny.

Kršák, B., Blistan, P., Pauliková, A., Puskarova, P., Kovanič, Ľ., Palková, J., & Zelizňaková, V. (2016). Use of low-cost UAV photogrammetry to analyse the accuracy of a digital elevation model in a case study. *Measurement*, *91*, 276–287. Advance online publication. doi:10.1016/j.measurement.2016.05.028

Kucharczyk, M., & Hugenholtz, C. H. (2018). *Pre-disaster mapping with drones: an urban case study in Victoria*. Nat. Hazards Earth Syst. Sci. Discuss.

Kucherov, D. P., & Kozub, A. N. (2015). Control System Objects with Multiple Stream of Information. *Proceedings IEEE 3rd International Conference Actual Problems of Unmanned Aerial Vehicles Developments (APUAVD)*, 290-293. 10.1109/APUAVD.2015.7346623

Kucherov, D. P., & Kozub, A. N. (2016). Group Behavior of UAVs in obstacle presence. *Proceeding IEEE 4th International Conference "Methods and Systems of Navigation and Motion Control (MSNMC)"*, 51-54. 10.1109/MSNMC.2016.7783104

Kucherov, D. P., Kozub, A. N., & Kostyna, A. N. (2017). Group of UAVs Moving on Smooth Control Law with Fixed Obstacles. *Advances in Science. Technology and Engineering Systems Journal*, *2*(3), 1034–1041. doi:10.25046/aj0203131

Kulik, A.S. (2016). *Elements of the theory of rational management of objects*. Kharkov: Nat. Aerospace. UN-T Them. N. Ye. Zhukovsky.

Kumar, A., Dudhia, J., Rotunno, R., Niyogi, D., & Mohanty, U. C. (2008). Analysis of the 26 July 2005 heavy rain event over Mumbai, India using the Weather Research and Forecasting (WRF) model. *Quarterly Journal of the Royal Meteorological Society*, *134*(636), 1897–1910. doi:10.1002/qj.325

Kumar, G. A., Kumar, A., Patil, R., Sill, S., & Chai, Y. (2017). A LiDAR and IMU Integrated Indoor Navigation System for UAVs and Its Application in Real-Time Pipeline Classification. *Sensors (Basel)*, *17*(6), 1268. doi:10.339017061268 PMID:28574474

Kumari, B., Tayyab, M., Mallick, J., Khan, M. F., & Rahman, A. (2018). Satellite-Driven Land Surface Temperature (LST) Using Landsat 5, 7 (TM/ETM+ SLC) and Landsat 8 (OLI/TIRS) Data and Its Association with Built-Up and Green Cover Over Urban Delhi, India. *Remote Sensing in Earth Systems Sciences*, *1*(3-4), 63–78. doi:10.100741976-018-0004-2

Kumar, P., Geneletti, D., & Nagendra, H. (2016). Spatial assessment of climate change vulnerability at city scale: A study in Bangalore, India. *Land Use Policy*, *58*, 514–532. doi:10.1016/j.landusepol.2016.08.018

Kumar, R., & Mishra, V. (2019). Decline in surface urban heat island intensity in India during heatwaves. *Environmental Research Communications*, *1*(3), 031001. doi:10.1088/2515-7620/ab121d

Kumar, S., Ghosh, S., Hooda, R. S., & Singh, S. (2019). Monitoring and Prediction of Land Use Land Cover Changes and its Impact on Land Surface Temperature in the Central Part of Hisar District, Haryana Under Semi-Arid Zone of India. *Journal of Landscape Ecology*, *12*(3), 117–140. doi:10.2478/jlecol-2019-0020

Kundu, A. (2006). Trends and patterns of urbanization and their economic implications. *India Infrastructure Report, 2006*, 27-41.

Kundu, A. (2011). *Trends and Processes of Urbanisation in India, Urbanisation and Emerging Population Issues 6*. International Institute for Environment and Development (IIED). https://pubs.iied.org/10597IIED/

Küng, O., Strecha, C., Beyeler, A., Zufferey, J.-C., Floreano, D., Fua, P., & Gervaix, F. (2011). The Accuracy of Automatic Photogrammetric Techniques on Ultra-Light UAV Imagery. *ISPRS International Archives of the Photogrammetry, Remote Sensing and Spatial Information Sciences*, *38*(1), 125–130.

Kuo, F. E., & Sullivan, W. C. (2001). Environment and crime in the inner city: Does vegetation reduce crime? *Environment and Behavior*, *33*(3), 343–367. doi:10.1177/0013916501333002

Kusaka, H., Kimura, F., Hirakuchi, H., & Mizutori, M. (2000). The effects of land-use alteration on the sea breeze and daytime heat island in the Tokyo metropolitan area. *Kisho Shushi. Dai2shu*, *78*(4), 405–420. doi:10.2151/jmsj1965.78.4_405

Kuzin, A. V. (2011). Computer networks. Moscow: FORUM: INFRA-M.

Labib, R. (2017). *What are the applications of drone technology in architecture and the built environment field?* [Commentary]. https://www.researchgate.net/post/What_are_the_applications_of_drone_technology_in_architectur_and_the_built_environment_field

Laganière, R. (2014). *OpenCV Computer Vision Application Programming Cookbook* (2nd ed.). Packt Publishing Ltd.

Lakes, T., & Kim, H.-O. (2012). The urban environmental indicator "Biotope Area Ratio"—An enhanced approach to assess and manage the urban ecosystem services using high resolution remote-sensing. *Ecological Indicators*, *13*(1), 93–103. doi:10.1016/j.ecolind.2011.05.016

Lambin, E. F., Turner, B. L., Geist, H. J., Agbola, S. B., Angelsen, A., Bruce, J. W., ... George, P. (2001). The causes of land-use and land-cover change: Moving beyond the myths. *Global Environmental Change*, *11*(4), 261–269. doi:10.1016/S0959-3780(01)00007-3

Landis, J. R., & Koch, G. G. (1977). The Measurement of Observer Agreement for Categorical Data. *Biometrics*, *33*(1), 159–174. doi:10.2307/2529310 PMID:843571

Landsberg, H. E. (1970). Man-Made Climatic Changes: Man's activities have altered the climate of urbanized areas and may affect global climate in the future. *Science*, *170*(3964), 1265–1274. doi:10.1126cience.170.3964.1265 PMID:17829423

Lantmäteriet. (2018). *Handbok i mät - och kartfagor Bilddata.* https://www.lantmateriet.se/sv/Om-Lantmateriet/Samverkan-med-andra/Handbok-i-mat--och-kartfragor-HMK/

Larondelle, N., Haase, D., & Kabisch, N. (2014). Mapping the diversity of regulating ecosystem services in European cities. *Global Environmental Change*, *26*, 119–129. doi:10.1016/j.gloenvcha.2014.04.008

Laureti, F., & Battisti, A. (2017). *A study of vegetation and cool materials effects on pedestrian comfort in Rome*. Paper presented at International Conference on Urban Confort and Environmental Quality: URBAN – CEQ, University of Genova, Italy.

Lazzarini, M., Marpu, P. R., & Ghedira, H. (2013). Temperature-land cover interactions: The inversion of urban heat island phenomenon in desert city areas. *Remote Sensing of Environment*, *130*, 136–152. doi:10.1016/j.rse.2012.11.007

Le Bris, A., & Chehata, N. (2019). Urban morpho-types classification from SPOT-6/7 imagery and Sentinel-2 time series. *2019 Joint Urban Remote Sensing Event*, 1–4. doi:10.1109/JURSE.2019.8808988

Leberl, F., Irschara, A., Pock, T., Meixner, P., Gruber, M., Scholz, S., & Wiechert, A. (2010). Point Clouds: LiDAR versus 3D Vision. *Photogrammetric Engineering and Remote Sensing*, *76*(10), 1123–1134. doi:10.14358/PERS.76.10.1123

Lee, D. O. (1979). The influence of atmospheric stability and the urban heat island on urban-rural wind speed differences. *Atmospheric Environment, 13*(8), 1175-1180.

Lee, A. C., & Maheswaran, R. (2011). The health benefits of urban green spaces: A review of the evidence. *Journal of Public Health*, *33*(2), 212–222. doi:10.1093/pubmed/fdq068 PMID:20833671

Lee, A., Mackenzie, A., Smith, G. J., & Box, P. (2020). Mapping platform urbanism: Charting the nuance of the platform pivot. *Urban Planning*, *5*(1), 116–128. doi:10.17645/up.v5i1.2545

Lefebvre, H. (1974). La production de l'espace. *Anthropos.*

Lehmann, S. (2014). Low carbon districts: Mitigating the urban heat island with green roof infrastructure. *City. Cultura e Scuola*, *5*(1), 1–8. doi:10.1016/j.ccs.2014.02.002

Lehner, A., Kraus, V., & Steinnocher, K. (2016). Urban growth scenarios of a future Mega City: Case study Ahmedabad. *ISPRS Annals of the Photogrammetry, Remote Sensing and Spatial Information Sciences*, *3*, 165.

Lei, M., Niyogi, D., Kishtawal, C., Pielke, R. A. Sr, Beltrán-Przekurat, A., Nobis, T. E., & Vaidya, S. S. (2008). Effect of explicit urban land surface representation on the simulation of the 26 July 2005 heavy rain event over Mumbai, India. *Atmospheric Chemistry and Physics*, *8*(20), 5975–5995. doi:10.5194/acp-8-5975-2008

Leitao, J. P., Moy de Vitry, M., Scheideffer, A., & Rieckermann, J. (2016). Assessing the quality of digital elevation models obtained from mini unmanned aerial vehicles for overland flow modelling in urban areas. *Hydrology and Earth System Sciences*, *20*(4), 1637–1653. doi:10.5194/hess-20-1637-2016

Leitão, J. P., Moy de Vitry, M., Scheidegger, A., & Rieckermann, J. (2015). Assessing the quality of Digital Elevation Models obtained from mini-Unmanned Aerial Vehicles for overland flow modelling in urban areas. *Hydrology and Earth System Sciences Discussions*, *12*(6), 5629–5670.

Lemmen, C.; van Oosterom, P.; Uitermark, H.; Thompson, R.; Hespanha, J. (2009). *Transforming the Land Administration Domain Model (LADM) into an ISO Standard (ISO19152)*. FIG Working Week 2009, Eilat, Israel.

Lemmen, C., & van Oosterom, P. (2006). *Version 1.0 of the FIG Core Cadastral Domain Model. XXIII FIG Congress*, Munich, Germany.

Li, X., & Gong, P. (2016). Urban growth models: progress and perspective. *Science Bulletin*, *61*(21), 1637–1650. doi:10.100711434-016-1111-1

Liang, S. (Ed.). (2018). *Comprehensive Remote Sensing*. Elsevier.

Liang, X., Liu, X., Li, D., Zhao, H., & Chen, G. (2018). Urban growth simulation by incorporating planning policies into a CA-based future land-use simulation model. *International Journal of Geographical Information Science*, *32*(11), 2294–2316. doi:10.1080/13658816.2018.1502441

Li, C. J., & Ling, H. (2015). Synthetic Aperture Radar Imaging Using a Small Consumer Drone [Unpublished Powerpoint Presentation]. 10.1109/APS.2015.7304729

Li, H., & Liu, Q. (2008). Comparison of NDBI and NDVI as indicators of surface urban heat island effect in MODIS imagery. In *International conference on earth observation data processing and analysis (ICEODPA)* (Vol. 7285, p. 728503). International Society for Optics and Photonics. 10.1117/12.815679

Lillesand, T., Kiefer, R. W., & Chipman, J. (2015). *Remote Sensing and Image Interpretation* (7th ed.). Wiley.

Li, M., Nan, L. L., Smith, N., & Wonka, P. (2016). Reconstructing Building Mass Models from UAV Images. *Computers & Graphics*, *54*(February), 84–93. doi:10.1016/j.cag.2015.07.004

Li, M., Zhang, Z., Lo Seen, D., Sun, J., & Zhao, X. (2016). Spatiotemporal characteristics of urban sprawl in Chinese port cities from 1979 to 2013. *Sustainability*, *8*(11), 1138. doi:10.3390u8111138

Lin, C. Y., Chen, F., Huang, J. C., Chen, W. C., Liou, Y. A., Chen, W. N., & Liu, S. C. (2008b). Urban heat island effect and its impact on boundary layer development and land–sea circulation over northern Taiwan. *Atmospheric Environment*, *42*(22), 5635–5649. doi:10.1016/j.atmosenv.2008.03.015

Lin, C. Y., Chen, W. C., Liu, S. C., Liou, Y. A., Liu, G. R., & Lin, T. H. (2008a). Numerical study of the impact of urbanization on the precipitation over Taiwan. *Atmospheric Environment*, *42*(13), 2934–2947. doi:10.1016/j.atmosenv.2007.12.054

Lindberg, F., Grimmond, C. S. B., Yogeswaran, N., Kotthaus, S., & Allen, L. (2013). Impact of city changes and weather on anthropogenic heat flux in Europe 1995–2015. *Urban Climate*, *4*, 1–15. doi:10.1016/j.uclim.2013.03.002

Linder, W. (2016). Digital photogrammetry: A practical course. In Digital Photogrammetry: A Practical Course. doi:10.1007/978-3-662-50463-5

Lin, P. (2009). Thermal perception, adaptation, and attendance in a public square in hot and humid regions. *Building and Environment*, *44*(10), 2017–2026. doi:10.1016/j.buildenv.2009.02.004

Lin, Y., Hyyppa, J., & Jaakkola, A. (2011). Mini-UAV-Borne LIDAR for Fine-Scale Mapping. *IEEE Geoscience and Remote Sensing Letters*, *8*(3), 426–430. doi:10.1109/LGRS.2010.2079913

Li, Q., Li, W., Si, P., Xiaorong, G., Dong, W., Jones, P., Huang, J., & Cao, L. (2010). Assessment of surface air warming in northeast China, with emphasis on the impacts of urbanization. *Theoretical and Applied Climatology*, *99*(3-4), 469–478. doi:10.100700704-009-0155-4

Little, K. (1960). West African urbanization as a social process. *Cahiers d'etudes Africaines*, *1*(3), 90-102.

Liu, Y. (2009). *Modelling Urban Development with Geographical Information Systems and Cellular Automata*. doi:10.1201/9781420059908

Liu, H., & Weng, Q. (2008). Seasonal variations in the relationship between landscape pattern and land surface temperature in Indianapolis, USA. *Environmental Monitoring and Assessment*, *144*(1-3), 199–219. doi:10.100710661-007-9979-5 PMID:17899413

Liu, J., & Niyogi, D. (2019). Meta-analysis of urbanization impact on rainfall modification. *Scientific Reports*, *9*(1), 1–14. doi:10.103841598-019-42494-2 PMID:31086196

Liu, L., Coops, N. C., Aven, N. W., & Pang, Y. (2017). Mapping urban tree species using integrated airborne hyperspectral and LiDAR remote sensing data. *Remote Sensing of Environment*, *200*, 170–182. doi:10.1016/j.rse.2017.08.010

Liu, L., & Zhang, Y. (2011). Urban heat island analysis using the Landsat TM data and ASTER data: A case study in Hong Kong. *Remote Sensing*, *3*(7), 1535–1552. doi:10.3390/rs3071535

Liu, P., Chen, A. Y., Huang, Y. N., Han, J. Y., Lai, J. S., Kang, S. C., ... Tsai, M. H. (2014). A review of rotorcraft unmanned aerial vehicle (UAV) developments and applications in civil engineering. *Smart Structures and Systems*, *13*(6), 1065–1094.

Liu, Y., Yue, T., Jiao, Y., Zhao, N., & Zhao, M. (2019). Temperature changes in the Heihe River Basin based on high accuracy surface modelling. *Meteorological Applications*, *26*(4), 720–732. doi:10.1002/met.1803

Li, X. X., Koh, T. Y., Entekhabi, D., Roth, M., Panda, J., & Norford, L. K. (2013). A multi-resolution ensemble study of a tropical urban environment and its interactions with the background regional atmosphere. *Journal of Geophysical Research, D, Atmospheres*, *118*(17), 9804–9818. doi:10.1002/jgrd.50795

Li, X. X., Koh, T. Y., Panda, J., & Norford, L. K. (2016). Impact of urbanization patterns on the local climate of a tropical city, Singapore: An ensemble study. *Journal of Geophysical Research, D, Atmospheres*, *121*(9), 4386–4403. doi:10.1002/2015JD024452

Li, X., Li, W., Meng, Q., Zhang, C., Jancso, T., & Wu, K. (2016). Modelling building proximity to greenery in a three-dimensional perspective using multi-source remotely sensed data. *Journal of Spatial Science*, *61*(2), 389–403. doi:10.1080/14498596.2015.1132642

Li, X., Mitra, C., Dong, L., & Yang, Q. (2018b). Understanding land use change impacts on microclimate using Weather Research and Forecasting (WRF) model. *Physics and Chemistry of the Earth Parts A/B/C*, *103*, 115–126. doi:10.1016/j.pce.2017.01.017

Li, X., & Yeh, A. G.-O. (2004). Analyzing spatial restructuring of land use patterns in a fast growing region using remote sensing and GIS. *Landscape and Urban Planning*, *69*(4), 335–354. doi:10.1016/j.landurbplan.2003.10.033

Li, X., Zhang, L., & Liang, C. (2010). A GIS-based buffer gradient analysis on spatiotemporal dynamics of urban expansion in Shanghai and its major satellite cities. *Procedia Environmental Sciences*, *2*, 1139–1156. doi:10.1016/j.proenv.2010.10.123

Li, Y., Zhao, C., Zhang, T., Wang, W., Duan, H., Liu, Y., Ren, Y., & Pu, Z. (2018a). Impacts of land-use data on the simulation of surface air temperature in Northwest China. *Journal of Meteorological Research*, *32*(6), 896–908. doi:10.100713351-018-7151-5

Li, Z. L., Tang, B. H., Wu, H., Huazhong, R., Guangjian, Y., Wan, Z., Trigo, I. F., & Sobrino, J. A. (2013). Satellite-derived land surface temperature: Current status and perspectives. *Remote Sensing of Environment*, *131*, 14–37. doi:10.1016/j.rse.2012.12.008

Loarie, S. R., Joppa, L. N., & Pimm, S. L. (2007). Satellites miss environmental priorities. *Trends in Ecology & Evolution*, *22*(12), 630–632. doi:10.1016/j.tree.2007.08.018 PMID:17996978

Lobaccaro, G., & Acero, J. A. (2015). Comparative analysis of green actions to improve outdoor thermal comfort inside typical urban street canyons. *Urban Climate*, *14*, 251–267. doi:10.1016/j.uclim.2015.10.002

Loitsyansky, L. (1983). Course of theoretical mechanics. Nauka.

Londe, L. R., Santos, L. B. L., Soriano, E., Tomás, L. R., & Carvalho, T. (2015). Urban mobility data to support the assessment of dynamic vulnerability to disasters. In *Proceedings of the Second International Conference on Information and Communication Technologies for Disaster Management*. Rennes, Paris: IEEE. 10.1109/ICT-DM.2015.7402036

Longo, F., Nicoletti, L., & Padovano, A. (2018). An interactive, interoperable and ubiquitous mixed reality application for a smart learning experience. *International Journal of Simulation and Process Modelling*, *13*(6), 589–603. doi:10.1504/IJSPM.2018.095864

Loose, T., & Bornstein, R. D. (1977). Observations of mesoscale effects on frontal movement through an urban area. *Monthly Weather Review*, *105*(5), 563–571. doi:10.1175/1520-0493(1977)105<0563:OOMEOF>2.0.CO;2

Lopes, J. M. de A. (2011). *Sobre arquitetos e sem-tetos: técnicas e arquitetura como prática política*. Rev. Brasileira Estudos Urbanos Regional. Universidade de São Paulo.

Lopes, J. M. de A., & Rizek, C. S. (2006). O mutirão autogerido como procedimento inovador na produção da moradia para os pobres: uma abordagem crítica. In A. L. Cardoso & A. K. Abiko (Eds.), *Procedimentos de gestão habitacional para população de baixa renda*. ANTAC.

Lord, J. M., & Norton, D. A. (1990). Scale and the spatial concept of fragmentation. *Conservation Biology*, *4*(2), 197–202. doi:10.1111/j.1523-1739.1990.tb00109.x

Lovell, S. T., & Taylor, J. R. (2013). Supplying urban ecosystem services through multifunctional green infrastructure in the United States. *Landscape Ecology*, *28*(8), 1447–1463. doi:10.100710980-013-9912-y

Lowry, W. P. (1998). Urban effects on precipitation amount. *Progress in Physical Geography*, *22*(4), 477–520. doi:10.1177/030913339802200403

Lutzoni, L. (2016). In-formalised urban space design. Rethinking the relationship between formal and informal. *City. Territory and Architecture*, *3*(20), 1–14.

Ma, W. (2018). *Aplicación de los UAV convencionales para la identificación y clasificación de los espacios verdes densos. Estudio de caso de la Parc Turó-Barcelona* (Master's thesis). Technical University of Catalonia. Upccommons. https://upcommons.upc.edu/handle/2117/330117

Madre, F., Vergnes, A., Machon, N., & Clergeau, P. (2014). Green roofs as habitats for wild plant species in urban landscapes: First insights from a large-scale sampling. *Landscape and Urban Planning*, *122*, 100–107. doi:10.1016/j.landurbplan.2013.11.012

Magarotto, M., Faria-De-Deus, R., Ferreira Costa, M., & Masanet, É. (2017). Green areas in coastal cities – conflict of interests or stakeholders' perspectives? *International Journal of Sustainable Development and Planning*, *12*(08), 1260–1271. doi:10.2495/SDP-V12-N8-1260-1271

Makinodan, D. Y., & Costa, S. M. F. (2004). Estudo das características socioeconômicas e ambientais da microbacia do Rio Comprido. In *Proceedings of XIV Encontro Nacional de Estudos Populacionais – ABEP*. Caxambú, Brazil: ABEP.

Maleki, A., & Mahdavi, A. (2016). Evaluation of urban heat islands mitigation strategies using 3dimentional urban micro-climate model envi-met. *Asian Journal of Civil Engineering*, *17*(3), 357–371.

Malik, M. S., Shukla, J. P., & Mishra, S. (2019). Relationship of LST, NDBI and NDVI using Landsat-8 data in Kandaihimmat Watershed, Hoshangabad, India. *Indian Journal of Geo-Marine Sciences*, *48*(1), 25–31.

Mallick, J., Kant, Y., & Bharath, B. (2008). Estimation of land surface temperature over Delhi using Landsat-7 ETM+. *The Journal of Indian Geophysical Union*, *12*, 131–140.

Mallick, J., Rahman, A., & Singh, C. K. (2013). Modeling urban heat islands in heterogeneous land surface and its correlation with impervious surface area by using night-time ASTER satellite data in highly urbanizing city, Delhi-India. *Advances in Space Research*, *52*(4), 639–655. doi:10.1016/j.asr.2013.04.025

Mallinis, G., Karteris, M., Theodoridou, I., Tsioukas, V., & Karteris, M. (2014). Development of a nationwide approach for large scale estimation of green roof retrofitting areas and roof-top solar energy potential using VHR natural colour orthoimagery and DSM data over Thessaloniki, Greece. *Remote Sensing Letters*, *5*(6), 548–557. doi:10.1080/2150704X.2014.933276

Manconi, A., Ziegler, M., & Blöchliger, T. (2019). Technical Note: Optimization of UAV flight planning in steep terrains. *International Journal of Remote Sensing*, *40*(7), 2483–2492. doi:10.1080/01431161.2019.1573334

Manyoky, M., Theiler, P., Steudler, D., & Eisenbeiss, H. (2011). Unmanned aerial vehicle in cadastral applications. In ISPRS. *The International Archives of the Photogrammetry, Remote Sensing and Spatial Information Sciences*, *38*(C22), 57–62.

Marandola, E. Jr. (2012). Mobilidades contemporâneas: distribuição espacial da população, vulnerabilidade e espaços de vida nas aglomerações urbanas. In J. M. P. C. Cunha (Ed.), *Mobilidade Espacial da População: Desafios teóricos e metodológicos para o seu estudo* (pp. 95–115). Nepo/Unicamp.

Maricato, E. (2000). As idéias fora do lugar e o lugar fora das idéias: planejamento urbano no Brasil. In O. Arantes, C. Vainer, & E. Maricato (Eds.), *A cidade do pensamento único - desmanchando consensos* (pp. 121–192). Vozes.

Maricato, E. (2003). *Metrópole, legislação e desigualdade*. Estudos Avançados.

Markham, B. L., & Barker, J. L. (1986). Landsat MSS and TM post-calibration dynamic ranges, exoatmospheric reflectances and at-satellite temperatures. *EOSAT Landsat Technical Notes, 1*, 3-8. http://landsathandbook.gsfc.nasa.gov/pdfs/L5_cal_document.pdf

Marques, L. (2017). *Augmented Valuation of Cultural Heritage through Digital Representation based upon Geographic Information Technologies (GIT)* (Unpublished doctoral dissertation). Universitat Politècnica de Catalunya. Departament de Tecnologia de l'Arquitectura, PhD thesis, 341. Retrieved from: http://luisfilipemarques.com/site/wp-content/uploads/2017/10/LM_PhDthesis_UPC_ETSAB_2017.pdf

Marques, L., Tenedório, J. A., Burns, M., Romão, T., Birra, F., Marques, J., & Pires, A. (2017). Cultural Heritage 3D Modelling and visualisation within an Augmented Reality Environment, based on Geographic Information Technologies and mobile platforms. *ACE: Architecture, City and Environment, 11*(33), 117–136.

Marques, L., Tenedório, J., Burns, M., Romão, T., Birra, F., Marques, J., & Pires, A. (2017). Cultural heritage 3D modelling and visualisation within an augmented reality environment, based on geographic information technologies and mobile platforms. *Architecture. City and Environment, 11*(33), 117–136. doi:10.5821/ace.11.33.4686

Marr, B. (2019). *Artificial Intelligence in Practice: How 50 Successful Companies Used AI and Machine Learning to Solve Problems*. John Wiley & Sons.

Marselle, M., Irvine, K., Lorenzo-Arribas, A., & Warber, S. (2015). Moving beyond green: Exploring the relationship of environment type and indicators of perceived environmental quality on emotional well-being following group walks. *International Journal of Environmental Research and Public Health, 12*(1), 106–130. doi:10.3390/ijerph120100106 PMID:25546275

Marshall, F., & Dolley, J. (2019). Transformative innovation in peri-urban Asia. *Research Policy, 48*(4), 983–992. doi:10.1016/j.respol.2018.10.007

Martinez, M. L., Silva, R., Lithgow, D., Mendoza, E., Flores, P., Martínez, R., & Cruz, C. (2017). Human Impact on Coastal Resilience along the Coast of Veracruz, Mexico. *Journal of Coastal Research, 77*(SI), 143–153. doi:10.2112/SI77-015.1

Martins, M. L. R. (1997). *São Paulo: movimentos populares urbanos e o direito à cidade*. FASE.

Masek, J. G., Lindsay, F. E., & Goward, S. N. (2000). Dynamics of urban growth in the Washington DC metropolitan area, 1973-1996, from Landsat observations. *International Journal of Remote Sensing, 21*(18), 3473–3486. doi:10.1080/014311600750037507

Mas, J. F., Kolb, M., Paegelow, M., Olmedo, M. T., & Houet, T. (2014). Inductive pattern-based land use/cover change models: A comparison of four software packages. *Environmental Modelling & Software, 51*, 94–111. doi:10.1016/j.envsoft.2013.09.010

Masoudi, M., Tan, P. Y., & Liew, S. C. (2019). Multi-city comparison of the relationships between spatial pattern and cooling effect of urban green spaces in four major Asian cities. *Ecological Indicators, 98*, 200–213. doi:10.1016/j.ecolind.2018.09.058

Masser, I. (2005). *GIS Worlds. Creating Spatial Data Infrastructures*. ESRI Press.

Matese, A., Toscano, P., Di Gennaro, S. F., Genesio, L., Vaccari, F. P., Primicerio, J., Belli, C., Zaldei, A., Bianconi, R., & Gioli, B. (2015). Intercomparison of UAV, Aircraft and Satellite Remote Sensing Platforms for Precision Viticulture. *Remote Sensing*, *7*(3), 2971–2990. doi:10.3390/rs70302971

Matheron, G. (1970). *La théorie des variables régionalisées, et ses applications*. Cahiers du Centre de Morphologie Mathématique de Fontainebleau. Fascicule 5. École des Mines de Paris.

Matheron, G. (1965). *Les variables régionalisées et leur estimation: Une application de la théorie des fonctions aléatoires aux sciences de la nature*. Masson.

Mathew, A., Khandelwal, S., & Kaul, N. (2016). Spatial and temporal variations of urban heat island effect and the effect of percentage impervious surface area and elevation on land surface temperature: Study of Chandigarh city, India. *Sustainable Cities and Society*, *26*, 264–277. doi:10.1016/j.scs.2016.06.018

Mathew, A., Khandelwal, S., & Kaul, N. (2018). Analysis of diurnal surface temperature variations for the assessment of surface urban heat island effect over Indian cities. *Energy and Building*, *159*, 271–295. doi:10.1016/j.enbuild.2017.10.062

Mathew, A., Sreekumar, S., Khandelwal, S., & Kumar, R. (2019). Prediction of land surface temperatures for surface urban heat island assessment over Chandigarh city using support vector regression model. *Solar Energy*, *186*, 404–415. doi:10.1016/j.solener.2019.04.001

McGarigal, Cushman, Neel, & Ene. (2002). *FRAGSTATS: spatial pattern analysis program for categorical maps*. Academic Press.

McIntyre, N. E. (1995). Effects of forest patch size on avian diversity. *Landscape Ecology*, *10*(2), 85–99. doi:10.1007/BF00153826

McLoughlin, L. (1991). Urban consolidation and urban sprawl: A question of density. *Urban Policy and Research*, *9*(3), 148–156. doi:10.1080/08111149108551499

McNeill, D. (2020). The volumetric city. *Progress in Human Geography*, *44*(5), 815–831. doi:10.1177/0309132519863486

Medeiros, A. (2019). *Evento gratuito mostra como fazer análise espacial com Mapas de Kernel em software livre*. https://mundogeo.com/blog/2015/07/02/evento-gratuito-mostra-como-fazer-analise-espacial-com-mapas-de-kernel-em-software-livre/

Medvedkov, Y. V. (1967). The concept of entropy in settlement pattern analysis. *Papers / Regional Science Association. Regional Science Association. Meeting*, *18*(1), 165–168. doi:10.1007/BF01940320

Mees, P. (2010a). *Density and sustainable transport in US, Canadian and Australian cities, another look at the data.* Compendium of Papers, the 12th WCTR conference, Lisbon, Portugal.

Mees, P. (2010b). *Transport for Suburbia: Beyond the Automobile Age*. Earthscan.

Meng, X., Shang, N., Zhang, X., Li, C., Zhao, K., Qiu, X., & Weeks, E. (2017). Photogrammetric UAV Mapping of Terrain under Dense Coastal Vegetation: An Object-Oriented Classification Ensemble Algorithm for Classification and Terrain Correction. *Remote Sensing*, *9*(11), 1187. doi:10.3390/rs9111187

Meola, A. (2017). *Drone market shows positive outlook with strong industry growth and trends.* https://www.businessinsider.com/drone-industry-analysis-market-trends-growth-forecasts-2017-7

Mesas-Carrascosa, F. J., Torres-Sánchez, J., Clavero-Rumbao, I., García-Ferrer, A., Peña, J. M., Borra-Serrano, I., & López-Granados, F. (2015). Assessing Optimal Flight Parameters for Generating Accurate Multispectral Orthomosaics by UAV to Support Site-Specific Crop Management. *Remote Sensing*, *7*(10), 12793–12814. doi:10.3390/rs71012793

Miao, S., Chen, F., LeMone, M. A., Tewari, M., Li, Q., & Wang, Y. (2009). An observational and modeling study of characteristics of urban heat island and boundary layer structures in Beijing. *Journal of Applied Meteorology and Climatology*, *48*(3), 484–501. doi:10.1175/2008JAMC1909.1

Miao, S., Chen, F., Li, Q., & Fan, S. (2011). Impacts of urban processes and urbanization on summer precipitation: A case study of heavy rainfall in Beijing on 1 August 2006. *Journal of Applied Meteorology and Climatology*, *50*(4), 806–825. doi:10.1175/2010JAMC2513.1

Micheletti, N., Chandler, J. H., & Lane, S. N. (2015). Structure from Motion (SfM) Photogrammetry. In Geomorphological Techniques. British Society for Geomorphology.

Mikhail, E., Bethel, J., & McGlone, J. (2001). *Introduction to modern photogrammetry*. John Wiley & Sons.

Mikrut, S. (2016). Classical Photogrammetry and UAV - Selected aspects. In *Proceedings of XXIII ISPRS Congress* (vol. XLI-B1, pp. 947-952). 10.5194/isprsarchives-XLI-B1-947-2016

Mindell, D. A., & Bingham, B. (2001). New Archaeological Uses of Autonomous Undersea Vehicles. *MTS/IEEE Oceans 2001. An Ocean Odyssey. Conference Proceedings*, 555-558. 10.1109/OCEANS.2001.968781

Ministério da Economia e do Emprego. (2013). *Diário da República n.º 159/2013, Série I de 2013-08-20*. Diário da República Eletrónico. https://dre.pt/pesquisa/-/search/499237/details/maximized?jp=true/en/en

Ministério das Cidades. (2004). *Política Nacional de Habitação*. Cadernos MCidades, Habitação 4. Espalhafato Comunicação.

Ministério das Obras Públicas. Transportes e Comunicações. (2006). *Decreto-Lei 80/2006, 2006-04-04*. Diário da República Eletrónico. https://dre.pt/pesquisa/-/search/672456/details/maximized?jp=true

Ministérios do Ambiente, Ordenamento do Território e Energia e da Solidariedade, Emprego e Segurança Social. (2013). *Portaria 349-D/2013 de 2013-12-02*. Diário da República Eletrónico. https://dre.pt/web/guest/pesquisa/-/search/263693/details/maximized?print_preview=print-preview

Mitra, C., Shepherd, J. M., & Jordan, T. (2012). On the relationship between the premonsoonal rainfall climatology and urban land cover dynamics in Kolkata city, India. *International Journal of Climatology*, *32*(9), 1443–1454. doi:10.1002/joc.2366

Mobility and Residential Costs Project – MORECO. (2014). *Why is MORECO of interest to you?* Retrieved from http://www.moreco-project.eu/

Mohammad, R., & Sidaway, J. D. (2012). Spectacular urbanization amidst variegated geographies of globalization: Learning from Abu Dhabi's trajectory through the lives of South Asian men. *International Journal of Urban and Regional Research*, *36*(3), 606–627. doi:10.1111/j.1468-2427.2011.01099.x

Mohammed, F., Idries, A., Mohamed, N., Al-Jaroodi, J., & Jawhar, I. (2014). UAVs for smart cities: Opportunities and challenges. In *2014 International Conference on Unmanned Aircraft Systems (ICUAS)* (pp. 267-273). IEEE.

Mohan, M., & Kandya, A. (2015). Impact of urbanization and land-use/land-cover change on diurnal temperature range: A case study of tropical urban airshed of India using remote sensing data. *The Science of the Total Environment*, *506*, 453–465. doi:10.1016/j.scitotenv.2014.11.006 PMID:25437763

Mohan, M., Kandya, A., & Battiprolu, A. (2011). Urban heat island effect over national capital region of India: A study using the temperature trends. *Journal of Environmental Protection*, *2*(04), 465–472. doi:10.4236/jep.2011.24054

Mohan, M., Kikegawa, Y., Gurjar, B. R., Bhati, S., Kandya, A., & Ogawa, K. (2012). Urban heat island assessment for a tropical urban airshed in India. *Atmospheric and Climate Science*, *2*(2), 127–138. doi:10.4236/acs.2012.22014

Mohan, M., Kikegawa, Y., Gurjar, B. R., Bhati, S., & Kolli, N. R. (2013). Assessment of urban heat island effect for different land use–land cover from micrometeorological measurements and remote sensing data for megacity Delhi. *Theoretical and Applied Climatology*, *112*(3-4), 647–658. doi:10.100700704-012-0758-z

Mohan, M., Pathan, S. K., Narendrareddy, K., Kandya, A., & Pandey, S. (2011). Dynamics of urbanization and its impact on land-use/land-cover: A case study of megacity Delhi. *Journal of Environmental Protection*, *2*(09), 1274–1283. doi:10.4236/jep.2011.29147

Mölders, N., & Olson, M. A. (2004). Impact of urban effects on precipitation in high latitudes. *Journal of Hydrometeorology*, *5*(3), 409–429. doi:10.1175/1525-7541(2004)005<0409:IOUEOP>2.0.CO;2

Mondal, D. M. S., Sharma, N., Kappas, M., & Garg, P. (2012). Modeling of spatio-temporal dynamics of land use land cover - a review and assessment. *Journal of Geomatics*, *6*.

Morrison, J. L. (1995). Spatial data quality. In S. C. Guptill & J. L. Morrison (Eds.), *Elements of spatial data quality. International Cartographic Association* (pp. 1–12). Elsevier Science. doi:10.1016/B978-0-08-042432-3.50008-2

Mosammam, H. M., Nia, J. T., Khani, H., Teymouri, A., & Kazemi, M. (2017). Monitoring land use change and measuring urban sprawl based on its spatial forms: The case of Qom city. *The Egyptian Journal of Remote Sensing and Space Sciences*, *20*(1), 103–116. doi:10.1016/j.ejrs.2016.08.002

Mukherjee, S., Joshi, P. K., & Garg, R. D. (2017). Analysis of urban built-up areas and surface urban heat island using downscaled MODIS derived land surface temperature data. *Geocarto International*, *32*(8), 900–918. doi:10.1080/10106049.2016.1222634

Muller, M. R., & Middleton, J. (1994). A Markov model of land-use change dynamics in the Niagara Region, Ontario, Canada. *Landscape Ecology*, *9*(2), 151–157. doi:10.1007/BF00124382

Mundia, C. N., & Aniya, M. (2005). Analysis of land use/cover changes and urban expansion of Nairobi city using remote sensing and GIS. *International Journal of Remote Sensing*, *26*(13), 2831–2849. doi:10.1080/01431160500117865

Município de Cuiabá. (2013a). *Composição dos bairros de Cuiabá.* https://www.cuiaba.mt.gov.br/upload/arquivo/composicao_bairros.pdf

Município de Cuiabá. (2013b). *Plano Local de Habitação de Interesse Social.* Unpublished.

Município de Cuiabá. (2015). *Lei Complementar nº 389 - Lei de Uso e Ocupação do Solo (LUOS).* https://www.legisweb.com.br/legislacao/?id=320141

Muñoz, F. (2003). Lock living: Urban sprawl in Mediterranean cities. *Cities (London, England)*, *20*(6), 381–385. doi:10.1016/j.cities.2003.08.003

Muñoz-Vallés, S., Cambrollé, J., Figueroa-Luque, E., Luque, T., Niell, F. X., & Figueroa, M. E. (2013). An approach to the evaluation and management of natural carbon sinks: From plant species to urban green systems. *Urban Forestry & Urban Greening*, *12*(4), 450–453. doi:10.1016/j.ufug.2013.06.007

Murakami, A., Zain, A. M., Takeuchi, K., Tsunekawa, A., & Yokota, S. (2005). Trends in urbanization and patterns of land use in the Asian mega cities Jakarta, Bangkok, and Metro Manila. *Landscape and Urban Planning*, *70*(3-4), 251–259. doi:10.1016/j.landurbplan.2003.10.021

Murison, M. (2018). *DJI cements drone industry dominance with new enterprise products.* https://internetofbusiness.com/dji-drone-industry-domination-enterprise-hardware-and-software-launches/

Murphy, J. T., & Carmody, P. R. (2019). Generative urbanization in Africa? A sociotechnical systems view of Tanzania's urban transition. *Urban Geography*, *40*(1), 128–157. doi:10.1080/02723638.2018.1500249

Mutani, G., & Todeschi, V. (2020). The Effects of Green Roofs on Outdoor Thermal Comfort, Urban Heat Island Mitigation and Energy Savings. *Atmosphere*, *11*(2), 123. doi:10.3390/atmos11020123

Naghibi, F., Delavar, M. R., & Pijanowski, B. (2016). Urban Growth Modeling Using Cellular Automata with Multi-Temporal Remote Sensing Images Calibrated by the Artificial Bee Colony Optimization Algorithm. *Sensors (Basel)*, *16*(12), 2122. Advance online publication. doi:10.339016122122 PMID:27983633

Nascimento, M. S. B., & Oliveira, M. E. (2005). *Diversidade e uso de plantas nativas*. Retrieved from https://www.embrapa.br/imprensa/artigos/2005/artigo.2005.05.8820647706/

Nathalia, D., Kumar, K. M., Kishore, N., & Krishnan, G. (2017). Environmental change detection using Geo-Spatial techniques in Aravalli hills and Environs (Faridabad District, Haryana). *International Journal of Applied Environmental Sciences*, *12*(5), 865–875.

National Academies of Sciences, Engineering, and Medicine. (2012). *Travel Demand Forecasting: Parameters and Techniques*. Washington, DC: The National Academies Press.. doi:10.17226/14665

Nazarnia, N., Harding, C., & Jaeger, J. A. (2019). How suitable is entropy as a measure of urban sprawl? *Landscape and Urban Planning*, *184*, 32–43. doi:10.1016/j.landurbplan.2018.09.025

Neirotti, P., deMarco, A., Cagliano, A. C., Mangano, G., & Scorrano, F. (2014). Current Trends in Smart City Initiatives: Some Stylised Facts. *Cities (London, England)*, *38*, 25–36. doi:10.1016/j.cities.2013.12.010

Newbold, K. B., & Scott, D. (2013). Migration, commuting distance, and urban sustainability in Ontario's Greater Golden Horseshoe: Implications of the Greenbelt and Places to Grow legislation. *The Canadian Geographer. Geographe Canadien*, *57*(4), 474–487. doi:10.1111/j.1541-0064.2013.12044.x

Newman, P. W., & Kenworthy, J. R. (1992). Is there a role for physical planners? *Journal of the American Planning Association*, *58*(3), 353–362. doi:10.1080/01944369208975814

Newman, P., & Kenworthy, J. (1999). *Sustainability and Cities: Overcoming Automobile Dependence*. Island Press.

Ng, E., Chen, L., Wang, Y., & Yuan, C. (2012). A study on the cooling effects of greening in a high-density city: An experience from Hong Kong. *Building and Environment*, *47*, 256–271. doi:10.1016/j.buildenv.2011.07.014

Nieuwenhuijsen, M. J., Khreis, H., Triguero-Mas, M., Gascon, M., & Dadvand, P. (2017). Fifty shades of green. *Epidemiology (Cambridge, Mass.)*, *28*(1), 63–71. doi:10.1097/EDE.0000000000000549 PMID:27525811

Nikiforov, A., & Munimaev, V. (2010). *Analysis of foreign unmanned aerial vehicles*. St. Petersburg Forestry Academy.

Niyogi, D., Pyle, P., Lei, M., Arya, S. P., Kishtawal, C. M., Shepherd, M., Chen, F., & Wolfe, B. (2011). Urban modification of thunderstorms: An observational storm climatology and model case study for the Indianapolis urban region. *Journal of Applied Meteorology and Climatology*, *50*(5), 1129–1144. doi:10.1175/2010JAMC1836.1

Norton, B. A., Coutts, A. M., Livesley, S. J., Harris, R. J., Hunter, A. M., & Williams, N. S. G. (2015). Planning for cooler cities: A framework to prioritise green infrastructure to mitigate high temperatures in urban landscapes. *Landscape and Urban Planning*, *134*, 127–138. doi:10.1016/j.landurbplan.2014.10.018

Ohashi, Y., & Kida, H. (2002). Local circulations developed in the vicinity of both coastal and inland urban areas: A numerical study with a mesoscale atmospheric model. *Journal of Applied Meteorology*, *41*(1), 30–45. doi:10.1175/1520-0450(2002)041<0030:LCDITV>2.0.CO;2

Oke, T. R. (1982). The energetic basis of the urban heat island. *Quarterly Journal of the Royal Meteorological Society*, *108*(455), 1–24. doi:10.1002/qj.49710845502

Oke, T. R., Johnson, G. T., Steyn, D. G., & Watson, I. D. (1991). Simulation of surface urban heat islands under 'ideal'conditions at night Part 2: Diagnosis of causation. *Boundary-Layer Meteorology*, *56*(4), 339–358. doi:10.1007/BF00119211

Oktay, D. (2012). Human sustainable urbanism: In pursuit of ecological and social-cultural sustainability. *Procedia: Social and Behavioral Sciences*, *36*, 16–27. doi:10.1016/j.sbspro.2012.03.003

Olgyay, V. (2015). *Design with Climate: Bioclimatic Approach to Architectural Regionalism - New and expanded Edition*. Princeton University Press. doi:10.2307/j.ctvc77kqb

Olifer, V. G., & Olifer, N. A. (2006). Computer networks: principles, technologies, protocols. Academic Press.

Olivieri, F., Olivieri, L., & Neila, J. (2014). Experimental study of the thermal-energy performance of an insulated vegetal facade under summer conditions. *Building and Environment*, *77*, 61–76. doi:10.1016/j.buildenv.2014.03.019

Olorunfemi, J. F. (1983). Monitoring urban land use in developing countries—An aerial photographic approach. *Environment International*, *9*(1), 27–32. doi:10.1016/0160-4120(83)90111-3

Oniga, V. E., Breaban, A. I., & Statescu, F. (2018). Determining the Optimum Number of Ground Control Points for Obtaining High Precision Results Based on UAS Images. *Proceedings*, *2*(7), 352. doi:10.3390/ecrs-2-05165

Önsoy, H., & Bocquillon, C. (2009). Interpolation de variables régionales et cartographie automathique: Modèle triangulaire hydrologique (MTH)/Interpolation of regional variables and automatic drawing of isoplets. *Hydrological Sciences Bulletin*, *25*(3), 297–306. doi:10.1080/02626668009491936

Onur, I., Maktav, D., Sari, M., & Kemal Sönmez, N. (2009). Change detection of land cover and land use using remote sensing and GIS: A case study in Kemer, Turkey. *International Journal of Remote Sensing*, *30*(7), 1749–1757. doi:10.1080/01431160802639665

OpenDroneMap. (2020). *Drone Mapping Software, Features.* https://www.opendronemap.org/webodm/

Opromolla, R., Inchingolo, G., & Fasano, G. (2019). Airborne Visual Detection and Tracking of Cooperative UAVs Exploiting Deep Learning. *Sensors (Basel)*, *19*(9), 4322. doi:10.339019194332 PMID:31591368

Ortiz, L. E., González, J. E., Horton, R., Lin, W., Wu, W., Ramamurthy, P., Arend, M., & Bornstein, R. D. (2019). High-resolution projections of extreme heat in New York City. *International Journal of Climatology*, *39*(12), 4721–4735. doi:10.1002/joc.6102

Paixão, S., Hespanha, J. P., Ghawana, T., Carneiro, A. F. T., Zevenbergen, J., & Frederico, L. N. (2015). Modeling indigenous tribes' land rights with ISO 19152 LADM: A case from Brazil. *Land Use Policy*, *49*, 587–597. doi:10.1016/j.landusepol.2014.12.001

Pal, S., & Ziaul, S. K. (2017). Detection of land use and land cover change and land surface temperature in English Bazar urban centre. *The Egyptian Journal of Remote Sensing and Space Sciences*, *20*(1), 125–145. doi:10.1016/j.ejrs.2016.11.003

Panchen, Z. A., Primack, R. B., Nordt, B., Ellwood, E., Stevens, A., Renner, S. S., Willis, C. G., Fahey, R., Whittemore, A., Du, Y., & Davis, C. C. (2014). Leaf out times of temperate woody plants are related to phylogeny, deciduousness, growth habit, and woody anatomy. *The New Phytologist*, *203*(4), 1208–1219. doi:10.1111/nph.12892 PMID:24942252

Pandey, B., & Seto, K. C. (2015). Urbanization and agricultural land loss in India: Comparing satellite estimates with census data. *Journal of Environmental Management*, *148*, 53–66. doi:10.1016/j.jenvman.2014.05.014 PMID:24958549

Pandey, P., Kumar, D., Prakash, A., Masih, J., Singh, M., Kumar, S., Jain, V. K., & Kumar, K. (2012). A study of urban heat island and its association with particulate matter during winter months over Delhi. *The Science of the Total Environment*, *414*, 494–507. doi:10.1016/j.scitotenv.2011.10.043 PMID:22154211

Panek, J., & Netek, R. (2019). Collaborative Mapping and Digital Participation: A Tool for Local Empowerment in Developing Countries. *Information (Basel)*, *10*(255), 1–14. doi:10.3390/info10080255

Pan, L., & Chu, L. M. (2016). Energy saving potential and life cycle environmental impacts of a vertical greenery system in Hong Kong: A case study. *Building and Environment*, *96*, 293–300. doi:10.1016/j.buildenv.2015.06.033

Pan, L., Gu, L., Ren, R., & Yang, S. (2020, August 20). Land cover classification based on machine learning using UAV multispectral images. *Proc. SPIE 11501. Earth Observing Systems*, *25*, 115011F.

Pannell, C. (1995). China's urban transition. *The Journal of Geography*, *94*(3), 394–403. doi:10.1080/00221349508979342 PMID:12178548

Pastor, J., & Johnson, C. A. (1992). *Using Simulation Models and Geographic Information Systems to Integrate Ecosystem and Landscape Ecology*. doi:10.1007/978-1-4612-4382-3_11

Pastore, L., Corrao, R., & Kvols Heiselberg, P. (2017). The effects of vegetation on indoor thermal comfort: The application of a multi-scale simulation methodology on a residential neighborhood renovation case study. *Energy and Building*, *146*, 1–11. doi:10.1016/j.enbuild.2017.04.022

Pathan, S. K., Sastry, S. V. C., Dhinwa, P. S., Rao, M., Majumdar, K. L., Sampat Kumar, D., Patkar, V. N., & Phatak, V. N. (1993). Urban growth trend analysis using GIS techniques—A case study of the Bombay metropolitan region. *International Journal of Remote Sensing*, *14*(17), 3169–3179. doi:10.1080/01431169308904431

Pauleit, S., Ennos, R., & Golding, Y. (2005). Modeling the environmental impacts of urban land use and land cover change—A study in Merseyside, UK. *Landscape and Urban Planning*, *71*(2-4), 295–310. doi:10.1016/S0169-2046(04)00083-0

Paul, M. J., & Meyer, J. L. (2001). Streams in the urban landscape. *Annual Review of Ecology, Evolution, and Systematics*, *32*(1), 333–365. doi:10.1146/annurev.ecolsys.32.081501.114040

Payne, G. (2001). Urban land tenure policy options: Titles or rights? *Habitat International*, *25*(3), 415–429. doi:10.1016/S0197-3975(01)00014-5

Peña, J. M., Nilsson, R., Björkegren, J., & Tegnér, J. (2007). Towards scalable and data efficient learning of Markov boundaries. *International Journal of Approximate Reasoning*, *45*(2), 211–232. doi:10.1016/j.ijar.2006.06.008

Peñalosa, J. (1985). Dinâmica de crescimento de Lianas. In A. Gomezpompa & R. S. Del Amo (Eds.), Investigaciones sobre la regeneración de selvas altas en Veracruz, México (pp. 147-169). México: Alhambra Mexicana.

Pérez, M. (2012). Estimación de incertidumbres Guía GUM. *Revista Española de Metrología*, 113-130.

Perez, L. P., Bischof dos Santos, R., Almeida, G., & Carvalho, G. C. (2017). Spatial data in the Global South: A case study of alternative land management tools for cities with limited resources. In *2017 IEEE Global Humanitarian Technology Conference (GHTC)* (pp. 1-10). 10.1109/GHTC.2017.8239224

Petersen, E. R., & Vovsha, P. (2008). Directions for Coordinated Improvement of Travel Surveys and Models. In *Proceedings of Innovations in Travel Demand Modeling* (Vol. 2, pp. 85–88). Transportation Research Board of the National Academies.

Petrov, L., Lavalle, C., & Kasanko, M. (2009). Urban land use scenarios for a tourist region in Europe: Applying the MOLAND model to Algarve, Portugal. *Landscape and Urban Planning* (Vol. 92). doi:10.1016/j.landurbplan.2009.01.011

Pevzner, A. (2013). *Living Architecture: Green Roofs for Public Buildings*. General Services Administration: Energy & Sustainability / Safety & Environmental Branch Facilities Management & Services Programs Division PBS, Region 9. https://sftool.gov/Content/attachments/Living_Architecture_Green_Roofs_for_Public_Buildings.pdf

Pichon, L., Ducanchez, A., Fonta, H., & Tisseyre, B. (2016). Quality of Digital Elevation Models obtained from Unmanned Aerial Vehicles for Precision Viticulture. *OENO One*, *50*(3). Advance online publication. doi:10.20870/oeno-one.2016.50.3.1177

Pielke, R. A., & Adegoke, J., BeltraáN-Przekurat, A., Hiemstra, C. A., Lin, J., Nair, U. S., ... Nobis, T. E. (2007). An overview of regional land-use and land-cover impacts on rainfall. Tellus B. *Chemical and Physical Meteorology*, *59*(3), 587–601.

Pielke, R. A. Sr, Pitman, A., Niyogi, D., Mahmood, R., McAlpine, C., Hossain, F., ... Reichstein, M. (2011). Land use/land cover changes and climate: Modeling analysis and observational evidence. *Wiley Interdisciplinary Reviews: Climate Change*, *2*(6), 828–850. doi:10.1002/wcc.144

Pinheiro, F. J. (2013). *Déficit habitacional municipal no Brasil*. Centro de Estatística e Informações., https://goo.gl/euvcpK

Pinto, N. N., Tenedório, J. A., Antunes, A. P., & Cladera, J. R. (2014). *Technologies for Urban and Spatial Planning: Virtual Cities and Territories*. IGI Global. doi:10.4018/978-1-4666-4349-9

Pinto, N., Antunes, A. P., & Roca, J. (2017). Applicability and calibration of an irregular cellular automata model for land use change. *Computers, Environment and Urban Systems*, *65*, 93–102. doi:10.1016/j.compenvurbsys.2017.05.005

Pires, A. (2016). *Combining paper maps and smartphones in the exploration of cultural heritage* (Unpublished master dissertation). Mestrado em engenharia informática, Faculdade de Ciências e Tecnologia da Universidade Nova de Lisboa. Retrieved from https://run.unl.pt/handle/10362/19962

Pix4D. (2017). *Do RTK/PPK drones give you better results than GCPs?* https://www.pix4d.com/blog/rtk-ppk-drones-gcp-comparison

Pix4D. (2018). *Do more GCPs equal more accurate drone maps?* https://www.pix4d.com/blog/GCP-accuracy-drone-maps

Pix4D. (2019). *How to verify that there is enough overlap between the images*. https://support.pix4d.com/hc/en-us/articles/203756125-How-to-verify-that-there-is-enough-overlap-between-the-images

Pix4D. (2020). *Number and distribution of ground control points (GCPs) in corridor mapping*. https://support.pix4d.com/hc/en-us/articles/202559299-Number-and-distribution-of-ground-control-points-GCPs-in-corridor-mapping

Polydoros, A., & Cartalis, C. (2014). Assessing thermal risk in urban areas–an application for the urban agglomeration of Athens. *Advances in Building Energy Research*, *8*(1), 74–83. doi:10.1080/17512549.2014.890536

Pontius, R. (2002). Statistical Methods to Partition Effects of Quantity and Location During Comparison of Categorical Maps at Multiple Resolutions. *Photogrammetric Engineering and Remote Sensing*, *68*.

Pontius, R. G. Jr, Shusas, E., & McEachern, M. (2004). Detecting important categorical land changes while accounting for persistence. *Agriculture, Ecosystems & Environment*, *101*(2), 251–268. doi:10.1016/j.agee.2003.09.008

Popescu, G., Iordan, D., & Păunescu, V. (2016). The Resultant Positional Accuracy for the Orthophotos Obtained with Unmanned Aerial Vehicles (UAVs). *Agriculture and Agricultural Science Procedia*, *10*, 458–464. doi:10.1016/j.aaspro.2016.09.016

Portugali, J. (2002). Self-Organization and the City. *J. Artificial Societies and Social Simulation* (Vol. 5). doi:10.1007/978-3-662-04099-7

Prashar, S., Shaw, R., & Takeuchi, Y. (2012). Assessing the resilience of Delhi to climate-related disasters: A comprehensive approach. *Natural Hazards*, *64*(2), 1609–1624. doi:10.100711069-012-0320-4

Prata, A. J., Caselles, V., Coll, C., Sobrino, J. A., & Ottlé, C. (1995). Thermal remote sensing of land surface temperature from satellites: Current status and future prospects. *Remote Sensing Reviews*, *12*(3-4), 175–224. doi:10.1080/02757259509532285

Prawjal, M., Jain, R., Srinivasa, V., & Karthik, K. S. (2016). Optimal Number of Ground Control Points for a UAV based Corridor Mapping. *International Journal of Innovative Research in Science, Engineering and Technology*, *5*(9), 28–32. doi:10.15680/IJIRSET.2016.0505505

Prefeitura de São José dos Campos - PSJC & Instituto de Pesquisa, Administração e Planejamento - IPPLAN. (2014). *Atlas da Pesquisa Origem e Destino: Panorama da Mobilidade em São José dos Campos*. São Carlos, Brazil: Editora Cubo Prefeitura de São José dos Campos - PSJC.

Prefeitura de São José dos Campos - PSJC. (2015). *Urban mobility report - Diagnosis and prognosis*. Retrieved from https://www.sjc.sp.gov.br/media/56153/diagnostico_planmob.pdf

Prefeitura de São José dos Campos - PSJC. (2016). *Evaluation of Risk Areas of São José dos Campos Municipality*. Retrieved from https://www.sjc.sp.gov.br/media/45004/pmrr_final_i_e_ii.pdf

Prefeitura de São José dos Campos - PSJC. (2019). *Public transport operational data (XLS table)*. Retrieved from https://www.sjc.sp.gov.br/servicos/governanca/portal-da-transparencia/dados-do-transporte-coletivo/dados-operacionais/

Pregnolato, M., Ford, A., Wilkinson, S. M., & Dawson, R. J. (2017). The impact of flooding on road transport: A depth disruption function. *Transportation Research Part D, Transport and Environment*, *55*, 67–81. doi:10.1016/j.trd.2017.06.020

Primack, R. B., & Rodrigues, E. (2001). *Biologia da conservação* (8th ed.). Planta Editora.

Propeller Aero. (2019). *What is Ground Sample Distance (GSD) and How Does it Affect Your Drone Data?* https://www.propelleraero.com/blog/ground-sample-distance-gsd-calculate-drone-data/

Pujals, A. (2011). *Ecologia de lianas e cipós em fragmentos florestais: Dualidade e propostas de manejo* (Unpublished master thesis). Universidade de Maringá.

Punia, M., & Singh, L. (2012). Entropy approach for assessment of urban growth: A case study of Jaipur, India. *Photonirvachak (Dehra Dun), 40*(2), 231–244. doi:10.100712524-011-0141-z

Qaid, A., Lamit, H. B., Ossen, D. R., & Shahminan, R. N. R. (2016). Urban heat island and thermal comfort conditions at micro-climate scale in a tropical planned city. *Energy and Building, 133*, 577–595. doi:10.1016/j.enbuild.2016.10.006

Qin, R., Tian, J., & Reinartz, P. (2016). 3D change detection - Approaches and applications. *ISPRS Journal of Photogrammetry and Remote Sensing, 122*, 41–56. doi:10.1016/j.isprsjprs.2016.09.013

Qin, Z., Karnieli, A., & Berliner, P. (2001). A mono-window algorithm for retrieving land surface temperature from Landsat TM data and its application to the Israel-Egypt border region. *International Journal of Remote Sensing, 22*(18), 3719–3746. doi:10.1080/01431160010006971

R Core Team. (2020). *R: A language and environment for statistical computing. R Foundation for Statistical Computing, Vienna, Austria.* https://www.R-project.org/

Rabinovitch, J. (1996). Innovative land use and public transport policy: The case of Curitiba, Brazil. *Land Use Policy, 13*(1), 51–67. doi:10.1016/0264-8377(95)00023-2

Raciti, S. M., Hutyra, L. R., & Newell, J. D. (2014). Mapping carbon storage in urban trees with multi-source remote sensing data: Relationships between biomass, land use, and demographics in Boston neighborhoods. *The Science of the Total Environment, 500–501*, 72–83. doi:10.1016/j.scitotenv.2014.08.070 PMID:25217746

Radjawali, I., Pye, O., & Flitner, M. (2017). Recognition through reconnaissance? Using drones for counter-mapping in Indonesia. *The Journal of Peasant Studies, 44*(4), 817–833. doi:10.1080/03066150.2016.1264937

Rai, B., & Nair, S. S. (2013). Change detection of Barkhal Lake in Faridabad District of Haryana using geo-informatic techniques. *International Journal of Remote Sensing & Geoscience, 2*(2), 38–41.

Rajasekar, U., Chakraborty, S., & Bhat, G. (2018). Climate resilient smart cities: opportunities for innovative solutions in India. In *Climate Change in Cities* (pp. 203–227). Springer. doi:10.1007/978-3-319-65003-6_11

Rakha, T., & Gorodetsky, A. (2020). Unmanned Aerial System (UAS) Applications in the Built Environment: Towards Automated Building Inspection Procedures Using Drones. In Unmanned Aerial Remote Sensing UAS for Environmental Applications. CRC Press.

Rakha, T., & Gorodetsky, A. (2018). Review of Unmanned Aerial System (UAS) applications in the built environment: Towards automated building inspection procedures using drones. *Automation in Construction, 93*, 252–264. doi:10.1016/j.autcon.2018.05.002

Ramachandra, T. V., & Kumar, U. (2010). Greater Bangalore: Emerging urban heat island. *GIS Development, 14*(1), 86-104.

Ramachandra, T. V., Aithal, B. H., & Sanna, D. D. (2012). Insights to urban dynamics through landscape spatial pattern analysis. *International Journal of Applied Earth Observation and Geoinformation, 18*, 329–343. doi:10.1016/j.jag.2012.03.005

Ramachandra, T. V., Bharath, A. H., & Sowmyashree, M. V. (2015). Monitoring urbanization and its implications in a mega city from space: Spatiotemporal patterns and its indicators. *Journal of Environmental Management, 148*, 67–81. doi:10.1016/j.jenvman.2014.02.015 PMID:24768450

Ramalho, C. L., & Proença, C. E. B. (2004). Trepadeiras ornamentais do cerrado. Brasília: Universidade de Brasília e Embrapa Cerrados.

Ranagalage, M., Estoque, R., Zhang, X., & Murayama, Y. (2018). Spatial changes of urban heat island formation in the Colombo District, Sri Lanka: Implications for sustainability planning. *Sustainability, 10*(5), 1367. doi:10.3390u10051367

Ranjan-Martha, T., Kerle, N., van Westen, C. J., & Jetten, V. G. (2010). Effect of Sun Elevation Angle on DSMs Derived from Cartosat-1 Data. *Photogrammetric Engineering and Remote Sensing, 76*(4), 429–438. doi:10.14358/PERS.76.4.429

Rao, B., Gopi, A., & Maione, R. (2016). The societal impact of commercial drones. *Technology in Society, 45*, 83–90. doi:10.1016/j.techsoc.2016.02.009

Raposo, I. (2017). Intervir nas margens urbanas: o papel da academia. *Revista Espaços vividos e espaços construídos. Estudos sobre a cidade, 5*, 29-39.

Rasul, A., Balzter, H., & Smith, C. (2015). Spatial variation of the daytime Surface Urban Cool Island during the dry season in Erbil, Iraqi Kurdistan, from Landsat 8. *Urban Climate, 14*, 176–186. doi:10.1016/j.uclim.2015.09.001

Rath, S. S., & Panda, J. (2019). A Study of Near-Surface Boundary Layer Characteristics During the 2015 Chennai Flood in the Context of Urban-Induced Land Use Changes. *Pure and Applied Geophysics, 176*(6), 2607–2629. doi:10.100700024-018-2069-5

Razzaghmanesh, M., Beecham, S., & Salemi, T. (2016). The role of green roofs in mitigating urban heat island effects in the metropolitan area of Adelaide, South Australia. *Urban Forestry & Urban Greening, 15*, 89–102. doi:10.1016/j.ufug.2015.11.013

Reagan, D., Sabato, A., & Niezrecki, C. (2017). Unmanned Aerial Vehicle Acquisition of Three-Dimensional Digital Image Correlation Measurements for Structural Health Monitoring of Bridges. Conference: SPIE Smart Structures and Materials + Nondestructive Evaluation and Health Monitoring.

Reagan, J. (2015). *How Low Will Drone Prices Go?* https://dronelife.com/2015/07/09/how-low-will-drone-prices-go/

Rebelo, C. (2016). *3D Point Clouds in Urban Planning: Developing and Releasing High-end Methodologies based on LiDAR and UAV Data for the Extraction of Building Parameters* (PhD Thesis). Portugal, Universidade NOVA de Lisboa, Faculdade de Ciências Sociais e Humanas. http://hdl.handle.net/10362/19005

Rebelo, C., Rodrigues, A. M., Tenedório, J. A., Gonçalves, J. A., & Marnoto, J. (2015). Building 3D City Models: Testing and Comparing Laser Scanning and Low-Cost UAV Data Using FOSS Technologies. In Computational Science and Its Applications – ICCSA 2015. ICCSA 2015. Lecture Notes in Computer Science, (vol. 9157). Springer. doi:10.1007/978-3-319-21470-2_26

Rebelo, C., Rodrigues, A., Tenedório, J., Goncalves, J., & Marnoto, J. (2015). Building 3D City Models: Testing and Comparing Laser Scanning and Low-Cost UAV Data Using FOSS Technologies. In Computational Science and Its Applications -- ICCSA 2015 (pp. 367–379). Springer International Publishing.

Rebelo, C., Rodrigues, A., & Tenedório, J. A. (2019). Dasymetric Mapping Using UAV High Resolution 3D Data Within Urban Areas. *Remote Sensing*, *11*(14), 1716. doi:10.3390/rs11141716

Redweik, P. (2007). *Fotogrametria Aérea*. Departamento de Engenharia Geográfica, Geofísica e Energia, Faculdade de Ciências da Universidade de Lisboa.

Reinhart, C. F., & Davila, C. C. (2016). Urban building energy modeling – A review of a nascent field. *Building and Environment*, *97*, 196–202. doi:10.1016/j.buildenv.2015.12.001

Revich, B. (2011). Heat-wave, air quality and mortality in European Russia in summer 2010: Preliminary assessment. *Yekologiya Cheloveka. Human Ecology*, *7*, 3–9.

Roca, J. (1988). *La estructura de los valores urbanos: análisis teórico empírico*. Instituto de Estudios de Administración Local.

Rocha, F. J. (2012). *Sistemas complexos, modelação e geossimulação da evolução de padrões de uso e ocupação do solo* [PhD Thesis]. IGOT, Universidade de Lisboa.

Rocha, J. (2012). *Sistemas complexos, modelação e geosimulação da evolução de padrões de uso e ocupação do solo*. Universidade de Lisboa. Retrieved from http://hdl.handle.net/10451/6772

Rocha, J. (2019). Simulation and Sustainability BT - Encyclopedia of Sustainability in Higher Education. doi:10.1007/978-3-319-63951-2_413-1

Rocha, J., Ferreira, J., Simões, J., & Tenedório, J. (2007). Modelling Coastal and Land Use Evolution Patterns through Neural Network and Cellular Automata Integration. *Journal of Coastal Research SI, 50.*

Rocha, J., Gutierres, F., Gomes, P., & Teodoro, A. C. (2018). *A Hybrid CA-ANN-Fuzzy Model for Simulating Coastal Changing Patterns BT - Beach Management Tools - Concepts, Methodologies and Case Studies* (C. M. Botero, O. Cervantes, & C. W. Finkl, Eds.). doi:10.1007/978-3-319-58304-4_10

Rock, G., Ries, J. B., & Udelhoven, T. (2011). Sensitivity analysis of UAV-Photogrammetry for creating Digital Elevation Models (DEM). In *Proceedings of Conference on Unmanned Aerial Vehicle in Geomatics* (vol. XXXVIII-1/C22, pp. 69-74). Academic Press.

Rogerson, P. (2001). *Statistical Methods for Geography*. doi:10.4135/9781849209953

Roggema, R. (2016). The Future of Sustainable Urbanism: A Redefinition. *City, Territory and Architecture*, *3*(22).

Rolnik, R., & Santoro, P. F. (2013). *Zonas Especiais de Interesse Social (ZEIS) em Cidades Brasileiras: Trajetória Recente de Implementação de um Instrumento de Política Fundiária*. Lincoln Institute of Land Policy. https://www.lincolninst.edu/sites/default/files/pubfiles/rolnik-wp14rr1po-full_0.pdf

Roman, C., & Mather, R. (2016). Autonomous Underwater Vehicles as Tools for Deep-Submergence Archaeology. *Journal of Engineering for the Maritime Environment, 224*(4), 327-340.

Roman, C., & Mather, R. (2016). Autonomous Underwater Vehicles as Tools for Deep-Submergence Archaeology. *Proceedings of the Institution of Mechanical Engineers, Part M: Journal of Engineering for the Maritime Environment*, *224*(4), 327-340.

Romero, A. B. (2013). Princípios bioclimáticos para o desenho urbano (2nd ed.). Brasília: Editora UnB. doi:10.7476/9788523013486

Ross, L. (2010). *Virtual 3D City Models in Urban Land Management—Technologies and Applications* (Ph.D. Thesis). Technische Universität Berlin, Berlin, Germany.

Rouse, J. J., Haas, R., Deering, D., Schell, J., & Harlan, J. (1974). Monitoring the Vernal Advancement and Retrogradation (Green Wave Effect) of Natural Vegetation. NASA/GSFC Type III Final Report, Greenbelt, MD.

Rousta, I., Sarif, M., Gupta, R., Olafsson, H., Ranagalage, M., Murayama, Y., Zhang, H., & Mushore, T. (2018). Spatiotemporal analysis of land use/land cover and its effects on surface urban heat island using Landsat data: A case study of Metropolitan City Tehran (1988–2018). *Sustainability*, *10*(12), 4433. doi:10.3390u10124433

Rozoff, C. M., Cotton, W. R., & Adegoke, J. O. (2003). Simulation of St. Louis, Missouri, land use impacts on thunderstorms. *Journal of Applied Meteorology*, *42*(6), 716–738. doi:10.1175/1520-0450(2003)042<0716:SOSLML>2.0.CO;2

Rudolf, C. (1867). *The Mechanical Theory of Heat: With its Applications to the Steam-engine and to the Physical Properties of Bodies*. Academic Press.

Rutkauskas, M., Asenov, M., Ramamoorthy, S., & Reid, D. T. (2019). Autonomous multi-species environmental gas sensing using drone-based Fourier-transform infrared spectroscopy. Optics Express, 27(7).

Saadatian, O., Sopian, K., Salleh, E., Lim, C. H., Riffat, S., Saadatian, E., Toudeshki, A., & Sulaiman, M. Y. (2013). A review of energy aspects of green roofs. *Renewable & Sustainable Energy Reviews*, *23*, 155–168. doi:10.1016/j.rser.2013.02.022

Sahana, M., Dutta, S., & Sajjad, H. (2019). Assessing land transformation and its relation with land surface temperature in Mumbai city, India using geospatial techniques. *International Journal of Urban Sciences*, *23*(2), 205–225. doi:10.1080/12265934.2018.1488604

Sahana, M., Hong, H., & Sajjad, H. (2018). Analyzing urban spatial patterns and trend of urban growth using urban sprawl matrix: A study on Kolkata urban agglomeration, India. *The Science of the Total Environment*, *628*, 1557–1566. doi:10.1016/j.scitotenv.2018.02.170 PMID:30045573

Saini, H. S., Kaura, L. P., Mujtaba, S. A. I., Verma, K. & Pant, N. C (2017). Assessment of the soils around Faridabad, NCR in the context of changing landuse. *Indian Journal of Geosciences, 70-71*(3&4), 219–226.

Salamanca, F., Martilli, A., & Yagüe, C. (2012). A numerical study of the Urban Heat Island over Madrid during the DESIREX (2008) campaign with WRF and an evaluation of simple mitigation strategies. *International Journal of Climatology*, *32*(15), 2372–2386. doi:10.1002/joc.3398

Salata, F., Golasi, I., Vollaro, A. D., & Vollaro, R. D. (2015). How high albedo and traditional buildings' materials and vegetation affect the quality of urban microclimate. A case study. *Energy and Building*, *99*, 32–49. doi:10.1016/j.enbuild.2015.04.010

Salvati, L., Serra, P., Bencardino, M., & Carlucci, M. (2019). Re-urbanizing the European city: A multivariate analysis of population dynamics during expansion and recession times. *European Journal of Population*, *35*(1), 1–28. doi:10.100710680-017-9462-0 PMID:30976266

Santana, V. D., Silva, J. S., Ribeiro, H. L. M., Arantes, A. J. R., Silva, O. C., & Alves, L. L. (2016). InfoDengue – Ferramenta Colaborativa de Combate e Prevenção à Dengue. *Computer on the Beach*, (9), 1-4. Retrieved from https://siaiap32.univali.br/seer/index.php/acotb/article/view/10787

Santos, A. P. (2010). Urbanização e Indústria: Interpretação Cartográfica da Evolução Urbana de São José dos Campos, São Paulo, 1920-2000. XI Seminário da História da Cidade e do Urbanismo - Seminário da História da Cidade e do Urbanismo, 11, 1-22.

Santos, J. D., Bacelar, R. B., & Santos, L. B. L. (2017). Desastres e mobilidade urbana - um recorte da literatura. In *Proceedings of XXII Simpósio Brasileiro de Recursos Hídricos (SBRH)*. Florianópolis, Brazil: ABRHidro.

Santos, C. (1981). *Movimentos urbanos no Rio de Janeiro*. Zahar Editores.

Santos, L. B. L., Carvalho, T., Anderson, L. O., Rudorff, C. M., Marchezini, V., Londe, L. R., & Saito, S. M. (2017). A RS-GIS-based Comprehensive Impact Assessment of Floods - a Case Study in Madeira River, Western Brazilian Amazon. *IEEE Geoscience and Remote Sensing Letters*, *14*(9), 1–4. doi:10.1109/LGRS.2017.2726524

Santos, L. B. L., Londe, L. R., Soriano, E., Souza, A., & Coelho, A. F. (2015). Potential flood-related daily urban mobility problems in Rio de Janeiro (Brazil). *Revista do Departamento de Geografia*, *29*, 175–190.

Santos, T., Silva, C., & Tenedório, J. A. (2019). Promoting Citizens' Quality of Life Through Green Urban Planning. In L. Ragia, R. Laurini, & J. G. Rocha (Eds.), *Geographical Information Systems Theory, Applications and Management* (pp. 153–175). Springer International Publishing. doi:10.1007/978-3-030-06010-7_10

Santos, T., Tenedório, J. A., & Gonçalves, J. A. (2016). Quantifying the City's Green Area Potential Gain Using Remote Sensing Data. *Sustainability*, *8*(12), 1247. doi:10.3390u8121247

Sanz-Ablanedo, E., Chandler, J. H., Rodríguez-Pérez, J. R., & Ordóñez, C. (2018). Accuracy of Unmanned Aerial Vehicle (UAV) and SfM Photogrammetry Survey as a Function of the Number and Location of Ground Control Points Used. *Remote Sensing*, *10*(10), 1606. doi:10.3390/rs10101606

Sarvestani, M. S., Ibrahim, A. L., & Kanaroglou, P. (2011). Three decades of urban growth in the city of Shiraz, Iran: A remote sensing and geographic information systems application. *Cities (London, England)*, *28*(4), 320–329. doi:10.1016/j.cities.2011.03.002

Satellite Imaging Corp. (n.d.). *Satellite Sensors and Specifications*. Retrieved March 23, 2020, from https://www.satimagingcorp.com/satellite-sensors/

Savard, J. P. L., Clergeau, P., & Mennechez, G. (2000). Biodiversity concepts and urban ecosystems. *Landscape and Urban Planning*, *48*(3-4), 131–142. doi:10.1016/S0169-2046(00)00037-2

Sawicki, D. S., & Flynn, P. (1996). Neighborhood Indicators: A Review of the Literature and an Assessment of Conceptual and Methodological Issues. *Journal of the American Planning Association*, *62*(2), 165–183. doi:10.1080/01944369608975683

Scannapieco, A. F., Renga, A., Fasano, G., & Moccia, A. (2017). Ultralight Radar for Small and Micro-UAV Navigation. *The International Archives of the Photogrammetry, Remote Sensing and Spatial Information Sciences, Volume XLII-2/W6, 2017. International Conference on Unmanned Aerial Vehicles in Geomatics*, 333-338. 10.5194/isprs-archives-XLII-2-W6-333-2017

Schatsky, D., & Ream, J. (2016). *Advanced software applications are driving commercial drone adoption.* https://www2.deloitte.com/insights/us/en/focus/signals-for-strategists/advanced-drone-software-applications.html

Schlünzen, K. H., Hoffmann, P., Rosenhagen, G., & Riecke, W. (2010). Long-term changes and regional differences in temperature and precipitation in the metropolitan area of Hamburg. *International Journal of Climatology*, *30*(8), 1121–1136. doi:10.1002/joc.1968

Schneider, A., & Woodcock, C. E. (2008). Compact, dispersed, fragmented, extensive? A comparison of urban growth in twenty-five global cities using remotely sensed data, pattern metrics and census information. *Urban Studies (Edinburgh, Scotland)*, *45*(3), 659–692. doi:10.1177/0042098007087340

Schreiber, E., Heinzel, A., Peichl, M., Engel, M., & Wiesbeck, W. (2019). Advanced Buried Object Detection by Multichannel, UAV/Drone Carried Synthetic Aperture Radar. In *2019 13th European Conference on Antennas and Propagation (EuCAP)* (vol. 1, pp. 1-15). Academic Press.

Schroth, L. (2018). *Drones and Artificial Intelligence.* https://www.droneii.com/drones-and-artificial-intelligence

Seifert, E., Seifert, S., Vogt, H., Drew, D., van Aardt, J., Kunneke, A., & Seifert, T. (2019). Influence of Drone Altitude, Image Overlap, and Optical Sensor Resolution on Multi-View Reconstruction of Forest Images. *Remote Sensing*, *11*(10), 1252. doi:10.3390/rs11101252

Semboloni, F. (2000). The Growth of an Urban Cluster into a Dynamic Self-Modifying Spatial Pattern. *Environment and Planning. B, Planning & Design*, *27*(4), 549–564. doi:10.1068/b2673

Senanayake, I. P., Welivitiya, W. D. D. P., & Nadeeka, P. M. (2013). Assessment of green space requirement and site analysis in Colombo, Sri Lanka–a remote sensing and GIS approach. *International Journal of Scientific and Engineering Research*, *4*(12), 29–34.

Seto, K. C., Güneralp, B., & Hutyra, L. R. (2012). Global forecasts of urban expansion to 2030 and direct impacts on biodiversity and carbon pools. *Proceedings of the National Academy of Sciences of the United States of America*, *109*(40), 16083–16088. doi:10.1073/pnas.1211658109 PMID:22988086

Sexton, J. O., Song, X. P., Huang, C., Channan, S., Baker, M. E., & Townshend, J. R. (2013). Urban growth of the Washington, DC–Baltimore, MD metropolitan region from 1984 to 2010 by annual, Landsat-based estimates of impervious cover. *Remote Sensing of Environment*, *129*, 42–53. doi:10.1016/j.rse.2012.10.025

Shang, D., Doulet, J.-F., & Keane, M. (2009). Urban Informatics in China: Exploring the Emergence of the Chinese City 2.0. In M. Foth (Ed.), *Handbook of Research on Urban Informatics: The Practice and Promise of the Real-Time City* (pp. 379–389). IGI Global. doi:10.4018/978-1-60566-152-0.ch026

Shannon, C. E. (1948). A mathematical theory of communication. *The Bell System Technical Journal*, *27*(3), 379–423. doi:10.1002/j.1538-7305.1948.tb01338.x

Sharifi, A. (2016). From Garden City to Eco-urbanism: The quest for sustainable neighborhood development. *Sustainable Cities and Society*, *20*, 1–16. doi:10.1016/j.scs.2015.09.002

Sharifi, E., & Lehmann, S. (2015). Correlation analysis of surface temperature of rooftops, streetscapes and urban heat island effect: Case study of central Sydney. *Journal of Urban and Environmental Engineering*, *9*(1), 3–11. doi:10.4090/juee.2015.v9n1.3-11

Sharma, D., & Tomar, S. (2010). Mainstreaming climate change adaptation in Indian cities. *Environment and Urbanization*, *22*(2), 451–465. doi:10.1177/0956247810377390

Sharma, M. P., Kaur, A., & Shalley, V. (2005). Satellite based study of land transformation in Ludhiana district, Punjab. *Photonirvachak (Dehra Dun)*, *33*(1), 63. doi:10.1007/BF02989991

Sharma, R., & Joshi, P. K. (2014). Identifying seasonal heat islands in urban settings of Delhi (India) using remotely sensed data–An anomaly based approach. *Urban Climate*, *9*, 19–34. doi:10.1016/j.uclim.2014.05.003

Shastri, H., Barik, B., Ghosh, S., Venkataraman, C., & Sadavarte, P. (2017). Flip flop of day-night and summer-winter surface urban heat island intensity in India. *Scientific Reports*, *7*(1), 40178. doi:10.1038rep40178 PMID:28067276

Shem, W., & Shepherd, M. (2009). On the impact of urbanization on summertime thunderstorms in Atlanta: Two numerical model case studies. *Atmospheric Research*, *92*(2), 172–189. doi:10.1016/j.atmosres.2008.09.013

Shen, G., Ibrahim Abdoul, N., Zhu, Y., Wang, Z., & Gong, J. (2017). Remote sensing of urban growth and landscape pattern changes in response to the expansion of Chongming Island in Shanghai, China. *Geocarto International*, *32*(5), 488–502. doi:10.1080/10106049.2016.1156166

Shen, L., Zhou, J., Skitmore, M., & Xia, B. (2015). Application of a hybrid Entropy–McKinsey Matrix method in evaluating sustainable urbanization: A China case study. *Cities (London, England)*, *42*, 186–194. doi:10.1016/j.cities.2014.06.006

Shepherd, J. M. (2005). A review of current investigations of urban-induced rainfall and recommendations for the future. *Earth Interactions*, *9*(12), 1–27. doi:10.1175/EI156.1

Shepherd, J. M., & Burian, S. J. (2003). Detection of urban-induced rainfall anomalies in a major coastal city. *Earth Interactions*, *7*(4), 1–17. doi:10.1175/1087-3562(2003)007<0001:DOUIRA>2.0.CO;2

Shepherd, J. M., Carter, M., Manyin, M., Messen, D., & Burian, S. (2010). The impact of urbanization on current and future coastal precipitation: A case study for Houston. *Environment and Planning. B, Planning & Design*, *37*(2), 284–304. doi:10.1068/b34102t

Shepherd, J. M., Pierce, H., & Negri, A. J. (2002). Rainfall modification by major urban areas: Observations from spaceborne rain radar on the TRMM satellite. *Journal of Applied Meteorology*, *41*(7), 689–701. doi:10.1175/1520-0450(2002)041<0689:RMBMUA>2.0.CO;2

Shi, Y., & Xiao, J. (2008). Evaluating landscape changing due to urbanization using remote sensing data: A case study of Shijiazhuang, China. In *Education Technology and Training, 2008. and 2008 International Workshop on Geoscience and Remote Sensing. ETT and GRS 2008. International Workshop on*, (vol. 1, pp. 559-562). IEEE. 10.1109/ETTandGRS.2008.183

Shiode, N. (2001). 3D urban models: Recent developments in the digital modelling of urban environments in three-dimensions. *GeoJournal*, *52*(3), 263–269. doi:10.1023/A:1014276309416

Shirowzhan, S., & Sepasgozar, S. M. E. (2019). Spatial Analysis Using Temporal Point Clouds in Advanced GIS: Methods for Ground Elevation Extraction in Slant Areas and Building Classifications. ISPRS International Journal of Geo-Information, 8(3), 120.

Shirowzhan, S., & Lim, S. (2014). Three-Dimensional Spatial Metrics for Compactness Assessment of Urban Forms. *Proceedings of the 31st International Symposium on Automation and Robotics in Construction and Mining*. 10.22260/ISARC2014/0119

Shirowzhan, S., & Lim, S. (2013). Development of Urban Sustainability Index using 3-D spatial metrics. In *Proceedings of the ICSDEC 2012 International Conference on Sustainable Design, Engineering, and Construction* (pp. 238-245). 10.1061/9780784412688.028

Shmelova, T., & Bondarev, D. (2015). Theory Applying for Quantitative Estimation of UAV's Group Flight. *IEEE 3D International Conference Actual Problems of Unmanned Aerial Vehicles Developments (APPUAVD),* 328-331.

Shmelova, T., Lazorenko, V., Bondarev, D., & Burlaka, O. (2019). Group flights of Unmanned Aviation Vehicles for Smart cities. *IEEE 9th International Conference on Advanced Computer Information Technologies (ACIT),* 230-233. 10.1109/ACITT.2019.8780071

Shmelova, T., Sterenharz, O., & Burlaka, O. (2019) Optimization of flows and flexible redistribution of autonomous UAV routes in multilevel airspace. *IEEE Proceedings of the 15th International Conference on ICT in Education, Research and Industrial Applications. Integration, Harmonization and Knowledge Transfer; 5th International Workshop on Theory of Reliability and Markov Modelling for Information Technologies (TheRMIT), 4*, 703-714.

Shmelova, T., Stratij, A. (2015). *Computer Program "Aggregation of heterogeneous information flows": certificate of registration of copyright in the product N° 59466*. Academic Press.

Shmelova, T., & Bondarev, D. (2016). Graph Theory Applying for Quantitative Estimation of UAVs Group Flight into Aerial Photography. *Electronics and Control Systems*, *4*(46), 128–133.

Shmelova, T., Bondarev, D., & Znakovska, Y. (2016). Modeling of the Decision Making by UAVs Operator in Emergency Situations. *4th International Conference on Methods and Systems of Navigation and Motion Control (MSNMC)*, 31-34. 10.1109/MSNMC.2016.7783099

Shmelova, T., Kovalyov, Yu., Dolgikh, S., & Burlaka, O. (2019) Geometry-modeling based flight optimization for autonomous groups of UAVs *5D International Conference Actual Problems of Unmanned Aerial Vehicles Developments (APUAVD)*, 79-82. 10.1109/APUAVD47061.2019.8943856

Shmelova, T., Sikirda, Y., Rizun, N., Kucherov, D., & Dergachov, K. (Eds.). (2019). *Automated Systems in the Aviation and Aerospace Industries*. Engineering Science Reference.

Shreffler, J. H. (1979). Urban-rural differences in tower-measured winds, St. Louis. *Journal of Applied Meteorology*, *18*(7), 829–835. doi:10.1175/1520-0450(1979)018<0829:URDITM>2.0.CO;2

Shrestha, R., Flacke, J., Martinez, J., & Van Maarseveen, M. (2016). Environmental health related socio-spatial inequalities: Identifying "hotspots" of environmental burdens and social vulnerability. *International Journal of Environmental Research and Public Health*, *13*(7), 691. doi:10.3390/ijerph13070691 PMID:27409625

Silva, B., Silva, C., & Santos, T. (2018). Protocolo de simulação computacional do microclima urbano para a cidade de Lisboa, Portugal. *Península Ibérica No Mundo: Problemas e Desafios Para Uma Intervenção Ativa Da Geografia*, 1066–1073.

Silva, F. D. S., & Blamires, D. (2007). Avifauna urbana no Lago Pôr do Sol, Iporá, Goiás, Brasil. *Lundiana: International Journal of Biodiversity*, *8*, 17–26.

Silva, M. F. da R. (2014). *Simulação energética de coberturas verdes*. https://fenix.tecnico.ulisboa.pt/cursos/mec/dissertacao/2353642480248

Silva, C. M., Flores-Colen, I., & Antunes, M. (2017). Step-by-step approach to ranking green roof retrofit potential in urban areas: A case study of Lisbon, Portugal. *Urban Forestry & Urban Greening*, *25*, 120–129. doi:10.1016/j.ufug.2017.04.018

Simon, H., Lindén, J., Hoffmann, D., Braun, P., Bruse, M., & Esper, J. (2018). Modeling transpiration and leaf temperature of urban trees - A case study evaluating the microclimate model ENVI-met against measurement data. *Landscape and Urban Planning*, *174*, 33–44. doi:10.1016/j.landurbplan.2018.03.003

Singh, P., Kikon, N., & Verma, P. (2017). Impact of land use change and urbanization on urban heat island in Lucknow city, Central India. A remote sensing based estimate. *Sustainable Cities and Society*, *32*, 100–114. doi:10.1016/j.scs.2017.02.018

Skayannis, P. (2010). Is urban underground space luxury? From infrastructures to subterranean structures. In *24th AESOP Annual Conference: Association of European Schools of Planning*. Aalto University.

Skitmore, A. (2018). *Launch and Recovery System for Improved Fixed-Wing UAV Deployment in Complex Environments* [Unpublished MSc Thesis]. Department of Maritime and Mechanical Engineering. Liverpool John Moores University.

Skitmore, A. (2020). Launch and Recovery System for Improved Fixed-Wing UAV Deployment in Complex Environments. In Unmanned Aerial Remote Sensing: UAS for Environmental Applications. CRC Press.

Sładkowski, A., & Wojciech, K. (2019). Cases on Modern Computer Systems in Aviation. In Using Unmanned Aerial Vehicles to Solve Some Civil Problems. Academic Press.

Sliuzas, R., Gevaert, C. M., Persello, C., & Vosselman, G. (2018). Evaluating the Societal Impact of Using Drones to Support Urban Upgrading Projects. *ISPRS International Journal of Geo-Information*, *7*(3). https://www.mdpi.com/2220-9964/7/3/91

Small, K. A., & Verhoef, E. T. (2007). *The Economics of Urban Transportation*. Routledge. doi:10.4324/9780203642306

Smart Cities Guide, G. S. M. A. Traffic Management, GSM Association. (2016) Available https://www.gsma.com/iot/gsma-smart-cities-guide-traffic-management/

Smeur, E. J. J., Bronz, M., & de Croon, G. C. H. E. (2018). *Incremental control and guidance of hybrid aircraft applied to the Cyclone tailsitter UAV.* https://arxiv.org/pdf/1802.00714.pdf

SMHI. (2015). *Vad är ett solbanediagram?* http://www.smhi.se/polopoly_fs/1.32097.1398236573!/image/stockholm_kontur.png_gen/derivatives/Original/stockholm_kontur.png

Snow, C. (2019). *Seven Trends That Will Shape the Commercial Drone Industry In 2019.* https://www.forbes.com/sites/colinsnow/2019/01/07/seven-trends-that-will-shape-the-commercial-drone-industry-in-2019/#31914dda7494

Sobrino, J. A., Jiménez-Muñoz, J. C., & Paolini, L. (2004). Land surface temperature retrieval from LANDSAT TM 5. *Remote Sensing of Environment, 90*(4), 434–440. doi:10.1016/j.rse.2004.02.003

Sodoudi, S., Shahmohamadi, P., Vollack, K., Cubasch, U., & Che-Ani, A. I. (2014). Mitigating the urban heat island effect in megacity Tehran. *Advances in Meteorology, 2014*, 1–19. doi:10.1155/2014/547974

Song, Y., & Knaap, G. J. (2004). Measuring urban form: Is Portland winning the war on sprawl? *Journal of the American Planning Association, 70*(2), 210–225. doi:10.1080/01944360408976371

Son, N. T., Chen, C. F., Chen, C. R., & Chiang, S. H. (2016). Mapping urban growth of the capital city of Honduras from Landsat data using the impervious surface fraction algorithm. *Geocarto International, 31*(3), 328–341. doi:10.1080/10106049.2015.1047469

Sood, A. (2015). Industrial townships and the policy facilitation of corporate urbanisation in India. *Urban Studies (Edinburgh, Scotland), 52*(8), 1359–1378. doi:10.1177/0042098014562318

Srinivasan, R. S., Lakshmanan, J., Srivastav, D., & Santosa, E. (2011). Benchmarking plug-load densities for k-12 schools. *Proceedings of Building Simulation, 2011*, 2746–2752.

Stanilov, K., & Batty, M. (2011). Exploring the Historical Determinants of Urban Growth Patterns through Cellular Automata. *Transactions in GIS, 15*(3), 253–271. doi:10.1111/j.1467-9671.2011.01254.x

Stathopoulou, M., & Cartalis, C. (2009). Downscaling AVHRR land surface temperatures for improved surface urban heat island intensity estimation. *Remote Sensing of Environment, 113*(12), 2592–2605. doi:10.1016/j.rse.2009.07.017

Stensson, L. N. (2016). *Spatial Accuracy in Orthophoto produced using UAV Photographic Images* [Unpublished master thesis]. Karlstads Universitet, Karlstad, Sweden.

Stern, M., Peck, S. W., & Joslin, J. (2019). *Green roof and wall policy in North America Regulations, Incentives, and Best Practices.* Green Roof for Healthy Cities. https://static1.squarespace.com/static/58e3eecf2994ca997dd56381/t/5d84dfc371cf0822bdf7dc29/1568989140101/Green_Roof_and_Wall_Policy_in_North_America.pdf

Stewart, I. D. (2011). A systematic review and scientific critique of methodology in modern urban heat island literature. *International Journal of Climatology, 31*(2), 200–217. doi:10.1002/joc.2141

Stöcker, C., Ho, S., Koeva, M., Nkerabigwi, P., Schmidt, C., Zevenbergen, J., & Bennett, R. (2018). Towards UAV-based Land Tenure Data Acquisition in Rwanda: Needs Assessment and Technology Response. *FIG Congress 2018: Embracing our smart world where the continents connect - Enhancing geospatial maturity of societies*. http://www.fig.net/resources/proceedings/fig_proceedings/fig2018/papers/ts08c/TS08C_stocker_koeva_et_al_9428.pdf

Stocker, C., Bennett, R., Nex, F., Gerke, M., & Zevenbergen, J. (2017). Review of the Current State of UAV. *Remote Sensing*, *9*(5), 459–486. doi:10.3390/rs9050459

Stone, B., Hess, J. J., & Frumkin, H. (2010). Urban form and extreme heat events: Are sprawling cities more vulnerable to climate change than compact cities? *Environmental Health Perspectives*, *118*(10), 1425–1428. doi:10.1289/ehp.0901879 PMID:21114000

Storch, H., & Downes, N. K. (2011). A scenario-based approach to assess Ho Chi Minh City's urban development strategies against the impact of climate change. *Cities (London, England)*, *28*(6), 517–526. doi:10.1016/j.cities.2011.07.002

Sudhira, H. S., Ramachandra, T. V., & Jagadish, K. S. (2004). Urban sprawl: Metrics, dynamics and modelling using GIS. *International Journal of Applied Earth Observation and Geoinformation*, *5*(1), 29–39. doi:10.1016/j.jag.2003.08.002

Sudhira, H. S., Ramachandra, T. V., Raj, K. S., & Jagadish, K. S. (2003). Urban growth analysis using spatial and temporal data. *Photonirvachak (Dehra Dun)*, *31*(4), 299–311. doi:10.1007/BF03007350

Sultana, S., & Satyanarayana, A. N. V. (2018). Urban heat island intensity during winter over metropolitan cities of India using remote-sensing techniques: Impact of urbanization. *International Journal of Remote Sensing*, *39*(20), 6692–6730. doi:10.1080/01431161.2018.1466072

Sun, C., Wu, Z. F., Lv, Z. Q., Yao, N., & Wei, J. B. (2013). Quantifying different types of urban growth and the change dynamic in Guangzhou using multi-temporal remote sensing data. *International Journal of Applied Earth Observation and Geoinformation*, *21*, 409–417. doi:10.1016/j.jag.2011.12.012

Surjan, A., & Shaw, R. (2009). Enhancing disaster resilience through local environment management: Case of Mumbai, India. *Disaster Prevention and Management: An International Journal*, *18*(4), 418–433. doi:10.1108/09653560910984474

Sutton, P. C. (2003). A scale-adjusted measure of "urban sprawl" using nighttime satellite imagery. *Remote Sensing of Environment*, *86*(3), 353–369. doi:10.1016/S0034-4257(03)00078-6

Takayama, N., Korpela, K., Lee, J., Morikawa, T., Tsunetsugu, Y., Park, B. J., & Kagawa, T. (2014). Emotional, restorative and vitalizing effects of forest and urban environments at four sites in Japan. *International Journal of Environmental Research and Public Health*, *11*(7), 7207–7230. doi:10.3390/ijerph110707207 PMID:25029496

Taleghani, M. (2018). Outdoor thermal comfort by different heat mitigantion strategies – A review. *Renewable & Sustainable Energy Reviews*, *81*, 2011–2018. doi:10.1016/j.rser.2017.06.010

Taleghani, M., Sailor, D. J., Tenpierik, M., & van den Dobblesteen, A. (2014). Thermal assessment of heat mitigation strategies: The case of Portland state university, Oregon, USA. *Building and Environment*, *73*, 138–150. doi:10.1016/j.buildenv.2013.12.006

Tan, J., Zheng, Y., Tang, X., Guo, C., Li, L., Song, G., Zhen, X., Yuan, D., Kalkstein, A. J., Li, F., & Chen, H. (2010). The urban heat island and its impact on heat waves and human health in Shanghai. *International Journal of Biometeorology*, *54*(1), 75–84. doi:10.100700484-009-0256-x PMID:19727842

Tao, X., Liang, S., Hea, T., & Jina, H. (2016). Estimation of fraction of absorbed photosynthetically active radiation from multiple satellite data: Model development and validation. *Remote Sensing of Environment*, *184*, 539–557. doi:10.1016/j.rse.2016.07.036

Tarabalka, Y., Chanussot, J., & Benediktsson, J. A. (2010). Segmentation and classification of hyperspectral images using watershed transformation. *Pattern Recognition*, *43*(7), 2367–2379. doi:10.1016/j.patcog.2010.01.016

Tarasov, V., & Gerasimov, B. (2007). Intelligent Decision Support Systems: Theory, Synthesis, Efficiency. Kyiv: International Academy of Computer Science and Systems.

Targ, S. M. (1986). *Short course of theoretical mechanics*. Vyssh. Shkola Publ.

Taubenböck, H., Esch, T., Felbier, A., Wiesner, M., Roth, A., & Dech, S. (2012). Monitoring urbanization in mega cities from space. *Remote Sensing of Environment*, *117*, 162–176. doi:10.1016/j.rse.2011.09.015

Taubenböck, H., Wegmann, M., Roth, A., Mehl, H., & Dech, S. (2009). Urbanization in India–Spatiotemporal analysis using remote sensing data. *Computers, Environment and Urban Systems*, *33*(3), 179–188. doi:10.1016/j.compenvurbsys.2008.09.003

Tayanç, M., Karaca, M., & Yenigün, O. (1997). Annual and seasonal air temperature trend patterns of climate change and urbanization effects in relation to air pollutants in Turkey. *Journal of Geophysical Research, D, Atmospheres*, *102*(D2), 1909–1919. doi:10.1029/96JD02108

Tehrany, M. S., Pradhan, B., & Jebuv, M. N. (2014). A comparative assessment between object and pixel-based classification approaches for land use/land cover mapping using SPOT 5 imagery. *Geocarto International*, *29*(4), 351–369. doi:10.1080/10106049.2013.768300

Templin, T., Brzezinski, G., & Rawa, M. (2019). Visualization of Spatio-Temporal Building Changes Using 3D Web GIS. *IOP Conference Series. Earth and Environmental Science*, *221*, 012084. doi:10.1088/1755-1315/221/1/012084

Tenedório, J. A., Estanqueiro, R., Lima, A. M., & Marques, J. (2016). Remote sensing from unmanned aerial vehicles for 3D urban modelling: case study of Loulé, Portugal. In Virtual City and Territory "Back to the Sense of the City: International Monograph Book". Barcelona: Centre de Política de Sòl i Valoracions.

Tenedório, J. A., Rebelo, C., Estanqueiro, R., Henriques, C. D., Marques, L., & Gonçalves, J. A. (2016). New Developments in Geographical Information Technology for Urban and Spatial Planning. In Geospatial Research: Concepts, Methodologies, Tools, and Applications, (pp. 1965-1997). Hershey, PA: IGI Global. doi:10.4018/978-1-4666-9845-1.ch094

Tenedório, J., & Marques, L. (2017). How can 3D models and augmented reality visualization based on mobile platforms enhance the value of urban heritage? *International Federation of Surveyors*. Retrieved from https://www.fig.net/resources/monthly_articles/2017/tenedorie_etal_december_2017.asp

Tenedório, J. A., Estanqueiro, R., Matos Lima, A., & Marques, J. (2016). Remote sensing from unmanned aerial vehicles for 3D urban modelling: case study of Loulé, Portugal. In *Back to the Sense of the City: International Monograph Book* (pp. 1118–1128). Centre de Política de Sòl i Valoracions.

Terfa, B. K., Chen, N., Liu, D., Zhang, X., & Niyogi, D. (2019). Urban Expansion in Ethiopia from 1987 to 2017: Characteristics, Spatial Patterns, and Driving Forces. *Sustainability*, *11*(10), 2973. doi:10.3390u11102973

Tewolde, M. G., & Cabral, P. (2011). Urban sprawl analysis and modeling in Asmara, Eritrea. *Remote Sensing*, *3*(10), 2148–2165. doi:10.3390/rs3102148

TGRTAG -Toronto Green Roof Technical Advisory Group. (2011). *Toronto Green Roof Construction Standard (TGRCS)*. Office of the Chief Building Official, Toronto Building, City of Toronto. Canada. http://www.toronto.ca/greenroofs/pdf/GreenRoof-supGuidelines.pdf

Thielen, J., Wobrock, W., Gadian, A., Mestayer, P. G., & Creutin, J. D. (2000). The possible influence of urban surfaces on rainfall development: A sensitivity study in 2D in the meso-γ-scale. *Atmospheric Research*, *54*(1), 15–39. doi:10.1016/S0169-8095(00)00041-7

Thillainayagi, R., & Kumar, K. S. (2019). Bi-dimensional Empirical Mode Decomposition Based Contrast Enhancement Technique for UAV Thermal Images. *Journal of the Institution of Electronics and Telecommunication Engineers*, 1–8. Advance online publication. doi:10.1080/03772063.2019.1619487

Thomas, G., Sherin, A. P., Ansar, S., & Zachariah, E. J. (2014). Analysis of urban heat island in Kochi, India, using a modified local climate zone classification. *Procedia Environmental Sciences*, *21*, 3–13. doi:10.1016/j.proenv.2014.09.002

Thomas, R. W. (1981). *Information statistics in geography, geo abstracts*. University of East Anglia.

Tian, Y., & Jim, C. Y. (2011). Factors influencing the spatial pattern of sky gardens in the compact city of Hong Kong. *Landscape and Urban Planning*, *101*(4), 299–309. doi:10.1016/j.landurbplan.2011.02.035

Tmuši, G. (2020). Current Practices in UAV-based Environmental Monitoring. *Remote Sensing*, *12*(6), 1001. doi:10.3390/rs12061001

Tobler, W. R. (1970). Computer Movie Simulating Urban Growth in the Detroit Region. *Economic Geography, 46*, 234-240. https://www.tandfonline.com/doi/abs/10.2307/143141

Todes, A., Kok, P., Wentzel, M., Van Zyl, J., & Cross, C. (2010). Contemporary South African urbanization dynamics. *Urban Forum, 21(3), 331-348*. doi:10.100712132-010-9094-5

Tooke, T. R., Coops, N. C., Goodwin, N. R., & Voogt, J. A. (2009). Extracting urban vegetation characteristics using spectral mixture analysis and decision tree classifications. *Remote Sensing of Environment, 113*(2), 398–407. doi:10.1016/j.rse.2008.10.005

Toosi, N. B., Fakheran, S., & Soffianian, A. (2012). Analysis of landscape pattern changes in Isfahan city during the last two decades. In *International Conference on Applied Life Sciences*. IntechOpen.

Torrens, P. M., & O'Sullivan, D. (2000). Cities, cells and complexity: developing a research agenda for urban geocomputation. *Geocomputation 2000*. Retrieved from http://www.geosimulation.org/geosim/gc044.htm

Torrens, P. M., & Benenson, I. (2005). Geographic Automata Systems. *International Journal of Geographical Information Science, 19*(4), 385–412. doi:10.1080/13658810512331325139

Tran, T. D.-B., Puissant, A., Badariotti, D., & Weber, C. (2011). Optimizing Spatial Resolution of Imagery for Urban Form Detection—The Cases of France and Vietnam. *Remote Sensing, 3*(10), 2128–2147. doi:10.3390/rs3102128

Tsai, Y.-H. (2005). Quantifying Urban Form: Compactness versus 'Sprawl'. *Urban Studies (Edinburgh, Scotland), 42*(1), 141–161. doi:10.1080/0042098042000309748

Tsou, J., Zhuang, J., Li, Y., & Zhang, Y. (2017). Urban heat island assessment using the Landsat 8 data: A case study in Shenzhen and Hong Kong. *Urban Science, 1*(1), 10. doi:10.3390/urbansci1010010

Tucci, C. E. M. (2007). *Inundações Urbanas*. ABRH.

Turner, J. (1977). *Vivienda: todo el poder para los usuarios*. H. Blume.

Turok, I., & McGranahan, G. (2013). Urbanization and economic growth: The arguments and evidence for Africa and Asia. *Environment and Urbanization, 25*(2), 465–482. doi:10.1177/0956247813490908

Tversky, B. (2005). Functional significance of visuospatial representations. In Handbook of Higher-Level Visuospatial Thinking, (pp. 1–34). Cambridge, UK: Cambridge University Press.

Tzoulas, K., Korpela, K., Venn, S., Yli-Pelkonen, V., Kaźmierczak, A., Niemela, J., & James, P. (2007). Promoting ecosystem and human health in urban areas using Green Infrastructure: A literature review. *Landscape and Urban Planning, 81*(3), 167–178. doi:10.1016/j.landurbplan.2007.02.001

U.S. Environmental Protection Agency. (2008). *Handbook for Developing Watershed Plans to Restore and Protect Our Waters. EPA 841-B-08-002*. Nonpoint Source Control Branch.

Ulam, S. (1952). Random processes and transformations. In *International Congress of Mathematicians*, (pp. 264–75). Providence, RI: American Mathematical Society.

UNDESA. (2015). *World urbanization prospects: The 2014 revision*. United Nations Department of Economics and Social Affairs, Population Division.

UNDESA. (2018). *Revision of world urbanization prospects*. United Nations Department of Economics and Social Affairs, Population Division.

UNDRR - United Nations Office for Disaster Risk Reduction. (2015). *Poorly planned and managed urban development. PreventionWeb, the knowledge plataform for disaster risk reduction*. UNDRR.

UNDRR - United Nations Office for Disaster Risk Reduction. (2019). *Terminology on Disaster Risk Reduction*. Retrieved from https://www.unisdr.org/we/inform/terminology

UN-FIG (1996). *The Bogor Declaration*. United Nations Interregional Meeting of Experts on the Cadastre, Bogor, Indonesia.

UN-FIG. (1999). *Executive Summary of the Bathurst Declaration. Workshop on Land Tenure and Cadastral Infrastructures for Sustainable Development*, Bathurst, Australia.

United Nations Development Programme. (2016). *Sustainable Urbanization Strategy. UNDP's support to sustainable, inclusive and resilient cities in the developing world.* https://www.undp.org/content/undp/en/home/librarypage/poverty-reduction/sustainable-urbanization-strategy.html

United Nations General Assembly. (1987). Report of the world commission on environment and development: Our common future. Oslo, Norway: United Nations General Assembly, Development, and International Co-operation. *Environment*.

United Nations Human Settlements Programme. (2008). *Secure Land Rights for All*. UN-Habitat/Global Land Tool Network.

United Nations Human Settlements Programme. (2014). *The Right to Adequate Housing*. Fact Sheet No.21, United Nations https://www.ohchr.org/Documents/Publications/FS21_rev_1_Housing_en.pdf

United Nations Human Settlements Programme. (2016). *Urbanization and development: emerging Futures. World Cities Report 2016*. https://unhabitat.org/sites/default/files/download-manager-files/WCR-2016-WEB.pdf

United Nations. (2015). Transforming our world: The 2030 agenda for sustainable development. *General Assembley 70 session*. United States Geological Survey. Landsat8 Operational Land Imager and Thermal Infrared Sensor Calibration Notices. Available online: https://landsat.usgs.gov/calibration_notices.php

United Nations. (2015). *Transforming our world: the 2030 Agenda for Sustainable Development*. https://sdgs.un.org/2030agenda

United Nations. (2016). *Sustainable Development Goals Report 2016*. UN.

United Nations. (2017). *New Urban Agenda*. https://www.habitat3.org/

United Nations. (2018). *World Urbanization Prospects, The 2018 Revision*. United Nations. Available online: https://www.un.org/development/desa/publications/2018-revision-of-world-urbanization-prospects.html

Ural, S., Hussain, E., & Shan, J. (2011). Building population mapping with aerial imagery and GIS data. *International Journal of Applied Earth Observation and Geoinformation, 13*(6), 841–852. doi:10.1016/j.jag.2011.06.004

Ustaoglu, E., Williams, B., Petrov, O. L., Shahumyan, H., & Van Delden, H. (2018). Developing and Assessing Alternative Land-Use Scenarios from the MOLAND Model: A Scenario-Based Impact Analysis Approach for the Evaluation of Rapid Rail Provisions and Urban Development in the Greater Dublin Region. *Sustainability, 10*(2), 61. Advance online publication. doi:10.3390u10010061

Uysal, M., Toprak, A. S., & Polat, N. (2015). DEM generation with UAV Photogrammetry and accuracy analysis in Sahitler hill. *Measurement, 73*, 539–543. doi:10.1016/j.measurement.2015.06.010

Valladares, L. P. (1980). *Passa-se uma casa: análise do programa de remoção de favelas do Rio de Janeiro*. Zahar Editores.

Vall-Casas, P., Koschinsky, J., & Mendoza, C. (2011). Retrofitting suburbia through pre-urban patterns: Introducing a European perspective. *URBAN DESIGN International, 16*(3), 171–187. doi:10.1057/udi.2011.9

Valor, E., & Caselles, V. (1996, September). Mapping land surface emissivity from NDVI: Application to European, African, and South American areas. *Remote Sensing of Environment, 57*(3), 167–184. doi:10.1016/0034-4257(96)00039-9

Valsson, S. & Bharat, A (2009). Urban Heat Island: Cause for Microclimate Variations. *Architecture-Time Space & People*, 21–25.

van den Bosch, M., & Nieuwenhuijsen, M. (2017). No time to lose-Green the cities now. *Environment International, 99*, 343–350. doi:10.1016/j.envint.2016.11.025 PMID:27923587

Van Dijk, T. (2011). Networks of urbanization in two Indian cities. *Environment & Urbanization Asia, 2*(2), 303–319. doi:10.1177/097542531100200211

van Rees, E. (2017). *Pix4D Extends Drone-based Imagery with Machine Learning Techniques*. https://www.commercialuavnews.com/energy/pix4d-extends-drone-based-imagery-machine-learning-techniques

Van, T. T., & Bao, H. D. X. (2010). Study of the impact of urban development on surface temperature using remote sensing in Ho Chi Minh City, northern Vietnam. *Geographical Research, 48*(1), 86–96. doi:10.1111/j.1745-5871.2009.00607.x

Vendrame, I. F., Mota, M. S., & Reis, D. (2016). Inundações no Córrego Senhorinha e suas Consequências na Mobilidade Urbana. In *Proceedings of II Encontro sobre Impactos Potenciais de Desastres Naturais em Infraestrutura de Transporte e Mobilidade Urbana (IPTMU)*. São José dos Campos, Brazil: UNESP.

Vendrame, I. F., Senaubar, M. A., & Scofield, G. B. (2000). Aspectos da Evolução da Urbanização e dos Problemas de Inundação em São José dos Campos. *Proceedings of XXVII Congresso Interamericano de Engenharia Sanitária e Ambiental, 27*, 1-9.

Verbesselt, J., Hyndman, R., Newnham, G., & Culvenor, D. (2010). Detecting trend and seasonal changes in satellite image time series. *Remote Sensing of Environment, 114*(1), 106–115. doi:10.1016/j.rse.2009.08.014

Verburg, P. H., Crossman, N., Ellis, E. C., Heinimann, A., Hostert, P., Mertz, O., Nagendra, H., Sikor, T., Erb, K.-H., Golubiewski, N., Grau, R., Grove, M., Konaté, S., Meyfroidt, P., Parker, D. C., Chowdhury, R. R., Shibata, H., Thomson, A., & Zhen, L. (2015). Land system science and sustainable development of the earth system: A global land project perspective. *Anthropocene, 12*, 29–41. doi:10.1016/j.ancene.2015.09.004

Verykokou, S., & Ioannidis, C. (2018). A photogrammetry-based structure from motion algorithm using robust iterative bundle adjustment techniques. *ISPRS Annals of the Photogrammetry, Remote Sensing and Spatial Information Sciences, 4*(4/W6), 73–80. 10.5194/isprs-annals-IV-4-W6-73-2018

Veyret, Y. (2007). *Os riscos: o homem como agressor e vítima do meio ambiente*. Contexto.

Viegas, C. V., Saldanha, D. L., Bond, A., Ribeiro, J. L. D., & Selig, P. M. (2013). Urban land planning: The role of a Master Plan in influencing local temperatures. *Cities (London, England), 35*, 1–13. doi:10.1016/j.cities.2013.05.006

Vijayaraghavan, K. (2016). Green roofs: A critical review on the role of components, benefits, limitations and trends. *Renewable & Sustainable Energy Reviews, 57*, 740–752. doi:10.1016/j.rser.2015.12.119

Villa, T. F., Gonzalez, F., Miljievi, B., Ristovsky, Z. D., & Morawska, L. (2016). An Overview of Small Unmanned Aerial Vehicles for Air Quality Measurements: Present Applications and Future Prospectives. *Sensors (Basel), 16*(7), 1072. doi:10.339016071072 PMID:27420065

Vitanova, L. L., Kusaka, H., Doan, V. Q., & Nishi, A. (2019). Numerical Study of the Urban Heat Island in Sendai City with Potential Natural Vegetation and the 1850s and 2000s Land-Use Data. *Journal of the Meteorological Society of Japan. Ser. II*.

Von Neumann, J. (1966). The theory of automata: Construction, reproduction, homogeneity. In J. Von Neumann & A. W. Burks (Eds.), *Theory of Self-Reproducing Automata* (pp. 89–250). University of Illinois Press.

Vučić, N. (2017). *LADM country profile Croatia*. The 6th Land Administration Domain Model (LADM) Workshop, Delft, The Netherlands.

Vu, T. T., Yamazaki, F., & Matsuoka, M. (2009). Multi-scale solution for building extraction from LiDAR and image data. *International Journal of Applied Earth Observation and Geoinformation*, *11*(4), 281–289. doi:10.1016/j.jag.2009.03.005

Vyrelkin, A., & Kucheryavy, A. (2014). The usage of unmanned aircraft to solve the tasks of "a smart city". SPbGUT.

Wadhawan, M., & Ahmad, S. (2010). *11th ESRI India User Conference 2010*. Academic Press.

Walter, H. (1971). *Ecology of tropical and subtropical vegetation*. Van Nostrand Reinhold.

Wang, J., Feng, J., Yan, Z., Hu, Y., & Jia, G. (2012). Nested high-resolution modeling of the impact of urbanization on regional climate in three vast urban agglomerations in China. *Journal of Geophysical Research, D, Atmospheres*, *117*(D21), n/a. doi:10.1029/2012JD018226

Wang, J., & Mountrakis, G. (2011). Developing a multi-network urbanization model: A case study of urban growth in Denver, Colorado. *International Journal of Geographical Information Science*, *25*(2), 229–253. doi:10.1080/13658810903473213

Wang, J., & Yan, Z. W. (2016). Urbanization-related warming in local temperature records: A review. *Atmospheric and Oceanic Science Letters*, *9*(2), 129–138. doi:10.1080/16742834.2016.1141658

Wang, L., & Li, D. (2019). Modulation of the urban boundary-layer heat budget by a heatwave. *Quarterly Journal of the Royal Meteorological Society*, *145*(722), 1814–1831. doi:10.1002/qj.3526

Wang, L., Lu, Y., & Yao, Y. (2019). Comparison of three algorithms for the retrieval of land surface temperature from Landsat 8 images. *Sensors (Basel)*, *19*(22), 5049. doi:10.339019225049 PMID:31752412

Wang, Y. C., Hu, B. K. H., Myint, S. W., Feng, C.-C., Chow, W. T. L., & Passy, P. F. (2018). Patterns of land change and their potential impacts on land surface temperature change in Yangon, Myanmar. *The Science of the Total Environment*, *643*, 738–750. doi:10.1016/j.scitotenv.2018.06.209 PMID:29957438

Wan, H., Zhong, Z., Yang, X., & Li, X. (2013). Impact of city belt in Yangtze River Delta in China on a precipitation process in summer: A case study. *Atmospheric Research*, *125*, 63–75. doi:10.1016/j.atmosres.2013.02.004

Wan, H., Zhong, Z., Yang, X., & Li, X. (2015). Ensembles to model the impact of urbanization for a summertime rainstorm process in Yangtze River Delta, China. *Meteorological Applications*, *22*(1), 105–112. doi:10.1002/met.1360

Watts, A. C., Ambrosia, V. G., & Hinkley, E. A. (2012). Unmanned Aircraft Systems in Remote Sensing and Scientific Research: Classification and Considerations of Use. *Remote Sensing*, *4*(6), 1671–1692. doi:10.3390/rs4061671

Weber, C., Petropoulou, C., & Hirsch, J. (2005). Urban development in the Athens metropolitan area using remote sensing data with supervised analysis and GIS. *International Journal of Remote Sensing*, *26*(4), 785–796. doi:10.1080/0143116051233l316856

Weier, J., & Herring, D. (2000, August 30). *Measuring Vegetation (NDVI and EVI)*. Retrieved 2013, from Earth Observatory: https://earthobservatory.nasa.gov/Features/MeasuringVegetation/

Weilenmann, B., Seidl, I., & Schulz, T. (2017). The socio-economic determinants of urban sprawl between 1980 and 2010 in Switzerland. *Landscape and Urban Planning*, *157*, 468–482. doi:10.1016/j.landurbplan.2016.08.002

Weiping, M., & Guangchuang, Y. (2018). *BasicTrendline: Add Trendline and Confidence Interval of Basic Regression Models to Plot. R package version 2.0.3.* https://CRAN.R-project.org/package=basicTrendline

Weng, Q. (2001). Modeling urban growth effects on surface runoff with the integration of remote sensing and GIS. *Environmental Management*, *28*(6), 737–748. doi:10.1007002670010258 PMID:11915963

Weng, Q. (2002). Land use change analysis in the Zhujiang Delta of China using satellite remote sensing, GIS and stochastic modelling. *Journal of Environmental Management*, *64*(3), 273–284. doi:10.1006/jema.2001.0509 PMID:12040960

Weng, Q. (2009). Thermal infrared remote sensing for urban climate and environmental studies: Methods, applications and trends. *ISPRS Journal of Photogrammetry and Remote Sensing*, *64*(4), 335–344. doi:10.1016/j.isprsjprs.2009.03.007

Weng, Q. (2012). Remote sensing of impervious surfaces in the urban areas: Requirements, methods, and trends. *Remote Sensing of Environment*, *117*, 34–49. doi:10.1016/j.rse.2011.02.030

Westoby, M. J., Brasington, J., Glasser, N. F., Hambre, M. J., & Reynolds, J. M. (2012). "Structure-from-Motion" photogrammetry: A low-cost, effective tool for geoscience applications. *Geomorphology*, *179*, 300–314. doi:10.1016/j.geomorph.2012.08.021

Whitehead, K., & Hugenholtz, C. H. (2015). Applying ASPRS accuracy standards to surveys from small unmanned aircraft systems (UAV). *Photogrammetric Engineering and Remote Sensing*, *81*(10), 787–793. doi:10.14358/PERS.81.10.787

White, R., & Engelen, G. (1994). Urban systems dynamics and cellular automata: Fractal structures between order and chaos. *Chaos, Solitons, and Fractals*, *4*(4), 563–583. doi:10.1016/0960-0779(94)90066-3

White, R., Engelen, G., & Uljee, I. (2015). *Modeling Cities and Regions as Complex Systems: From Theory to Planning Applications*. MIT Press. doi:10.7551/mitpress/9780262029568.001.0001

Wickham, H., François, R., Henry, L., & Müller, K. (2020). *dplyr: A Grammar of Data Manipulation. R package version 0.8.5.* https://CRAN.R-project.org/package=dplyr

Wierzbicki, D., Kedzierski, M., & Fryskowska, A. (2015). Assessment of the influence of UAV image quality on the orthophoto production. In *Proceedings of International Conference on Unmanned Aerial Vehicles in Geomatics* (vol. XL-1/W4, pp. 1-8). 10.5194/isprsarchives-XL-1-W4-1-2015

Wihbey, J. (2017). The Drone Revolution: UAV-Generated Geodata Drives Policy Innovation. In Land Lines. Lincoln Institute of Land Policy.

Wilby, R. L. (2008). Constructing climate change scenarios of urban heat island intensity and air quality. *Environment and Planning. B, Planning & Design*, *35*(5), 902–919. doi:10.1068/b33066t

Williamson, I., Enemark, S., Wallace, J., & Rajabifard, A. (2009). *Land Administration for Sustainable Development*. ESRI Press.

Willis, C. G., Ruhfel, B. R., Primack, R. B., Miller-Rushing, A. J., Losos, J. B., & Davis, C. C. (2010). Favorable climate change response explains non-native success in Thoreau's woods. *PLoS One*, *5*(1), e8878. doi:10.1371/journal.pone.0008878 PMID:20126652

Wilson, A. (2010). Entropy in Urban and Regional Modelling: Retrospect and Prospect. *Geographical Analysis*, *42*(4), 364–394. doi:10.1111/j.1538-4632.2010.00799.x

Wilson, A. G. (1967). 1967: A statistical theory of spatial distribution models. *Transportation Research*, *1*(3), 253–269. doi:10.1016/0041-1647(67)90035-4

Wolf, P. R., Dewitt, B. A., & Wilkinson, B. (2014). *Elements of Photogrammetry with Applications in GIS* (4th ed.). McGraw-Hill.

Wolfram, S. (2002). *A New Kind of Science*. Wolfrom Media.

Wolfram, S. (1983). Statistical mechanics of cellular automata. *Reviews of Modern Physics*, *55*(3), 601–644. doi:10.1103/RevModPhys.55.601

Wong, I., & Baldwin, A. N. (2016). Investigating the potential of applying vertical green walls to high-rise residential buildings for energy-saving in sub-tropical region. *Building and Environment*, *97*, 34–39. doi:10.1016/j.buildenv.2015.11.028

Wong, J. K. W., & Lau, L. S.-K. (2013). From the 'urban heat island' to the 'green island'? A preliminary investigation into the potential of retrofitting green roofs in Mongkok district of Hong Kong. *Habitat International*, *39*, 25–35. doi:10.1016/j.habitatint.2012.10.005

Wong, N., Kwang, T., Chen, Y., Sekar, K., Tan, P., Chan, D., & Chiang, K. (2010). Thermal evaluation of vertical greenery systemns for building walls. *Building and Environment*, *45*(3), 663–672. doi:10.1016/j.buildenv.2009.08.005

Wong, N., Tay, S., Wong, R., Ong, C., & Sia, A. (2014). Life cycle cost analysis of rooftop gardens in Singapore. *Building and Environment*, *38*(3), 499–509. doi:10.1016/S0360-1323(02)00131-2

Wu, D., Arkhipov, D. I., Kim, M., Talcott, C. L., Regan, A. C., McCann, J. A., & Venkatasubramanian, N. (2017). ADDSEN: Adaptive Data Processing and Dissemination for Drone Swarms in Urban Sensing. *IEEE Transactions on Computers*, *66*(2), 183–198.

Wulder, M. A., Loveland, T. R., Roy, D. P., Crawford, C. J., Masek, J. G., Woodcock, C. E., Allen, R. G., Anderson, M. C., Belward, A. S., Cohen, W. B., Dwyer, J., Erb, A., Gao, F., Griffiths, P., Helder, D., Hermosilla, T., Hipple, J. D., Hostert, P., Hughes, M. J., ... Zhu, Z. (2019). Current status of Landsat program, science, and applications. *Remote Sensing of Environment*, *225*, 127–147. doi:10.1016/j.rse.2019.02.015

Xiao, J., Shen, Y., Ge, J., Tateishi, R., Tang, C., Liang, Y., & Huang, Z. (2006). Evaluating urban expansion and land use change in Shijiazhuang, China, by using GIS and remote sensing. *Landscape and Urban Planning*, *75*(1-2), 69–80. doi:10.1016/j.landurbplan.2004.12.005

Xu, G., Dong, T., Cobbinah, P. B., Jiao, L., Sumari, N. S., Chai, B., & Liu, Y. (2019). Urban expansion and form changes across African cities with a global outlook: Spatiotemporal analysis of urban land densities. *Journal of Cleaner Production*, *224*, 802–810. doi:10.1016/j.jclepro.2019.03.276

Yacef, F., Rizoug, O., & Hamerlain, M. (2017). Optimization of Energy Consumption for Quadrotor UAV. In *Proceedings of 9th International Micro Air Vehicle Conference and Flight Competition (IMAV) 2017* (*vol. 1*, pp. 215-222). Academic Press.

Yang, J., & Wang, Z.-H. (2017). Planning for a sustainable desert city: The potential water buffering capacity of urban green infrastructure. *Landscape and Urban Planning*, *167*, 339–347. doi:10.1016/j.landurbplan.2017.07.014

Yang, L., Smith, J. A., Baeck, M. L., Bou-Zeid, E., Jessup, S. M., Tian, F., & Hu, H. (2014). Impact of urbanization on heavy convective precipitation under strong large-scale forcing: A case study over the Milwaukee–Lake Michigan region. *Journal of Hydrometeorology*, *15*(1), 261–278. doi:10.1175/JHM-D-13-020.1

Yang, X., Hou, Y., & Chen, B. (2011). Observed surface warming induced by urbanization in east China. *Journal of Geophysical Research, D, Atmospheres*, *116*(D14), D14113. doi:10.1029/2010JD015452

Yang, X., & Lo, C. P. (2002). Using a time series of satellite imagery to detect land use and land cover changes in the Atlanta, Georgia metropolitan area. *International Journal of Remote Sensing*, *23*(9), 1775–1798. doi:10.1080/01431160110075802

Yang, Y., Zhang, X., Lu, X., Hu, J., Pan, X., Zhu, Q., & Su, W. (2018). Effects of Building Design Elements on Residential Thermal Environment. *Sustainability*, *10*(2), 57. doi:10.3390u10010057

Yao, H., Quin, R., & Chen, X. (2019). Unmanned Aerial Vehicle for Remote Sensing Applications – A review. *Remote Sensing*, *11*(12), 1443. doi:10.3390/rs11121443

Yaroslav, K. (2019). *Remote expert air traffic management system "Decision making in a collaborative environment " FF-ICE" obese* [Unpublished Master's thesis]. National Aviation University, Ukraine.

Yin, L., & Shiode, N. (2014). 3D spatial-temporal GIS modeling of urban environments to support design and planning processes. *Journal of Urbanism*, *7*(2), 152–169. doi:10.1080/17549175.2013.879452

Yow, D. M. (2007). Urban heat islands: Observations, impacts, and adaptation. *Geography Compass*, *1*(6), 1227–1251. doi:10.1111/j.1749-8198.2007.00063.x

Yue, W., Xu, J., Tan, W., & Xu, L. (2007). The relationship between land surface temperature and NDVI with remote sensing: Application to Shanghai Landsat 7 ETM+ data. *International Journal of Remote Sensing*, *28*(15), 3205–3226. doi:10.1080/01431160500306906

Yu, X., Guo, X., & Wu, Z. (2014). Land surface temperature retrieval from Landsat 8 TIRS—Comparison between radiative transfer equation-based method, split window algorithm and single channel method. *Remote Sensing*, *6*(10), 9829–9852. doi:10.3390/rs6109829

Zadeh, L. A. (1965). Fuzzy sets. *Information and Control*, *8*(3), 338–353. doi:10.1016/S0019-9958(65)90241-X

Zeng, C., Liu, Y., Stein, A., & Jiao, L. (2015). Characterization and spatial modeling of urban sprawl in the Wuhan Metropolitan Area, China. *International Journal of Applied Earth Observation and Geoinformation*, *34*, 10–24. doi:10.1016/j.jag.2014.06.012

Zeng, H., Sui, D. Z., & Li, S. (2005). Linking urban field theory with GIS and remote sensing to detect signatures of rapid urbanization on the landscape: Toward a new approach for characterizing urban sprawl. *Urban Geography*, *26*(5), 410–434. doi:10.2747/0272-3638.26.5.410

Zhang, X., Zhu, Q., & Wang, J. (2004). 3D City Models Based Spatial Analysis to Urban Design. *Geographic Information Sciences: A Journal of the Association of Chinese Professionals in Geographic Information Systems*, *10*(1), 82–86. doi:10.1080/10824000409480658

Zhang, K., Wang, R., Shen, C., & Da, L. (2010). Temporal and spatial characteristics of the urban heat island during rapid urbanization in Shanghai, China. *Environmental Monitoring and Assessment*, *169*(1-4), 101–112. doi:10.100710661-009-1154-8 PMID:19795217

Zhang, T. (2001). Community features and urban sprawl: The case of the Chicago metropolitan region. *Land Use Policy*, *18*(3), 221–232. doi:10.1016/S0264-8377(01)00018-7

Zhang, Y., & Guindon, B. (2006). Using satellite remote sensing to survey transport-related urban sustainability: Part 1: Methodologies for indicator quantification. *International Journal of Applied Earth Observation and Geoinformation*, *8*(3), 149–164. doi:10.1016/j.jag.2005.08.005

Zhang, Y., Smith, J. A., Luo, L., Wang, Z., & Baeck, M. L. (2014). Urbanization and rainfall variability in the Beijing metropolitan region. *Journal of Hydrometeorology*, *15*(6), 2219–2235. doi:10.1175/JHM-D-13-0180.1

Zhan, W., Zhou, J., Ju, W., Li, M., Sandholt, I., Voogt, J., & Yu, C. (2014). Remotely sensed soil temperatures beneath snow-free skin-surface using thermal observations from tandem polar-orbiting satellites: An analytical three-time-scale model. *Remote Sensing of Environment*, *143*, 1–14. doi:10.1016/j.rse.2013.12.004

Zheng, G., & Moskal, L. M. (2007). Retrieving Leaf Area Index (LAI) Using Remote Sensing: Theories, Methods and Sensors. *Sensors (Basel)*, *9*(4), 2719–2745. doi:10.339090402719 PMID:22574042

Zheng, Y., Chen, T., Cai, J., & Liu, S. (2009). Regional concentration and region-based urban transition: China's mega-urban region formation in the 1990s. *Urban Geography*, *30*(3), 312–333. doi:10.2747/0272-3638.30.3.312

Zhihui, G., Yan, Z., Yu, C., & Xiaomeng, C. (2016). Analysis of Attraction Features of Tourism Destinations in a Mega-City Based on Check-in Data Mining - A Case Study of Shenzhen, China. *ISPRS, International Journal of Geo-Information*, *5*(11), 1–13. doi:10.3390/ijgi5110210

Zhong, S., Qian, Y., Zhao, C., Leung, R., Wang, H., Yang, B., Fan, J., Yan, H., Yang, X.-Q., & Liu, D. (2017). Urbanization-induced urban heat island and aerosol effects on climate extremes in the Yangtze River Delta region of China. *Atmospheric Chemistry and Physics*, *17*(8), 5439–5457. doi:10.5194/acp-17-5439-2017

Zhong, S., Qian, Y., Zhao, C., Leung, R., & Yang, X. Q. (2015). A case study of urbanization impact on summer precipitation in the Greater Beijing Metropolitan Area: Urban heat island versus aerosol effects. *Journal of Geophysical Research, D, Atmospheres*, *120*(20), 10,903–10,914. doi:10.1002/2015JD023753

Zhong, S., & Yang, X. Q. (2015a). Ensemble simulations of the urban effect on a summer rainfall event in the Great Beijing Metropolitan Area. *Atmospheric Research*, *153*, 318–334. doi:10.1016/j.atmosres.2014.09.005

Zhong, S., & Yang, X. Q. (2015b). Mechanism of urbanization impact on a summer cold-frontal rainfall process in the greater Beijing metropolitan area. *Journal of Applied Meteorology and Climatology*, *54*(6), 1234–1247. doi:10.1175/JAMC-D-14-0264.1

Zhou, W., Jiao, M., Yu, W., & Wang, J. (2019). Urban sprawl in a megaregion: A multiple spatial and temporal perspective. *Ecological Indicators*, *96*, 54–66. doi:10.1016/j.ecolind.2017.10.035

Zhu, Z., Zhou, Y., Seto, K. C., Stokes, E. C., Deng, C., Pickett, S. T. A., & Taubenböck, H. (2019). Understanding an urbanizing planet: Strategic directions for remote sensing. *Remote Sensing of Environment*, *228*, 164–182. doi:10.1016/j.rse.2019.04.020

Zimmermann, F., Eling, C., Klingbeil, L., & Kuhlmann, H. (2017). Precise positioning of UAVs - Dealing with challenging RTK-GPS measurement conditions during automated UAV flights. In *Proceedings of International Conference on Unmanned Aerial Vehicles in Geomatics* (vol. IV-2/W3, pp. 95-103). 10.5194/isprs-annals-IV-2-W3-95-2017

Zink, J., & Lovelace, B. (2015). *Unmanned Aerial Vehicle Bridge Inspection Demonstration Project.* Research Project Final Report 2015-40. Minnesota Department of Transportation.

Zölch, T., Maderspacher, J., Wamsler, C., & Pauleit, S. (2016). Using green infrastructure for urban climate-proofing: An evaluation of heat mitigation measures at the micro-scale. *Urban Forestry & Urban Greening*, *20*, 305–316. doi:10.1016/j.ufug.2016.09.011

Zonneveld, I. S. (1995). *Land ecology: An introduction to landscape ecology as a base for land evaluation. Land management and conservation.* SPB, Academic Publishing.

About the Contributors

José António Tenedório is a Portuguese geographer, urbanist and remote sensing scientist. He his Associate Professor at Universidade NOVA de Lisboa, School of Social Sciences and Humanities (NOVA FCSH), Lisbon, and Integrated Researcher at CICS.NOVA. He has been Visiting Professor at Polytechnic University of Catalonia, BarcelonaTech (UPC), CPSV, Spain, since 2005. His research is focused on geographic analysis, urban remote sensing and modelling of urban environments. In 2013 and 2015 (the latter, ex aequo) he received the SANTANDER Award (first prize) for the Internationalisation of the NOVA FCSH Scientific Production.

Rossana Estanqueiro is a Geographer, Assistant Professor at Universidade NOVA de Lisboa, School of Social Sciences and Humanities (NOVA FCSH), Lisbon, where since 2012 she teaches aggregate subjects in the area of Geographic analysis, Remote Sensing and Geographic Information Systems for territorial and urban planning. Integrated Researcher at Interdisciplinary Centre of Social Sciences (CICS.NOVA). Graduated in Geography and Regional Planning (2002), Master in Territorial Management (2007) and PhD in Geography and Territorial Planning, specialization area in Remote Sensing and Geographic Information Systems (2012).

Cristina Delgado Henriques is a Geographer and Urban and Regional Planner, Assistant Professor at the Lisbon School of Architecture, Universidade de Lisboa, since 1997. PhD in Urban Planning from the Technical University of Lisbon. Graduated in Geography and Regional Planning from Universidade Nova de Lisboa. Holder of a scholarship from the Institute of Tropical Research, for the development of Cabo Verde Atlas, and of a scholarship, from FCT (Portuguese national funding agency, for science, research and technology) in the Military Geographical Institute, for the working up of the international project Vector Smart Map.Post-graduated in Land Information Surveying, from Erasmus Program (Ghent University and Roskilde University), and in GIS from Instituto Superior Técnico, (Universidade de Lisboa). Author of the book "Maputo. Cinco décadas de mudança territorial. O uso do solo observado por tecnologias de informação geográfica", published in 2008 by Instituto

Português para o Desenvolvimento (IPAD). Her main scientific areas of research are Applied Geography, GIS, Spatial Analysis, Urban Modeling, Urban Planning, African Studies.

* * *

Amilton Amorim has a graduate degree at Engenharia Cartográfica from Universidade Estadual Paulista Júlio de Mesquita Filho (1988), master's at Civil Engineering from Universidade Federal de Santa Catarina (1993) and Doctorate at Transportes from Escola de Engenharia de São Carlos (2000). Has experience in Urban and Regional Planning, focusing on Information, Register and Mapping.

Blanca Arellano is an Architect (Graduate with Honours) M.Sc. Master's Degree in Urban Management and Valuations Director of the UPC Master in Urban Planning and Sustainability and Postgraduate program of Smart Cities: Urbanism and Technologies. Assistant lecturer, Department of Architectural Technology (TA) UPC's Barcelona School of Architecture (ETSAB). Researcher in the Centre of Land Policy and Valuations (CPSV). Head of the Department of Housing and Land Reserves of the Secretariat of Infrastructure and Urban Development of the State of Baja California (2005-10) Founding Member and Coordinator of Urban Planning within the Municipal Institute of Research and Urban Planning of Mexicali, and President of the Mexicali Association of Architects. She has undertaken research in the areas of urban planning, land and housing policy, analysis of urban growth processes: Sprawl vs. Compact City, sustainable urban planning, mobility, public space, urban regeneration and models of city management.

Liliana Raquel Simões Azevedo is Superior Geography Technician, since 2004, at Coimbra City Council (Division of Geographic and Cadastral Information). In 2015 she concluded his Master's dissertation in Geographic Information Technologies, at the University of Coimbra, with a work entitled "Geolocation, Geocolaboration and Social Networks. Case study applied to tourism in the Historic Center of Coimbra", supervised by José Gomes dos Santos.

Guilherme Barros de Souza has a Bachelor's at Cartographic Engineering from São Paulo State University (UNESP) (2004) and master's and doctorate degree on Cartographic Sciences from São Paulo State University (UNESP) (2011). Has experience in Urban and Regional Planning, focusing on Information, Register and Mapping, acting on the following subjects: Multipurpose Cadastre, Cadastral Systems, Database and Cartography.

Oleksandr Burlaka is a Master of Department of Air Navigation Systems at National Aviation University (Ukraine). Areas of Scientific Interests: Research of Air Navigation systems; Information Technology; Decision-making; Data Science; Computer Science. Thesis in the field of UAVs Air Navigation, Neural Networks Application to UAVs.

Abner Luis Calixter is a PhD candidate at the University of Brasília (UnB) and held a Master of Arts (M.A) degree in Sustainable International Development from the Heller School for Social Policy and Management - Brandeis University focusing on sustainable development planning, design and public policies. His expertise in Resilient and Sustainable Urbanism has been further enriched through his time with the MIT (Massachusetts Institute of Technology) DUSP – Department of Urban Studies and Planning. At the Harvard Graduate School of Design (GSD), besides contributing as research assistant to the project "A Sustainable Future for Exuma" he has earned Executive Education Certificates in Climate Adapted Design and Master Planning for Sustainable Cities. Currently, Calixter investigates potential influences of urban heat and human aggressive behaviour. He also coordinates projects to tackle excessive heat in public schools with green infrastructure with the aim of leveraging education and alleviating violence in Brasília.

Andrea Figueiredo Arruda Canavarros is an Architect and Urbanist from the Faculty of Architecture and Urbanism of the Pontifical Catholic University of Campinas (2001). Master degree from the Graduate Program of the Faculty of Architecture and Urbanism of the University of São Paulo (2007). PhD in Urbanism from the Faculty of Architecture of the University of Lisbon (2016). She was an effective professor at the Department of Architecture and Urbanism at the State University of Mato Grosso (UNEMAT), between February 2006 and February 2008, a visiting professor at the Department of Architecture and Urbanism at the Federal University of Mato Grosso (UFMT), between June 2006 to June 2008. Since 2013 she has been an effective professor in the same Department of UFMT and researcher at the Research and Extension Group "Study of Urban and Regional Planning - EPURA", where she has been developing local and regional actions concerning Housing Policy, Land Regularization and Technical Assistance for Housing of Social Interest (ATHIS). In 2019 she joined the Socio-Territorial Studies Group for Urban and Local Action (GESTUAL-CIAUD / Lisbon School of Architecture - University of Lisbon.

Kostyantin Dergachov is Head of the Department of Aircraft Control Systems of the National Aerospace University KhAI, Kharkov Ukraine.

Rafael Destro graduated at Engenharia Cartográfica by Universidade Estadual Paulista Júlio de Mesquita Filho (2017). Currently is of Universidade Estadual Paulista Júlio de Mesquita Filho. Has experience in the area of Geoscience, with emphasis on Geodésia.

Janaina dos Santos is an Administrative Technique, Undergraduate in Surveying and Cartographic Engineering at Federal Rural University of Rio de Janeiro Former Member in the Academic Directory of Surveying and Cartographic Engineering, Secretary General - Management 2015 - 2016 Former Member of the Junior Association of Surveying and Cartographic Engineering - Junior Measure as: Advisor of Human Resources, Director of Human Resources (2016) and Diretora Presidente Gestão (2017); Member of the Board of Engineering of Surveying and Cartographic Engineering 2016 Faperj Researcher and Fellow with the Climate Susceptibility Project for Desertification in Regions of the State of Rio de Janeiro, DCA / IF (Department of Environmental Sciences - Forest Institute of UFRRJ), project delivered in 2016. Faperj Researcher and Fellow with the Climate Susceptibility Project to Desertification in the Southeast Region of Brazil, Has experience in the area of Geosciences and Topography with emphasis in Geophysics.

Raquel Faria de Deus is a Geographer. She holds an MS in Geographic Information Systems from Universidade de Lisboa, and a PhD degree in Geography and Territorial Planning, specialization area in Remote Sensing and Geographic Information Systems from Universidade NOVA de Lisboa. In 2017 she was awarded the National Prize of Geography Orlando Ribeiro for the best Portuguese PhD thesis (2015-2016) from the Portuguese Association of Geographers. In 2019 she was hired as a PhD researcher by the Faculty of Social Sciences and Humanities at the Universidade NOVA de Lisboa. She writes on several issues including Land-use and Land-cover change, Geographic Information Systems, Spatial analysis, Modelling and Planning.

Swagata Ghosh is currently working as Assistant Professor in the Amity Institute of Geoinformatics and Remote Sensing (AIGIRS), Amity University, Noida, India since 2014. Prior to join in current job, she has worked as Postdoctoral Research Fellow in IIT Bombay, India for more than a year (2013-14). She has completed her M.Tech in 2009 and Ph.D in Remote Sensing from BIT Mesra Ranchi India in 2013. She was awarded the CSIR-SRF fellowship to pursue her Ph.D. studies. Her earlier qualification includes M.Sc. in Geography from University of Calcutta, India. She has published 13 research papers in International/National journals and conference. Her research area includes urban growth modelling, pollution studies, glacier retreat analysis, climate change, remote sensing. She is running a project in Urban Remote Sensing funded by SERB, DST, Govt. of India.

José Gomes dos Santos is Associate Professor at the Department of Geography and Tourism of the Faculty of Arts of the University of Coimbra since August 2019. In 2008, he became an integrated member of the Center for Studies in Geography and Spatial Planning (CEGOT). In 2018 he was invited to collaborate with the Center for Applied Cartography, University of Brasília (UNB) and, in 2019, with the Cartography Laboratory - GeoCart, Department of Geography, Federal University of Rio de Janeiro (UFRJ); in both. cases, with the status of Invited External Investigator. Entered the university teaching career in June 1992, after having completed his Degree in Geography, in 1991, and the Pedagogical Internship held at the José Falcão Secondary School, in Coimbra, in 1992. In 1996, he completed his Master's Degree in Physical Geography and, in 2005, a PhD in the same scientific area, both from FLUC. Between 2009 and 2011, he was Mentor and CEO of the company iNovmapping LDA, elected partner of Google, in 2010 and Mozilla, in 2011, a company that can be defined as Start-up and Spin-off of UC that was awarded in international competitions (Google My Maps - 2009) and national (Arrisca Coimbra 2009). That same year, it was incubated at the Pedro Nunes Institute - UC. In 2011, he was mentor and PEP for accreditation of the Master's Degree in Geographic Information Technologies (MGIT) created, in consortium, by FLUC and FCTUC, of which he was also the first Director between 2012 and 2015. In 2014 he was mentor and Chair of the 1st event of the cycle of International Lusophone Congresses on Geographic Information Sciences and Technologies which, in 2017, had the organization of the 2nd event at UNESP, Presidente Prudente - Brazil and, in 2022, will have a sequence in its 3rd edition, in Cape Verde. His main areas and research, interests are applied geomorphology, natural hazards, risk and catastrophes, GIS and WebGIS, GIS open source software, GIS, innovation and entrepreneurship, and geo-law, environment and spatial planning.

David R. Green is Director of the Aberdeen Institute for Coastal Science and Management (AICSM), Director of the M.Sc. Degree Programme in Geographical Information Systems, and Director of the UAV Centre for Environmental Monitoring and Mapping (UCEMM) (www.abdn.ac.uk/research/ucemm) at the Department of Geography and Environment, University of Aberdeen, Scotland, United Kingdom. He is past Chairman of the Association for Geographic Information (AGI); past Editor in Chief of The Cartographic Journal.

Billy Gregory is Director of DroneLIte and an affiliate of UCEMM at the University of Aberdeen.

Jason Hagon studied Geography and Geographical Information Systems at the University of Aberdeen between 2011 and 2017. Working alongside the Universities UAS Centre for Environmental Monitoring and Mapping (UCEMM), he is also a director of GeoDrone Survey Ltd, a company specialising in aerial image data processing. Since 2017 he has worked as a data processor for Andrews Survey, a hydrographic survey company based in Aberdeen.

Camilo Hernández-Mejía is a Mechatronic Engineer - University EIA, Colombia. Developed the first full scale concrete 3D printer and the first 3D printed house in Latin America at Constructora Conconcreto SA. Currently Ph.D. Student at the Integrated Actuators Laboratory (LAI) at EPFL, Switzerland.

João Paulo Hespanha, PhD by TU Delft, currently works at the Águeda School of Technology and Management, University of Aveiro. João Paulo does research in Geoinformatics (GIS) and Land Administration (LIS). Their most recent publication is 'Applying the Knowledge Innovation Value Chain Model to Land Administration Research'.

Rui Pedro Julião is a PhD in Geography and Regional Planning – New Technologies in Geography, Associate Professor at New University of Lisbon, Department of Geography and Regional Planning, and researcher at CICS.NOVA – Interdisciplinary Centre for Social Sciences, coordinating the Laboratory of Ideas (LabID). Main research and competencies are in the field of Geographic Information Science and Systems, namely spatial data modeling, Geocomputation, Spatial Data Infrastructures, Cadastre, and GIS in Organizations. Coordinator of the "Smart Territories" task force from Association for the Promotion and Development of the Information Society (APDSI) since 2018. Recently (from 2002 till 2012) was Deputy Director-General of the Portuguese Geographic Institute (in charge of research, cadastre, IT and National Spatial Data Infrastructure), national representative at the INSPIRE Committee, and also elected member of the Executive Committee of EUROGI (European Umbrella Association for Geographic Information) and of the Management Board of Eurogeographics (European Association of National Mapping and Cadastre Agencies). From 2012 till 2016 was President of the Portuguese Geographers Association. He is the author of several publications including books, chapters in books, articles and communications, edited nationally and internationally, and regularly participates in major national and international conferences in the field of science and geographic information systems as a speaker and session moderator as well as at the organization.

Maria Carolina Jurema is an Environmental Engineering student at the State University of São Paulo Júlio de Mesquita Filho - UNESP, Brazil. Currently internship at the company Imagem Geosistemas in the Marketing Area Esri Portfolio.

Alex Karachok has many hobbies, the most significant being art, geology and photography. Alex acquired his first mini-camera in 2011 and his first drone in 2014. He currently owns a DJI Spark and Osmo Pocket, and uses them to create video projects. Alex is hoping to become a filmmaker and to use drones in the field of geology. Alex is also an affiliate of UCEMM (UAV Centre for Environmental Monitoring and Mapping) at the Department of Geography and Environment, University of Aberdeen, Scotland, United Kingdom, focusing on raising awareness and educating school age pupils in safe, legal and responsible drone flying. He is also an Honorary Pilot for DroneLite.

Anatolii Kulik is Professor of the Department of Aircraft Control Systems at the National Aerospace University KhAI, Kharkov Ukraine.

Sunil Kumar is currently working as JRF in SERB, DST funded project in the Amity Institute of Geoinformatics and Remote Sensing (AIGIRS), Amity University, Noida, India since 2018. He is also pursuing his Ph.D. in Impact of urban expansion in environment. Prior to join Ph.D. he has completed M Tech in Geo-informatics in 2016 from Guru Jambheshwar University of Science and Technology, Hisar.

Jakob Larsen is a planning student at the University of Aalborg, and drone pilot.

Vitalii Lazorenko is Senior Lecturer and Instructor on ATCO simulator of the National Aviation University of Ukraine. Graduated from the State Flight Academy of Ukraine in 2004. Examiner on Phraseology of Radiotelephony communication in the "AeroLingua" company. Crew Resource Management Instructor (CRM I) in "Ukrainian Helicopters" aviation company. Ground Instructor (GI) on general aviation disciplines. Areas of Scientific Researches: Air Traffic Management, Safety, Phraseology of radiotelephony, Human Factors, Decision Making, Practical preparation of ATCOs.

Luís Carlos Roseiro Leitão is Senior Technician at the Institute for the Conservation of Nature and Forests, since 2011, and is also the responsible technical for the Paul de Arzila Nature Reserve (RNPA). In 2012 he concluded the Master's Dissertation in Physical Geography, Environment and Spatial Planning, supervised by José Gomes dos Santos, at the University of Coimbra, with the work entitled The Special Protection Zone (SPA) Ria de Aveiro. Methodological Essay for the

Construction of a Letter of Vulnerability to Support Ecological Landscape Management. In September 2015, he formally enrolled in the PhD course in Physical Geography, under the guidance of José Gomes dos Santos, at the University of Coimbra, and began his research career at the Center for Geography and Spatial Planning Studies (CEGOT).

João Marques is CEO at GEODRONE. Graduated in Geography and Regional Planning, Master degree in Territorial Management at NOVA FCSH, School of Social Science and Humanities. PhD student at NOVA FCSH, School of Social Science and Humanities in Geography and Territorial Planning.

Luis Marques, PhD (maximum grade, cum laude in 2017) in Barcelona School of Architecture of the Polytechnic University Of Catalonia – BarcelonaTech (ETSAB/UPC). Master (pre-Bologna, 2010) and Post-Graduate (pre-Bologna, 2002) in "Land Use Management and Environmental Planning" at the Faculty of Science and Technology of the New University of Lisbon (FCT/UNL), Portugal. Degree (pre-Bologna – four years) in Geography and Regional Planning at the Faculty of Human and Social Sciences of the New University of Lisbon (FCSH/UNL), Portugal (1997-2001). Image Analyst at European Union Satellite Centre. Has been collaborating with several organizations of the public administration (at central, regional and local level) and private companies, mainly in research projects related to the themes of spatial planning and environment (commonly with the use of Geographic Information Technologies). www.luisfilipemarques.com.

Ronaldo Messias Correia is Bachelor of Computer Science from Universidade do Oeste Paulista (1993), Master in Computer Science and Computational Mathematics from the Universidade de São Paulo (1998) and PhD in Electronic Engineering and Computing from Instituto Tecnológico de Aeronáutica (2005). He is currently assistant professor at UNESP. Has experience in Computer Science, acting on the following subjects: database, distributed environments and augmented and virtual reality.

Kousik Midya is currently working as Assistant Professor in the Dept. of ECE, Ajay Kumar Garg Engineering College, Ghaziabad (Affiliated to Dr. APJ Abdul Kalam Technical University), India. He has more than 5 years teaching experience. He has completed his Ph.D. in Nanotechnology from IIT Bombay. His earlier qualification includes M.Tech in Electronics & Telecom. Engg. from Jadavpur University. He has published research papers in International journals and conferences.

Claudio Miranda is an Architect and Urbanist graduated from the Federal University of Rio de Janeiro, PhD in Urbanism from the Postgraduate Program in Urbanism FAU-UFRJ and Master in Environmental Engineering from the same University. His interests are focused on the themes of social housing, land tenure, conservation of historical heritage and urban planning. He worked in the Urban Development area for 30 years, as an Architect at BNH / Caixa Econômica Federal, he was President of the Mato Grosso Architecture and Urbanism Council in the period 2012 - 2014 and Deputy Secretary of Urban Policies of the State of Mato Grosso in the period from 2015 to 2017. He is a Researcher at the Faculty of Architecture, Engineering and Technology - FAET of the Federal University of Mato Grosso.

Srutisudha Mohanty is now doing her PhD and completed a Masters in Environmental Sciences and a post-graduate diploma in remote sensing and GIS.

Thiago Montenegro Góes is an Architect and Urban Planner by Federal University of Santa Catarina (2011) and holds a MSc (2018) in Architecture and Urban Planning by the University of Brasília and a Specialization in Architecture, Construction and Sustainable Building Management. He is currently a PhD candidate in University of Brasília. His research interests are in sustainability with emphasis on urban planning and natural drainage, energy efficiency and resource savings in architectural design and sustainable certification in civil construction. Has professional experience in management of works, in architectural and interior projects in the commercial and residential areas.

Krishna Vidhata N. completed his M.Tech. in Geoinformatics and Remote Sensing from Amity Institute of Geoinformatics and Remote Sensing (AIGIRS), Amity University. He has done his B.Tech. in Computer Science Engineering from Aurora Engineering College, Bhongir, Hyderabad during 2013-2017. His area of research is LULC Change, Forestry, Climate Change and Remote Sensing.

Jagabandhu Panda is working as an Associate Professor. He has more than 10 years of research and teaching experience. He did his PhD from Indian Institute of Technology Delhi and Masters from Utkal University, Bhubaneswar.

Sudhansu Rath is currently doing his PhD in Atmospheric Science. Prior to joining the PhD program, he completed a Masters degree in remote sensing and GIS.

Regina Reani, PhD in geography. Specialist in social and environmental disasters at National Center for Monitoring and Early Warning of Natural Disasters - CEMADEN, Brazil.

Carla Rebelo holds a Ph.D. in Geography and Territorial Planning (specialization in Remote Sensing and Geographic Information Systems) from the Faculty of Social Sciences and Humanities (FCSH) at the Universidade NOVA de Lisboa. Lecturer with a vast experience in teaching Geomatics-related courses, such as Geographic Information Systems (GIS), FOSS GIS, and Photogrammetry in different higher education institutions. Presently holds the position of Deputy Technical Director at the National Land Registry System Project of East-Timor. Also, she is research collaborator at the Interdisciplinary Centre of Social Sciences - UNL (CICS.NOVA). Current research interests include GIS, Remote Sensing and Photogrammetry, 3D Modelling and Free and Open Source Software (FOSS) GIS. She has been the Deputy Technical Director in the National Cadastral Survey and land registry System Project of East-Timor since 2014.

Josep Roca is a Ph.D. Architect and Founding Director of Centre of Land Policy and Valuations (CPSV) (1986); Chair of Legal Architecture, Planning Law and Valuations and member of the Department of Architectural Technology of the UPC's School of Architecture (ETSAB). Director of UPC's Master's Degree in Urban Land Management and Valuations, MBArch), Doctorate Program in Urban and Architectural Management and Valuation, and Master's and Postgraduate Programs in Urban Land and Property Management; Spatial Planning, Sustainability and Climate Change, Smart Cities and Geographical Information Systems. Director of more than 40 doctoral theses.

Jorge Rocha, geographer. He is MSc in Geographic Information Systems (2003) and in Spatial Planning (2013), and PhD in Geographic Information Science (2012). Currently is an Assistant Professor of the Institute of Geography and Spatial Planning and member of the Modelling, Urban and Regional Planning, and Environmental Hazard and Risk Assessment and Management research groups of the Centre of Geographical Studies, all from the University of Lisbon. His field of expertise is Geosimulation and Geocomputation involving Artificial Neural Networks, Graphs Theory, Cellular Automata, and Multi-agent Systems. Jorge's works are quite diverse, focusing mainly on, but not exclusively, in Urban Morphology, Remote Sensing, Epidemiology, Health Geography, Smart Cities and Big Data.

Marta Romero is a Full Professor at the University of Brasília - Brazil. Coordinator at Laboratory of Sustainability Applied on Architecture and Urban Design - LaSUS (www.lasus.unb.br). Coordinator of research group named "Reabilita" (Environmental and Sustainable Rehabilitation in Buildings and Cities).

José Gomes dos Santos is Associate Professor at the Department of Geography and Tourism of the Faculty of Arts of the University of Coimbra since August 2019. In 2008, he became an integrated member of the Center for Studies in Geography and Spatial Planning (CEGOT). In 2018 he was invited to collaborate with the Center for Applied Cartography, University of Brasília (UNB) and, in 2019, with the Cartography Laboratory - GeoCart, Department of Geography, Federal University of Rio de Janeiro (UFRJ); in both. cases, with the status of Invited External Investigator. Entered the university teaching career in June 1992, after having completed his Degree in Geography, in 1991, and the Pedagogical Internship held at the José Falcão Secondary School, in Coimbra, in 1992. In 1996, he completed his Master's Degree in Physical Geography and, in 2005, a PhD in the same scientific area, both from FLUC. Between 2009 and 2011, he was Mentor and CEO of the company iNovmapping LDA, elected partner of Google, in 2010 and Mozilla, in 2011, a company that can be defined as Start-up and Spin-off of UC that was awarded in international competitions (Google My Maps - 2009) and national (Arrisca Coimbra 2009). That same year, it was incubated at the Pedro Nunes Institute - UC. In 2011, he was mentor and PEP for accreditation of the Master's Degree in Geographic Information Technologies (MGIT) created, in consortium, by FLUC and FCTUC, of which he was also the first Director between 2012 and 2015. In 2014 he was mentor and Chair of the 1st event of the cycle of International Lusophone Congresses on Geographic Information Sciences and Technologies which, in 2017, had the organization of the 2nd event at UNESP, Presidente Prudente - Brazil and, in 2022, will have a sequence in its 3rd edition, in Cape Verde. His main areas and research interests are applied geomorphology; natural hazards, risk and catastrophes; GIS and WebGIS; GIS open source software; GIS, innovation and entrepreneurship, and geo-law, environment and spatial planning.

Teresa Santos is a researcher at CICS.NOVA, Interdisciplinary Centre of Social Sciences, Universidade NOVA de Lisboa, Portugal. She has a MSc in Geographic Information Systems - GIS (2004), and a PhD in Geography and Regional Planning (2011). Her academic path has been focused on using geographic modelling to produce relevant information to support urban planners and decision makers, based on data retrieved from Remote Sensing, GIS and 3D modelling. Current research activities include the development of planning tools to promote environmentally sustainable and liveable cities. Her main lines of investigation are: solar potential in the urban environment; urban regeneration and building rehabilitation; outdoor thermal comfort; green urban areas and environmental quality indicators, among others. She is involved in research projects and is the author of several international journal articles.

Tetiana Shmelova, Doctor of Science, Professor of Air Navigation Systems Department at National Aviation University (Ukraine). Thesis in the field of Scientific and Methodological Basis of Decision-Making in the Air Navigation System (2012). Areas of Scientific Interests: Mathematical Models of Decision-Making by Human operator, especially in Emergency Situation; Decision Support System; Research of Air Navigation system as Socio-technical system; Aviation Sociometry and Socionics; Management and Marketing in aviation; Information technology and Informatics of Decision-Making; Problems of Human factors in aviation. Teaching courses: Theory of decision-making, Mathematical Programming, Effectiveness of Air Traffic Management and UAVs, Artificial Intelligence.

Caio Silva is an Architect and Urban Planner by Federal University of Piauí (2006), MSc (2009) and PhD by University of Brasília / Polytechnic University of Catalunya, Barcelona - Spain. He is a Professor at the FAU- UnB Technology. Department and researcher at the Sustainability Laboratory Applied to Architecture and Urbanism- LaSUS/ UnB and a researcher at the Environmental Control Laboratory- Lacam / UnB, and a collaborator researcher at Universidade NOVA de Lisboa (UNL). Currently, he is a Visiting Scholar at Harvard University, and he coordinates the research group about computational simulation in buildings and cities (www.sicac.unb.br).

Alastair Skitmore is a graduate of Liverpool John Moores University with a Master of Science in Drone Technology and Applications. He is an affiliate of UCEMM.

Pedro Sorte graduated in Biological Science at the University de Brasília (UnB), Graduated in Architecture and Urbanism (UnB), Msc in Environmental and Sustainable Rehabilitation in Buildings and Cities (UnB) and MPhil in Architecture and Urbanism (UnB). He is interested in architectural, theoretical and built expressions, derived from the relationship between the built environment and the natural environment. Currently, he works as an Architect and Urban Planner at the VIV office, he is a professor in the Landscape Architecture course at JK Faculty and the School of Landscape Architecture of Brasília, and he is a student on the Msc in Teaching on Higher Education at the Catholic University (UCB), student of the undergraduate course in Philosophy at the UnB and is a member of the research group on Computational Simulation in the Built Environment (SICAC - UnB).

Engelberth Soto-Estrada: civil engineer, MEng and Ph. D. in environmental engineering. Lecturer at the Civil Engineering programme at the EIA University.

Lívia Tomás is graduated in Architecture and Urbanism from Federal University of Uberlandia - UFU, Brazil (2002), owns a Master Degree in Geography with emphasis in Regional and Urban Planning from UFU (2005), accomplished part of her PhD in Remote Sensing at the Electronic Department of University of Pavia, Italy, (2009) as an affiliate researcher and achieved a PhD in Remote Sensing from Brazilian National Institute for Space Research - INPE (2010). She worked six years at Institute of Research, Administration and Planning – IPPLAN in São José dos Campos, Brazil, developing projects on urban mobility, city planning, sustainable city development, geoprocessing, and others. She is currently a research fellow at Brazilian Center for Monitoring and Early Warning of Natural Disasters - CEMADEN, where she investigates the impact of disasters on transport infrastructure and people mobility.

Ann Wellens: chemical engineer, MEng. Lecturer at the Systems Engineering programme at the Facultad de Ingeniería, UNAM.

Index

D

E

F

G

H

I

K

L

M

N

O

P

R

S

T

U

V

W

www.ingramcontent.com/pod-product-compliance
Lightning Source LLC
LaVergne TN
LVHW080616300926
849162LV00005B/1180

* 9 7 8 1 7 9 9 8 2 2 4 9 3 *